ŬÑŦÅÑG£ĨÑG
ŦHË
W̊Ëß

A Guide To Internet Research

(original cover)

Untangling the Web was published by the Center for Digital Content of the National Security Agency in 2007. It was obtained by the Freedom of Informaion Act in 2013.

~~This product has been produced for official use only. This product is not approved for public release. The information contained herein is for the exclusive use of the original recipient and is not for further distribution outside the recipient's agency or organization. The original recipient may make copies for distribution only within the recipient's agency or organization.~~

~~Ultimate responsibility for the protection of this product from public disclosure resides with the user, with severe penalties for non-compliance.~~

~~For additional information, please contact:~~

[redacted]
9800 Savage Road
Suite 6324
Fort Meade, MD 20755-6324

(b)(3)-P.L. 86-36

This Page Intentionally Left Blank

Table of Contents

Preface: The Clew to the Labyrinth

One of the most famous stories about libraries tells of the tenth century Grand Vizier of Persia, Abdul Kassem Ismael who, "in order not to part with his collection of 117,000 volumes when traveling, had them carried by a caravan of 400 camels trained to walk in alphabetical order."[1] However charming this tale may be, the actual event upon which it is based is subtly different. According to the original manuscript, now in the British Museum, the great scholar and literary patron Sahib Isma'il b. 'Abbad so loved his books that he excused himself from an invitation by King Nuh II to become his prime minister at least in part on the grounds that four hundred camels would be required for the transport of his library alone.[2]

A 21st Century version of the story might feature any number of portable electronic devices—a laptop, a PDA, or even a mobile phone—designed to overcome this difficulty. Today, 1000 years later, the Persian scholar/statesman would have to find a new excuse for declining the job offer. Abdul Kassem Ismael (aka Sahib Isma'il b. 'Abbad) would be hard pressed to explain why he couldn't just find what he needed on the Internet. The message seems to be that books are passé, replaced by ones and zeroes, the real world replaced by a virtual one, knowledge supplanted by information at best and chaotic data at worst. Have we shrunk the world or expanded it? Or have we in some way replaced it?

Untangling the Web for 2007 is the twelfth edition of a book that started as a small handout. After more than a decade of researching, reading about, using, and trying to understand the Internet, I have come to accept that it is indeed a Sisyphean task. Sometimes I feel that all I can do is to push the rock up to the top of that virtual hill, then stand back and watch as it rolls down again. The Internet—in all its glory of information and misinformation—is for all practical purposes limitless, which of course means we can never know it all, see it all, understand it all, or even imagine all it is and will be. The more we know about the Internet, the more acute is our

---

[1] Alberto Manguel, *A History of Reading*, New York: Penguin, 1997, 19. Manguel cites as his source Edward G. Browne's *A Literary History of Persia*, 4 vols., London: T. Fisher Unwin, 1902-24. I found the specific reference to this story on pages 374-375 of Vol. 1, Book IV, "Decline of the Caliphate." There is, sadly, no mention of the alphabetical arrangement of the library. This entire masterpiece is available online at The Packard Humanities Institute, Persian Texts in Translation, 23 February 2006, <http://persian.packhum.org/persian/pf?file=90001011&ct=0> (15 November 2006).

[2] Edward G. Browne. Vol. 1, Book IV, "Decline of the Caliphate," *A Literary History of Persia*," 4 vols., London: T. Fisher Unwin, 1902-24, 374-375. Available online at The Packard Humanities Institute, Persian Texts in Translation, 23 February 2006, <http://persian.packhum.org/persian/pf?file=90001011&ct=0> (15 November 2006).

awareness of what we do not know. The Internet emphasizes the depth of our ignorance because "our knowledge can only be finite, while our ignorance must necessarily be infinite."[3] My hope is that *Untangling the Web* will add to our knowledge of the Internet and the world while recognizing that the rock will always roll back down the hill at the end of the day.

I will end this beginning with another story and a word of warning. "Tlön, Uqbar, Orbis Tertius" describes the discovery of an encyclopedia of an unknown planet. This unreal world is the creation of a secret society of scientists, and gradually, the imaginary world of Tlön replaces and obliterates the real world. Substitute "the Internet" for Tlön and listen. Does this sound familiar?

> "Almost immediately, reality yielded on more than one account. The truth is that it longed to yield...The contact and the habit of Tlön have disintegrated this world. Enchanted by its rigor, humanity forgets over and again that it is a rigor of chess masters, not of angels...A scattered dynasty of solitary men has changed the face of the world. Their task continues. If our forecasts are not in error, a hundred [or a thousand] years from now someone will discover the hundred volumes of the Second Encyclopedia of Tlön. Then English and French and mere Spanish will disappear from the globe. The world will be Tlön."[4]

As we enjoy, employ, and embrace the Internet, it is vital we not succumb to the chauvinism of novelty, that is, the belief that somehow whatever is new is inherently good, is better than what came before, and is the best way to go or best tool to use. I am reminded of Freud's comment about the "added factor of disappointment" that has occurred despite mankind's extraordinary scientific and technical advances. Mankind, claims Freud, seems "to have observed that this newly-won power over space and time, this subjugation of the forces of nature, which is the fulfillment of a longing that goes back thousands of years, has not increased the amount of pleasurable satisfaction which they may expect from life and has not made them feel happier."[5] Indeed, most of the satisfactions derived from technology are analogous to the "cheap enjoyment...obtained by putting a bare leg from under the bedclothes on a cold winter night and drawing it in again."[6] What good is all this technology and information if, instead of improving our lot, it only adds to our confusion and suffering? We are continually tempted to treat all technology as an end in itself instead of a means to some end. The Internet is no exception: it has in large

---

[3] Karl Popper, *Conjectures and Refutation: The Growth of Scientific Knowledge,* London & New York: Routledge, 2002, p. 38.

[4] Jorge Luis Borges, "Tlön, Uqbar, Orbis Tertius," in *Labyrinths*, ed. Donald A. Yates and James E. Irby, New York: New Directions Books, 1962, 17-18.

[5] Sigmund Freud, "Civilization and Its Discontents," tr. James Strachey, New York: Norton, 1962, 34-35.

[6] Freud, 35.

measure become the thing itself instead of a means of discovery, understanding, and knowledge.

Like Tlön, the Internet, "is surely a labyrinth, but it is a labyrinth devised by men, a labyrinth destined to be deciphered by men." We must avoid getting lost in the labyrinth without a clew. My hope is that *Untangling the Web* will be something akin to Ariadne's clew,[7] so that as you unravel it, you can wind your way through the web while avoiding some of its dangers. Remember also that those who use the Internet to do harm, to spread fear, and to carry out crimes are like the mythical Minotaur who, as well as being the monster in the Minoan maze, was also its prisoner.

8

[7]Daedalus, the architect of the infamous labyrinth on Crete, purportedly gave King Minos' daughter Ariadne the clew, a ball of thread or yarn, to use to find a way out of the maze. Ariadne in turn gave the clew to Theseus, who slew the Minotaur and found his way out of the labyrinth. Theseus repaid Ariadne's kindness by leaving her on an island on their way back to Athens.

[8] "Minotaurus," WikiMedia Commons, <http://commons.wikimedia.org/wiki/Image:Minotaurus.gif> (6 February 2007). This image is in the public domain because its copyright has expired.

Notes

"Every Angle of the Universe"

One wag has suggested that the Internet is an "electronic Boswell," the chronicler of our age. It is that and more because the Internet chronicles not only a time and place but all times and all places, known and unknown, real and imaginary. The Internet is the closest thing to the fantastical "Aleph" imagined by the great Argentine story-teller Jorge Luis Borges, an object whose diameter is "little more than an inch" but which nonetheless contains all space, "actual and undiminished," and in which one can see "every angle of the universe."

While the comparison with the mythical Aleph may strike you as a bit whimsical, it is in fact not an altogether unfair metaphor. There has never been anything that approaches the Internet's reach (to almost every part of the globe in less than thirty years), its size (estimated at 532,897 *terabytes* way back in 2003[9]), and its ability to link us together in a new kind of world community (words, pictures, sounds, ideas beyond imagining). But, as with all new technologies, it comes at a cost—many costs, in fact. ***We pay for the benefits of the Internet less in terms of money and more in terms of the currencies of our age: time, energy, and privacy.***

The goal of this book is to help you save some of each of these valuable resources: time, by making your searches more efficient; energy, by reducing the frustration using the Internet often entails; and privacy, by pointing out some simple measures to take to lower your cyber-profile and enhance your security.

I cannot emphasize strongly enough that this book was already out of date by the time it was published. Even though I have checked and rechecked every link in this book, some addresses are bound to have changed, some sites will have shut down, and some tips and techniques—such as search engine rules and syntax—will no longer be accurate. This is a testament to the changeable nature of the Internet and I must beg your forbearance for any such errors. Writing about the Internet is much like trying to catch Proteus[10]—as with the mythical prophet, it keeps changing and escaping our grasp.

---

[9] School of Information Management and Systems, University of California at Berkeley, "How Much Information? 2003," 27 October 2003, <http://www.sims.berkeley.edu/research/projects/how-much-info-2003/internet.htm> (October 2005), Executive Summary.

[10] "*Proteus—i.e.* full of shifts, aliases, disguises, etc. Proteus was Neptune's herdsman, an old man and a prophet...There was no way of catching him but by stealing upon him during sleep and binding him; if not so captured, he would elude anyone who came to consult him by changing his shape, for he had the power of changing it in an instant into any form he chose." "Proteus," *Brewer's Dictionary of Phrase and Fable*, 1898, <http://www.bartleby.com/81/13723.html> (14 November 2006).

> **"The Internet has often been called the world's largest library with all of the books on the floor."**
>
> Curtin, M., Ellison, G., Monroe, D., "What's Related? Everything But Your Privacy," 7 October 1998, Revision: 1.5, <http://www.interhack.net/pubs/whatsrelated/> (14 November 2006).

What Will I Learn?

To achieve these goals, this book will:

- help you understand how to use the Internet more efficiently to find useful information and, in so doing...

- make clear why the Internet is an invaluable resource.

This year I have reorganized the book to make it more logical and easier to use. The first part of the book still focuses on the ***ins and outs of searching***: how search engines work, types of search tools, how to handle different types of searches. The next section has expanded to offer in-depth tutorials on six major search engines. Next, the book covers ***specialized search tools and techniques***, including a new section devoted to Wikipedia. I have also moved the discussion of maps and mapping to this section. This is followed by ***"invisible" web research*** to include the changes to A9 and Amazon's search inside the book option. Next is the ***international search and language tools*** section, followed by specialized research tools, including new sections on video, audio, and podcast searches. The next section covers ***specific topical research***, such as news, telecommunications, blogs, and RSS feeds. This is followed by a series of "how to" guides, culminating with ***tips and techniques*** for more effective searching. The book then delves into ***using the Internet to research the Internet***, with the final section still addressing crucial ***privacy and security issues***.

Why Do I Need Help?

There are no Internet research experts.

There are people who make a living using the Internet for research and who know more than others about what is on the Internet, how to find what they want on the Internet, and how to do this with relative efficiency. But no one knows what is truly "out there" for two fundamental reasons:

- The Internet changes constantly. By that I mean daily, hourly, minute-to-minute, incessantly.
- It's too darned big! If we can't accurately size the Internet (which we can't), you can be sure we don't know what is available via this resource with any degree of accuracy or completeness.

This doesn't mean you can't ever hope to find anything on the Internet. You often can find what you're looking for (and usually a lot more) with comparative ease, but no one should be deluded into believing he has a good grasp of the entire world of information available on the Internet. Realistically, the best search engines index ***only a fraction of all webpages*** and keyword searching is at best an art that routinely misses relevant sites while loading you down with dross.

Are you discouraged? Don't be...novices often have more luck finding something arcane than seasoned researchers because of the power of creative thinking and serendipity. I've learned never to underrate luck and intuition when doing Internet research, but I think the two most important tools for successful Internet research are:

> ***1. a good set of bookmarks***
>
> ***2. other people with experience searching the Web***

Never assume others are already aware of some website, tool, or technique you find particularly useful. The sheer quantity of data, information, and knowledge associated with the Internet is so enormous that ***no one*** can know more than a fraction of what's on it. While we're talking size, let me mention an important distinction. The **Internet** and the **web** are not one and the same, though the web is what most people think of when you say "Internet."

In fact, as huge as it is, the ***Worldwide Web is actually a subset of the Internet***. The Internet is the network of networks, all the world's servers connected by routers, to put it in semi-technical terms. The web is that portion of the Internet that uses a browser (typically Netscape or Firefox—browsers built upon Mozilla—or Microsoft's Internet Explorer) and some type of hypertext language (usually HTML) to move around. This book focuses primarily on the web because tackling the web by itself is a big enough challenge.

As you have no doubt guessed by now, the Worldwide Web does not come with an instruction manual or users guide, which means much if not most of what you learn about researching using the Internet will come from hard-won experience. On the up side, you probably will not be able to break anything on the Internet. More than likely, no matter how lost or hopelessly confused you become, you will only damage your own computer and/or network—and perhaps your good humor and sanity. However, because of the almost astronomical growth of malicious activity, the Internet has become a dangerous place, and users have discovered that they have inadvertently spread malicious software (malware) such as viruses, worms, and Trojan horses. That is why I have devoted the last section of the book to personal computer security and privacy. We are all at risk from the rising tide of bad and in some cases criminal behavior, so we must take responsibility for protecting ourselves and our computers from the ruses and attacks that grow in number and sophistication each year.

This book will expand on simple "rules" of Internet research, rules that are really more in the nature of friendly suggestions. These rules are the fruit of my own experiences as an Internet user and may prevent you from repeating all the mistakes I made that gave rise to the rules in the first place. Some of these suggestions may at first strike you as odd or inconsistent, but the rationale for each I hope will become clear as we go along.[11] The fact is that today we are drowning in information and starving for knowledge. The goal of *Untangling the Web* is to help rescue users from the ocean of information (and misinformation) by throwing them a virtual lifeline.

What's New This Year

Most people probably have not thought about or been very much affected by the changing search landscape because, as is only natural, most people have one or two sites they routinely use for search and research, regardless of the nature of the inquiry. However, virtually all search professionals will agree that ***knowing where to look for information is the key to successful searching***. Yet few venture beyond

[11] If you are using the hypertext version of this book on line, the links in the paper may not load correctly. Try the refresh button, copy and paste the url, or type in the url directly.

the comfortable confines of the familiar search engine. While the major search engines continue to improve each year, they are far from the be all and end all of search. The problem with general search tools is that they cannot provide targeted or tailored results, certainly not without a lot of work on the part of the user. For this reason, a large part of *Untangling the Web* is devoted to other ways to uncover information, be it subject guides, "deep web" resources, targeted search tools, or unusual tips and techniques for revealing what is hidden.

Again this year, I have included detailed information on how to use Google, Yahoo, Gigablast, and Live Search (formerly MSN Search) to find very specific data. I have also updated and expanded the section on Exalead and added Ask to the major search engines. However, unless you spend a fair amount of time using each of these search tools, you will probably find their many options too complicated and cumbersome for everyday use. A different approach is to use specialized search tools, which begs the question of how to find these tools. *Untangling the Web* maps a number of the Internet's less-traveled roads, i.e., excellent but unheralded specialty search tools such as Fagan Finder, Amazon's A9 multipurpose search, and ThomasGlobal's business search. Also, the section on international search is substantially larger than before.

In recognition of the growing importance and influence of collaborative websites, there are several new sections in this year's book. One is a separate section devoted to Wikipedia, contributed in part by my colleague Diane White. Video and audio search exploded during 2006, and this year's edition contains a new and extensive examination of video search sites as well as a new section on audio search and podcasting. Two other new sections are devoted to custom search engines and book search, neither of which is an entirely new technology but both of which spread in popularity and improved in quality in the past year. Custom search is fast becoming a replacement for web directories, which continue their slide into irrelevance.

The section on researching and understanding the Internet now begins with a new section on "internetworking." This tutorial is a response to a number of requests from people such as myself who need basic knowledge and understanding of how the Internet works without too much technical jargon or expertise. I hope you find that it falls in a comfortable middle ground between simplistic and abstruse.

Once again, the section on privacy and security grows in proportion to concerns about protecting our privacy and security on the Internet. Fortunately, as the problems increase and the malicious users become more enterprising, so do the ways and means for protecting ourselves. However, home computer security is a personal responsibility few people take seriously until it is too late. *Untangling the Web*'s privacy and security information is designed to help users avoid becoming victims and instead take the offense in protecting themselves, their families, and by extension, the Internet community from the Internet's evil-doers. The 2007 edition includes new sections on clearing private data in Firefox, encrypting files in

Windows, pretexting, protecting yourself from search engine leaks, whether or not you can really opt out of online directories, and a brief discussion of wireless Internet use.

I have also reorganized Untangling the Web to make it easier to use. The new section on "Specialized Search Tools & Techniques" brings together some already existing topics, such as Google hacking, with the new sections on Wikipedia and Custom Search Engines. I also moved maps up to this section because they have become integral to basic search. Specialized Research Tools now include the video and audio search sections as well as telephone and email search. Basically, all types of search comprise the first two-thirds of the book, while the remainder focuses on the Internet itself and privacy and security..

As was true of last year's edition, I can again say with confidence that the 2007 *Untangling the Web* was already out of date before it reached your desk. Experienced Internet users understand the Internet is truly a river of information that is impossible to step into twice. And the basic concepts for using the Internet to research topics of interest to our community of readers are sound despite changes in websites, links, and technology.

Web Tip

Web links often change. In case of a bad link for a news article, use the site's search facility and search by the headline, author, or date. In the case of a bad link inside a website, try going to the site's homepage and working your way down to the page, which may still be there, only in a different location.

Introduction to Searching

Search Fundamentals

The September-October 1997 issue of *IEEE Internet Computing* estimated the Worldwide Web contained over 150 million pages of information. At the end of 1998, the web's size had grown to more than 500 million pages. By early 2000, the best estimates put the number over 1 billion and by mid-2000 there was a study showing that there are over ***550 billion unique documents on the web***.[12] Netcraft, which has been running Internet surveys since 1995, reported in its November 2006 survey that there are now ***more than 100 million websites***. "The 100 million site milestone caps an extraordinary year in which the Internet has already added 27.4 million sites, easily topping the previous full-year growth record of 17 million from 2005. The Internet has doubled in size since May 2004, when the survey hit 50 million."[13] The major factors driving this boom are free blogging sites, small businesses, and the relative and lower cost of setting up a website. Another recent survey found:

- *The World Wide Web* contains about 170 terabytes of information on its surface; in volume this is seventeen times the size of the Library of Congress print collections.
- *Instant messaging* generates five billion messages a day (750GB), or 274 Terabytes a year.
- *Email* generates about 400,000 terabytes of new information each year worldwide.[14]

The numbers hardly matter anymore. The enormous size of the Internet means we simply must use search tools of some sort to find information. Otherwise, we are voyagers lost on a vast uncharted ocean.

---

[12] Michael K. Bergman, "The Deep Web: Surfacing Hidden Value," *BrightPlanet*, August 2001, <http://www.brightplanet.com/technology/deepweb.asp> (14 November 2006).

[13] "November 2006 Web Server Survey," Netcraft.com, 1 November 2006, <http://news.netcraft.com/archives/2006/11/01/november_2006_web_server_survey.html> (15 November 2006).

[14] School of Information Management and Systems, University of California at Berkeley, "How Much Information? 2003," 27 October 2003, <http://www.sims.berkeley.edu/research/projects/how-much-info-2003/execsum.htm#summary > (14 November 2006) Executive Summary.

> **Consider this:**
>
> **When you do a search, you are going through more information in less than 30 seconds than a librarian probably could scan in an entire career 30 years ago.**

All the major search engines now index well over a billion pages of information. The problem generally isn't lack of data but finding that one tiny needle in a virtual haystack of almost limitless size (much like looking for a needle in a stack of needles).

Any serious researcher needs to know more about search engines than the average person using the Net for fun or even for very specialized searches associated with a hobby or perhaps a certain topic, e.g., cancer research. How do you learn the ins and outs of search?

The Past, Present, and Future of Search

"Search has become the most hotly contested field in the world of technology."[15]

Remember Northern Light? How about Excite, Galaxy, Lycos, HotBot, Magellan, InfoSpace, Go, Webcrawler, iWon, Netfind, or Webtop? If so, you've been searching the Internet a long time because many of these search engines are long gone and forgotten. However many changes in search and search engines have taken place in recent years, nothing has been quite so dramatic as what has occurred in the past two years with the appearance of the new Yahoo and Live Search engines.

While many smaller, focused search tools still exist, the sad fact is that, in terms of large, powerful, world-encompassing search engines, Internet searchers at this moment have fewer major search engines from which to choose.[16] What happened to get us to this point and what does the future portend?

---

[15] Terry McCarthy, "On the Frontier of Search," *Time.com*, 28 August 2005, <http://www.time.com/time/magazine/article/0,9171,1098955-1,00.html> (14 November 2006).

[16] Of course there are many non-US search engines beyond those run by Google, Yahoo, and Microsoft, but they generally target a particular part of the world and are not serious competitors with Google, Yahoo, or Live Search at this time.

In the early years of the Internet, there was enormous competition in the search market among a large number of search engines vying not only for users but, more importantly, for investors. The "dot bomb" crash in mid-2000 began the shakeout of search companies that continues to this day. The biggest change wrought by the failure of so many Internet-based investments was the growth of pay-per-click advertising in search results. Pioneered by Overture, these so-called sponsored results began to show up at the top of search result lists: the more an advertiser was willing to pay, the higher his result on the list. Then, in 2002 the big search engine consolidation began: first, Yahoo purchased Inktomi, a little known but major player in the search engine world. In 2005, Overture bought AltaVista, one of the oldest and most venerable search engines on the Internet, then quickly acquired AlltheWeb, another major search engine. To top it off, in July 2003, Yahoo bought Overture, thus acquiring three huge search properties at one time.

All this was done publicly. The real revolution was what was happening behind the scenes: with a remarkable degree of secrecy, Yahoo gave the engineers it had acquired from AltaVista, AlltheWeb, and Inktomi a new task—create a whole new search engine to compete with Google. On February 18, 2004, Yahoo unveiled its new search engine, which has a database and search features to rival Google's. Shortly thereafter, Yahoo began killing off the "parents" of its new progeny: first Inktomi, then AlltheWeb and AltaVista. While users can still go to the AlltheWeb and AltaVista websites and run searches, the results are pulled from the Yahoo database and many of the unique search options and features of both search engines are no longer available. However, Yahoo continues to add new features and options that are improving its capabilities.

During 2006, two major search engines unveiled major changes that make them serious contenders: Ask and Exalead. During 2006 Teoma and Ask Jeeves ceased to exist as separate search sites and merged under the Ask.com umbrella. The French search engine Exalead came out of beta for a new look and major overhaul during 2006 and continues to offer a number of important and unique search features. MSN Search became Live Search, which left beta status in September 2006 and increased the much-needed competition from a company that knows how to make successful (if imperfect) products. **Amazon.com** still offers its own search engine, A9, although during 2006, Amazon eliminated some of A9's unique functions, switched from Google to Live Search to power web searches, and appeared to be if not abandoning A9 then certainly scaling it back.

All the major search sites are still trying to be the "Swiss army knife" of search engines. Google, Yahoo, Live Search, Ask, and Exalead all competed hotly with each other to roll out new, better, faster, fancier, more powerful tools to do everything from search the contents of your computer in a heartbeat to letting you "fly" around the world with a bird's (or satellite's) eye view of the planet. Among the new search engine-based tools and programs arriving this past year were vastly improved maps and mapping technologies, enhanced multimedia search, desktop search utilities, toolbars integrated into the browser, and application programming interfaces (APIs) for use by individual developers.

If 2004 was the year of the new search engine and 2005 the year of tailored search, 2006 seems to have been the first year of Web 2.0. Interactive, participatory Internet activities such as blogging, podcasts, online video sharing, and wikis dominated the discourse.

Podcasting finally came into its own last year. Podcasting is recording and broadcasting any non-musical information—be it news, radio shows, sporting events, audio tours, or personal opinions— usually in MP3 format for playback using a digital audio player. Many websites now serve as directories to help users find podcasts of every variety anywhere in the world. Podcasting has caught on because it is easy, inexpensive, mobile, flexible, and powerful. Yahoo got out in front of the podcasting trend with its new Podcasts Search site after a study the search giant published with Ipsos Insight, which disclosed that most of the people who are using RSS do so without even knowing it.[17] RSS, which either stands for Rich Site Summary or Really Simple Syndication, is an XML format for news and content syndication. News aggregators are programs designed to read RSS formatted content, which is very popular in the blogging community. Many if not most blogs make their content available in RSS.

Although there is no agreed upon definition of what Web 2.0 means, in general terms most people believe it involves at a minimum users collaborating to share information online, i.e., an interactive, participatory web in contrast to what is now being called the static web (or Web 1.0). I think the Wikipedia article on Web 2.0 sums the current state of affairs up nicely when it says "To some extent **Web 2.0** is a buzzword, incorporating whatever is newly popular on the Web (such as tags and podcasts), and its meaning is still in flux."[18]

Another important aspect of Web 2.0 is that it organizes information differently from traditional web and other news and knowledge models. So reports a *Time* article on the frontiers of search in its 5 September 2005 issue. There is good reason to believe this claim, given a major investment firm's assessment that "by 2010, search-engine advertising will be a $22 billion industry worldwide, up from an estimated $8 billion today."[19]

One casualty of Web 2.0 appears to be directories. Directories are hierarchical guides to a subset of what are presumably the best, most relevant (or at least most popular) websites on a specific topic. Yahoo was always the king of directories, but

---

[17] Yahoo! and Ipsos Insight, "RSS: Crossing into the Mainstream," October 2005 [PDF], <http://publisher.yahoo.com/rss/RSS_whitePaper1004.pdf> (14 November 2006).

[18] "Web 2.0," Wikipedia, <http://en.wikipedia.org/wiki/Web_2.0> (15 November 2006).

[19] McCarthy.

several years ago, I noted a marked decline in both the quantity and quality of the Yahoo directory. The other major directory was and remains the Open Directory Project, which has always powered the Google Directory and, ironically, now powers the Yahoo Directory. What distinguished the Open Directory from Yahoo was that, while Yahoo was heavily commercial, the Open Directory has always relied upon volunteers to populate and maintain it. Now that most of users' creative energy seems to have moved to wikis, the ODP is in what may be a permanent and ultimately fatal decline. Today, the most successful directories tend to be specialty directories such as NewsDirectory.com or yourDictionary.com, and vertical search engines, such as Business.com or MedlinePlus, which focus on a particular topic instead of trying to catalog the entire Internet.

Directories were almost always a part of the **portal concept**. Portals were all the rage for a few years, while search was considered the Internet boondocks—no one was terribly interested in the boring (and unprofitable) technology of search. So where are portals now—those one-stop handy-dandy Swiss army knife websites that tried to do and be all things to all people? Most of them are gone, thanks in large part to Google's ascendancy. With its clean, spare look, Google changed the face of Internet search by moving away from the portal concept to pure search. While it is true that Google offers a directory as well as other types of searches—Image, news, shopping, groups—Google's focus has always been on web search. Google's new look, which debuted in April 2004, included removing the directory tab from the Google home page, further evidence of the decreasing importance of directories. Although there is growing criticism of the "googlization" of websites, Google continues to be the standard by which most sites are judged.

The rapid and dramatic decline in web directories is only partially attributable to Google's success. The other explanation for the waning of directories is the Tristram Shandy paradox. *The Life and Opinions of Tristram Shandy, Gentleman* is a nine-volume 18th century novel in which Tristram Shandy tries to record every detail of his life but discovers his task is hopeless because it takes him one year to document only one day. As Shandy writes an additional day, it takes him an additional year to complete the events of that day. Such is the fate, to a somewhat lesser degree, of those who seek to compile an Internet directory. By the time the information in the directory is researched, compiled, and published, the Internet has changed and made much of that information obsolete.

I believe Yahoo's decision to metamorphose from directory to search engine was in part a result of a tacit recognition of the Tristram Shandy paradox. Yahoo just couldn't keep up with the Internet's changes and it became too costly to try. ***Creating and maintaining a directory is an extremely manpower intensive endeavor, which flies in the face of the Internet model of relying on automation and technology.*** Undoubtedly, Yahoo's changes were largely driven by Google's enormous financial success. Yahoo sat by for years and watched as Google's popularity (and revenues) increased as Yahoo's stagnated. "By the late '90s much of [Yahoo's] focus was actually diametrically opposed to search, which is supposed to send you to other sites. The Yahoo portal strategy was to keep the eyeballs on its turf, where they viewed more ad units, shopped, and bought premium services. Only when a third of online ad spending moved to search within a few short years did Yahoo decide to buy in big."[20]

Again in 2006 Yahoo changed the look of its homepage, but I believe Yahoo is making a fundamental error by still presenting its busy, messy portal face to the world. Although savvy Internet searchers know to go directly to http://search.yahoo.com in order to avoid the confusion and get a clean interface, most users are still going to the main Yahoo page where they are confronted with this:

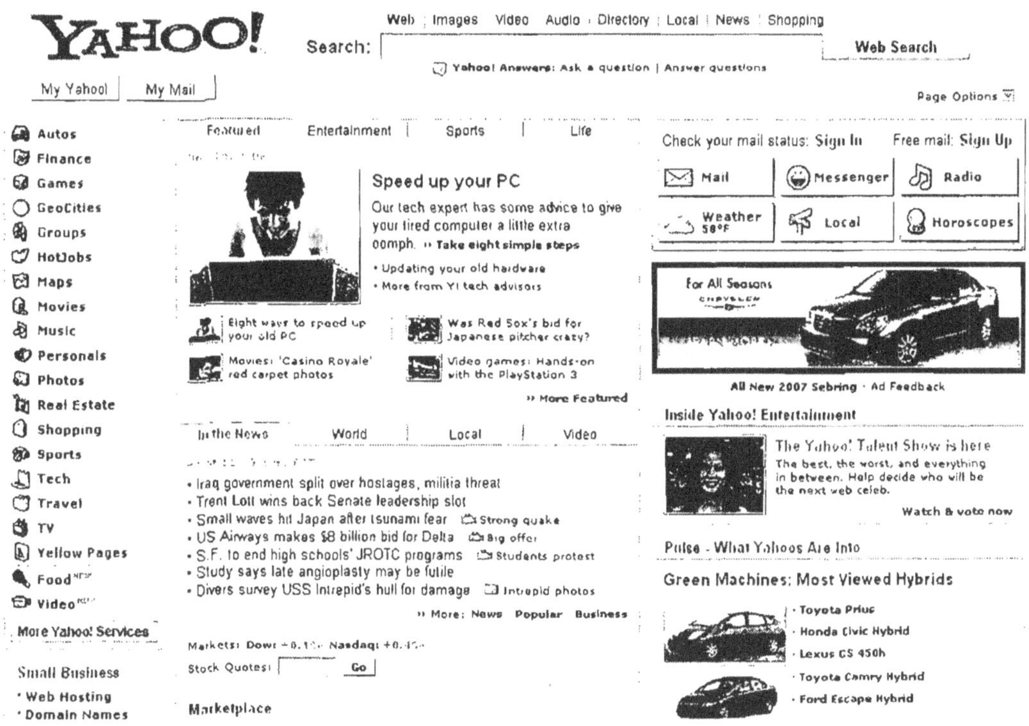
YAHOO!
Web Images Video Audio Directory Local News Shopping
Search:
Web Search
Yahoo! Answers: Ask a question | Answer questions
My Yahoo!
My Mail
Page Options
Autos
Finance
Games
GeoCities
Groups
HotJobs
Maps
Movies
Music
Personals
Photos
Real Estate
Shopping
Sports
Tech
Travel
TV
Yellow Pages
Food
Video
More Yahoo! Services
Small Business
Web Hosting
Domain Names
Featured
Entertainment
Sports
Life
Speed up your PC
Our tech expert has some advice to give your tired computer a little extra oomph. Take eight simple steps
Updating your old hardware
More from Y! tech advisors
Eight ways to speed up your old PC
Was Red Sox's bid for Japanese pitcher crazy?
Movies: 'Casino Royale' red carpet photos
Video games: Hands-on with the PlayStation 3
More Featured
In the News
World
Local
Video
Iraq government split over hostages, militia threat
Trent Lott wins back Senate leadership slot
Small waves hit Japan after tsunami fear
Strong quake
US Airways makes $8 billion bid for Delta
Big offer
S.F. to end high schools' JROTC programs
Students protest
Study says late angioplasty may be futile
Divers survey USS Intrepid's hull for damage
Intrepid photos
More: News Popular Business
Stock Quotes:
Go
Marketplace
Check your mail status: Sign In
Free mail: Sign Up
Mail
Messenger
Radio
Weather
Local
Horoscopes
For All Seasons
All New 2007 Sebring · Ad Feedback
Inside Yahoo! Entertainment
The Yahoo! Talent Show is here
The best, the worst, and everything in between. Help decide who will be the next web celeb.
Watch & vote now
Pulse - What Yahoos Are Into
Green Machines: Most Viewed Hybrids
Toyota Prius
Honda Civic Hybrid
Lexus GS 450h
Toyota Camry Hybrid
Ford Escape Hybrid

Here's Yahoo's dilemma: how does it compete with Google for searchers seeking a simple, clean interface while simultaneously retaining and attracting users who want "one stop shopping"? Thus far, more searchers are still going to Google first rather than muddling their way through the kind of mess you see above. Where Yahoo excels—and in my opinion beats Google—is in shopping and in finding local information. This is a fact Yahoo not only recognizes but also embraces. Says Ted Meisel, head of Yahoo's Overture division, "We never claimed it [Yahoo] was a better approach for doing research on 18th century Spain. But if you are trying to buy a power washer for your back deck, it's a pretty good way to find what you

---

[20] Steve Smith, "Search Wars: Google vs. Yahoo!," *MediaPost.com*, April 2004 Issue, <http://www.mediapost.com/dtls_dsp_mediamag.cfm?magID=245868> (registration required).

need."[21] That's fine for personal searches, but it does not help the searcher who is using the Internet for work-related, academic, or other types of research.

The future of search seems to be in fewer but more experienced and more commercially driven hands now than a decade ago. Certainly both the quantity and quality of search results are much better today. And there are other trends in search that are going to have a major impact on users, love them or hate them. Among these are greater ***personalization*** of search, an area in which Google, Yahoo, and Live Search are all vying for your attention. Then there is the concept of ***social networking***, through which Internet users with similar interests share their web knowledge and experience. ***Social bookmarking*** sites such as del.icio.us or digg and sharing software such as Stumbleupon are growing in popularity as individual users seek ways to help each other discover and propagate information.

There has also been a strong impetus towards more ***localized search*** for shopping, news, map directions, services, telephone lookups, and more. Yahoo initially outpaced Google in this area because it already owns an enormous warehouse of information about where its users live and work, shop and play. However, Google, Yahoo, Ask, and Live Search all moved strongly into the local and personalized search arena during 2006. Add to the mix all the other services search companies offer or plan to offer, such as Google's much ballyhooed and controversial foray into email with Gmail. The move toward greater personalization (likes and dislikes/interests/shopping/travel) and more services (especially email and tailored news) brings increased concerns about privacy and security. The more Yahoo, Google, Amazon, Microsoft, et al. know about us, the more they can serve up what we want.

But the more they know, the less control we have over our privacy and computer security. I am reminded of a scene from the film *Minority Report* in which the main character walks into a clothing store and, after his eye scan, the computer welcomes him by name, asks if he was happy with his previous purchase (which it details) and what he would like now. It doesn't take a lot of imagination to see how this technology can be abused. Everyone wants convenience but it is a virtual axiom of technology that every increase in convenience brings with it some decrease in privacy and, most likely, security. Now more than ever, the future of search is one that appears to be heading towards more personalization, more features, more options and, inevitably it seems, less privacy, less security, and fewer companies with the will, technological know-how, and financial resources to build and maintain search engines.

---

[21] Steven Levy, "All Eyes on Google," *Newsweek*, 29 March 2005, p. 54, <http://www.msnbc.msn.com/id/4570868/> (14 November 2006).

Understanding Search Engines

The best way to keep up to date with search engines in the US is to visit websites devoted to search and to read their newsletters. One of the oldest sites about search is Search Engine Watch. Although Search Engine Watch was originally designed for webmasters (by webmaster Danny Sullivan), it is a good resource for researchers who want and need in-depth information about the major English-language search services and some country specific engines. Search Engine Watch is also home to **Search Day**, noted search maven Chris Sherman's daily newsletter. While Search Day is kept current, Search Engine Watch now has many out of date pages.

Stepping into the breach is the superb **Pandia Search Central**, which offers current search news and an almost endless number of tips, tutorials, guides, and even its own search tools. Pandia has emerged as the premier site for news about and help with search.

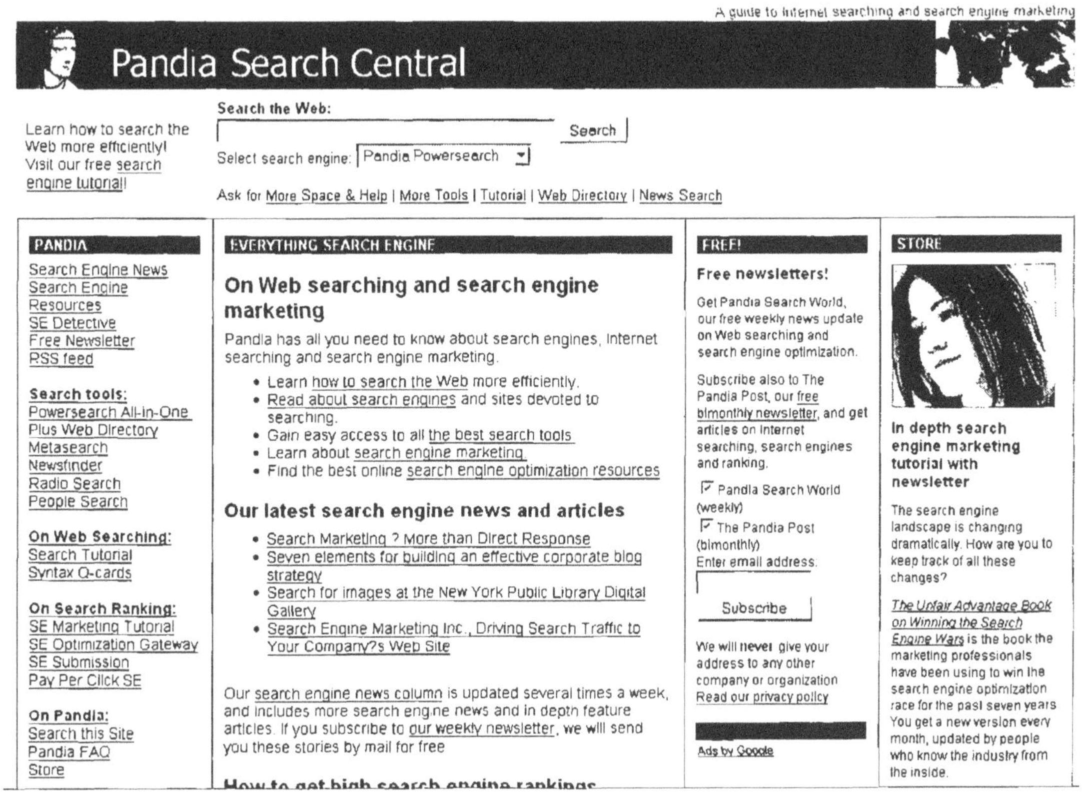
A guide to internet searching and search engine marketing
Pandia Search Central
Search the Web:
Search
Learn how to search the Web more efficiently! Visit our free search engine tutorial!
Select search engine: Pandia Powersearch
Ask for More Space & Help | More Tools | Tutorial | Web Directory | News Search
PANDIA
Search Engine News
Search Engine Resources
SE Detective
Free Newsletter
RSS feed
Search tools:
Powersearch All-in-One
Plus Web Directory
Metasearch
Newsfinder
Radio Search
People Search
On Web Searching:
Search Tutorial
Syntax Q-cards
On Search Ranking:
SE Marketing Tutorial
SE Optimization Gateway
SE Submission
Pay Per Click SE
On Pandia:
Search this Site
Pandia FAQ
Store
EVERYTHING SEARCH ENGINE
On Web searching and search engine marketing
Pandia has all you need to know about search engines, internet searching and search engine marketing.
• Learn how to search the Web more efficiently.
• Read about search engines and sites devoted to searching.
• Gain easy access to all the best search tools
• Learn about search engine marketing
• Find the best online search engine optimization resources
Our latest search engine news and articles
• Search Marketing ? More than Direct Response
• Seven elements for building an effective corporate blog strategy
• Search for images at the New York Public Library Digital Gallery
• Search Engine Marketing Inc., Driving Search Traffic to Your Company?s Web Site
Our search engine news column is updated several times a week, and includes more search engine news and in depth feature articles. If you subscribe to our weekly newsletter, we will send you these stories by mail for free
FREE!
Free newsletters!
Get Pandia Search World, our free weekly news update on Web searching and search engine optimization.
Subscribe also to The Pandia Post, our free bimonthly newsletter, and get articles on internet searching, search engines and ranking.
Pandia Search World (weekly)
The Pandia Post (bimonthly)
Enter email address:
Subscribe
We will never give your address to any other company or organization. Read our privacy policy
Ads by Google
STORE
In depth search engine marketing tutorial with newsletter
The search engine landscape is changing dramatically. How are you to keep track of all these changes?
The Unfair Advantage Book on Winning the Search Engine Wars is the book the marketing professionals have been using to win the search engine optimization race for the past seven years. You get a new version every month, updated by people who know the industry from the inside.

Other good web search sites include John Battelle's Searchblog, Philipp Lenssen's Google Blogoscoped (which covers much more than just Google), Gary Price's

Resource Shelf, Phil Bradley's Weblog, Greg Notess's Search Engine Showdown, as well as Web Master World and Web Search Guide. Among the best search engine-specific blogs are the Yahoo Search Blog, the Official Google Blog, Google Operating System, and Live Search Weblog.

The only thing predictable about search engines is how quickly and frequently they change not only their content but also their features. Because there are websites devoted to keeping up with the myriad changes, they are your best bet for staying on top of the ever-changing world of search tools.

Search News and Blogs

Google Operating System	http://googlesystem.blogspot.com/
John Battelle's Searchblog	http://battellemedia.com/
Live Search Weblog	http://blogs.msdn.com/msnsearch/default.aspx
Official Google Blog	http://googleblog.blogspot.com/
Pandia Search Central	http://pandia.com/
Philipp Lenssen's Google Blogoscoped	http://blog.outer-court.com/
Phil Bradley's Weblog	http://philbradley.typepad.com/phil_bradleys_weblog/
Research Buzz	http://www.researchbuzz.com/
Resource Shelf	http://www.resourceshelf.com/
Search Day	http://searchenginewatch.com/searchday/
Search Engine Showdown	http://www.searchengineshowdown.com/
Search Engine Showdown Reviews	http://www.searchengineshowdown.com/reviews/
Search Engine Watch	http://searchenginewatch.com/
Search Engine Watch Web Searching Tips	http://www.searchenginewatch.com/facts/index.html
Web Master World	http://www.webmasterworld.com/
Web Search Guide	http://www.websearchguide.ca/
Search Engine Watch Blog	http://blog.searchenginewatch.com/blog/
Yahoo Search Blog	http://www.ysearchblog.com/

Web Tip

Browsers assume the prefix "http://" unless you tell them otherwise, which means you do not need to type "http://"— just type the url (address).

Search Engine Basics

A search engine comprises three basic parts:

1. The **spider/robot/crawler** is software that "visits" sites on the Internet (each search engine does this differently). The spider reads what is there, follows links at the site, and ultimately brings all that data back to:

2. The **search engine index**, catalog, or database, where everything the spider found is stored;

3. The **search engine software** that actually sifts through everything in the index to find matches and then ranks or sorts them into a list of results or hits.

Important points to consider about search engines:

- Spiders are programmed to return to websites on a regular basis, but the ***time interval varies widely*** from engine to engine. Monthly or better is considered "fresh."
- When you use a search engine, you are searching the index or database, not the web pages themselves. This is important to remember because ***no search engine operates in "real time."***
- Spiders do not index all web pages they find, including pages that employ the "**Robots Exclusion Protocol**" or the "**Robots META tag**." The first of these mechanisms is a special file website administrators use to indicate which parts of the site should not be visited by the robot or spider. The second is a special HTML metatag that may be inserted by a web page author to indicate if the page may be indexed or analyzed for links. Not every robot/spider

respects these mechanisms. Password protection, firewalls, and other measures will generally keep spiders from crawling a website and indexing it.

The Web Robots Pages ◂

Robots Exclusion

Sometimes people find they have been indexed by an indexing robot, or that a resource discovery robot has visited part of a site that for some reason shouldn't be visited by robots.

In recognition of this problem, many Web Robots offer facilities for Web site administrators and content providers to limit what the robot does. This is achieved through two mechanisms:

The Robots Exclusion Protocol	A Web site administrator can indicate which parts of the site should not be vistsed by a robot, by providing a specially formatted file on their site, in http://.../robots.txt.
The Robots META tag	A Web author can indicate if a page may or may not be indexed, or analysed for links, through the use of a special HTML META tag.

The remainder of this pages provides full details on these facilities.

Note that these methods rely on cooperation from the Robot, and are by no means guaranteed to work for every Robot. If you need stronger protection from robots and other agents, you should use alternative methods such as password protection.

Robots Exclusion Page http://www.robotstxt.org/wc/exclusion.html

Not every search engine has its own proprietary search program but instead relies upon another company's search service for its results. Most of these ***strategic alliances*** now involve Yahoo, Google, and Windows Live Search. All these partnerships are subject to change without notice; for more on these strategic alliances, see:

Search Engine Alliances http://searchenginewatch.com/reports/alliances.html

Knowing that Yahoo, for example, is the search tool behind a search engine can save you time because you can be pretty sure that using AltaVista will get you similar (although not identical) results to the other search engines also powered by Yahoo. It is critical to remember that ***each service powered by a particular search engine produces different results*** even though they may all use the same core database. Why is this? Because the search interfaces have their own algorithms that decide how queries are run, how results are returned, or even if they query the entire database (most do not). In short, go to the primary search engine—Google, Yahoo, or Live Search for best results.

A Word About Browsers: Internet Explorer and Mozilla Firefox

Two years ago I declared that the "browser wars" were for all intents and purposes over and Microsoft's Internet Explorer (IE) had won. IE still commands more than 90 percent of the world's browser market, and AOL abandoned Netscape's Navigator/Communicator in mid-2003. However, during 2004, Mozilla browsers experienced a resurrection thanks largely to user frustration.

> ### *Caveat Browser*
>
> Alexa **and Smart Browsing technology are very controversial because of their invasion of privacy implications. For more information, take a look at the article:**
> **"What's Related? Everything But Your Privacy"**
>
> Curtin, M., Ellison, G., Monroe, D., "What's Related? Everything But Your Privacy," 7 Oct 1998, Revision: 1.5, <http://www.interhack.net/pubs/whatsrelated/> (24 October 2006)

Because of Internet Explorer's continued dominance of the browser market and, more importantly, because it is the standard browser for many *Untangling the Web* readers, I will focus my attention on Internet Explorer.

Nonetheless, all browsers have advantages and drawbacks. ***I still recommend you configure two browsers, both Internet Explorer and Mozilla Firefox.*** Both types of browsers allow you to make a number of decisions that affect your privacy and security while browsing. Also, both browsers have become much more customizable with each new release, allowing every user to select and save his own preferences for everything from fonts to what will appear on the toolbar. Be sure to familiarize yourself with the many evolving features of your browser(s). The Microsoft and Mozilla websites have extensive information and documentation about their browsers. At the Mozilla site you can download and install the highly regarded Firefox browser as well as other free software, such as the Thunderbird email program.

In October 2006, both Microsoft and Mozilla introduced new versions of their browsers: Internet Explorer 7 (IE7) and Firefox 2. Microsoft, which had owned upwards of 90 percent of the browser market until Firefox took off a couple of years ago, recognized it has a genuine competitor on its hands and made significant changes and improvements to its browser to try to bring some Firefox users back into the fold. Will it work? *PC World* offered an excellent comparison of IE7 and Firefox 2.[22] While Firefox 2's changes are mostly refinements of already existing features with no change in the browser's look and feel, IE7 marks a major overhaul since IE6 was released way back in 2001.

Among the changes to **Internet Explorer 7** are tabbed browsing, integrated searching, RSS newsfeed support, and an antiphishing tool. The most noticeable change is IE7's look and feel, which is designed to resemble Microsoft's new operating system, Vista. Probably the most obvious and popular addition to IE7 is tabbed browsing, something Firefox already offered. Also, IE7 has a built-in search box, which lets users search from anywhere without having to go to the search engine's home page. Google and other search engines had successfully lobbied Microsoft not to make Live Search the default search service, so you can pick your search engine.

The other major change is invisible: improved security features designed to cope with the almost endless number of vulnerabilities that have afflicted IE6.[23] The most prominent of these security upgrades is one shared with Firefox: an "antiphishing" tool that works by warning users that a website they are about to visit may be fake and redirects them away from the page unless they actively choose to go to it. The other major new IE7 security feature is something called Protected Mode, which prevents a website from changing a computer's files or settings. However, Protected Mode will not work with any Windows operating system except Vista, which is due out next year. Also, one of IE's major appeals had been its universality, that is, it would work with most websites. The security features in IE7 mean that some sites that could be viewed in earlier versions of IE cannot be viewed in IE7, undermining one reason many people still continued to use the Microsoft browser.

Firefox 2 is another in a long line of gradual updates. This version adds a spell checker, a system for suggesting popular search terms, and an option to pick up where you left off after a crash. Firefox 2 also upgrades the RSS newsfeed so that now, if you click on the feed itself, instead of seeing the usual XML gibberish, Firefox 2 will parse the raw feed into something readable and also subscribe to a feed using one of a numerous (but not all) newsreaders.

What is the bottom line? Firefox users should upgrade to version 2; it will be easy and pain free. IE6 users probably should wait a while before downloading IE7 to let

---

[22] Erik Larkin, "Radically New IE7 or Updated Mozilla Firefox 2--Which Browser is Better?" PC World, 18 October 2006, <http://www.pcworld.com/printable/article/id,127309/printable.html> (24 October 2006).

[23] Not 24 hours after its release and the first vulnerability was detected in IE7. Of course, it also affects IE6, but this is embarrassing for Microsoft given that the company has touted the security of IE7 over its predecessor.
<http://secunia.com/Internet_Explorer_Arbitrary_Content_Disclosure_Vulnerability_Test/>

early adopters find the inevitable bugs that Microsoft will have to fix. Frankly, after five years, you would think Microsoft could do better than come up with a browser that basically mimics the best features of Firefox and its other (much smaller) competitors. This looks mostly like catch-up and very little like innovation.

If you are going to use Netscape, another Mozilla-based browser, I do not recommend using Netscape 8x because it has many reported problems. Stick with either Netscape 7.1x or 7.2x. Also, if you prefer a streamlined version of Netscape 7x without all the annoying "extras," I can recommend one from Sillydog (silly name, great tool). "Netscape 7.1 is based on Mozilla 1.4. Both applications share almost identical features, such as tabbed browsing, custom keywords, and Sidebar. Exceptions are additions of proprietary features such as the support for Netscape WebMail and AOL mail."[24] Netscape 7.2 is based on Mozilla 1.7.2. "In addition to the technologies that Netscape 7.2 shares with Mozilla 1.7.2, it includes additional features such as a number of installed plugins, support for Windows Media Player Active X control which are not available in Mozilla."[25]

Microsoft Internet Explorer	http://www.microsoft.com/windows/ie/default.htm
Mozilla Firefox	http://www.mozilla.com/firefox/
Netscape 7.1 Streamline	http://sillydog.org/narchive/sd/71.html
Netscape Archive (7.1 or 7.2)	http://browser.netscape.com/ns8/download/archive.jsp

[24] Mozilla FAQ, <http://www.mozilla.org/start/1.4/faq/general.html#ns7> (14 November 2006).

[25] Sillydog.org Browser Archive, 31 October 2005, <http://sillydog.org/narchive/full67.php> (24 October 2006).

What the heck are "cookies"?

Cookies are text placed on your computer's hard disk (yes!) by a website in order to remember something about you. For example, a site may set a cookie that enables you to reenter without logging in or customize its pages based on the type of browser you're using. Cookies remain controversial (more later).

The Great Internet Search-Offs

Over the last decade, the inevitable "search offs" have become commonplace (both Internet vs. traditional researcher and Internet researchers against each other). Some of the findings of these "contests" provide insight into how search engines work.

1. Most search-offs and wide-ranging studies continue to find surprisingly little overlap among major search engines, so ***use more than one search engine*** as a general rule.

2. The Internet is now being widely used for "serious" research, which means higher quality, more reliable information on the web. But, as with any research source, you must ***weigh the validity, accuracy, currency, and overall quality*** of the information before using it.

3. Search engines rely on statistical interfaces, concept-based search mechanisms, or link analysis to return and rank hits; using **boolean expressions**[26] usually interferes with or defeats these statistical approaches. In general, ***do not use boolean queries*** unless you know exactly what you are looking for and are very comfortable with that search engine's boolean rules (no, they are not all the same; for example, you may have to use CAPS for all operators). Also, many search engines do not correctly process nested boolean queries (boolean searches with parentheses).

4. Be aware that search engines are giving ***more weight to popular and/or pay-for-placement web pages.*** In fact, most search engines use services to determine which are the most visited, and therefore most popular, websites and return them at the top of the results list. This is a strategic move away from the traditional "words on a page" ranking system. Trustworthy search engines will clearly indicate which hits are paid entries.

[26] The term "boolean," often encountered when doing searchers on the web (and frequently spelled "Boolean"), refers to a system of logical thought developed by the English mathematician and computer pioneer George Boole (1815-64). In boolean searching, an "and" operator between two words or other values (for example, "pear AND apple") means one is searching for documents containing both the words or values, not just one of them. An "or" operator between two words or other values (for example, "pear OR apple") means one is searching for documents containing either of the words. "Boolean," SearchSMB.com, <http://searchsmb.techtarget.com/sDefinition/0,290660,sid44_gci211695,00.html> (14 November 2006).

5. ***Learn the search syntax*** of the search engines you use (never assume). Most search engines use double quotes ("") to enclose a phrase and the plus + and minus - keys to indicate "must include" and "must exclude" respectively. But these are by no means universal rules (especially when using *international* or *metasearch* engines).

6. ***The default operator for all major US search engines is now AND.*** As of February 2002, no major search engine used OR as its default operator. However, most search engines will let you use an OR in the simple search box: Yahoo and Google permit OR searches in the simple search box, but you must capitalize the OR.

7. Keep in mind that because ***HTML does not have a "date" tag***, ***"date"*** can mean many things: creation date; the last modified date for the page; or the date search engine found the page. *I do not recommend searching by date except when using weblog, news, or newsgroup search engines.*

Understanding ***statistical interfaces*** is important, especially for researchers used to boolean and other non-statistical query languages. Most search engines use statistical interfaces. The search engine assigns relative weights to each **search term**, depending on:

- its rarity in their database
- how frequently the term occurs on the webpage
- whether or not the term appears in the url
- how close to the top of the page the term appears
- (sometimes) whether or not the term appears in the metatags.

When you query the database, the search engine adds up all the weights that match your query terms and returns the documents with the highest weight first. ***Each search engine has its own algorithm*** for assigning weights, and they tweak these frequently. In general, rare, unusual terms are easier to find than common ones because of the weighting system.

However, remember that "popularity" measured by various means often trumps any statistical interface.

Types of Search Tools

Before delving into the intricacies of search engines, let's look at some other ways of finding information on the web. Search engines are not the only and often not even the best way to access information on the Internet.

Web Directories/Subject Guides/Portals

Web directories are organized subject catalogues that allow the web searcher to browse through lists of sites by subject in search of relevant information. **Yahoo**, **Galaxy**, **Google Directory, Lycos**, and the **Open Directory** are select lists of links to pertinent websites. Directories were once viewed as the future of the Internet because they could sift through the mountains of information and millions of websites to offer only the best and most relevant. However, directories have truly fallen by the wayside over the past several years with the rise of Google and, even more importantly, wikis in general and Wikipedia in particular. Directories continue to recede in importance and value to researchers as they are increasingly replaced by better alternatives, including Custom Search, by which a voluntary community of searchers shares expertise to create more focused searches with more relevant results. The reason for the decline of directories is obvious: directories are simply too manpower intensive and expensive to keep up with the ever-changing and expanding web. I would say at this point directories, while not dead, are probably moribund.

Directories rely on people to create their listings

Obviously, this is a much more labor-intensive business than operating a search engine robot. Websites indexed in a directory are either described/evaluated by editors/reviewers or rely on descriptions provided by web page owners who may pay for placement in a directory. When you search a directory, the only retrievals will come from those descriptions, so keep this in mind. Although directories give you a much more limited view of the web, directories do have their own utility. Most directories also have a backup search that provides responses to queries that don't match anything in the directory listings.

Directories may produce more relevant results

Subject guide databases are always smaller than those of search engines, which means that the number of hits returned tends to be smaller as well. On the bright side, this means the results directories produce are often more relevant. For example, while a search engine typically indexes every page of a given website, a subject guide is more likely to provide a link only to the site's home page. For this reason, they ***lend themselves best to searching for information about a general subject***, rather than for a specific piece of information.

Yahoo still has the best-known subject guide/directory and can be a good starting place for research, even on technical subjects. Yahoo used to list links alphabetically, but once Google came along with its ranked list of sites, Yahoo started offering most popular sites first before going to its alphabetical list. However, Yahoo's directory has suffered in recent years as the **Google Directory** has steadily improved. Google gets its directory data free in the form of the **Open Directory Project**.

You may not recognize the Open Directory Project by this name, but you have probably used it. The ODP is the directory behind the Google Directory, AOL Search, Yahoo Directory, and many others. The ODP "is the largest, most comprehensive human-edited directory of the Web. It is constructed and maintained by a vast, global community of volunteer editors."

Galaxy is definitely worth a look because it was designed for and by "professionals," so it has a bent toward business, technology, and science that other directories lack. You may search either the Galaxy collection or the web using their proprietary search engine. **Best of the Web** started life in 1994 as a web awards site and is now a full-fledged directory.

Many more specialized directories are discussed under the "Invisible" Internet.

Best of the Web	http://botw.org/default.aspx
Galaxy	http://www.galaxy.com/
Google Directory	http://directory.google.com/
Open Directory	http://dmoz.org/
Yahoo Directory	http://dir.yahoo.com/

Metasearch Sites

The growth in the number of search engines has led to the creation of "meta" search sites. These services allow you to ***invoke several or even many search engines simultaneously***. These metasearchers may do a more thorough job of sifting through the net for your topic than any single search engine. If you are new to using search engines, these are a great way to do a very broad search, while familiarizing yourself with the popular engines and how they respond. But metasearch engines inevitably ***lack the flexibility*** of individual search tools.

It is important to note that many metasearch engines do not employ some of the best search engines, such as Google and Yahoo. Also, my biggest complaint about metasearch engines is that ***they perform shallow searches***, usually only retrieving the top ten or so hits from a site, which is far too few to be comprehensive or truly representative of what is "out there."

However, metasearch engines do serve a purpose. If you are unsure if a term will be found anywhere on the web, try a metasearch engine first to ***"size" the problem***: you may get zero hits with a dozen search engines (you've got a problem) or you may get a half-dozen right-on-the-money hits right off the bat.

Clusty http://clusty.com/

Vivisimo, in my opinion the best free metasearch tool available, opened a new search site—Clusty—in 2004 and then made Clusty its search home in 2006. Fundamentally, Vivisimo and Clusty are the same, but Clusty adds options for news, image, Wikipedia, government, and blog searches.

The Vivisimo technology behind Clusty is unique because it employs its own ***clustering engine***, software that organizes unstructured information into hierarchical folders. Clusty offers clustered results of web, news, and certain specialty searches. The Clusty default is to search the web using Live Search, Gigablast, Ask, Wikipedia, and the Open Directory.

Clusty is especially useful for searching ambiguous terms, such as *cardinal*, because it clusters them by logical categories, as shown below.

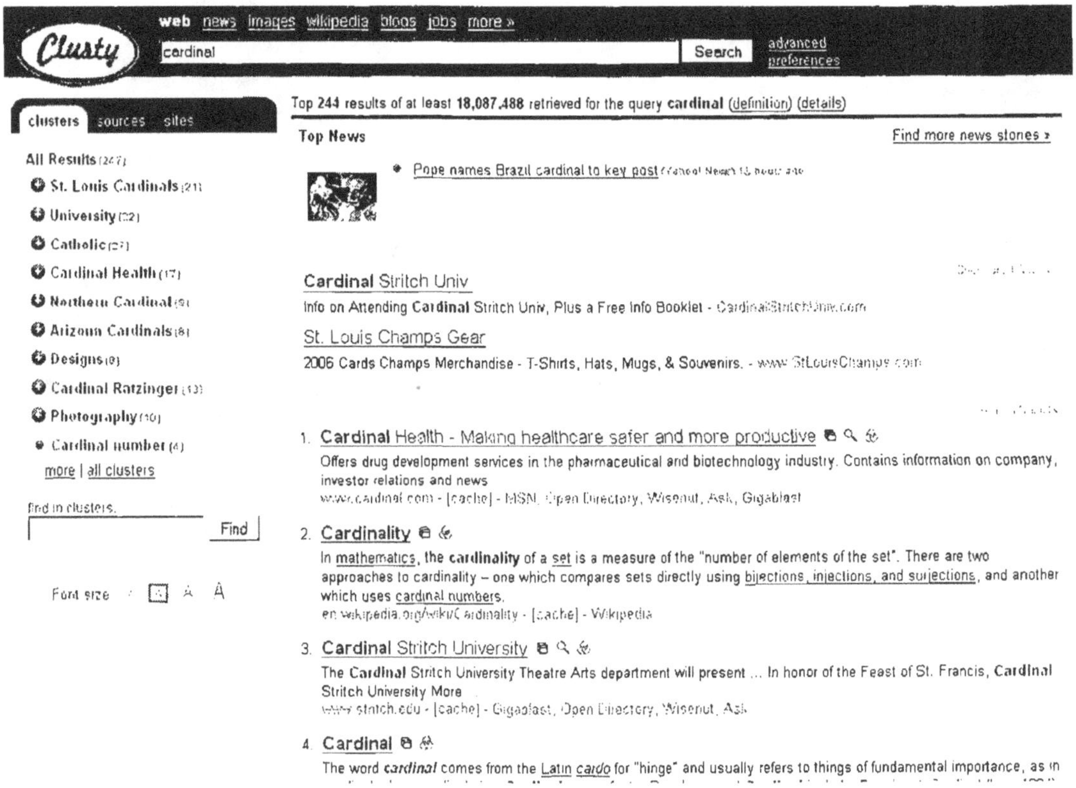
Clusty
web news images wikipedia blogs jobs more »
cardinal
Search
advanced
preferences
clusters sources sites
Top 244 results of at least 18,087,488 retrieved for the query cardinal (definition) (details)
Top News
Find more news stories »
Pope names Brazil cardinal to key post
All Results
St. Louis Cardinals
University
Catholic
Cardinal Health
Northern Cardinal
Arizona Cardinals
Designs
Cardinal Ratzinger
Photography
Cardinal number
more | all clusters
Find in clusters:
Find
Font size
Cardinal Stritch Univ
Info on Attending Cardinal Stritch Univ, Plus a Free Info Booklet
St. Louis Champs Gear
2006 Cards Champs Merchandise - T-Shirts, Hats, Mugs, & Souvenirs.
1. Cardinal Health - Making healthcare safer and more productive
Offers drug development services in the pharmaceutical and biotechnology industry. Contains information on company, investor relations and news
2. Cardinality
In mathematics, the cardinality of a set is a measure of the "number of elements of the set". There are two approaches to cardinality – one which compares sets directly using bijections, injections, and surjections, and another which uses cardinal numbers.
en.wikipedia.org/wiki/Cardinality - [cache] - Wikipedia
3. Cardinal Stritch University
The Cardinal Stritch University Theatre Arts department will present ... In honor of the Feast of St. Francis, Cardinal Stritch University More
4. Cardinal
The word cardinal comes from the Latin cardo for "hinge" and usually refers to things of fundamental importance, as in

Also, Clusty lets users look at the sources of the search results and types of sites (e.g., .com, .gov). Clusty has a unique feature that allows users to search inside clusters. In this example, the original search was [iran] and the "find in clusters" search was [nuclear]. Here are the results of this recursive search looking at the sources of data:

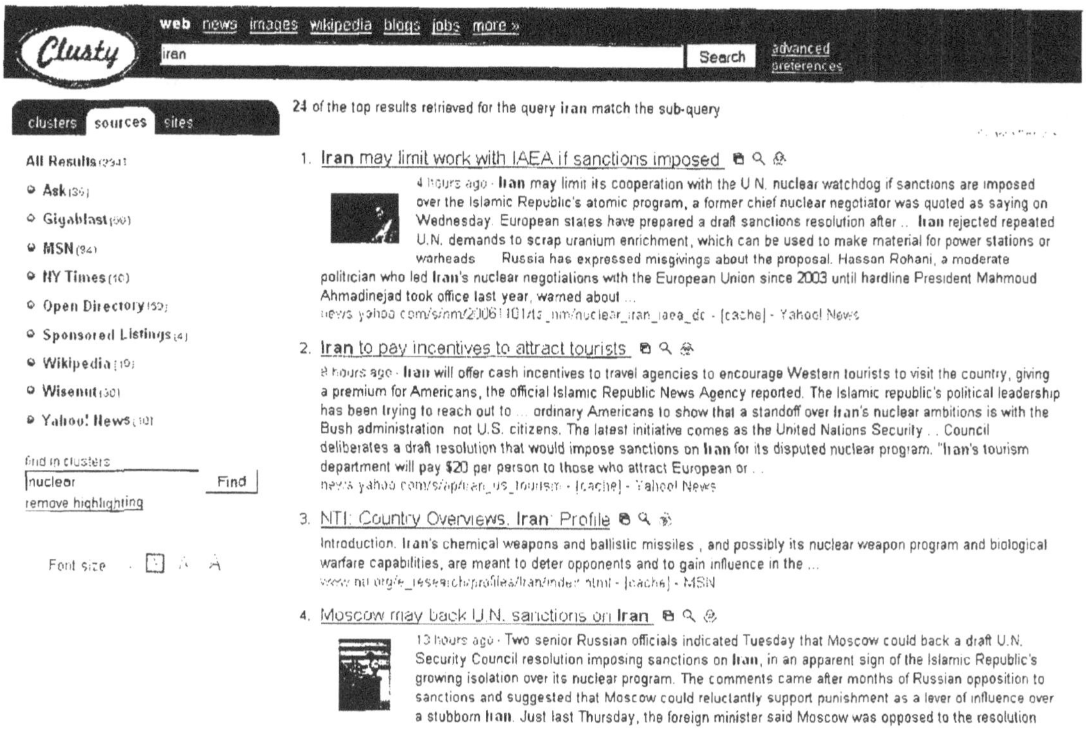
Clusty
web news images wikipedia blogs jobs more »
iran
Search
advanced
preferences
clusters sources sites
24 of the top results retrieved for the query iran match the sub-query
All Results (234)
Ask (36)
Gigablast (90)
MSN (94)
NY Times (10)
Open Directory (52)
Sponsored Listings (4)
Wikipedia (10)
Wisenut (30)
Yahoo! News (10)
find in clusters
nuclear
Find
remove highlighting
Font size
1. Iran may limit work with IAEA if sanctions imposed
4 hours ago · Iran may limit its cooperation with the U.N. nuclear watchdog if sanctions are imposed over the Islamic Republic's atomic program, a former chief nuclear negotiator was quoted as saying on Wednesday. European states have prepared a draft sanctions resolution after ... Iran rejected repeated U.N. demands to scrap uranium enrichment, which can be used to make material for power stations or warheads. Russia has expressed misgivings about the proposal. Hassan Rohani, a moderate politician who led Iran's nuclear negotiations with the European Union since 2003 until hardline President Mahmoud Ahmadinejad took office last year, warned about ...
2. Iran to pay incentives to attract tourists
8 hours ago · Iran will offer cash incentives to travel agencies to encourage Western tourists to visit the country, giving a premium for Americans, the official Islamic Republic News Agency reported. The Islamic republic's political leadership has been trying to reach out to ... ordinary Americans to show that a standoff over Iran's nuclear ambitions is with the Bush administration not U.S. citizens. The latest initiative comes as the United Nations Security ... Council deliberates a draft resolution that would impose sanctions on Iran for its disputed nuclear program. "Iran's tourism department will pay $20 per person to those who attract European or ...
3. NTI: Country Overviews: Iran: Profile
Introduction. Iran's chemical weapons and ballistic missiles, and possibly its nuclear weapon program and biological warfare capabilities, are meant to deter opponents and to gain influence in the ...
4. Moscow may back U.N. sanctions on Iran
13 hours ago · Two senior Russian officials indicated Tuesday that Moscow could back a draft U.N. Security Council resolution imposing sanctions on Iran, in an apparent sign of the Islamic Republic's growing isolation over its nuclear program. The comments came after months of Russian opposition to sanctions and suggested that Moscow could reluctantly support punishment as a lever of influence over a stubborn Iran. Just last Thursday, the foreign minister said Moscow was opposed to the resolution

For news, Clusty searches the *New York Times*, Associated Press, Reuters, and Yahoo News (which subsumes a huge number of sources). One of the best features of Clusty news search is the ability to toggle among clustered results, sources, and sections (such as business, health, tech, science).

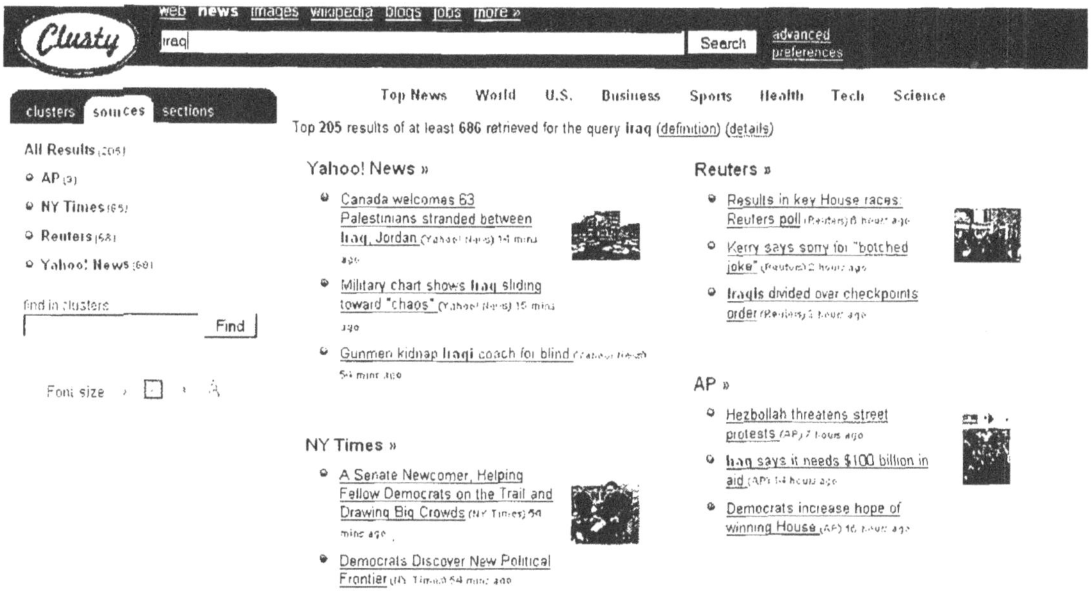
Clusty
web news images wikipedia blogs jobs more »
iraq
Search
advanced
preferences
clusters sources sections
Top News World U.S. Business Sports Health Tech Science
Top 205 results of at least 686 retrieved for the query iraq (definition) (details)
All Results (205)
AP (3)
NY Times (65)
Reuters (68)
Yahoo! News (69)
find in clusters
Find
Font size
Yahoo! News »
Canada welcomes 63 Palestinians stranded between Iraq, Jordan
Military chart shows Iraq sliding toward "chaos"
Gunmen kidnap Iraqi coach for blind
NY Times »
A Senate Newcomer, Helping Fellow Democrats on the Trail and Drawing Big Crowds
Democrats Discover New Political Frontier
Reuters »
Results in key House races: Reuters poll
Kerry says sorry for "botched joke"
Iraqis divided over checkpoints order
AP »
Hezbollah threatens street protests
Iraq says it needs $100 billion in aid
Democrats increase hope of winning House

Clusty also provides a number of advanced search options and preferences, including the option to add your own customized tabs to the main search page.

Clusty stands out as one of—if not indeed—the best metasearch tools available for free and without registration on the Internet. When clustering works (and the Vivisimo technology was independently rated as accurate 90 percent of the time), it offers advantages for automatically grouping huge amounts of information logically. Because there is no human intervention, Vivisimo's clustering algorithm "also helps in discovering new areas of subject development, avoiding the 'mummy's curse,' in which human catalogers have to recognize a term before approving it for usage and then leaving the earlier material using the term un-indexed and irretrievable by that term as an authorized descriptor or metatag."[27]

Jux2 http://www.jux2.com/

Jux2 lets users query three search engines—Google, Yahoo, and Live Search (still referred to as MSN Search)—and then shows you:

1. The Best Results from all three search engines and the total hits for each.
2. What only Google found and what is missing from Google.
3. What only Yahoo found and what is missing from Yahoo.
4. What only Live/MSN Search found and what is missing from Live/MSN Search.

---

[27] Barbara Quint, "Vivísimo Clustering Chosen to Enhance Searching at Institute of Physics Publishing Site," *Infotoday*, 25 March 2002, <http://www.infotoday.com/newsbreaks/nb020325-2.htm> (14 November 2006).

jux
A Better Way to Search

| atropos | | search | advanced search
preferences |

| | Best Results | Compare Google's
Results | Compare Yahoo's
Results | Compare MSN's
Results |

What only **Google** found

ATROPOS : The deity from Greek Mythology

Greek Mythology. Meet the Classical Gods of Ancient Greece. ATROPOS: Oldest of the three FATES. She's the one who cuts the Thread of Life with her fatal ...

http://www.godchecker.com/pantheon/greek-mythology.php?deity=ATROPOS [#6 on Google]

www.katsudon.net/gwfic/**atropos**.html

http://www.katsudon.net/gwfic/atropos.html [#7 on Google]

Amazon.com: Hornblower and the **Atropos** (Hornblower Saga (Paperback ...

Amazon.com: Hornblower and the **Atropos** (Hornblower Saga (Paperback)): Books: CS Forester by CS Forester.

http://www.amazon.com/exec/obidos/tg/detail/-/0316289299?v=glance [#9 on Google]

What's missing from **Google**

http://en.wikipedia.org/wiki/**Atropos**

From Wikipedia, the free encyclopedia. **Atropos** is also a British entomological journal - see **Atropos** (journal). In Greek mythology, **Atropos** was the third of the Moirae. ... It was **Atropos** who chose the mechanism of death and ended the life of each mortal by cutting their thread ...

http://en.wikipedia.org/wiki/atropos [#3 on Yahoo!, #3 on MSN]

Blogger: User Profile: **Atropos**

Atropos Gender: female Industry: Fashion Occupation: Model Location: Ohio : United States About Me I'm a 22 year old female who enjoys exploring deviant fashion as a form of self expression

http://www.blogger.com/profile/4664126 [#5 on MSN]

I believe you will be as surprised as I was to see how little overlap there often is among the "big three" search engines.

Dogpile   http://www.dogpile.com/

Dogpile, despite its name, is a good metasearch engine. Dogpile includes Live Search results, along with those from Google, Yahoo, and Ask Jeeves. This is, of course, very good news because Dogpile is now drawing from all the major US-based search engines with the exception of Gigablast. It also searches smaller or lesser-known search engines and directories, including the MIVA (formally FindWhat), LookSmart, Ask, About, and more. Interestingly, the European version's name is Webfetch because of "unfortunate associations" between Dogpile and manure.

Mamma   http://www.mamma.com/

Mamma, the "Mother of All Search Engines," might just be exaggerating a wee bit. Mamma offers web, news, image, and yellow and white page search options.

Search engines queried are Ask, Wisenut, Gigablast, and Entireweb (a serious misnomer) and directories queried are Open Directory, About, Business.com, and two pay-per-click sources.

The Pandia Metasearch Engine http://www.pandia.com/metasearch/index.html

The famed search guide site, Pandia, offers its own excellent metasearch engine. The Pandia metasearch engine "collects and sorts the hits, takes out duplicates, and presents the end result in a simple format. "The first results you'll see are from what Pandia describes as the "essential search engines and directories," which include Google, Yahoo, HotBot and Wisenut. Strangely, Pandia continues to list AlltheWeb (Fast) and AltaVista as search engines while they acknowledge at other places on their site that Yahoo subsumed both engines. Still, this is a very good metasearch site.

More metasearch sites:

Ithaki http://www.ithaki.net/indexu.htm

IxQuick http://www.ixquick.com/

Metacrawler http://www.metacrawler.com/

Search.com http://www.search.com/

Surfwax http://www.surfwax.com/

Open Directory's List of Metasearch Sites
http://dmoz.org/Computers/Internet/Searching/Metasearch/

Megasearch Sites

Megasearch sites simply store several search engines under one roof, but you have to do the searches one search engine at a time. They are becoming more sophisticated and better as time passes, serving as good entry points for finding and evaluating search engines. They are especially useful for ***locating international search engines***.

All Search Engines http://www.allsearchengines.com/

Find It Quick http://www.quickfindit.com/Search_Engines/

Search—22 http://www.search-22.com/

SearchEzee http://www.searchezee.com/search.shtml

Types of Searches and the Best Ways to Handle Them

The first thing to ask yourself is the one question a lot of people never consider: is the Internet the best place to start? In general, the Internet has become so good at ***answering factual questions***—the kinds of things you find in an almanac, an encyclopedia, or a phone book—that it is now usually better in terms of speed, timeliness, and accuracy than other resources. For example, if I need to know the world's largest hydroelectric plants, I can open an almanac and look up this information or I can type [world's largest hydroelectric plants] into Google, Yahoo, or Live Search, where the first result links me to a page at **Information Please.com** that contains the answer to the question.

Still, compared to traditional library-type resources, the Internet may be:

- slower (though this is changing with new technologies).
- less reliable (large amounts of bad data in among the good).
- disorganized (a library with all the books on the floor).
- frustrating (lots of "broken" links).
- hard to use (generally poor search tools and too much data to sift through).
- risky because of growing privacy and security threats.

This being said, why do we need to use to the Internet? Because:

- it has almost **unlimited amount** of data (also a minus...too much of a good thing and way too much of the bad).
- the data tend to be **current.**
- it offers **multimedia** (video, audio, charts, tables, illustrations).
- it allows the individual to do much more of his **own research.**
- it is relatively **inexpensive** (at least in some countries).
- most importantly, it contains a vast amount of **unique** information.

You've thought through your research question and decided to use the Internet to find information either because you've already tried traditional sources without success or you believe the Internet is your best option. You're sitting in front of your terminal, you've logged onto the Internet and you're staring at a blank screen. Now what? Let's start with a (relatively) easy type of search. You need to find **general information about a fairly broad topic.**

Let's say you need to research a broad topic unfamiliar to you, for example, Java. The best approach may not be to type *java* into a search engine. Why? Because you'll probably get millions of hits, and the first ones may be to commercial sites trying to sell you something relating to Java and will undoubtedly also include other meanings of Java, such as Indonesia and coffee. If you are looking for general information on a topic, **wikis, specialized (vertical) search engines**, and **virtual libraries** are often better starting points for researching general or broad topics than big search engines.

The single biggest mistake searchers make is using the wrong search tool. For example, search engines are generally not the best tools for finding current news (use a news search engine), for researching broad topics (use a specialty directory or virtual library), or for performing specialized searches such as scientific research (use a specialty search engine). That's why the number one rule for web research is:

Rule One

Use the right tool for the job.

Let's go back to the Java example where you want to find general information on the web about Java programming. Start with the **Yahoo** directory and see what categories it offers on Java. You can ignore the sponsored results and the categories about Indonesia, classic arcade games, and commercial Java services. Instead, your best bet is Programming Languages > Java:

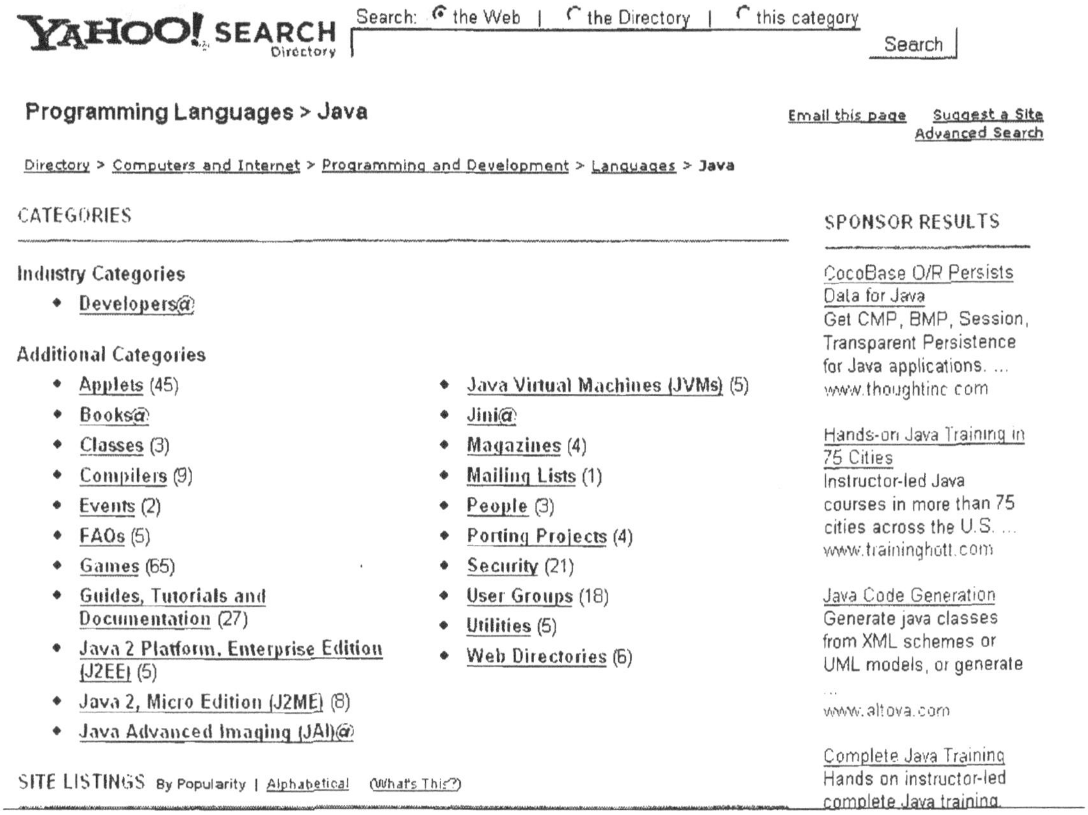
YAHOO! SEARCH Directory
Search: the Web | the Directory | this category
Search
Programming Languages > Java
Email this page Suggest a Site
Advanced Search
Directory > Computers and Internet > Programming and Development > Languages > Java
CATEGORIES
Industry Categories
Developers@
Additional Categories
Applets (45)
Books@
Classes (3)
Compilers (9)
Events (2)
FAQs (5)
Games (65)
Guides, Tutorials and Documentation (27)
Java 2 Platform, Enterprise Edition (J2EE) (5)
Java 2, Micro Edition (J2ME) (8)
Java Advanced Imaging (JAI)@
Java Virtual Machines (JVMs) (5)
Jini@
Magazines (4)
Mailing Lists (1)
People (3)
Porting Projects (4)
Security (21)
User Groups (18)
Utilities (5)
Web Directories (6)
SITE LISTINGS By Popularity | Alphabetical (What's This?)
SPONSOR RESULTS
CocoBase O/R Persists Data for Java
Get CMP, BMP, Session, Transparent Persistence for Java applications. ...
www.thoughtinc.com
Hands-on Java Training in 75 Cities
Instructor-led Java courses in more than 75 cities across the U.S. ...
www.traininghott.com
Java Code Generation
Generate java classes from XML schemes or UML models, or generate ...
www.altova.com
Complete Java Training
Hands on instructor-led complete Java training.

Computers_and_Internet/Programming_and_Development/Languages/Java/

Right there on one page is a wealth of promising links to documentation, reference, tutorials, news, downloads, articles, etc., and to the most lucrative resource of all, the **metaguide**. In this case, take a look at Java Boutique, which is a collection of useful Java information, news, forums, and more collected in one convenient location.

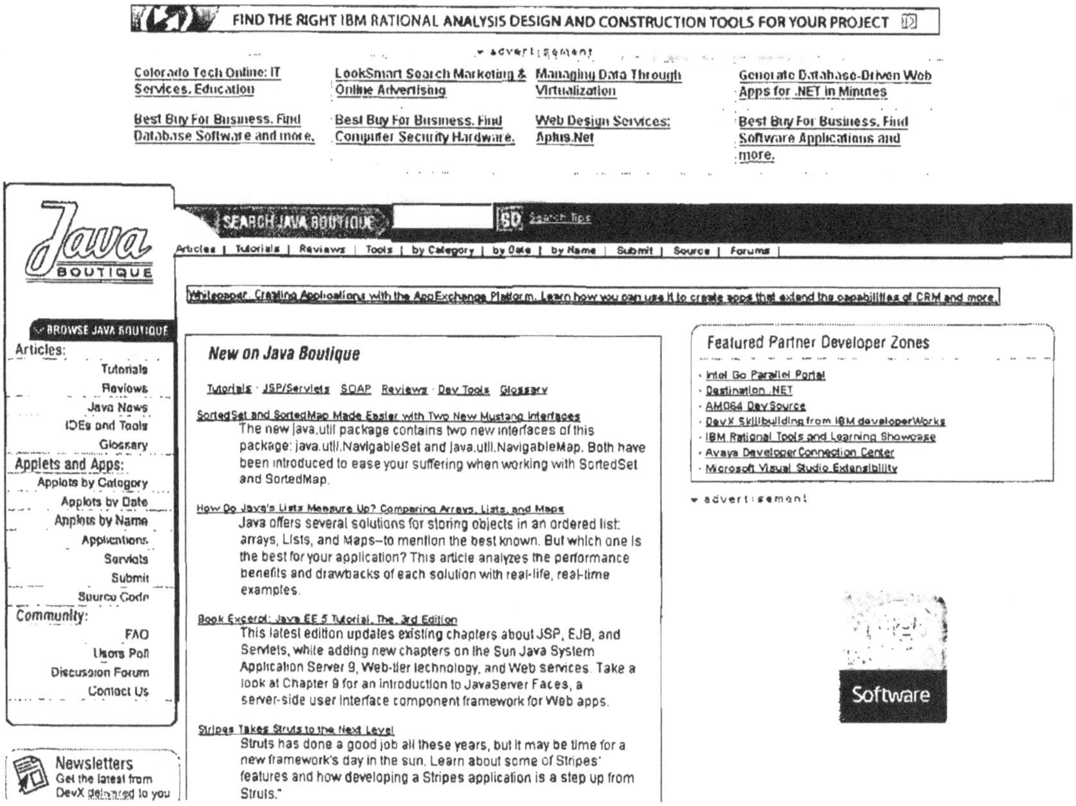
FIND THE RIGHT IBM RATIONAL ANALYSIS DESIGN AND CONSTRUCTION TOOLS FOR YOUR PROJECT
Java BOUTIQUE
SEARCH JAVA BOUTIQUE
Articles | Tutorials | Reviews | Tools | by Category | by Date | by Name | Submit | Source | Forums
BROWSE JAVA BOUTIQUE
Articles:
Tutorials
Reviews
Java News
IDEs and Tools
Glossary
Applets and Apps:
Applets by Category
Applets by Date
Applets by Name
Applications
Servlets
Submit
Source Code
Community:
FAQ
Users Poll
Discussion Forum
Contact Us
Newsletters
New on Java Boutique
The new java.util package contains two new interfaces of this package: java.util.NavigableSet and java.util.NavigableMap. Both have been introduced to ease your suffering when working with SortedSet and SortedMap.
Java offers several solutions for storing objects in an ordered list: arrays, Lists, and Maps—to mention the best known. But which one is the best for your application? This article analyzes the performance benefits and drawbacks of each solution with real-life, real-time examples.
This latest edition updates existing chapters about JSP, EJB, and Servlets, while adding new chapters on the Sun Java System Application Server 9, Web-tier technology, and Web services. Take a look at Chapter 9 for an introduction to JavaServer Faces, a server-side user interface component framework for Web apps.
Struts has done a good job all these years, but it may be time for a new framework's day in the sun. Learn about some of Stripes' features and how developing a Stripes application is a step up from Struts.
Featured Partner Developer Zones
Software

Thanks to thousands of individuals, corporations, and organizations, the Internet offers countless such metaguide sites on a huge variety of subjects. Which brings us directly to...

Rule Two

Let other people do as much work for you as possible (use their metaguides, their FAQs, their expertise to your advantage).

Directories are not the only good sources of general information. A number of **virtual libraries** and **reference desks** have sprung up on the web and they tend to be terrific starting places for all types of general information because they have thousands of pre-selected links to sources of data the researchers know to be good.

Let's continue with the Java example. If we go to the **Intute Science, Engineering, and Technology** page (formerly EEVL, the Internet Guide to Engineering,

Mathematics and Computing) and search on *java*, we get back a list of highly relevant and carefully evaluated websites:

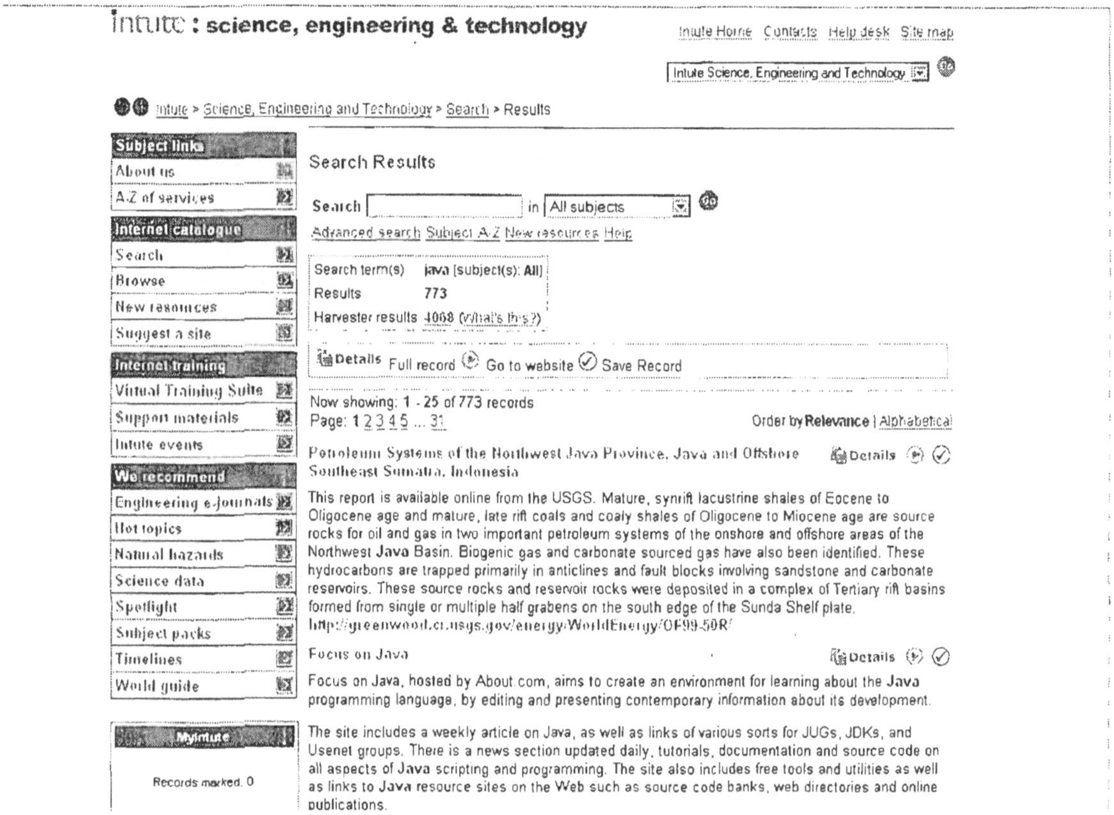
intute : science, engineering & technology
Intute Home Contacts Help desk Site map
Intute Science, Engineering and Technology
Intute > Science, Engineering and Technology > Search > Results
Subject links
About us
A-Z of services
Internet catalogue
Search
Browse
New resources
Suggest a site
Internet training
Virtual Training Suite
Support materials
Intute events
We recommend
Engineering e-Journals
Hot topics
Natural hazards
Science data
Spotlight
Subject packs
Timelines
World guide
MyIntute
Records marked: 0
Search Results
Search in All subjects
Advanced search Subject A-Z New resources Help
Search term(s) java [subject(s): All]
Results 773
Harvester results 4068 (What's this?)
Details Full record Go to website Save Record
Now showing: 1 - 25 of 773 records
Page: 1 2 3 4 5 ... 31
Order by Relevance | Alphabetical
Petroleum Systems of the Northwest Java Province, Java and Offshore Southeast Sumatra, Indonesia
Details
This report is available online from the USGS. Mature, synrift lacustrine shales of Eocene to Oligocene age and mature, late rift coals and coaly shales of Oligocene to Miocene age are source rocks for oil and gas in two important petroleum systems of the onshore and offshore areas of the Northwest Java Basin. Biogenic gas and carbonate sourced gas have also been identified. These hydrocarbons are trapped primarily in anticlines and fault blocks involving sandstone and carbonate reservoirs. These source rocks and reservoir rocks were deposited in a complex of Tertiary rift basins formed from single or multiple half grabens on the south edge of the Sunda Shelf plate.
http://greenwood.cr.usgs.gov/energy/WorldEnergy/OF99-50R/
Focus on Java
Details
Focus on Java, hosted by About.com, aims to create an environment for learning about the Java programming language, by editing and presenting contemporary information about its development.
The site includes a weekly article on Java, as well as links of various sorts for JUGs, JDKs, and Usenet groups. There is a news section updated daily, tutorials, documentation and source code on all aspects of Java scripting and programming. The site also includes free tools and utilities as well as links to Java resource sites on the Web such as source code banks, web directories and online publications.

In addition to the obvious SUN sites about Java, there are many others, such as links to Java FAQs, news, tutorials, course notes, seminar slides, articles, development tools, users' groups, mailing lists, books, conferences, links to web-based courses, and other resources.

Now you have a new resource for future Java-related research. Naturally, the first thing to do is bookmark the page.

Rule Three

Bookmark constantly, organize your bookmarks, and back them up as though your life depends on it.

One of the biggest and most influential entries into the reference/research world on the Internet is **Wikipedia**, a self-described free encyclopedia that anyone can edit. Because of its growth and importance, Wikipedia has earned a separate section in this year's edition. According to the Wikipedia, the term "wiki" describes "a group of

Web pages that allows users to add content, as on an Internet forum, but also allows others (often completely unrestricted) to edit the content. The term *wiki* also refers to the collaborative software (wiki engine) used to create such a website (*see wiki software*). In essence, the wiki is a vast simplification of the process of creating HTML pages, and thus is a very effective way to exchange information through collaborative effort. **Wiki** is sometimes interpreted as the acronym for 'what I know, is', which describes the knowledge contribution, storage and exchange up to some point."[28] The most obvious potential problem with an encyclopedia that "anyone can edit" is quality control, and in fact, one of the Wikipedia's co-founders admitted serious problems with the quality and accuracy of some (perhaps a lot) of the Wikipedia content.[29] While there is a tremendous amount of good information in Wikipedia, it should not be relied upon as a sole source. Neither should it be ignored as this example of a "disambiguation" page on "java" shows:

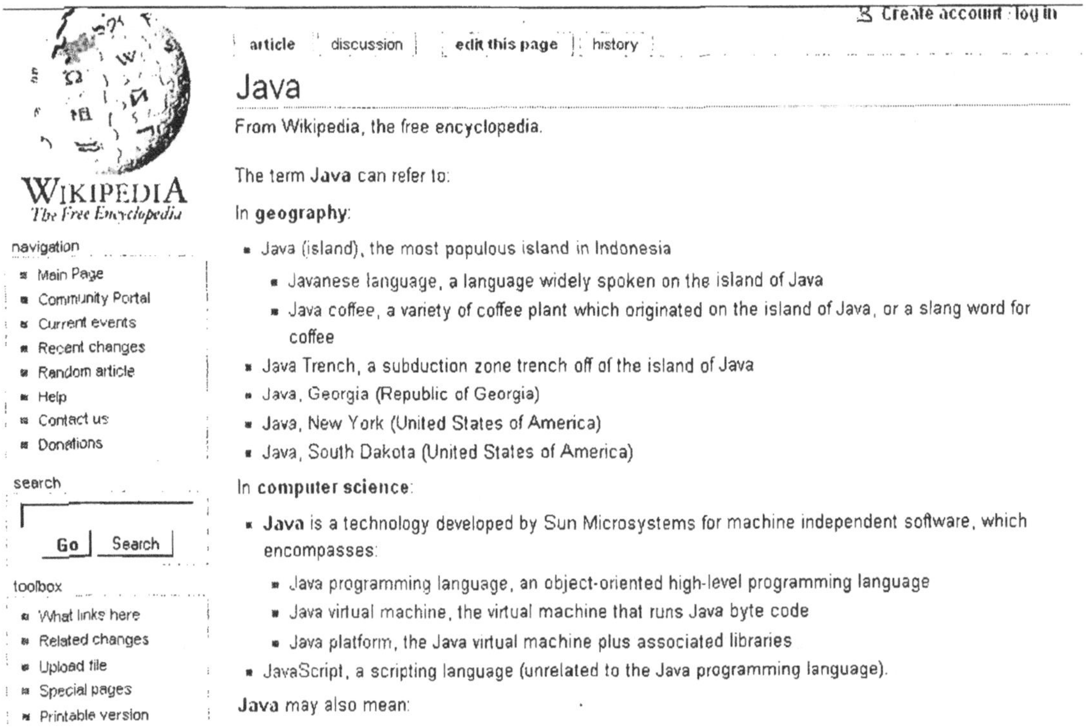

Create account / log in

article | discussion | edit this page | history

WIKIPEDIA
The Free Encyclopedia

navigation
- Main Page
- Community Portal
- Current events
- Recent changes
- Random article
- Help
- Contact us
- Donations

search

Go | Search

toolbox
- What links here
- Related changes
- Upload file
- Special pages
- Printable version

Java

From Wikipedia, the free encyclopedia.

The term **Java** can refer to:

In **geography**:

- Java (island), the most populous island in Indonesia
  - Javanese language, a language widely spoken on the island of Java
  - Java coffee, a variety of coffee plant which originated on the island of Java, or a slang word for coffee
- Java Trench, a subduction zone trench off of the island of Java
- Java, Georgia (Republic of Georgia)
- Java, New York (United States of America)
- Java, South Dakota (United States of America)

In **computer science**:

- **Java** is a technology developed by Sun Microsystems for machine independent software, which encompasses:
  - Java programming language, an object-oriented high-level programming language
  - Java virtual machine, the virtual machine that runs Java byte code
  - Java platform, the Java virtual machine plus associated libraries
- JavaScript, a scripting language (unrelated to the Java programming language).

Java may also mean:

Wikipedia also has the advantage of offering a free encyclopedia in a number of languages besides English, including French, Polish, Portuguese, Spanish, Dutch, Swedish, Italian, German, and Japanese.

---

[28] "Wiki," *Wikipedia*, Wikipedia, 2005. Answers.com <http://www.answers.com/topic/wiki> (14 November 2006).

[29] Andrew Orlowski, "Wikipedia Founder Admits to Serious Quality Problems," *The Register*, 18 October 2005, <http://www.theregister.co.uk/2005/10/18/wikipedia_quality_problem/> (14 November 2006).

To review, the best starting places for general information on broad topics are web directories/subject guides, virtual libraries, and reference desks. There are hundreds of such websites, but I've selected a few of the best.

About	http://www.about.com/
Encyclopedia.com	http://www.encyclopedia.com/
Encyclopedia Britannica[30]	http://www.britannica.com/
Hotsheet	http://www.hotsheet.com/
INFOMINE	http://infomine.ucr.edu/
Information Please	http://www.infoplease.com/
Internet Library for Librarians	http://www.itcompany.com/inforetriever/index.htm
Intute (formerly RDN)	http://www.intute.ac.uk/
The Internet Public Library	http://www.ipl.org/
Librarians' Index to the Internet	http://lii.org/
The Library Spot	http://www.libraryspot.com/
Martindale's The Reference Desk	http://www.martindalecenter.com/
My Virtual Reference Desk	http://www.refdesk.com/
Pinakes Subject Gateway[31]	http://www.hw.ac.uk/libWWW/irn/pinakes/pinakes.html
Wikipedia	http://en.wikipedia.org/
WWW Virtual Library	http://vlib.org/Overview.html
Yahoo Reference	http://education.yahoo.com/reference/

Web Tip

Think of search engine databases as *huge warehouses in which everything from diamonds to debris is stored*. Your job is to find the jewels amid the muck.

[30] Although full-text articles require a paid subscription to *Encyclopedia Britannica*, the site is still a useful starting place for research and includes free access to the *Britannica Concise Encyclopedia*.

[31] Pinakes is the gateway to Intute and dozens of other equally valuable specialized research sites.

Search Savvy—Mastering the Art of Search

While directories and virtual libraries contain information selected by people, search engine databases are mostly unfiltered, that is, no human being is looking at the data being indexed to determine its value, authenticity, and reliability. Search engines are where the researcher's experience, knowledge, judgment, and intuition really come into play. Because of their vast scope and size, search engines are the heart and soul of Internet search and research. No other resource reaches as far or wide or quickly as a search engine. A researcher must learn to use search engines to their fullest extent despite their limitations.

Individual search engines have some very important advantages over directories, metasearch, and megasearch sites. Foremost among these is the fact that they have much larger databases of indexed sites. However, ***no single search engine is best.*** Each has its own advantages and drawbacks. Furthermore, ***there is a remarkable lack of overlap among search engines databases***, so it is vital that you train yourself to use more than one search engine.

Greg Notess ran an interesting little experiment that demonstrated the need to use more than one search engine.[32] He was looking for the real name for an AOL screen name, a piece of information that is often hard to find. One only search engine—in his example, Yahoo, found the name—while Google, Live, Gigablast, Ask, and Exalead all failed to locate the information. It could have been any search engine, not just Yahoo, that provided the data, but the point is clear: you must try multiple search engines, especially when looking for obscure or hard to find information.

On a larger scale, the metasearch engine Dogpile touted the results of a 2005 study they did in collaboration with researchers from the University of Pittsburgh and Pennsylvania State University showing a lack of duplication in the top results of the major search engines.

> "When the researchers ran 12,570 different queries through search engines at Yahoo, Google, MSN and Ask Jeeves, they found that only 1.1 percent of the results appeared on all four engines, while 84.9 percent of the top results were

---

[32] Greg Notess, "Overlap Showdown: Only 1 of 6," Search Engine Showdown, 28 December 2006, <http://www.searchengineshowdown.com/blog/2006/12/overlap_showdown_only_at_1_of_1.shtml> (16 January 2007).

unique to one engine. Only 2.6 percent of the results were shared by three search providers, and 11.4 percent were delivered by two search engines." [33]

I am not surprised by the results, although I doubt the lack of overlap is quite as significant as the study indicates. The researchers used a relatively small sample, they only looked at the top ten results, they included paid results, and—probably most significantly—Dogpile sponsored the study. If a study to test metasearch engine results compared to individual search engine results concludes that metasearch engines do a better job, it is hardly surprising (and not necessarily convincing).

You can read the Dogpile/University of Pittsburgh and Penn State University study and take a look at Dogpile's "Compare Search Engines" page to see how the comparison works.

Dogpile's Compare Search Engines http://comparesearchengines.dogpile.com/

"Different Engines, Different Results"
http://comparesearchengines.dogpile.com/OverlapAnalysis.pdf

Rule Four

Use more than one search engine.

All search engines have their own way of doing things, which means there is no set of rules or guidelines that users can apply to them all. It helps, however, to familiarize yourself with the kinds of features available so you will at least know what to look for.

Often research involves the ***search for specific information***: a telephone number, a name or title, a specific company or product, a piece of equipment, etc. Even researching a general subject may require a broader data set than a virtual library or subject guide offers if you need to find out as much as possible about a subject. For example, if I need to know everything available on the web about Mexico and NAFTA, I cannot limit myself to someone else's edited list. Besides, there won't be

[33] Dogpile.com in collaboration with researchers from the University of Pittsburgh and Pennsylvania State University, "Different Engines, Different Results, A Research Study," July 2005, <http://comparesearchengines.dogpile.com/OverlapAnalysis.pdf> (14 November 2006). I have serious doubts about the accuracy of this claim, but the general conclusion of the study that there is a lack of overlap among search engine results is valid, if exaggerated.

much, if any information because I am looking for specialized information (Mexico) within a big topic (NAFTA).

Compared to directories and metasearch services, individual search engines offer much greater flexibility and many more options for searching, not the least of which is the ability to search using ***boolean* expressions.** Search engine companies have concluded (probably rightly) that boolean searches are beyond the ken of most users, although you may find the boolean queries permitted by the best search engines are inferior to what you've used before.

One of the hottest areas of contention surrounding search engines has always been and continues to be ***search engine index size***. I recommend you take size claims with a grain of salt. Search engine index sizes are self-reported and not validated by any objective third party. This old contest came to a head in 2005. First Yahoo claimed to have indexed over 20 billion "items" in its index. These items included "just over 19.2 billion web documents, 1.6 billion images, and over 50 million audio and video files."[34] Yahoo's claim at first appeared to mark the beginning of another competition to retain the "honor" of having the biggest search engine database, something Google had prided itself on for years. This time, however, instead of fighting back with bigger number counts on its homepage, Google dropped those numbers entirely as part of its seventh birthday celebration in September 2005. At the same time, Google announced a "newly expanded web search index that is 1,000 times the size of our original index...which makes Google more than 3 times larger than any other search engine."[35] Google did not offer any specific number but insisted it offers the most comprehensive collection of websites and documents on the Internet. Yahoo makes a similar claim. The answer? There is no one "best" search engine or site; researchers need a good toolkit of many resources when looking for rare information.

Determining search engine database size is something more akin to alchemy than arithmetic, so I suggest you take all such estimates of size with a large dose of skepticism. Besides, numbers are one thing and good search results are quite another. What good do 20 billion web documents do if not one of them provides the results you are seeking? **Relevant results** are the best measure of a search engine's value, but from my experience, having a larger pool in which to fish for these answers really does make it more likely that a search engine will retrieve the results users seek in the case of obscure information, which is after all the kind of information we are often seeking. Search engine size wars are almost always a good thing for researchers because it keeps the big players on their toes and

---

[34] "Our Blog is Growing Up—And So Has *[sic]* Our Index," *Yahoo! Search Blog*, 8 August 2005, <http://www.ysearchblog.com/archives/000172.html> (15 November 2006).

[35] "We Wanted Something Special for Our Birthday," *Google Blogspot*, 26 September 2005, <http://googleblog.blogspot.com/2005/09/we-wanted-something-special-for-our.html> (15 November 2006).

motivates them to improve their services. This past year's competition was no exception.

Another important fact to remember is that most search engines do not index entire websites or documents. It is no longer clear exactly how much of a webpage the major search engines index. For example, Google used to only index approximately the first 100KB of HTML, and reportedly the first megabyte of PDF documents, but in October 2005, Google dramatically increased the size of its cache limit. Yahoo indexes at least the first 500KB of HTML and PDF documents. As for Microsoft files types, my experimentation with them indicates that, in most cases, Yahoo indexes virtually the entire file, even in the case of very large documents.

The following is an overview of the major search engines in terms of their features, how to use them effectively, and what makes each one distinctive. It is important to remember there is no such thing as a perfect search engine. Each one has its advantages and drawbacks. The only way to fully exploit a search engine is to take the time to learn to use it, which means you must read the instructions.

Rule Five

Read the instructions.

Google

Google first gained fame and widespread use because of its single-minded focus on search, exemplified by its "clean" interface, and its **PageRank™** "weighted link popularity." In simple terms, Google gives each webpage a rank based on the number of other pages linking to it and the "importance" of those pages, where importance is derived from an overall link count. While PageRank is imperfect, it works better than most other approaches to ranking search results and, indeed, is one of the primary reasons for Google's success.

Some of Google's features that helped to create this very successful and powerful search tool are:

- **cached versions** of webpages; Google was the first search engine to offer this option, which let users peek into its vast database.
- **automatic conversion of non-HTML filetypes** to HTML is available; Google was not the first to do this, but certainly has been the most successful.
- **backlinks** (the link: syntax); unfortunately, Google now limits the number of backlinks it shows, greatly reducing the utility of this option.
- Google seems to have **increased its limits on the size of indexed pages**. I found an indexed PDF document over 764K, a text file over 1000K, and a webpage over 366K. Very few webpages are larger than 500K. Google does not offer HTML versions of very large PDF or Word documents, e.g., the complete 9/11 Commission Report, but exactly what their cut-off size is, I do not know.
- Google **refreshes its index** continuously, not on a schedule (this is a good thing); Google's Matt Cutts explains Google's refresh rate: "It's true that when an event happens on the web, our index can often pick it up in 1-2 days, and usually even faster. But a typical page in Google's main web index is updated every 2-3 weeks or faster; it's not the case that the entire main web index is updated every 2-3 days."[36]
- Google stopped advertising the **size of its database** in 2005, but Google is one of the largest if not the largest search database.

In determining the overall size of its index, Google also ***includes urls of pages that it has not crawled and for which it has not indexed the text.*** These "orphan"

---

[36] Matt Cutts, "Google Update Speed," Google Blogoscoped, 26 July 2006, <http://blog.outer-court.com/archive/2006-07-26.html#n28> (14 November 2006).

pages may be any number of things, including pages with robots.txt command or tag. Unindexed pages are identifiable by what they lack: no summary, no page size, and no cached copy.

www.stat.vcu.edu/robots.txt
Similar pages

Google Orphans—no cached copy, no summary, no page size

javangelist.snipsnap.org/space/SnipSnap/config/rob...
Similar pages

www.atmos.washington.edu/robots.txt
Similar pages

fichier indiquant aux robots les endroits interdits # # voir http ... - [Translate this page]
fichier indiquant aux robots les endroits interdits # # voir
http://info.webcrawler.com/mak/projects/robots/norobots.html User-agent: * Disallow: ...
www.ann.jussieu.fr/robots.txt - 1k - Cached - Similar pages

www.pamarys.ku.lt/robots.txt
Similar pages

Indexed page—cached copy, summary, page size

sagittarius.student.utwente.nl/robots.txt
Similar pages

Google no longer displays the number of pages searched on its home page, but a search on **[the]** returns an estimated 4.8 billion pages, so that represents the *minimum* number of pages in the Google database (in fact, it probably is far larger). Remember, all size claims are "self validated," so take them with a grain of salt. Still, for the types of Internet research we perform, bigger really is better because we have a much better chance of finding obscure information in billions of webpages than in millions.

Customizing Google Preferences

Google offers five basic **Preferences** settings:

1. Interface language: if you are more comfortable working in another language, Google can display in dozens.
2. Search language: generally, most searchers choose to search in any language, but there are occasions when it makes sense to limit your search to one or more specific languages. Google supports 35 languages, including non-Latin languages such as Arabic, Simplified and Traditional Chinese, Greek, Hebrew, Japanese, Korean, Russian, and Turkish.
3. Filtering of pages containing explicit sexual content.
4. Number of results: a purely individual preference, but 10 results per page is simply frustrating; Google lets you see up to 100 results per page.
5. Results window: opens results in a new browser window.

The Google Results Page

Once you've entered your search terms and selected the *Google Search* button, Google will present you with a list of results (hits). For each result returned you may see:

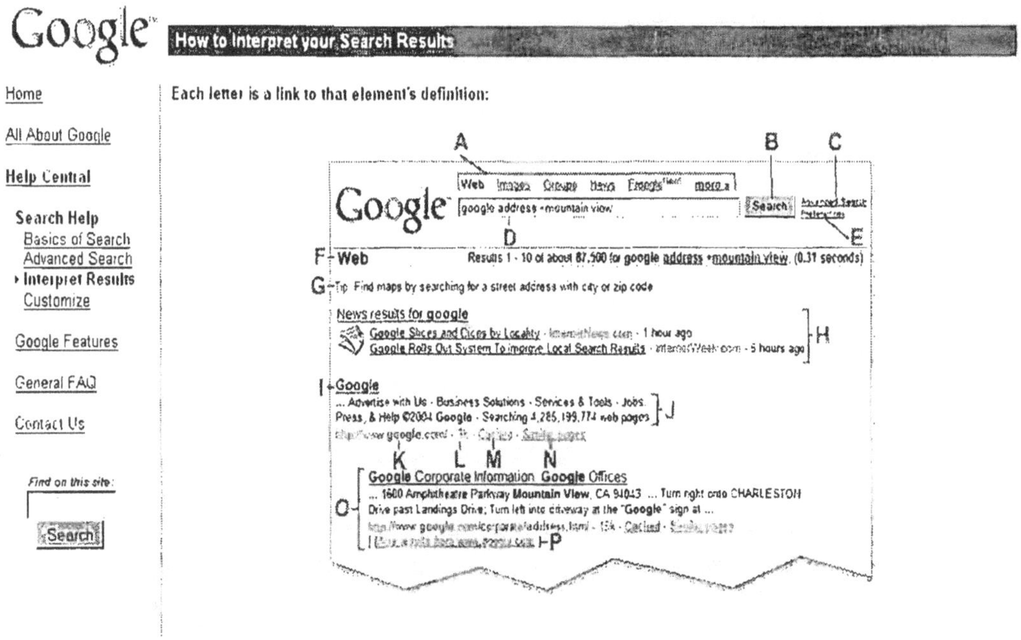
Google
How to Interpret your Search Results
Home
All About Google
Help Central
Search Help
Basics of Search
Advanced Search
Interpret Results
Customize
Google Features
General FAQ
Contact Us
Find on this site:
Search
Each letter is a link to that element's definition:
A
B
C
D
E
F
G
H
I
J
K
L
M
N
O
P
Web
News results for google
Google
Google Corporate Information Google Offices

- ➢ the **statistics bar (F)** describes the type of search, e.g., web, and shows the number of results returned as well as the amount of time it took to complete your search
- ➢ a **tip (G)** may appear here, but not all queries generate tips. A typical tip might be "Try removing quotes from your search to get more results."
- ➢ **"OneBox Results" (H)** typically include news, stock quotes, maps, weather and local websites related to your search
- ➢ the **title (I)** of the webpage found
- ➢ an **excerpt (J)** from the webpage with your search terms **bolded**
- ➢ the **url (K)** of the webpage
- ➢ you will see **Supplemental Result** only if Google retrieved the result from its supplemental and not its main index
- ➢ the **size (L)** of the text portion of the webpage (omitted if page is not yet indexed)
- ➢ a **cached (M)** link to the version of the site stored by Google if it is indexed
- ➢ a **similar pages (N)** link for pages related to this result

- **indented results (O)** indicate Google has found more than one result from the same website; the most relevant page is listed first
- a **more results from (P)** link if there are more than two results from the same site

Google Basic Search http://www.google.com/

Google assumes as its default that multiple search terms are joined by the **AND** operator, so that a search on the keywords [windows explorer][37] will find all the webpages that contain ***both*** search terms. Furthermore, Google will first try to find all the webpages that contain the ***phrase*** ["windows explorer"]. Google will search:

- first, for ***phrases*** (keywords as one long phrase)
- second, for webpages containing all the keywords with the ***greatest adjacency*** (closest together),
- third, for webpages containing ***all the keywords***, regardless of where they appear on the webpage

Google ***will not return any results*** if there is no webpage containing all the search terms. Try this query to see what I mean:

[kong spektioneer synecdoche]

There is an exception to this rule. Google often returns results when a keyword is not actually on a webpage but is in a ***link to a website***, usually as text in a link anchor.

---

[37] Matt Cutts, one of Google's software engineers who also writes a blog mostly about Google, let his readers in on a little bit of Google insider information. "At Google, we use [and] to mark the beginning and end of queries. So ["scorpio submarine"] means to do a phrase search for "scorpio submarine", while [scorpio submarine] means just to type in those words without the quotes–you leave the brackets out when you actually do the search." That's an interesting and useful bit of trivia. I have tried various schemes to distinguish queries and ended up using italics (not a very elegant solution). From now on, I will be using brackets to set off queries in UTW. *Matt Cutt's Blog*, 11 August 2005, <http://www.mattcutts.com/blog/writing-google-queries/> (14 November 2006).

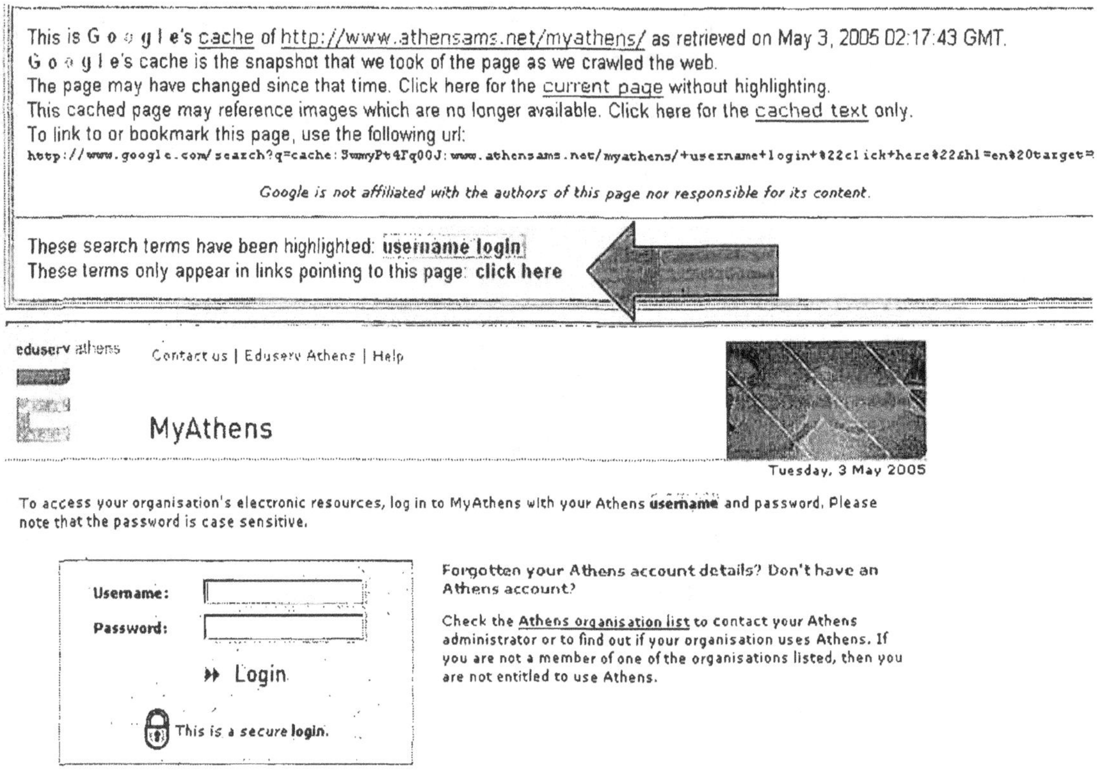
This is Google's cache of http://www.athensams.net/myathens/ as retrieved on May 3, 2005 02:17:43 GMT.
Google's cache is the snapshot that we took of the page as we crawled the web.
The page may have changed since that time. Click here for the current page without highlighting.
This cached page may reference images which are no longer available. Click here for the cached text only.
To link to or bookmark this page, use the following url:
Google is not affiliated with the authors of this page nor responsible for its content.
These search terms have been highlighted: username login
These terms only appear in links pointing to this page: click here
Contact us | Eduserv Athens | Help
MyAthens
Tuesday, 3 May 2005
To access your organisation's electronic resources, log in to MyAthens with your Athens username and password. Please note that the password is case sensitive.
Username:
Password:
Login
This is a secure login.
Forgotten your Athens account details? Don't have an Athens account?
Check the Athens organisation list to contact your Athens administrator or to find out if your organisation uses Athens. If you are not a member of one of the organisations listed, then you are not entitled to use Athens.

Google ***limits the number of search terms to 32 keywords***. It ignores any terms beyond that number. However, there are ways to force Google to search for more than 32 keywords.

Google is ***not case sensitive***. There does not appear to be anything you can do about this.

In late 2003 Google introduced automatic ***word stemming*** or ***truncation***, i.e., searching for variations of search terms. Normally, word stemming involves searching for plurals and verb conjugations such as *drink*, *drank*, *drunk*. However, Google's word stemming is not consistent and somewhat confusing. For example, stemming does not work either with single word or phrase searches, i.e., a search on [child] will not find *children*. Yet a search on [child health] will find child, *childhood*, *children*, and *children's*. Google will also find some variations of verbs, e.g., a search on [drink water] will find *drinking water*. Users should still search on all variations of a term, including plurals. There is a Google hack for disabling word stemming.

Google automatically ***clusters search results***. Multiple hits from the same site are indented and there is usually an option to see more results from a specific site.

Google ***permits the use of the OR operator*** in simple search. The OR must be capitalized.

Beyond the use of the OR operator in its simple search, ***Google does not support <u>boolean</u> search***.

While Google ***assumes that multiple keywords are a phrase***, searchers can delimit phrases using double-quotes. For example, if I search on:

[the last king of france]

without double-quotes, Google will ignore the "the" and the "of" in its search. The results I get include many irrelevant hits, such as music from a group called "The Last King" and an article about Lance Armstrong. However, if I enclose the same query in ***double-quotes***, Google will search on exactly the phrase ["the last king of france"], and return a result with the name of the last king of France. Enclosing searches in double-quotes is much more effective for finding precise results than relying on automatic phrase searching.

Google no longer routinely ignores ***stop words*** outside double quotes. Each of these searches will now return different results:

[the last king of france] [last king france] ["the last king of france"]

Stop words are English words that are so commonplace they are not included in a search unless the searcher forces Google to do so. The stop words Google recognizes include: *a, an, about, and, are, as, at, be, by, com, from, how, I, in, is, it, of, on, or, that, the, this, to, we, what, when, where, which, with, why*. There probably are others!

However, Google's handling of stop words is inconsistent. For example, in the query [to be or not to be], Google ignores OR because it may be a logical operator, and it also appears to ignore TO and BE, only searching for NOT. Therefore, you may need to force Google to search for a stop word on occasion. There is a <u>Google hack for forcing Google to search for stop words</u>.

It is unnecessary to use the plus sign (+) with any terms except stop words because by default Google searches for all keywords. However, there are many times when searchers need to exclude certain terms that are commonly associated with a keyword but irrelevant to their search. That's where the ***minus sign (-)*** comes in. Using the minus sign in front of a keyword ensures that Google excludes that term from the search. For example, the results for the search ["pearl harbor" –movie] are very different from the results for ["pearl harbor"].

Google's handling of words with diacritical marks such as accents or umlauts is inconsistent. ***By default, Google will search for terms matching those with and without the diacritic***. As Google's Vanessa Fox explains, "When a searcher enters a query that includes a word with accented characters, our algorithms consider web pages that contain versions of that word both with and without the accent. For instance, if a searcher enters [México], we'll return results for pages about both "Mexico" and "México.""[38]

For example, a search on [façade] will return pages containing both facade and façade. To force Google to search only for the term with the diacritic, put a ***plus sign in front of the term: [+façade]***. You may see a few pages that do not appear to have the diacritic, but that is probably because that term appears in anchor text or an inbound link that is pointing to the page but not actually on the page in question.

However, Fox goes on to explain that results also vary depending upon whether you are searching at Google.com or a Google international site (e.g., Google.fr), whether your preferred language at Google is English or another language, and from where you are coming to the Google site as indicated by your IP address. If Google detects that your IP address geolocates to Peru, your search results will be different from those provided to someone coming to Google from Norway, regardless of the preferred language or the site you search. Also, users who have registered with Google and set up personalized search will find that their results are affected by their previous searches. In other words, while there are ways to *manipulate* the results Google provides, there is no way to *control* them.

Google treats most ***punctuation marks*** the same way, as links in a search string. For example, Google handles a search for [c-span], [c.span], ["c span"], and [c?span] basically the same way. However, a search for [cspan] with no space or mark is treated differently.

---

[38]Vanessa Fox, "How search results may differ based on accented characters and interface languages," Official Google Blog, 32 August 31 2006, <http://googlewebmastercentral.blogspot.com/2006/08/how-search-results-may-differ-based-on.html> (November 27, 2006).

Google will search for several punctuation marks and special characters:

- the ***ampersand*** **[&]**: Google will search for [barnes&noble] or [barnes & noble]
- the ***underscore*** **[_]**: Google will search for a phrase such as [public_records.doc] or even more specifically [public_records.html]
- the dollar sign **[$]**, used with a number, for example [$100]
- the sharp **[#]**, for example [F#]
- the **slash**, but only when used in the search [I/O]
- While Google will not actually search on a plus sign, the search engine does recognize the difference between searches for **[c]**, **[c+]**, and **[c++]**

Google Advanced Search

Google has a number of "query modifiers" to restrict searches and make them more effective in many cases. These query modifiers can be used in simple search in the following syntax or on the advanced web search page using the appropriate menu options. The query modifiers Google supports are:

- **site:** restricts results to websites in a given domain. *This syntax no longer requires you to add a keyword.* Google's site: syntax will also ***search within folders***, e.g., [site:jpl.nasa.gov/technology]. Remember you *can* add keywords to the site: search. [site:jpl.nasa.gov/technology "jpl spacecraft"]

  Advanced Web Search > Domains

  Examples of how to use the **site:** command:

  [shuttle site:www.nasa.gov] finds pages about the space shuttle at the NASA website.

  [site:info] finds all the pages in the Google database in the .info top-level domain

  ["bulletin officiel" site:fr] finds pages in the French top-level domain about official bulletins

  [cirrus -site:mastercard.com] finds pages about the keyword cirrus that are not at the Mastercard.com site

  [site:jpl.nasa.gov/technology]

[site:jpl.nasa.gov/technology "jpl spacecraft"]

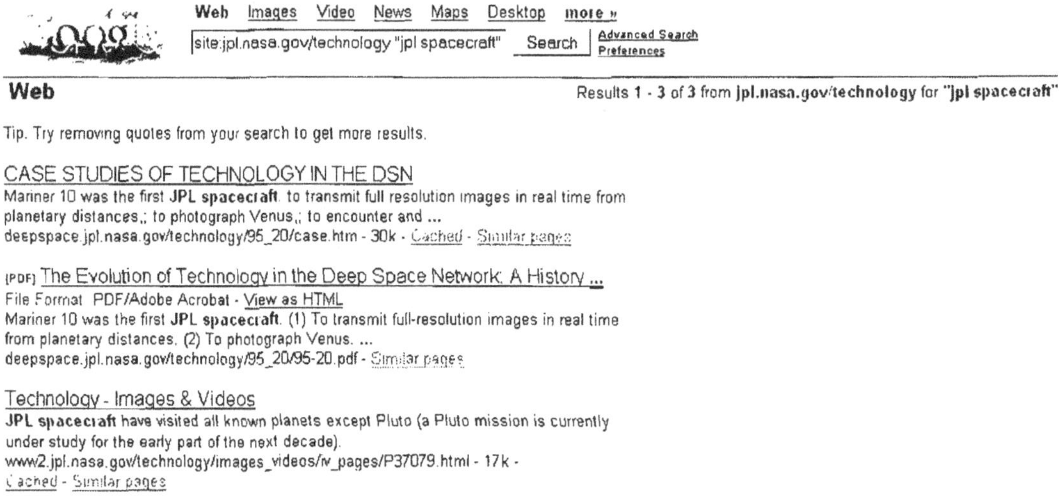

Web Images Video News Maps Desktop more »

site:jpl.nasa.gov/technology "jpl spacecraft" Search | Advanced Search Preferences

Web Results 1 - 3 of 3 from **jpl.nasa.gov/technology** for **"jpl spacecraft"**

Tip. Try removing quotes from your search to get more results.

CASE STUDIES OF TECHNOLOGY IN THE DSN
Mariner 10 was the first **JPL spacecraft** to transmit full resolution images in real time from planetary distances,; to photograph Venus,; to encounter and ...
deepspace.jpl.nasa.gov/technology/95_20/case.htm - 30k - Cached - Similar pages

[PDF] The Evolution of Technology in the Deep Space Network: A History ...
File Format: PDF/Adobe Acrobat - View as HTML
Mariner 10 was the first **JPL spacecraft**. (1) To transmit full-resolution images in real time from planetary distances. (2) To photograph Venus. ...
deepspace.jpl.nasa.gov/technology/95_20/95-20.pdf - Similar pages

Technology - Images & Videos
JPL spacecraft have visited all known planets except Pluto (a Pluto mission is currently under study for the early part of the next decade).
www2.jpl.nasa.gov/technology/images_videos/iv_pages/P37079.html - 17k - Cached - Similar pages

- **intitle:** restricts the results to documents containing the keyword in the title.

  Advanced Web Search > Occurrences

  Examples of how to use the **intitle:** command:

  [intitle:amazon] finds all pages that include the word *amazon* in their title

  [intitle:amazon "rain forest"] finds all pages that include the word *amazon* in their title and mention the phrase *"rain forest"* anywhere in the document (title or text or anywhere in the document)

- **allintitle:** restricts the results to documents containing all the keywords in the title of the document.

  Advanced Web Search > Occurrences

  Example of how to use the **allintitle:** command:

  [allintitle:amazon jungle "rain forest"] finds pages that include *all* the words in the *title* (not the text) of the document, e.g.,

  <title>Amazon Rain Forest Jungle Tours</title>

- **inurl:** restricts the results to documents containing the keyword in the url.

  Advanced Web Search > Occurrences

<u>Examples of how to use the **inurl:** command</u>:

[inurl:nasa] finds all pages that include *nasa* anywhere in the url (address)

[inurl:nasa -site:gov] finds all pages that include *nasa* anywhere in the url of sites that are *not* in the *.gov* top-level domain

[inurl:nasa shuttle] finds all pages that include *nasa* anywhere in the url of the site and *shuttle* anywhere in the document (url or anywhere else).

- **allinurl:** restricts the results to documents containing all the keywords in the url.

  Advanced Web Search > Occurrences

  <u>Example of how to use the **allinurl:** command</u>:

  [allinurl:nasa shuttle] finds all pages that include *both* nasa and shuttle in the url of the site.

- **link:** restricts the results to documents that have links to a specific webpage. [39] ***Cannot use with keyword search terms.***

  Advanced Web Search > Page Specific Search > Links

  <u>Example of how to use the **link:** command</u>:

  [link:www.noaa.gov] finds all pages linking to the NOAA homepage.

  [link:www.noaa.gov/wx.html] finds all pages linking to a specific page at the NOAA site.

- **info:** presents information Google has about a webpage. This option is only available via the main Google search.

  <u>Example of how to use the **info:** command</u>:

  [info:www.noaa.gov] provides links to Google's cache of the page, pages that are similar to *www.noaa.gov*, pages that link to *www.noaa.gov*, and pages that contain the term *www.noaa.gov*.

---

[39] The Google *link:* command no longer shows all links as it once did in order to cut down on the amount of webspam created by hidden links on webpages. Therefore, the Google *link:* command is not nearly as useful as it used to be.

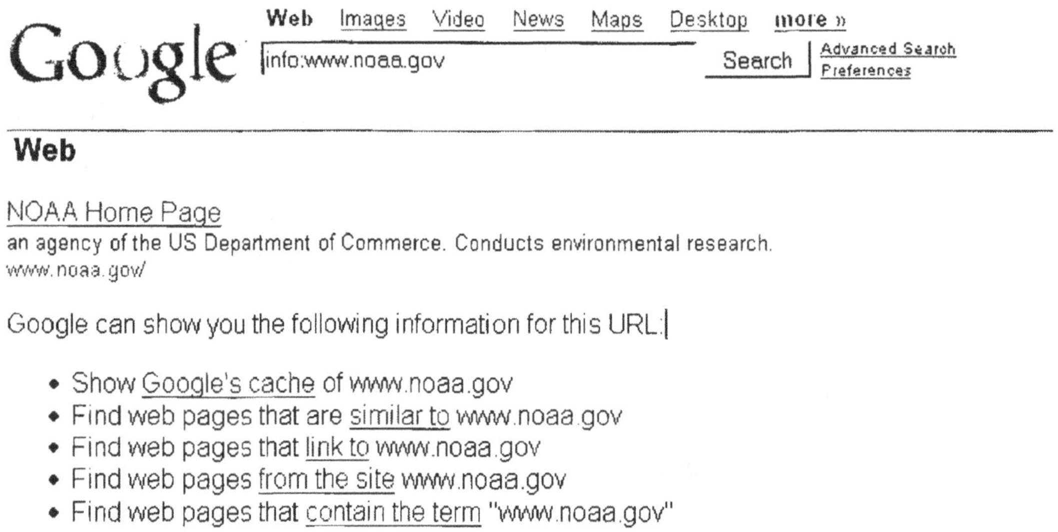
Google
Web Images Video News Maps Desktop more »
info:www.noaa.gov
Search
Advanced Search
Preferences
Web
NOAA Home Page
an agency of the US Department of Commerce. Conducts environmental research.
www.noaa.gov/
Google can show you the following information for this URL:
• Show Google's cache of www.noaa.gov
• Find web pages that are similar to www.noaa.gov
• Find web pages that link to www.noaa.gov
• Find web pages from the site www.noaa.gov
• Find web pages that contain the term "www.noaa.gov"

➢ **related:** restricts the results to documents Google has determined are similar to a specific webpage.

Advanced Web Search > Page Specific Search > Similar

Example of how to use the **related:** command:

[related:www.nasa.gov] finds other US government homepages.

➢ **cache:** presents the version of the webpage Google has stored. This option can also be accessed by clicking on the Cached link on the main results page.

Examples of how to use the **cache:** command:

[cache:www.noaa.gov] shows the stored version of the NASA homepage

[cache:www.noaa.gov hurricane] shows the stored version of the NOAA homepage with the keyword *hurricane* highlighted.

With no fanfare, Google once again began showing the date and time when a webpage was cached. This is not a new feature in Google; the Google cache option showed date/time until mid-2000 when date and time unceremoniously disappeared. Now it's back. Who knows why? Who cares? It's a good thing. Here's what you'll see:

> This is Go gle's cache of http://www.lib.virginia.edu/science/guides/s-clim.htm as retrieved on Oct 30, 2005 03:11:51 GMT.
> Go gle's cache is the snapshot that we took of the page as we crawled the web.
> The page may have changed since that time. Click here for the current page without highlighting.
> This cached page may reference images which are no longer available. Click here for the cached text only.
> To link to or bookmark this page, use the following url:
> http://www.google.com/search?q=cache:JGBNAKmoBeUJ:www.lib.virginia.edu/science/guides/s-clim.htm+&hl=en
>
> *Google is neither affiliated with the authors of this page nor responsible for its content.*

Important: there is a Google hack that lets you view the cached text only version without first viewing the cached page containing images and other non-text data that could send information back to the original website. Gigablast also offers a "stripped" cache option.

- **filetype:** Google will search the content of many file types and must be used with keyword(s). However, there is a Google hack that lets you get around the keyword requirement. ***Microsoft filetypes are potentially dangerous*** to open in their native formats. Please follow these instructions for handling Microsoft files on the Internet safely.

Warning: use Google option to "view as html" instead of opening certain file types (mainly Microsoft Word and Excel) that could contain macro viruses.

Google will search the content of these file types:

HTML

Corel WordPerfect (wp)

Lotus 1-2-3 (wk1, wk2, wk3, wk4, wk5, wki, wks, wku)

Lotus WordPro (lwp)

MacWrite (mw)

Microsoft Excel (xls)

Microsoft PowerPoint (ppt)

Microsoft Word (doc)

Microsoft Works (wks, wps, wdb)

Microsoft Write (wri)

Portable Document Format (pdf)

Postscript (ps)

Rich Text Format (rtf)

Text (ans, txt)

Macromedia Shockwave Flash (swf)

Example of a **filetype:** search:

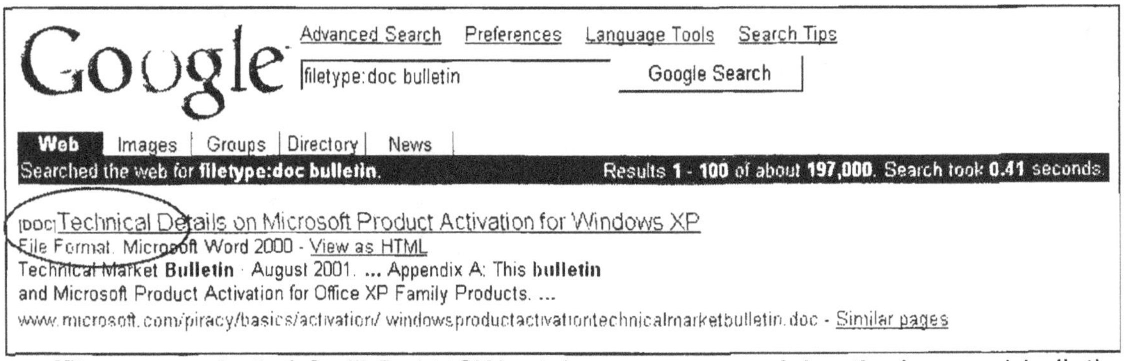
Google
Advanced Search
Preferences
Language Tools
Search Tips
filetype:doc bulletin
Google Search
Web
Images
Groups
Directory
News
Searched the web for filetype:doc bulletin.
Results 1 - 100 of about 197,000. Search took 0.41 seconds.
[DOC] Technical Details on Microsoft Product Activation for Windows XP
File Format: Microsoft Word 2000 - View as HTML
Technical Market Bulletin · August 2001. ... Appendix A: This bulletin
and Microsoft Product Activation for Office XP Family Products. ...
www.microsoft.com/piracy/basics/activation/ windowsproductactivationtechnicalmarketbulletin.doc - Similar pages

[filetype:doc bulletin] will find MSWord documents containing the keyword *bulletin*

For details on how to use the Google filetype option, please refer to the Google Filetype help FAQ. Also, there are a number of undocumented file type searches available using Google.

Google Filetype FAQ http://www.google.com/help/faq_filetypes.html

Google Special Search Features

Spell Checker: Google has a very good spell check option. When you input a query, Google checks to see if you are using the most common spelling of the keyword. If not, Google nicely asks, "Did you mean: x?" where x is the most common spelling. I really love this because Google doesn't presume. Sometimes you are intentionally misspelling a term. The classic example is [http referrer]. This computer term is almost always misspelled, so searching on [http referrer] won't yield nearly as many results as searching on the misspelled term. Google's dictionary also includes ***proper names***.

Calculator: the Google calculator will evaluate basic and complicated mathematical expressions as well as convert units of measurement and physical constants. Soople makes the Google calculator extremely easy to use. For detailed help in using the calculator, see:

Google Calculator Help http://www.google.com/help/calculator.html

Dictionary: Google has integrated dictionary definitions into its search options. Nothing could be easier to use. Underlined keywords appearing at the top of the results page are linked to Answers.com.

Define: new to Google is the *define* option. To use it, enter *define* then a word or phrase. This feature augments the dictionary option by searching a wider variety of sources. For example, the query [define blog] will return a web definition as the first result. *The advantage of the define option is that the definition appears at the top of* *the Google results list*, whereas using the dictionary option (clicking on *define*) requires the user to click on the link and go to that site to read the definition.

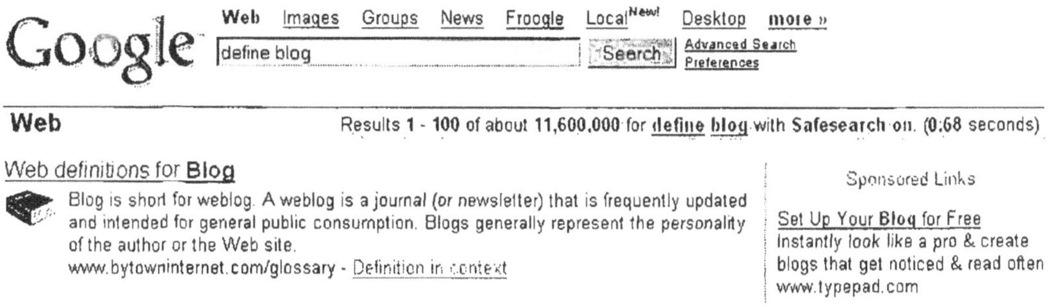
Google
Web Images Groups News Froogle Local New! Desktop more »
define blog
Search
Advanced Search
Preferences
Web
Results 1 - 100 of about 11,600,000 for define blog with Safesearch on. (0.68 seconds)
Web definitions for Blog
Blog is short for weblog. A weblog is a journal (or newsletter) that is frequently updated and intended for general public consumption. Blogs generally represent the personality of the author or the Web site.
www.bytowninternet.com/glossary - Definition in context
Sponsored Links
Set Up Your Blog for Free
Instantly look like a pro & create blogs that get noticed & read often
www.typepad.com

Translations: Currently, Google offers webpage translations to/from English and Arabic, simplified Chinese, French, German, Italian, Japanese, Korean, Portuguese, Russian, and Spanish. If a page appears in the results list in one of the languages Google translation supports, you will see [**Translate this page**] after the page title. All the newest additions to Google's translation list use statistical machine translation software developed by Google and the quality of these translations is far superior to that provided by Systran. These languages include at present Arabic, simplified Chinese, Japanese, Korean, and Russian.

Number Search: The numbers Google will search for include:

- US Patent numbers: syntax is [patent 5521308]
- **UPS** tracking: enter the tracking number with or without spaces [1Z9999X99999999]
- **USPS** tracking: enter the tracking number with or without spaces [9999999999999999999999]
- **FedEx** tracking: enter the tracking number with or without spaces [999999999999999]
- **DHL and Airborne Express** tracking: enter DHL plus the tracking number [DHL 9999999999]; DHL queries are the least reliable on Google
- **ZIP codes**: enter a US ZIP code, either five or nine digits
- **ISBN**: enter any International Standard Book Number
- **VIN** Information: to find information about a vehicle's history, search on its 17-character Vehicle Identification Number (VIN)
- **FAA airplane registration numbers**: [n158ua] (simply enter the FAA registration number; no special syntax is required)

> - **FCC equipment Ids**: syntax is [fcc EJM386S303]

Weather: The Google weather search is for US locations only. Simply enter the keyword *weather* followed by a city and/or state or even just a zip code (which works just fine by itself) and Google will present you with an attractive, succinct weather chart:

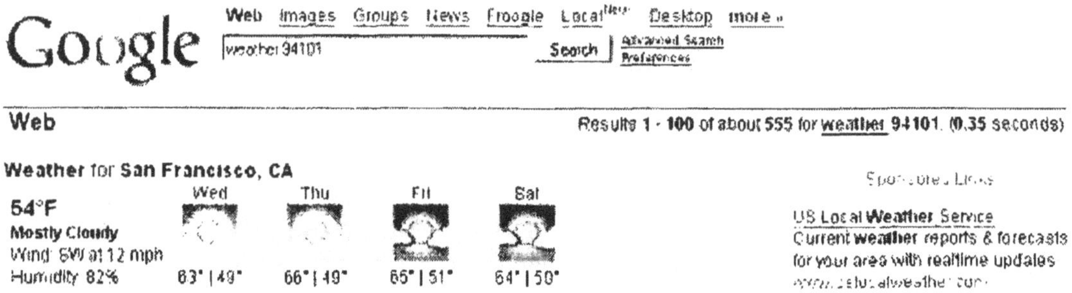
Google
Web Images Groups News Froogle Local Desktop more »
weather 94101
Search
Web
Results 1 - 100 of about 555 for weather 94101. (0.35 seconds)
Weather for San Francisco, CA
54°F
Mostly Cloudy
Wind: SW at 12 mph
Humidity: 82%
Wed
Thu
Fri
Sat
63° | 49°
66° | 49°
65° | 51°
64° | 50°
US Local Weather Service
Current weather reports & forecasts for your area with realtime updates

Airport Delays and Weather: To see delays and weather conditions at a US airport, enter the airport's three-letter code and the word *airport*. For example, to see delays and weather conditions at Baltimore-Washington International, enter [bwi airport]. At the top of the Google results page you will see the image of an airplane followed by a link to "View conditions at Baltimore-Washington International Airport (BWI), Baltimore, Maryland." The link takes you directly to the FAA's Air Traffic Control System Command Center's real-time status information page for BWI.

Phonebook, Street Maps, and Stock Quotes: US residential and business phonebook lookups, US addresses, and US stock exchange data. Please see Google Help for information on using these features:

Google Help http://www.google.com/help/features.html

Google Guides: Did you know that Google publishes "a variety of reviewer's guides to selected Google products on the Google Press Center" designed for journalists who are reviewing these products? However, the guides are very well done and include a lot of useful set-up and user instructions.

Google Guides http://www.google.com/press/guides.html

Google Services

Google has many services hidden behind its spare homepage. By selecting "more" and then "even more" on the homepage…

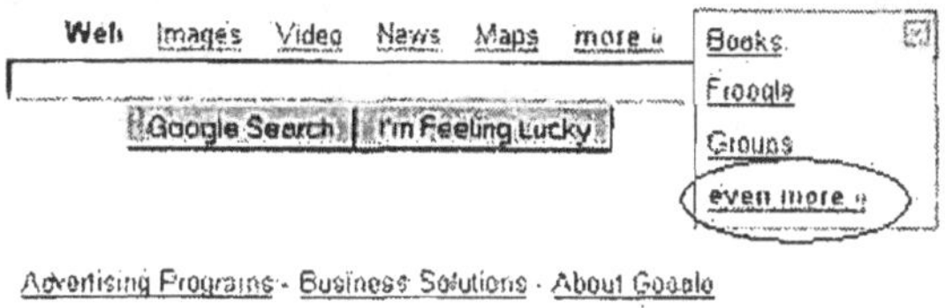
Web Images Video News Maps more »
Books
Froogle
Groups
even more »
Google Search I'm Feeling Lucky
Advertising Programs - Business Solutions - About Google

You will see the many products and services Google offers that are not reflected on the "googlized" homepage.

Froogle
Shop for items to buy online and at local stores

Images
Search for images on the web

Local
Find local businesses and get directions

Maps
View maps and get directions

News - now with archive search New!
Search thousands of news stories

Scholar
Search scholarly papers

Specialized Searches
Search within specific topics

Toolbar
Add a search box to your browser

Video
Search TV programs and videos

Web Search
Search over billions of web pages

Web Search Features
Find movies, music, stocks, books, and more

Search

Alerts
Get email updates on the topics of your choice

Blog Search
Find blogs on your favorite topics

Book Search
Search the full text of books

Catalogs
Search and browse mail-order catalogs

Checkout
Complete online purchases more quickly and securely

Desktop
Search and personalize your computer

Directory
Browse the web by topic

Earth
Explore the world from your PC

Finance
Business info, news, and interactive charts

Picasa
Find, edit and share your photos

SketchUp
Create 3D models for Google Earth

Talk
IM and call your friends through your computer

Translate
View web pages in other languages

Go mobile

Maps for mobile
View maps and get directions on your phone

Mobile
Use Google on your mobile phone

SMS
Use text messaging for quick info

Make your computer work better

Pack
A free collection of essential software

Web Accelerator
Speed up the web

Explore and innovate

Code
Download APIs and open source code

Co-op - now with Custom Search Engine New!
Contribute your expertise and customize the web search experience

Labs
Explore Google's technology playground

Communicate, show & share

Blogger
Share your life online with a blog -- it's fast, easy, and free

Calendar
Organize your schedule and share events with friends

Docs & Spreadsheets
Create and share your projects online and access them from anywhere

Gmail
Fast, searchable email with less spam

Groups
Create mailing lists and discussion groups

News: Google News headlines are entirely generated using a computer algorithm that scours more than 4500 worldwide news sources. Google News also offers ***international editions*** for France, Germany, India, Italy, Spain, and several other countries. For details on Google News, see the news search engine section below.

Images: Google Images indexes more than a billion images in JPEG, GIF, and PNG (Portable Network Graphics) formats. The Advanced Image Search lets users limit searches by filetype, size of image, coloration, and site or domain.

Google Image Search http://images.google.com/

Book Search: To use the book search, simply preface any search with the keyword *book* or *books*. The first three results, if there are any matches, will come from Google Book Search.

Groups: Google acquired Deja, the last remaining Usenet newsgroup search engine, in 2001. Even before that, Google began its own Usenet archive in August 2000. The complete Usenet archive, more than a billion messages dating back to 1995, is searchable via Google Groups. In 2004 Google introduced a new version of Google Groups that includes a mailing list and discussion forum creation option to rival Yahoo Groups' similar service. Also new is the ability for registered users to keep track of their favorite topics using the star (☆) feature. By clicking the star next to a favorite topic, that topic is added to the user's "My starred topics" page. Postings now appear in minutes in Google Groups rather than the hours it used to take.

Google Groups still offers both a simple and advanced interface to search the newsgroup postings. Both interfaces are extremely easy to use. Google Groups not only returns results (sorted by relevance or date), it also shows you the most relevant groups for your topic. So a search on the term *oceanography* suggests I might want to take a look at the related groups sci.geo.oceanography and bionet.biology.deepsea. Newsgroup searching in general and Google Groups are discussed in greater detail in a later section.

Google Groups http://groups.google.com/

Mobile SMS Search: This service is different from the SMS text messaging that has been available at AOL, Yahoo, Live Search, etc., for some time. The new Google SMS permits queries using mobile technology. Google's SMS service offers similar services but with different shortcuts. It is open to all US subscribers using a "major" US cell phone provider and also to most UK mobile subscribers. The US number is 46645 (GOOGL on most phones) and for the UK it is 64664 (6GOOG on most phones). Google explains how to use the SMS search service and offers a number of sample queries at its new SMS webpage.

Google SMS http://www.google.com/sms/

Patent Search (beta): New for 2007, Google Patent Search now has its own discrete page. Users could always search for US patents by number, but Google decided to create a separate page for these searches. The new site offers many advanced search options, including options to search by patent number, title, inventor's name, assignee's name, US and international classifications, and issue or filing date range. Even more valuable than the search options are the view choices.

The new Google patent search has the ability to show you the patent itself, complete with zoom, page scrolling, drawings, internal search, and other patents that reference the current one. Here is the drawing of H. K. Markey (aka the actress Hedy Lamarr) and George Antheil's 1941 patent for a "secret communication system" that used a new concept: frequency hopping.

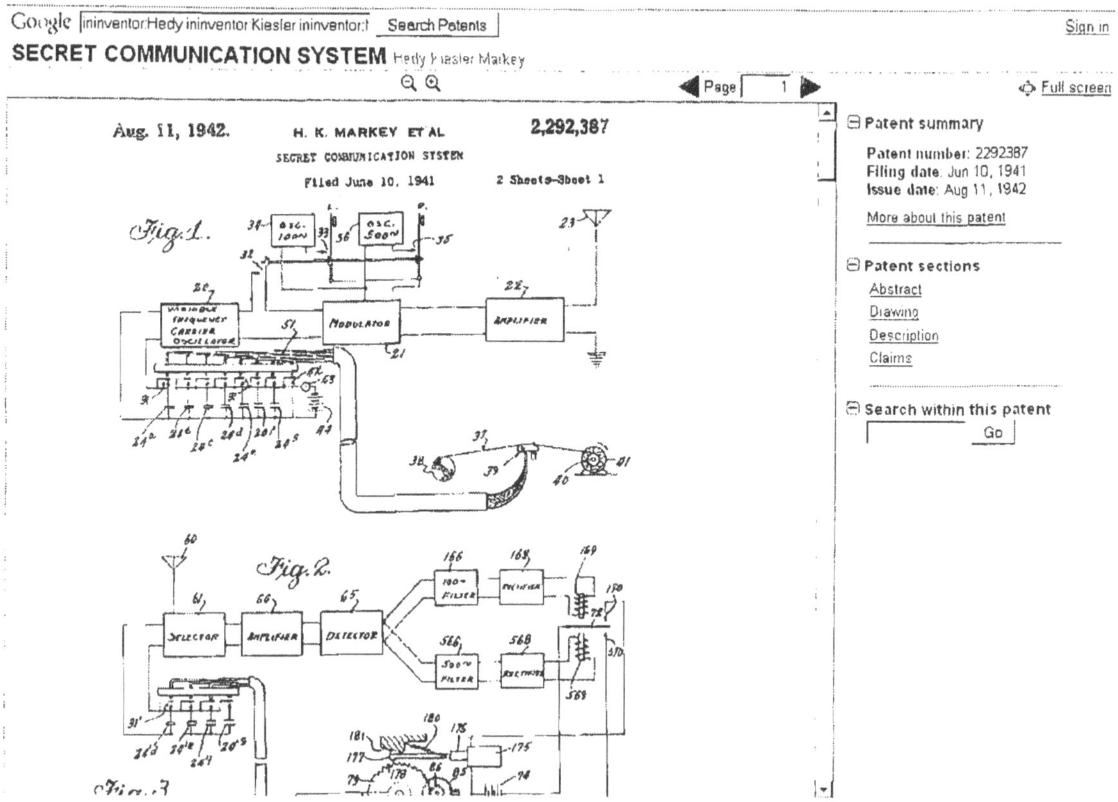
Google ininventor:Hedy ininventor:Kiesler ininventor:I Search Patents
Sign in
SECRET COMMUNICATION SYSTEM Hedy Kiesler Markey
Page 1
Full screen
Aug. 11, 1942. H. K. MARKEY ET AL 2,292,387
SECRET COMMUNICATION SYSTEM
Filed June 10, 1941 2 Sheets-Sheet 1
Fig. 1.
Fig. 2.
Patent summary
Patent number: 2292387
Filing date: Jun 10, 1941
Issue date: Aug 11, 1942
More about this patent
Patent sections
Abstract
Drawing
Description
Claims
Search within this patent
Go

Google Patent Search http://www.google.com/patents

Blog Search: Google Blog Search is a direct competitor with Technorati, until now the big dog on the blog search block. Despite all the chatter about them, blogs are still kind of mysterious and confusing in part because there are lots of things called blogs that don't fit the earlier definition. Blogs originally referred to on-line personal journals, often updated daily, but now everything using RSS or Atom (XML formats for distributing newsfeeds) seems to be considered a blog. Therefore, blog search engines generally restrict themselves to indexing and searching for anything that uses a site feed. Google Blog Search is no exception. The new Google Blog Search FAQ says, "The goal of Blog Search is to *include every blog that publishes a site feed* (either RSS or Atom)." [emphasis added] This means that Google Blog Search defines "blog" as any site with an XML site feed, and that is fine as long as we know what we're getting. However, Google Blog Search is apparently excluding feeds from news sources to try to prevent overlap between Google News and Blog Search.

Keep in mind that ***Google Blog Search only indexes the site feed, not the full content at the website*** that originated the feed.

Google Blog Search indexes feeds dating back to January 2000. Also, one of the big advantages of the XML format is that it, unlike HTML, includes date/time data, which means you can use Google Blog Search to find information from a specific day or a range of dates. Google Blog Search will also enable users to search entire blogs or specific posts.

Some of the Google Search operators work in Google Blog Search and it has its own unique operators, too, as the About Google Blog Search page explains:

All of the standard Google Search operators are supported in Blog Search. These include:

- link: [very useful in finding who's linking to whom]
- site:
- intitle:

Additionally, Blog Search supports the following new operators of its own:

- inblogtitle:
- inposttitle:
- inpostauthor:
- blogurl:

For example, a search such as [mandolin inpostauthor:Graham] will show you posts about mandolins written by people named Graham. Note that you can also use the Advanced Search option to achieve the same effect.

In addition, you can restrict your results to any one or any combination of 35 languages using the Advanced Search option. Google Blog Search will also give users the option of subscribing to the blogs in the news aggregator of your choice.

The main drawback of Google Blog Search seems to be that it indexes only the content of *feeds* and often what is syndicated in a newsfeed is very sparse. Technorati wins hands down on this point because it does do full text searching. Google may eventually decide it needs to do so as well. After all, this is a beta version.

Google Blog Search http://blogsearch.google.com/

About Google Blog Search http://www.google.com/help/about_blogsearch.html

Directory: Google's web directory uses the Open Directory Project's collection and its own search technology to rank the sites based upon "importance" (which usually means popularity). Google Directory lets users limit searches to a specific directory category. For example, if I search the directory for the keyword [afghan], Google presents two categories:

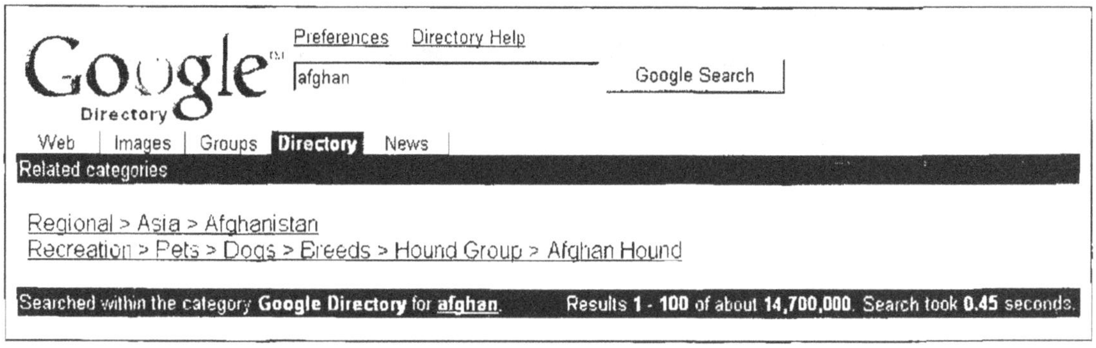
Google Directory
Preferences Directory Help
afghan
Google Search
Web Images Groups Directory News
Related categories
Regional > Asia > Afghanistan
Recreation > Pets > Dogs > Breeds > Hound Group > Afghan Hound
Searched within the category Google Directory for afghan.
Results 1 - 100 of about 14,700,000. Search took 0.45 seconds.

By selecting the *Regional > Asia > Afghanistan* category, I have the option to search only in this category, thus focusing my search and avoiding irrelevant results. The Google Directory contains over 1.5 million urls.

Google Directory http://directory.google.com/

Video Search: As of the first of the year 2007, Google Video began to include results from YouTube, which it purchased in October 2006. For now, when users click on the YouTube results, they are taken to the YouTube website. YouTube videos do not appear on the Google video homepage, only in search results where they are recognizable from their address.

Google Video Search is a way to ***search and view*** TV shows, including news, entertainment, and more, for free. The search includes not only national networks, such as ABC, CBS, NBC, and CNN, but also local programming and shows from around the world. How does this work? Google Video indexes the closed captioning of TV shows so that when you search for a keyword, it finds that word in the captioning transcripts and displays a list of the shows with that keyword. In most cases, you can only see still shots of the show, but in a few instances, you will be able to view the entire broadcast. This is a very new Google option, and it is sure to expand and improve over time.

One of the best features of Google Video is that you do not need any special software to view the videos, only Macromedia Flash Player, which is a free browser plug-in that most users already have on their computers.

Google Video offers users the opportunity to rank video using a system of one to five stars for users to rate videos as well as the ability to add labels (tags to describe a video) and comments.

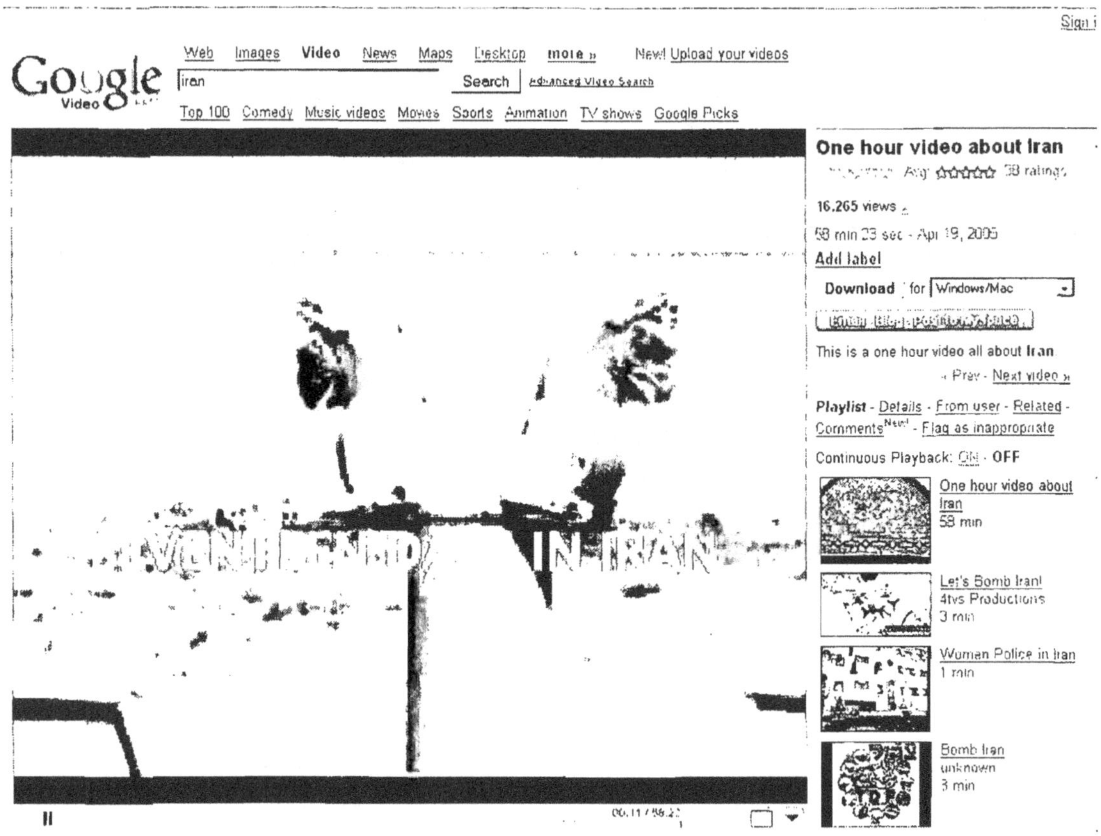
Google Video
Web
Images
Video
News
Maps
Desktop
more »
New! Upload your videos
iran
Search
Top 100
Comedy
Music videos
Movies
Sports
Animation
TV shows
Google Picks
One hour video about Iran
16,265 views
Add label
Download
Windows/Mac
This is a one hour video all about Iran
Playlist - Details - From user - Related -
Comments - Flag as inappropriate
Continuous Playback: ON - OFF
One hour video about Iran
58 min
Let's Bomb Iran!
4tvs Productions
3 min
Women Police in Iran
1 min
Bomb Iran
unknown
3 min

There are some of the operators that work in Google Video:

title: enter the name of a TV show on one of the stations Google Video indexes, with or without keywords. For example:

[title:nature] or [title:newshour robotics]

genre: enter news, comedy, music, animation with or without a keyword; I recommend referring to the complete list of genres on the Advanced Search page.

type: enter sports, music_video, movie with or without a keyword.

duration: short (< 4 min.), medium (4-20 min.), or long (> 20 min.) with or without a keyword.

is:free or **is:forsale** determines whether or not the search finds free or for purchase videos.

language: limits the search to videos in a specific language using the same digraphs as main Google search.

Google Video http://video.google.com/

Google Scholar

In November 2004, Google introduced a new tool called Google Scholar. Here's how the Google site describes it: "Google Scholar enables you to search specifically for scholarly literature, including peer-reviewed papers, theses, books, preprints, abstracts and technical reports from all broad areas of research. Use Google Scholar to find articles from a wide variety of academic publishers, professional societies, preprint repositories and universities, as well as scholarly articles available across the web." Moreover, "Google Scholar...automatically analyzes and extracts citations and presents them as separate results, even if the documents they refer to are not online. This means your search results may include citations of older works and seminal articles that appear only in books or other offline publications." Google Scholar not only indexes journal articles, dissertations, and technical reports, it also indexes books, which means you can use Google's new Library Search (OCLC's WorldCat search) to locate the book in a local library or find a place to purchase the book online.

Although a number of scholarly search sites and tools already exist—e.g., CiteSeer, DOAJ, ArXiv, and even Google's own partnership with IEEE—the fact that the premier search engine has branched off into scholarly search is obviously significant. Google Scholar searches across a far wider range of sources than any other publicly available scholarly search tool currently available. Users should be able to read at least an abstract of articles that require registration and access the full text if they or their institution have a subscription for the content. The best thing about Google Scholar is that it gives users the range, power, and flexibility of Google. As far as I can tell, all the types of Google syntax—*site:*, *inurl:*, *filetype:*, etc.—work with Google Scholar. You can limit your search to file type using either the *filetype:* or *ext:* syntax, e.g., [ext:pdf] (*filetype:* and *ext:* work interchangeably). The most useful addition to Google Scholar is probably the new *author:* syntax (which, by the way, already existed in Google Groups search).

As you can see from this query, Google Scholar searches and retrieves scholarly references from many types of sources and also provides a handy "Cited by" link that shows all the pages referring to the original work.

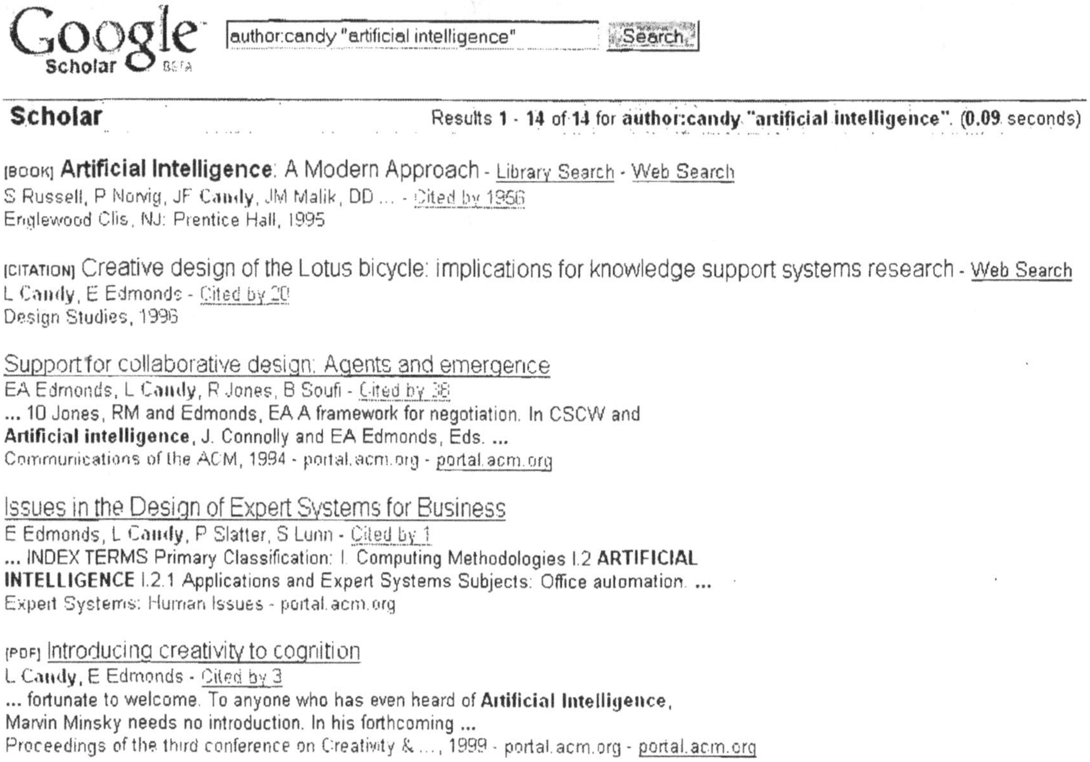

Google Scholar BETA | author:candy "artificial intelligence" | Search

Scholar — Results 1 - 14 of 14 for author:candy "artificial intelligence". (0.09 seconds)

[BOOK] **Artificial Intelligence**: A Modern Approach - Library Search - Web Search
S Russell, P Norvig, JF **Candy**, JM Malik, DD ... - Cited by 1956
Englewood Clis, NJ: Prentice Hall, 1995

[CITATION] Creative design of the Lotus bicycle: implications for knowledge support systems research - Web Search
L **Candy**, E Edmonds - Cited by 20
Design Studies, 1996

Support for collaborative design: Agents and emergence
EA Edmonds, L **Candy**, R Jones, B Soufi - Cited by 38
... 10 Jones, RM and Edmonds, EA A framework for negotiation. In CSCW and
Artificial intelligence, J. Connolly and EA Edmonds, Eds. ...
Communications of the ACM, 1994 - portal.acm.org - portal.acm.org

Issues in the Design of Expert Systems for Business
E Edmonds, L **Candy**, P Slatter, S Lunn - Cited by 1
... INDEX TERMS Primary Classification: I. Computing Methodologies I.2 **ARTIFICIAL INTELLIGENCE** I.2.1 Applications and Expert Systems Subjects: Office automation. ...
Expert Systems: Human Issues - portal.acm.org

[PDF] Introducing creativity to cognition
L **Candy**, E Edmonds - Cited by 3
... fortunate to welcome. To anyone who has even heard of **Artificial Intelligence**,
Marvin Minsky needs no introduction. In his forthcoming ...
Proceedings of the third conference on Creativity & ..., 1999 - portal.acm.org - portal.acm.org

Google Scholar also offers an **advanced search option**. It certainly simplifies searching for articles by author, articles published in a specific publication, and words in the articles' title. However, as with most date searches, forget it. I searched for articles about chemistry published in the year 2020 and found three. Either Google knows something about the future that we don't or their software is misreading some number as a year. The advanced Google Scholar search also let users limit their search by publication. This is somewhat misleading because the "publication" can be a citation, article, or book, although there is no way to tell Google Scholar to distinguish among these choices. Also, the publication searches are imperfect; a search limited to the publication *Nature* also returns results from *Nature Medicine*, for example.

During 2006, Google Scholar added a new feature that "will make it easier for researchers to keep up with recent research...It's not just a plain sort by date, but rather we try to rank recent papers the way researchers do, by looking at the prominence of the author's and journal's previous papers, how many citations it already has, when it was written, and so on. Look for the new link on the upper right for 'Recent articles'—or switch to 'All articles' for the full list."[40]

---

[40] Dejan Perkovic, "Keeping up with recent research," Google Blogspot, 20 April 2006, <http://googleblog.blogspot.com/2006/04/keeping-up-with-recent-research.html> (31 October 2006).

Also new for 2006 was a related search option: "For every Google Scholar search result, we try to automatically determine which articles in our repository are most closely related to it. You can see a list of these articles by clicking the 'Related Articles' link that appears next to each result. The list of related articles is ranked primarily by how similar these articles are to the original result, but also takes into account the relevance of each paper."[41]

Péter Jacsó has called Google Scholar's quality into question in his excellent and thorough analysis of Google Scholar's citation ability. Jacsó, Professor of Library and Information Science, University of Hawaii, concluded that "Google Scholar (GS) does a really horrible job matching cited and citing references."[42] There are numerous other scholarly citation search options (CiteSeer, ISI Highly Cited, and Scirus) that, for now at least, are superior to Google Scholar.

However, I would not count Google Scholar out in the long run. Google Scholar is yet another example of what are called "vertical search engines," that is, search services that focus on indexing and searching specialized data sources. Vertical search has fundamentally replaced the portal concept as a more targeted, less manpower-intensive, and more cost effective means of getting the right information to the right people at the right time.

Google Scholar http://scholar.google.com/

Advanced Google Scholar Search http://scholar.google.com/advanced_scholar_search

Google Trends

Google unveiled Google Trends in May 2006 and set a lot of people thinking about its potential utility. Google Trends is a new technology that lets users see how many searches have been performed on one to five terms and where those searches originate.

> "Google Trends analyzes a portion of Google web searches to compute how many searches have been done for the terms you enter relative to the total number of searches done on Google over time. We then show you a graph with the results—our search-volume graph—plotted on a linear scale.

---

[41] Luiz Barroso, Distinguished Engineer, "Exploring the scholarly neighborhood," Google Blogspot, 22 August 2006, <http://googleblog.blogspot.com/2006/08/exploring-scholarly-neighborhood.html> (10 October 2006).

[42] Péter Jacsó, "Google Scholar and *The Scientist*," Péter Jacsó's Review Extras, October 2005, <http://www2.hawaii.edu/~jacso/extra/gs/> (31 October 2006).

Located just beneath our search-volume graph is our news-reference-volume graph. This graph shows you the number of times your topic appeared in Google News stories. When Google Trends detects a spike in the volume of news stories for a particular term, it labels the graph and displays the headline of an automatically selected Google News story written near the time of that spike. Currently, only English-language headlines are displayed, but we hope to support non-English headlines in the future. Below the search and news volume graphs, Google Trends displays the top cities, regions, and languages for the first term you entered."[43]

There are some very important limitations to Google Trends, however. First of all, the feedback provided by Google Trends is based on ***a portion of Google's searches***, not all of them. Google Trends seeks to provide "insights into broad search patterns," not detailed and verifiable data about searches. Second, "as a Google Labs product, it is still in the early stages of development," meaning it is prone to error because "several approximations are used when computing your results," but Google does not say what these are.[44]

Here's a look at Google Trends' results for the query comparing search terms ["north korea",dprk] for all regions and all years. Note that in Google Trends, you can ***compare terms by using a comma to separate them***.

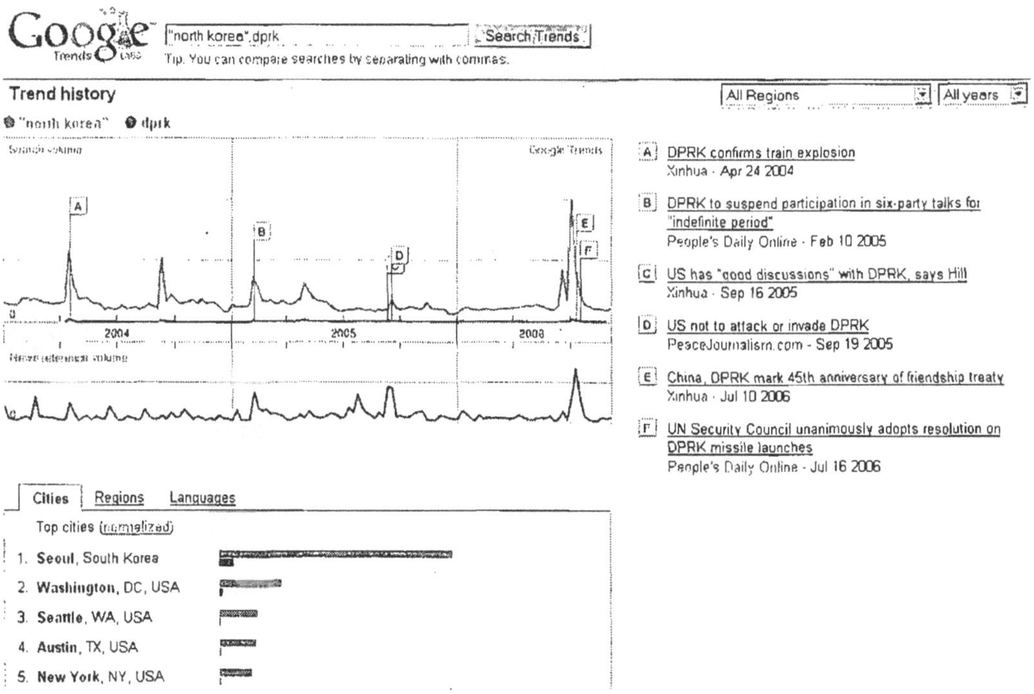
"north korea",dprk
Search Trends
Tip: You can compare searches by separating with commas.
Trend history
All Regions
All years
"north korea"
dprk
2004
2005
2006
A DPRK confirms train explosion
Xinhua - Apr 24 2004
B DPRK to suspend participation in six-party talks for "indefinite period"
People's Daily Online - Feb 10 2005
C US has "good discussions" with DPRK, says Hill
Xinhua - Sep 16 2005
D US not to attack or invade DPRK
PeaceJournalism.com - Sep 19 2005
E China, DPRK mark 45th anniversary of friendship treaty
Xinhua - Jul 10 2006
F UN Security Council unanimously adopts resolution on DPRK missile launches
People's Daily Online - Jul 16 2006
Cities
Regions
Languages
Top cities (normalized)
1. Seoul, South Korea
2. Washington, DC, USA
3. Seattle, WA, USA
4. Austin, TX, USA
5. New York, NY, USA

[43] "About Google Trends," Google Trends, 2006, <http://www.google.com/intl/en/trends/about.html> (31 October 2006).

[44] Google Trends.

Despite all its limitations, I am intrigued by the potential this tool offers. For example, if I were looking back over the past couple of years for a very obscure term, I would definitely use Google Trends to see if I could discern anything useful from this data.

Google Trends http://www.google.com/trends

Google Guides

Did you know that Google publishes "a variety of reviewer's guides to selected Google products on the Google Press Center" designed for journalists who are reviewing these products? However, the guides are very well done and include a lot of useful set-up and user instructions.

Google Guides http://www.google.com/press/guides.html

Web Tip

Everything isn't on the Internet (or it's not free)!

Contrary to popular opinion, everything is not on the Internet. In fact, much of the kind of information you are used to working with is not and never will be on the Internet. Unrealistic expectations about the kinds of information you may find on the Internet can lead to frustration and wasted time and effort. A general rule of thumb: the more sensitive, rare, or expensive the information, the less likely it is to be on the Internet. Also, much valuable data on the Internet requires payment.

Google Hacks

Google hacks—a term usually associated with the book of the same name published by O'Reilly Publishing—are tips, tricks, techniques, and scripts that make Google more powerful and useful. Some are extraordinarily simple, such as being conscious of word order, while others are either so complicated ("scraping the code") or trivial (Googlisms) that I doubt many of you will ever need them.

Because there is so much interest in Google hacks, I thought I would catalog links to the best sites on the web for finding more information about Google hacks as well as bring the best Google hack techniques together in one place.

First, a word about **Google APIs**, which are used to create many Google hacks. API stands for Application Programming Interface. Google offers its own free APIs developers' kit, which provides documentation and example code for using the Google Web APIs service.[45] That's fine if you are in a situation that permits downloading, installing, and running code from the web, but that is not always an option. However, many Google hacks either do not require an API key or, if they do, are available, thanks to the kindness of strangers, on websites.

Best Google Hack Websites

The Official (but not the best): Google hacks from O'Reilly. Taken directly from the book of the same name by Tara Calishain and Rael Dornfest, the complete list of 100 hacks is here, but only a few have details about how to use them.

Google Hacks http://hacks.oreilly.com/pub/ht/2

Researchbuzz.com: Google Hacks Archive. Much better source of Google hacks from Tara Calishain's website. Google hacks are listed by date and you can search the site.

Researchbuzz http://www.researchbuzz.org/archives/google_hacks.shtml

Google API FAQs. Everything you need to know about Google's API service. Remember, this process involves registration, downloading software, and other interaction with Google, so it's not for everyone.

Google API FAQS http://www.google.com/apis/api_faq.html

---

45 "As of December 5, 2006, we are no longer issuing new API keys for the SOAP Search API. Developers with existing SOAP Search API keys will not be affected." Google SOAP Search API <http://code.google.com/apis/soapsearch/> (20 February 2007).

<u>Staggernation: Three Google APIs</u>.

- <u>API Proximity Search (GAPS)</u> "uses the Google API to search Google for two search terms that appear within a certain distance from each other on a page. It does this by using a seldom-discussed Google feature: within a quoted phrase, * can be used as a wildcard meaning 'any word.'" ***This is a very useful tool***; it gives users the option of searching for two terms within one, two, or three words of each other in any order or a specific order. <u>http://www.staggernation.com/cgi-bin/gaps.cgi</u>

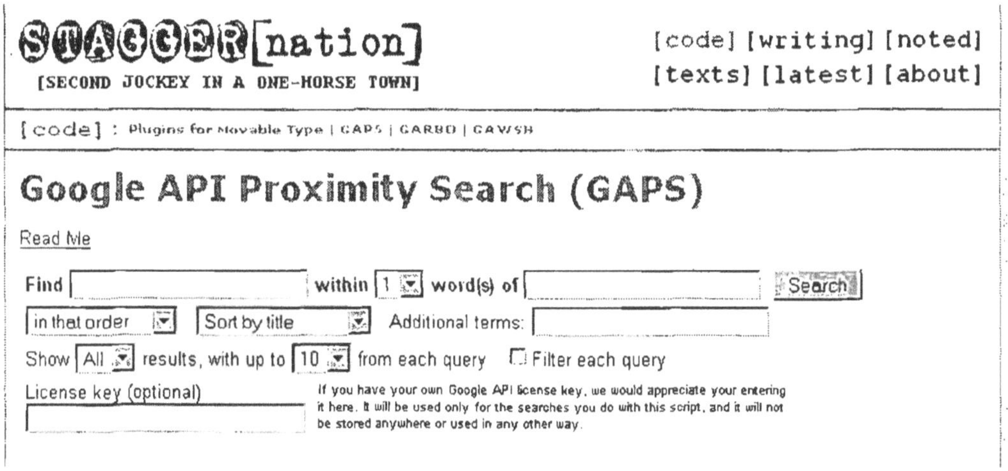
STAGGER[nation]
[SECOND JOCKEY IN A ONE-HORSE TOWN]
[code] [writing] [noted]
[texts] [latest] [about]
[code] : Plugins for Movable Type | GAPS | GARBO | GAWSH
Google API Proximity Search (GAPS)
Read Me
Find
within 1 word(s) of
Search
in that order
Sort by title
Additional terms:
Show All results, with up to 10 from each query
Filter each query
License key (optional)
If you have your own Google API license key, we would appreciate your entering it here. It will be used only for the searches you do with this script, and it will not be stored anywhere or used in any other way.

- <u>API Web Search By Host (GAWSH)</u> "uses the Google API to search Google for a query string, and returns a list of the web hosts found in the set of results. You can then expand any of these hosts and display only the results from that host...Clicking on the triangle to the left of a host will perform the same query again, but restricted to that host (using Google's "**site:www.foo.com**" query syntax), and expand the listing to display the first 10 results." I have to admit I find this particular script more confusing than useful. I prefer to use Google's **site:** syntax, but others may find this API to their liking.
<u>http://www.staggernation.com/gawsh/</u>

- <u>API Relation Browsing Outliner (GARBO)</u> uses the Google API to search Google for pages that are either related to (using the related: keyword) or linked to (using the **link:** keyword) that URL. ***Does not add a great deal to what Google can do already***, beyond offering the option to view the results as snippets or urls.
<u>http://www.staggernation.com/garbo/</u>

<u>Google Rankings</u>. This site contains a number of different options created as search engine optimization tools for website creators and maintainers. However, some of these Google tools are what you might call "dual use technologies."

➢ Keyword Density. Probably not something everyone needs everyday, but a pretty neat tool. It lets you enter a url, then see which are the words and phrases that address uses most in the form of a detailed report on their numbers and density. Although it was designed as a search engine optimization tool, I can see its utility as a rudimentary traffic analysis tool.

Google Rankings http://www.googlerankings.com/kdindex.php

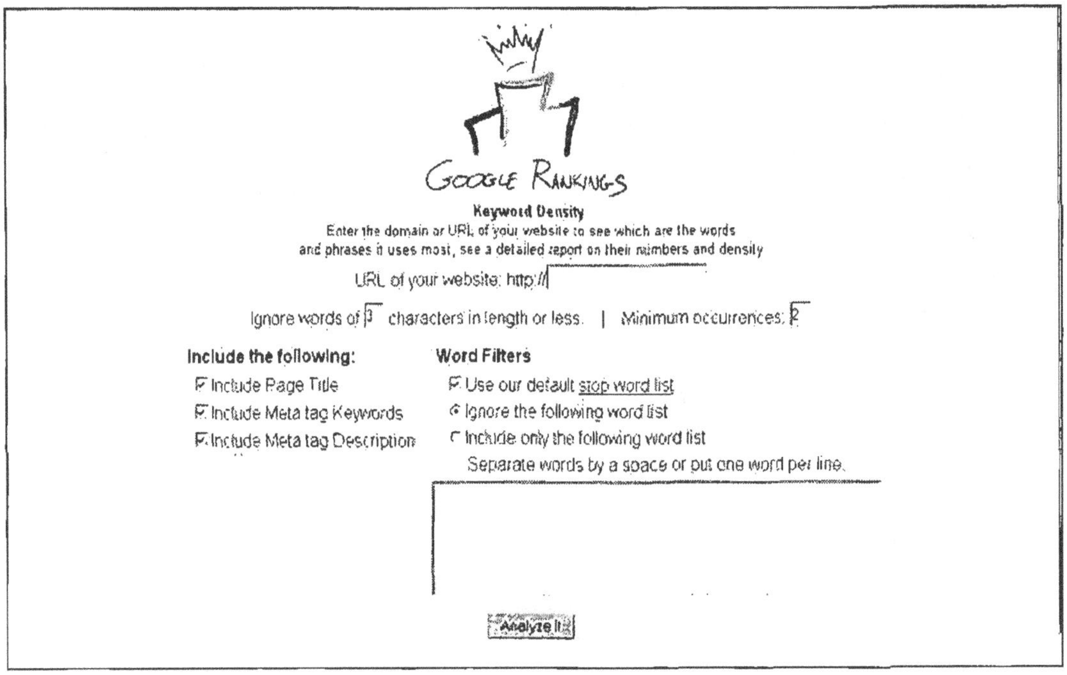
Google Rankings
Keyword Density
Enter the domain or URL of your website to see which are the words
and phrases it uses most, see a detailed report on their numbers and density
URL of your website: http://
Ignore words of 3 characters in length or less. | Minimum occurrences: 2
Include the following:
Include Page Title
Include Meta tag Keywords
Include Meta tag Description
Word Filters
Use our default stop word list
Ignore the following word list
Include only the following word list
Separate words by a space or put one word per line.
Analyze It

➢ Mass Keyword Search: This tool lets you enter from two to ten keywords and a url to see where (if anywhere) the site ranks in Google *vis-à-vis* those keywords. The tool only searches up to the top 1000 Google sites. This could be useful to see if an unusual term is found at a specific site. http://www.googlerankings.com/mkindex.php

➢ Mass Domain Search: With this tool, you can enter up to 10 different urls (domains) for the same keyword and check the position of the websites in Google up to the top 1000 sites. This could be useful to see where an unusual term ranks in a number of different sites at once. http://www.googlerankings.com/mdindex.php

FindForward. If you love Google, you will probably find Philipp Lenssen's creation very useful. Lenssen has done Google one better by creating a simple way to power search Google with the help of the Google Web APIs. All users have to do is to enter a query (FindForward supports all basic and advanced Google search options and syntax) and select the type of search desired from the pulldown list. As you can see from the list below, there are many possible types of searches available. Try the **Search Grid** option, in which you can enter up to five keywords to create a grid showing inter-relationships of the terms. Try [iran korea nuclear terror], for example. Then there is the potentially dangerous **Just Files** option for searching website directories that were (probably) not meant to be browsed.

Find Forward http://www.findforward.com/

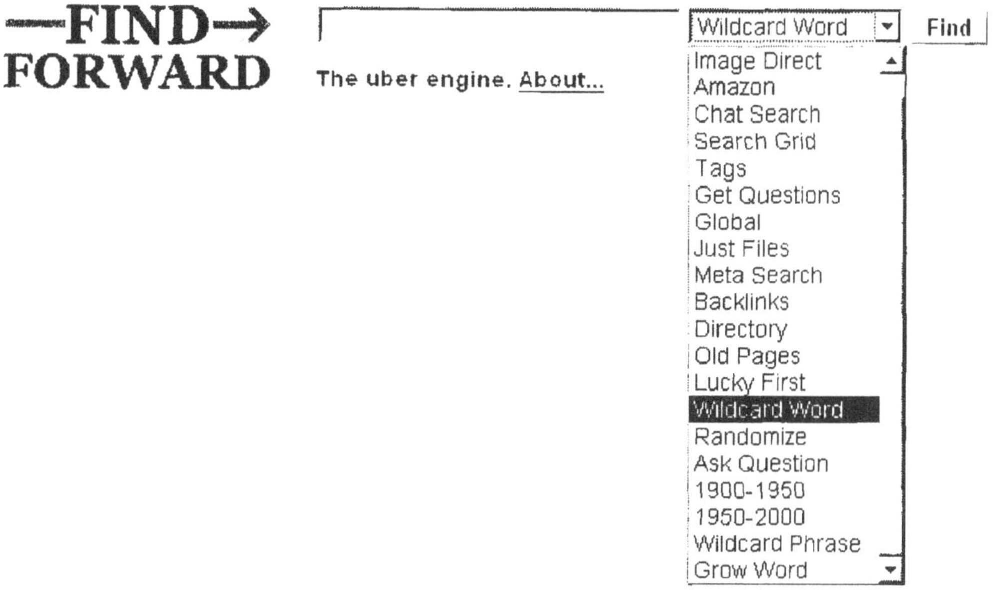
FIND
FORWARD
The uber engine. About...
Wildcard Word
Find
Image Direct
Amazon
Chat Search
Search Grid
Tags
Get Questions
Global
Just Files
Meta Search
Backlinks
Directory
Old Pages
Lucky First
Wildcard Word
Randomize
Ask Question
1900-1950
1950-2000
Wildcard Phrase
Grow Word

Fagan Finder's Google Ultimate Interface. A useful and friendly Google interface that allows users to maximize Google power without learning the syntax that it deserves to be on this list.

http://www.faganfinder.com/google.html (for Internet Explorer)
http://www.faganfinder.com/google2.html (for Mozilla browsers)

Soople. A Google interface that makes Google's advanced features so easy to use it's amazing. The main page offers lots of query boxes that enable users to run complex Google queries without knowing anything about Google. But I think the best Soople tool is the calculator page. Did you know Google has a very powerful calculator function as a basic part of the search engine? Most people don't, and even if they did, they wouldn't use it because it is not what I would call intuitive. Therefore, the Soople calculator interface to the Google calculator is a godsend for the mathematically impaired such as yours truly.

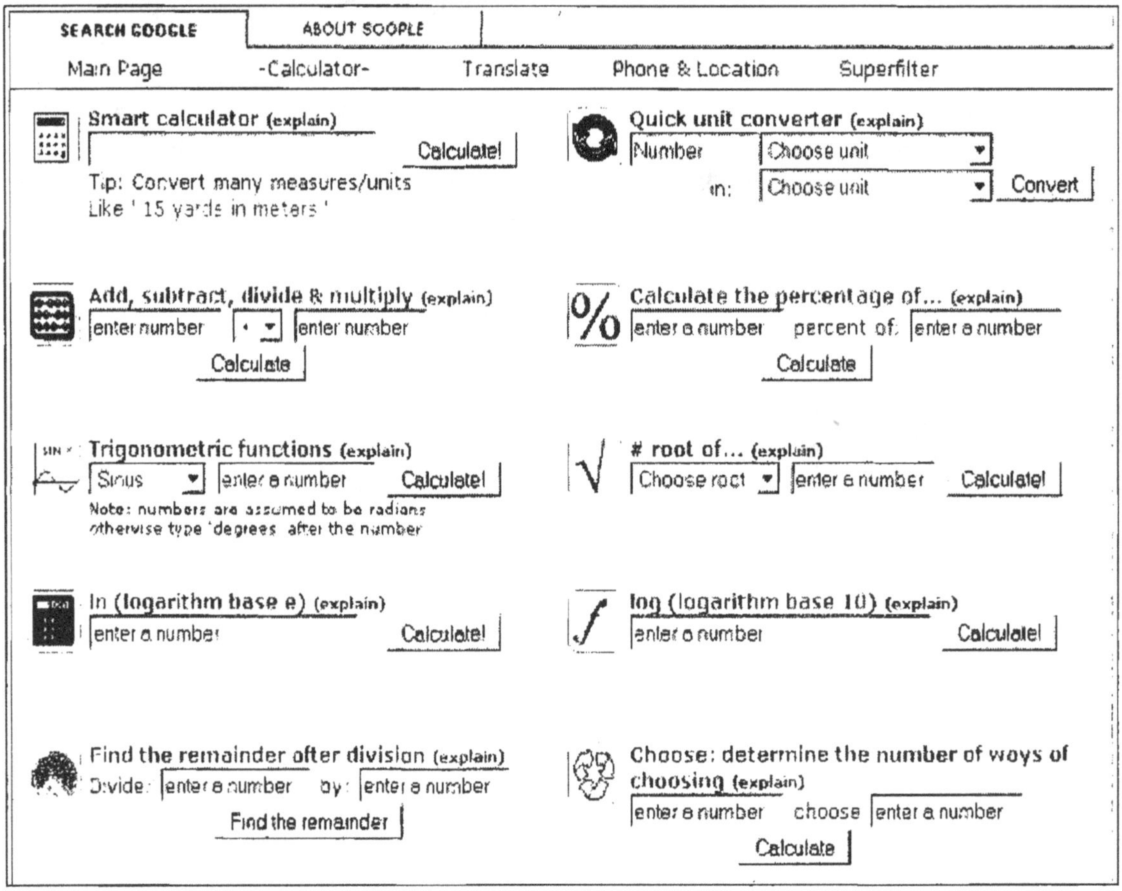
SEARCH GOOGLE
ABOUT SOOPLE
Main Page
-Calculator-
Translate
Phone & Location
Superfilter
Smart calculator (explain)
Calculate!
Tip: Convert many measures/units
Like ' 15 yards in meters '
Quick unit converter (explain)
Number
Choose unit
in:
Choose unit
Convert
Add, subtract, divide & multiply (explain)
enter number
enter number
Calculate
Calculate the percentage of... (explain)
enter a number
percent of:
enter a number
Calculate
Trigonometric functions (explain)
Sinus
enter a number
Calculate!
Note: numbers are assumed to be radians
otherwise type 'degrees' after the number
root of... (explain)
Choose root
enter a number
Calculate!
ln (logarithm base e) (explain)
enter a number
Calculate!
log (logarithm base 10) (explain)
enter a number
Calculate!
Find the remainder after division (explain)
Divide:
enter a number
by:
enter a number
Find the remainder
Choose: determine the number of ways of choosing (explain)
enter a number
choose
enter a number
Calculate

Soople Calculator http://www.soople.com/index.php?sub=calculator

Compare Google Results From Different Countries. This is a valuable and much needed new tool. This site lets you run one query simultaneously against two different Google locations. What is more, it also lets you choose between two different Google datacenters. For example, if you select "www.google.com" as a datacenter, the query could go to a number of different access points. It is probably better to pick a specific Google datacenter. The site also lists many but not all local Google domains. If you select a local domain, the search and results' language will default to that country's language. However, if you specify a local Google domain, the tool may use any existing datacenter.

It is not really as hard to use this tool as I've made it sound. You can easily play around with this tool and see what works for you. Here's an example of a search for a Spanish term in the US using the generic "www.google.com" and the local Google and datacenter for Spain. The results are quite different.

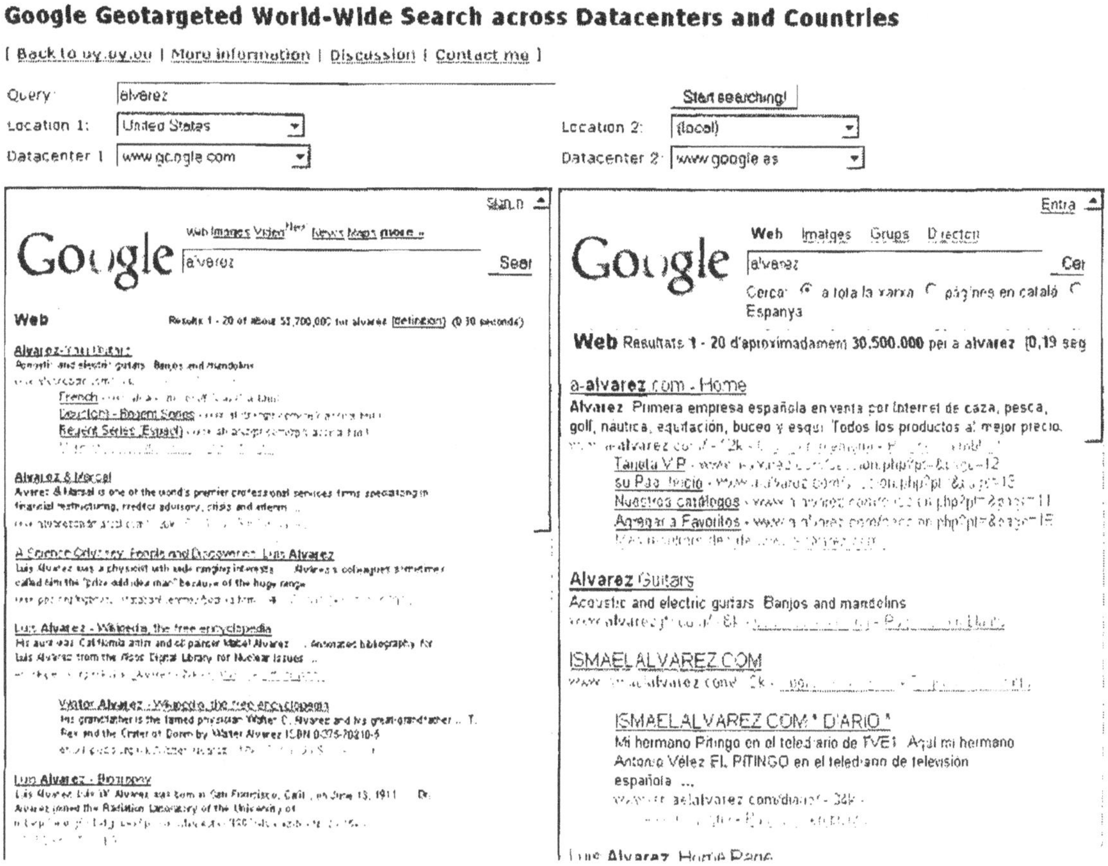
Google Geotargeted World-Wide Search across Datacenters and Countries
[Back to oy-oy.eu | More information | Discussion | Contact me]
Query: alvarez
Start searching!
Location 1: United States
Location 2: (local)
Datacenter 1: www.google.com
Datacenter 2: www.google.es
Google
Google
Web
Web
Alvarez Guitars
Acoustic and electric guitars. Banjos and mandolins

Using Google to search for terms using non-Latin character and/or diacritical marks remains problematic. I recommend that you ***put a plus sign in front of any non-English term*** when using this tool. While the Google geographic comparison tool is imperfect, it is a useful and interesting way to see how Google handles queries differently depending upon geographic region.

Compare Google results for different geographic locations
http://oy-oy.eu/google/world/

Simply Google. On one page you can see and use pretty much every Google search, find most Google sites, link to all the Google blogs, and even locate some other search blogs and sites. This really is a useful site. Notice the live bookmarks for all the blogs. Google search syntax works fine here (it's just an interface to the "real thing").

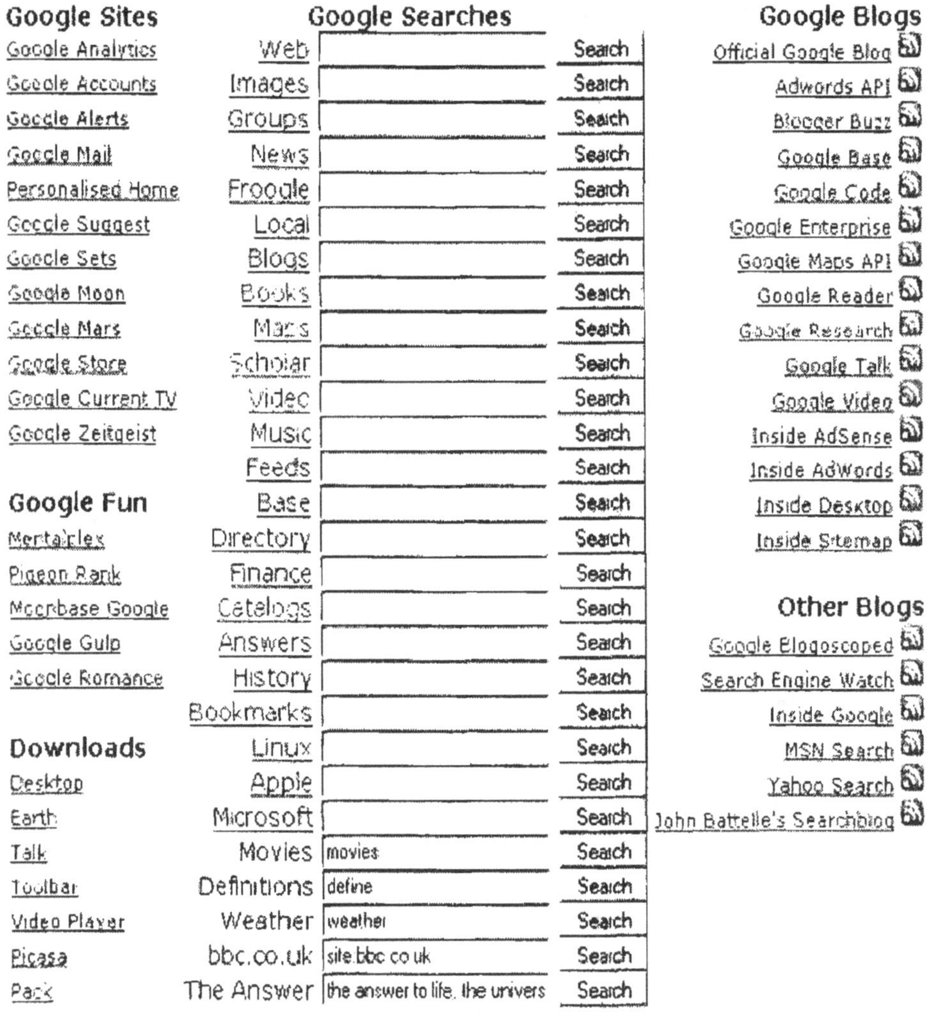
Google Sites
Google Analytics
Google Accounts
Google Alerts
Google Mail
Personalised Home
Google Suggest
Google Sets
Google Moon
Google Mars
Google Store
Google Current TV
Google Zeitgeist
Google Fun
Pigeon Rank
Moonbase Google
Google Gulp
Google Romance
Downloads
Desktop
Earth
Talk
Toolbar
Video Player
Picasa
Pack
Google Searches
Web
Images
Groups
News
Froogle
Local
Blogs
Books
Maps
Scholar
Video
Music
Feeds
Base
Directory
Finance
Catalogs
Answers
History
Bookmarks
Linux
Apple
Microsoft
Movies
Definitions
Weather
bbc.co.uk
The Answer
movies
define
weather
site:bbc.co.uk
the answer to life, the univers
Search
Google Blogs
Official Google Blog
Adwords API
Blogger Buzz
Google Base
Google Code
Google Enterprise
Google Maps API
Google Reader
Google Research
Google Talk
Google Video
Inside AdSense
Inside AdWords
Inside Desktop
Inside Sitemap
Other Blogs
Google Blogoscoped
Search Engine Watch
Inside Google
MSN Search
Yahoo Search
John Battelle's Searchblog

One thing that I found really interesting behind the concept of this page is that it has been "de-googlized." What does that mean? Well, "googlizing" a home page means stripping it down basically to one thing, in Google's case, that one thing is search. The critics of googlization point out that while that worked great for Google, it rarely works most of the time for the simple reason that most websites are trying to do something more than just one thing. In fact, "the experience of using Google is not the experience of using any other site. People go to Google to search the entire web."[46] People go to other sites for other reasons and to do other things, sometimes many other things. Furthermore, Google is not simple! It just hides its complexity behind that plain homepage.

Simply Google http://www.usabilityviews.com/simply_google.htm

---

[46] Jared Spool, "Home Page Googlization," User Interface Engineering Brainsparks.com, April 6, 2006, <http://www.uie.com/brainsparks/2006/04/05/home-page-googlization/> (30 October 2006).

Google Related Image Search. The ever-creative Philipp Lenssen has developed a new tool that looks for related images. It is a small PHP5 script that is "first screenscraping Google Sets, and then it's screenscraping Google Images (once per term found in the set, which is why this takes some seconds)." <http://blog.outer-court.com/forum/68971.html> I have to say it is mostly a novelty at this point, but the results are interesting and could potentially be useful. Here are the results for the query [solar system, milky way]:

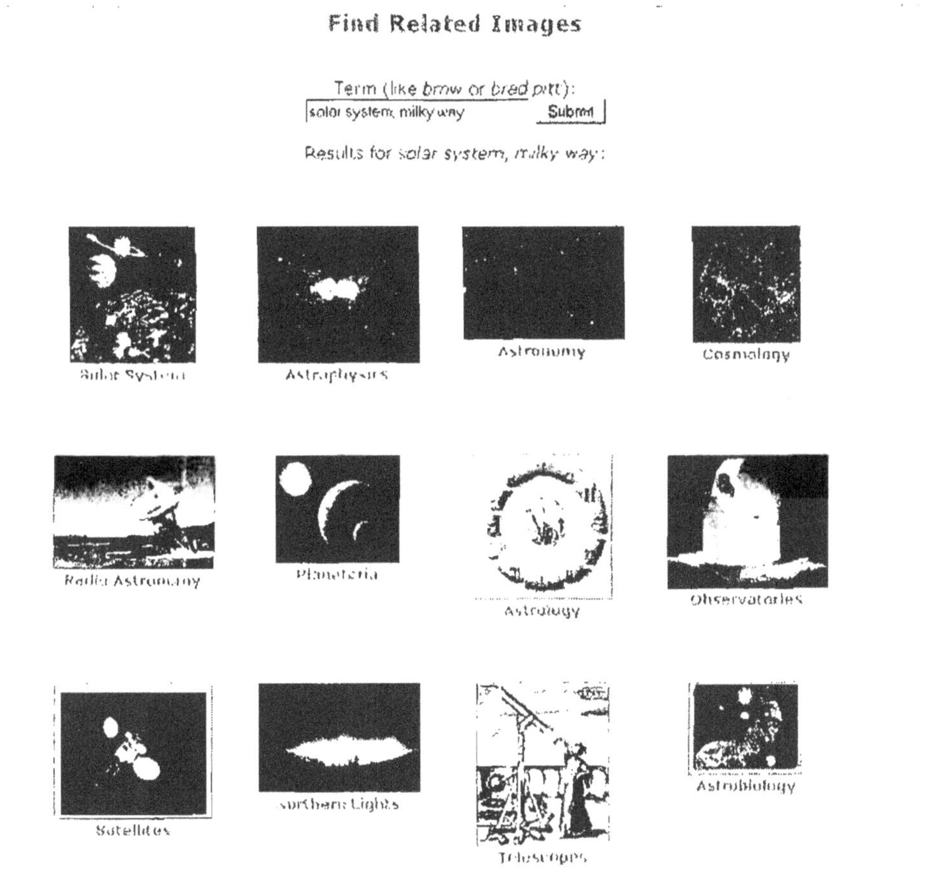
Find Related Images
Term (like bmw or brad pitt):
solar system, milky way
Submit
Results for solar system, milky way:
Astronomy
Cosmology
Radio Astronomy
Planetaria
Astrology
Observatories
Satellites
Northern Lights
Telescopes
Astrobiology

Notice that you can search for up to two terms separated by a comma. That means that you can search for [solar system, milky way] but a search for three single terms, e.g., [mercury, pluto, mars] may return strange results at this time. In any event, this is an interesting and unique tool, so if you need to search for related images, this is the place to start.

Google Find Related Images http://blog.outer-court.com/related/

The Best Individual Google Hacks

While not the only useful and interesting Google hacks, these are among the most valuable and less well-known techniques every serious searcher needs to know.

How Google Parses Queries. Google is sensitive to the number and order of query terms. If you enter the query [windows explorer] without quotation marks, Google will first try to find all the webpages that contain the phrase ["windows explorer"] Google will search:

- First, for keywords as one long phrase.
- Second, for webpages containing all the keywords with the *greatest adjacency* (closest together).
- Third, for webpages containing *all the keywords*, regardless of where they appear on the webpage.

Word Order Matters. Google gives more weight to the first term in a query, so ***put the most important search term(s) first***. Try these two queries and you'll see how different the results are: [new york city] vs. [city york new]

Repetition, Repetition, Repetition. If you keep getting irrelevant hits, you can try repeating a keyword that will be emphasized by Google, such as [java coffee coffee coffee], which cuts down considerably on the number of results about the programming language.

Boilerplate Words or Phrases Yield Gold. Used in combination with keywords, standardized words or phrases can produce very useful results from Google. Whether it's "company proprietary," "not for distribution," or a copyright disclaimer, these are the kinds of identifying query terms that searchers need to look for. Tara Calishain sites the example of using "copyright * the new york times company" plus keywords to locate not only articles at *The New York Times* website but those reprinted elsewhere.

Disabling Word Stemming. The problem with Google's word stemming is that Google does not give users the option to turn it off, which can frustrate users trying to perform precise searches. However, ***if you put a plus (+) sign in front of a term, this will disable word stemming***.

Searching on Stop Words. There are two ways to force Google to search on stop words such as *the, a, an, I*. The first is to ***include stop words in phrases enclosed by double-quotes***, e.g., ["to be or not to be"]. The second way to force Google to search for stop words is to put a ***plus sign (+)*** in front of them, e.g., [+who +what +when].

Google Wildcard. Google has tinkered with its wildcard and it is more useful but still woefully inadequate. The Google wildcard (*) can only be used to *replace a term or terms in a query*. It cannot be used to truncate search terms, e.g., [child*] to find *children*, *childhood*, etc., and it cannot be used to find alternate spellings of terms, e.g., [kazak*stan]. That being said, the Google wildcard is not useless. The wildcard search is very good at helping you do "fill in the blank" searches, and it works great not only in English but in other languages as well. Here are a few examples of "fill in the blank" searches using the wildcard (there is no need to capitalize because Google ignores uppercase):

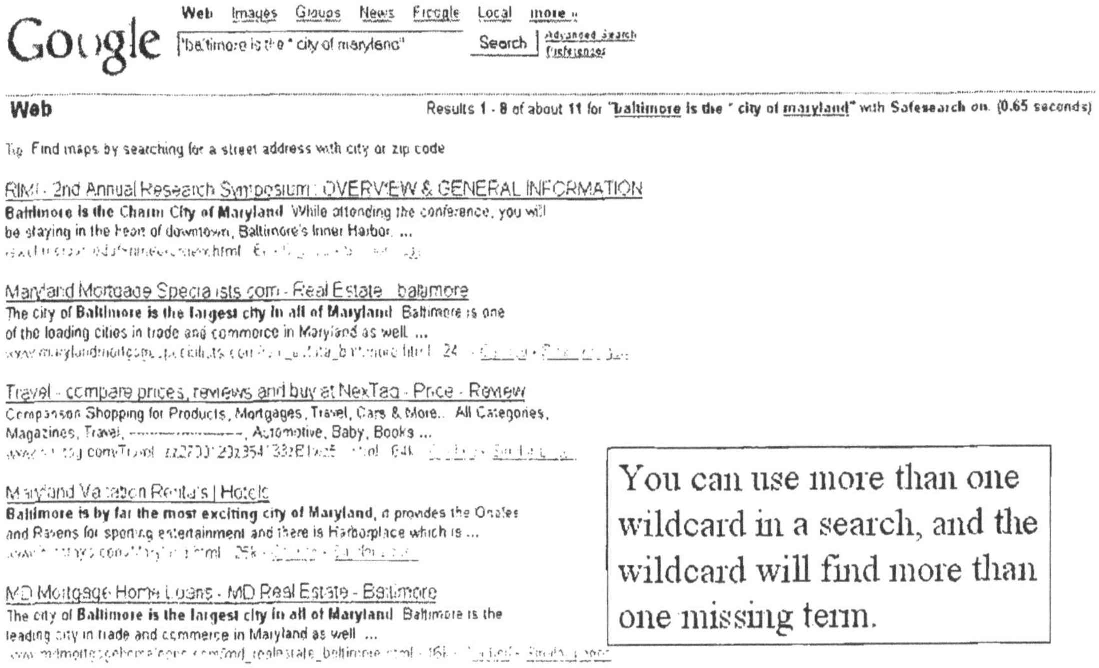
Google
Web Images Groups News Froogle Local more »
"baltimore is the * city of maryland"
Search
Web
Results 1 - 8 of about 11 for "baltimore is the * city of maryland" with Safesearch on. (0.65 seconds)
Tip: Find maps by searching for a street address with city or zip code
Baltimore is the Charm City of Maryland. While attending the conference, you will be staying in the heart of downtown, Baltimore's Inner Harbor. ...
The city of Baltimore is the largest city in all of Maryland. Baltimore is one of the leading cities in trade and commerce in Maryland as well. ...
Comparison Shopping for Products, Mortgages, Travel, Cars & More.. All Categories, Magazines, Travel, Automotive, Baby, Books ...
Baltimore is by far the most exciting city of Maryland, it provides the Orioles and Ravens for sporting entertainment and there is Harborplace which is ...
The city of Baltimore is the largest city in all of Maryland. Baltimore is the leading city in trade and commerce in Maryland as well. ...

You can use more than one wildcard in a search, and the wildcard will find more than one missing term.

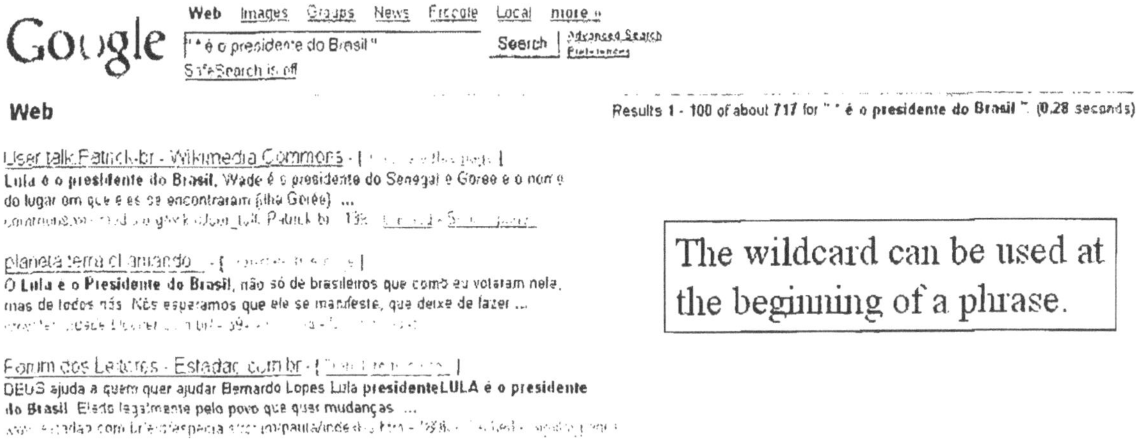
Google
Web Images Groups News Froogle Local more »
"* é o presidente do Brasil"
Search
SafeSearch is off
Web
Results 1 - 100 of about 717 for "* é o presidente do Brasil". (0.28 seconds)
Lula é o presidente do Brasil, Wade é o presidente do Senegal e Goree é o nome do lugar em que eles se encontraram (ilha Gorée) ...
O Lula é o Presidente do Brasil, não só de brasileiros que como eu votaram nele, mas de todos nós. Nós esperamos que ele se manifeste, que deixe de fazer ...
DEUS ajuda a quem quer ajudar Bernardo Lopes Lula presidenteLULA é o presidente do Brasil. Eleito legalmente pelo povo que quer mudanças ...

The wildcard can be used at the beginning of a phrase.

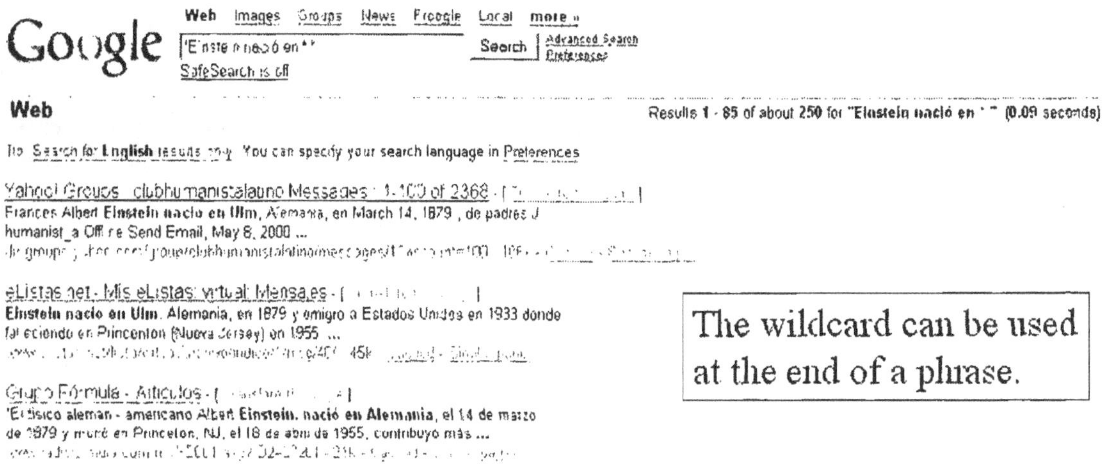
Google
Web Images Groups News Froogle Local more »
Search
Advanced Search
Preferences
SafeSearch is off
Web
Results 1 - 85 of about 250 for "Einstein nació en * *" (0.09 seconds)
Frances Albert Einstein nacio en Ulm, Alemania, en March 14, 1879 , de padres J
humanist_a Offline Send Email, May 8, 2000 ...
Einstein nacio en Ulm. Alemania, en 1879 y emigro a Estados Unidos en 1933 donde
Grupo Fórmula - Artículos

The wildcard can be used at the end of a phrase.

Stripping the Cached Copy: while Google offers the option to view the cached text version of a webpage from the full cached page, there is a Google hack that lets you view the **cached text version** only without having to open the cached page that contains images or other non-text data that might be sent back to the original website or that might redirect you to another page simply by adding **&strip=1** onto the end of the url the cached page as follows:

1. **Right-click** on the Cached link [in red below] and select **Copy link location** in Mozilla or **Copy Shortcut** in Internet Explorer.

2. **Paste** that location (url) into your browser's address bar [do not hit return yet!].

3. Add **&strip=1** to the end of the url and hit return.

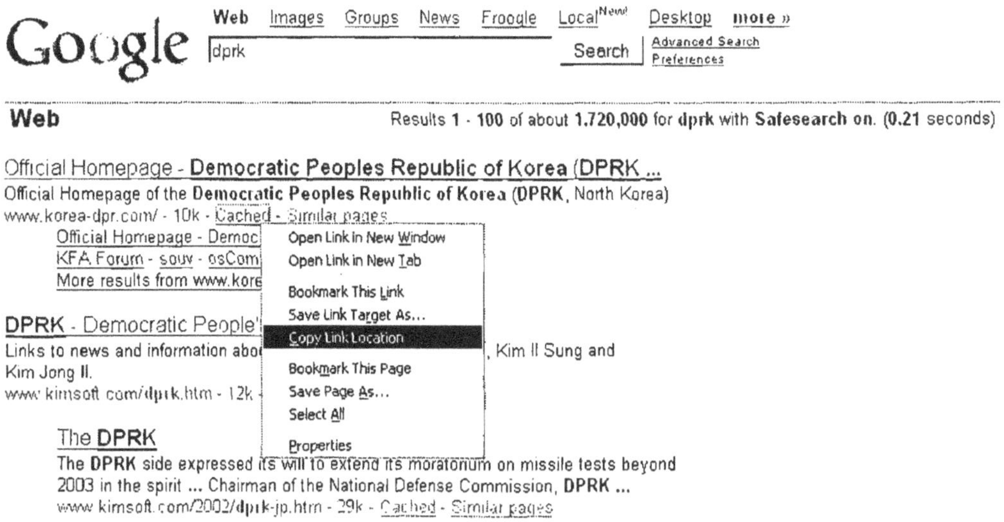
Google
Web Images Groups News Froogle Local New! Desktop more »
dprk
Search
Advanced Search
Preferences
Web
Results 1 - 100 of about 1,720,000 for dprk with Safesearch on. (0.21 seconds)
Official Homepage - Democratic Peoples Republic of Korea (DPRK ...
Official Homepage of the Democratic Peoples Republic of Korea (DPRK, North Korea)
www.korea-dpr.com/ - 10k - Cached - Similar pages
Open Link in New Window
Open Link in New Tab
Bookmark This Link
Save Link Target As...
Copy Link Location
Bookmark This Page
Save Page As...
Select All
Properties
DPRK - Democratic People's
Kim Il Sung and
Kim Jong Il.
www.kimsoft.com/dprk.htm - 12k
The DPRK
The DPRK side expressed its will to extend its moratorium on missile tests beyond
2003 in the spirit ... Chairman of the National Defense Commission, DPRK ...
www.kimsoft.com/2002/dprk-jp.htm - 29k - Cached - Similar pages

This will show the cached version of the page that contains only text:

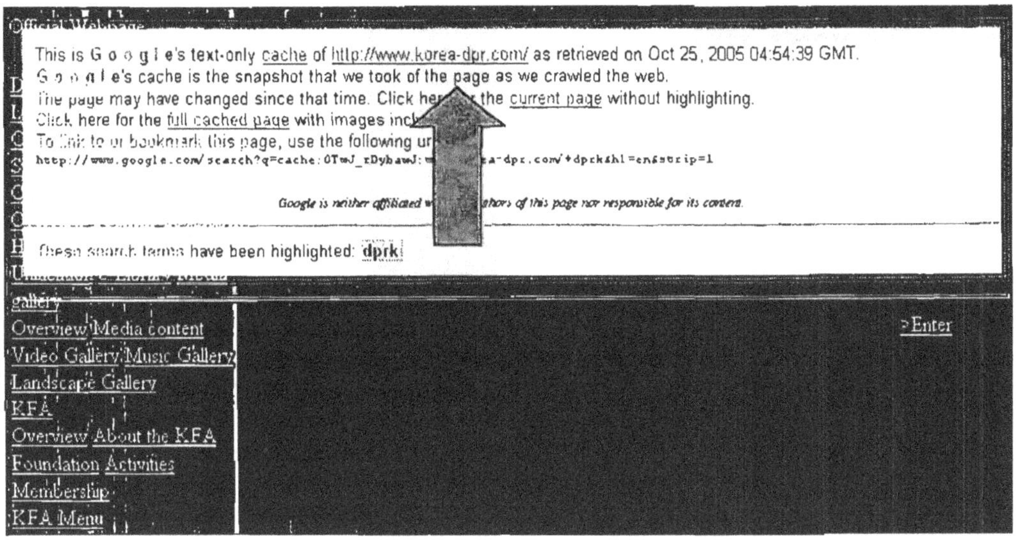
This is Google's text-only cache of http://www.korea-dpr.com/ as retrieved on Oct 25, 2005 04:54:39 GMT.
Google's cache is the snapshot that we took of the page as we crawled the web.
These search terms have been highlighted: dprk
Overview Media content
Video Gallery Music Gallery
Landscape Gallery
KFA
Overview About the KFA
Foundation Activities
Membership
KFA Menu
>Enter

Getting around the 32-word limit. For years Google had a 10-word limit for search queries, meaning that anything more than that, and Google would drop those terms from your query. However, Google expanded the number of terms searched to a 32-word limit. While the casual Google searcher will probably never notice the difference, professional researchers certainly will. There are many times when researchers need to search for long phrases (error codes, for example), exclude large numbers of terms to avoid unwanted results, run complex Google API searches, run queries of multiple sites, etc., and that darned 10-word interfered with the search. While there are a number of work-arounds all were unsatisfactory. Allowing more search terms is a big improvement, but I am sorry to report that the new 32-word limit only applies at present to main Google search, Google Images, Froogle and the Google Web API, while the 10-word limit is still in effect for Google Groups and Google News. This is especially disappointing vis-à-vis Google Groups because it has long been one of the best sources of information about complicated computer error codes and other computer arcana. Perhaps the folks at Google will see fit to expand the 32-word limit to include Google Groups.

You can, however, still ***use the wildcard to trick Google Groups into searching more than 10 keywords.*** Google will not count wildcards as search terms, so inserting a wildcard into a phrase will let you search for more than 10 terms. I have found this most useful when searching for a long phrase such as a computer error message, which may frequently run well over 10 words. By simply removing the "little words" such as *an*, you can easily search for the entire error message.

Here's an example of an error message containing more than 10 terms: Windows Socket Error: An Invalid Argument was supplied (10022), on API 'connect'

It can be written using wildcards to run the complete message as a Google Groups query:

["Windows Socket Error: * Invalid Argument * supplied (10022), * API 'connect'"]

Undocumented Google Filetype Searches. Google can search for many more file types than those documented on the Google FAQ page. Here are some—but not all—of the file types users can search for using ***filetype: plus keyword or another special syntax***, *e.g., site:* (try [filetype:cgi bin] or [filetype:js inurl:login]).

Undocumented	Filetype	Searches
bak	system	backup file
back	system	backupfile
bat	system	batch file
bin	system	binary file
gz	UNIX zip	binary
hlp	text	help files
ico	graphic	icon
ini	system	initialization file
js	script	Javascript
log	text	log files
php	script	HTML
pls	script	PERL script
sql	language	database
tmp	system	Windows temporary file
uu	script	encoding
vbs	script	Microsoft's Visual Basic Script

Google Hack: Create Your Own Google Video RSS Feed. Users can easily create their own personal RSS feed of Google videos using a simple Google hack. Let's say you want a feed of Google videos about Iraq. Here's how to create it:

1. Go to Google Video<http://video.google.com/>and enter a search term, e.g., [iraq]

2. The result will be http://video.google.com/videosearch?q=iraq

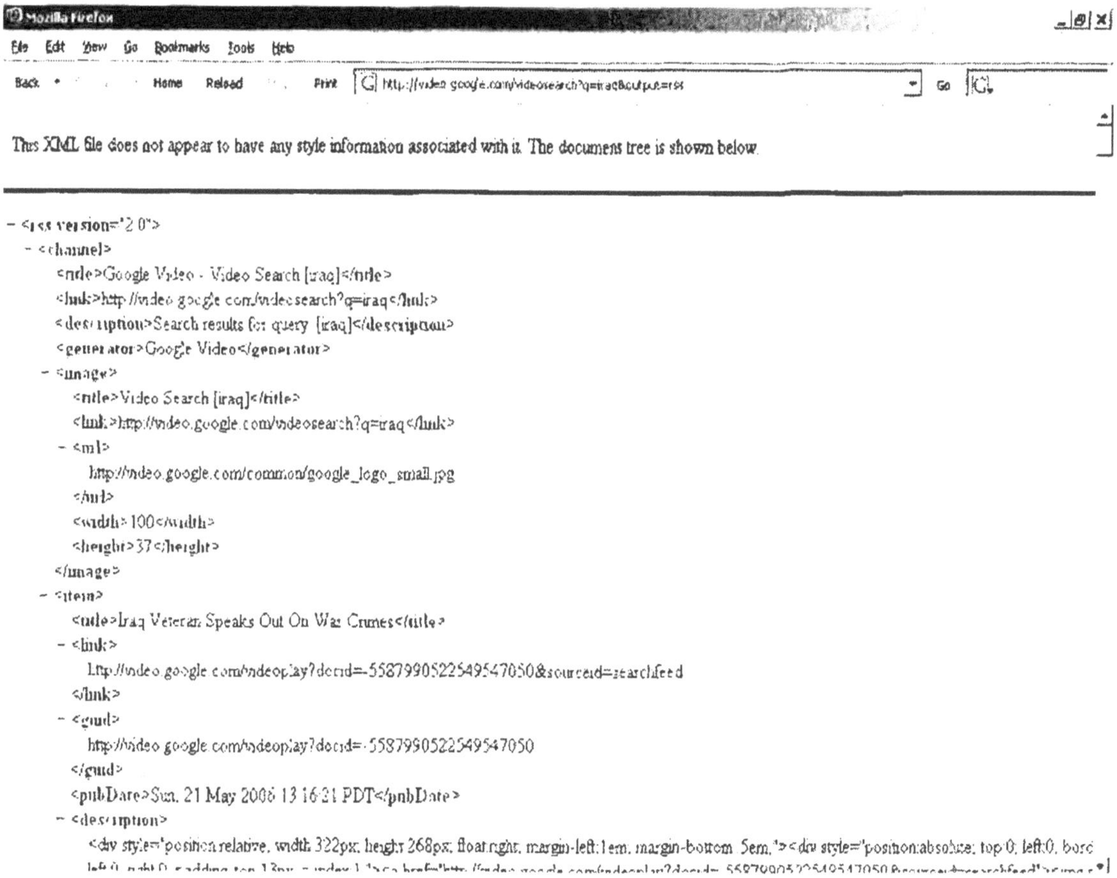
Mozilla Firefox
File Edit View Go Bookmarks Tools Help
Back Home Reload Print http://video.google.com/videosearch?q=iraq&output=rss Go
This XML file does not appear to have any style information associated with it. The document tree is shown below.
- <rss version="2.0">
- <channel>
<title>Google Video - Video Search [iraq]</title>
<link>http://video.google.com/videosearch?q=iraq</link>
<description>Search results for query [iraq]</description>
<generator>Google Video</generator>
- <image>
<title>Video Search [iraq]</title>
<link>http://video.google.com/videosearch?q=iraq</link>
- <url>
http://video.google.com/common/google_logo_small.jpg
</url>
<width>100</width>
<height>37</height>
</image>
- <item>
<title>Iraq Veteran Speaks Out On War Crimes</title>
- <link>
http://video.google.com/videoplay?docid=-5587990522549547050&sourceid=searchfeed
</link>
- <guid>
http://video.google.com/videoplay?docid=-5587990522549547050
</guid>
<pubDate>Sun, 21 May 2006 13:16:21 PDT</pubDate>
- <description>

3. Now add &output=rss so that your new query string http://video.google.com/videosearch?q=iraq&output=rss

4. Now you can add this XML output to your favorite newsreader, e.g., Bloglines to create a Google Video feed:

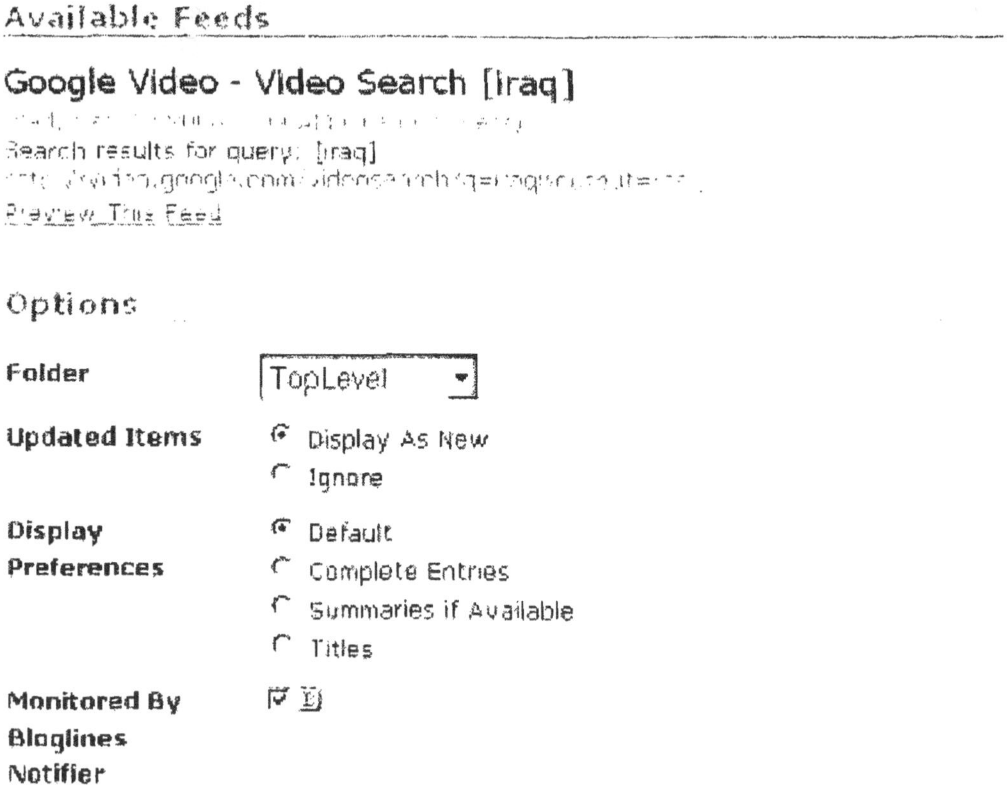
Available Feeds
Google Video - Video Search [iraq]
Search results for query: [iraq]
Preview This Feed
Options
Folder
TopLevel
Updated Items
Display As New
Ignore
Display Preferences
Default
Complete Entries
Summaries if Available
Titles
Monitored By Bloglines Notifier

Every time a new video containing the tag "iraq" is added to Google video, it will be automatically added to your blog feed. Unfortunately, this hack does not work with Google search; that would be a very nice feature. Google Blogsearch and News have RSS/Atom feed options already built into them.[47]

Getting around Google's keyword restriction for *filetype:* searches. The Google *filetype*: syntax requires a keyword. To get around the requirement for a keyword, use the filetype extension as the keyword, e.g., [filetype:pdf pdf].

Google's "synonym" (related term) search. If you place a ***tilde*** *(~)* in front of a keyword, Google will search for the keyword and for its synonyms. For example, a search for ["computer ~security"] will find not only *security* but also *vulnerability*, *encryption*, *secure*, *firewall*. As you can see, this is not a search for synonyms but for related terms.

---

[47] Ionut Alex. Chitu, "Secret Feeds in Google Video," Google Operating System, 12 September 2006, <http://googlesystem.blogspot.com/2006/09/secret-feeds-in-google-video.html> (5 October 2006).

Google Search Sinker. This query option comes from Tara Calishain's website, where she explains how it works: "Search Sinker asks for two things: a query, and then a 'sinker"—a word that you want to emphasize as much as possible in the search, e.g., [java] as the query and [coffee] as the "sinker." The script counts your query words, then fills up any remaining space in the query—remember Google's query limit is ten words—with your 'sinker.'" No API required.

http://www.researchbuzz.org/2004/01/happy_google_hacks_week_2004_2.shtml

Use Google to "Search This Site." Have you ever noticed how bad many internal site searches are? Maybe it's just me, but I can't ever seem to find what I want using a site's internal search. However, Google can perform internal site searches for you, and generally my experience is that Google will do it better. All you have to do is use the *site:* search syntax. So, if I want to search the French Ministry of Foreign Affairs' website for [afrique], all I need to do at Google is search on:

[site:www.france.diplomatie.fr afrique]

Number Range Search Option. Google has a powerful *numrange* search, which uses two number separated by two periods (dots) and no spaces. Numrange has proven invaluable to malicious types who use it to harvest credit card numbers. But it has other legitimate uses as well. It may be important to indicate what the numbers mean, e.g., weight, money, pixels, etc. Google does recognize the almighty dollar sign (but results searching on the Euro symbol are inconsistent). I'm sure you can think of many uses for the Google *numrange* search, such as searching on phone number ranges, dates, address ranges, etc. Try a search such as [amman telephone 617..680] to see how this search works. *Numrange* can be used with other Google search options, such as [site:www.jordanislamicbank.com 617..780]. Also, you don't have to use two numbers: this search finds DVD players up to $150: ["dvd players" .. $150].

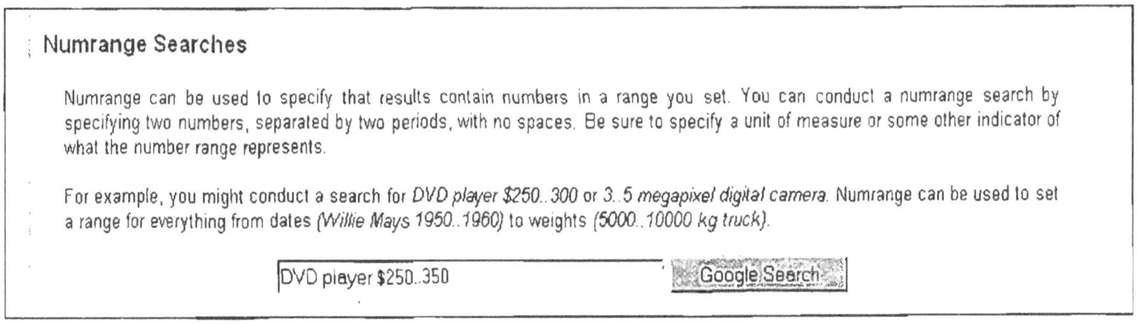
Numrange Searches

Numrange can be used to specify that results contain numbers in a range you set. You can conduct a numrange search by specifying two numbers, separated by two periods, with no spaces. Be sure to specify a unit of measure or some other indicator of what the number range represents.

For example, you might conduct a search for *DVD player $250..300* or *3..5 megapixel digital camera*. Numrange can be used to set a range for everything from dates *(Willie Mays 1950..1960)* to weights *(5000..10000 kg truck)*.

DVD player $250..350 | Google Search

Numrange has other uses that I discuss under Google hacking.

Yahoo Search

In February 2004 Yahoo did what has been expected for about a year: it dropped Google. However, what had not been correctly predicted was the new search engine used by Yahoo in place of Google. Pretty much everyone expected Yahoo to go with the Inktomi search engine it purchased the previous year. Instead, ***Yahoo introduced a new search engine*** that "draws on" the technology of Inktomi and other search engines such as AlltheWeb and AltaVista, which are owned by Overture, a company Yahoo also acquired in 2003. After many years, ***Yahoo is a legitimate search engine*** and not just a directory or a pale copy of Google.

Yahoo has been and remains one of the most visited and most popular websites on the Internet. It was one of the first web portals, debuting in 1994 as a relatively simple web guide and expanding rapidly to over 25 countries in 13 languages. Along with Google, Yahoo is one of the most recognizable brand names on the Internet. At the core of Yahoo was its directory of websites, a hierarchical list of sites organized by subject. Until October 2002, that list was compiled and maintained by humans. At that point, Yahoo switched to Google search results, thereby undermining, in my opinion, its uniqueness and value.

Yahoo announced in mid-2006 what it claims is the "most significant redesign of the www.yahoo.com home page ever...[because] we're also on a mission to empower people to find information and turn it into knowledge, play, and meaningful communication."[48] In my opinion, this was sorely needed. What are you going to find on the new Yahoo homepage? A great deal of personalization choices: links to Yahoo email and Yahoo Messenger; local weather, traffic, events; and the most popular trends, primarily in entertainment and pop culture. The most significant addition to search is Yahoo Answers, Yahoo's version of "ask a question."

The current Yahoo Search Technology (YST) combines the technologies of the various Yahoo search properties—Inktomi, AlltheWeb, and AltaVista—as well as certain Google-like features to create a very powerful search tool. Among these features are:

- cached versions of webpages
- automatic conversion of non-HTML filetypes to HTML is available

---

[48] Yahoo! Search Blog, 15 May 2006, <http://www.ysearchblog.com/archives/000304.html> (31 October 2006).

- **<u>all backlinks</u> are shown** (Google has started limiting the number of backlinks it shows)
- Yahoo refreshes its index continuously, not on a schedule (this is a good thing)
- Yahoo limits the size of indexed pages to the about the first 500KB
- Yahoo claims its database contains over 20 billion "items" (webpages, images, audio and video files); if true, it would be the largest search engine database
- default operator is AND; searchers may use OR
- search by filetype, but can only be used via the Advanced Search option (there is no special syntax)
- it appears there is no upper limit on the number of search terms (Google ignores anything beyond 32)
- an option to open link in new window (small double-window icon at end of url)

At the same time it was introducing its proprietary search engine, Yahoo streamlined its search page interface so that it is resembles the Google homepage.

YAHOO! SEARCH
Web Images Video Audio Directory Local News Shopping Jobs More »
Web Search
Advanced Search
Tip. Get directions fast. Type: [city] map >>Try it
Search Preferences - Search Services

One feature Yahoo Search offers that Google would do well to imitate is the ability to edit the search options. Here are the options Yahoo offers; you can select the ones you want to appear on your Yahoo Search page. With Google, you get what they give you.

Customize Yahoo! Search Homepage Tabs

Select the tabs you want to appear on your Search Homepage and click "Save".

☐	Tabs	Description
	Web	The most relevant results from across the Web. This tab cannot be removed.
☑	Images	Find photos and illustrations from all over the Web.
☑	Directory	Search Yahoo!'s categorized guide to the web.
☑	Local New!	Find local businesses serving your area.
☑	News	Search for news stories, pictures, and audio/video.
☑	Products	Find millions of product reviews and prices on the Web.
☐	Maps	Find maps and directions to anywhere you'd like to go.
☐	People Search	Search for people using Yahoo!'s telephone and email directories.
☐	Travel	Search for airplane tickets, hotel reservations, car rentals, and more.

Yahoo also introduced a number of international versions as well as "local" Yahoos for the US. These international versions can be very useful for locating information about a specific region or country. The pages are in the native language, so if Catalan is unfamiliar, you are probably better off sticking with the main Yahoo search page.

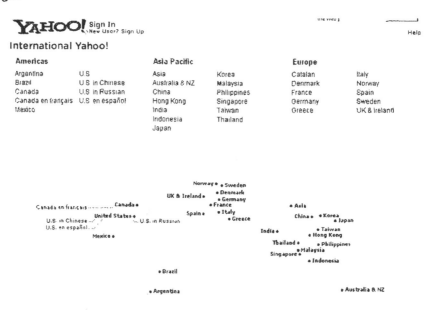

YAHOO! Sign In
New User? Sign Up

Help

International Yahoo!

Americas
Argentina
Brazil
Canada
Canada en français
Mexico
U.S.
U.S. in Chinese
U.S. in Russian
U.S. en español

Asia Pacific
Asia
Australia & NZ
China
Hong Kong
India
Indonesia
Japan
Korea
Malaysia
Philippines
Singapore
Taiwan
Thailand

Europe
Catalan
Denmark
France
Germany
Greece
Italy
Norway
Spain
Sweden
UK & Ireland

Norway
Sweden
Denmark
UK & Ireland
Germany
France
Canada en français
Canada
Spain
Italy
Greece
Asia
United States
China
Korea
U.S. in Chinese
U.S. in Russian
Japan
U.S. en español
India
Taiwan
Mexico
Hong Kong
Thailand
Philippines
Malaysia
Singapore
Indonesia
Brazil
Argentina
Australia & NZ

Yahoo is systematically adding "local" searches outside the US. During 2006, Yahoo added Germany and UK/Ireland to its "local" page. See the map section for more information.

Yahoo International http://world.yahoo.com/

Customizing Yahoo Preferences

Yahoo currently offers four basic ***Search Preferences*** settings:

1. SafeSearch Filter: choose among Strict, Moderate, Off. The Filter lock option requires registration.
2. Languages: search in any language or one or more of 35+ languages, including Arabic, Chinese, Greek, Hebrew, Korean, etc.
3. Display and Layout:

   Open search results in a new browser window when you click on the link. Note: Yahoo also offers an "Open this result in a new window" option following the link for each result if you do not select this option.

   Results per page: choose to display 10, 15, 20, 30, 40, or 100 results at a time.

   Show Save and Block links (to easily save or block pages to Yahoo's My Web service directly from search results)

   Show Instant Search results

4. Subscriptions: search Yahoo's premium content sources such as LexisNexis, Factiva, *The Wall Street Journal*, *Consumer Reports*. Access to these sources requires a paid subscription.
5. Multimedia Search: allows you to search and access songs already available from existing audio service provider.

The Yahoo Results Page

Once you have entered your search term(s) and clicked the Yahoo Search button, Yahoo will present you with a list of results or hits. Depending on the search you are running, you will see some or all of the following for a web search:

DOCID: 4046925

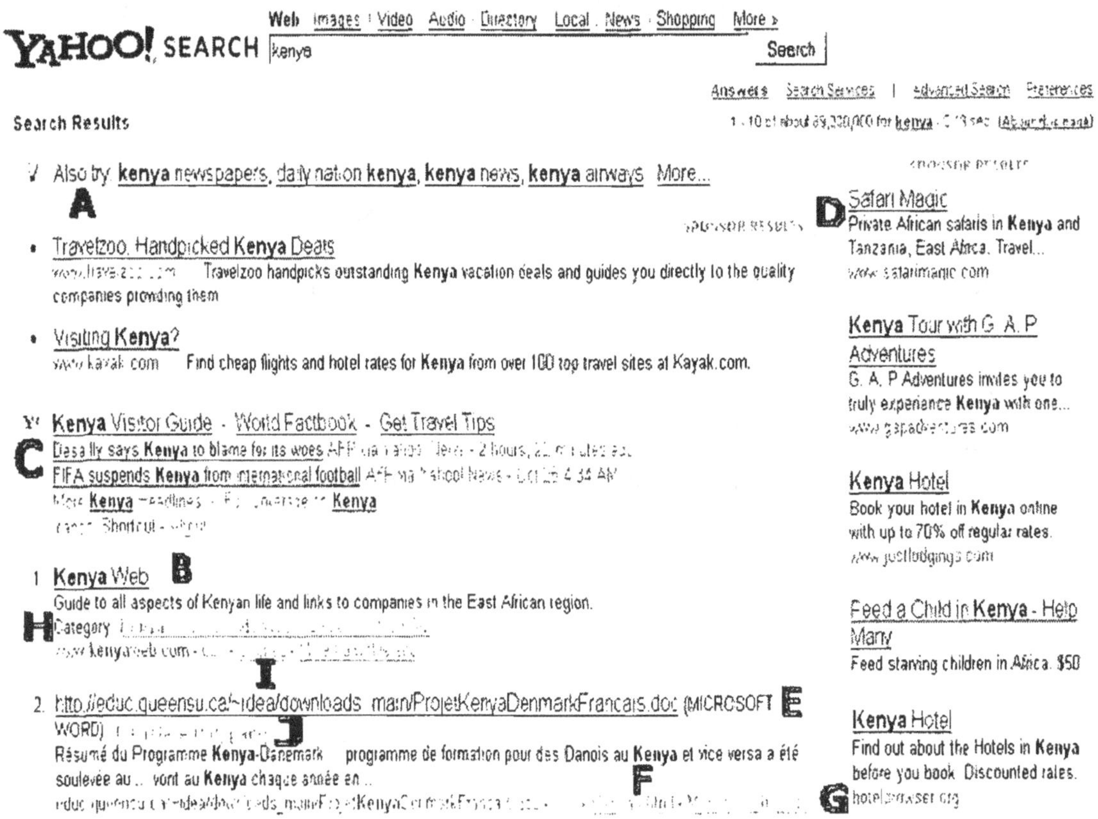
YAHOO! SEARCH
Web Images Video Audio Directory Local News Shopping More »
kenya
Search
Answers Search Services | Advanced Search Preferences
Search Results
Also try: kenya newspapers, daily nation kenya, kenya news, kenya airways More...
A
Travelzoo: Handpicked Kenya Deals
Travelzoo handpicks outstanding Kenya vacation deals and guides you directly to the quality companies providing them
Visiting Kenya?
www.kayak.com Find cheap flights and hotel rates for Kenya from over 100 top travel sites at Kayak.com.
Kenya Visitor Guide - World Factbook - Get Travel Tips
C
FIFA suspends Kenya from international football
1 Kenya Web
B
Guide to all aspects of Kenyan life and links to companies in the East African region.
H
I
2. http://educ.queensu.ca/~idea/downloads_main/ProjetKenyaDenmarkFrancais.doc (MICROSOFT WORD)
E
J
Résumé du Programme Kenya-Danemark programme de formation pour des Danois au Kenya et vice versa a été soulevée au ... vont au Kenya chaque année en ...
F
G
D
Safari Magic
Private African safaris in Kenya and Tanzania, East Africa. Travel...
Kenya Tour with G. A. P Adventures
G. A. P Adventures invites you to truly experience Kenya with one...
Kenya Hotel
Book your hotel in Kenya online with up to 70% off regular rates.
Feed a Child in Kenya - Help Many
Feed starving children in Africa. $50
Kenya Hotel
Find out about the Hotels in Kenya before you book. Discounted rates.

- A Also try: queries to help you refine your search.
- B Results: 1-x out of total number of results.
- C ***YAHOO Shortcuts*** Indicated by Y! next to the first result, Yahoo Shortcuts provide "links to useful content from Yahoo, its partners, or across the web."
- D Sponsor Results are pay for placement results provided by Overture, a Yahoo company.
- E File Type: if the document found is not HTML, the file type will follow the link, e.g., (MICROSOFT WORD).
- F View as HTML: for non-HTML file types, select this link to see the page in HTML format. ***Microsoft file types are potentially dangerous*** to open in their native formats. Use view as HTML to open the document as a webpage instead of downloading it. This is a safe procedure. For more information, please review these instructions for ***handling Microsoft files on the Internet safely***.

- G More pages from this site: select this link to search for other pages within this particular website that *include the search term(s)*.
- H Category: links to the appropriate category in the Yahoo Directory.
- I Cached: links to a copy of the page as saved by Yahoo's search engine; search terms are highlighted on the cached page.
- J Translation: if Yahoo offers a translation, you will see **[Translate this page]** following the link.

Yahoo Basic Search http://search.yahoo.com/

Yahoo assumes as its default that multiple search terms are joined by the **AND** operator, so that a search on the keywords [windows explorer] will find all the webpages that contain ***both*** search terms.

Yahoo ***will not return any results*** if there is no webpage containing all the search terms. Try this query to see what I mean:

rollerskate handshake buckyball

Unlike Google, Yahoo ***does not limit the number of search terms to 32 keywords***. Yahoo will try to match all the keywords you enter.

Yahoo is ***not case sensitive***.

Yahoo does not have true ***word stemming*** or ***truncation***, i.e., searching for variations of search terms. Normally, word stemming involves searching for plurals and verb conjugations such as *drink*, *drank*, *drunk*. However, Yahoo's word stemming is not consistent and somewhat confusing. For example, stemming works with some words but not with others, i.e., a search on [child] will not find *children*; a search on [drink] will find *drinks* but not *drinking* or *drunk*. Users should search on all variations of a term, including plurals, and not rely upon the automatic but inconsistent stemming feature. However, there is a Yahoo hack to disable word stemming.

Yahoo automatically ***clusters search results***. If you want to ***see more pages*** from a specific site, simply select that link following the url of the result.

Undocumented Yahoo Search Feature:

Yahoo ***permits the use of boolean operators*** in simple search. The operators must be capitalized. Yahoo Search will run full nested boolean queries (those using parentheses), such as:

[cardinals AND (bird OR catholic) AND NOT (baseball OR football)]

Yahoo recognizes ***double-quotes*** as enclosing a phrase.

Yahoo does not have any ***stop words***, i.e., commonplace words. In fact, you can search on any single letter or number.

It is unnecessary to use the plus sign (+) with any terms because by default Yahoo searches for all keywords. However, there are many times when searchers need to exclude certain terms that are commonly associated with a keyword but irrelevant to their search. That's where the ***minus sign (-)*** comes in. Using the minus sign in front of a keyword ensures that Yahoo excludes that term from the search. For example, the results for the search ["pearl harbor" –movie] are very different from the results for ["pearl harbor"].

Yahoo will search for the ***ampersand*** **[&]**. Yahoo will search for [barnes&noble], but if you insert spaces—[barnes & noble]—Yahoo ignores the ampersand. Also, while Yahoo will not actually search on a plus sign, the search engine will search for **[c+]** and **[c++]**, although it does not recognize the difference between one and two pluses.

Yahoo Advanced Search

Yahoo has many advanced search features that can be accessed from the Advanced Search page or, in many cases, employed in the simple search screen by using the correct syntax.

Yahoo has incorporated from AlltheWeb and AltaVista most of the ***languages*** in which users may search; like its predecessors, Yahoo is superb at searching for non-English and non-Latin languages. Using either the language preference settings or the advanced search page, users can select from over 35 languages and encodings in which to search and see results.

Keep in mind that if you search on a word using diacritical marks such as accents or umlauts, ***Yahoo will only search for terms matching those with the diacritic***. However, if you search for the unaccented term, Yahoo will find the term with and without the diacritic. For example, a search on [façade] will not return pages containing only the term *façade*, but a search on [facade] will find both *facade* and *façade*.

Yahoo offers many "search meta words," i.e., special search terms to restrict searches and make them more effective. These special operators can be used in both simple and advanced search.

- **site/domain:** restricts results to a specific website or domain, including a specific top-level domain. May be used with or without keywords.

Advanced Web Search > Site/Domain > Filter results from specific domains (com, gov, dell.com, etc.)

Examples of how to use the **site/domain**: command:

[site:amazon.com] finds *www.amazon.com, cards.amazon.com, www.amazon.com/dvd/.* However, it will not find *www.amazon.com.br.*

[domain:ir] finds all the pages from the Iranian (.ir) top-level domain indexed by Yahoo.

- **url:** use to find a specific document in the Yahoo index. This command is very limited in its usefulness and requires the full url (address), including the *http://* to work. This command works best using the Yahoo Advanced Search to find all the words in the url. May be used with or without additional keywords.

  Advanced Web Search > Show Results > **all of these words > In the URL of the page**

  Examples of how to use the **url:** command:

  [url:http://impact.arc.nasa.gov/intro.cfm] finds this specific page as it is stored in the Yahoo index.

- **inurl:** restricts results to any page with a term in its url (address). May be used with or without additional keywords.

  Advanced Web Search > Show Results > In the URL of the page

  Examples of how to use the **inurl:** command:

  [inurl:nasa] finds any site containing *nasa* anywhere in the url. Will find webpages at *www.nasa.gov* as well as pages at *www.beeville.net/NASA/*

  [inurl:nasa.gov columbia] finds any site at *nasa.gov* that contains the keyword *columbia* anywhere on the website.

- **title or intitle:** restricts results to pages containing a specific word or phrase anywhere in the webpage's title, which usually appears in the browser's title bar and is the HTML *<title>* tag. ***Title and intitle appear to work identically***. May be used with or without additional keywords.

  Advanced Web Search > Show Results > In the title of the page

  Examples of how to use the **title:** or **intitle:** command:

[title:amazon] finds all pages that include the word *amazon* in their title

[title:amazon "rain forest"] finds all pages that include the word *amazon* in their title and mention the phrase *"rain forest"* anywhere in the document (title or text or anywhere in the document)

- **link:** restricts results to pages containing links to a specific url. Yahoo has the unfortunate ***requirement to enter the full url including the http://*** to use the link command, which adversely affects its usefulness. May be used with or without additional keywords.

No Advanced Web Search Option

Examples of how to use the **link:** command:

[link:http://jpl.nasa.gov] finds all pages containing links to any page at *jpl.nsa.gov*

[link:http://jpl.nasa.gov asteroid] finds all pages containing links to any page on the *jpl.nasa.gov* site and the keyword *asteroid* anywhere on the page.

- **linkdomain:** the Yahoo *link* command finds every other page that links to a specific webpage, but *linkdomain* finds every page that links to an entire domain. Simply put, the *linkdomain* command should not be used with the full address while the *link* command requires the full address to work properly. May be used with or without additional keywords.

No Advanced Web Search Option

Examples of how to use the **linkdomain:** command vs. the **link:** command:

[linkdomain:amazon.com]
use to find links to an entire domain

[link:http://www.amazon.com] or [link:http://help.yahoo.com/help/us/ysearch]
use to find links to a specific webpage

Here is an interesting twist on link searching, that is, finding sites that link to a specific address. This search, which works with Yahoo and to a lesser extent Live Search, finds pages that link to a specific domain or domains but not to another specific domain or domains. An example would help. Let's say I start by finding the sites that link to the Iranian Ministry of Defense. Here is the query I would use:

[linkdomain:mod.ir]

This query returns 545 hits. Now, suppose I want to see which sites link to both the Iranian MOD and the Iranian Electronics Industries. I can do that easily with this query:

[linkdomain:mod.ir linkdomain:ieimil.com]

However, I see lots of sites that also link to the ever-present CIA World Factbook, which, while a wonderful resource, isn't want I want. I would really like to see the sites that link to both the Iranian MOD and IEI sites but *not* to the CIA Factbook. Can I do this? Sure:

[linkdomain:mod.ir linkdomain:ieimil.com -linkdomain:cia.gov]

While this technique has obvious applicability for search engine optimization ("who is linking to my competitors but not linking to me?"), I think it is worth knowing about because you may come up with some creative ways to use it. Just as an interesting example, try these two queries in both Live Search and Yahoo. It's interesting to see what drops from the results' list on the second query.

[linkdomain:cia.gov linkdomain:nsa.gov]

[linkdomain:cia.gov linkdomain:nsa.gov -linkdomain:fbi.gov]

I believe you will consistently find that Yahoo provides more results than Live Search for the linkdomain: searches. However, the results will vary, so it's worth using both search engines. Google does not offer a linkdomain: search, and its link: search has been hobbled.

- **search by file type:** restricts results to PDF, MS Word, XML, and other filetypes.

  Yahoo does not offer a **filetype:** syntax. However, you can search by file type using the advanced search option:

  Advanced Web Search > File Format > Only find results that are: > All formats or one format

  There is also a Yahoo Hack that lets you search by file type.

- Yahoo's video search now includes an option to search for **MacroMedia Flash** files. To limit your Yahoo video search to Flash files, go to Yahoo Video Search, select Advanced Search, uncheck all formats except Flash and run your query. A simpler approach is to enter the query [filetype:swf keyword]. ***This query only works in Yahoo Video Search***, not Web Search.

Yahoo Video Search http://video.search.yahoo.com/

- Yahoo now links its cached copies of webpages to the **Internet Archive's Wayback Machine**. To use the link, select the Cached copy of a result, then click on "check for previous versions at the Internet Archive." That link takes you to the results of a Wayback Machine search of that precise url.

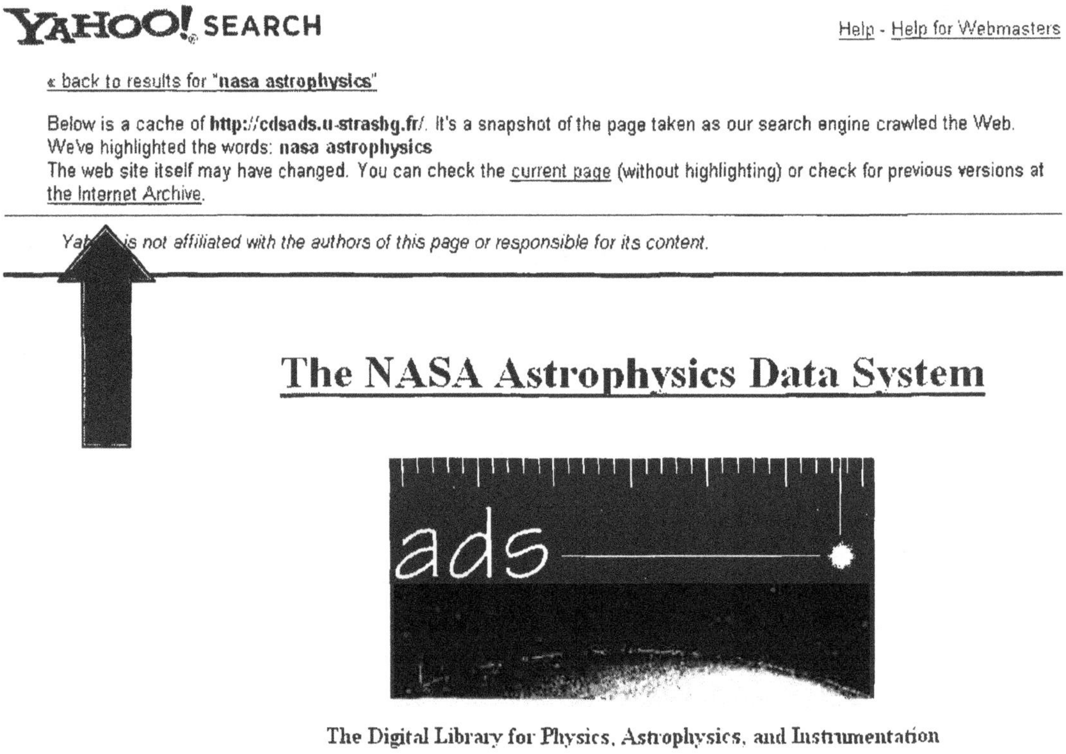
YAHOO! SEARCH
Help - Help for Webmasters
« back to results for "nasa astrophysics"
Below is a cache of http://cdsads.u-strasbg.fr/. It's a snapshot of the page taken as our search engine crawled the Web.
We've highlighted the words: nasa astrophysics
The web site itself may have changed. You can check the current page (without highlighting) or check for previous versions at the Internet Archive.
is not affiliated with the authors of this page or responsible for its content.
The NASA Astrophysics Data System
ads
The Digital Library for Physics, Astrophysics, and Instrumentation

Using Yahoo's More from this site option, you can select other specific pages to view using the Wayback Machine's access to the huge Internet Archive database.

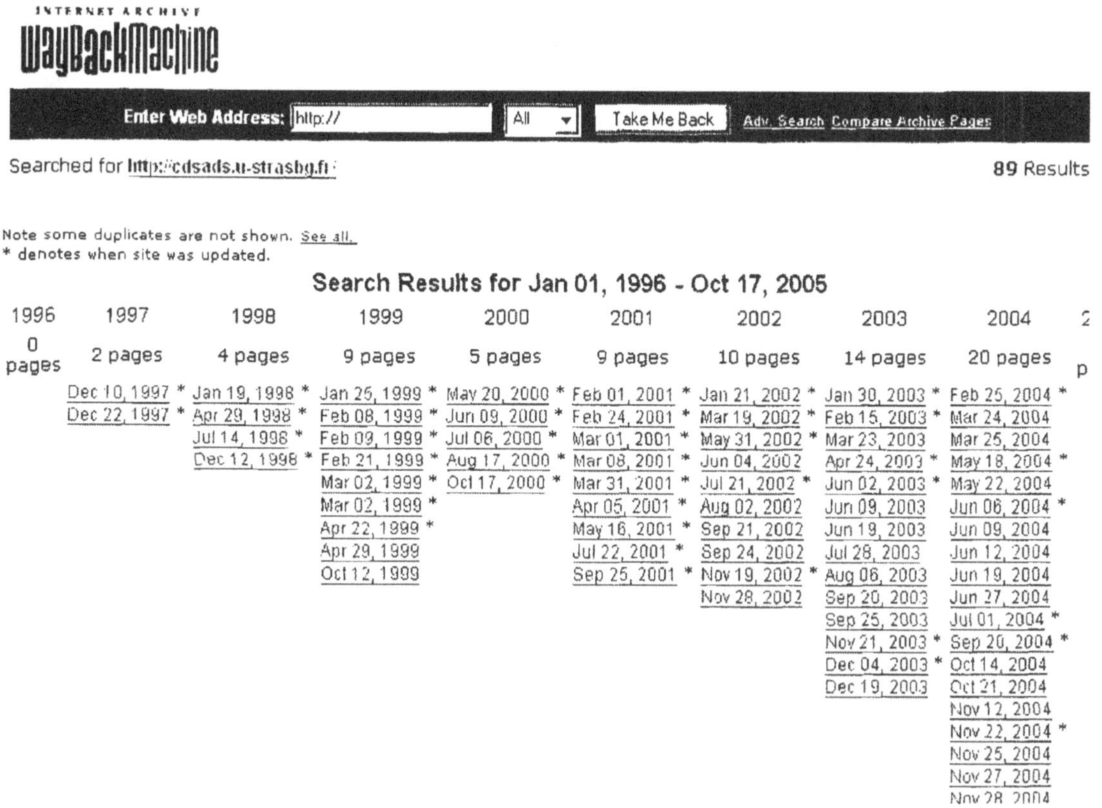

INTERNET ARCHIVE
WaybackMachine

Enter Web Address: http:// | All | Take Me Back | Adv. Search | Compare Archive Pages

Searched for http://cdsads.u-strasbg.fr | 89 Results

Note some duplicates are not shown. See all.
* denotes when site was updated.

Search Results for Jan 01, 1996 - Oct 17, 2005

1996	1997	1998	1999	2000	2001	2002	2003	2004	2
0 pages	2 pages	4 pages	9 pages	5 pages	9 pages	10 pages	14 pages	20 pages	p
	Dec 10, 1997 *	Jan 19, 1998 *	Jan 25, 1999 *	May 20, 2000 *	Feb 01, 2001 *	Jan 21, 2002 *	Jan 30, 2003 *	Feb 25, 2004 *	
	Dec 22, 1997 *	Apr 29, 1998 *	Feb 08, 1999 *	Jun 09, 2000 *	Feb 24, 2001 *	Mar 19, 2002 *	Feb 15, 2003 *	Mar 24, 2004	
		Jul 14, 1998 *	Feb 09, 1999 *	Jul 06, 2000 *	Mar 01, 2001 *	May 31, 2002 *	Mar 23, 2003	Mar 25, 2004	
		Dec 12, 1998 *	Feb 21, 1999 *	Aug 17, 2000 *	Mar 08, 2001 *	Jun 04, 2002	Apr 24, 2003 *	May 18, 2004 *	
			Mar 02, 1999 *	Oct 17, 2000 *	Mar 31, 2001 *	Jul 21, 2002 *	Jun 02, 2003 *	May 22, 2004	
			Mar 02, 1999 *		Apr 05, 2001 *	Aug 02, 2002	Jun 09, 2003	Jun 06, 2004 *	
			Apr 22, 1999 *		May 16, 2001 *	Sep 21, 2002	Jun 19, 2003	Jun 09, 2004	
			Apr 29, 1999		Jul 22, 2001 *	Sep 24, 2002	Jul 28, 2003	Jun 12, 2004	
			Oct 12, 1999		Sep 25, 2001 *	Nov 19, 2002 *	Aug 06, 2003	Jun 19, 2004	
						Nov 28, 2002	Sep 20, 2003	Jun 27, 2004	
							Sep 25, 2003	Jul 01, 2004 *	
							Nov 21, 2003 *	Sep 20, 2004 *	
							Dec 04, 2003 *	Oct 14, 2004	
							Dec 19, 2003	Oct 21, 2004	
								Nov 12, 2004	
								Nov 22, 2004 *	
								Nov 25, 2004	
								Nov 27, 2004	
								Nov 28, 2004	

This is an excellent way to use the Wayback Machine to its fullest because it integrates the power and speed of a huge search engine such as Yahoo with the powerful, though sometimes lumbering and hard to use, Internet Archive query tool.

Yahoo Shortcuts

Yahoo Shortcuts are special features and syntax designed to help users find specific kinds of information faster and more easily. The Shortcuts, indicated by the Y! include:

Local Shortcuts, including zip and area codes, weather, traffic reports, gas prices, and

> **Maps:** to map any US location, search on the street address, city and state or the word *map* and a location. Some international maps are now available.

News & Information links when search term matches current news stories to include video and images.

Travel Shortcuts: airport information, hotel reservations, exchange rates, and flight tracker. To find directions, terminal maps, flight delays, and weather conditions at a

US airport, enter the airport's three-letter code and the word *airport*. For example, to information about Baltimore-Washington International, enter [bwi airport]. At the top of the Yahoo results page you will see a Yahoo Shortcut link marked by the . Clicking on the Shortcut link takes you to the Yahoo BWI resource page:

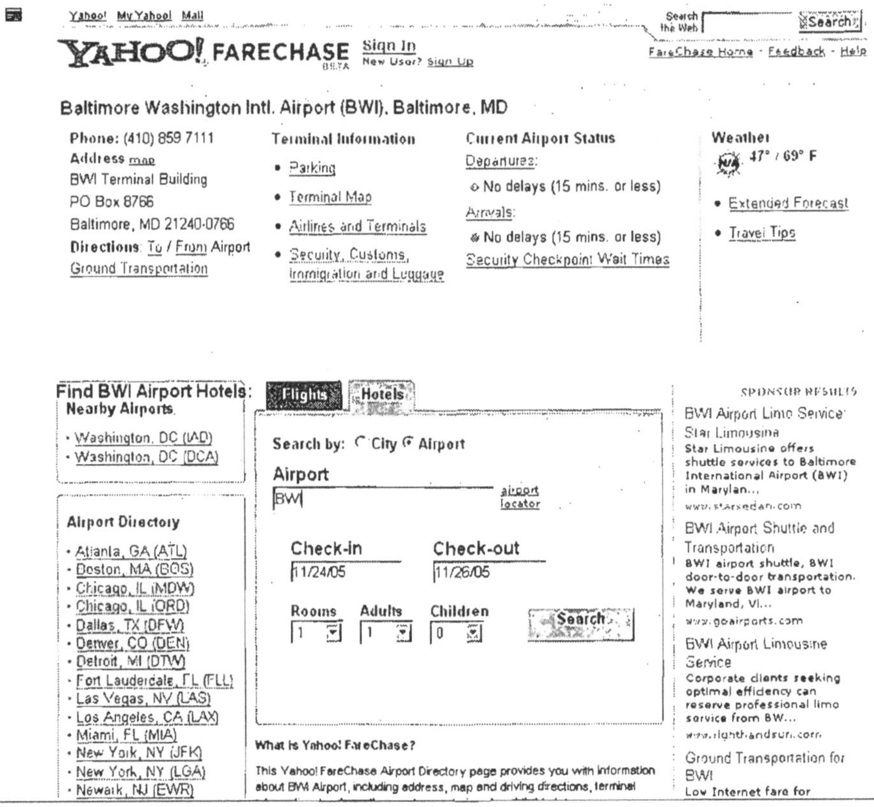
Yahoo! My Yahoo! Mail
YAHOO! FARECHASE
Sign In
New User? Sign Up
FareChase Home - Feedback - Help
Baltimore Washington Intl. Airport (BWI), Baltimore, MD
Phone: (410) 859 7111
Address map
BWI Terminal Building
PO Box 8766
Baltimore, MD 21240-0766
Directions: To / From Airport
Ground Transportation
Terminal Information
Parking
Terminal Map
Airlines and Terminals
Security, Customs, Immigration and Luggage
Current Airport Status
Departures:
No delays (15 mins. or less)
Arrivals:
No delays (15 mins. or less)
Security Checkpoint Wait Times
Weather
47° / 69° F
Extended Forecast
Travel Tips
Find BWI Airport Hotels:
Nearby Airports
Washington, DC (IAD)
Washington, DC (DCA)
Airport Directory
Atlanta, GA (ATL)
Boston, MA (BOS)
Chicago, IL (MDW)
Chicago, IL (ORD)
Dallas, TX (DFW)
Denver, CO (DEN)
Detroit, MI (DTW)
Fort Lauderdale, FL (FLL)
Las Vegas, NV (LAS)
Los Angeles, CA (LAX)
Miami, FL (MIA)
New York, NY (JFK)
New York, NY (LGA)
Newark, NJ (EWR)
Flights
Hotels
Search by: City Airport
Airport
BWI
airport locator
Check-in
11/24/05
Check-out
11/26/05
Rooms
Adults
Children
Search
What is Yahoo! FareChase?
This Yahoo! FareChase Airport Directory page provides you with information about BWI Airport, including address, map and driving directions, terminal
SPONSOR RESULTS
BWI Airport Limo Service
Star Limousine
Star Limousine offers shuttle services to Baltimore International Airport (BWI) in Marylan...
BWI Airport Shuttle and Transportation
BWI airport shuttle, BWI door-to-door transportation. We serve BWI airport to Maryland, VI...
www.goairports.com
BWI Airport Limousine Service
Corporate clients seeking optimal efficiency can reserve professional limo service from BW...
Ground Transportation for BWI
Low Internet fare for

Reference Shortcuts such as:

Dictionary Definitions: as with Google, Yahoo offers the *define* option. To use it, enter *define* then a word or phrase. Yahoo only refers to the *American Heritage Dictionary*; for definitions, so Yahoo's *define* option is more limited than Google's.

Synonym Finder: similarly, *Roget's Thesaurus* provides synonyms for Yahoo Search. They syntax is [synonym keyword].

Encyclopedia: the *Columbia Encyclopedia* supplies the facts about a topic. The syntax is: ["facts about" keyword]

Number Search: Yahoo offers many types of number searches. The numbers Yahoo will search for are:

- **US Patent** numbers: syntax is [patent 5521308].

- **UPS** tracking numbers: [1Z9999X99999999] (simply enter the UPS tracking number; no special syntax is required).
- **USPS** tracking numbers: search on [usps 9999999999999999999999].
- **FedEx** tracking numbers; syntax is [fedex 9999999999999999].
- **FAA** airplane registration numbers; [n158ua] (simply enter the FAA registration number; no special syntax is required).
- **ZIP** codes: enter a US ZIP code, either five or nine digits.
- **ISBN**: enter any International Standard Book Number.
- **UPC** codes: to find information such as the manufacturer for any product, search on the UPC bar code.
- **VIN** Information: to find information about a vehicle's history, search on its 17-character Vehicle Identification Number (VIN).

Calculator Shortcut: a calculator, time zone calculator, and weights and measure converter.

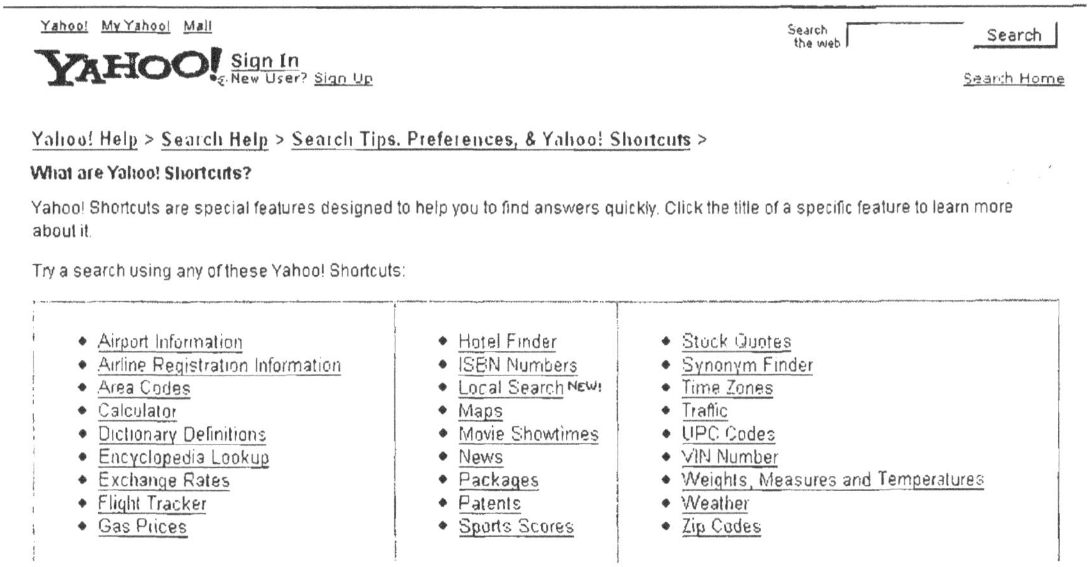
Yahoo! My Yahoo! Mail

YAHOO! Sign In
New User? Sign Up

Search the web | Search

Search Home

Yahoo! Help > Search Help > Search Tips, Preferences, & Yahoo! Shortcuts >

What are Yahoo! Shortcuts?

Yahoo! Shortcuts are special features designed to help you to find answers quickly. Click the title of a specific feature to learn more about it.

Try a search using any of these Yahoo! Shortcuts:

Airport Information	Hotel Finder	Stock Quotes
Airline Registration Information	ISBN Numbers	Synonym Finder
Area Codes	Local Search NEW!	Time Zones
Calculator	Maps	Traffic
Dictionary Definitions	Movie Showtimes	UPC Codes
Encyclopedia Lookup	News	VIN Number
Exchange Rates	Packages	Weights, Measures and Temperatures
Flight Tracker	Patents	Weather
Gas Prices	Sports Scores	Zip Codes

Yahoo Shortcut Help http://help.yahoo.com/help/us/ysearch/tips/tips-01.html

Yahoo Shortcuts http://tools.search.yahoo.com/shortcuts/

Yahoo Services and Specialty Searches

News: Yahoo News headlines are entirely generated using a computer algorithm that scours approximately 4500 worldwide news sources. Yahoo News also offers ***international editions*** for France, Germany, India, Italy, Spain, and several other countries. For details on Yahoo News, see the news search engine section below.

Images: the Yahoo image database contains more than a billion images and includes content from Yahoo news and movies. The advanced search options for images permit users to limit the search by size, color, and image type. Users of Yahoo Search can add the *Image* tab to the left-hand side of the main search page.

Video: Yahoo Video search is extremely powerful. I address this option below in the video search section.

Translations: Yahoo absorbed AltaVista's ***Systran*** translation page (Babelfish) but without the excellent virtual keyboard. Systran provides reasonably decent ***machine translations*** of web pages from many European and Asian languages. The translation page lets you automatically translate a search result, enter any url you like, or copy/type text directly onto the translation page. If a page appears in the results list in one of the languages Yahoo translation supports, you will see **[Translate this page]** after the page title.

Yahoo Babelfish	http://babelfish.yahoo.com/
Yahoo Language Tools	http://tools.search.yahoo.com/language/

Mobile SMS Search: This is a new service that is different from the SMS text messaging that has been available at Yahoo for some time, i.e., Yahoo Messenger. The new SMS at Yahoo permits queries using mobile technology. Yahoo's SMS number is 92466 (which spells Yahoo on most phones). Here is Yahoo's explanation of their service, which is presently only available to US Cingular, Sprint and Verizon subscribers:

> Right now you can search for any local information by sending a query with your location or zip code like: "pizza 94025," you can get a stock quote with: "s yhoo," weather information: "w 94025," dictionary definitions: "d garrulous," horoscopes: "h aquarius," WiFi hotspots: "wifi 94123," and more are coming. Hence typing "w" and the ZIP code get you a short weather forecast.'

Yahoo Mobile Search	http://mobile.yahoo.com/search

Directory: As with Google, Yahoo's web directory uses the Open Directory Project's collection but the two versions are not identical. For example, if I search the directory for the keyword *java*, Yahoo presents multiple directory categories:

YAHOO! search
java
Yahoo! Search
Advanced
Preferences
Web
Images
Directory
Yellow Pages
News
Products
RELATED DIRECTORY CATEGORIES
Programming Languages > Java
Indonesia > Java
More...
Sponsor Results
Hands-On Java Training
Nationwide
Learn hands-on from experts in more

Then I can select an appropriate category where Yahoo Directory lets me limit my search to a specific directory category.

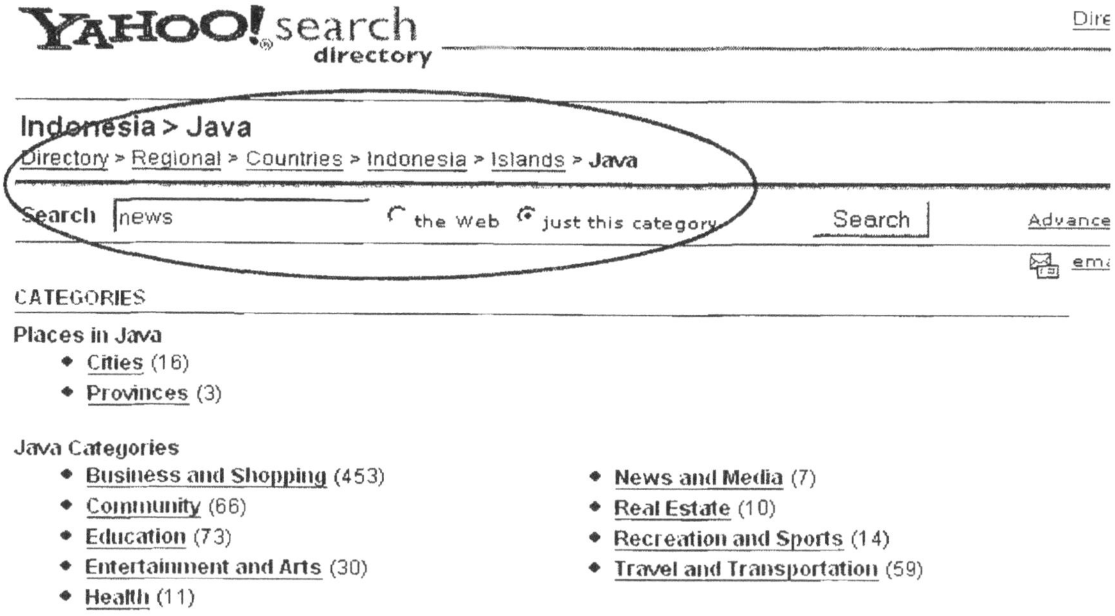
YAHOO! search
directory
Indonesia > Java
Directory > Regional > Countries > Indonesia > Islands > Java
Search
news
the Web
just this category
Search
CATEGORIES
Places in Java
Cities (16)
Provinces (3)
Java Categories
Business and Shopping (453)
Community (66)
Education (73)
Entertainment and Arts (30)
Health (11)
News and Media (7)
Real Estate (10)
Recreation and Sports (14)
Travel and Transportation (59)

Yahoo Answers: the biggest change to Yahoo search during 2006 was the integration of "ask a question" or Yahoo Answers into core search. Yahoo began integrating responses from Answers into its main search results in mid-2006. Although you can read answers at the Yahoo Answers site, in order to ask and answer questions users must register with Yahoo. If a question has already appeared at the Answers site, it will now turn up in the results from the main Yahoo search. Don't expect to get answers to esoteric or difficult questions (you are much more likely to get opinions about the best DVD player than an answer to a question about which languages are spoken in Afghanistan). But the reservoir of questions and answers will continue to grow, so perhaps we will see a richer set of answers than exist thus far.

Yahoo Answers http://answers.yahoo.com/

Yahoo Site Explorer: The Yahoo Site Explorer website is still in beta; its goal is to help users learn detailed information about a specific website:

> The Yahoo search database contains detailed information about the structure of the web. In addition to the web pages themselves, the database stores information about links among pages, and uses that information (as well as additional algorithms) to gauge the popularity of a given page.
>
> Site Explorer gives you access to this information so you can learn about a site. To explore a site, you submit a URL using a search box, just as you would for a normal web search. You can then click links on the results page to see detailed information.

The Yahoo Site Explorer will reveal all the pages in a specific domain, all the pages in any subdirectory of a domain, and all the links to a domain. The main purpose of the Site Explorer is to help webmasters improve the rankings of their sites, as evidenced by the capability for sites to submit missing urls, the fact that Site Explorer provides 50 results by default, its web services APIs, and its ability to export the data to a tab separated (TSV) file for further analysis. The initial response from the search community has been lukewarm, but I like this new tool because it simplifies learning about a site and, unlike Google, Site Explorer provides all the links to a site (which Yahoo calls "inlinks") instead of a limited subset of links.

Let's examine Site Explorer from the point of view of a researcher instead of a developer. Here's the Site Explorer result page for the url [http://www.who.int] showing all the pages in all subdomains of that website. The order is by the most visited pages at the domain according to Yahoo's records about the page:

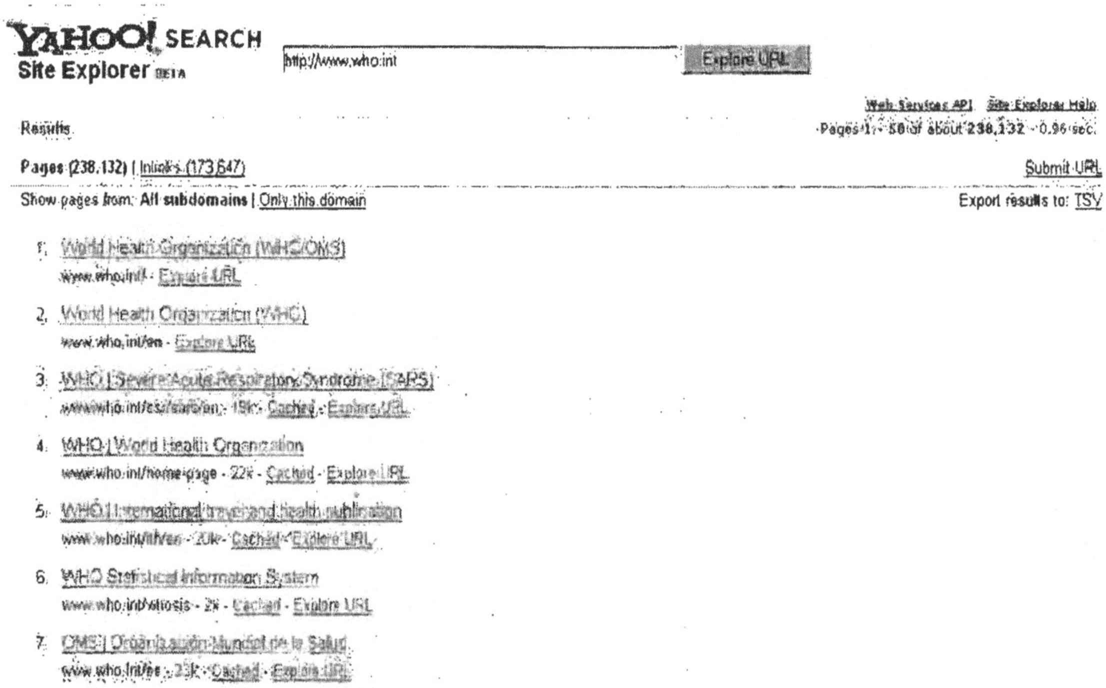
YAHOO! SEARCH
Site Explorer BETA
http://www.who.int
Explore URL
Results
Pages (238,132) | Inlinks (173,647)
Submit URL
Show pages from: All subdomains | Only this domain
Export results to: TSV
World Health Organization (WHO)
www.who.int/en - Explore URL
WHO | World Health Organization

Now look at the "inlinks" (links to the WHO website) Site Explorer found that show the link to this url (http://www.who.int) only:

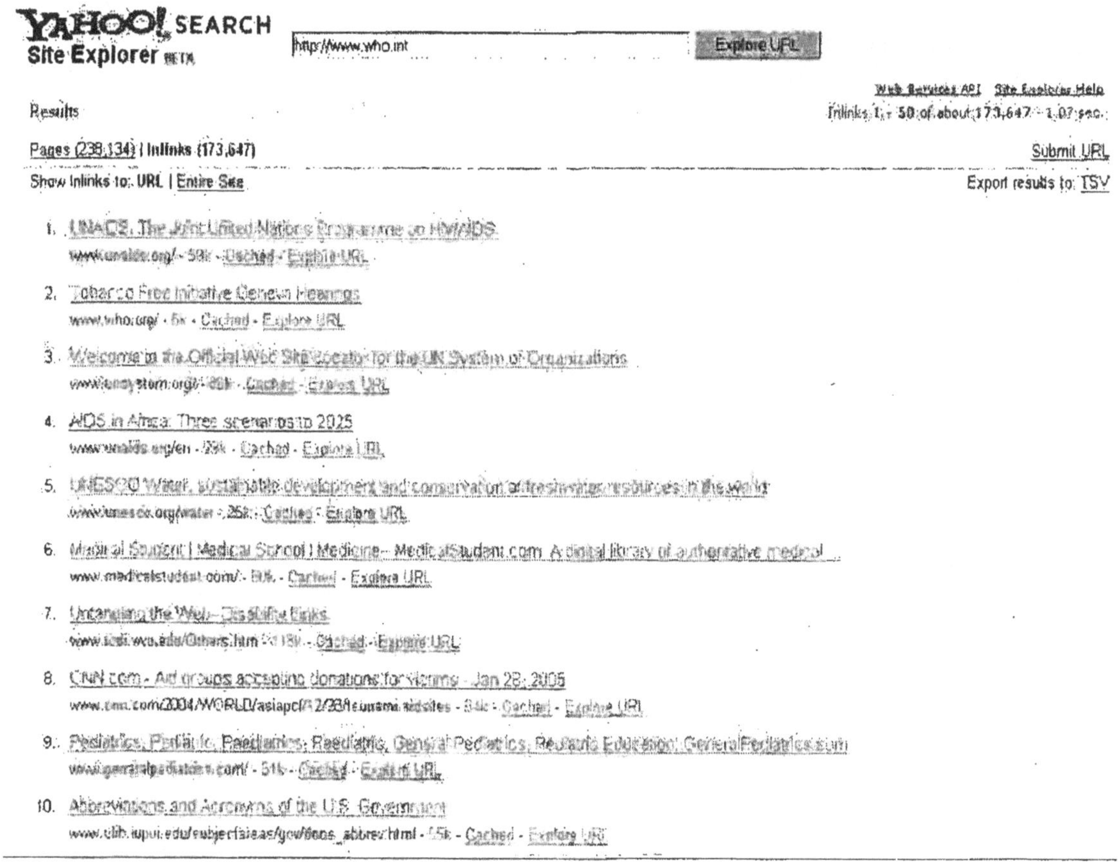
YAHOO! SEARCH
Site Explorer BETA
http://www.who.int
Explore URL
Web Services API
Site Explorer Help
Results
Inlinks 1 - 50 of about 173,647 - 1.07 sec.
Pages (238,134) | Inlinks (173,647)
Submit URL
Show Inlinks to: URL | Entire Site
Export results to: TSV
2. Tobacco Free Initiative Geneva Hearings
4. AIDS in Africa: Three scenarios to 2025
8. CNN.com - Aid groups accepting donations for victims - Jan 28, 2005
10. Abbreviations and Acronyms of the U.S. Government

By way of comparison, the **site:** search in Google—[site:www.who.int]—returned a whopping 2.2 million pages for this domain but only about 29,000 inlinks to the url, whereas Yahoo Site Search returned nearly 174,000 inlinks. However, I find it easier to explore a domain or subdomain using Site Explorer. Note the Explore URL link for every page on the results' list. This effectively lets you "dig" deep into specific pages, directories, or subdirectories in a very orderly way. Keep in mind that you cannot use any keywords or other special syntax with the Yahoo Site Explorer whereas both the Google and Yahoo **site:** command lets users include both other special syntax and keywords, e.g., [site:www.who.int inurl:sars vaccine].

The main value of Site Explorer is the powerful **inlinks** command. Look at the big increase in the number of inlinks (917,704) when you look at inlinks to the Entire Site:

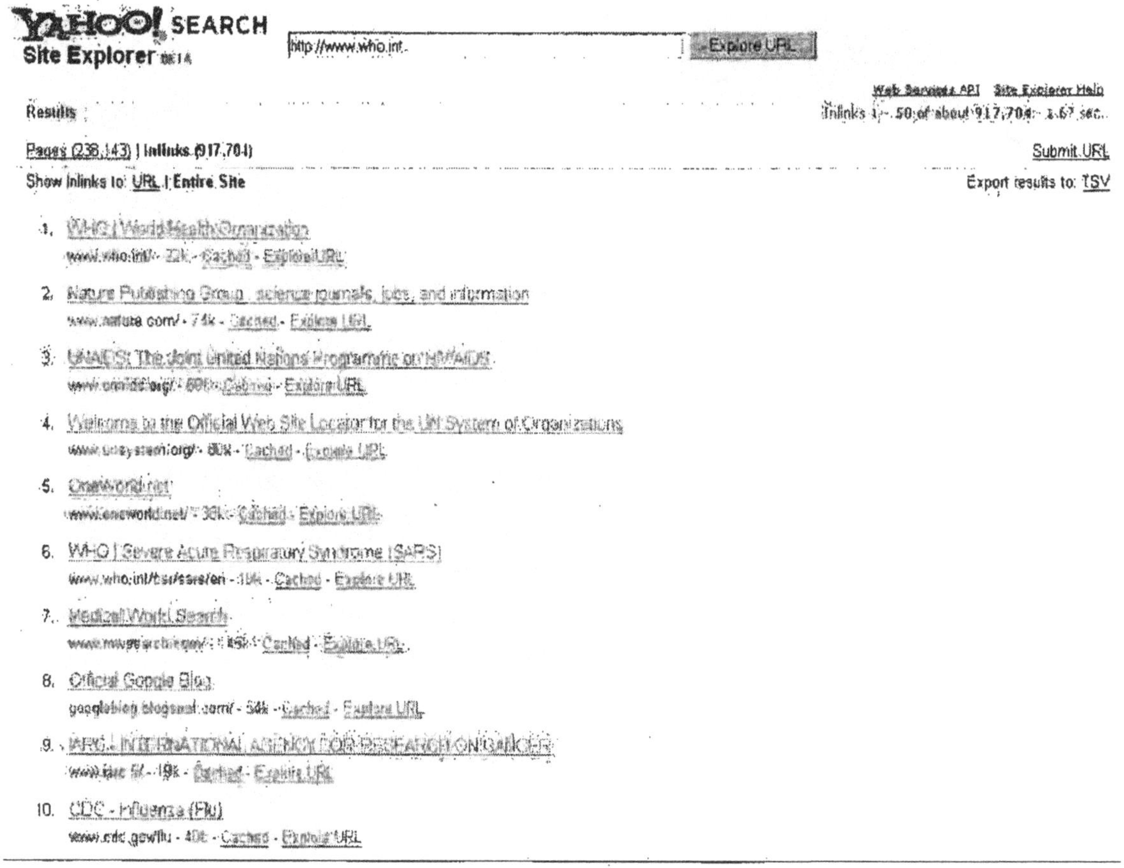
YAHOO! SEARCH
Site Explorer
Explore URL
Results
Submit URL
Show Inlinks to: URL | Entire Site
Export results to: TSV

Remember: Google will not let you use any keywords or other syntax with its **link:** command, and it purposely limits the number of inlinks as a way of trying to control webspam. However, Yahoo's **linkdomain:** syntax will let you use keywords:, e.g., [linkdomain:www.who.int sars]. The bottom line is that the Yahoo Site Explorer does not add any genuinely new functionality for researchers, while it does offer new capabilities for developers. You can do everything (and in some cases more) with the old Yahoo search syntax. So why use Yahoo Site Explorer? Because:

- it provides an extremely orderly and easy to use way of digging deep into a site,
- it provides if not a complete then a huge set of inlinks to a specific url, and
- it ranks the pages of a site by their popularity in Yahoo's statistical records of the site.

Yahoo Site Explorer (beta) http://siteexplorer.search.yahoo.com/

Yahoo Mindset: This new search tool from Yahoo Research Labs is worth a look, especially for those queries that turn up a lot of commercial/shopping hits at the top of the list when you are trying to find "academic, non-commercial, or research-

DOCID: 4046925

oriented sources." Mindset uses a single slider with two options: at the far left is *Shopping* and at the far right is *Researching*. You can move the slider anywhere along the continuum to minimize commercial results and maximize research results or vice versa. Here's a good example of the two different sets of results you'll get for the query ["windows xp"]:

"Researching" Windows XP (ignore the sponsored results)

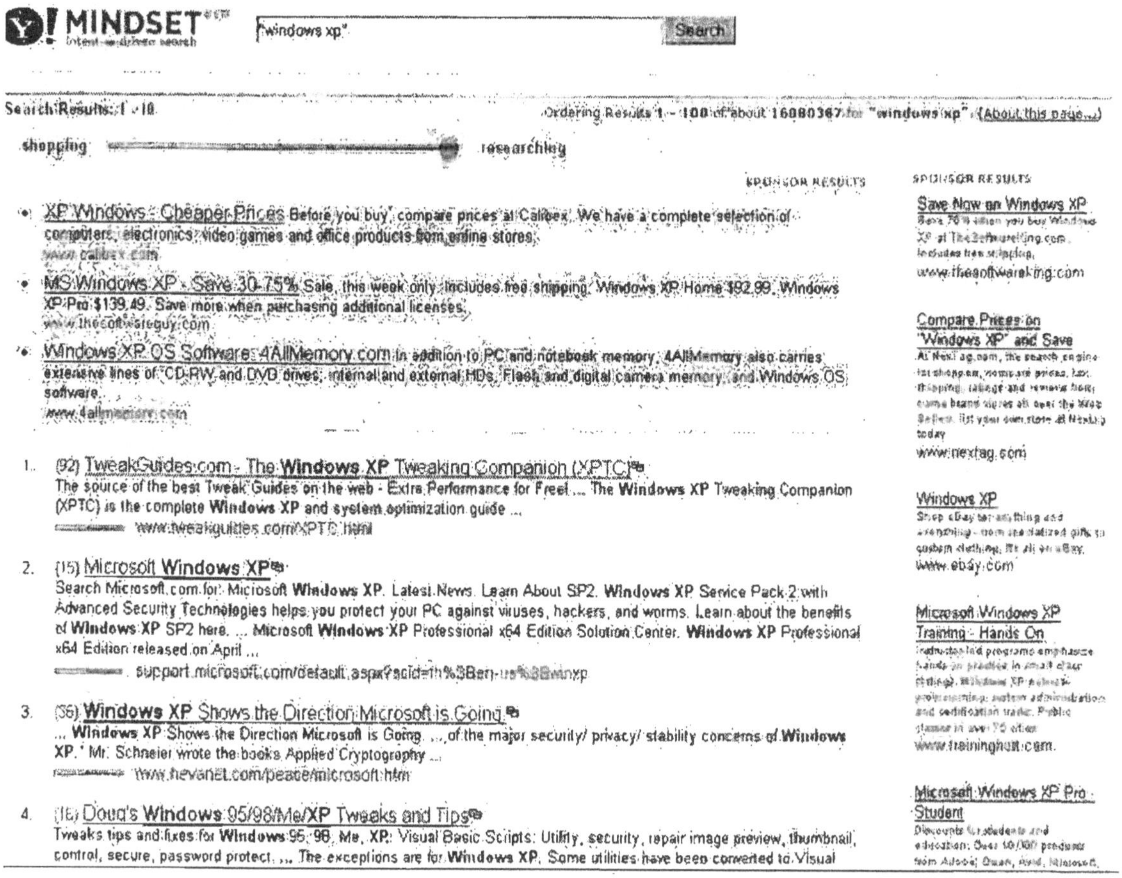
MINDSET
"windows xp"
Search
shopping
researching
SPONSOR RESULTS
XP Windows - Cheaper Prices
MS Windows XP - Save 30-75%
Windows XP OS Software: 4AllMemory.com
1. (92) TweakGuides.com - The Windows XP Tweaking Companion (XPTC)
2. (15) Microsoft Windows XP
3. (56) Windows XP Shows the Direction Microsoft is Going
4. (16) Doug's Windows 95/98/Me/XP Tweaks and Tips
Save Now on Windows XP
Compare Prices on "Windows XP" and Save
Windows XP
Microsoft Windows XP Training - Hands On
Microsoft Windows XP Pro - Student

DOCID: 4046925

"Shopping for" Windows XP

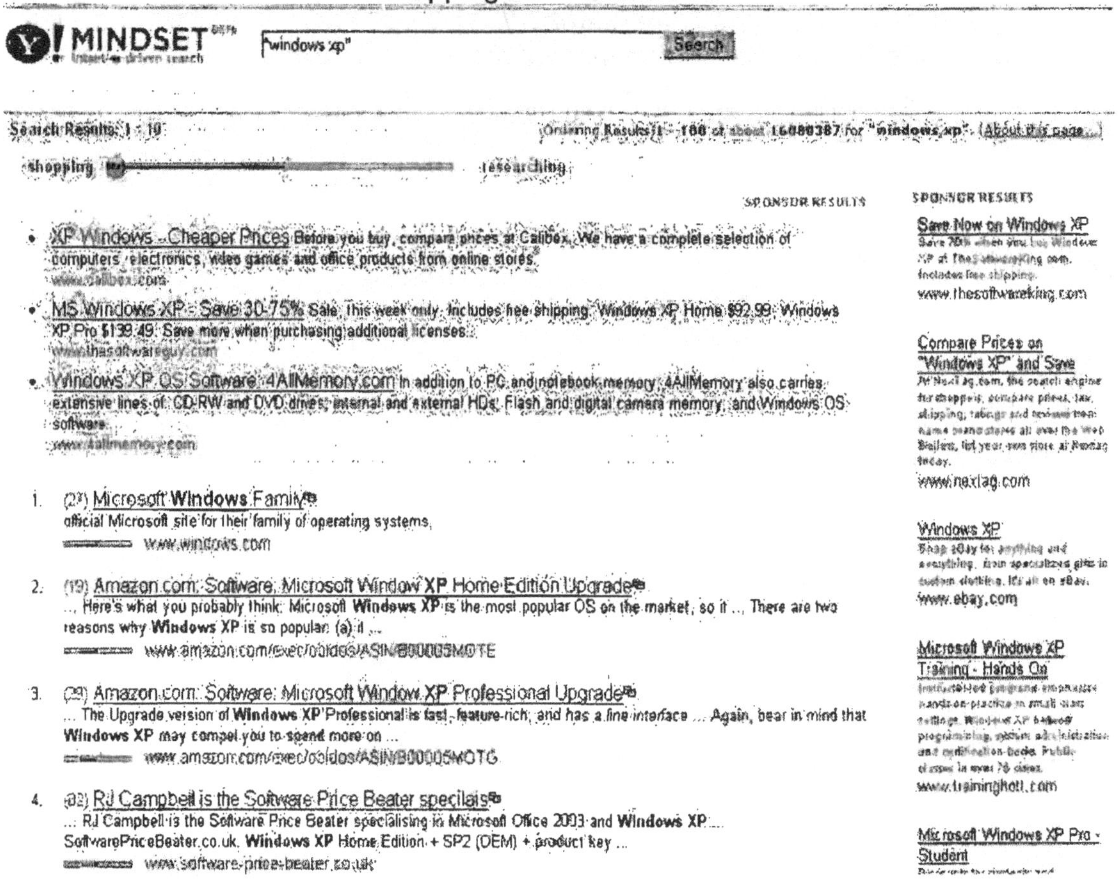
MINDSET
windows xp"
Search
shopping
researching
SPONSOR RESULTS
XP Windows - Cheaper Prices Before you buy, compare prices at Calibex. We have a complete selection of computers, electronics, video games and office products from online stores.
MS Windows XP - Save 30-75% Sale, This week only. Includes free shipping. Windows XP Home $92.99; Windows XP Pro $139.49. Save more when purchasing additional licenses.
Windows XP OS Software - 4AllMemory.com In addition to PC and notebook memory, 4AllMemory also carries extensive lines of CD-RW and DVD drives, internal and external HDs, Flash and digital camera memory, and Windows OS software.
1. Microsoft Windows Family
official Microsoft site for their family of operating systems.
www.windows.com
2. Amazon.com: Software: Microsoft Window XP Home Edition Upgrade
3. Amazon.com: Software: Microsoft Window XP Professional Upgrade
4. RJ Campbell is the Software Price Beater specilais
Save Now on Windows XP
www.thesoftwareking.com
Compare Prices on "Windows XP" and Save
www.nextag.com
Windows XP
www.ebay.com
Microsoft Windows XP Training - Hands On
www.traininghott.com
Microsoft Windows XP Pro - Student

Yahoo Mindset http://mindset.research.yahoo.com/

Yahoo Instant Search: This is yet another attempt by a major search engine to provide answers instead of lists of search results, Yahoo's Instant Search tries to give you a single, relevant response to a query. As their blog explains,

> If search engines are so smart, why do they give you millions of results when you type in "boston weather"? Why even ten, for that matter? Why not just **one**? Or better yet, why not just tell me what the weather is?...

Of course, the answer is that often there isn't one obvious "answer" to a query and you may want a number of possible sites to explore. However, sometimes you just want a straightforward response, and that is what Instant Search is trying to achieve. Instant Search works best (in fact, only works) with simple requests, e.g., [weather london] or [bwi] or [artificial intelligence] or [convert 150 dollars to pounds]. Keep in mind, Instant Search is not trying to be another Answers.com, which "researches" answers to specific questions. Rather, Instant Search tries to "guess" what *most users* would want to know if they entered a query such as [ravens]; Instant Search guesses most users are looking for information about the Baltimore Ravens and not about birds. Sometimes the guesses are good, sometimes not.

Instant Search employs the AJAX[49] web development technique that is starting to make a big splash. **AJAX** uses asynchronous JavaScript and XML (thus the name "AJaX") to allow interactive web browsing. You can see AJAX in action with one of the most intriguing features of Instant Search. You don't need to hit the Search button to get your answer. As you type, the answer (or best guess) will appear in a "speech bubble" below the query box:

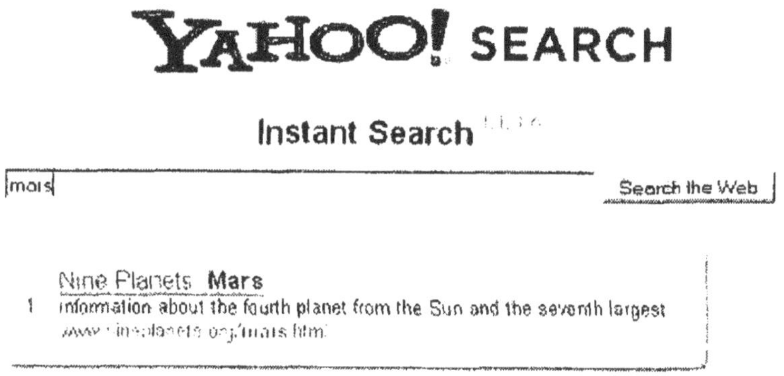
YAHOO! SEARCH
Instant Search
mars
Search the Web
Nine Planets: Mars
1 information about the fourth planet from the Sun and the seventh largest

You can also add Instant Search to your Yahoo Search page by clicking on a link on the Instant Search webpage. I actually think Instant Search is closer to Google's "I Feel Lucky" option than to Answers.com, something the Yahoo press release announcing Instant Search alludes to: "Why feel lucky when you can be right?" Instant Search doesn't just link to a webpage but actually tries to figure out what you want and give it to you (fast). It will be interesting to see if this tool catches on. I suspect that people will find the magically appearing "speech bubble" irresistible for a while, but whether or not Instant Search has staying power will depend on the quantity, quality, and reliability of the responses it provides.

Yahoo Instant Search http://instant.search.yahoo.com/

Yahoo <u>Podcasts</u>: Yahoo's new search site is designed not only to find podcasts on topics of interest but also let users search podcasts by keywords, categories or user-generated topic tags. The new site is a variation on the traditional Yahoo directory, offering a category list by topics, lists of "what other people like" and "what we like," and a search box that lets users choose to search either series, episodes or both. A search on [spyware] returned 5 series results and 202 episode results. It is clear from the highlighted terms in the results that Yahoo's Podcast search looks not only at tags but at the content as well:

---

[49] "AJAX." Computer Desktop Encyclopedia. Computer Language Company Inc., 2005. Answers.com, <http://www.answers.com/ajax> (15 November 2006).

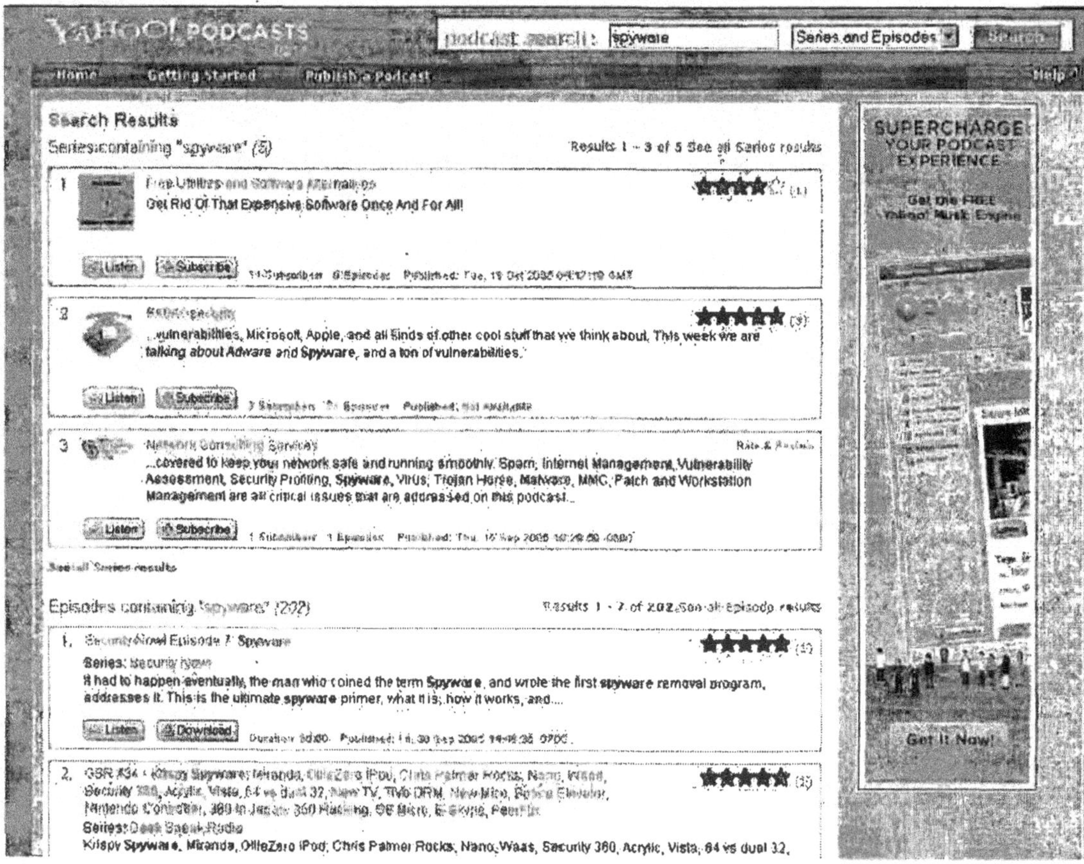
podcast search:
spyware
Series and Episodes
Search Results
Series containing "spyware" (5)
Get Rid Of That Expensive Software Once And For All!
Listen
Subscribe
...vulnerabilities, Microsoft, Apple, and all kinds of other cool stuff that we think about. This week we are talking about Adware and Spyware, and a ton of vulnerabilities.
...covered to keep your network safe and running smoothly. Spam, Internet Management, Vulnerability Assessment, Security Profiling, Spyware, Virus, Trojan Horse, Malware, MMC, Patch and Workstation Management are all critical issues that are addressed on this podcast...
Episodes containing "spyware" (202)
It had to happen eventually, the man who coined the term Spyware, and wrote the first spyware removal program, addresses it. This is the ultimate spyware primer, what it is, how it works, and....
Download
SUPERCHARGE YOUR PODCAST EXPERIENCE
Get It Now!

Yahoo Podcasts includes a player that does not require any installation so that users can simply click on the  button to hear the podcast in MP3 format. If you prefer, Yahoo Podcasts even lets users download a copy of a podcast for later listening. All this is free and does not require registration. The site is in beta as of now.

Why is Yahoo breaking into podcasting in such a big way? Perhaps this is in response to the new study from Yahoo and Ipsos that shows that while **RSS feeds** are gaining ground among the technology elite, they have made little headway with most Internet users. At least that is what most users believe, but the reality is somewhat different. How's that? "The survey found that 31 percent of respondents used RSS, but only 4 percent were aware of it. A full 96 percent of people participating in the survey told Ipsos Insight that they did not use RSS. Their obliviousness came from the fact that most people used browser-based feed-reading software." In short, folks are using RSS and don't know it." When users go to My Yahoo or Firefox's live bookmarks, they are using RSS technology. Moreover, the study found that even the most tech-savvy users prefer to use browser or web-based RSS readers than stand-alone software. That's easy to understand because people are so deluged with technology that unless it is easy to use or brings some indispensable new capability, people simply won't use it. The study concluded:

> Internet users do not understand how to use the XML button, how to actively seek out RSS feeds, or even what the term RSS means. Instead, they need a simple interface where they can choose the information and content that interests them. This is where personalized start pages and browser-based experiences can help move RSS into the mainstream.

Yahoo's new Podcasts Search tries to simplify and demystify not only podcasting but also RSS feeds so that users will feel comfortable enough to try a new technology and, Yahoo hopes, get hooked on it. I know I have pretty much given up on email newsletters in favor of RSS feeds at Bloglines because it is just so much more convenient.[50]

Yahoo Podcasts Search http://podcasts.yahoo.com/

RSS: Crossing into the Mainstream, by Yahoo and Ipsos Insight, October 2005 [PDF] http://publisher.yahoo.com/rss/RSS_whitePaper1004.pdf

---

[50] Chris Sherman reviews eight RSS readers (for parsing primarily XML formatted news and blogs); some are integrated into a browser while others are standalone products that must be downloaded and installed. Chris Sherman, "Choosing an RSS Reader," SearchDay, 1 September 2005, <http://searchenginewatch.com/searchday/article.php/3531486> (14 November 2006).

Yahoo Hacks

While Google hacks—tips, tricks, techniques, and scripts that make Google more powerful and useful—are plentiful and fairly well documented, the same cannot be said (yet) for Yahoo Hacks, despite the fact that O'Reilly published a *Yahoo Hacks* book in late 2005. Part of the reason for this was the absence of Yahoo APIs, a problem Yahoo recognized and rectified with its Developer site.

Yahoo Developer Network http://developer.yahoo.net/

Yahoo Developer Network Blog http://developer.yahoo.net/blog/

While many of the hacks, mostly employing some form of API, are geared toward maps, Yahoo launched a webpage devoted exclusively to Yahoo and "mixed" API applications.

Yahoo Search Application Gallery
http://developer.yahoo.net/search/applications.html

I recommend you pay special attention to the following applications that use Yahoo APIs, although you may find others even more useful to you:

Link Harvester http://www.linkhounds.com/link-harvester/

This is a very powerful—but very slow—tool for examining links to a domain or a specific url. The example below shows the links to [www.mfa.gov.cn]. Link Harvester does the following:

- quickly finds almost every single site linking into a domain or page.
- scrapes past the 1,000 search result limit by making domain filtering a snap.
- grabs number of pages indexed.
- grabs links to any page.
- grabs total inbound links, home page links, and deep link ratio.
- tool is fast and free. which is great considering all it does.
- grabs C block IP address information.
- tool provides links to Wayback Machine and Whois Source (now Domain Tools) next to each domain.
- free & open source

- uses the Yahoo API so it complies with their TOS [terms of service].[51]

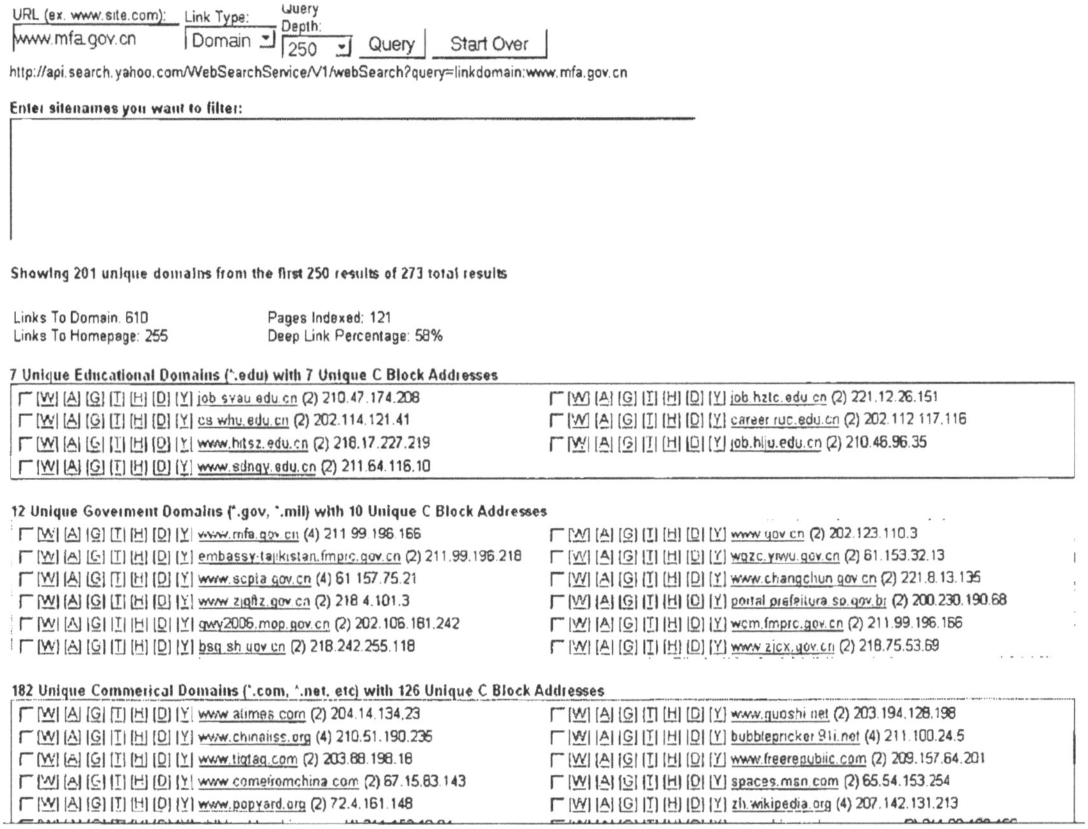
URL (ex. www.site.com):
www.mfa.gov.cn
Link Type:
Domain
Query Depth:
250
Query
Start Over
http://api.search.yahoo.com/WebSearchService/V1/webSearch?query=linkdomain:www.mfa.gov.cn
Enter sitenames you want to filter:
Showing 201 unique domains from the first 250 results of 273 total results
Links To Domain. 610
Links To Homepage: 255
Pages Indexed: 121
Deep Link Percentage: 58%
7 Unique Educational Domains (*.edu) with 7 Unique C Block Addresses
[W] [A] [G] [T] [H] [D] [Y] job svau edu.cn (2) 210.47.174.208
[W] [A] [G] [T] [H] [D] [Y] job.hztc.edu cn (2) 221.12.26.151
[W] [A] [G] [T] [H] [D] [Y] cs whu.edu.cn (2) 202.114.121.41
[W] [A] [G] [T] [H] [D] [Y] career ruc.edu.cn (2) 202.112 117.116
[W] [A] [G] [T] [H] [D] [Y] www.hitsz.edu.cn (2) 218.17.227.219
[W] [A] [G] [T] [H] [D] [Y] job.hlju.edu.cn (2) 210.46.96.35
[W] [A] [G] [T] [H] [D] [Y] www.sdngy.edu.cn (2) 211.64.116.10
12 Unique Goverment Domains (*.gov, *.mil) with 10 Unique C Block Addresses
[W] [A] [G] [T] [H] [D] [Y] www.mfa.gov.cn (4) 211 99.196.166
[W] [A] [G] [T] [H] [D] [Y] www gov.cn (2) 202.123.110.3
[W] [A] [G] [T] [H] [D] [Y] embassy-tajikistan.fmprc.gov.cn (2) 211.99.196.218
[W] [A] [G] [T] [H] [D] [Y] wgzc.ywu.gov.cn (2) 61.153.32.13
[W] [A] [G] [T] [H] [D] [Y] www.scpta gov.cn (4) 61 157.75.21
[W] [A] [G] [T] [H] [D] [Y] www.changchun gov cn (2) 221.8.13.135
[W] [A] [G] [T] [H] [D] [Y] www zjqftz.gov.cn (2) 218 4.101.3
[W] [A] [G] [T] [H] [D] [Y] portal prefeitura sp.gov.br (2) 200.230.190.68
[W] [A] [G] [T] [H] [D] [Y] gwy2006.mop.gov.cn (2) 202.106.181.242
[W] [A] [G] [T] [H] [D] [Y] wcm.fmprc.gov.cn (2) 211.99.196.166
[W] [A] [G] [T] [H] [D] [Y] bsq sh gov cn (2) 218.242.255.118
[W] [A] [G] [T] [H] [D] [Y] www zjcx.gov.cn (2) 218.75.53.69
182 Unique Commerical Domains (*.com, *.net, etc) with 126 Unique C Block Addresses
[W] [A] [G] [T] [H] [D] [Y] www atimes com (2) 204.14.134.23
[W] [A] [G] [T] [H] [D] [Y] www.quoshi net (2) 203.194.128.198
[W] [A] [G] [T] [H] [D] [Y] www.chinaiiss.org (4) 210.51.190.236
[W] [A] [G] [T] [H] [D] [Y] bubblepricker 91i.net (4) 211.100.24.5
[W] [A] [G] [T] [H] [D] [Y] www.tigtag.com (2) 203.88.198.18
[W] [A] [G] [T] [H] [D] [Y] www.freerepublic.com (2) 209.157.64.201
[W] [A] [G] [T] [H] [D] [Y] www comefromchina com (2) 67.15.83.143
[W] [A] [G] [T] [H] [D] [Y] spaces.msn.com (2) 65.54.153.254
[W] [A] [G] [T] [H] [D] [Y] www.popyard.org (2) 72.4.161.148
[W] [A] [G] [T] [H] [D] [Y] zh.wikipedia.org (4) 207.142.131.213

For each unique domain, Link Harvester provides [W]=Whois Source data for domain; [A]=Internet Archive data for domain; [G]=Google cache of actual webpage; [T]=Google's text only cache of actual webpage; [H]=Google's text only cache of domain; [D]=Whois Source's information about the domain from the Open Directory; [Y]=Yahoo's Directory Listing of Whois Source data about the domain.

Hub Finder http://www.linkhounds.com/hub-finder/

"Hub Finder looks for sites which have co-occurring links to related authoritative websites on a particular topic." Basically, Hub Finder locates authoritative websites on a particular subject, as in the example below, for *java*. In this case, the top sites (most authoritative resources) for *java* are shown. Hub Finder also permits users to download the data in CSV (Comma Separated Value) format that can be easily merged into a spreadsheet or database.

---

[51] Link Harvester, *Linkhounds*, <http://www.linkhounds.com/link-harvester/> (14 November 2006).

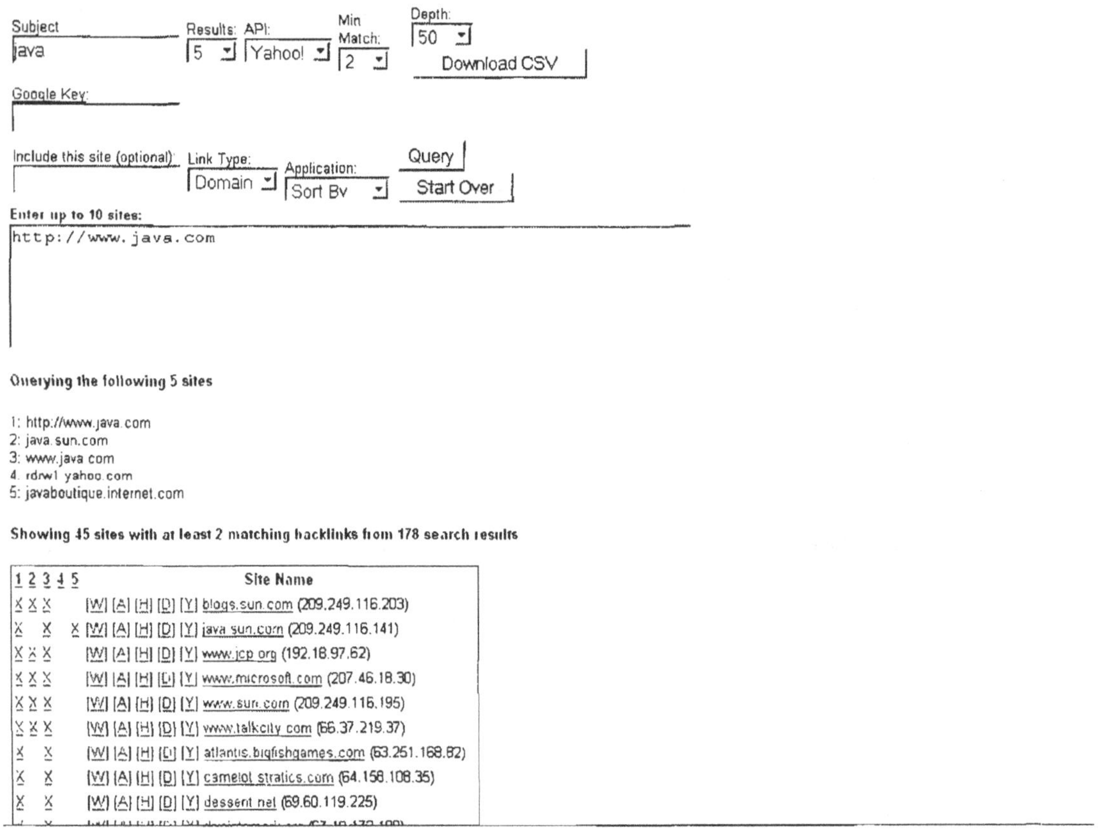
Subject
java
Results: API:
5
Yahoo!
Min Match:
2
Depth:
50
Download CSV
Google Key:
Include this site (optional):
Link Type:
Domain
Application:
Sort By
Query
Start Over
Enter up to 10 sites:
http://www.java.com
Querying the following 5 sites
1: http://www.java.com
2: java.sun.com
3: www.java.com
4. rdrw1 yahoo.com
5: javaboutique.internet.com
Showing 45 sites with at least 2 matching backlinks from 178 search results
1 2 3 4 5 Site Name
X X X [W] [A] [H] [D] [Y] blogs.sun.com (209.249.116.203)
X X X [W] [A] [H] [D] [Y] java.sun.com (209.249.116.141)
X X X [W] [A] [H] [D] [Y] www.jcp.org (192.18.97.62)
X X X [W] [A] [H] [D] [Y] www.microsoft.com (207.46.18.30)
X X X [W] [A] [H] [D] [Y] www.sun.com (209.249.116.195)
X X X [W] [A] [H] [D] [Y] www.talkcity.com (66.37.219.37)
X X [W] [A] [H] [D] [Y] atlantis.bigfishgames.com (63.251.168.82)
X X [W] [A] [H] [D] [Y] camelot.stratics.com (64.158.108.35)
X X [W] [A] [H] [D] [Y] dessent.net (69.60.119.225)

The following Yahoo Hacks generally mirror certain Google hacks, with the exception of the **originurlextension:** syntax, which is unique to Yahoo and very powerful.

- Disabling Word Stemming. Yahoo does not give users the option to turn off word stemming, which can frustrate users trying to perform precise searches. To run a precise search, enclose the term in double-quotes, e.g., ["drink"] will not find *drinks* (except in sponsored results).

- Searching by Filetype. Despite the fact Yahoo mysteriously disabled its *filetype* syntax, you can use **originurlextension:** to search by file type, but this syntax is imperfect.

  Examples of how to use the **originurlextension:** command:

- [originurlextension:pdf "white paper"] finds pages indexed by Yahoo that are in PDF format and contain the phrase ["white paper"] anywhere in the text, title, or url.

To search by specific type of file, use the syntax *originurlextension*: plus one of these or ***any file extension***, such as *cgi*, *log*, *zip*, etc. Because this workaround is not a true filetype search, you can search on any file extension.

- htm or html—standard webpage
- pdf —Adobe Acrobat
- xls—MS Excel
- ppt—MS PowerPoint
- doc—MS Word
- txt—text
- xml, rdf, rss—RSS or XML feeds[52]

Searchroller. Searchroller uses a JavaScript to let you create a neat little search query bookmarklet[53] for your future use. The bookmarklet comprises a set of domains you like to search on routinely but don't want to type in each time. For example, perhaps you'd like to search simultaneously on a whole group of news sites. Tara Calishain's script lets you input the urls for the news' sites once, then save them to your Favorites or Bookmarks. Each time you click on the bookmarklet, a screen will appear asking you to enter a query term or terms, then the bookmarklet will automatically go to Yahoo and run that query against all the urls you have previously selected. It's a great timesaver when you consider this is a typical Searchroller bookmarklet query, although it could be much longer:

[iraq (site:cnn.com OR site:msnbc.com OR site:usatoday.com] OR [site:nytimes.com OR site:washingtonpost.com OR site:bbc.co.uk)]

Searchroller
http://www.researchbuzz.org/2004/10/new_yahoo_hack_searchroller_fo.shtml

Artificial Proximity Search. Since Yahoo's APIs are so new and as yet not fully exploited, clever folks like Tara Calishain have come up with ways to force Yahoo to perform new types of searches. The proximity search lets you input one search term and look for it from 1 to 5 "spaces" (really, words) from a second search term. For example, I can search for *henry* within two words of *thoreau* and find many instances

---

[52] In order to read RSS or XML feeds, you need a reader or aggregator to parse this type of data.

[53] A bookmarklet is a tiny JavaScript application contained in a bookmark that can be saved and used the same way you use normal bookmarks. Bookmarklets do not require users to download and install software. For more on bookmarklets, visit <http://www.bookmarklets.com/>.

of *Henry David Thoreau*. This tool is very good for finding names with the last name listed first, e.g., *Thoreau, Henry David*.

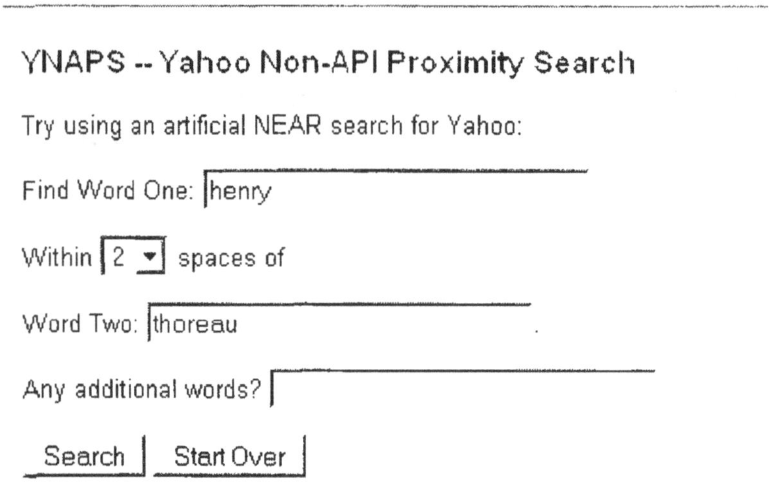
YNAPS -- Yahoo Non-API Proximity Search
Try using an artificial NEAR search for Yahoo:
Find Word One: henry
Within 2 spaces of
Word Two: thoreau
Any additional words?
Search
Start Over

Yahoo Proximity Search
http://www.researchbuzz.org/2004/10/ynaps_yahoo_nonapi_proximity_s.shtml

Boilerplate Words or Phrases Yield Gold. Used in combination with keywords, standardized words or phrases can produce very useful results from Yahoo as well as Google. Whether it's "company proprietary," "not for distribution," or a copyright disclaimer, these are the kinds of identifying query terms that searchers need to look for.

Windows Live Search

MSN Search is no more. As of mid-September 2006, Windows Live Search was out of beta and officially supplanted MSN Search. It came as a surprise to no one that the new Live Search has the familiar clean, uncluttered look popularized by Google. Live marks a clear change in Microsoft's overall direction from a multipurpose portal to a search service: "Live.com is now first and foremost a search destination," according to Christopher Payne, Microsoft's corporate vice president.[54]

The question on everyone's mind is whether or not Live Search is any better than MSN Search or Google or Yahoo or any number of other search engines. Thus far, Live is not noticeably superior to MSN Search, but it is a one of the top three largest and most powerful US-based search engines.

The new Windows Live Search:

- uses its own database for web search.
- indexes at least 5 billion pages.
- offers cached links with the date Microsoft estimates the page was last updated (usually the date the Microsoft spider last crawled the page); sometimes a date will appear next to the cached link on the results' page if that page has recently been updated.
- has a "Near Me" search option that only works in the US; it uses your IP address to determine your location; users can override this location by changing it on the Options page. Note that you cannot leave the default location empty. ***If you do not enter a location, Live Search will default to what it reads as your IP address's geolocation***.
- offers web, news, image, local, Q&A, academic, feeds, video, products, and new "build your own" searches.
- offers preference control via "options."

The "Search Builder" query customization tool has been replaced by the "Advanced" option; as with "Search Builder" the Advanced option opens a little window beneath the search form.

---

[54] Chris Sherman, "Microsoft Upgrades Live Search Offerings," SearchEngineWatch, 12 September 2006, <http://searchenginewatch.com/showPage.html?page=3623401> (5 October 2006).

Customizing Live Search Settings ("Options")

Live Search currently offers these user-defined options (preferences):

- Display: display the site in a specific language (most major languages with some notable exceptions, e.g., Arabic, Thai).
- Number of Results: choose to display 10, 15, 30, or 50 results at a time.
- Group results from the same site: Show the first 1, 2, or 3 results.
- Open Links in New Browser Window: yes or no.
- SafeSearch Filter: choose among Strict, Moderate, Off.
- Location: set a default location; Microsoft detects your physical location from your IP address, but you may enter a new geographical location in its place. Remember: you cannot leave the default location empty. If you do not enter a location, Live Search will default to what it reads as your IP address's geolocation.
- Search Language: search in any language or search in one or more of 38 languages including Arabic, Japanese, Chinese, Korean, and Hebrew.

The Live Search Results Page

The clean look continues on the results' page. Once you have entered your search term(s) and clicked the Live Search button, Live will present you with a list of results. Depending on the search you are running, you will see some or all of the following for a web search:

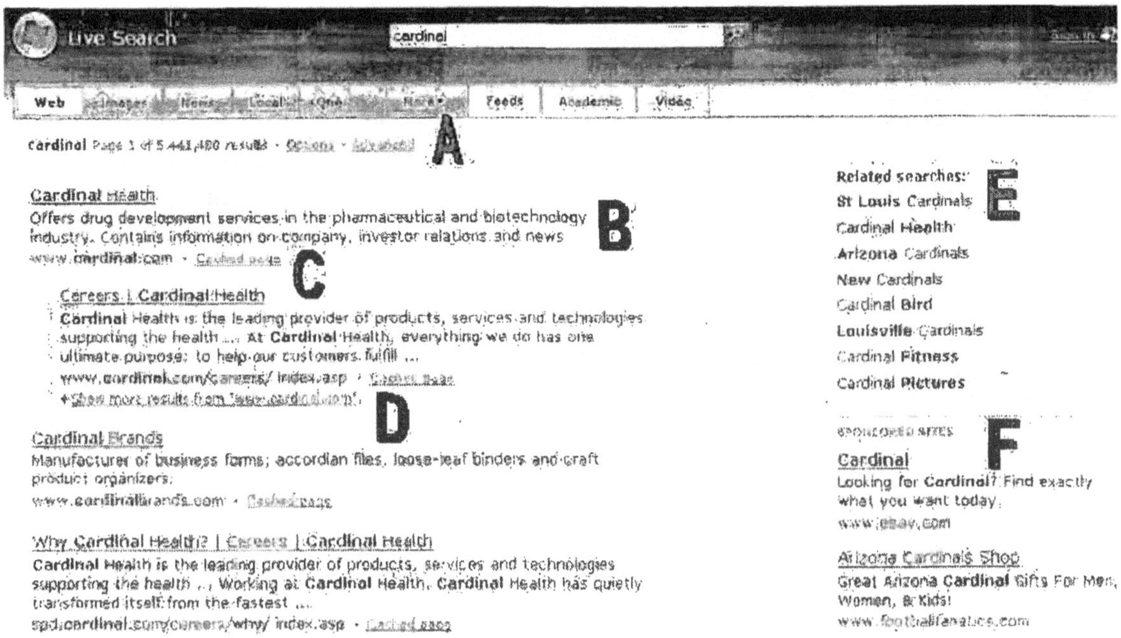
Live Search
cardinal
Web
Feeds
Academic
Video
A
Cardinal Health
Offers drug development services in the pharmaceutical and biotechnology industry. Contains information on company, investor relations and news
www.cardinal.com
B
C
Careers | Cardinal Health
Cardinal Health is the leading provider of products, services and technologies supporting the health .... At Cardinal Health, everything we do has one ultimate purpose: to help our customers fulfill ...
D
Cardinal Brands
Manufacturer of business forms; accordian files, loose-leaf binders and craft product organizers.
www.cardinalbrands.com
Why Cardinal Health? | Careers | Cardinal Health
Cardinal Health is the leading provider of products, services and technologies supporting the health ... Working at Cardinal Health, Cardinal Health has quietly transformed itself from the fastest ...
Related searches:
E
St Louis Cardinals
Cardinal Health
Arizona Cardinals
New Cardinals
Cardinal Bird
Louisville Cardinals
Cardinal Fitness
Cardinal Pictures
F
Cardinal
Looking for Cardinal? Find exactly what you want today.
www.ebay.com
Arizona Cardinals Shop
Great Arizona Cardinal Gifts For Men, Women, & Kids!

- **A** Type of results (web, image, etc.): the number of resulting pages and estimated total number of results.

- **B** The title of the webpage found, an excerpt from the webpage with the search terms bolded, the url of the webpage.

- **C** Cached page: links to a copy of the page as saved by the Live Search engine; Live Search shows the last date the page was examined by its spider; search terms are not highlighted on the cached page. **Important Note**: the cached copy of Microsoft file types are safe to view.

- **D** Additional results from the same site; clicking on "Show more results from..." will bring up the pages from that site that match the keyword(s).

- **E** Related Searches offer options either for similar terms or terms with multiple meanings, e.g., "cardinal."

- **F** Sponsored Sites are paid results.

Live Basic Search http://www.live.com/

Live Search has changed little about its basic search internals in terms of how it handles queries and the number of basic search options.

Live Search assumes as its default that multiple search terms are joined by the **AND** operator, so that a search on the keywords [windows explorer] will find all the webpages that contain both search terms.

Live Search recognizes **double quotes as enclosing a phrase**.

Live Search **will not return any results** if there is no webpage containing all the search terms. Try this query to see what I mean:

[rollerskate handshake specktioneer]

Unlike Google, Live Search **does not limit the number of search terms** to 10 keywords. Live will try to match all the keywords you enter.

Live Search is **not case sensitive**.

Live Search does **not offer any word stemming or truncation**, i.e., searching for variations of search terms. A search for [child] will not find [children].

Live Search **automatically clusters search results**. If you want to see more pages from a specific site, simply select the link following the url of the result.

Live permits the use of nested **boolean** queries in simple search. The operators must be **capitalized**. Live Search will run nested boolean queries (those using parentheses), such as:

[cardinals AND ("st louis" OR arizona) NOT (bird OR catholic)]

Live Search will **ignore stop words**, i.e., commonplace words, **if the query contains non-stop words**; the query [to be or not to be] will only search for the term "not." However, you can search on any single letter or number by itself, e.g., [1]. You can also force Live Search to look for stop words either by enclosing the query in double quotes ["to be or not to be"] or by placing a plus sign in front of the stop word, e.g., [+1 number] or [+to +be +or +not].

Otherwise, it is unnecessary to use the plus sign (+) with any terms because by default Live Search searches for all keywords. However, many times searchers need to exclude certain terms that are commonly associated with a keyword but irrelevant to their search. That's where the minus sign (-) comes in. Using the **minus sign** in front of a keyword ensures that Live Search excludes that term from the search. For example, the results for the search ["pearl harbor" –movie] are very different from the results for ["pearl harbor"]. You may use the boolean operator NOT instead of the minus sign.

Live Search interprets the **ampersand** [&] as a space, so these searches are virtually identical: [at&t], [at & t], ["at t"]. Also, while Live Search will not actually search on a plus sign, the search engine will search for **[c++]**, although it does not recognize [c+].

Live Search Advanced Search

Thus far, while Live Search added more advanced search options, it still falls behind Yahoo and Google in the number and type of advanced search options it offers. Nonetheless, Live Search has several advanced search features that are accessible by clicking on the "Advanced" link, which opens a small window that used to be labeled "Search Builder" in MSN Search and is still called that on the Help pages. The advanced search options may also be employed directly by using the correct syntax in the query box. Live Search's web search help is accessible from a link on the Live Search home page, but I prefer the old MSN Search, which is still available and at this point still accurate.

Windows Live Search Help http://search.msn.com/docs/help.aspx

Live Search now offers as many **languages** in which users may search as Yahoo and Google. Using either the language preference settings or the advanced search window, users can select from nearly 40 languages in which to search and see results. There are three ways to specify a search language:

1. in the Advanced search window, select Language, then pull down and click on a specific language.

2. type your search terms into the search box, and then add language: followed immediately by the two-character language code. For example, to search only for sites in French: [language:fr keyword]

3. a more permanent change would be to go to the Options page and change your primary search language.

Live Search does not distinguish words using **diacritical marks** such as accents or umlauts. Live Search finds terms matching those with and without the diacritic. The term [façade] finds façade and facade, and vice versa.

Live Search offers several special search terms to restrict searches and make them more effective.

- **site/domain:** restricts results to a specific website or domain, including a specific top-level domain. Can be used with or without keywords.

  Advanced Search > Site/Domain returns results from specific domains (com, gov, dell.com, a country digraph, etc.)

  Examples of how to use the site: command:

  [site:amazon.com] finds www.amazon.com, auction.amazon.com, www.amazon.com/dvd/. However, it will not find www.amazon.com.br.

  [books -site:amazon.com] finds pages containing the keyword "books" that are not at any amazon.com website.

  [site:ir] finds all the pages from the Iranian (.ir) top-level domain indexed by Live Search.

- **country/region:** on the Advanced menu; it is identical to the site/domain search for a country digraph. However, if you do not know a country's top-level domain, you can use the Country/Region pull-down menu to select the country, and Live Search will automatically enter the correct country digraph for you.

- **language:** restricts results to pages in a specific language. Users must specify a language using the two-letter code or use Advanced Search. Can be used with or without additional keywords.

  Advanced Search > Language uses pull-down menu to select languages.

  Examples of how to use the language: command:

  [language:ro] restricts results to sites written in Romanian.

  [language:es domain:mx méxico] restricts results to sites written in Spanish in the Mexican top-level domain that contain the term "méxico."

- **url:** unlike Google's url query, Live's url query checks to see if the domain or web address is in the Live Search index. This query is not really intended to be used with other search terms.

  Examples of how to use the url: command:

  [url:nasa.gov] or [url:education.jpl.nasa.gov] will check to see if a site is indexed by Live Search.

- **inurl:** restricts results to pages that contain search terms within the url of a site. Multiple terms can be used, but all must appear in the url (this query is similar to Google's allinurl: query).

  Examples of how to use the inurl: command:

[inurl:microsoft] finds all pages containing "microsoft" anywhere in the url

[inurl:microsoft downloads] finds all pages containing both the terms "microsoft" and "downloads" anywhere in the url.

- **inbody:** restricts results to pages containing search term(s) in the body of a webpage. Can be used with or without other search terms.

  Example of how to use the inbody: command:

  [inbody:amazon -inurl:amazon] finds all pages containing the term "amazon" anywhere in the body (text) of a webpage but which do not contain the term "amazon" in the url of the page.

- **intitle:** restricts results to pages containing search term(s) in the webpage's title. Can be used with or without other search terms.

  Examples of how to use the intitle: command:

  [intitle:amazon inbody:brazil] will find pages that contain "amazon" in the title of the webpage and "brazil" in the body text of the webpage.

- **contains:** restricts results to pages that have links to specific the file type(s). Can be used with or without other search terms.

  Examples of how to the contains: command:

  [music contains:mp3] finds webpages that contain links to MP3 files and have the keyword "music" in them.

  ["final report" contains:pdf] finds webpages that contain links to PDF files that have the phrase "final report" in them.

- **link:** Restricts results to pages containing links to a specific url. Can be used with or without additional keywords.

  Advanced Search > Links to returns results for pages that currently link to a specific url.

  Examples of how to use the link: command:

  [link:jpl.nasa.gov] finds all pages containing links to the specific domain jpl.nasa.gov.

  [link:jpl.nasa.gov asteroid] finds all pages containing links to any page in the jpl.nasa.gov domain and the keyword "asteroid" anywhere on the linking webpage.

- **linkdomain:** Restricts results to pages that link to any page within the specified domain. This is a broader search than the link: query. You can use this option to determine how many links there are to a specific page from sites indexed by Live Search. Can be used with or without additional keywords.

  Examples of how to use linkdomain:

  [linkdomain:jpl.nasa.gov] finds all pages containing links to any page at jpl.nasa.gov, including echo.jpl.nasa.gov, voyager.jpl.nasa.gov, etc.

  [linkdomain:jpl.nasa.gov cassini] finds all pages containing links to any page at jpl.nasa.gov and that also include the term "cassini" at the linking website.

  star [linkdomain:jpl.nasa.gov -site:jpl.nasa.gov] will find all pages containing links to any page at jpl.nasa.gov from sites other than jpl.nasa.gov (this eliminates internal links from the overall results).

- **linkfromdomain:** Restricts results to pages that are linked from the specified domain. This query only works with second-level domains, e.g., [domain.com]. You can use this option to determine how many links there are from a specific page. Can be used with or without additional keywords.

  Examples of how to use linkfromdomain:

  [linkfromdomain:nasa.gov] finds all the pages the nasa.gov domain links to, i.e., links from nasa.gov to site x.

  [linkfromdomain:nasa.gov standards] finds all pages the nasa.gov domain links to that contain the term "standards" on their webpage, i.e., links from nasa.gov to site x where site x contains the keyword "standards."

- **Results ranking:** allows users to emphasize different factors to get a different set of results for the same search.

  1. Type your search terms into the search text box, and then click Advanced Search.

  2. Select Results ranking, and then move the equalizer slider(s) in the direction you want.

  Live Search Help explains Results ranking in this way:

  "You can put emphasis on different factors to get a different set of results for the same search. The sliders control:

  - Updated recently: To modify your search to add emphasis to sites that have been recently added to the search index, move the left slider up.

- Very popular: To add emphasis to sites by the number of other sites that link to them, move the middle slider up.

- Approximate match: To put the most emphasis on the match between your exact search words and your results, move the right slider down.

Notes

- Approximate match overrides the first two slider rankings.

- Results ranking applies to web searches only."

It is easier to visualize how to use results ranking by looking at an example. In this case, the search on ["saudi arabia"] has been reranked to emphasize pages that have been recently updated {frsh=100} means the "freshness" ranking of these pages is 100 or the most recently updated pages in the Live Search database:

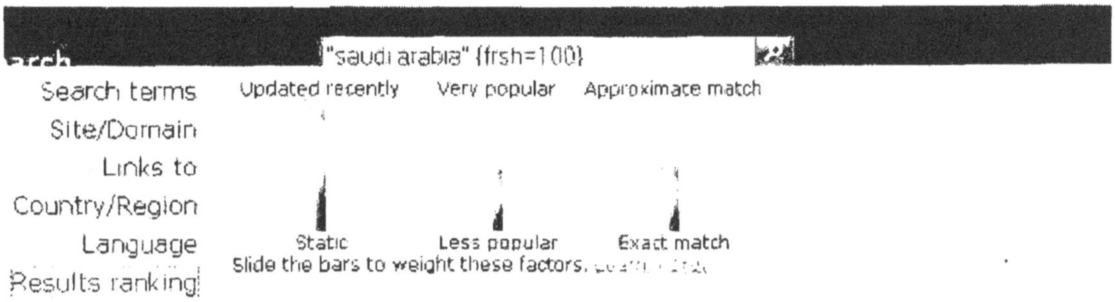
"saudi arabia" {frsh=100}
Search terms
Site/Domain
Links to
Country/Region
Language
Results ranking
Updated recently
Very popular
Approximate match
Static
Less popular
Exact match
Slide the bars to weight these factors.

➢ **filetype:** restricts results to a specific filetype. Can be used with or without additional keywords. The file types Live Search will search for include the major Microsoft file types and a few others:

Microsoft Excel (xls)

Microsoft PowerPoint (ppt)

Microsoft Word (doc)

Portable Document Format (pdf)

Rich Text Format (rtf)

Text (txt)

Examples of how to use the filetype: command:

[filetype:doc domain:nasa.gov] finds all Word files at the NASA domain in Word format.

[filetype:xls "financial data"] finds all Excel spreadsheets that contain the phrase "financial data."

Live Search ***does offer safe previewing of non-HTML file types, and this is especially useful for Microsoft file types, such as Word documents and PowerPoint slides***. In order to access the safe HTML versions, users must select the "Cached page" on the results page:

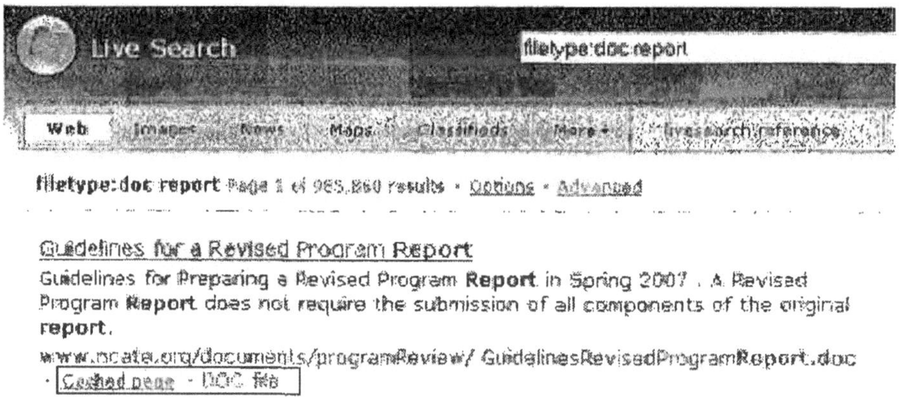
Live Search
filetype:doc report
Web
filetype:doc report Page 1 of 985,860 results · Options · Advanced
Guidelines for a Revised Program Report
Guidelines for Preparing a Revised Program Report in Spring 2007 . A Revised Program Report does not require the submission of all components of the original report.
Cached page · DOC file

- **IP: finds all the sites on a specific host computer.** Can be used with or without additional keywords.

  Examples of how to use the ip: command:

  [ip:66.218.77.68] finds all the sites on this specific host computer.

  [ip:66.218.77.68 "computer security"] finds all the sites on this specific host computer containing the phrase "computer security."

- **feed: one of two RSS search options; similar to the filetype: command.** It limits searches to text within a feed. Feeds are specially formatted brief descriptions of content with a link to the full version of that content. RSS (and the competing Atom) feeds are in XML format. These feeds are usually used for syndicating web content such as blogs and news. The feed: command only searches the text of the feed, which is often a very condensed description of the full web content.

  Example of how to use the feed: command:

  [feed:"trojan horse"]

  Each of the results represents an XML feed that includes the phrase "trojan horse." There is no point in clicking on the link in a browser because that brings up the XML page that most browsers are not designed to parse. The cached copy shows the search terms as they appeared in the feed.

- **hasfeed:** shows the pages that offer feed links and, if you add a keyword (something I'm pretty sure Live intended you to do), the pages with feed links and that also have that keyword somewhere on the page.

  Examples of how to use the hasfeed: command:

  [hasfeed:"trojan horses"]

  The results are webpages that offer news feeds and contain the phrase "trojan horses" on the webpage. This does not guarantee, however, that the news feed will be about Trojan horses, but the chances are good that if you are looking for sites with newsfeeds about this topic, you can find them using this query.

  [hasfeed:encryption site:microsoft.com]

  This query should find the pages at the Microsoft website with feeds about encryption. What this query actually finds are pages at the Microsoft site that contain both XML feeds and the word encryption in the text, so a little research will reveal which of these Microsoft newsfeeds are the most appropriate to the topic of encryption.

  This command is listed at the Live.com but is not working properly:

- **inanchor:** restricts results to pages containing search term(s) in the webpage anchor.

Live Search Special Features

Spell Checker: Live Search has a very good spell check option. When you input a query, Live checks to see if you are using the most common spelling of the keyword. If not, just like Google, Live nicely asks, * Were you looking for x, where x is the most common spelling. The Live Search dictionary also includes some proper names.

Dictionary Definitions: as with Google and Yahoo, Live Search offers the define option. To use it, type [define] then a word or brief phrase, e.g., [define king cobra]. Live's define option is more limited than some others because it only refers to Encarta.

Encarta: Microsoft's encyclopedia and general reference source Encarta provides answers to questions and facts about a topic. Users can type questions and (sometimes) get direct answers to them by simply entering a question and clicking on Search. Live Search does a much better job of correctly answering questions than MSN Search did (unlike its predecessor, it correctly identified Chirac as the

French president). Live Search can also directly answer certain specific questions, such as [how tall is the empire state building]:

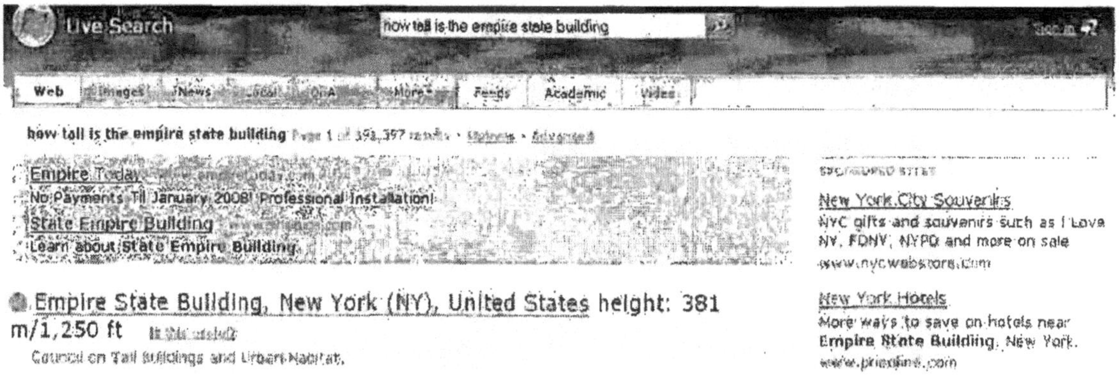
Live Search
how tall is the empire state building
Web
Feeds
Academic
Empire State Building, New York (NY), United States height: 381 m/1,250 ft
New York City Souvenirs
NYC gifts and souvenirs such as I Love NY, FDNY, NYPD and more on sale
New York Hotels
More ways to save on hotels near Empire State Building, New York.

Live Search no longer has the Encarta option that used to exist on MSN Search. For now, the easiest way I have found to invoke Encarta from Live Search and to take advantage of the Encarta Free Pass is to limit your search to Encarta using the site: syntax, e.g., [site:encarta.msn.com keyword]. This will give you two hours of free Encarta research.

Measurement Conversions: Live Search uses Encarta Answers to convert distance, weight, time, volume, and temperature. The conversions may be stated as questions, e.g., [how many seconds in a year?], or as a simple phrase:

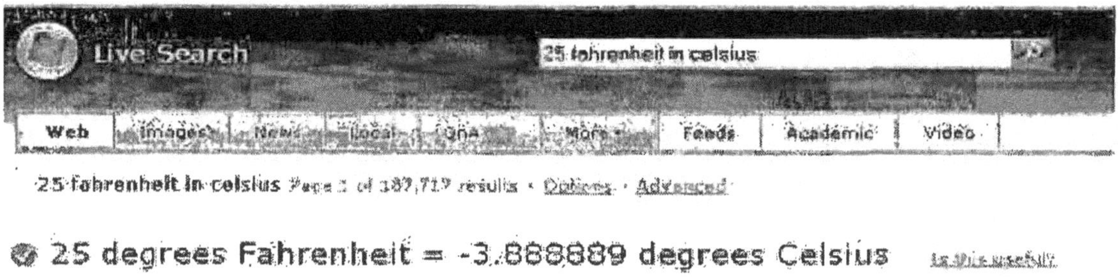
Live Search
25 fahrenheit in celsius
Web
Feeds
Academic
Video
25 fahrenheit in celsius
25 degrees Fahrenheit = -3.888889 degrees Celsius

<u>RSS</u> **Results:** when added to the end of any search result, the **&format=rss** parameter will provide users those search results via RSS. "When you subscribe to this RSS feed from Live Search, you'll get the top ten search results for this query delivered to your RSS Reader or personalized site. You can subscribe to any number of RSS feeds of Live Search results and view them all in your RSS Reader without re-running your search queries." To use this option, first search for your terms, e.g., [tsunami relief]. On the results' page, add &format=rss to the end of the url in the address bar and hit return:

http://search.live.com/results.aspx?q=tsunami&mkt=en-US&form=QBRE&go.x=0&go.y=0&go=Search&format=rss

The resulting page will look something like this; from here, follow the instructions on the webpage:

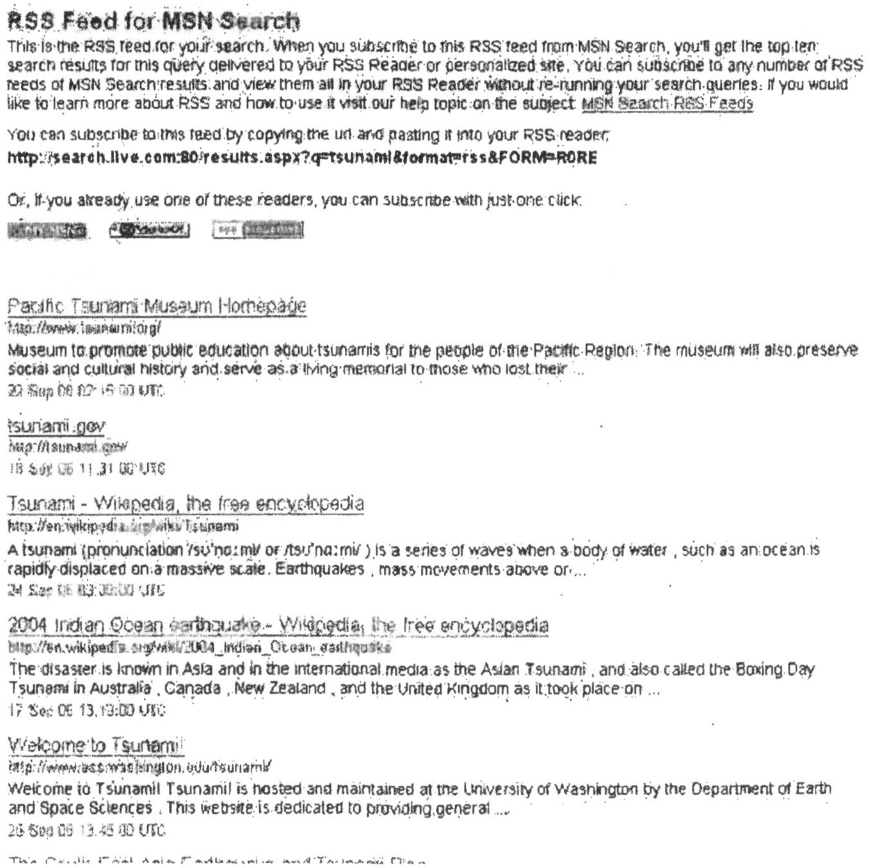
RSS Feed for MSN Search

This is the RSS feed for your search. When you subscribe to this RSS feed from MSN Search, you'll get the top ten search results for this query delivered to your RSS Reader or personalized site. You can subscribe to any number of RSS feeds of MSN Search results and view them all in your RSS Reader without re-running your search queries. If you would like to learn more about RSS and how to use it visit our help topic on the subject MSN Search RSS Feeds

You can subscribe to this feed by copying the url and pasting it into your RSS reader:
http://search.live.com:80/results.aspx?q=tsunami&format=rss&FORM=RORE

Or, if you already use one of these readers, you can subscribe with just one click:

Pacific Tsunami Museum Homepage
http://www.tsunami.org/
Museum to promote public education about tsunamis for the people of the Pacific Region. The museum will also preserve social and cultural history and serve as a living memorial to those who lost their ...
20 Sep 06 02:16:00 UTC

tsunami.gov
http://tsunami.gov/
18 Sep 06 11:31:00 UTC

Tsunami - Wikipedia, the free encyclopedia
http://en.wikipedia.org/wiki/Tsunami
A tsunami (pronunciation /suːˈnɑːmi/ or /tsuˈnɑːmi/) is a series of waves when a body of water , such as an ocean is rapidly displaced on a massive scale. Earthquakes , mass movements above or ...
24 Sep 06 03:00:00 UTC

2004 Indian Ocean earthquake - Wikipedia, the free encyclopedia
http://en.wikipedia.org/wiki/2004_Indian_Ocean_earthquake
The disaster is known in Asia and in the international media as the Asian Tsunami , and also called the Boxing Day Tsunami in Australia , Canada , New Zealand , and the United Kingdom as it took place on ...
17 Sep 06 13:13:00 UTC

Welcome to Tsunami!
http://www.ess.washington.edu/tsunami/
Welcome to Tsunami! Tsunami! is hosted and maintained at the University of Washington by the Department of Earth and Space Sciences . This website is dedicated to providing general ...
26 Sep 06 13:45:00 UTC

Number Search: Live Search offers many types of number searches, including:

- UPS tracking: enter the UPS tracking number [1Z9999X99999999] or [ups 1Z9999X99999999].
- USPS tracking: enter the tracking number or USPS plus the tracking number with or without spaces [usps 9999999999999999999999].
- FedEx tracking: enter the tracking number or FEDEX plus the tracking number [fedex 9999999999999999].
- DHL and Airborne Express tracking: enter DHL plus the tracking number [DHL 9999999999]; a DHL tracking search must include DHL in the query.

ISBN: enter any International Standard Book Number or [isbn 9999999999].

Calculator: Live Search uses the Encarta Calculator and Equation Solver to perform mathematical functions using "operators, exponents, and roots, factorials, modulo, percentages, logarithms, trig functions, and mathematical constants." The Encarta calculator appears to be the most sophisticated of all those offered by major search engines because it will even solve complex algebraic equations, such as 4x^3-2x+.9=0

The Live Search calculator uses the following symbols:

Add	+
Subtract	-
Multiply	*
Divide	/
Raise a number to an exponent (For example, 3^2 is 3 squared)	^
Specify the order of operation	( )
Find a percent of a number	% of
Find the square root of a number	sqrt
Find the sine of an angle	sin
Find the cosine of an angle	cos
Find the cosine of an angle	!

http://search.world.msn.com/docs/help.aspx?t=SEARCH_PROC_FindFactsNStatistics.htm

Live Search Services

Images: the Live Search image database is no longer Picsearch. Instead, Live Image Search uses Microsoft's own proprietary image database. Images are displayed as thumbnails (small versions of the original images), and the user can resize the thumbnails either using the slider or the dropdown "all image size" menu. One of the other changes to image search is the addition of a Scratchpad, which lets users drag and drop images onto a collection of images on the right-hand size of the screen. At this time, you do not have to have an account with Live in order to retrieve your image collections (they are retrieved based upon a cookie set by Live). When you mouse over an image, it zooms to a slightly larger size and moves toward the center and a box appears that shows the image source and size, and a link to the page where the image resides. If you click on the link, the linked page appears on the right with a "show image" in the top left corner. At present there are no advanced search options for images.

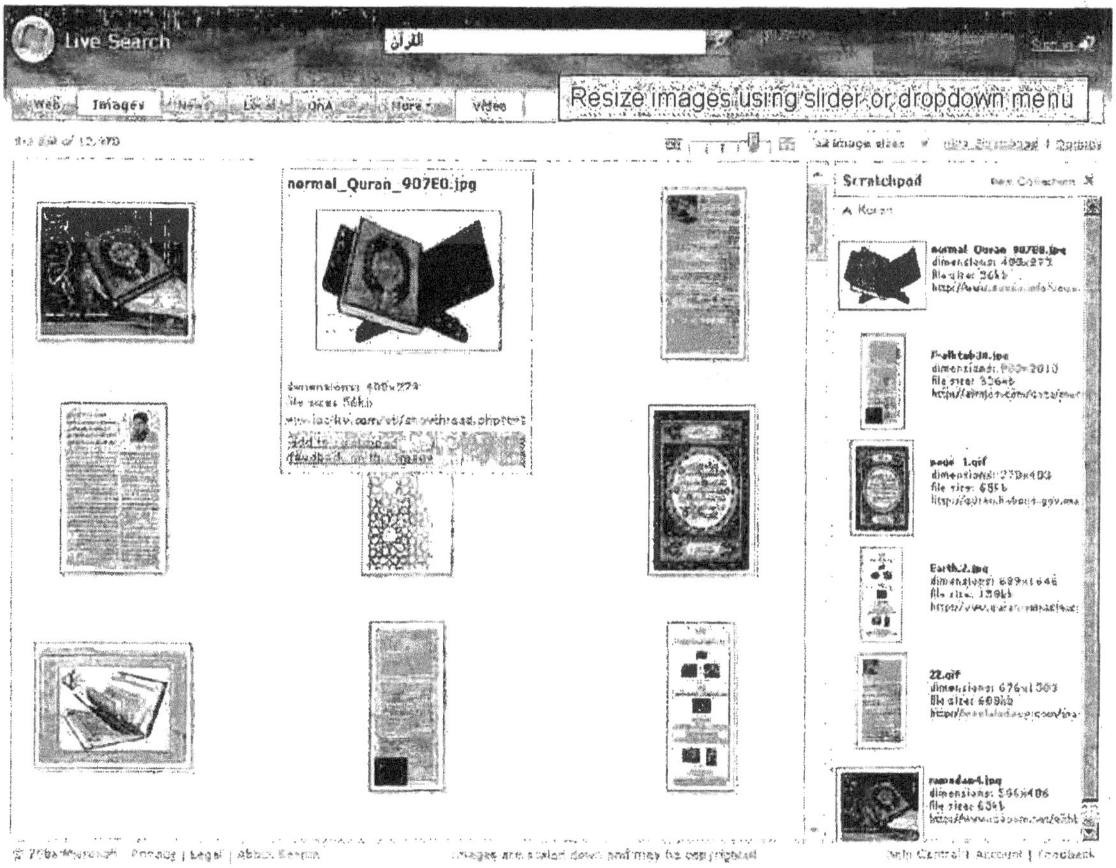
Live Search
القرآن
Web
Images
Video
Resize images using slider or dropdown menu
Scratchpad
normal_Quran_907E0.jpg

Also, when you search for a famous person using Live image search, look for the "Related People" window to appear on the right side of the screen. This can be an extremely useful tool in finding relationships between people in the news or historical figures.

The Live image search respects some but not all of the web search syntax and some of it is not really very useful for image search:

- **site/domain:** restricts results to images from a specific website or domain, including a specific top-level domain (com, gov, dell.com, a country digraph, etc.). May be used with or without keywords.

  Examples of how to use the site: command in image search:

  [site:amazon.com "twelfth night"] finds images of "twelfth night" that are from amazon.com; note that the images from amazon.com may reside on another website (amazon.com is in the image's url).

  [site:ir] finds all the image pages from the Iranian (.ir) top-level domain indexed by Live Search.

- **inurl:** restricts results to images that contain the term in the url of the image itself. Can be used with or without other search terms.

  Examples of how to use the inurl: command in image search:

  [inurl:amazon "rain forest"] finds all pages containing "amazon" in the url of the image and "rain forest" anywhere on the webpage.

- **intitle:** restricts results to images that appear on pages containing search term(s) in the title of the webpage. Can be used with or without other search terms.

  Examples of how to use the intitle: command in image search:

  [intitle:amazon brazil] will find pages that contain "amazon" in the title of the webpage and "brazil" anywhere on the webpage.

  [intitle:amazon inurl:brazil] will find pages that contain "amazon" in the title of the webpage and "brazil" in the image's url.

Video Search: Live video search is clearly trying to be competitive in the video search market. In October, Microsoft announced a new partnership with Blinkx to power its video search. This looks like a very good move for Microsoft. "Blinkx already powers video search on sites ranging from AOL to ITN, Lycos and Times Online. It also indexes video from the likes of BCC, Fox, MTV, Sky News, Reuters and YouTube and makes and makes videos on those sites searchable on Blinkx or partner sites. To date, the company has indexed more than six million hours of audio, video, and TV programming to make it searchable."[55] However, as of this writing, ***the Live video search has not yet been updated to reflect this partnership.***

---

[55] Eric Auchard, "Blinkx Signs Microsoft Pact," Reuters via Yahoo, 9 October 2006, <http://news.yahoo.com/s/nm/20061009/wr_nm/media_blinkx_dc_3> (17 October 2006).

As of now, the Live video search results include a thumbnail image from the video with the title, source, length, and format. All videos are viewed at the originating site, as shown below with the Newsweek On Air interview with Iranian President Ahmadinejad.

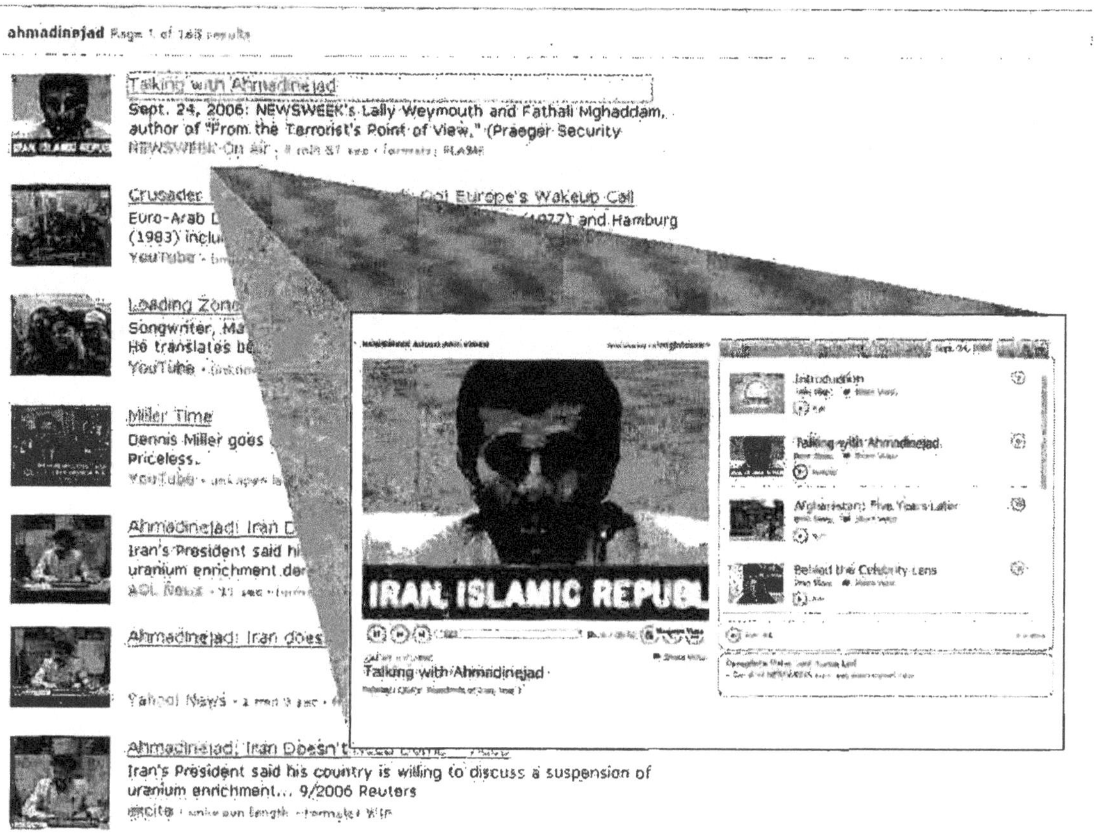
ahmadinejad
Talking with Ahmadinejad
Sept. 24, 2006: NEWSWEEK's Lally Weymouth and Fathali Mghaddam,
author of "From the Terrorist's Point of View," (Praeger Security
and Hamburg
Miller Time
Dennis Miller goes
Priceless.
Talking with Ahmadinejad
Iran's President said his country is willing to discuss a suspension of
uranium enrichment... 9/2006 Reuters

You can use some of the web search syntax for video search. Note the difference between these two searches:

[site:reuters.com iran]

[reuters iran]

The first query returns only those videos on Iran from the Reuters website; the second query returns queries from any site that includes the keywords "reuters" and "iran." We will have to wait and see how these query options change once the results come from Blinkx.

News Search: as of now, the Live news search is only a list of stories listed by relevance. MSN Newsbot <http://newsbot.msnbc.msn.com/> remains Microsoft's premier news page. However, if you want to search for news stories, MSN Newsbot takes you directly to the new Live news search. ***Most of the web search commands work for news search.*** Especially useful is the site/domain: syntax, which lets users limit a news query to a specific source:

[site:washingtonpost.com iran] finds pages at the Washington Post website that contain the keyword "iran." One big drawback of the Live news search is its inability to list the results by date.

Feed Search (Beta): This search is virtually identical to the feed: websearch. It limits searches to text within a feed. Feeds are specially formatted brief descriptions of content with a link to the full version of that content. RSS (and the competing Atom) feeds are in XML format. These feeds are usually used for syndicating web content such as blogs and news. The feed search only searches the text of the feed, which is often a very condensed description of the full web content.

Example of how to use the feed: command:

[feed:"trojan horse"]

Each of the results represents an XML feed that includes the phrase "trojan horse." There is no point in clicking on the link in a browser because that brings up the XML page that most browsers are not designed to parse. The cached copy shows the search terms as they appeared in the feed.

Live Book Search (beta): Microsoft added its own proprietary book search in late 2006. Details are in the Book Search section below.

Academic (Beta): Microsoft introduced Academic Search Beta for scholarly search earlier this year, and it is now also a Live search option. Academic search still has a separate website at the Windows Academic Live Beta Homepage. Clearly, Academic search is intended to compete with Google Scholar and other scholarly search sites. Unlike Google Scholar, Academic search focuses on computer science, physics, medical, and electrical engineering publications. As with Amazon and Google Scholar, Academic search has partnered with the **Online Computer Library Center (OCLC)**. "OCLC's involvement in Windows Live Academic is part of the Open WorldCat Find in a Library program,"[56] and also provides metadata from WorldCat to Academic search to give researchers access to the resources in library collections around the world.

As with almost anything, Academic search has good features and weaknesses. Here is a snapshot of the first page of results on the search [neural network]. When you execute a query, you will be presented with an interface that looks like this. One of the first things you notice is the split screen, which I actually like.

---

[56] "WorldCat live in Windows Live Academic search tool," OCLC Newsletter, Issue 2, 2006, <http://www.oclc.org/nextspace/002/updates.htm> (17 October 2006).

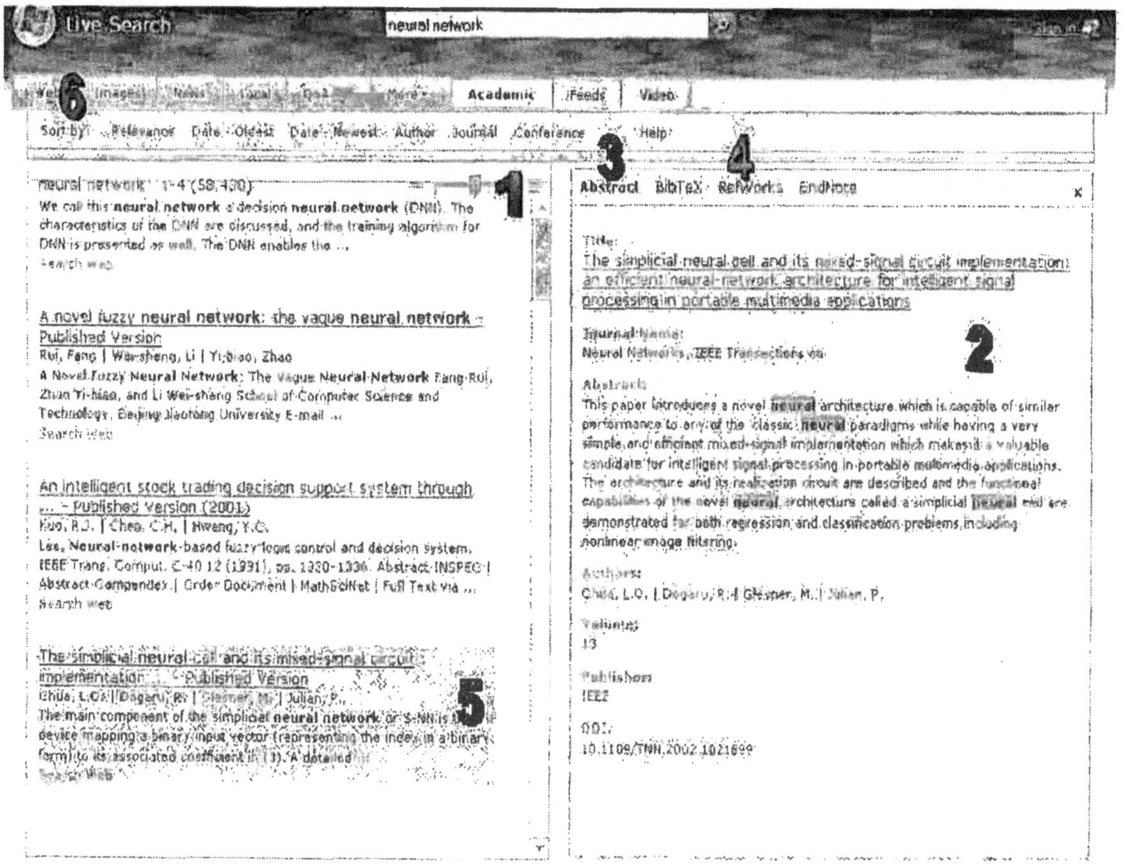
Live Search
neural network
Academic
Feeds
Abstract
BibTeX
RefWorks
EndNote
1
2
3
4
5
6

On the left you see the results; on the right-hand side of the screen is more detailed information that appears automatically as you click on different results. You have the option to view the abstract or properly formatted citations:

1. Slider bar: This allows you to expand or contract the amount of information contained in the search result
2. Preview pane: This pane allows you to obtain more information on the result that you are hovering over with your mouse on the results pane
3. Abstract: one of the options in the preview pane - choosing this option will allow you to see the abstract of the article that you are hovering over with your mouse on the results pane
4. BibTeX/RefWorks/EndNote: citation options in the preview pane - choosing one of these options will allow you to see the formatted citation (BibTeX, RefWorks, or EndNote format) on the preview pane for the search result that you are hovering over with your mouse on the results pane. BibTeX, RefWorks, and EndNote are different formats that allow users to create citations automatically. The EndNote RIS

format is compatible with EndNote, Reference Manager, and ProCite programs.

5. Search result: the actual search result; this includes links to the full text of the paper, link to search the web for that paper and potentially links that allow you to search your library for access to the full text from their subscription
6. Sort by options: allows you to sort the search results by relevance (default), oldest or newest date, author of paper, journal, or conference.

The best things about Academic search are:

- a list of journals it searches (something Google Scholar sorely needs); still, the list is too general (for example, IEEE Computer Society encompasses a huge number of journals and publications): <http://academic.live.com/AcademicJournals.htm>
- the preview pane is a good idea—no need to open new windows.
- the slider to view more or less information.
- the ability to extract citations (if you need to cite the information, this is a big benefit).
- the "find it in a library near you" search: [worldcatlibraries keyword].

Academic search needs to improve:

- lack of citation search (everyone seems to agree this is the biggest problem that simply must be rectified).
- no advanced search (may come later).
- not enough content.

Edit Macros: http://search.live.com/macros/default.aspx

This new feature allows users to "create their own search engine," so to speak. Of course, you are not really making a new search engine. In fact, what you are really doing with a basic macros' search is automatically generating a "site:" search. A basic macros search for ["north korea" "nuclear test"] on CNN, Reuters, and USA Today is equivalent to:

["north korea" "nuclear test" (site:www.cnn.com OR site:www.reuters.com OR site:www.usatoday.com)]

The advantage of the macros is that they are much simpler to create, especially if you want to search 30 sites, and you can easily save and retrieve your macros, but ***you must sign in to Live.com in order to save and retrieve your macros.***

Find Macros: http://gallery.live.com/default.aspx?l=4

Using other people's macros may be a much more fruitful activity. Microsoft has created a special page at its Gallery website to help users find already existing macros to add to a Live search page. The macros are by type: top categories, top downloads, and what's new, as well as new macros highlighted at the top of the page. Users can add any macros to their Live Search homepage by clicking on the "Add to Live.com" button. However, be careful. As of now the only way to remove/delete a macro that you have added to Live.com in your browser is to delete the browser's cookies (Microsoft, this needs to be fixed!). Also, if you try to add an uncertified macro to Live.com, you will get this: "This third-party application could include code that is unsafe." While the danger from these simple macros is probably very small to non-existent, this message does not exactly instill confidence, so caveat quaesitor.

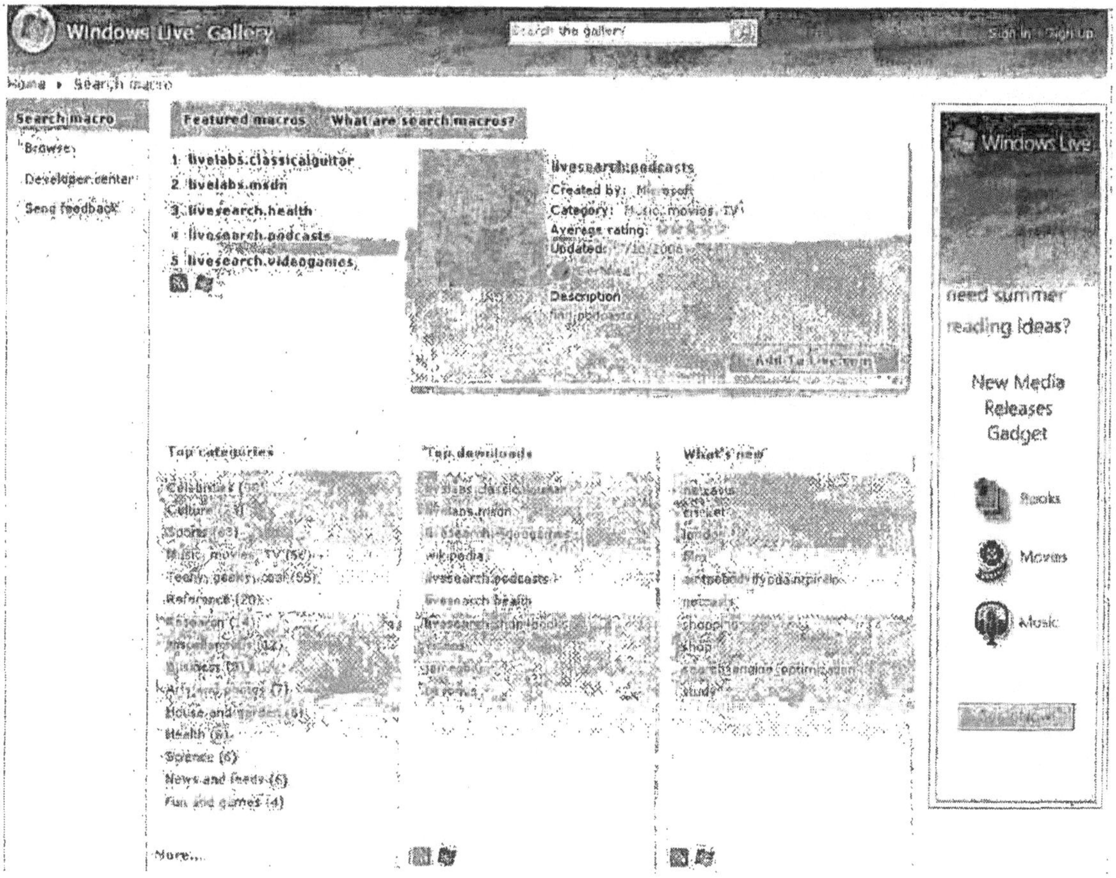
Windows Live Gallery
Search macro
Featured macros
What are search macros?
1 livelabs.classicalguitar
2 livelabs.msdn
3 livesearch.health
4 livesearch.podcasts
5 livesearch.videogames
livesearch.podcasts
need summer reading ideas?
New Media Releases Gadget
Books
Movies
Music

Given the newness of Live Macros, there are not very many to choose from yet; however, I expect to see this list grow and there are already some useful macros, such as the "reference" macros. Here is the reference macro added to the Live Search main menu with the results from querying the reference macro only.

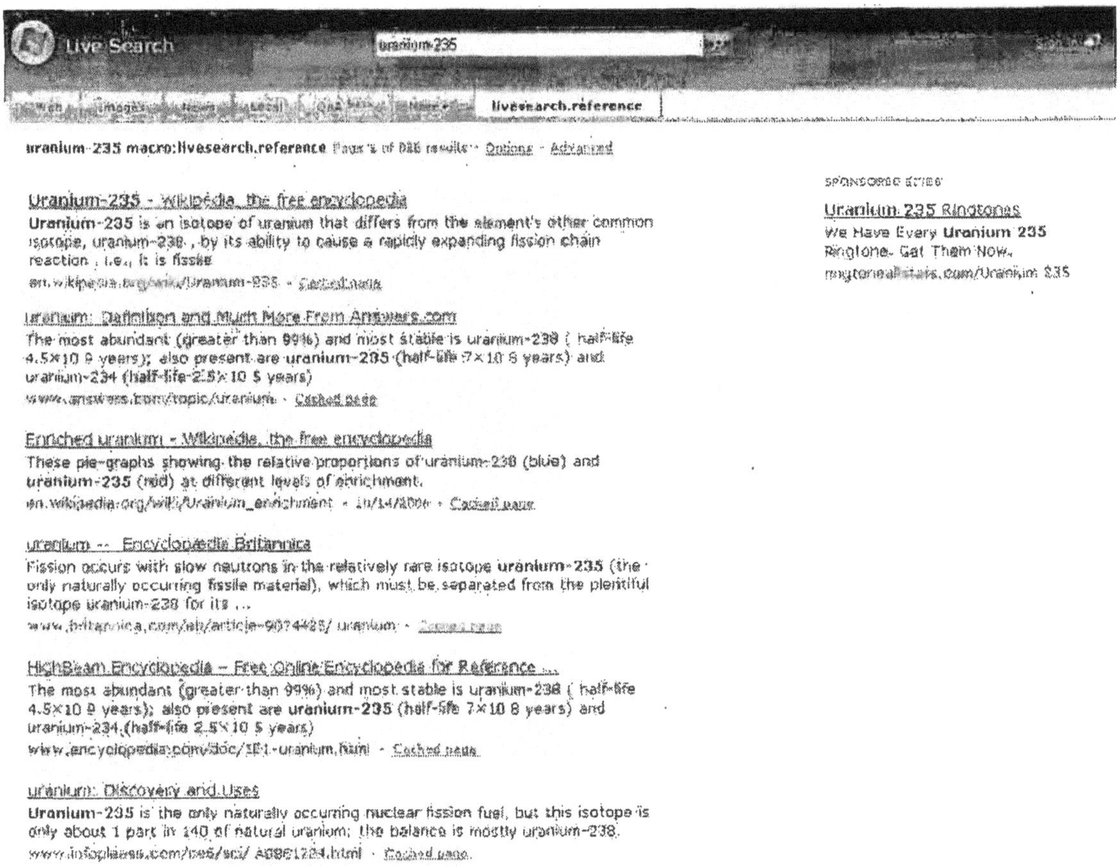
Live Search
uranium-235
livesearch.reference
uranium-235 macro:livesearch.reference
Options · Advanced
Uranium-235 - Wikipedia, the free encyclopedia
Uranium-235 is an isotope of uranium that differs from the element's other common isotope, uranium-238, by its ability to cause a rapidly expanding fission chain reaction, i.e., it is fissile
uranium: Definition and Much More From Answers.com
The most abundant (greater than 99%) and most stable is uranium-238 (half-life 4.5×10 9 years); also present are uranium-235 (half-life 7×10 8 years) and uranium-234 (half-life 2.5×10 5 years)
Enriched uranium - Wikipedia, the free encyclopedia
These pie-graphs showing the relative proportions of uranium-238 (blue) and uranium-235 (red) at different levels of enrichment.
en.wikipedia.org/wiki/Uranium_enrichment
uranium -- Encyclopædia Britannica
Fission occurs with slow neutrons in the relatively rare isotope uranium-235 (the only naturally occurring fissile material), which must be separated from the plentiful isotope uranium-238 for its ...
HighBeam Encyclopedia - Free Online Encyclopedia for Reference ...
The most abundant (greater than 99%) and most stable is uranium-238 (half-life 4.5×10 9 years); also present are uranium-235 (half-life 7×10 8 years) and uranium-234 (half-life 2.5×10 5 years)
Cached page
uranium: Discovery and Uses
Uranium-235 is the only naturally occurring nuclear fission fuel, but this isotope is only about 1 part in 140 of natural uranium; the balance is mostly uranium-238.
Cached page
Uranium 235 Ringtones
We Have Every Uranium 235 Ringtone. Get Them Now.

I believe there are too many results from Wikipedia in the reference search, but you can easily eliminate the Wikipedia results by adding [-site:wikipedia.org] to any query (conversely, you could limit your search to Wikipedia by adding [site:wikipedia.org] to your query. Live Search Macros are only the latest in a number of "create your own search engine" options, all of which are variations on complex queries of already existing search engines. For comparison, see the section on <u>Custom Search Engines</u> below.

QnA: Live Search's new QnA (question and answer) search is mostly fluff, at least for now. You can look at the questions and responses to see what I mean (typical questions: "How can i get my Space Cadet Pinball that was preinstalled in Windows XP back in Windows Vista?" "Do you think the Internet is contributing to 'Intellectual Laziness'?"). Lots of opinion, not a lot of fact. Let us hope this is not all that "Web 2.0" portends.

Live Platform: In September 2005 Microsoft announced it would begin offering APIs for Live Search, Virtual Earth, Spaces (weblogs), Messenger, Gadgets, and Expo classified ads database. These have begun to rival Google in terms of innovation and shared technology. To keep abreast of these changes, I recommend the MSN Developer Center.

MSN Developer Center's Windows Live Platform and Services for Web Mashups http://msdn.microsoft.com/live/default.aspx

Microsoft subsequently opened **Windows Live Dev (Beta)**, a "one-stop shop for the Windows Live Platform, including information on getting started with Windows Live services, latest documentation and APIs, samples, access to community areas and relevant blogs, and announcements of future releases and innovations."[57] Microsoft is trying to make it easy for users to integrate their products with Live regardless of platform, browser, or language. Certainly the first two are a departure for Microsoft, which in the past had made the requirement of a Windows platform and an Internet Explorer browser a necessity in most cases in order to "play ball" with the software giant. A further example of Microsoft's reluctant openness is the fact that Microsoft's Internet Explorer 7+ browser will not default to Live Search, something other search engines had objected to.

Windows Live Dev (Beta) http://dev.live.com/

Microsoft is working very hard to improve and expand its search properties, so much so that at times one feels as if we can see them working under the hood as we watch. Clearly, there are many things that need improvement and many things that are very good about Live.com. It will continue to be one of the top search sites on the Internet. If you are interested in keeping up with news about and changes to Live Search, there is a blog devoted to it; the blog offers RSS and Atom syndication. Also, all the Windows Live Beta projects are accessible through one webpage if you want to see what Microsoft is planning.

Windows Live Ideas Beta http://ideas.live.com/

Live Search Weblog http://blogs.msdn.com/livesearch/

---

[57] Windows Live Dev, Live Dev News, 8 June 2006, <http://dev.live.com/blogs/devlive/archive/2006/05/19/15.aspx> (17 October 2006).

Gigablast

The Gigablast search engine, which has been around since 2002, is still not quite in the same league as powerhouses Google, Yahoo, and Live Search, but it is well on its way to becoming one of the best search engines. That's something of a surprise given Gigablast's humble origins and unique status among major search engines. In case you're not familiar with Gigablast, it is different from its major competitors most notably because it is still owned and largely run by the guy who first wrote its C++ code in 2000. Matt Wells is still the very hands-on proprietor of Gigablast. Its database ***now indexes over 2 billion pages***, up from 650 million in late 2004. While this falls short of the size of the Google, Yahoo, and Live Search databases, it's not bad, especially considering a lot of the "stuff" in those databases is dross and the numbers are not verified independently.

How does Gigablast stack up to the big boys? Gigablast has some very nice features, some of which are unique to it, such as the IP range search (something AlltheWeb once offered).

Gigablast http://www.gigablast.com/

Strengths

- simple interface
- cached copies with date indexed [archived copies]
- cached copies of webpages without images [stripped]
- links to Internet Archives [older copies]
- clusters results by default (can be turned off)
- no limit on number of search terms
- file types indexed include Microsoft Word, Excel, and PowerPoint, as well as PDF, PostScript, HTML, and text; syntax is:
    - **type:pdf** for Adobe Acrobat PDFs
    - **type:doc** for Microsoft Word documents
    - **type:ppt** for PowerPoint presentations
    - **type:xls** for Excel spreadsheets
    - **type:ps** for PostScript files
    - **type:text** for ASCII text files
    - **type:html** for HTML Web pages

- **unique feature: IP range**; Gigablast adds the ability (unique as far as I know) to search on an IP address range. [ip:216.239.41] will find all IP addresses that begin with 216.239.41

  This query finds all the sites in the Gigablast database that begin with the IP address 66.218.77:

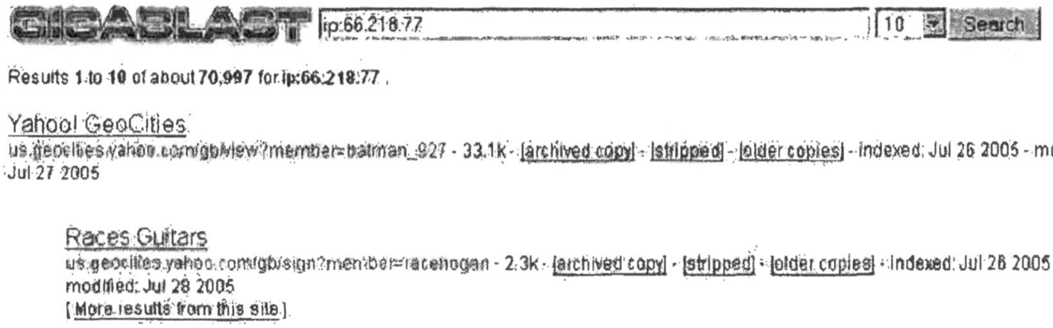
GIGABLAST
ip:66.218.77
10
Search
Results 1 to 10 of about 70,997 for ip:66.218.77 .
Yahoo! GeoCities
us.geocities.yahoo.com/gb/view?member=batman_927 - 33.1k - [archived copy] - [stripped] - [older copies] - indexed: Jul 26 2005 - modified: Jul 27 2005
Races Guitars
us.geocities.yahoo.com/gb/sign?member=racehogan - 2.3k - [archived copy] - [stripped] - [older copies] - indexed: Jul 28 2005 - modified: Jul 28 2005
[More results from this site]

  This query finds all the sites in the Gigablast database residing on the specific host whose IP address is 66.218.77.68:

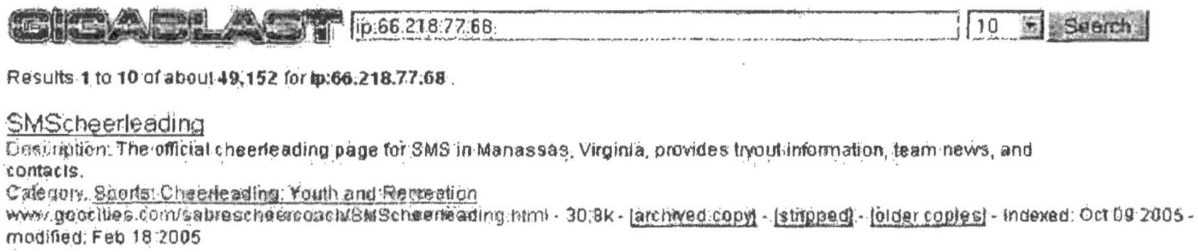
GIGABLAST
ip:66.218.77.68
10
Search
Results 1 to 10 of about 49,152 for ip:66.218.77.68 .
SMScheerleading
Description: The official cheerleading page for SMS in Manassas, Virginia, provides tryout information, team news, and contacts.
Category: Sports: Cheerleading: Youth and Recreation
www.geocities.com/sabrescheercoach/SMScheerleading.html - 30.8k - [archived copy] - [stripped] - [older copies] - indexed: Oct 09 2005 - modified: Feb 18 2005

- other special syntax includes **link:**, **site:**, **title:**, and **suburl:**, which searches for webpages that have the keyword anywhere in the url
- although Gigablast will ignore stop words in a long query, users can search on any word or number by itself
- default operator is AND; OR and AND NOT also work; nested queries (with parentheses) are supported
- **unique feature: indexes and displays of generic meta tags**; only search engine that will display the metatags in the results list, but the syntax for this query is very complex. Please see the Gigablast review at Search Engine Showdown for details on this type of query:

  "**Meta Tag Searching and Display:** Gigablast is the only search engine indexing meta tags beyond just the meta description and meta keywords that some others index. It is the only search engine that can also display meta tags in the results list. Gigablast claims to be indexing all "generic" meta tags. In addition, it can display the meta tags in the results list. Doing this requires adding commands to the URL of the results list. At the end of the url, add a &dt= followed by the word(s) for the meta tags, followed by a colon, and then a number to represent how many characters from each meta tag should be displayed. So, for example, adding &dt=keywords+author+generator+description:30 will display the meta tag content for meta keywords, meta author, meta generator, and meta description tags for any records retrieved. Use a + between meta tag words. It seems that this "generic" meta tag approach excludes more complex meta tags like Dublin Core, which use a syntax like DC.Creator. The dot syntax will not work for the display command, although Gigablast does index some of the content of these tags."[58]

Sample Output of Meta Tag Search

Back
Reload
http://www.gigablast.com/search?k1z=134827&q=dublin+core&dt=k
add string to the end of resulting url in address
DC-dot
..DC-dot now conforms with the Expressing Dublin Core in HTML/XHTML meta and l
..Now you can click on the DC-dot button, wherever you are, to create Dublin Core me
about.. ..This service will retrieve a Web page and automatically generate Dublin Core
metadata, either as..
Description: Give DC-dot a URL and see the Dublin Core it generates.
keywords: Dublin Core, DC; generator; editor; Warwick Framework; SOIF; TEI; USMARC; XML; GILS; ROADS; RDF; IMS
generator: HTML Tidy, see www.w3.org
description: A CGI based Dublin Core
Category: Reference: Libraries: Library and Information Science: Technical Services: Cataloguing: Metadata: Dublin Core
www.ukoln.ac.uk/metadata/dcdot/ - 8.8k - [archived copy] - [stripped] - [older copies] - indexed: Oct 05 2005 - modified: Dec 11 2001
Dublin Core Metadata Template
..When the list of Qualifiers for Dublin Core elements is finally decided upon, this template
will.. ..You may include my name and email-address in a list of those using Dublin Core.
Additional DC.. ..Dublin Core Metadata Template.. This service is provided by the "Nordic
Metadata Project" in..
Description: from the Nordic Metadata Project
Category: Reference: Libraries: Library and Information Science: Technical Services: Cataloguing: Metadata: Dublin Core
www.lub.lu.se/cgi-bin/nmdc.pl - 40.5k - [archived copy] - [stripped] - [older copies] - indexed: Oct 05 2005
Dublin Core/MARC/GILS Crosswalk
..For conversion of MARC 21 into Dublin Core, many fields may be mapped into a single
Dublin Core.. ..In the Dublin Core to MARC mapping, two mappings are provided,
one for unqualified Dublin Core.. ..The following is a crosswalk between the fifteen elements
in the Dublin Core Element Set and MARC..
Description: Library of Congress
keywords: MARC Dublin Core GILS Crosswalk
author: Library of Congress Network Development and MARC Standards Office
description: Crosswalk from Dublin Core
Category: Reference: Libraries: Library and Information Science: Technical Services: Cataloguing: Metadata: Crosswalks
lcweb.loc.gov/marc/dccross.html - 18.6k - [archived copy] - [stripped] - [older copies] - indexed: Oct 06 2005 - modified: Dec 31 2002

- clearly displays date webpage was indexed and, in some cases, modified
- search query spellchecker (**Did you mean?** option)

---

[58] Greg R. Notess, "Review of Gigablast," Searchengineshowdown, 17 September 2006, http://www.searchengineshowdown.com/features/gigablast/review.html> (14 November 2006).

- **undocumented feature:** will search in some specific languages, but I don't know how many; use ***language:de*** to search for webpages in German, for example.

Weaknesses

- most obviously, the Gigablast index is still smaller than those of Google, Yahoo, or Live Search
- no truncation
- is not case sensitive
- no wildcard
- limited file type searches
- limited language options
- poor documentation

Gigablast Options & Services

Custom Topic Search: Gigablast offers some special options, the most important of which is a Custom Topic Search, which I discuss in detail under the Custom Search Engines section below. ***If you don't read anything else about Gigablast, please take a look at this innovation.***

Directory: As with Google and Yahoo, Gigablast's web directory uses the Open Directory Project's collection but Gigablast use a "hypertechnology for searching the directory that allows its users to perform searches over websites, not just the actual pages, under any topic in the directory, in effect, instantly creating over 500,000 vertical search engines Additionally, all directory searches are enhanced by the massive amount of link information from Gigablast's multi-billion page index." So a Gigablast directory search returns not only DMOZ categories but "Giga Bits" and website listings as well.

XML Search Feed: Gigablast also offers an XML Search Feed that will run up to 1000 queries per day with a maximum of ten results each. But remember, you must have XML parsing software to read XML feeds, so this new feature isn't an option for all users.

XML Search Feed http://www.gigablast.com/searchfeed.html

Giga Bits: Gigablast has its own *refine* option called "Giga Bits." Giga Bits are terms that appear in a blue box at the top of a results page to help refine and focus your search.

Related Pages: Gigablast's Related Pages were introduced in March 2005. Related Pages are "relevant search results which do not necessarily contain the searcher's query terms." Related Pages are results that are contextually related to the query terms without having a direct connection to them. The Related Pages appear in the yellow box on the results page.

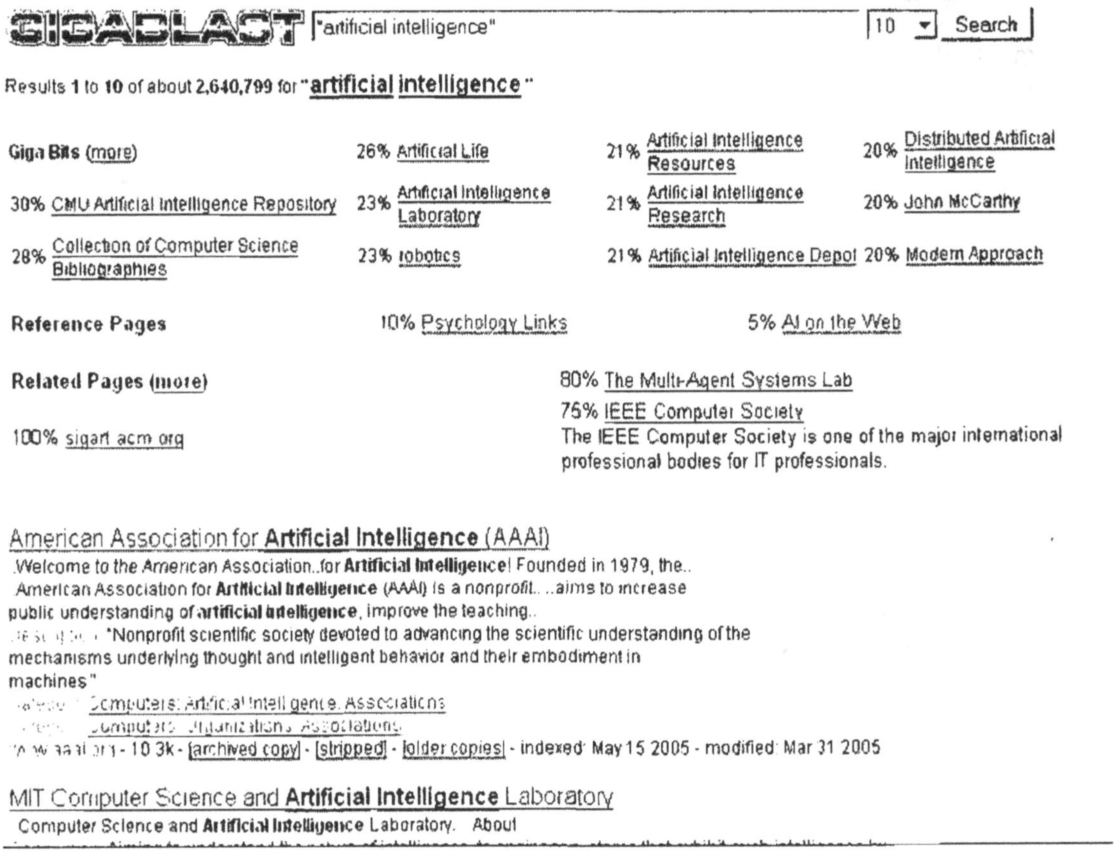
GIGABLAST
"artificial intelligence"
10
Search
Results 1 to 10 of about 2,640,799 for "artificial intelligence"
Giga Bits (more)
26% Artificial Life
21% Artificial Intelligence Resources
20% Distributed Artificial Intelligence
30% CMU Artificial Intelligence Repository
23% Artificial Intelligence Laboratory
21% Artificial Intelligence Research
20% John McCarthy
28% Collection of Computer Science Bibliographies
23% robotics
21% Artificial Intelligence Depot
20% Modern Approach
Reference Pages
10% Psychology Links
5% AI on the Web
Related Pages (more)
100% sigart acm org
80% The Multi-Agent Systems Lab
75% IEEE Computer Society
The IEEE Computer Society is one of the major international professional bodies for IT professionals.
American Association for Artificial Intelligence (AAAI)
Welcome to the American Association..for Artificial Intelligence! Founded in 1979, the..
American Association for Artificial Intelligence (AAAI) is a nonprofit.. ..aims to increase
public understanding of artificial intelligence, improve the teaching..
"Nonprofit scientific society devoted to advancing the scientific understanding of the
mechanisms underlying thought and intelligent behavior and their embodiment in
machines"
Computers: Artificial Intelligence: Associations
10 3k - [archived copy] - [stripped] - [older copies] - indexed: May 15 2005 - modified: Mar 31 2005
MIT Computer Science and Artificial Intelligence Laboratory
Computer Science and Artificial Intelligence Laboratory. About

Gigablast still "runs on eight desktop machines, each with four 160-GB IDE hard drives, two gigs of RAM, and one 2.6-GHz Intel processor. It can hold up to 320 million Web pages (on 5 TB), handle about 40 queries per second and spider about eight million pages per day. Currently it serves half a million queries per day to various clients, including some metasearch engines and some pay-per-click engines." We are not talking about a huge "server farm" here. Interestingly, despite keeping his search engine "small," Gigablast creator/proprietor Matt Wells says "I am a firm believer that bigger is better," and toward that end he is hoping to get the Gigablast index up to 5 billion pages. For more on Wells and Gigablast, read his interview with his former boss at Infoseek in the April 2004 edition of *AMC Queue*:

"A Conversation with Matt Wells: Steve Kirsh of Propel Software Interviews Gigablast Designer," *ACM Queue*, vol. 2, no. 2, April 2004, http://www.acmqueue.com/modules.php?name=Content&pa=showpage&pid=135 (15 November 2006).

Exalead

The French search engine Exalead, which introduced a new look in 2006, has features that make it worth special mention. Exalead offers both proximity searches and truncation, two options no other major search engine offers anymore. In addition, Exalead presents thumbnail images of websites in the results list (if you want them) and related search terms, directory categories, website locations, and filetypes. Exalead now claims to index more than eight billion pages. Although this is far smaller than some major search engines, it is a respectable number and one that is sure to increase.

While the new version of Exalead did away with one of its best features—the safe page preview—Exalead offers a number of other unusual or unique features designed to create a very powerful search tool:

- Exalead refreshes its index continuously, not on a schedule (this is a good thing).
- default operator is AND; users may use OR.
- Exalead does not publish a search term limit; it handled some very long searches perfectly while it had trouble with others.
- truncation, proximity, phonetic, and true wildcard searches.
- as of now, Exalead has no sponsored links.

exalead
Web | Images
Web Search
Advanced search
Welcome to your exalead homepage. Add a shortcut to personalize it.
Edit
Edit
Edit
Add a shortcut
Yellow Pages
weather.com
Wikipedia
Add more shortcuts - Hide edit buttons

Notice the images below the query box. Exalead lets users put "shortcuts" here by entering a title and url for your favorite websites.

Exalead is in the process of updating its help pages; thus far, you can find various types of help at these pages:

Exalead http://www.exalead.com/search

Exalead Refine Your Search http://www.exalead.com/search/?action=kourou&id=49

Exalead Advanced Search Help
http://www.exalead.com/search/?action=kourou&id=24

Exalead Search Syntax Help
http://www.exalead.com/search/C?definition=querySyntaxReference

Customizing Exalead Preferences

Exalead currently offers these ***Search*** Preferences settings:

1. Interface Language: English, French, or German.
2. Search language: any or any combination of most languages.
3. Adult content Filtering: on or off.
4. Display: Open results and shortcuts in new window?
5. Number of search results: up to 100 for web and up to 60 for image.
6. Number of shortcuts per row: 4 up to 12.
7. Display view on results page: text only; text and thumbnail; text thumbnail and extra

The Exalead Results Page

Once you have entered your search term(s) and clicked the Exalead search button, Exalead will present you with a complex results screen. Depending on the search you are running, you will see some or all of the following for a Web search:

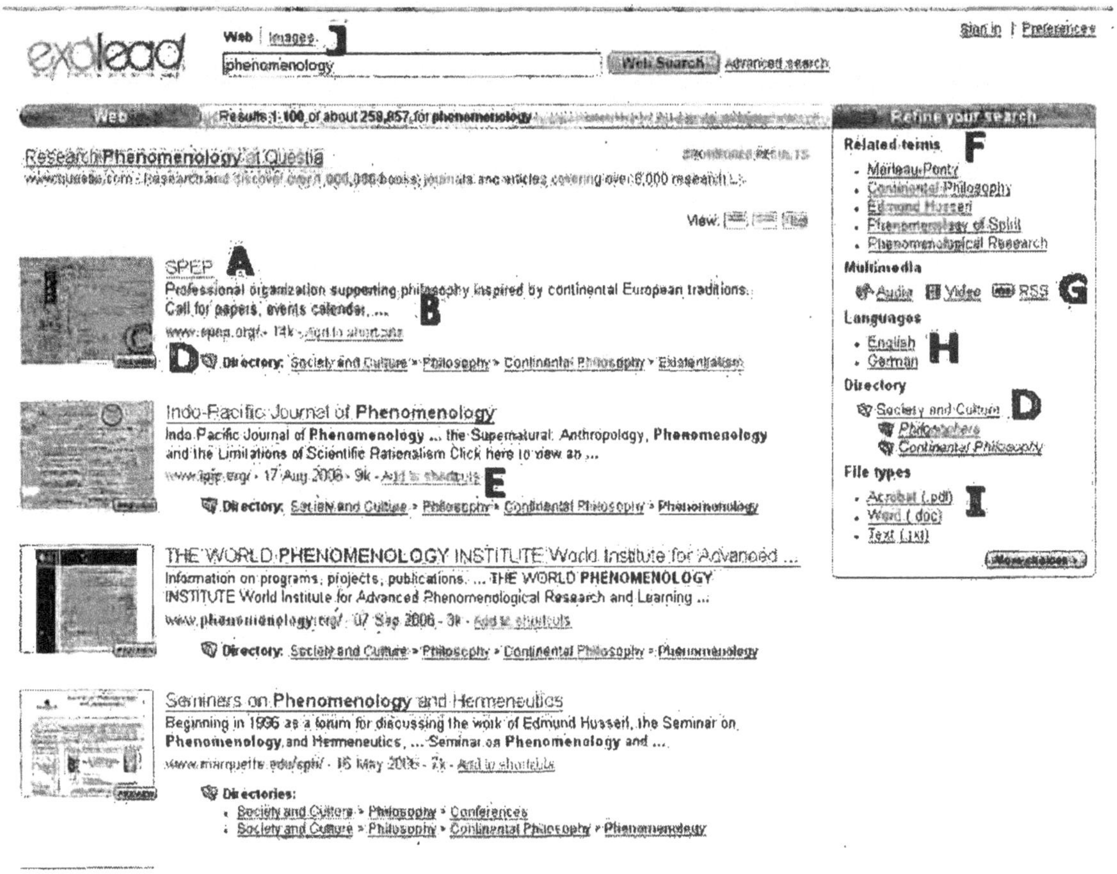
exalead
Web | Images
phenomenology
Web Search
Advanced search
Sign in | Preferences
Refine your search
Related terms
Multimedia
Languages
English
German
Directory
File types
SPEP
Indo-Pacific Journal of Phenomenology
THE WORLD PHENOMENOLOGY INSTITUTE World Institute for Advanced ...
Seminars on Phenomenology and Hermeneutics
A
B
C
D
E
F
G
H
I
J

- A Matching Documents: the best results for the query with the page title listed first; Exalead clusters results, showing only the "best" page for each website.

- B Webpage Description: a brief summary of the website.

- C Page preview and thumbnail image: The biggest disappointment of the new Exalead is that it no longer offers the safe page preview option for webpages. Instead it has chosen to give a thumbnail image of the cached copy of the webpage; users can click on "Preview" to see the cached copy, complete with highlighted search terms and the date cached. Fortunately, ***Exalead does offer safe previewing of non-HTML file types, and this is especially***

useful for Microsoft file types, such as Word documents and PowerPoint slides.

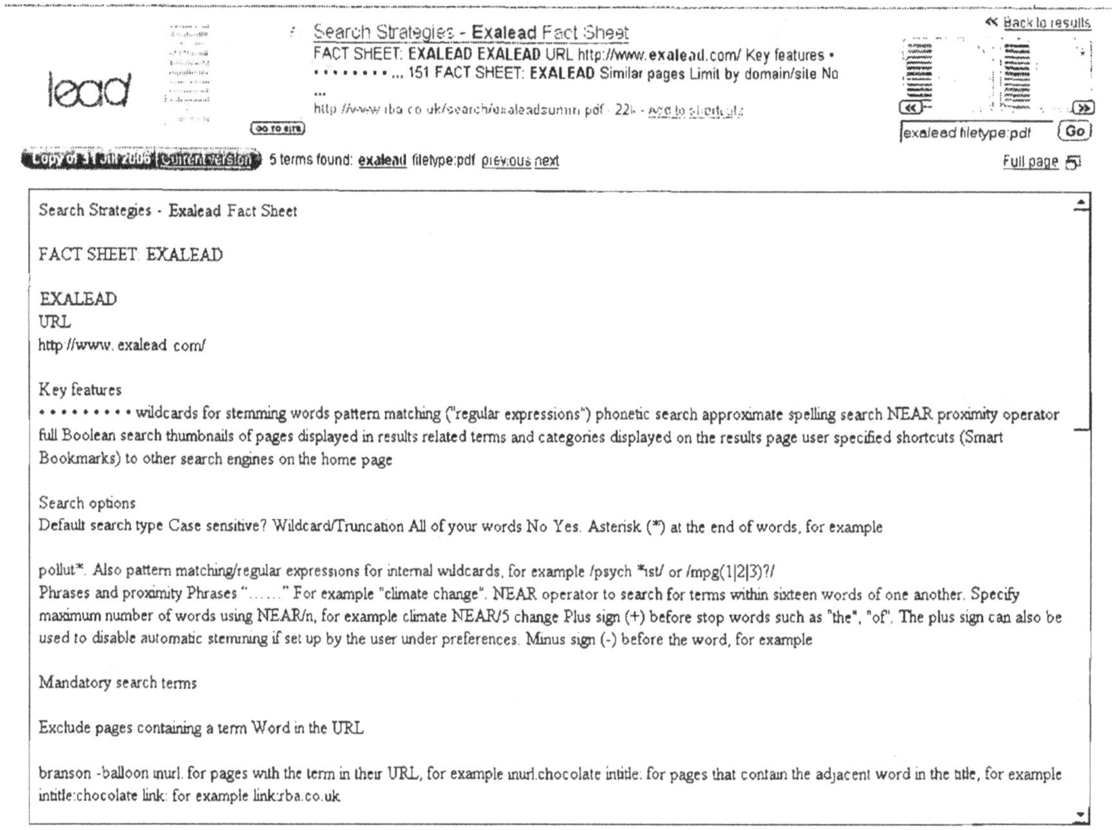
Search Strategies - Exalead Fact Sheet
FACT SHEET: EXALEAD EXALEAD URL http://www.exalead.com/ Key features • • • • • • • • • ... 151 FACT SHEET: EXALEAD Similar pages Limit by domain/site No
« Back to results
exalead filetype:pdf
Go
5 terms found: exalead filetype:pdf previous next
Full page
Search Strategies - Exalead Fact Sheet
FACT SHEET EXALEAD
EXALEAD
URL
http://www.exalead.com/
Key features
• • • • • • • • • wildcards for stemming words pattern matching ("regular expressions") phonetic search approximate spelling search NEAR proximity operator full Boolean search thumbnails of pages displayed in results related terms and categories displayed on the results page user specified shortcuts (Smart Bookmarks) to other search engines on the home page
Search options
Default search type Case sensitive? Wildcard/Truncation All of your words No Yes. Asterisk (*) at the end of words, for example
pollut*. Also pattern matching/regular expressions for internal wildcards, for example /psych *ist/ or /mpg(1|2|3)?/
Phrases and proximity Phrases "......" For example "climate change". NEAR operator to search for terms within sixteen words of one another. Specify maximum number of words using NEAR/n, for example climate NEAR/5 change Plus sign (+) before stop words such as "the", "of". The plus sign can also be used to disable automatic stemming if set up by the user under preferences. Minus sign (-) before the word, for example
Mandatory search terms
Exclude pages containing a term Word in the URL
branson -balloon inurl: for pages with the term in their URL, for example inurl:chocolate intitle: for pages that contain the adjacent word in the title, for example intitle:chocolate link: for example link:rba.co.uk

- **D** Directory link: opens the related categories folders from The Open Directory Project, which are also listed to the right. You can completely alter the results by selecting a different related category, e.g., in this example, *continental philosophy* instead of *phenomenology*. Clicking on "More choices" will greatly expand the related terms and related categories lists.

- **E** Add to shortcuts: selecting this link will make the current site one your shortcuts that appears on the Exalead homepage.

- **F** Related Terms: clicking on a related term runs a new search on that term and displays a new results page with new and different related terms, related categories, etc. Clicking on "More choices" will greatly expand the related terms and related categories lists.

- **G** Multimedia: selecting this option causes Exalead to restrict the search to webpages that have links to audio, video RSS content. You can select one, two, or all three multimedia options. If you click on RSS, any feeds available at any of the sites in your results' list will become visible.

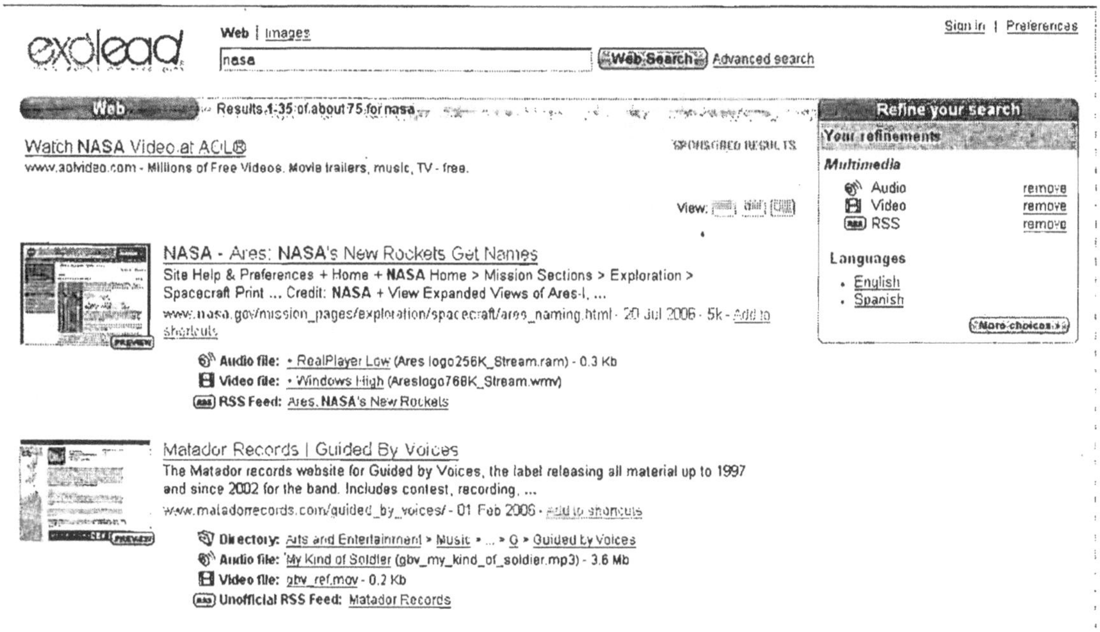
exalead
Web | Images
nasa
Web Search
Advanced search
Sign in | Preferences
Web
Results 1-35 of about 75 for nasa
Watch NASA Video at AOL®
www.aolvideo.com - Millions of Free Videos. Movie trailers, music, TV - free.
SPONSORED RESULTS
View:
NASA - Ares: NASA's New Rockets Get Names
Site Help & Preferences + Home + NASA Home > Mission Sections > Exploration > Spacecraft Print ... Credit: NASA + View Expanded Views of Ares-I, ...
www.nasa.gov/mission_pages/exploration/spacecraft/ares_naming.html - 20 Jul 2006 - 5k - Add to shortcuts
Audio file: RealPlayer Low (Ares logo256K_Stream.ram) - 0.3 Kb
Video file: Windows High (Areslogo768K_Stream.wmv)
RSS Feed: Ares, NASA's New Rockets
Matador Records | Guided By Voices
The Matador records website for Guided by Voices, the label releasing all material up to 1997 and since 2002 for the band. Includes contest, recording, ...
www.matadorrecords.com/guided_by_voices/ - 01 Feb 2006 - Add to shortcuts
Directory: Arts and Entertainment > Music > ... > G > Guided by Voices
Audio file: My Kind of Soldier (gbv_my_kind_of_soldier.mp3) - 3.6 Mb
Video file: gbv_ref.mov - 0.2 Kb
Unofficial RSS Feed: Matador Records
Refine your search
Your refinements
Multimedia
Audio remove
Video remove
RSS remove
Languages
English
Spanish
More choices

- **H** Languages: limit results to a specific language.

- **I** Document Type: clicking on a specific file type will only return matching documents in that specific file type, e.g., PDF, TXT, DOC, PPT, RTF, and XLS (remember: *do not open the Microsoft file types on the Internet; use the page preview option in the thumbnail image to view these files*).

- **J** Image Search: Clicking on image search will automatically run the web search against the image database.

Exalead Basic Search

Exalead assumes as its default that multiple search terms are joined by the **AND** operator, so that a search on the keywords [windows explorer] will find all the webpages that contain ***both*** search terms. However, unlike Google, Exalead does not search first for phrases, then the terms anywhere on a webpage.

Exalead ***will not return any results*** if there is no webpage containing all the search terms. Try this query to see what I mean:

[rollerskate handshake buckyball]

However, remember you can use the OPT (optional) operator to make a term desirable but not required.

Unlike Google, Exalead ***does not limit the number of search terms to 32 keywords***. Exalead will try to match all the keywords you enter.

Exalead is ***not case sensitive***.

Exalead ***automatically clusters*** search results. If you want to see more pages from a specific site, the only way I know to do so now is to run a site search. For example, to see the pages at Amazon UK search for [site:amazon.co.uk].

Exalead permits the use of ***the OR operator*** in simple search. The OR must be capitalized.

Exalead recognizes ***double-quotes*** as enclosing a phrase.

Exalead ignores certain ***stop words***, i.e., when searched alone or with other stop words. If you include a stop word such as *a*, *an*, *the*, *in*, or *be* in a search, Exalead searches for it. If you need to search for stop words by themselves or with other stop words, you must either enclose them in double-quotes or put the plus sign (+) in front of them. Compare [to be or not] to ["to be or not to be"] and compare [fire and ice] to ["fire and ice"].

Using the ***minus sign (-)*** in front of a keyword ensures that Exalead excludes that term from the search. For example, the results for the search [phenomenology –philosophy] are very different from the results for [phenomenology].

Exalead Advanced Search

Exalead has a unique and very appealing way of presenting advanced search features. Clicking on the "Advanced search" link on the main page brings up a window that displays and explains the advanced search options. In every case, these options work in the simple search screen by using the correct syntax.

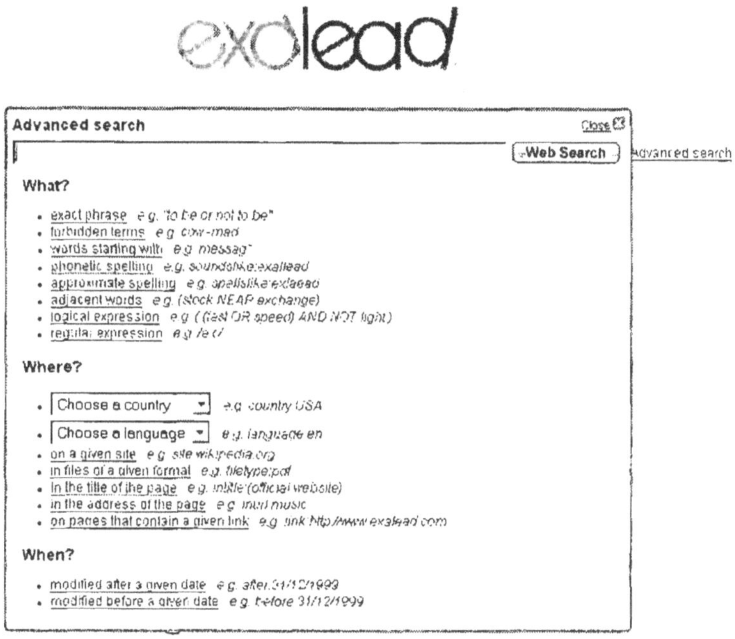
exalead
Advanced search
Close
Web Search
Advanced search
What?
exact phrase
forbidden terms
words starting with
phonetic spelling
approximate spelling
adjacent words
logical expression
regular expression
Where?
Choose a country
Choose a language
on a given site
in files of a given format
in the title of the page
in the address of the page
on pages that contain a given link
When?
modified after a given date
modified before a given date

Two features Exalead offers that have almost vanished from search elsewhere are ***proximity searches*** and ***truncation/wildcards***.

Exalead's proximity search uses **NEAR**. The default setting is for ***terms that are within sixteen terms of each other***, but users can change the proximity by adding a number, e.g., [empire NEAR/5 building]. With the NEAR operator, order is almost irrelevant as this query demonstrates. A query using the name of an 18th Century French foreign minister, Charles Jean-Baptiste Fleuriau, comte de Morville, shows how the NEAR operator works: the query [comte de Morville NEAR Fleuriau NEAR Charles NEAR Jean-Baptiste] finds any indexed page containing all these terms within sixteen words of each other, regardless of the order in which they appear either in the query or in the text.

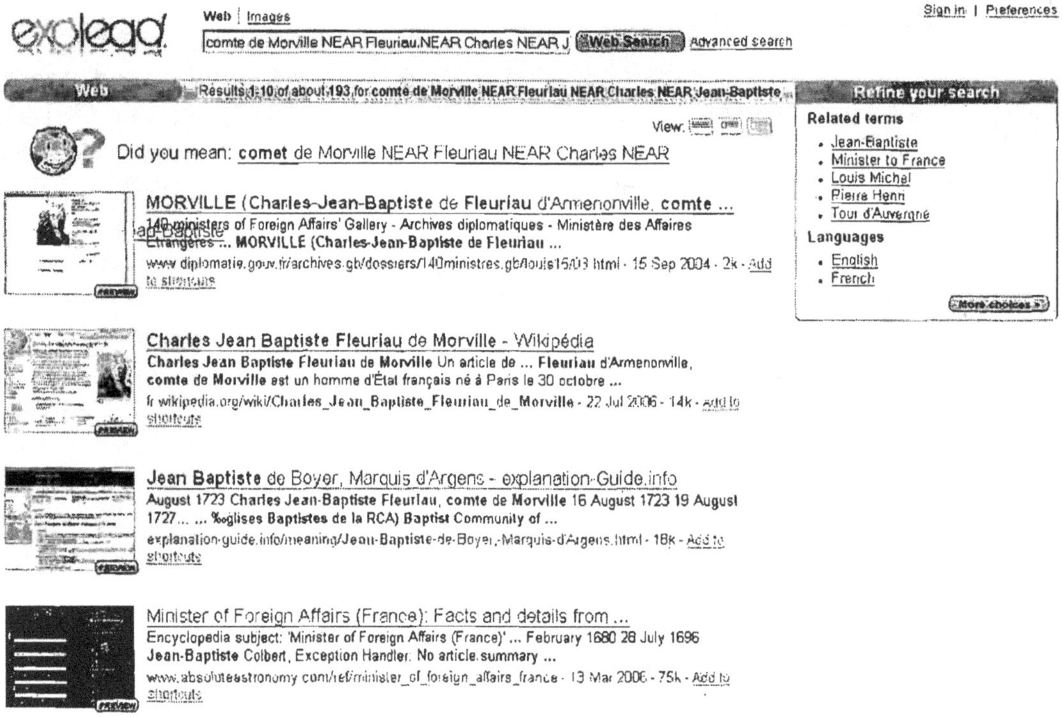
exalead
Web | Images
comte de Morville NEAR Fleuriau NEAR Charles NEAR J
Web Search
Advanced search
Sign in | Preferences
Web
Results 1-10 of about 193 for comte de Morville NEAR Fleuriau NEAR Charles NEAR Jean-Baptiste
Refine your search
Related terms
Jean-Baptiste
Minister to France
Louis Michel
Pierre Henri
Tour d'Auvergne
Languages
English
French
More choices
View:
Did you mean: comet de Morville NEAR Fleuriau NEAR Charles NEAR
MORVILLE (Charles-Jean-Baptiste de Fleuriau d'Armenonville, comte ...
Charles Jean Baptiste Fleuriau de Morville - Wikipédia
Charles Jean Baptiste Fleuriau de Morville Un article de ... Fleuriau d'Armenonville, comte de Morville est un homme d'État français né à Paris le 30 octobre ...
Jean Baptiste de Boyer, Marquis d'Argens - explanation-Guide.info
August 1723 Charles Jean-Baptiste Fleuriau, comte de Morville 16 August 1723 19 August 1727... ... Églises Baptistes de la RCA) Baptist Community of ...
Minister of Foreign Affairs (France): Facts and details from ...
Encyclopedia subject: 'Minister of Foreign Affairs (France)' ... February 1680 28 July 1696 Jean-Baptiste Colbert, Exception Handler. No article.summary ...

Also, the presence or absence of parentheses does not appear to affect the NEAR search. ***Proximity operators can be extremely useful in finding pages with search terms that may not be in a precise order while excluding a lot of irrelevant hits.***

Exalead ***supports both limited and true wildcard*** searching.

Exalead supposedly offers both ***automatic truncation*** (word stemming) and the ***wildcard***, which are welcome features discarded by other search engines. As of now, Exalead is the only major search engine to offer truncation or a wildcard. On a search with two or more words, stemming is supposed to be automatic. However, I find that the automatic truncation feature is so capricious as to be useless: sometimes it works, usually it doesn't. In a search for [child play toy], Exalead does not find *children, plays/played/playing*, or *toys*.

However, when I search on [child*], Exalead will return pages with *children* highlighted as a search result. The wildcard also can be ***used inside a search term***, e.g., [kazak*stan]. However, this search will also find *kazakh* and *kazak* as well as *kazakstan* and *kazakhstan*. The wildcard option is listed in the Advanced search window as ***words starting with***, but keep in mind the asterisk can be used inside words as well.

Exalead has a number of other interesting features. For example, in the advanced search window, users can choose among these search method options: ***exact***

search, ***forbidden terms***, ***phonetic search***, and ***approximate spelling***. Exact search is what you would expect, i.e., phrase searching inside double-quotes. "Forbidden terms" is a different way of saying NOT or using the minus sign.

The ***phonetic search*** sounds great, but I am often frustrated by it because so many websites misspell so many words, Exalead is going to find those misspelled words first (try: [geneology] to see what I mean). However, the phonetic search successfully figured out that [criptografy] meant [cryptography]. The phonetic search has genuine utility.

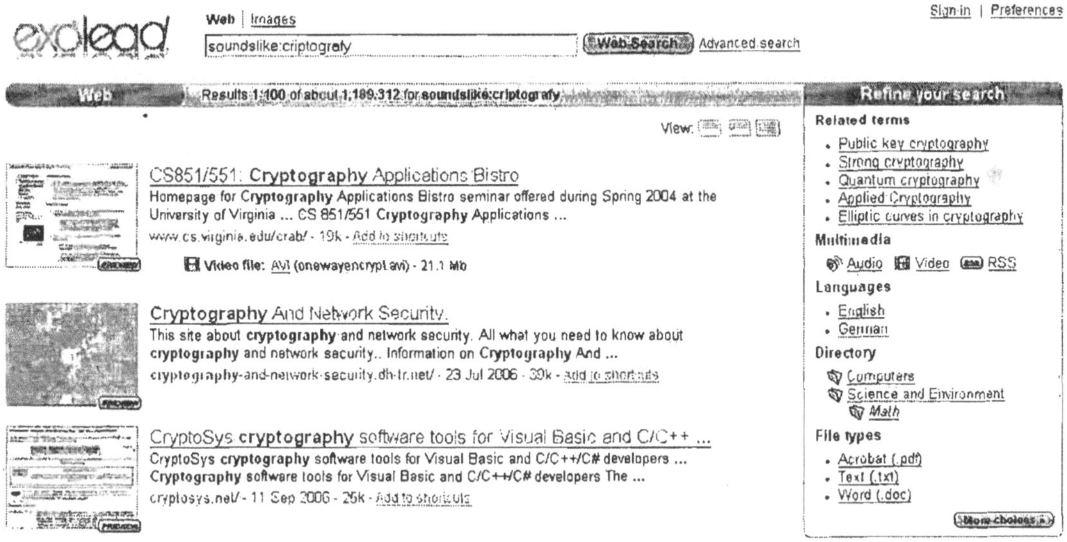
exalead
Web | Images
soundslike:criptografy
Web Search
Advanced search
Sign in | Preferences
Web
Results 1-100 of about 1,189,312 for soundslike:criptografy
Refine your search
View:
Related terms
Public key cryptography
Strong cryptography
Quantum cryptography
Applied Cryptography
Elliptic curves in cryptography
Multimedia
Audio
Video
RSS
Languages
English
German
Directory
Computers
Science and Environment
Math
File types
Acrobat (.pdf)
Text (.txt)
Word (.doc)
More choices
CS851/551: Cryptography Applications Bistro
Homepage for Cryptography Applications Bistro seminar offered during Spring 2004 at the University of Virginia ... CS 851/551 Cryptography Applications ...
www.cs.virginia.edu/crab/ - 19k - Add to shortcuts
Video file: AVI (onewayencrypt.avi) - 21.1 Mb
Cryptography And Network Security.
This site about cryptography and network security. All what you need to know about cryptography and network security.. Information on Cryptography And ...
cryptography-and-network-security.dh-tr.net/ - 23 Jul 2006 - 30k - Add to shortcuts
CryptoSys cryptography software tools for Visual Basic and C/C++ ...
CryptoSys cryptography software tools for Visual Basic and C/C++/C# developers ... Cryptography software tools for Visual Basic and C/C++/C# developers The ...
cryptosys.net/ - 11 Sep 2006 - 25k - Add to shortcuts

The ***approximate spelling*** option can be similarly frustrating. A search on [programme] will find a few sites containing *program, programmen,* or *programs,* but usually the results are for the actual term searched, in this case [programme]. However, it worked very well with [colour], finding a good mix of *color* and *colour* and the approximate search on [geneology] found *genealogy*.

What I like much, much better is Exalead's ***regular expression patterns*** option, which amounts to a ***true wildcard search***. Here's how it works:

> Use a forward slash (/) at the beginning and end of the term; use a period (.) to indicate one missing term; if you are not sure how many letters are missing, use the wildcard (*) after the period. For example, the query [/crypt.*c/] will find *cryptographic* and *cryptologic*:

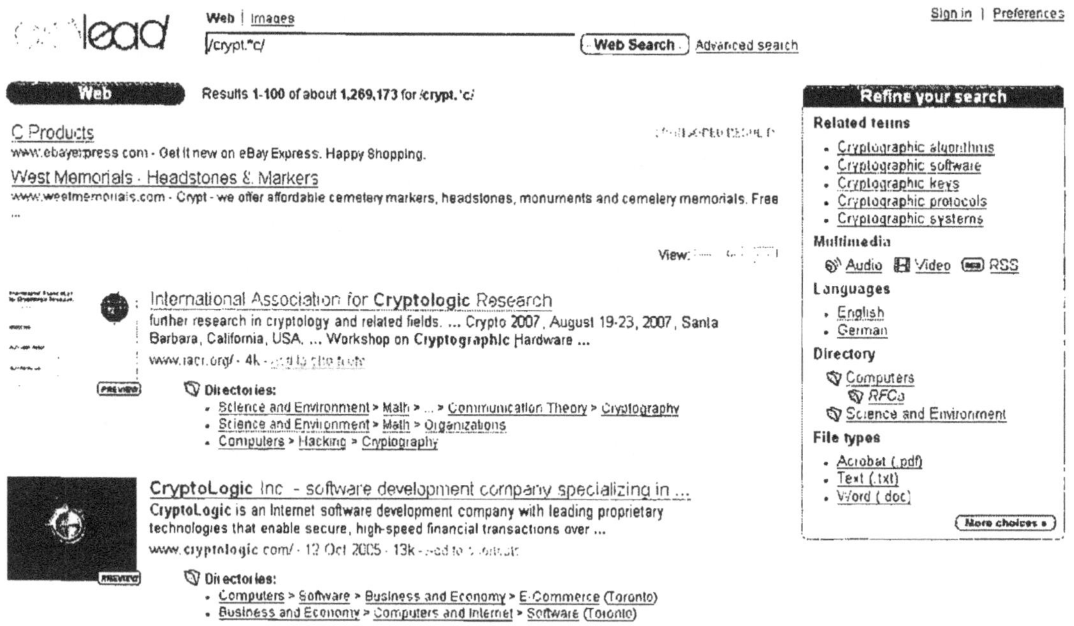
Web | Images
/crypt.*c/
Web Search
Advanced search
Sign in | Preferences
Web
Results 1-100 of about 1,269,173 for /crypt.*c/
C Products
www.ebayexpress.com - Get it new on eBay Express. Happy Shopping.
West Memorials - Headstones & Markers
www.westmemorials.com - Crypt - we offer affordable cemetery markers, headstones, monuments and cemetery memorials. Free ...
View:
International Association for Cryptologic Research
further research in cryptology and related fields. ... Crypto 2007, August 19-23, 2007, Santa Barbara, California, USA. ... Workshop on Cryptographic Hardware ...
Directories:
Science and Environment > Math > ... > Communication Theory > Cryptography
Science and Environment > Math > Organizations
Computers > Hacking > Cryptography
CryptoLogic Inc. - software development company specializing in ...
CryptoLogic is an Internet software development company with leading proprietary technologies that enable secure, high-speed financial transactions over ...
Directories:
Computers > Software > Business and Economy > E-Commerce (Toronto)
Business and Economy > Computers and Internet > Software (Toronto)
Refine your search
Related terms
Cryptographic algorithms
Cryptographic software
Cryptographic keys
Cryptographic protocols
Cryptographic systems
Multimedia
Audio
Video
RSS
Languages
English
German
Directory
Computers
RFCs
Science and Environment
File types
Acrobat (.pdf)
Text (.txt)
Word (.doc)
More choices

Here are the basic rules for pattern matching (wildcard) searches:

The first character is always a **slash** (/). This tells Exalead a special pattern will follow.

Within the pattern, the **period** (.) is a special character that can represent any character.

The **asterisk** (*) stands for character repetition, i.e., any number of characters.

The **pipe** (|) stands for OR, and **parentheses** are used to group characters.

A **question mark** (?) is placed at the end of a character group to make that group optional.

The last character is always a **slash (/)**. This tells Exalead this is the end of the query.

In this example—[/**mpg(1|2|3)?**/]—Exalead will search for any page containing the query term *mpg* and 1 or 2 or 3. It will also find pages containing only *mpg* because the ? makes the 1, 2, and 3 optional. Without the ? Exalead will only find pages containing *mpg1*, *mpg2*, or *mpg3*.

Exalead will handle ***complex boolean queries*** in the simple search screen or from the Advanced search window. The boolean operators Exalead supports are AND, OR, NOT or AND NOT (in caps). A typical boolean query would be:

[(baseball OR football) NOT cardinals]

In addition, there are two other operators that can be used in a boolean query: NEAR and OPT. NEAR finds search terms within 16 words of each other and OPT makes a query term preferable but does not require it. For example:

[(football NEAR cardinals) OPT "st louis"]

This is nice to know because most search engines use AND as their default, and will not return results unless all terms are found. Check the difference between the results for these two searches in Exalead: [buckyball skateboard OPT flyswatter] and [buckyball skateboard flyswatter].

Exalead will search in all or one of most ***languages***. Use either the syntax *language:* followed by the language digraph or the pulldown menu in the Advanced search window. Also, Exalead offers a country search option either from the Advanced search window or using the syntax *country:* followed by the country digraph.

Exalead does not recognize ***diacritical marks at this time***. This means that a search on [façade] finds both *façade* and *facade*. However, Exalead will handle some ***non-Latin character sets***. Exalead officially supports Unicode (UTF), Windows encodings, and miscellaneous encodings (Arabic, Chinese, Korean, Japanese, and Russian).

Exalead offers ***limited field searching***, i.e., special search terms to restrict searches and make them more effective. These special operators can be used in both simple search and in the Advanced search window.

- **language:** restricts results to pages in a specific language. The language syntax uses the obsolete two-letter ISO language codes (639-1). Must be used with additional keywords.

  Advanced Search > Where? > Choose a language

  Example of how to use the **language:** command:

  [language:de welt] finds all the pages indexed by Exalead that are written in

German and contain the keyword "welt," which has a very different meaning in German than in English.

- **country:** restricts results to pages in a specific country. The country syntax uses the two-letter ISO country codes. Must be used with additional keywords.

  Advanced Search > Where? > Choose a country

  Example of how to use the **country:** command:

  [country:de wissenschaft] finds all the pages indexed by Exalead that are purportedly in Germany and contain the term "wissenschaft." It will not limit the search to the German TLD "de."

- **site:** restricts results to a specific website or domain, *excluding* specific top-level domains. You must search on a second-level domain for site to work. May be used with or without keywords.

  Advanced Search > Where? > on a given site

  Examples of how to use the **site:** command:

  [site:amazon.com] finds www.amazon.com, cards.amazon.com, www.amazon.com/dvd/. However, it will not find www.amazon.com.br.

  [site:ir] *does not* find the pages from the Iranian (.ir) top-level domain. However, [site:gov.ir] does find all the pages from the Iranian government domain indexed by Exalead.

  [site:federalreserve.gov "statistical data"] finds all the pages at the Federal Reserve website indexed by Exalead containing the phrase statistical data.

- **filetype:** restricts results to PDF, MS Word, and other filetypes. May be used with or without keywords. **Exalead converts these other types of files to HTML, making them safe to view.** Select PREVIEW to see the HTML version.

  Advanced Search > Where? > in files of a given format

  To search by specific type of file, use the syntax **filetype:** plus one of these abbreviations:

html or htm	standard webpage
pdf	Adobe Acrobat
xls	MS Excel Spreadsheet
ppt	MS PowerPoint
doc	MS Word
wpd	Corel WordPerfect versions 6 & 7
rtf	Rich Text Format
swf	MacroMedia Flash text & hypertext link
txt	text

<u>Examples of how to use the **filetype:** command:</u>

[filetype:xls] finds all pages indexed by Exalead that are in Excel spreadsheet format.

[filetype:pdf "white paper"] finds all pages indexed by Exalead that are in PDF format and contain the phrase *"white paper"* anywhere in the text, title, or url.

- **intitle:** restricts results to pages containing a specific word or phrase anywhere in the webpage's title, which usually appears in the browser's title bar and is the HTML <title> tag. May be used with or without additional keywords.

  Advanced Search > Where? > in the title of the page

  <u>Examples of how to use the **intitle:** command:</u>

  [intitle:amazon] finds all pages that include the word *amazon* in their title

  ["rain forest" intitle:amazon] finds all pages that include the word *amazon* in their title and mention the phrase *"rain forest"* anywhere in the document (title or text or anywhere in the document)

- **inurl:** restricts results to pages containing a specific word or phrase *anywhere* in the webpage's url, that is, the webpage address. May be used with or without additional keywords.

  Advanced Search > Where? > in the address of the page

Examples of how to use the **inurl:** command:

[inurl:amazon] finds all pages that include the word *amazon* anywhere in their url.

["cosmic ray" inurl:spacecraft] finds all pages that include the exact phrase *"cosmic ray"* anywhere in the document (title or text or anywhere in the document) and include *spacecraft* anywhere in the site's url.

- **link:** restricts the results to documents that have links to a specific website. Will work without the full url (absent the http://) but the preferred syntax is [link:http://www.domain.com]. Also, the link: command does not work beyond the top level of a site, so the query [link:www.noaa.gov/wx.html] treats the "wx.html" as a keyword. May be used with or without keywords.

  Advanced Search > Where? > on pages that contain a given link

  Example of how to use the **link:** command:

  [link:http://www.noaa.gov] finds all pages linking to the NOAA homepage.

  [link:http://www.noaa.gov drought] finds all pages linking to the NOAA site that contain the keyword *drought*.

Exalead Search Services and Tools

Exalead does not offer any special services or tools such as news, maps, reference tools, except for a browser toolbar that works with both Internet Explorer and Firefox. At present, the two types of specialized Exalead search are the multimedia (audio, video, and RSS) refinement option and image search.

Image Search: Exalead offers some nice options with its image search. You can look for images of specific sizes (small, medium, large), computer wallpaper by resolution, image color, layout, or filetype. Exalead's advanced search options work in image search as well.

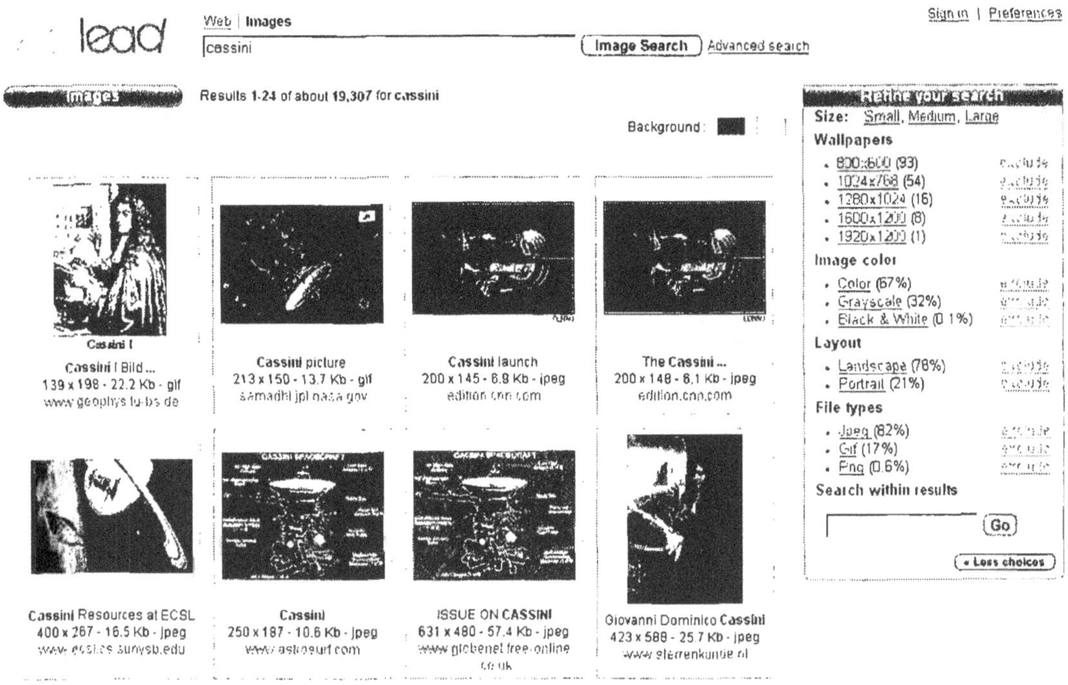
lead
Web | Images
cassini
Image Search
Advanced search
Sign in | Preferences
Images
Results 1-24 of about 19,307 for cassini
Background:
Refine your search
Size: Small, Medium, Large
Wallpapers
800x600 (93)
1024x768 (54)
1280x1024 (16)
1600x1200 (8)
1920x1200 (1)
Image color
Color (67%)
Grayscale (32%)
Black & White (0.1%)
Layout
Landscape (78%)
Portrait (21%)
File types
Jpeg (82%)
Gif (17%)
Png (0.6%)
Search within results
Go
« Less choices
Cassini I Bild ...
139 x 198 - 22.2 Kb - gif
www.geophys.tu-bs.de
Cassini picture
213 x 150 - 13.7 Kb - gif
samadhi.jpl.nasa.gov
Cassini launch
200 x 145 - 6.8 Kb - jpeg
edition.cnn.com
The Cassini ...
200 x 148 - 6.1 Kb - jpeg
edition.cnn.com
Cassini Resources at ECSL
400 x 267 - 16.5 Kb - jpeg
www.ecsl.cs.sunysb.edu
Cassini
250 x 187 - 10.6 Kb - jpeg
www.astrosurf.com
ISSUE ON CASSINI
631 x 480 - 57.4 Kb - jpeg
www.globenet.free-online.co.uk
Giovanni Dominico Cassini
423 x 588 - 25.7 Kb - jpeg
www.sterrenkunde.nl

The Bottom Line

Exalead is not in the Google and Yahoo class yet, but because it offers unique and important features dealing with truncation, wildcards, proximity searching, etc., **it is one of the top-tier search services**. In addition, Exalead offers the option to preview non-html files (e.g., Microsoft file types) safely, which is extremely important given the security dangers that plague Internet users. Exalead is a valuable addition to the world of Internet search.

Ask

During 2006 Teoma and Ask Jeeves ceased to exist as separate search sites and merged under the Ask.com umbrella. I had never been impressed with Ask Jeeves, which was one of the few sites that continued to try to respond to users questions, though not very successfully. Teoma was always an "also ran" in the world of search. However, when Barry Diller, former Chairman and CEO of Paramount Pictures and Fox, Inc.'s, and his IAC/Interactive Corp. acquired Ask Jeeves this year, things changed dramatically. The name was shortened to Ask, the annoying butler icon was gone, along with the ubiquitous ads and usually unfulfilled promise of answers to natural language queries. Ask incorporated Teoma's search algorithm, ExpertRank, and the Teoma site went away. Now, Ask.com has become a major player.

One of the most striking differences is obvious as soon as you run a search. Instead of a list of sponsored links, which Google, Live Search, and Yahoo all display, Ask shows "zoom related search" links, designed to help users either narrow or expand a search. Of course, Ask still serves up ads with its search results, but the search company is putting the primary focus on free search results and not on sponsored results.

Customizing Ask's Settings

Ask offers six general **Settings**:

1. Locations: you may enter a specific location, including a street address or a city, state, and zip code for the US. This is an optional feature and you can sign up for an account if you want to enter multiple locations. This information is used to provide tailored search results relevant to your location.
2. Displaying results: Ask lets you see as few as 10 and up to 100 results per page. There is also an option to open results in a new window.
3. Content filtering: Unlike most search engines, Ask automatically filters adult content; the two options are to alert the user when content is filtered and provide a link to it or to minimize adult content and not link to it.
4. Interface language: if you are more comfortable working in another language, Ask can display in dozens.
5. Make Ask your Default Search Engine: In this case, you are telling your browser to use Ask as the default search engine from the browser address bar.

6. Default Ask Site: You can chose one location from a list including the US, France, Germany, Italy, Netherlands, Spain, and UK, or no default site. Your results will vary depending on the default site.

The other setting option is similar to Yahoo's feature that lets users edit the search tools. Here are the options Ask offers; you can select only the ones you want to appear on your Ask main search page.

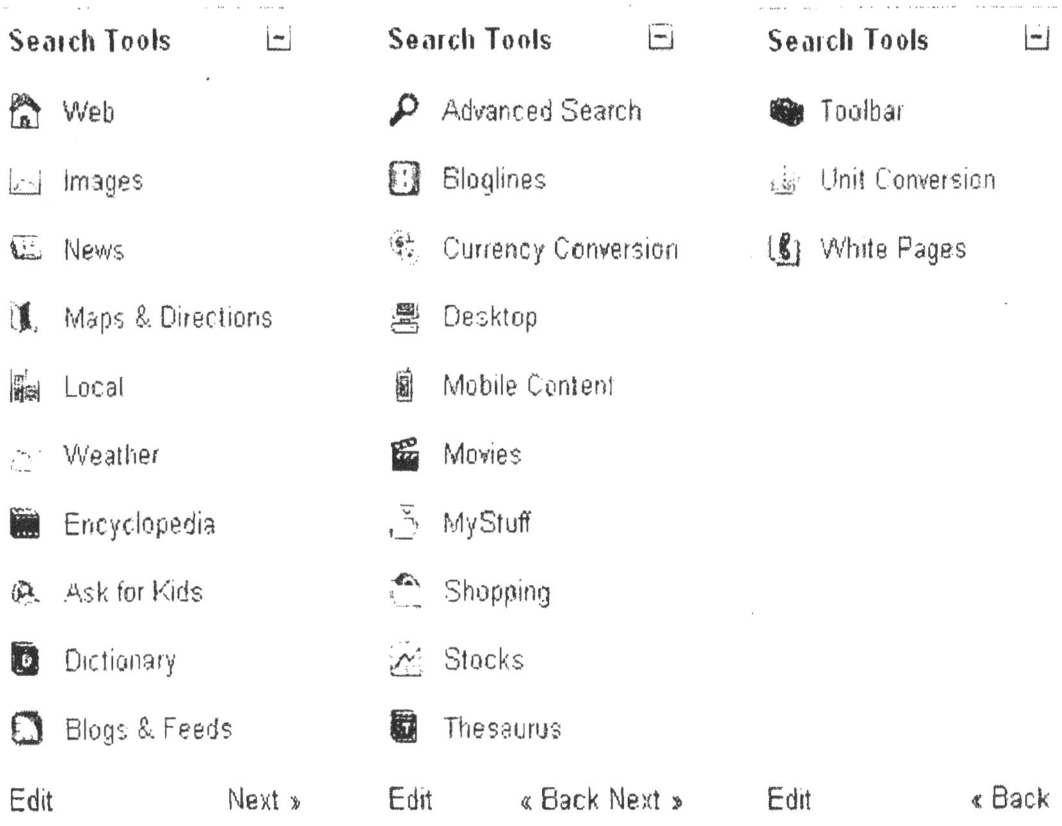
Search Tools
Web
Images
News
Maps & Directions
Local
Weather
Encyclopedia
Ask for Kids
Dictionary
Blogs & Feeds
Edit
Next »
Search Tools
Advanced Search
Bloglines
Currency Conversion
Desktop
Mobile Content
Movies
MyStuff
Shopping
Stocks
Thesaurus
Edit
« Back Next »
Search Tools
Toolbar
Unit Conversion
White Pages
Edit
« Back

The Ask Results Page

Once you've entered your search terms and selected the *Ask Search* button, Ask will present you with a list of results (hits). For each result returned you may see:

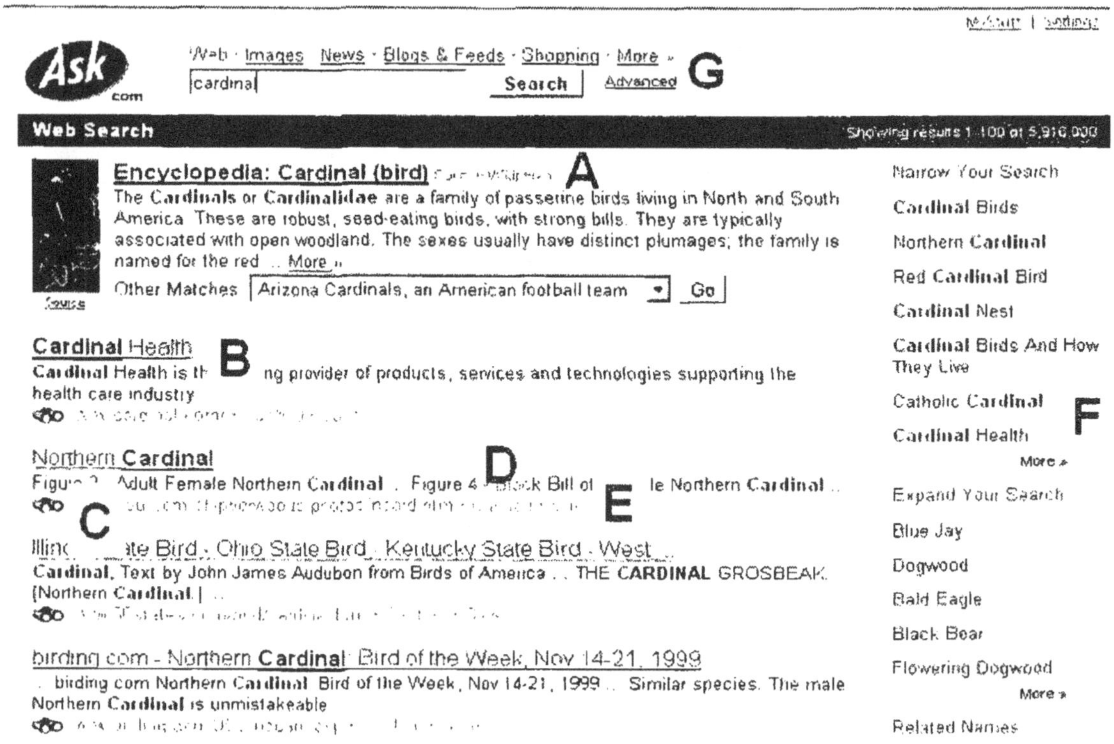
Web · Images · News · Blogs & Feeds · Shopping · More
cardinal
Search
Advanced
G
Web Search
Encyclopedia: Cardinal (bird)
A
The Cardinals or Cardinalidae are a family of passerine birds living in North and South America. These are robust, seed-eating birds, with strong bills. They are typically associated with open woodland. The sexes usually have distinct plumages; the family is named for the red ... More »
Other Matches
Arizona Cardinals, an American football team
Go
Cardinal Health
B
health care industry
Northern Cardinal
D
C
E
Cardinal. Text by John James Audubon from Birds of America ... THE CARDINAL GROSBEAK. (Northern Cardinal.) ...
birding.com - Northern Cardinal: Bird of the Week, Nov 14-21, 1999
Northern Cardinal is unmistakeable
Narrow Your Search
Cardinal Birds
Northern Cardinal
Red Cardinal Bird
Cardinal Nest
Cardinal Birds And How They Live
Catholic Cardinal
Cardinal Health
F
More »
Expand Your Search
Blue Jay
Dogwood
Bald Eagle
Black Bear
Flowering Dogwood
More »
Related Names

- **A** Smart Answers: Ask's best guess about what you want, Smart Answers provides quick access to encyclopedias (Wikipedia, Houghton Mifflin, or Columbia), weather, dictionary results, translations, conversions, etc. Note that "other matches" will try to disambiguate a search term with multiple meanings such as [cardinal]. This is an extremely useful way to find information about commonplace topics, such as [Rwanda]:

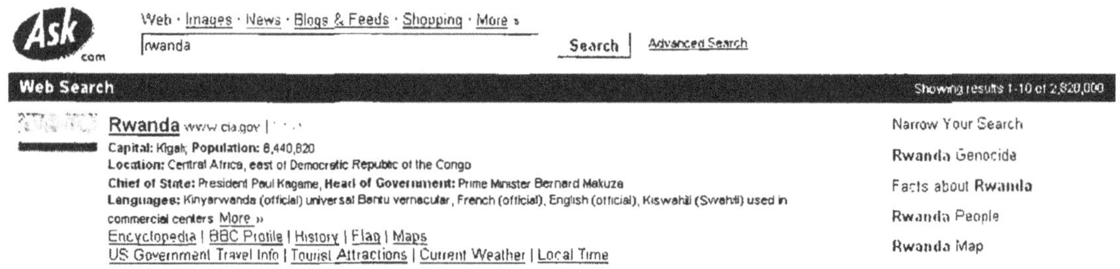
Web · Images · News · Blogs & Feeds · Shopping · More »
rwanda
Search
Advanced Search
Web Search
Rwanda
Capital: Kigali; Population: 8,440,820
Location: Central Africa, east of Democratic Republic of the Congo
Chief of State: President Paul Kagame, Head of Government: Prime Minister Bernard Makuza
Languages: Kinyarwanda (official) universal Bantu vernacular, French (official), English (official), Kiswahili (Swahili) used in commercial centers More »
Encyclopedia | BBC Profile | History | Flag | Maps
US Government Travel Info | Tourist Attractions | Current Weather | Local Time
Narrow Your Search
Rwanda Genocide
Facts about Rwanda
Rwanda People
Rwanda Map

- B Webpage Title & Description: the title and a brief summary of the website.

- C Binoculars Site Preview: Ask's Binoculars Site Preview are periodic screen captures of the browser navigating a page. To view the site preview, users should only move the mouse over the binoculars because clicking on the binoculars takes you to the site. The mouseover is of a static image, so it is safe to view, but I find it too small to be very useful beyond revealing the general nature of a site.

- D Cached: a link to the version of the site stored by Ask with the date and time the page was indexed.

- E Save: Ask offers this service for web and image searches. When users click on a "save" link on either a web or image search, for web searches Ask will save the title of the result, the url, the description, the binoculars icon, and the query used to find that result. For image searches, Ask will save the name and location of the picture, as well as the query used to find the image. Also, everything saved is fully searchable so all saved content is easy to find again later. However, for the save feature to work properly, users need to allow search history to be enabled (the default). *If you do not want Ask to save your search history, go to My Stuff | Settings and uncheck "Record all my searches into my 'Search History.'"*

- F Zoom Related Search: This is a popular feature retained from Teoma that helps users either narrow or broaden a search "with possible alternative search terms which appear on the right hand side of the Ask results page.

  - **Narrow Your Search**: helps you to drill down into topics that are specifically related to your search

  - **Expand Your Search**: allows you to explore topics that are conceptually related to your search

  - **Related Names**: presents a list of names that are conceptually tied to topic options within the 'Narrow Your Search' and 'Expand Your Search' lists."[59]

- G More Search Types: Selecting any of these other search options causes Ask to search automatically for images, news stories, blog entries, etc., with your search term(s).

---

[59] Ask.com Site Features, "Zoom Related Search," <http://help.ask.com/en/docs/about/site_features.shtml#relatedsearch> (14 November 2006).

Ask Basic Search

http://www.ask.com/

Ask assumes as its default that multiple search terms are joined by the **AND** operator, so that a search on the keywords [windows explorer] will find all the webpages that contain ***both*** search terms.

Ask ***will not return any results*** if there is no webpage containing all the search terms. Try this query to see what I mean:

[kong spektioneer synecdoche]

Ask ***does not appear to limit the number of search terms.***

Ask is ***not case sensitive.*** There does not appear to be anything you can do about this.

Ask does not offer ***word stemming*** or ***truncation***, i.e., searching for variations of search terms. Ask searches for exactly the term as you enter it, e.g., a search for [window] will not search for [windows].

Ask automatically ***clusters search results.*** Multiple hits from the same site are indented and there is usually an option to see more results from a specific site.

Ask ***permits the use of the OR operator*** in simple search. The OR needs to be capitalized.

Beyond the use of the OR operator in its simple search, ***Ask does not support <u>boolean</u> search.***

Searchers can delimit phrases using double-quotes. For example, if I search on:

[the last king of france]

without double-quotes, Ask will ignore the "the" and the "of" in its search. I noticed that the results from this search are more relevant than the ones I received from Google for the same query. If I enclose the same query in ***double-quotes***, Ask will search on exactly the phrase ["the last king of france"], and the first hit links to a site that lists all the Kings of France, where Louis XVIII can be readily identified. Enclosing searches in double-quotes is much more effective for finding precise results than relying on automatic phrase searching.

Ask appears to ignore ***stop words*** outside double quotes only when other search terms are used. These two searches will return identical results:

[the last king of france] [last king france]

However, if I search for [the], Ask returns over 2 billion hits. If I add another search term, e.g., [the france], that query is identical to searching for [france], so the stop word is ignored. Nonetheless, it appears that if you search only for stop words, Ask will find pages containing them all, e.g., [i a an the].

Ask does not seem to like the plus sign (+) because it returns an error message when I try to use it. By default Ask searches for all keywords except stop words. However, there are many times when searchers need to exclude certain terms that are commonly associated with a keyword but irrelevant to their search. That's where the ***minus sign (-)*** comes in. Using the minus sign in front of a keyword ensures that Ask excludes that term from the search. For example, the results for the search ["pearl harbor" -movie] are very different from the results for ["pearl harbor"].

Ask treats most ***punctuation marks*** the same way, as links in a search string. For example, Ask handles a search for [c-span], [c.span], ["c span"], and [c?span] basically the same way. However, a search for [cspan] with no space or mark is treated differently.

Ask Advanced Search

Ask has a number of "query modifiers" to restrict searches and make them more effective in many cases. These query modifiers can be used in simple search using the following syntax or on the advanced web search page using the appropriate menu options. Interestingly, Ask using the "must exclude" minus sign differently from other search engines: the minus sign goes after the command syntax, for example, [inurl:nasa site:-gov]

The query modifiers Ask supports are:

- **site:** restricts results to websites in a given domain. *This syntax requires a keyword.*

Advanced Web Search > Domain or Site

Examples of how to use the **site:** command:

[shuttle site:www.nasa.gov] finds pages about the space shuttle at the NASA website.

["bulletin officiel" site:fr] finds pages in the French top-level domain about official bulletins.

["bulletin officiel" site:-fr] finds pages containing the phrase "bulletin officiel" that are not in the French top-level domain. Note that the minus sign goes after the site: syntax.

- **title:** or **intitle:** restricts the results to documents containing the keyword in the title.

  Advanced Web Search > Location of words or phrases > In page title

  Examples of how to use the **title:** command:

  [title:amazon] finds all pages that include the word *amazon* in their title

  [intitle:amazon jungle rainforest] finds all pages that include the words *amazon*, *jungle*, and *rainforest* in their title. Using **intitle:** makes this search function the same as Google's allintitle: query. Note: use a hyphen to search for phrases using the intitle: syntax because the double-quotes do not work.

  [-books title:amazon] finds all pages that contain amazon in the title and do not contain the term *books* anywhere on the page. Note that you must put the excluded term before the intitle: syntax.

  [title:galileo site:-nasa.gov] finds all pages that contain the term *galileo* in the title but are not at any *nasa.gov* website.

- **inurl:** restricts the results to documents containing the keyword in the url.

  Advanced Web Search > Occurrences

  Examples of how to use the **inurl:** command:

  [inurl:nasa] finds all pages that include *nasa* anywhere in the url (address)

  [inurl:nasa site:-gov] finds all pages that include *nasa* anywhere in the url of sites that are *not* in the *.gov* top-level domain. Note that the minus sign goes after the site: syntax.

  [inurl:shuttle inurl:-nasa] finds all pages that include *shuttle* in the url but exclude *nasa* from the url. Note that the minus sign goes after the site: syntax.

  [inurl:nasa shuttle sts-90] finds all pages that include both *nasa* and *shuttle* in the url of a site. Used this way, Ask's inurl: command functions the same as Google's allinurl: command, that is, all terms must be in the url.

  [-shuttle inurl:nasa] finds all pages with *nasa* in the url but do not include the term *shuttle* anywhere on the page. Note that you must put the excluded term before the intitle: syntax.

Ask's Services and Specialty Searches

Ask offers a number of special features designed to help users find specific kinds of information faster and more easily.

Blog Search: Ask is partnered with Bloglines, the most popular (and my favorite) RSS feed reader, to create blog and RSS feed search. The blog search options are:

- sort by date, popularity, or relevance (which combines date and popularity).
- sort by posts, feeds, or news.
- binoculars preview last five posts from a feed.
- options to subscribe and/or post to a feed using several different applications.

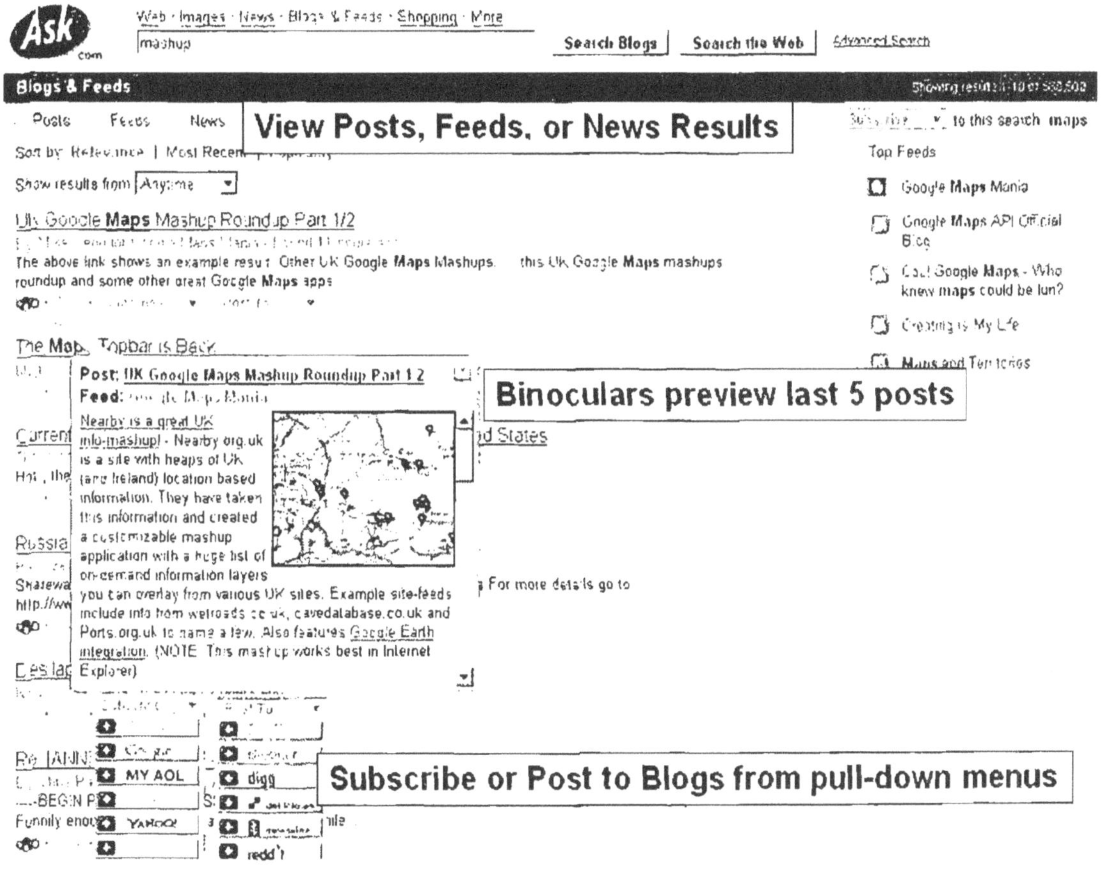
View Posts, Feeds, or News Results
Binoculars preview last 5 posts
Subscribe or Post to Blogs from pull-down menus
Search Blogs
Search the Web
Blogs & Feeds
Top Feeds

- **RSS Answers** will display the three most recent entries at a blog. Obviously, only a limited number of blogs work in RSS Answers, but it is a quick way to see what is new at your favorite blog site. Here is an example of an RSS Answers for John Battelle's Searchblog:

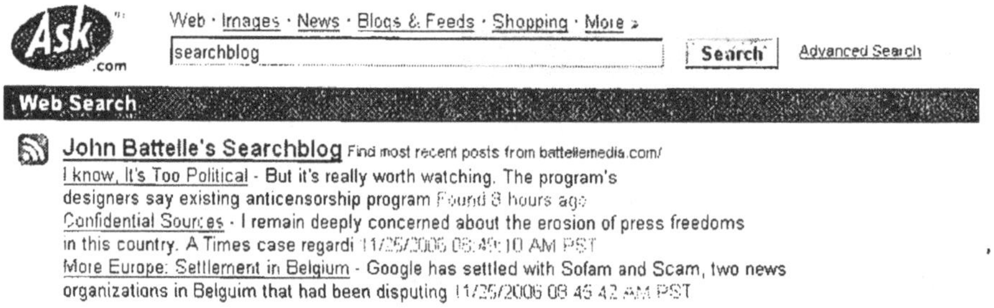
Web · Images · News · Blogs & Feeds · Shopping · More »
searchblog
Search
Advanced Search
Web Search
John Battelle's Searchblog Find most recent posts from battellemedia.com/
I know, It's Too Political - But it's really worth watching. The program's designers say existing anticensorship program
Confidential Sources - I remain deeply concerned about the erosion of press freedoms in this country. A Times case regardi
More Europe: Settlement in Belgium - Google has settled with Sofam and Scam, two news organizations in Belguim that had been disputing

Definitions: Ask will present a dictionary or encyclopedia definition of a term if you phrase the query as [define keyword], [definition of keyword], [the meaning of keyword], or [dictionary], which brings up the Dictionary Search option:

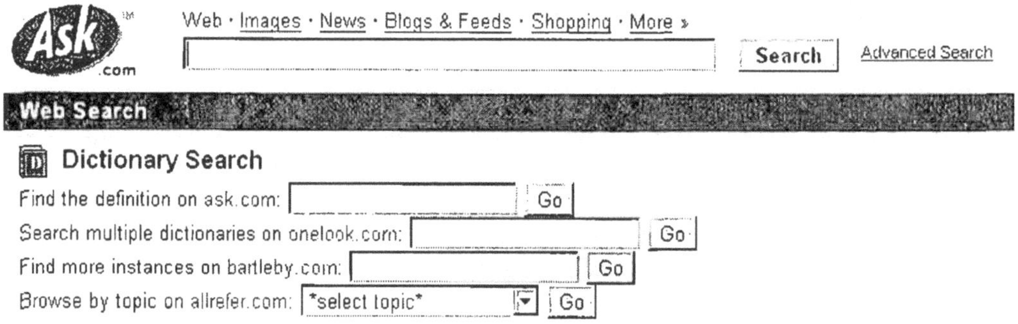
Web · Images · News · Blogs & Feeds · Shopping · More »
Search
Advanced Search
Web Search
Dictionary Search
Find the definition on ask.com:
Go
Search multiple dictionaries on onelook.com:
Go
Find more instances on bartleby.com:
Go
Browse by topic on allrefer.com:
select topic
Go

Local Search: search for services or businesses by US zip code or city.

Maps: to map a US or Canadian location, search on the street address, city and state or the word *map* and a location. Some international maps are now available. See the <u>section on maps for details</u>.

News: links to news stories appear when a search term matches current news stories. Sort news by date or relevance. A separate Ask News page is available at <u>http://news.ask.com/</u>

Travel Shortcuts: To find arrival and departure information, flight delays, airport status, and weather conditions at a US or Canadian airport, enter the airport's three-letter code and the word *airport*. For example, to information about Baltimore-Washington International, enter [bwi airport].

White Page Search: search for US phone numbers and addresses for people, businesses, government offices, doctors, and schools in the U.S.

Web Answers: This option is the remnant of Ask Jeeves, that is, Ask's attempt to provide direct answers to questions. Users may write a natural language question or, in this example, if an answer exists to a commonly asked question, such as the meaning of 'ontology,' the Web Answers will appear under the definition.

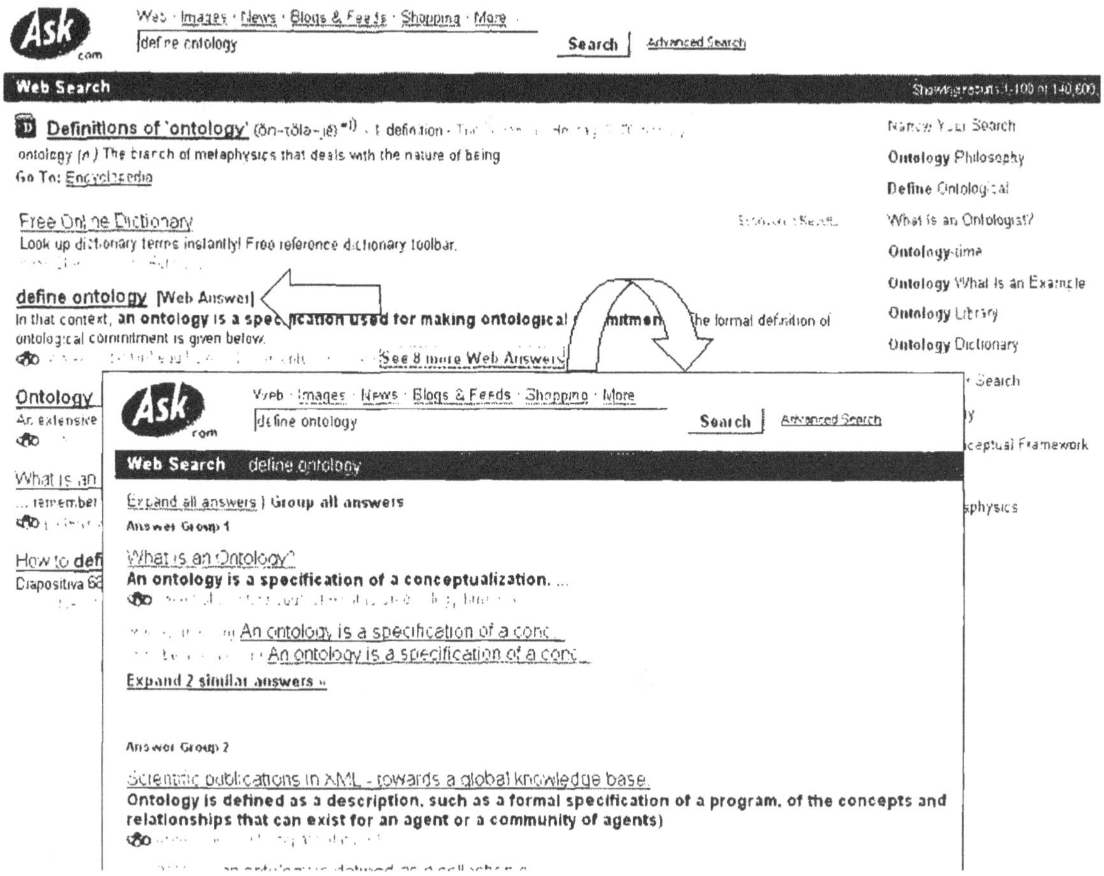
Web · Images · News · Blogs & Feeds · Shopping · More
define ontology
Search
Advanced Search
Web Search
Definitions of 'ontology'
ontology (n.) The branch of metaphysics that deals with the nature of being
Go To: Encyclopedia
Free Online Dictionary
Look up dictionary terms instantly! Free reference dictionary toolbar.
define ontology [Web Answer]
See 8 more Web Answers
Narrow Your Search
Ontology Philosophy
Define Ontological
What is an Ontologist?
Ontology What Is an Example
Ontology Library
Ontology Dictionary
Web · Images · News · Blogs & Feeds · Shopping · More
define ontology
Search
Advanced Search
Web Search define ontology
Expand all answers | Group all answers
Answer Group 1
What is an Ontology?
An ontology is a specification of a conceptualization. ...
An ontology is a specification of a conc...
An ontology is a specification of a conc...
Expand 2 similar answers »
Answer Group 2
Scientific publications in XML - towards a global knowledge base.
Ontology is defined as a description, such as a formal specification of a program, of the concepts and relationships that can exist for an agent or a community of agents)

Conversions: The Ask conversion tool will automatically convert world currency, temperature, weight, length, area, and cooking/volume. Users can use the query [convert amount x to amount y], e.g., [convert 200 iraqi dinars to pound sterling] or try a natural language query such as [how many kilometers are in a nautical mile]. The conversion tool is very easy to use and impressive.

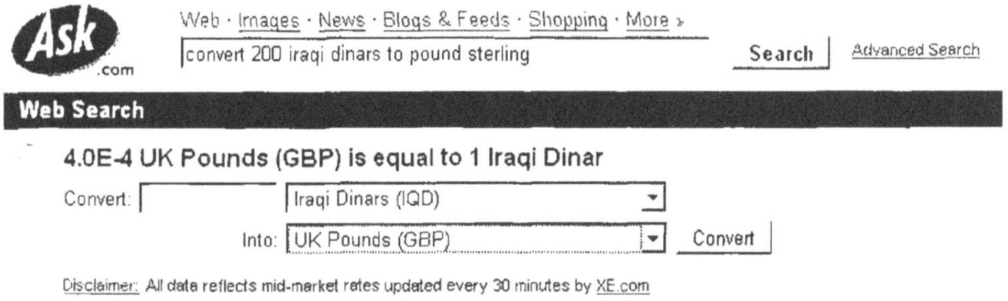
Web · Images · News · Blogs & Feeds · Shopping · More »
convert 200 iraqi dinars to pound sterling
Search
Advanced Search
Web Search
4.0E-4 UK Pounds (GBP) is equal to 1 Iraqi Dinar
Convert:
Iraqi Dinars (IQD)
Into:
UK Pounds (GBP)
Convert
Disclaimer: All data reflects mid-market rates updated every 30 minutes by XE.com

Image Search: the Ask image search uses "authoritativeness" to rank its results and also accesses a proprietary image index. ***It is one of the best image search tools available.*** The image search appears as one of the default search tools on the right-hand side of the main search page. There is no advanced image search and no special image search options. However, when you search for an image, zoom related search terms to expand or narrow the search appear. If you select the "save" option, this link will save the image to your personal "stuff," which can later be accessed via http://mystuff.ask.com/. If you select "info" about an image, you will then see detailed information about the image, including copyright information, and its source homepage will appear in a frame in the bottom portion of the screen.

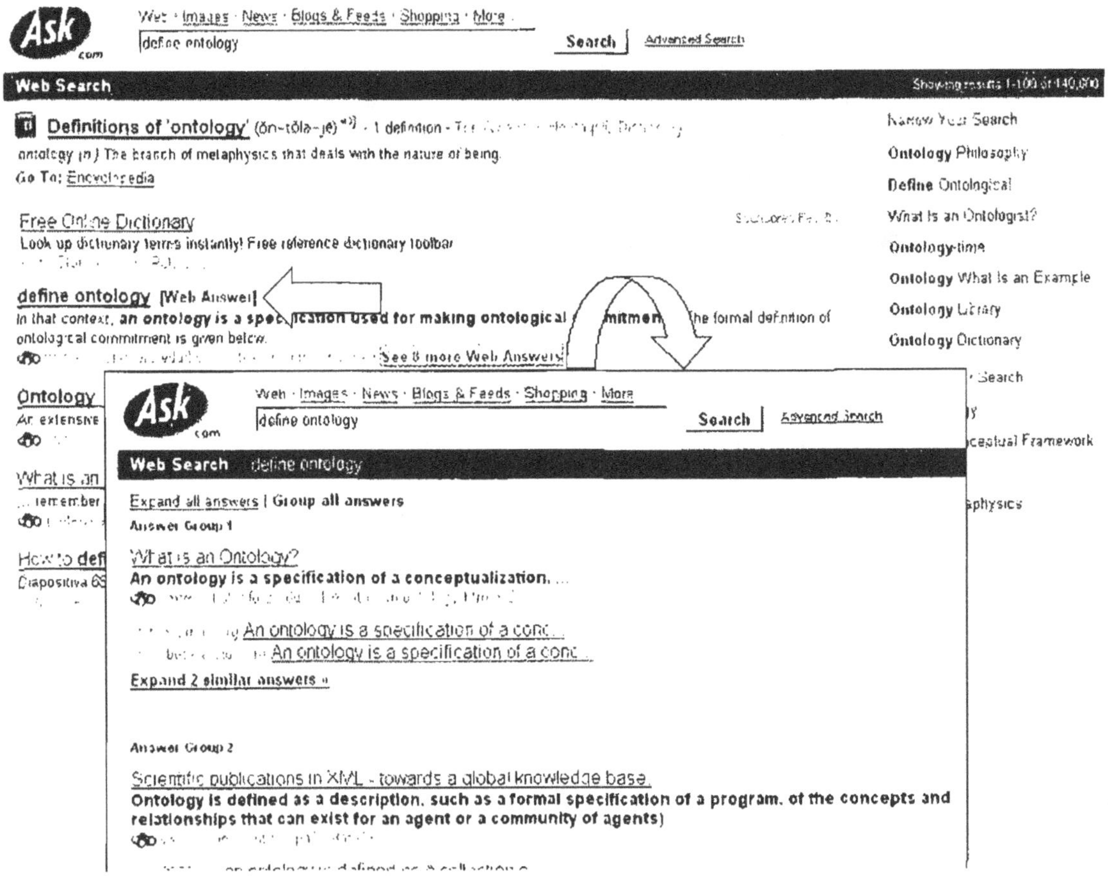
Ask
Web · Images · News · Blogs & Feeds · Shopping · More
define ontology
Search
Advanced Search
Web Search
Definitions of 'ontology'
ontology (n) The branch of metaphysics that deals with the nature of being.
Go To: Encyclopedia
Free Online Dictionary
Look up dictionary terms instantly! Free reference dictionary toolbar
define ontology [Web Answer]
See 8 more Web Answers
Narrow Your Search
Ontology Philosophy
Define Ontological
What Is an Ontologist?
Ontology-time
Ontology What Is an Example
Ontology Library
Ontology Dictionary
Web Search define ontology
Expand all answers | Group all answers
Answer Group 1
What is an Ontology?
An ontology is a specification of a conceptualization. ...
An ontology is a specification of a conc...
Expand 2 similar answers »
Answer Group 2
Scientific publications in XML - towards a global knowledge base
Ontology is defined as a description, such as a formal specification of a program, of the concepts and relationships that can exist for an agent or a community of agents)

Ask Image Search "Info" Page http://pictures.ask.com/

Number Search: Ask offers many types of number searches. The numbers Ask will search for are:

- **UPS** tracking: enter the UPS tracking number [1Z9999X99999999], or enter [ups tracking] to bring up the UPS tracking query option.

- **USPS** tracking: enter USPS plus the tracking number with or without spaces [usps 9999999999999999999999], or enter [usps tracking] to bring up the USPS tracking query option.
- **FEDEX** tracking: enter FEDEX plus the tracking number [fedex 9999999999999999], or enter [fedex tracking] to bring up the FEDEX tracking query option.
- **DHL and Airborne Express** tracking: enter DHL plus the tracking number [DHL 9999999999], or enter [DHL tracking] to bring up the DHL tracking query option.
- **ZIP codes**: enter a US ZIP code, either five or nine digits
- **ISBN**: enter any International Standard Book Number
- **VIN** Information: to find information about a vehicle's history, search on its 17-character Vehicle Identification Number (VIN)

More Help: Internet Guides and Tutorials

For anyone who wants additional help in learning how to use the Internet more effectively, many excellent resources are available for free via the Internet. Also, there are more and more sites appearing to help new Internet users get started with searching the web. Some help you choose the right search engine, others how to formulate a query, and others are step-by-step tutorials.

The Internet Detective Tutorial is a free online tutorial that is part of the Intute: Virtual Training Suite, a set of "free Internet tutorials to help you learn how to get the best from the Web for your education and research...[created by] a national team of subject specialists based in universities and colleges across the UK."[60] Not familiar with Intute? It is the newly evolved face of the Resource Discovery Network, a carefully selected and evaluated set of academic research resources. The Internet Detective tutorial focuses on how to evaluate Internet sources for quality and authoritativeness, how to avoid wasting time on questionable websites and searches, and how to avoid violating copyright laws and plagiarism. The tutorial includes a set of practical exercises to try your Internet research skills. Although the tutorial is aimed at university research, I highly recommend it for all readers. The tutorial requires about an hour to complete, but it is designed so you can do it in more than one sitting.

The Internet Detective Tutorial http://www.vts.intute.ac.uk/detective/index.html

All the Intute tutorials are available at:

Intute: Virtual Training Suite http://www.vts.intute.ac.uk/

The following are tutorials, guides, and search-oriented sites available on the Internet:

BrightPlanet's Guide to Effective Searching of the Internet
http://www.brightplanet.com/deepcontent/tutorials/search/index.asp

Finding Information on the Internet: A Tutorial
http://www.lib.berkeley.edu/TeachingLib/Guides/Internet/FindInfo.html

Internet Tutorials from University of Albany Libraries http://www.internettutorials.net/

Internet Scout Report
http://scout.wisc.edu/Projects/PastProjects/toolkit/searching/index.html

[60] Intute: Virtual Training Suite, <http://www.vts.intute.ac.uk/> (12 September 2006).

Pandia's Goalgetter http://www.pandia.com/goalgetter/index.html

Phil Bradley's Searching the Internet http://www.philb.com/searchindex.htm

Search Engine Watch Tutorials (old but still useful) http://www.searchenginewatch.com/resources/article.php/2156611

Web Search Guide http://www.websearchguide.ca/tutorials/tocfram.htm

Specialized Search Tools & Techniques

This section, which first appeared in the 2006 edition, was born of the rapid growth of both unconventional search techniques such as Google hacking and the wildfire spreading of such tools as online maps. This year, I have added a new section on Wikipedia and expanded the maps and mapping section.

"Google Hacking"

This topic has received a great deal of attention in the world of Internet search in the past few years. While this activity is generically referred to as "Google hacking,"[61] this is a double misnomer. First, to limit this practice to "Google" is a mistake because many of these kinds of searches can be run using any search engine, though they are clearly going to be most effective with a large, powerful search tool that offers many search options, such as Google. Second, this is not hacking in the sense that most people use the term, i.e., gaining access to a computer or data on a computer illegally or without authorization. Nothing I am going to describe to you is illegal, nor does it in any way involve accessing unauthorized data. ***"Google (or search engine) hacking" involves using publicly available search engines to access publicly available information that almost certainly was not intended for public distribution***. In short, it's using clever but legal techniques to find information that doesn't belong on the public Internet.

To understand how this information has found its way into search engine databases, we need a quick overview of how search engines work. Very simply, search engines deploy "spiders" (aka crawlers or bots), which is actually software that "crawls" websites looking for new sites, updating old ones, following links, and dumping all that data into search engine databases where it is stored, sorted, and eventually accessed by users. There is nothing illegal, immoral, or even fattening about search

---

[61] Let's talk about the term *hacking* for a minute. A hacker is someone who is proficient at using or programming a computer; in short, a computer expert. While there is no universal agreement on a preferred term for someone engaged in illegal/illicit computer or network activity, I will call these "black hat" hackers "malicious hackers" to distinguish them from "white hat" or neutral "hackers," meaning proficient or expert computer users.

engine spiders. Indeed, without them, we would have little or no idea what is "out there" and available to us. The problem for webmasters is that it is their responsibility to keep the search engine spiders out of any parts of their websites they do not want to be accessed and indexed by a search engine. The spider is not smart; it simply knows that if a "door" is open, it can—and will—go in and crawl around. Webmasters must tell spiders "do not enter" (primarily) by the use of the Robots Exclusion Protocol.

Robots Exclusion[62] comes in two basic flavors: either a metatag that can be inserted into the HTML of a web page (usually used by an individual) or a Robots Exclusion Protocol (robots.txt) file, a specially formatted file inserted by the website administrator to tell the spider which parts of the website may and may not be indexed by the spider. If a robots exclusion is missing or improperly configured, the spider will index pages that the website owner may not have wished to have been accessed.

The whole problem of keeping information on the Internet private dramatically worsened almost overnight a couple of years ago when Google quietly started indexing whole new types of data. Originally, most of what got spidered and indexed was HTML webpages and documents, with some plain text thrown in for good measure. However, the ever-innovative Google decided this wasn't good enough and started to index PDF, PostScript, and—most importantly—a whole range of Microsoft file types: Word, Excel, PowerPoint, and Access. Problem was, lots of folks had assumed these file types were "immune" to spidering not because it couldn't be done but because no one had yet done it. As a result, many companies, organizations, and even governments had quite a lot of egg on their faces when sensitive documents began turning up in the Google database.

That was then, this is now. You might think people would have learned, but judging by the amount of "sensitive" information still available, many have not. Even though search engines now routinely index many non-HTML file types, many individuals and organizations still do not protect these files from the long reach of search engine spiders. Furthermore, there are many ways for sensitive information to end up in search engine databases. An improperly configured server, security holes, and unpatched software can give search engine spiders unintended access. Quite frankly, most of the problems boil down to one thing: human error, either through ignorance or neglect.

What kinds of sensitive information can routinely be found using search engines? The types of data most commonly discovered by Google hackers usually falls into one of these categories:

---

[62] For additional information, see: <http://www.robotstxt.org/wc/exclusion.html> (14 November 2006).

- personal and/or financial information
- userids, computer or account logins, passwords
- private, confidential, or proprietary company data
- sensitive government information
- vulnerabilities in websites and servers that could facilitate breaking into the site

Now, you may be thinking to yourself, "I use Google all the time and I've never encountered this type of information." That's not surprising. It's not usually the kind of thing you would stumble across inadvertently. Normally, one would have to be actively looking for this type of information. Of course, many of the documents Google hackers find using these techniques are not sensitive and indeed are intended for the public Internet. Only a tiny fraction of the over eight billion pages in the Google index were not meant to be made available to the public. *And, it so happens, these techniques are excellent unconventional ways of finding useful information that might not be discovered using routine search engine queries.* Here are some of the typical techniques used in Google hacking:

- search by file type[63], site type, and keyword: many organizations store financial, inventory, personnel, etc., data in *Excel spreadsheet format* and often mark the information "Confidential," so a Google hacker looking for sensitive information about a company in South Africa might use a query such as:

  [filetype:xls site:za confidential]

  a similar but more specific search could involve use of a keyword such as *budget* to search for Excel spreadsheets at Indian websites; for example: [filetype:xls site:in budget]

- one of the most popular Google hacking technique is to employ ***stock words and phrases*** such as *proprietary, confidential, not for distribution, do not distribute,* along with a search for specific file types, especially Excel spreadsheets, Word documents, and PowerPoint briefings.

- search for files containing ***login, userid, and password*** information; note, even at international sites, these terms usually appear in English. This type of information is typically stored in spreadsheet format, so a typical search might be:
  [filetype:xls site:ru login]

---

[63] It is critical that you handle all Microsoft file types on the Internet with extreme care. Never open a Microsoft file type on the Internet. Instead, use one of the techniques described here.

- ***misconfigured web servers*** that list the content of directories not intended to be on the web often offer a rich load of information to Google hackers; a typical command to exploit this error is:

  [intitle:"index of" site:kr password]

- <u>numrange search</u>: this is one of the least known and (formerly) one of the scariest searches available through Google. Numrange uses two number separated by two periods (dots) and no spaces. While "***legitimate***" numrange users probably will want to indicate what the numbers mean, e.g., weight, money, pixels, etc. Google does not require any special words or symbols to run a successful numrange search; hence its power. *Numrange* can be used with keywords and other Google search options, such as:

  [site:www.jordanislamicbank.com 617..780]

How is numrange typically used in Google hacking? It used to be extremely effective in finding credit card numbers and social security numbers. Because of the publicity about criminals using Google to look for private data, this particular search no longer works for credit card and Social Security numbers, which is not a bad thing.

The disabled "hack" was:

[numrange:4567000000000000..4567999999999999 visa] or

[numrange:222000000..250999999 ssn]

Now if you try these searches, you will see this message:

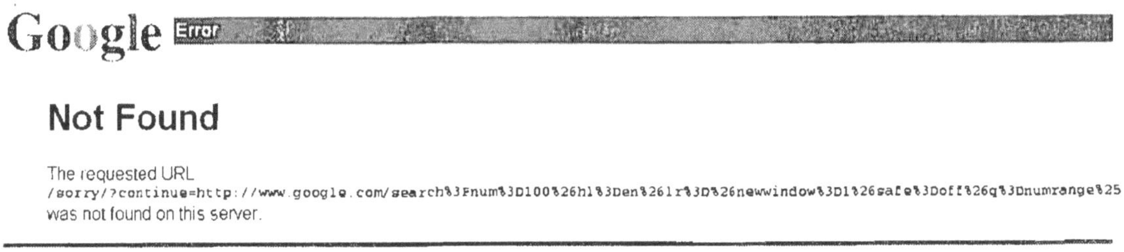
Google
Error
Not Found
The requested URL
/sorry/?continue=http://www.google.com/search%3Fnum%3D100%26hl%3Den%26lr%3D%26newwindow%3D1%26safe%3Doff%26q%3Dnumrange%25
was not found on this server.

Lest you think I am spilling the beans here, I assure you I am not revealing anything that is not already widely known and used on the Internet both by legitimate and illicit Google hackers. I am fully indebted to Johnny (johnnyihackstuff) Long for many of the "Google hacking" techniques[64] I have learned. Please use the information he provides judiciously because many of the Google *hacking* techniques he discusses are really designed for *cracking*, i.e., breaking into websites and servers. That is not

[64] Johnny Long, *Google Hacking for Penetration Testers*, Syngress: Rockland, MA, 2004.

something I encourage or advocate. I do encourage you to "hack" your own website to see what kinds of information is being revealed inadvertently via Google and other search engines.

Also, a lot of the best information Johnny offers is for his site members only, and I do not want to suggest you register there. Nonetheless, Johnny's briefing slides from the 2004 Black Hat and Defcon12 conferences are available at the official Black Hat Briefings website and elsewhere (so much for registration). I have also found his excellent white paper "The Google Hacker's Guide" at other sites that do not require registration; there is another very good briefing on the dangers of Google by Sebastian Wolfgarten.

There was a fair amount of sniping following Long's talks at Black Hat and Defcon, mostly of the "big deal" variety, i.e., it is not "real" hacking and therefore not worthy of presenting at Defcon. However, this is a very shortsighted point of view when one considers the kinds of information that is so very easily available via Google, et al. How would you like to see your Social Security Number, credit card number, and that very handy little three digit number on the back of your credit card used for "verification," bank routing information, mother's maiden name, etc., in the next Google hacking briefing? Yes, all this kind of information is readily available (I know...I've uncovered quite a bit of it myself). And this doesn't even take into consideration all the other website weaknesses, such as multiple vulnerabilities with IIS 6.0 Web-based administration, that can be exposed using Google.

Johnny Long's Googledorks Page http://johnny.ihackstuff.com/ghdb.php

Johnny Long's "The Google Hacker's Guide"
http://www.securitymanagement.com/library/Google_Hacker0704.pdf

Johnny Long, "You Got That With Google?" Black Hat Briefings and Defcon12, July 2004.
http://www.blackhat.com/html/bh-media-archives/bh-archives-2004.html#USA-2004

Johnny Long, "Google Hacking Mini-Guide," *Informit.com*, 7 May 2004
http://www.informit.com/articles/printerfriendly.asp?p=170880

Sebastian Wolfgarten, "Watch Out Google"
http://www.wolfgarten.com/downloads/Watch_out_google.pdf

Joe Barr, "Google Hacks are for Real," *Newsforge.com*, 6 August 2004
http://www.newsforge.com/article.pl?sid=04/08/05/1236234

Taken all together, the information Johnny Long has found using Google (he sticks with this one search engine), combined with the techniques he details at his website, provide an excellent tutorial on using Google to find stuff that really should not be on the public Internet or easily accessible via a search query. Furthermore, the greatest value of his efforts may not be in finding useful information but in demonstrating the vulnerabilities of any given website and the necessity of taking strong measures to ensure the information that gets into Google (as well as other search engine databases and the Internet Archive) is only that which is intended.

Given the large amount of "sensitive" or private data readily available via Internet search engines, people naturally wonder why companies and individuals do not actively try to remove this information. Sometimes they do, but much still remains accessible. Why? ***Getting private information "back" is harder than preventing its disclosure in the first place.*** There are steps you can take to remove your data, but as hacker Adrian Lamo says, "removing links after the fact isn't a very elegant solution." Nor is it likely to be terribly effective. There are a number of reasons for this, but what it boils down to is: it's very hard to put the genie back in the bottle.

First of all, you have to find out if your data is "out there" in order to ask search engines to remove it and, clearly, many people and organizations are not playing defense, that is, they are not routinely checking to see what is indexed from their websites. Let's say you find something on Google that shouldn't be on the public Internet. The first thing you have to do is to protect the sensitive pages on your site or remove them entirely. However, even when you have removed those pages from your website, this doesn't mean they can't be accessed. Once documents are indexed in a search engine database, a publicly available copy of those documents (usually referred to as the cache copy) may remain behind for days, weeks, even months.

The next step is to ask Google to remove your sensitive pages from its database. However, even when Google removes your data, there are literally hundreds of other search engines around the world, and who knows what they have indexed from your site. It will not be an easy task finding out. And I'll hazard a guess that not all of them will be quite so accommodating as Google in removing pages.

To make matters worse, if something really "juicy" shows up in a search engine, chances are someone will find it and copy it to another website. Once this happens, you can forget about removing that information from the Internet. To further complicate matters, even if no individual comes across your sensitive data, the Internet Archive[65] spider is almost certainly going to find that webpage and index it in the Archive, and there it will remain until and unless you find it first and ask the Archive to remove it. As you can see, the genie is running amuck! Prevention is much easier (though certainly not easy) than curing this particular disease, so it's vital to pay close attention to anything you put on a website, especially something you do not want the whole world to see.

---

[65] The Internet Archive is a non-profit organization that was founded to "build an 'Internet library,' with the purpose of offering permanent access for researchers, historians, and scholars to historical collections that exist in digital format. Based in San Francisco, the Internet Archive has been harvesting the World Wide Web since 1996, to create one of the largest data collections in the world. The Internet Archive's web archive contains over 100 terabytes of data, and the collection is growing at a rate of 12 terabytes per month." <http://www.archive.org/> (14 November 2006).

Because of the vast amount of information available using public search engines, it's relatively easy to find lots of interesting, amusing, shocking examples of sensitive information. While this is all fine and good for entertaining yourself and impressing your friends, what we are really after is useful, meaningful, and actionable information. Put succinctly:

It's Easier to Find Anything Than It Is to Find Something

So how do you find "something" useful? While it isn't easy to do so, I can make some suggestions that might help. The most valuable assets you have are your subject matter knowledge and your creativity. Add these to a few search engine strategies, and you can probably find many relevant and genuinely useful pieces of information. The strategies I recommend for finding "something" rather than just "anything" are:

<u>Limit the search by site</u>

This can be as broad as a county [site:fr] or as specific as an individual server on a company website [site:office.microsoft.com].

<u>Try to be as specific as possible</u>

You will have a lot more success searching for information within the Chinese Ministry of Foreign Affairs [site:fmprc.cn.gov] than looking at all the sites indexed for China [site:cn] or even for the government of China [site:gov.cn]

<u>Add keywords</u>

Here's where your subject matter knowledge and creativity really help. You are the best source of information about what words are most likely to yield the best quality and quantity of useful information. As a general rule, more uncommon words work best (consider using unusual proper names).

<u>Limit the search by file type</u>

Most of the best information found by Google hackers is not on webpages (HTML) but in other types of files. Try all or most of the file types one at a time (these are not the only searchable file types; check the particular search engine's documentation (*Help* page) for others):

filetype:pdf—good for large documents of all types; widely used in academia, government, and business; many PowerPoint briefings are also made available in PDF at the same website

filteype:doc—good for internal working documents, reports, etc.

filetype:xls—good for personnel data, computer records, financial information filetype:ppt—good for briefings, which often contain company or government plans for the future

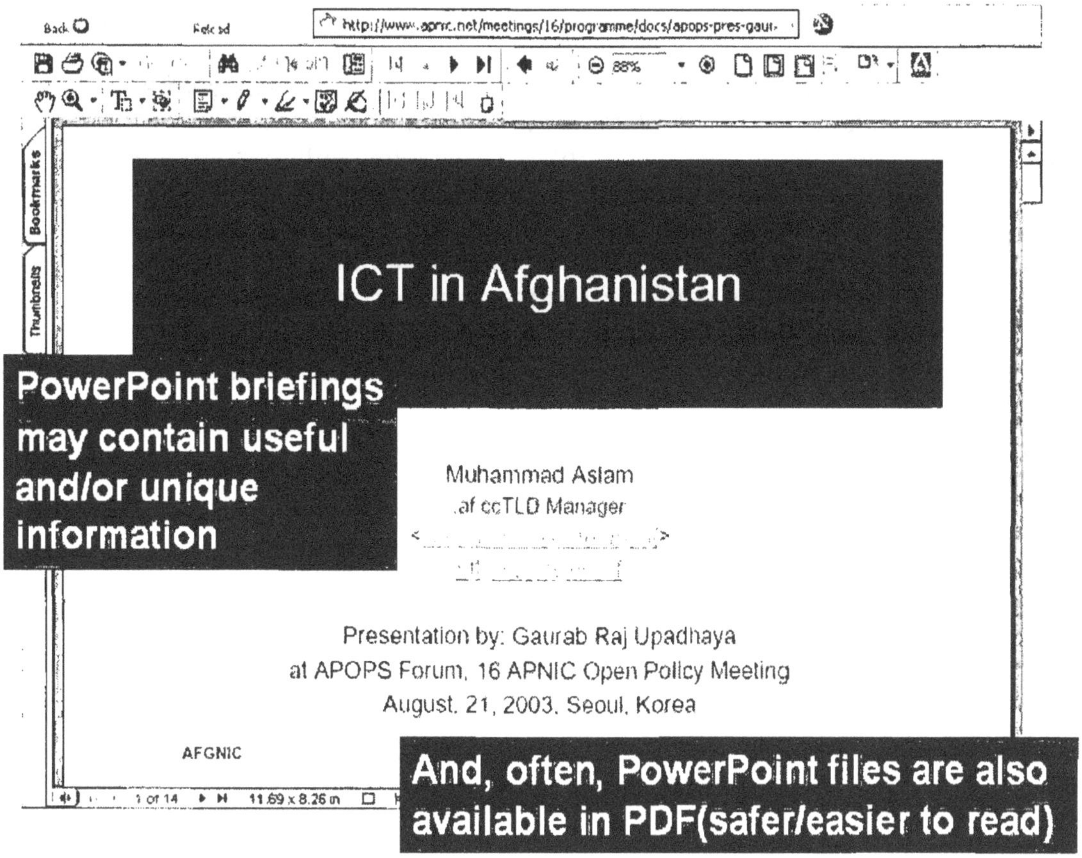
ICT in Afghanistan
PowerPoint briefings may contain useful and/or unique information
Muhammad Aslam
.af ccTLD Manager
Presentation by: Gaurab Raj Upadhaya
at APOPS Forum, 16 APNIC Open Policy Meeting
August, 21, 2003, Seoul, Korea
AFGNIC
And, often, PowerPoint files are also available in PDF(safer/easier to read)

<u>Use Google hacking techniques to search inside websites requiring registration</u>

You will frequently encounter a website, perhaps a database, that requires registration to view its contents. On occasion, you can use Google to get at that data without registering. For example, let's say you find a database of international companies that requires *free* registration. Without registering, you may be able to use Google to list all the companies and even get a look at the individual entries. Try this series of queries or something similar:

[site:www.companyname.com inurl:database] or

[site:www.companyname.com inurl:directory] or

[site:www.companyname.com inurl:index]

Then, look for keywords, such as *companies*, and move to the next level query:

[site:www.companyname.com inurl:companies]

You may be able to browse through the list of companies and get names, addresses, phone numbers, etc.

Search in the native language

I cannot emphasize strongly enough how important it is to use keyword search terms that are in the native language of the entity you are researching. The Internet is becoming much less dependent upon English, and sites written in languages that do not use the Latin alphabet are growing by leaps and bounds. For example, a search term written in the native language and encoding is far more likely to yield interesting, useful results than the same word transliterated into English. Most good quality search engines now correctly render non-Latin search terms regardless of how the term is transliterated in English. A search on the Arabic محمد returns very different results than searching on [muhammad], [mohamet], [mohammed], etc.

Google
Web Images Video News Maps Desktop more »
محمد
Search
Advanced Search
Preferences
Web
Results 1 - 100 of about 25,800,000 for محمد. (0.06 seconds)
Sheikh Muhammad Jebril | الشيخ محمد جبريل - [Translate this page]
... by sheikh Muhammad Jebril's spectacular voice.
www.jebril.com/ - 6k - Cached - Similar pages
www.jebril.com/ar/index.htm - 33k - Cached - Similar pages
Mohammad Esfahani Official Web Site
Iranian singer. Profile, discography, and pictures.
www.mohammad-esfahani.com/ - 10k - Cached - Similar pages
photographing,landscape,Portrait. ... All works of art copyright © MOHAMED MANNAI. All rights reserved Copyright © 2000-2006 MOHAMED MANNAI.
www.mmannai.com/ - 7k - Cached - Similar pages
MUHAMMAD ALI - The Greatest Of All Time
This is the Official website of Muhammad Ali, the greatest of all time.
www.ali.com/ - 22k - Cached - Similar pages
Welcome to His Highness Sheikh Mohammed bin Rashid Al Maktoum's ...
Official site of the Ruler of Dubai, who is also the UAE Vice President and Prime Minister. Contains news, his poetry and other information in Arabic and ...
www.sheikhmohammed.co.ae/ - 2k - Cached - Similar pages
www.sheikhmohammed.co.ae/arabic/index.asp - 2k - Cached - Similar pages
Al-Hammadi.com - A Website for All
Al-Hammadi.com is a website with information on Qatar, Islam, Arabic music, and more. Come on in and enjoy what we have to offer
www.al-hammadi.com/ - 12k - Cached - Similar pages

Remember that Diacritics Also Affect Searches

Most search engine algorithms are now set up to "read" accented search terms differently from those without accents. It's easy to test this by searching first for a term without any diacritical marks and then the same word with the marks, e.g., resume vs. résumé.

Types

Some common types of diacritical marks:

- acute accent (´)
- ring[1] above (°) used for angstrom (Å), aka krouzek
- breve (˘)
- caron or háček (ˇ)
- cedilla (¸)
- circumflex (^)
- umlaut[1] or diaeresis (¨)
- double acute accent (˝)
- grave accent (`)
- macron (¯)
- ogonek (˛)
- spiritus asper
- spiritus lenis

[1]/ Strictly taken not diacritics but parts of the character.

[66]

Look for Misspellings (Intentional or Accidental)

I am constantly amazed by the frequency of misspelled words, urls, file names, etc., I encounter on the Internet. By far, most appear to be simple mistakes, often made by non-English speakers trying to cope with our confusing language. These mistakes tend to propagate as users copy and paste them again and again, which is what I believe happened here:

[66] Fact Index, <http://www.fact-index.com/d/di/diacritic.html>

Web Images Groups News Froogle more »

enlgish site:www.chinadaily.com.cn Search Advanced Search Preferences

Web Results **1** - **10** of **10** from **www.chinadaily.com.cn** for **enlgish**. (**0.30** seconds)

jobs
Chinadaily.com.cn Recruitment 中国日报网站招聘. 网站设计部-
高级制作师(1名) (北京市). 要求 有良好的美术基础 ...
www.chinadaily.com.cn/**enlgish**/ doc/2004-03/16/content_315314.htm - 23k - Cached - Similar pages

jobs
Chinadaily.com.cn Recruitment 中国日报网站招聘. 网站设计部-
设计师(1名) (北京市). 要求 有专业美术设计基础 ...
www.chinadaily.com.cn/**enlgish**/ doc/2004-03/16/content_315316.htm - 23k - Cached - Similar pages

jobs
Chinadaily.com.cn Recruitment 中国日报网站招聘. 市场部-项目经理（
发行推广业务）(2名) (北京市). 要求： 30 ...
www.chinadaily.com.cn/**enlgish**/ doc/2004-03/16/content_315317.htm - 23k - Cached - Similar pages

jobs
Chinadaily.com.cn Recruitment 中国日报网站招聘. 《21
世纪少年英文报》诚聘各地发行代理. 《21世纪少年英文报 ...
www.chinadaily.com.cn/**enlgish**/ doc/2004-04/06/content_321050.htm - 9k - Cached - Similar pages

jobs
Chinadaily.com.cn Recruitment 中国日报网站招聘. 英语学习栏目编辑(
2-3名) (北京市). 工作所在地：北京 职责 ...
www.chinadaily.com.cn/**enlgish**/ doc/2004-03/16/content_315312.htm - 30k - Cached - Similar pages

Finally, the enormity of the task of finding meaningful and useful information on the Internet is both daunting and comforting: daunting because we know we can only scratch the surface of all the data and comforting because there is an almost limitless pool of possibilities. I find it useful to keep the challenge in perspective by recalling that a study published in 2000 showed "*the sixty known, largest deep Web sites* contain data of about 750 terabytes (HTML-included basis) or *roughly forty times the size of the known surface Web*."[67] In short, there is just so much data and information available via the Internet that no institution, no government, no computer, and certainly no individual can possibly grasp more than a small portion of all there is.

---

[67] Michael K. Bergman, "The Deep Web: Surfacing Hidden Value," *BrightPlanet.com*, July 2001, <http://www.brightplanet.com/technology/deepweb.asp> (14 November 2006), Introduction.

Custom Search Engines

This topic is new this year and expands upon the entries on Rollyo and Gigablast's Custom Topic Search from last year's edition. During 2006 there was an explosion in the number of custom search engines, including entries from Google, Yahoo, and Live Search, so you know the powerhouses think this is worth a try. Whether this trend catches on remains to be seen.

The phrase "custom search engine" is very misleading. None of these sites permits users to create a new search engine. What each site does in its own way is to let users customize an existing search engine to search specific sites in specific ways and return results in a personalized fashion. Thus, a better name for these services would be customizable searching, but that moniker is clearly unappealing. Just remember that you are not creating a new search engine any more than customizing a car is building a new automobile from the tires up.

Most of the custom search sites operate on a simple principle: they automate a long "site" search, e.g., the search is equivalent to [keyword(s) AND (site 1 OR site 2 OR site 3...OR site n)], where n stands for the maximum number of sites you are allowed to search.

In short, the proliferation of customizable search means that companies, educational institutions, government agencies, and individuals can easily put the power of the big search engines such as Google, Yahoo, and Live Search with its search Macros to work creating tailored and specialized search services in a way that has never before been possible. Customizable search may be "the next big thing," and I believe it is one of the most positive examples of that vague but ubiquitous concept called Web 2.0.

Gigablast's Custom Topic Search — http://www.gigablast.com/cts.html

Gigablast's Custom Topic Search was one of the first "create your own search engines" to appear, although Gigablast's creator Matt Wells never claimed it was anything other than a way to customize Gigablast. The beauty of the Gigablast CTS is that it requires no software installation but is very, very simple HTML code, so simple anyone can edit and understand it. No registration is required.

Many of Gigablast's features were primarily designed for webmasters instead of users, but this one is potentially valuable to both: **"Build Your Own Topic Search Engine."** Gigablast "allows you to create a list of up to 200 web sites (or subsites) and a search box that searches just those sites." **Custom Topic Search** even lets you decide if you want Gigablast to cluster the results for you. The concept behind topic search is that you, and not some anonymous marketer, choose the sites you want to search. This "tool" (for want of a better word) is amazingly easy to use and powerful. As someone whose eyes glaze over at the mere sight of code, let me put this in "user" language. If you are familiar with Google's **site:** syntax, imagine being able to have a "canned" query that runs against up to 200 websites of your own choosing and lets you run it whenever you like and use whatever keyword(s) you want at any time. The query on Google would look something like this:

[keyword site:cnn.com OR site:dmoz.org OR site:amazon.com OR site:usatoday.com OR site:cia.gov (etc.)]

The problem with Google is that multiple site/domain searches are cumbersome at best, and they quickly run up against Google's 32-word limit. Enter Matt Wells and Gigablast. As the creator and sole proprietor of his own search engine, Matt has the luxury of being able to add new options easily. I think CTS is his best innovation yet. Even if you are as HTML-averse as I am, this code is so easy to edit that it's a piece of cake. To make things even easier, I have done the basics for you. First, however, I highly recommend you read through the Gigablast pages below on the concepts behind CTS.

Build Your Own Topic Search Engine of Custom Topic Search
http://www.gigablast.com/byose.html

http://www.gigablast.com/cts.html

Now you're ready to take a look at, edit, and try the CTS. Copy and paste this HTML code into an application such as Notepad.

```
<head>
 <title>Gigablast Custom Search</title>
</head>
<body>
  Search News Websites
<form method="post" action="http://www.gigablast.com/search">
<input type="text" name="q" size="60">
<input type="submit" value="search" border="0">
<input type="hidden" name="sc" value="1">
<input type="hidden" name="sites" value="cnn.com news.yahoo.com
news.google.com usatoday.com foxnews.com">
</form>
</body>
```

This is a bare bones version of the CTS code. Now you can play with the code and make it into your own custom topic search page. I should mention that I set the "site clustering" option to ON <input type="hidden" name="sc" value="1"> but you can reset it to OFF by changing 1 to 0. Once you ***save as an HTML file***, all you have to do to use it is to ***open the file in your browser***, insert keyword(s), and go. Obviously, you will want to add more sites to search (I only put in a few) and change the topic to something of interest to you (I chose the rather bland News topic for demonstration purposes). Also, you can enter sub-sites or more specific sites, such as cnn.com/WORLD or dir.yahoo.com.

One thing to keep in mind that is ***you are searching Gigablast's database of pages from these websites, not the sites themselves***. The "work" that goes into creating a CTS is mostly up front because once you create your list of sites, it is not a complicated matter to add to or subtract from it. I can easily imagine creating a set of these search forms on a variety of topics using existing bookmarks.

Rollyo http://rollyo.com/

Rollyo stands for "Roll your own" search engine, meaning that you select the sources you want to search. Rollyo is powered by Yahoo, so results will come from Yahoo only. Rollyo lets users search up to 25 sites (not a huge number) and also try out and use other people's "Searchrolls." In order to save, share, and use your Searchrolls on other computers, you must register with an email address and a user-created name and password.

Rollyo has some unusual features. For example, Rollyo permits users to upload their bookmarks to create Searchrolls, edit someone else's Searchroll to make it your own, keep your Searchrolls private or share them. Rollyo searches entire sites or you can limit your search to a subdomain; however, you cannot limit your search to directories within a site, e.g., in this case, everything after the slash is ignored: *security.news.com/library*.

Rollyo has a nice little bookmarklet called Rollbar that "gives you access to all of your Searchrolls wherever you are.

- Search any site you visit, from the same spot on your browser, without having to dig around for every site's search page.
- Add sites to your Searchrolls on the fly.
- Create a new Searchroll from anywhere." <http://rollyo.com/bookmarklet.html>

One of the most attractive features of Rollyo is the ability to share Searchrolls. Here is an example of a Searchroll named "Muslim World Views." The sources searched are on the left side:

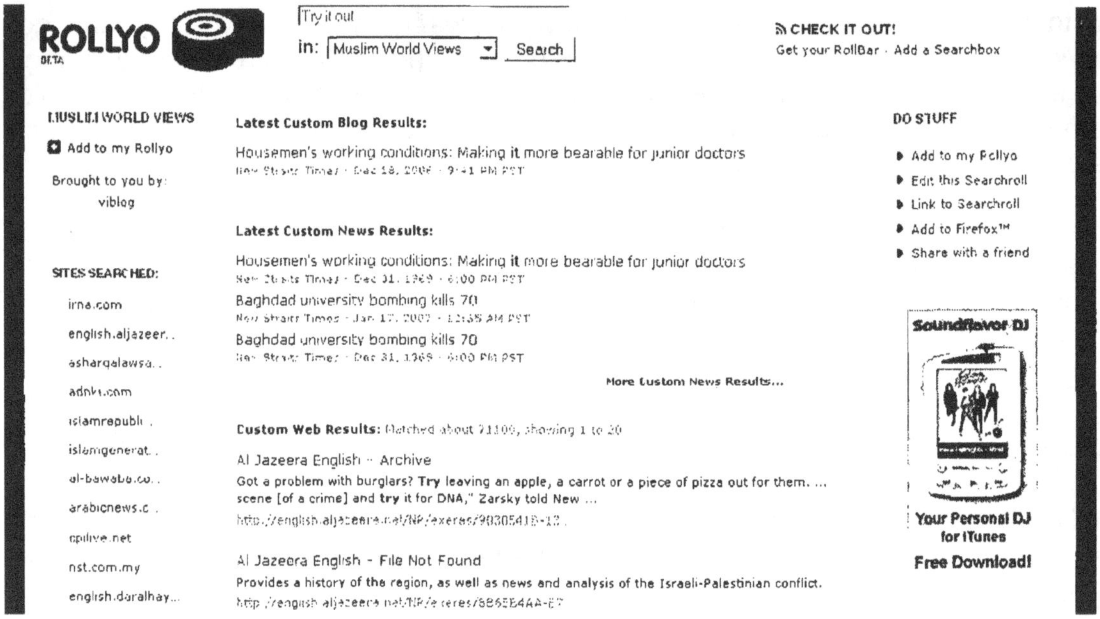
ROLLYO
Try it out
in: Muslim World Views
Search
CHECK IT OUT!
Get your RollBar · Add a Searchbox
MUSLIM WORLD VIEWS
Add to my Rollyo
Brought to you by:
viblog
SITES SEARCHED:
irna.com
english.aljazeer..
asharqalawsa..
islamrepubl..
islamgenerat..
arabicnews.c..
nst.com.my
english.daralhay...
Latest Custom Blog Results:
Housemen's working conditions: Making it more bearable for junior doctors
Latest Custom News Results:
Housemen's working conditions: Making it more bearable for junior doctors
Baghdad university bombing kills 70
Baghdad university bombing kills 70
More Custom News Results...
Custom Web Results:
Al Jazeera English - Archive
Got a problem with burglars? Try leaving an apple, a carrot or a piece of pizza out for them. ... scene [of a crime] and try it for DNA," Zarsky told New ...
Al Jazeera English - File Not Found
Provides a history of the region, as well as news and analysis of the Israeli-Palestinian conflict.
DO STUFF
Add to my Rollyo
Edit this Searchroll
Link to Searchroll
Add to Firefox™
Share with a friend
Your Personal DJ for iTunes
Free Download!

Rollyo has added blog and news searches (again, from Yahoo) to the results. Rollyo makes it very easy to create, save, and edit custom searches.

Google Custom Search Engine http://www.google.com/coop/cse/overview

Google got into the custom search game rather late. In October 2006 Google announced its own version of a custom search engine. In the announcement, Google said,

> "When we say we're letting people build a custom search engine, we mean the whole thing: choosing which pages they want to include in their index, how the content should be prioritized, whether others can contribute to the index, and what the search results page will look like...Here's how a Custom Search Engine works: organizations or individuals simply go to www.google.com/coop/cse and select the websites or pages they'd like to include in their search index. Users can choose to restrict their search results to include only those pages and sites, or they can give those pages and sites higher priority and ranking within the larger Google index when people search their site. Users can then customize the look, feel and functionality of their search engine."[68]

---

[68] Google Press Release, "The Power of Google Search is Now Customizable," 23 October 2006, <http://www.google.com/press/annc/custom_search.html> (17 January 2007).

After a telephone conference with Google's Marissa Mayer and the Google product managers, search and Google expert John Battelle shared his comments, which I think are excellent insights:

> "While similar to Rollyo's innovative custom roll, the Google CSE adds the benefit of allowing users to roll an unlimited number of sites together and display the results on their own site, with personalized presentation. Someone on the call described this as the fragmentation of search. The ability to build verticals will allow experts to build specialized engines. But while the engines will be individual, the collaborative element of tagging the domains encourages communities of knowledge to create together. So while each will stand apart from the amazing all-in-one answer box, the Custom Search will also allow a thickening or deepening of intelligent tags in Co-op, which feeds the one box that unites them all." <http://battellemedia.com/archives/003006.php>

Not surprisingly, ***you must have a Google account to use this service***. Also, Google Custom Search includes AdSense sponsored links alongside search results, but government sites, non-profits, and educational institutions are exempt from the advertising requirement. To see the Google Custom Search in action, take a look at Real Climate.org's internal search: <http://www.realclimate.org/> Even better, check out **Customsearchguide**, a directory of Google Custom Search Engines that others have created but you can use. Here is an example of general science and technology custom searches.

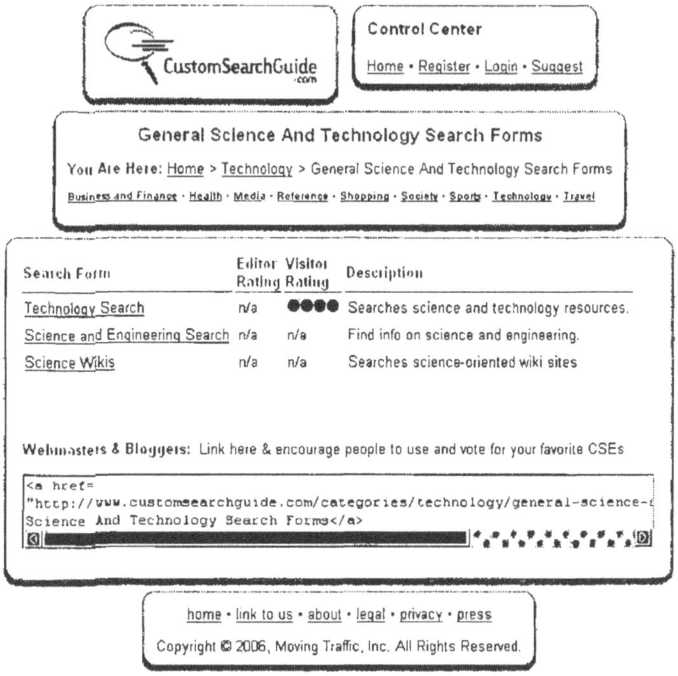
CustomSearchGuide.com
Control Center
Home • Register • Login • Suggest
General Science And Technology Search Forms
You Are Here: Home > Technology > General Science And Technology Search Forms
Business and Finance • Health • Media • Reference • Shopping • Society • Sports • Technology • Travel
Search Form
Editor Rating
Visitor Rating
Description
Technology Search
n/a
Searches science and technology resources.
Science and Engineering Search
n/a
n/a
Find info on science and engineering.
Science Wikis
n/a
n/a
Searches science-oriented wiki sites
Webmasters & Bloggers: Link here & encourage people to use and vote for your favorite CSEs
<a href=
"http://www.customsearchguide.com/categories/technology/general-science-
Science And Technology Search Forms</a>
home • link to us • about • legal • privacy • press
Copyright © 2006, Moving Traffic, Inc. All Rights Reserved.

Customsearchguide http://www.customsearchguide.com/

Yahoo Search Builder http://builder.search.yahoo.com/

Yahoo's custom search option requires registration and is very similar to others create your own search sites. Rollyo predates the Yahoo Search Builder and also searches the Yahoo database, giving you a good idea of what you can do with this tool.

Live Search Macros http://search.live.com/macros/default.aspx

I discuss creating and finding search macros in the section devoted to Microsoft's Live Search.

Alexa Web Search Platform http://websearch.alexa.com/welcome.html

What has been for many the holy grail of search is now a big step closer to reality. With little fanfare, Amazon's Alexa subsidiary announced in December 2005 that it was opening up its search tools and index to the world in a new project named the Alexa Web Search Platform (AWSP)—and for a very modest price.

According to its website, "The Alexa Web Search Platform provides public access to the vast web crawl collected by Alexa Internet. Users can search and process billions of documents and even create their own search engines using Alexa's search and publication tools. Alexa provides compute and storage resources that allow users to quickly process and store large amounts of web data. Users can view the results of their processes interactively, transfer the results to their home machine, or publish them as a new web service."

What exactly is Alexa offering to the user? In essence, Alexa gives the user, whether an individual or organization, access to the same kind of powerful technology used by Google, Yahoo, and Live Search. "Alexa spiders 4 billion to 5 billion pages a month and archives 1 terabyte of data a day. The new platform will allow developers to build their own search engines." The goal? To democratize web search by taking it out of the hands of giants like Google and putting it into the hands of literally anyone and everyone. The implications are enormous. And it appears it is a hit. In fact, within a very short time of its initial opening, Alexa had to cut off new applications temporarily because it was overloaded with customers wanting to sign up for the new service, but the site soon reopened registration.

The Alexa Web Search Platform (AWSP) offers the user the capability to:

➢ define (search): AWSP has a much more robust set of search options, syntax, and APIs than other search engines and also permits the use of stored (canned) queries; the AWSP "data store" contains text, html, music, video, images, and more types of files.

➢ process: users can search the entire Alexa data store and "are able to process both the raw content and the metadata extracted by Alexa's internal processes."

➢ publish: the output of the search can be anything from one result to an entirely new vertical search engine, for example a new video search engine or a new search engine for automotive parts. Quite literally, "by making use of these utilities, a user might introduce a great new search service to the world with nothing more than a home computer."[69]

The costs are modest and are based on consumption (you pay for what you use and not for a subscription or service contract):

$1 per cpu hour ($0.50 for reserved but unused hours)

$1 per GB/year of user storage

$1 per 50 GB processed

$1 per GB uploaded/downloaded

$1 for every 4,000 user-published web service requests

In case you're curious, Alexa has a long history. Now owned by Amazon, Alexa was created by Bruce Gilliat and Brewster Kahle (of Internet Archive fame), and until now has been both famous and infamous as the technology behind the controversial web traffic and website statistics "What's Related" toolbar feature in both Netscape and Internet Explorer. The new AWSP is actually integrated into Amazon's web services platform, something no one has done before.[70]

Simply stated, Alexa/Amazon are "renting" their huge database ("data store") to any and all takers for a remarkably reasonable price and, what is more, offering detailed

---

[69] Alexa Web Search Platform User Guide, Introduction: What Can I Do with the Platform? <http://pages.alexa.com/awsp/docs/WebHelp/AWSP_User_Guide.htm> (17 January 2007).

[70] There is one example of something similar, which came to my and some others' minds. If you are familiar with IBM's WebFountain and its proprietary implementations for specific customers, you may see some similarities. WebFountain also spidered the web and then let IBM's customers run queries against that data set in more sophisticated ways than simple querying (something akin to datamining). However, the problem with WebFountain and its progeny was that IBM had to write the programs, and thereby hangs a tale of woe. For more, I recommend Jeff Dalton's blog entry on this topic (I think he nails it). Jeff Dalton, "Alexa Web Search Platform: IBM WebFountain 2.0," Jeff's Search Cafe, <http://searchcafe.blogspot.com/2005/12/alexa-web-search-platform-ibm.html>

user support on how to maximize the effectiveness of this data to get the most out of it. The customer is empowered to write his own program to run against the Alexa/Amazon data, download the results (metadata), and even create his own private search engine on their platform. Perhaps I am wrong, but this could be a huge development, perhaps even a major change in the way we use the web.

Amazon Web Services Platform
http://www.amazon.com/gp/browse.html/104-1308416-9976726?node=3435361&

Alexa Web Search Platform (beta) http://websearch.alexa.com/welcome.html

Alexa Web Search Platform Users Guide
http://pages.alexa.com/awsp/docs/WebHelp/AWSP_User_Guide.htm

More Custom Search Sites

There are other sites offering customized search that you may want to experiment with to find one that best suits your needs. Search expert Phil Bradley reviews some of these custom search sites in a two-part article on Searchenginewatch.com:

"Search Your Own Way," Part I,
http://searchenginewatch.com/showPage.html?page=3623434 and Part 2,
http://searchenginewatch.com/showPage.html?page=3623482

Eurekster's Swicki http://swicki.eurekster.com/

PSS http://www.pssdir.com/

Fagan Finder

The Fagan Finder site has been a boon to searchers for some time not so much because of its basic interface, which is a good but unexceptional megasearch tool, but because of the many other "useful tools" site creator Michael Fagan has made available.

Fagan Finder File Format Search

Instead of having to visit a number of different search engines to search for files in a variety of formats, users can now go to the Fagan Finder "search by File Format" page, which is still in beta testing but appears to be running just fine. By selecting a specific file format, e.g., Microsoft PowerPoint, Fagan Finder automatically shows which of the search engines is capable of searching for that particular type of file. Not every search engine on the list searches for every file type.

Also, keep in mind that the Fagan Finder file type search for XML is less precise than going directly to Google or Yahoo and searching by *filetype:* in Google and by *originurlextension:* in Yahoo. If you use one of these search engines, you can specify that you only want to search for, say, those files that are *.rss* by entering the query [filetype:rss] or [originurlextension:rss]. These queries will return only those documents in RSS format, not those in XML or RDF. So I recommend using the Fagan Finder search by file type for files types other than XML, RSS, or RDF..

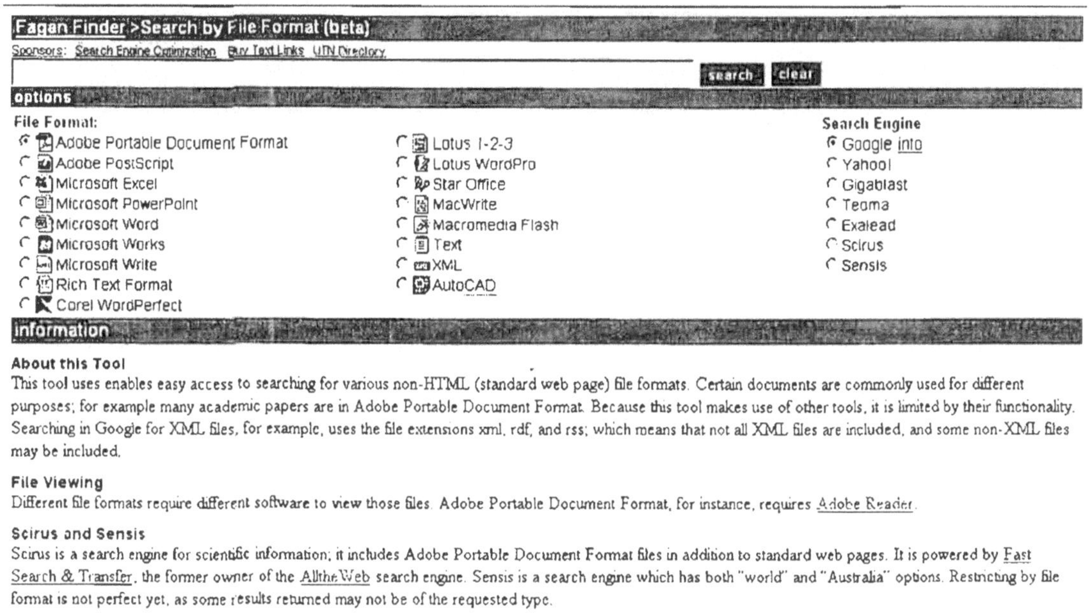
Fagan Finder >Search by File Format (beta)
search
clear
options
File Format:
Adobe Portable Document Format
Adobe PostScript
Microsoft Excel
Microsoft PowerPoint
Microsoft Word
Microsoft Works
Microsoft Write
Rich Text Format
Corel WordPerfect
Lotus 1-2-3
Lotus WordPro
Star Office
MacWrite
Macromedia Flash
Text
XML
AutoCAD
Search Engine
Google info
Yahoo!
Gigablast
Teoma
Exalead
Scirus
Sensis
information
About this Tool
This tool uses enables easy access to searching for various non-HTML (standard web page) file formats. Certain documents are commonly used for different purposes; for example many academic papers are in Adobe Portable Document Format. Because this tool makes use of other tools, it is limited by their functionality. Searching in Google for XML files, for example, uses the file extensions xml, rdf, and rss; which means that not all XML files are included, and some non-XML files may be included.
File Viewing
Different file formats require different software to view those files. Adobe Portable Document Format, for instance, requires Adobe Reader.
Scirus and Sensis
Scirus is a search engine for scientific information; it includes Adobe Portable Document Format files in addition to standard web pages. It is powered by Fast Search & Transfer, the former owner of the AlltheWeb search engine. Sensis is a search engine which has both "world" and "Australia" options. Restricting by file format is not perfect yet, as some results returned may not be of the requested type.

Fagan Finder Search by File Type http://www.faganfinder.com/filetype/

URLinfo http://www.faganfinder.com/urlinfo/

The indefatigable Michael Fagan also introduced a beta version of a new tool, URLinfo, in mid-2004. URLinfo fills a void created when AlltheWeb effectively shut down and took with it the useful "url investigator." While Yahoo now offers Site Explorer and Google a lame version [info:domain.com], Fagan's URLinfo provides many more options for exploring a site. As with everything he does, Fagan has gone all out with URLinfo, almost to the point of providing too many options! However, he has done a smart thing in keeping the main URLinfo page simple, "hiding" the nearly 85 investigative tools in his toolkit behind a variety of tabs. I think URLinfo is important and valuable enough to spend time looking at most of the options in some detail.

Here is a snapshot of the URLinfo main page.

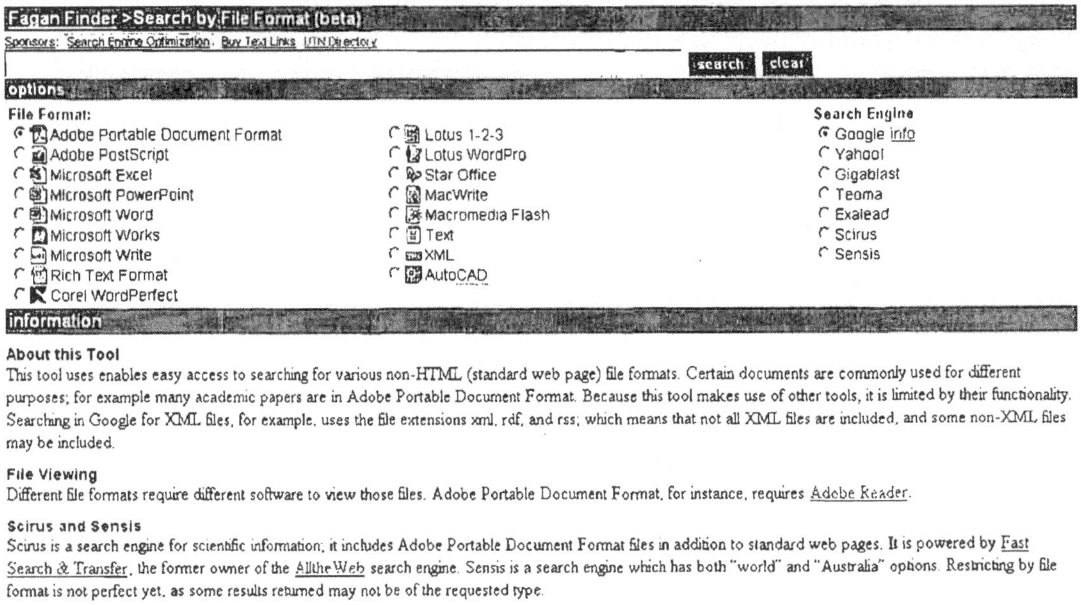
Fagan Finder >Search by File Format (beta)
Sponsors: Search Engine Optimization · Buy Text Links · LTN Directory
search clear
options
File Format:
Adobe Portable Document Format
Adobe PostScript
Microsoft Excel
Microsoft PowerPoint
Microsoft Word
Microsoft Works
Microsoft Write
Rich Text Format
Corel WordPerfect
Lotus 1-2-3
Lotus WordPro
Star Office
MacWrite
Macromedia Flash
Text
XML
AutoCAD
Search Engine
Google info
Yahoo!
Gigablast
Teoma
Exalead
Scirus
Sensis
information
About this Tool
This tool uses enables easy access to searching for various non-HTML (standard web page) file formats. Certain documents are commonly used for different purposes; for example many academic papers are in Adobe Portable Document Format. Because this tool makes use of other tools, it is limited by their functionality. Searching in Google for XML files, for example, uses the file extensions xml, rdf, and rss; which means that not all XML files are included, and some non-XML files may be included.
File Viewing
Different file formats require different software to view those files. Adobe Portable Document Format, for instance, requires Adobe Reader.
Scirus and Sensis
Scirus is a search engine for scientific information; it includes Adobe Portable Document Format files in addition to standard web pages. It is powered by Fast Search & Transfer, the former owner of the AlltheWeb search engine. Sensis is a search engine which has both "world" and "Australia" options. Restricting by file format is not perfect yet, as some results returned may not be of the requested type.

Note the eleven tabs at the top, behind each of which is a range of investigatory options. For help using URLinfo simply click on the dark blue **[info]** link on the far right. The first step in using URLinfo is to enter a url (address) in the search box at the top of the page. Keep in mind that ***if you enter a url in the search box and simply hit return, you will be taken to that webpage, not to information about it***.

Entering a url can prove to be more problematic than you might think because not every URLinfo tool can handle the same format. For example, in the General tab, the one most users are likely to use most frequently, you will get very different results depending on the type of url entered. For basic .com, .org, .net, .info, .biz, and .us domains, Domain Tools is great. However, for any other top-level domain, you must use Global Whois, and it will not search on anything but first-level domain names. This means that neither Domain Tools nor Global Whois can look up [www.duma.gov.ru]. Global Whois, however, will find first-level domains such as [www.feb-web.ru]. This does not mean you cannot find information about [www.duma.gov.ru].

Take a look at the results from the first tab, Alexa.

As you can see, you get lots of data about the Russian Duma website. Note that there are many additional useful links from the Alexa page, including one to the Internet Archive's Wayback machine.

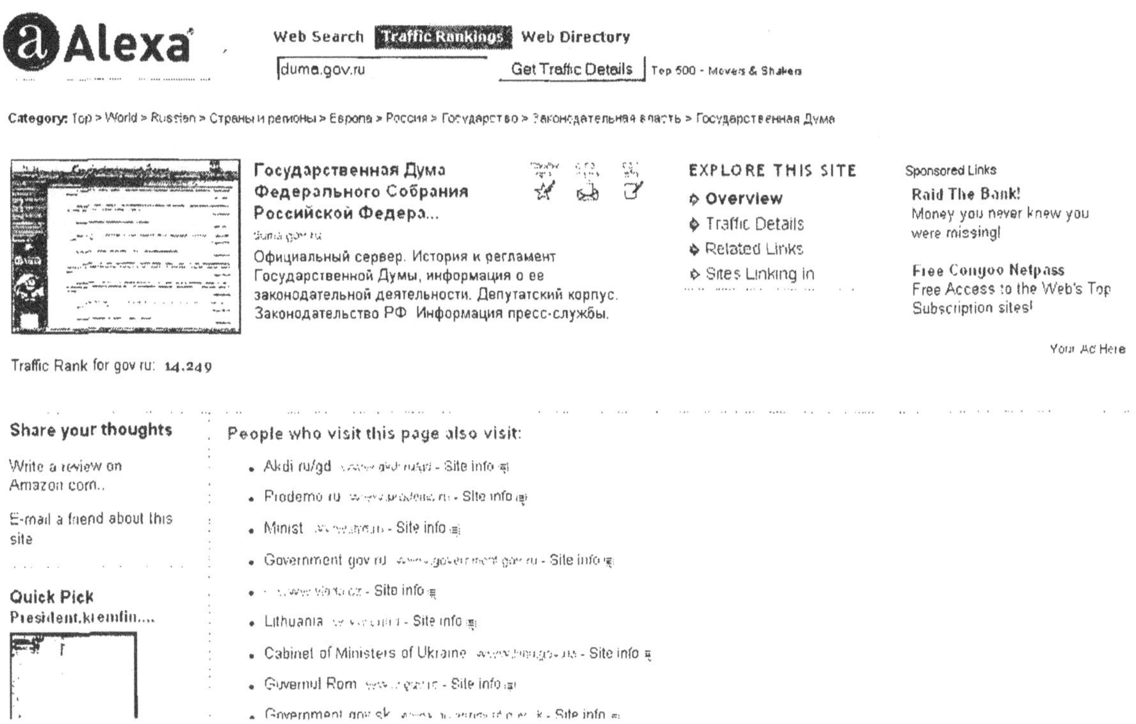
Alexa
Web Search
Traffic Rankings
Web Directory
duma.gov.ru
Get Traffic Details
Top 500 - Movers & Shakers
Category: Top > World > Russian > Страны и регионы > Европа > Россия > Государство > Законодательная власть > Государственная Дума
Государственная Дума Федерального Собрания Российской Федера...
Официальный сервер. История и регламент Государственной Думы, информация о ее законодательной деятельности. Депутатский корпус. Законодательство РФ Информация пресс-службы.
EXPLORE THIS SITE
Overview
Traffic Details
Related Links
Sites Linking in
Sponsored Links
Raid The Bank!
Money you never knew you were missing!
Free Congoo Netpass
Free Access to the Web's Top Subscription sites!
Your Ad Here
Traffic Rank for gov.ru: 14,249
Share your thoughts
Write a review on Amazon.com..
E-mail a friend about this site
Quick Pick
President.kremlin....
People who visit this page also visit:
Akdi ru/gd - Site info
Prodemo ru - Site info
Minist - Site info
Government gov ru - Site info
- Site info
Lithuania - Site info
Cabinet of Ministers of Ukraine - Site info
Guvernul Rom - Site info

The Alexa database contains site statistics, contact information, similar pages, and more.

> **What is Alexa?** Many things, but most interesting and useful is Alexa's Site Information:
>
> "Alexa has built an unparalleled database of information about sites that includes statistics, related links and more. All of this information can be found on Alexa's Site Overview pages, Traffic Detail pages and Related Links pages. To access these pages, simply type the URL of any site into the Alexa Search box."
>
> Alexa Site Information, http://www.alexa.com/site/company

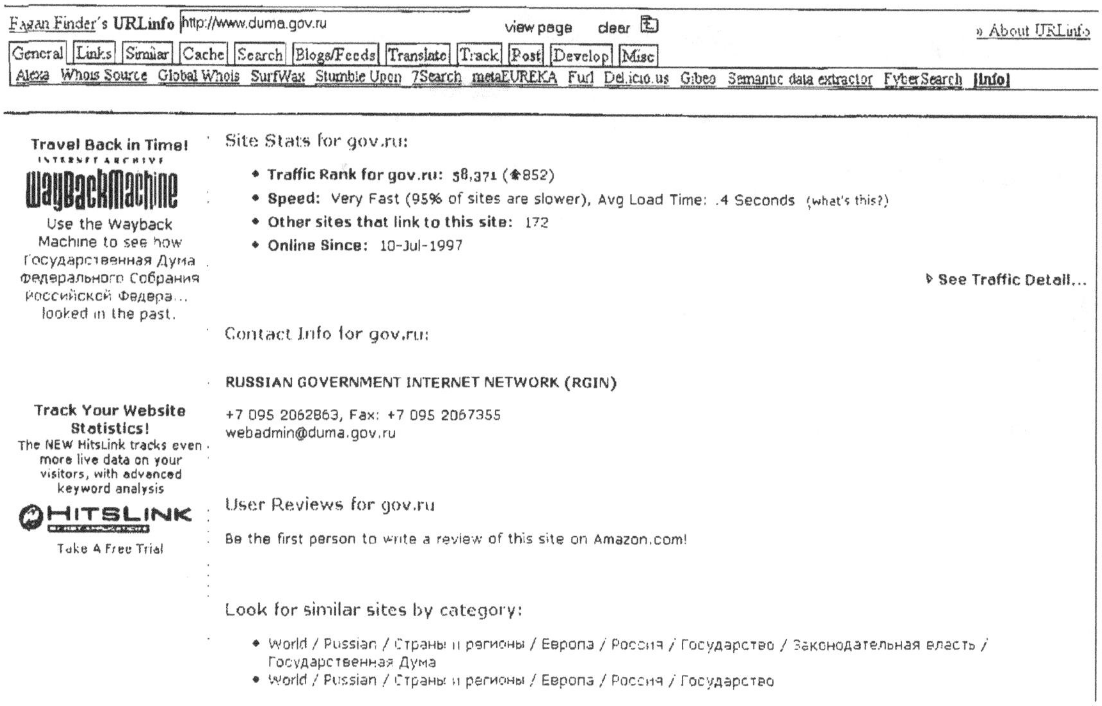

Fagan Finder's URLinfo http://www.duma.gov.ru view page clear » About URLinfo

General | Links | Similar | Cache | Search | Blogs/Feeds | Translate | Track | Post | Develop | Misc

Alexa Whois Source Global Whois SurfWax Stumble Upon 7Search metaEUREKA Furl Del.icio.us Gibeo Semantic data extractor FyberSearch [info]

Travel Back in Time!
INTERNET ARCHIVE
WaybackMachine
Use the Wayback Machine to see how Государственная Дума Федерального Собрания Российской Федера... looked in the past.

Track Your Website Statistics!
The NEW HitsLink tracks even more live data on your visitors, with advanced keyword analysis
HITSLINK
Take A Free Trial

Site Stats for gov.ru:

- Traffic Rank for gov.ru: 58,371 (▲852)
- Speed: Very Fast (95% of sites are slower), Avg Load Time: .4 Seconds (what's this?)
- Other sites that link to this site: 172
- Online Since: 10-Jul-1997

▸ See Traffic Detail...

Contact Info for gov.ru:

RUSSIAN GOVERNMENT INTERNET NETWORK (RGIN)

+7 095 2062863, Fax: +7 095 2067355
webadmin@duma.gov.ru

User Reviews for gov.ru

Be the first person to write a review of this site on Amazon.com!

Look for similar sites by category:

- World / Russian / Страны и регионы / Европа / Россия / Государство / Законодательная власть / Государственная Дума
- World / Russian / Страны и регионы / Европа / Россия / Государство

Let's look at a different url for the **SurfWax** results. What you are seeing are "SurfWax SiteSnaps™, [which] count the number of links, images, words, and forms on a page, shows the meta description tag, and extracts 'key points' and 'FocusWords.'" This is a very useful way to analyze a website without actually visiting it, though the amount of information is considerably less for some sites than others, *cf.*, *www.fateh.net*.

Fagan Finder's URLinfo http://www.nla.gov.au view page clear » About URLinfo

General | Links | Similar | Cache | Search | Blogs/Feeds | Translate | Track | Post | Develop | Misc

Alexa Whois Source Global Whois SurfWax Stumble Upon 7Search metaEUREKA Furl Del.icio.us Gibeo Semantic data extractor FyberSearch [info]

SurfWax SiteSnaps™ (patent pending) [Help]

National Library of Australia
http://www.nla.gov.au
Links: 32 • Images: 28 • Words: 383 • Forms: 0

Author Summary

Our collections and services underpin Australian cultural life and intellectual pursuits. We are the preeminent source for the documentary record of Australia and its place in the world. This site provides access to our catalogue and links to other information resources, details of our collections and services, upcoming events at the Library; the NLA Shop and information on our initiatives in the field of information management.

Site's FocusWords

- Australian Libraries Gateway
- call 0410 627 684
- collections of Australia's national
- combined catalogues of Australian libraries
- curiosities and masterpieces
- exhibitions and publications
- Exhibitions Displays
- GUIDES INDEXES DATABASES
- HOME CATALOGUE ASK
- information journey
- Kinetica
- Libraries Australia
- Library
- Library's
- million items
- MusicAustralia
- National Library of Australia
- phone services
- phone system. the main
- PictureAustralia
- state and territory libraries
- trove of Australia's musical gems
- watercolours

The next tab is a link to **Stumble Upon**, a collaborative bookmarking service. If people who belong to the service have recommended a site, Stumble Upon will show who they are, any comments, the category it is in, and the page size. The tab for 7Search brings up some interesting results, including the website's traffic ranking (how many people visit it), number of links to the site, link popularity, language, area of service, contact information, Open Directory category, and "TrustGauge." **TrustGauge** is a commercial program that measures how "trustworthy" a site is in terms of such things as the amount and quality of contact information, secure billing, third party validations (e.g., Truste seal), and what people think of the site.

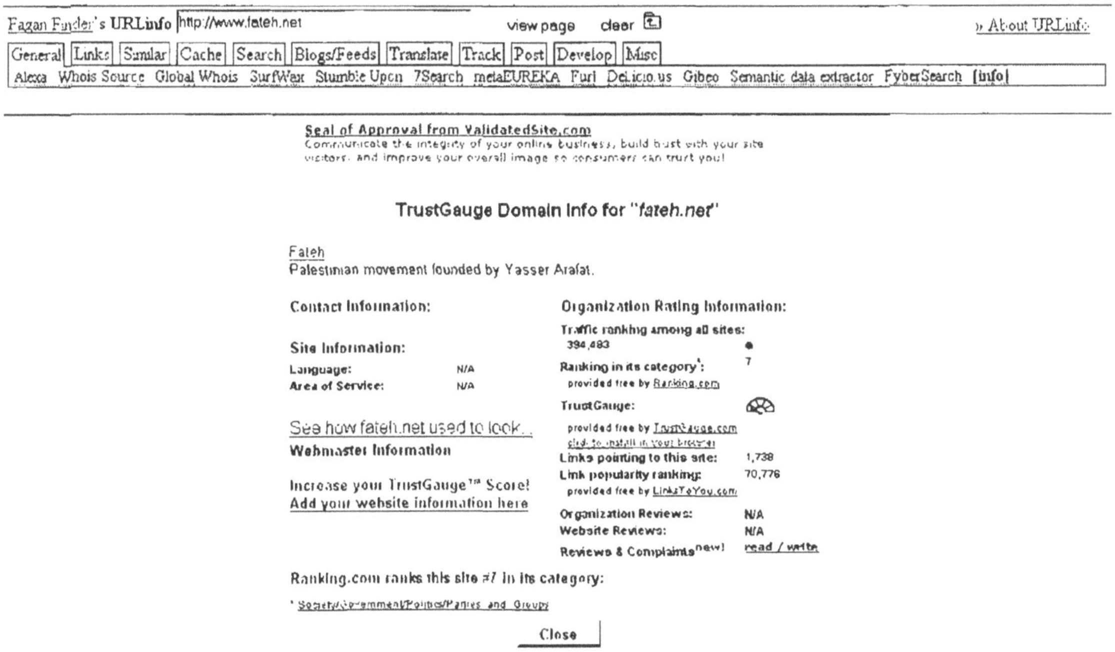
Fagan Finder's URLinfo
http://www.fateh.net
view page
clear
» About URLinfo
General
Links
Similar
Cache
Search
Blogs/Feeds
Translate
Track
Post
Develop
Misc
Alexa Whois Source Global Whois SurfWax Stumble Upon 7Search metaEUREKA Furl Del.icio.us Gibeo Semantic data extractor FyberSearch [info]
Seal of Approval from ValidatedSite.com
TrustGauge Domain Info for "fateh.net"
Fateh
Palestinian movement founded by Yasser Arafat.
Contact Information:
Organization Rating Information:
Traffic ranking among all sites:
394,483
Site Information:
Language:
N/A
Area of Service:
N/A
Ranking in its category:
7
TrustGauge:
See how fateh.net used to look...
Webmaster Information
Links pointing to this site:
1,738
Link popularity ranking:
70,776
Increase your TrustGauge™ Score!
Add your website information here
Organization Reviews:
N/A
Website Reviews:
N/A
Reviews & Complaints
read / write
Ranking.com ranks this site #7 in its category:
Close

The remainder of the **General** tab links are:

metaEUREKA

metaEUREKA shows information about the page (last modified date, page size), meta information (description, keywords, author), web server information, and the number of backlinks

Furl

Furl is a collaborative bookmarking system. This tool allows you to see the comments others have written about a webpage.

Del.icio.us

Del.icio.us is a collaborative bookmarking system. This tool allows you to see the comments others have written about a webpage.

Gibeo

Gibeo allows anyone to annotate any part of a web page, and others can comment on the annotation. Gibeo requires registration.

Semantic data extractor

The Semantic data extractor finds information about a page (metadata, page outline) by looking at its HTML code.

The next tab is for **Links**. This is pretty straightforward. The first two links are to Yahoo, the first for the *link:* command (links to a specific page) and the second for the Yahoo Site Explorer or alternately *linkdomain:* command (links to a website). The next is the Live Search (MSN) *link:* search, and then the Google *link:* command, which no longer shows all links as it once did. Gigablast does not show all links to a page, either.

The links from **blogs** is a very useful service because it lets you check to see if a website is mentioned in a number of weblogs very quickly (I expect Technorati to give the best results).

Blogpulse
Intelliseek's Blog search (was not working when I tried it)

Bloglines
Backlinks from blogs known to Bloglines, an online RSS/Atom aggregator.

Blogdex is defunct.

Technorati
Backlinks from blogs known to the Technorati blog indexer. Each result is shown with an extract containing the link.

Feedster
Backlinks from blogs known to the Feedster RSS/Atom search engine.

BlogDigger
Backlinks from blogs known to the BlogDigger RSS/Atom search engine.

Waypath
Backlinks from blogs, known to the Waypath blog indexer, each is listed with the date that the link was first seen and an extract from the page Unlike some other backlinks tools, Waypath lists the permalinks rather than blog home pages.

Daypop
Backlinks from blogs and news websites known to the Daypop search engine.

BlogRolling
BlogRolling is a service for bloggers to include blogrolls (lists of blogs) on their own blogs. This shows what users include the given site on their blogroll.

Popdex
Backlinks from blogs (as well as the date of linkage) known to the Popdex blog indexer.

The **Similar** tab is not entirely self-explanatory. Alexa, UCmore, Furl, and Google all try to show related or similar websites, though not in the same way. **Alexa** shows 'people who visit this page also visit...'; **UCmore** clusters related pages by topic; **Furl** is a collaborative bookmarking tool, so it only shows pages bookmarked by the same person (of dubious use); and Google's related pages is, in Fagan's and my opinion, of poor quality. Google News will show related news articles, but only if the original article has been indexed by Google News. The **Waypath** tool looks for blog entries about a website, and Waypath is showing no links to *http://www.google.com* and two hits on *http://www.microsoft.com*. There is obviously a problem with this specific search.

The **Cache** tab is much more useful at this time. Fagan has done us all the great service of bringing the search tools that cache webpages together so they can be searched from one convenient interface. Also, ***URLinfo makes it possible to see Google's cached pages without images, style sheets, or forms with Google (plain).*** Openfind is an Asian search engine and does not yet have an English version. I was unable to figure out how their caching works because of the language barrier. For news and blogs Daypop caches each page it crawls. "Its cache is often the most up-to-date copy of the page, and it shows the exact time that the copy was made."

Here's the low-down on the other general cache tools at Fagan Finder:

Internet Archive

The Internet Archive has been crawling the web and caching pages since 1996. The Wayback Machine allows you to view the copies made during any of those crawls, and also to compare any two versions of the same page.

Google

When Google crawls the web, it stores a copy of each web page. This is the most recent copy. This can also be used as a means of viewing some non-HTML files converted to HTML.

Google (plain)

Google's stripped cache, with images, styles (style sheets), and forms removed.

Gigablast

Gigablast does not provide direct access to its cache. You must follow the link labeled [archived copy]. Gigablast's cache shows the date on which the copy was made.

Openfind

Openfind is an Asian search engine; their English version is under construction.

Spurl

Spurl is a collaborative online bookmarking tool. Whenever someone using Spurl bookmarks a page, a cached copy is stored. So Spurl may contain many different copies of the same page on different dates and times, which can be accessed from a selection box at the top of any Spurl cached page.

IncyWincy

This is the cached version of a web page from when it was last crawled by IncyWincy. That date is shown at the top of the page.

Scrub The Web

Cached version of the page from the Scrub The web search engine.

Ay-Up

Cached version of the page from the Ay-Up search engine.

Objects Search

Cached version of the page from the Objects Search engine. Objects Search has a small index, so don't expected every page to be cached. After using this tool, follow the link below the page you want labeled 'cached.'

SearchSpider

This is the cached version of a web page from when it was last crawled by SearchSpider. Most pages appear to have been last cached during July 2003.

The **Search** section is pretty much self-explanatory, except that MSN searches Live and Teoma searches Ask. Fagan explains the **Blogs/Feeds** tab very well for those who are interested in searching weblogs and RSS or Atom news feeds. The **Translate** tab simply sends your request to Fagan Finder's superb Translation Wizard discussed in the online dictionary and translators' section. The **Track** and **Post** tabs are in general not going to be useful for most of you in your work environment. The **Develop** tab offers an excellent selection of web authoring resources such as validation, editing, spelling, cacheability, and keyword analysis tools. One tool users may not recognize and which could prove quite useful is **Traffic** from Alexa. Here's Fagan's description:

> 'Shows a (logarithmic) graph of a website's (not a web page) popularity over time, as determined by Alexa. Alexa gathers this data from users of their toolbar. The six-month graph is shown by default. You can also use this tool to compare the popularity of a second website. Also shown are popular subdomains, reach per million users, average page views per user, etc.'

I find the graphic representation is so much clearer than the results from a tool such as Google PageRank (which is not a Google product, by the way). Traffic also lets users compare two sites and shows you "Where do people go" on the site. It's a gold mine of data about the sites in Alexa's top 100,000; unfortunately, most of the sites I wanted to research were not in that top group, so no statistics were available when a site fell below the 100,000 threshold.

In case the Google PageRank tool confuses you, it normally requires users to download and install the Google Toolbar. However, you can access the Google PageRank option from URLinfo without the Google Toolbar. The results look rather mysterious, but the PageRank is there. In the following example, AOL's home page has a page rank of 8 (where 10 is the highest...and Google gets a 10 ranking, by the way):

> http://www.aol.com
>
> PR Toolbar: 9
>
> PR Actual: 9

Finally, under **Misc** you'll find the tools that didn't quite fit anywhere else. One word of caution about **BugMeNot**: this is a service for sharing login information for websites that require user registration and, as such, its ethics is questionable. I do not recommend using it. It may also violate organizational Internet usage rules.

I think URLinfo will prove to be a very useful if not indispensable tool for researchers, but I also think the key to using it effectively is not using every bell and whistle.

Fagan Finder's URLinfo beta http://www.faganfinder.com/urlinfo/

Wikipedia

Wikipedia http://en.wikipedia.org/

The 2007 edition is the first to include a separate section on and discussion of Wikipedia and the entire "wiki" phenomenon. The extraordinary growth and success of Wikipedia demand recognition and comment. Although the numbers change constantly, in mid-2006, Wikipedia sites were the twelfth most visited Internet sites among US properties, *up over 300 percent from the previous year.*[71] On March 1, 2006, Wikipedia reached one million articles, and "the site receives as many as

[71] Safa Rashtchy, et al., "Silk Road: Solid Search Results Could Boost the Sector," PiperJaffray Industry Note, 10 July 2006, available at John Battelle's Searchblog, <http://battellemedia.com/archives/Rashtchy%20-%20Silk%20Road%200710.pdf> [PDF] (14 November 2006).

fourteen thousand hits per second."[72] Just what is the Wikipedia itself and the wiki concept in general that have led to a level of success that is nothing short of astounding? For an excellent overview, I turn to my colleague Diane White's article from an internal publication many of you read, *The WorthWhile Web*. In the May 2006 edition, Diane wrote:

"In true Ouroborosian fashion, the Wikipedia defines itself as a 'multilingual Web-based free-content encyclopedia...written collaboratively by volunteers, allowing most articles to be changed by anyone with access to a web browser and an Internet connection.' It exists as a wiki, which again Wikipedia self-defines as 'a type of website that allows anyone visiting the site to add, remove or otherwise edit all content very quickly and easily, often without the need for registration.' Truly collaboration to the extreme, wikis are the latest trend in open-ended community involvement and public debate. But it also conjures fears of authority and validity run amok, and general mischief and vandalism. Wikis are popping up everywhere; but just what are they, and how did they become so ubiquitous? More to the point, can they be trusted, or are they just the work of a few people with big egos and lots of time?...The term wiki is a shortened form of the Hawaiian language term *wiki wiki*, which is commonly used as an adjective to denote something quick or fast. It is also sometimes interpreted as the backronym for *What I Know Is*. The invention of the wiki is credited to Ward Cunningham, author of the book, *The Wiki Way* (Addison-Wesley Longman, March 2001, ISBN 0-201-71499-X). The first wiki, WikiWikiWeb, was created in 1994 and installed on the web by Cunningham in 1995.[73]

"Once begun, almost anyone can edit a wiki, often without actually registering to do so. Wikis can be on any subject, on every subject, and in multiple languages. The most famous wiki, Wikipedia, was begun in 2001, initially as part of a broader, peer-reviewed project and later as a stand-alone, 'neutral point of view' product. Guided from the beginning by Larry Sanger and Jimmy Wales, today it is available in over 100 languages, with over 1 million articles in the English edition alone...

"Questions of Validity and Reliability

"But can Wikis be trusted? From almost the beginning, people have questioned the wiki's seemingly radical departure from traditional methods of scholarship; that is, the use of a community of interested parties instead of the work of appointed experts. In the December 2005 issue of *Nature*, there began a major debate over which site was more 'right,' Wikipedia or the fee-based ($85/year) Britannica Online; with the conclusion being that 'Wikipedia comes close to Britannica in terms of the

---

[72] Stacy Shiff, "Can Wikipedia Conquer Expertise?" *The New Yorker*, 24 July 2006, <http://www.newyorker.com/fact/content/articles/060731fa_fact> (14 November 2006).

[73] "Wikipedia," Wikipedia: The Free Encyclopedia, <http://en.wikipedia.org/wiki/Wikipedia > (23 August 2006).

accuracy of its science entries.'[74] From there it has escalated, with refutations and calls for retraction from *Encylopaedia Britannica* and heated responses from *Nature*. Wikipedia itself has steered clear of this particular fray; however, it does attempt to respond to criticism and has a page on its site for common criticisms. It also addresses issues such as copyright, vandalism, and authorship.

"So what's the bottom line? The same as it's always been. When performing thorough research, be it Internet-based or otherwise, *the onus is always on the researcher to check sources, validity, and authority*. The speed and relative ease at which changes can be made to a wiki, while good for consensus correction and corroboration, are not so good for measured and thoughtful debate. A number of articles in Wikipedia are sourced, but many are not, and just because it's on the Internet, does not mean it is true. In addition, merely because it's free does not mean Wikipedia is more suspect and Britannica is more reliable. There is an argument to be made for being so passionate about a topic that you feel the need to share that passion with the world. But one man's passion is also another's conceit. There is a counter to every argument, a rebuttal to every claim.

"Like it or not, wikis and wiki behaviors have entered the mainstream, just like blogs and MySpace and the iPod. Love it or hate it, if you are involved in open source research you need to know about wikis."[75]

The Wikipedia Itself: The Good, the Bad, and the Dubious

As Diane White clearly indicates, there are many, many wikis now available on the Internet, and their numbers continue to increase at present. I want to focus on Wikipedia itself because it remains the center of the wiki universe and thus far shows no signs of decline. Many Wikipedia critics mourn the decline of traditional encyclopedias because they are thinking of an encyclopedia such as *Britannica* in its current form, that is, "the most authoritative source of...information and ideas," the "definitive source of knowledge."[76] According to Tom Panelas, *Britannica*'s Director of Corporate Communications, "We can't cover as many things as they [Wikipedia] do but we wouldn't even try to. What they do is very different from what we do. We don't have an article on extreme ironing, and we shouldn't."[77]

---

[74] Jim Giles, "Internet Encyclopaedias Go Head to Head," *Nature*, 14 December 2005 (last updated 28 March 2006), <http://www.nature.com/news/2005/051212/full/438900a.html> (14 November 2006).

[75] Diane White, "Wikis and the Wikipedia," *The WorthWhile Web*, May 2006, <http://www.fggm.osis.gov/Worthwhile/archive/20060501.html>.

[76] Paula Berinstein, "Wikipedia and Britannica: The Kids Are All Right (And So's the Old Man)," *Information Today*, March 2006, <http://www.infotoday.com/searcher/mar06/berinstein.shtml> (11 September 2006).

[77] Berinstein.

However, *Britannica* today (and by extension any other encyclopedia) is very different from Britannicas of the past. In thinking about this controversy, I was reminded of a passage in *The Fatal Shore*, Robert Hughes' masterpiece about the founding of modern Australia. Hughes writes about one transported convict, Thomas Palmer, who finished his sentence and went into business with his close friend John Boston.[78] Neither man had much business experience, "but they possessed a singular advantage: ***the only encyclopedia in the colony***. With it, they taught themselves to make beer. Then they learned how to make soap. Next they looked up 'ship' and, after some trial and error, contrived to build a somewhat cranky but adequate small vessel for trading stores to Norfolk Island." [emphasis added][79] Their lone encyclopedia probably made it possible for these men not merely to survive but to thrive in this perilous new world.

The modern encyclopedia is very different from the encyclopedias of earlier centuries, which bear rather more resemblance to the Wikipedia than to the current *Britannica* in that the older encyclopedias were not only "sources of knowledge" but also "practical" how-to guides and almanacs. In other ways, however, encyclopedias are and always were quite different from the Wikipedia. They have always relied upon paid experts whose work is reviewed and edited. And they have always been for-profit enterprises.

Wikipedia *relies almost entirely upon individual users* to create, edit, maintain, and often argue about its entries. It is free and carries no advertising; it is a nonprofit and has a tiny staff.

- For practical purposes, Wikipedia has *no physical limits*: it could conceivably continue to expand indefinitely, something no print encyclopedia could ever do.
- Its content is "open," that is, *almost any topic can be included*; traditional encyclopedias generally do not include "how-to" instructions ("How to draw a diagram with Microsoft Word"), new or transient popular culture ("24: The TV Series"), or breaking stories ("JonBenét Ramsey").
- Wikipedia's heavy emphasis on current events and popular culture bespeak a prejudice of the present at the expense of the important: it favors *the fashionable over the important*.

---

[78] In what must be one of the most profound examples of friendship since Damon and Pythias, Boston actually traveled voluntarily with his wife to New South Wales to "keep Palmer company." Anyone who has read about a sea voyage from England to Australia at that time knows the trip in and of itself was a major sacrifice. Robert Hughes, *The Fatal Shore* (New York: Vintage Books, 1988), 180.

[79] Hughes, 180.

- Wikipedias are available in *229 languages*. These are not always just translations of the English language Wikipedia but often contain their own content.

- Wikipedia's eight-word self-description—"*neutral and unbiased compilation of previously written, verifiable facts*"—usually keeps out articles about "my funniest dreams and what they mean" (no original "research" allowed), but firefights over controversial topics and outright vandalism occur on a regular basis.

- In 2006 comedian Steven Colbert's amusing rant against "wikiality" and "truthiness," i.e., that *reality and truth are what the most people say they are*, and his charge to his viewers to change a Wikipedia article on African elephants caused the entire site to go down temporarily. His point is well taken: if enough Wikipedians agree that the earth is flat, then the Wikipedia will reflect that "wikiality." While that is an absurd example, people vehemently (and often violently) disagree over the most basic topics (try to think of anything that isn't controversial).

- Wikipedia "does not favor the Ph.D. over the well-read fifteen year old."[80] While the *democratization of knowledge and information* has a certain appeal, the fact that Wikipedia pages dealing with policies, rules, administration, coordination, and other metadata now comprise thirty percent of Wikipedia indicates that the free-for-all nature of Wikipedia is giving ground to the harsh reality of the need for "crowd control." There is a fine line between democracy and mob rule.

- There is *no "weighting" of the relative significance* of any topic: compare the Wikipedia entries on the Beatles v. Boethius. Judged by sheer quantity, articles on popular culture far exceed those of traditional scholarly topics. Given its potentially limitless size, this may not be a drawback, but if everything from "The Simpsons" to "The Nicomachean Ethics" is on an equal footing, then aren't we back to the Colbert criticism that all objective standards are obliterated?

- Diane White correctly identified Wikipedia's "ouroborosian" nature: it is *fiercely self-referential* in that all the works cited in this creature of the Internet are also on the Internet.

- Some critics maintain that emergent enterprises such as Wikipedia reflect an "online collectivism" that lead to a kind of group think and produce poor quality results that both appeal to and are a product of the lowest common

---

[80] Shiff, "Can Wikipedia Conquer Expertise?"

denominator. For more on this topic, read Jaron Lanier's now famous think piece "Digital Maoism" and the many responses to it on Edge.org.[81]

- Finally, Wikipedia has *no editorial quality review*. Traditional encyclopedias do not guarantee zero mistakes; what they do promise are "strong scholarship, sound judgment, and disciplined editorial review."[82]

All this being said, nothing is going to stop people from using Wikipedia as a reference, in many cases, their primary source for information. Some search engines—for example, Ask—now proudly display Wikipedia responses at the top of the results list. Most will return Wikipedia links near the top. The best advice I can give you vis-à-vis Wikipedia and related community generated resources is as follows:

- <u>Use multiple sources</u>: Do not as a rule rely on Wikipedia as your sole reference or source of information. Any Wikipedia entry that is not well sourced should raise a red flag.

- <u>Trust but Verify</u>: Look for verification of Wikipedia information from sources such as traditional references that have been through editorial review: encyclopedias, dictionaries, scholarly (peer-reviewed) publications, university websites, books, etc.

- <u>Follow those links</u>: The best thing about Wikipedia in my opinion are the *external links* from entries; with the virtual demise of web directories, Wikipedia fills that void by supplying excellent links to what are often the best websites on a topic.

- <u>Be skeptical</u>: The more controversial the topic, the more skepticism you need to apply to the Wikipedia entry. For example, the article "Asteroid" is quite well done, but there isn't quite the controversy about that topic that there is about, say, Hezbollah, an article that was locked because of vandalism.

Wikipedia has an internal search option, but as any Wikipedia user knows, it is not the best way to search Wikipedia. First, unlike virtually every search engine on the web, its default is OR not AND, meaning it searches for ANY of the terms you enter. To search Wikipedia content you are better off using a separate search engine, either one of the major search engines or a specialty search tool designed to search Wikipedia.

---

[81] Jaron Lanier, "Digital Maoism," Edge.org, June 2006, <http://www.edge.org/3rd_culture/lanier06/lanier06_index.html> (14 November 2006).

[82] "Britannica Rips Nature Magazine on Accuracy Study," Encyclopedia Britannica Corporate Website, 24 March 2006, <http://corporate.britannica.com/press/releases/nature.html> (14 November 2006).

How to Check Links from Wikipedia

Wikipedia has a special page that lets users easily check to see which Wikipedia pages have links to a specific webpage. It even includes a wildcard function. Here's how it works. You can search for a specific link or a very general one. Here are examples of both, starting with a general search using a wildcard [*.nasa.gov]:

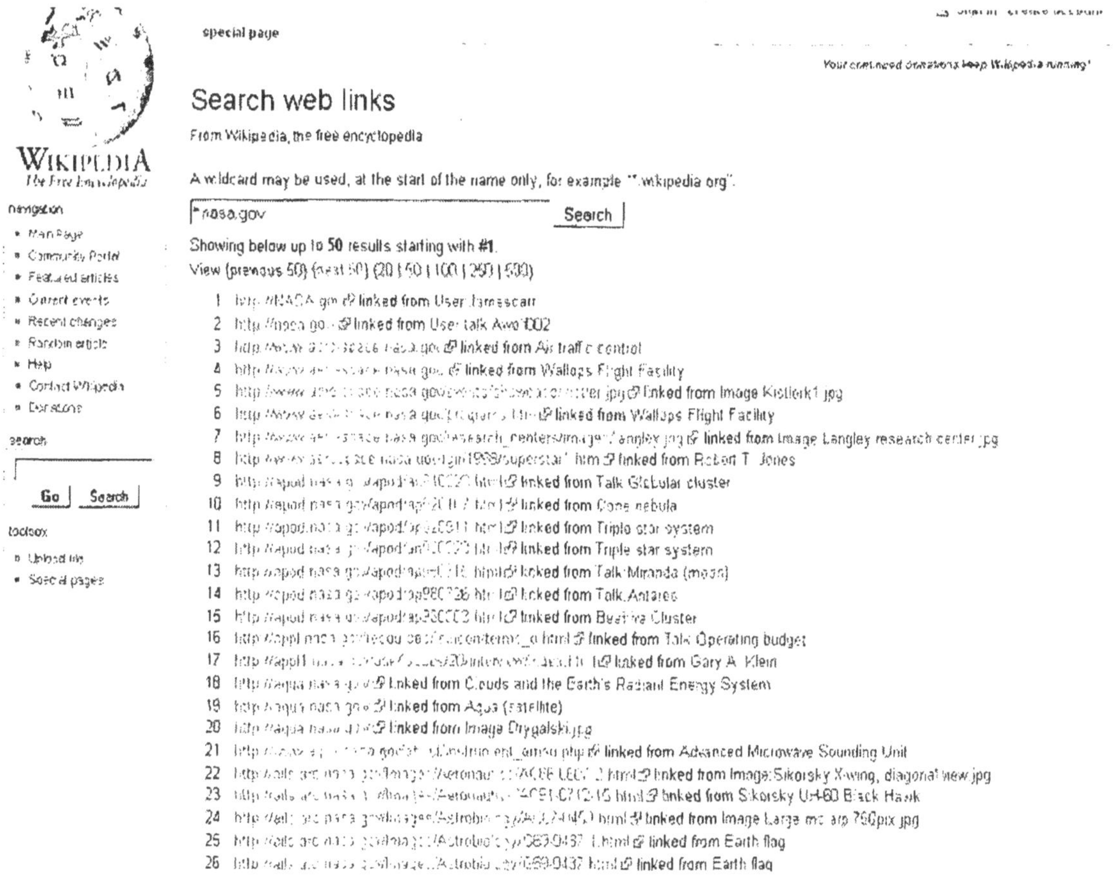
special page
Search web links
From Wikipedia, the free encyclopedia
A wildcard may be used, at the start of the name only, for example "*.wikipedia.org".
*nasa.gov
Search
Showing below up to 50 results starting with #1.
WIKIPEDIA
Go
Search

Search Web Links at Wikipedia
http://en.wikipedia.org/w/index.php?title=Special%3ALinksearch

There are literally hundreds of results for this query. However, you can limit your search to a specific page [leonid.arc.nasa.gov/meteor.html], which in this case returns one result:

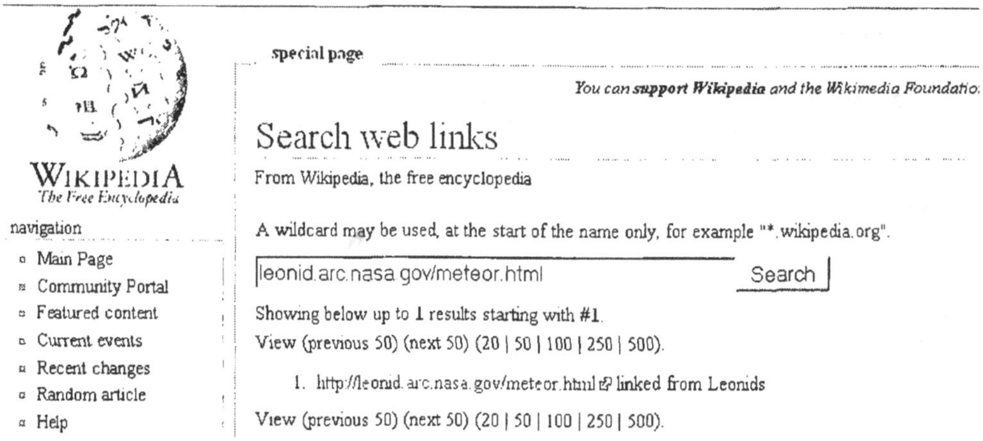
WIKIPEDIA
The Free Encyclopedia
navigation
Main Page
Community Portal
Featured content
Current events
Recent changes
Random article
Help
special page
You can support Wikipedia and the Wikimedia Foundatio
Search web links
From Wikipedia, the free encyclopedia
A wildcard may be used, at the start of the name only, for example "*.wikipedia.org".
leonid.arc.nasa.gov/meteor.html
Search
Showing below up to 1 results starting with #1.
View (previous 50) (next 50) (20 | 50 | 100 | 250 | 500).
1. http://leonid.arc.nasa.gov/meteor.html linked from Leonids
View (previous 50) (next 50) (20 | 50 | 100 | 250 | 500).

This is a very useful tool if you need to find out what pages in Wikipedia link to a specific site. Be sure to follow these basic rules for using this feature:

1. a full domain name, e.g., [www.nasa.gov] (this will only find links to this specific domain) OR

2. a partial domain name with a wildcard, e.g., [*.nasa.gov] (this will find links to any site at nasa.gov, such as ase.arc.nasa.gov) OR

3. a full domain name plus directory and/or webpage, e.g., [leonid.arc.nasa.gov/meteor.html] (this will find links only to this specific webpage)

Some Wikipedias other than the English language version have a similar page. For example, the German language Wikipedia link search page is: <http://de.wikipedia.org/wiki/Spezial:Linksearch>.

If you use the English Wikipedia link below and substitute the appropriate language digraph for the "en," you can find these non-English language link search pages. See this page <http://meta.wikimedia.org/wiki/List_of_Wikipedias> for all the Wikipedias and the appropriate digraph.

Search English Wikipedia Web Links
http://en.wikipedia.org/w/index.php?title=Special%3ALinksearch

Wiki Search Engines

FUTEF (Beta) http://futef.com/

FUTEF, which uses is own proprietary search engine, provides both a list of relevant articles but also a list of related categories that can be used to further refine a search. ***FUTEF handles non-Latin searches, something not every Wikipedia search engine can do.*** Try a search on Σμύρνη and you will see that FUTEF finds this term in the English language Wikipedia.

Qwika http://www.qwika.com/

Qwika indexes English, German, French, Japanese, Italian, Dutch, Portuguese, Spanish, Greek, Korean, Chinese and Russian wikis; the original content is combined with machine translated content to/from English. However, when searching for a non-Latin term Qwika will only find that term in the international Wikipedia not the English language Wikipedia even if it is, e.g., Σμύρνη.

LuMriX http://wiki.lumrix.net/

LuMirX uses AJAX technology and searches English, German, Japanese, French, Polish, Italian, Swedish, Dutch, Portuguese, Russian, Danish, Spanish, Finnish, Norwegian, Hungarian, Turkish, and Chinese Wikipedias. However, when searching for a non-Latin term LuMriX will only find that term in the international Wikipedia not the English language Wikipedia even if it is there, e.g., Çeşme.

Clusty's Wikipedia Search (English only) http://wiki.clusty.com/

One of the best Wikipedia search engines, Clusty not only searches the Wikipedia, it clusters the results into easy to understand categories that make it possible to zero in on the appropriate subtopic. Its main drawback is that the search is limited to the English-language Wikipedia.

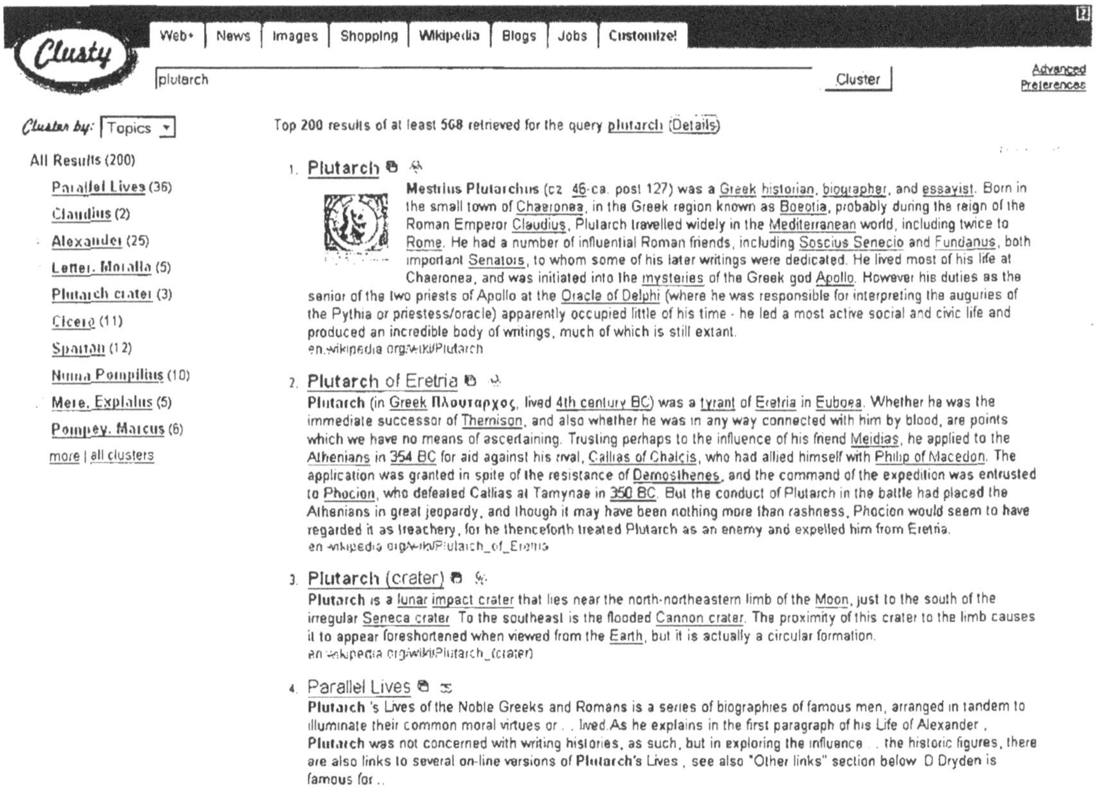
Clusty
Web+ News Images Shopping Wikipedia Blogs Jobs Customize!
plutarch
Cluster
Advanced
Preferences
Cluster by: Topics
Top 200 results of at least 568 retrieved for the query plutarch (Details)
All Results (200)
Parallel Lives (36)
Claudius (2)
Alexander (25)
Letter, Moralia (5)
Plutarch crater (3)
Cicero (11)
Spartan (12)
Numa Pompilius (10)
Mere, Explains (5)
Pompey, Marcus (6)
more | all clusters
1. Plutarch
Mestrius Plutarchus (ca. 46-ca. post 127) was a Greek historian, biographer, and essayist. Born in the small town of Chaeronea, in the Greek region known as Boeotia, probably during the reign of the Roman Emperor Claudius, Plutarch travelled widely in the Mediterranean world, including twice to Rome. He had a number of influential Roman friends, including Soscius Senecio and Fundanus, both important Senators, to whom some of his later writings were dedicated. He lived most of his life at Chaeronea, and was initiated into the mysteries of the Greek god Apollo. However his duties as the senior of the two priests of Apollo at the Oracle of Delphi (where he was responsible for interpreting the auguries of the Pythia or priestess/oracle) apparently occupied little of his time - he led a most active social and civic life and produced an incredible body of writings, much of which is still extant.
en.wikipedia.org/wiki/Plutarch
2. Plutarch of Eretria
Plutarch (in Greek Πλουταρχος, lived 4th century BC) was a tyrant of Eretria in Euboea. Whether he was the immediate successor of Themison, and also whether he was in any way connected with him by blood, are points which we have no means of ascertaining. Trusting perhaps to the influence of his friend Meidias, he applied to the Athenians in 354 BC for aid against his rival, Callias of Chalcis, who had allied himself with Philip of Macedon. The application was granted in spite of the resistance of Demosthenes, and the command of the expedition was entrusted to Phocion, who defeated Callias at Tamynae in 350 BC. But the conduct of Plutarch in the battle had placed the Athenians in great jeopardy, and though it may have been nothing more than rashness, Phocion would seem to have regarded it as treachery, for he thenceforth treated Plutarch as an enemy and expelled him from Eretria.
en.wikipedia.org/wiki/Plutarch_of_Eretria
3. Plutarch (crater)
Plutarch is a lunar impact crater that lies near the north-northeastern limb of the Moon, just to the south of the irregular Seneca crater. To the southeast is the flooded Cannon crater. The proximity of this crater to the limb causes it to appear foreshortened when viewed from the Earth, but it is actually a circular formation.
en.wikipedia.org/wiki/Plutarch_(crater)
4. Parallel Lives
Plutarch 's Lives of the Noble Greeks and Romans is a series of biographies of famous men, arranged in tandem to illuminate their common moral virtues or . . lived.As he explains in the first paragraph of his Life of Alexander , Plutarch was not concerned with writing histories, as such, but in exploring the influence . the historic figures, there are also links to several on-line versions of Plutarch's Lives , see also "Other links" section below D Dryden is famous for ..

Wikiseek http://wikiseek.com/

Launched in early 2007, Wikiseek was created with the assistance of Wikipedia, although it is not a part of Wikipedia. "The contents of Wikiseek are restricted to Wikipedia pages and only those sites which are referenced within Wikipedia, making it an authoritative source of information less subject to spam and SEO schemes. Wikiseek utilizes Searchme's category refinement technology, providing suggested search refinements based on user tagging and categorization within Wikipedia, making results more relevant than conventional search engines." <http://www.wikiseek.com/> Wikiseek uses AJAX technology to create changing "tag clouds" of possible terms as you type.

This is a good way to find articles within English language Wikipedia and to search sites referenced in Wikipedia, but it is by no means a substitute for a general search engine. Results from Wikipedia are identified by the W icon. The tag cloud that appears at the top of each successful search is designed to show related categories to help users either narrow or broaden a search. Keep in mind these are user-generated tags, so many of them, e.g., "Japanese terms," do not correspond to Wikipedia categories.

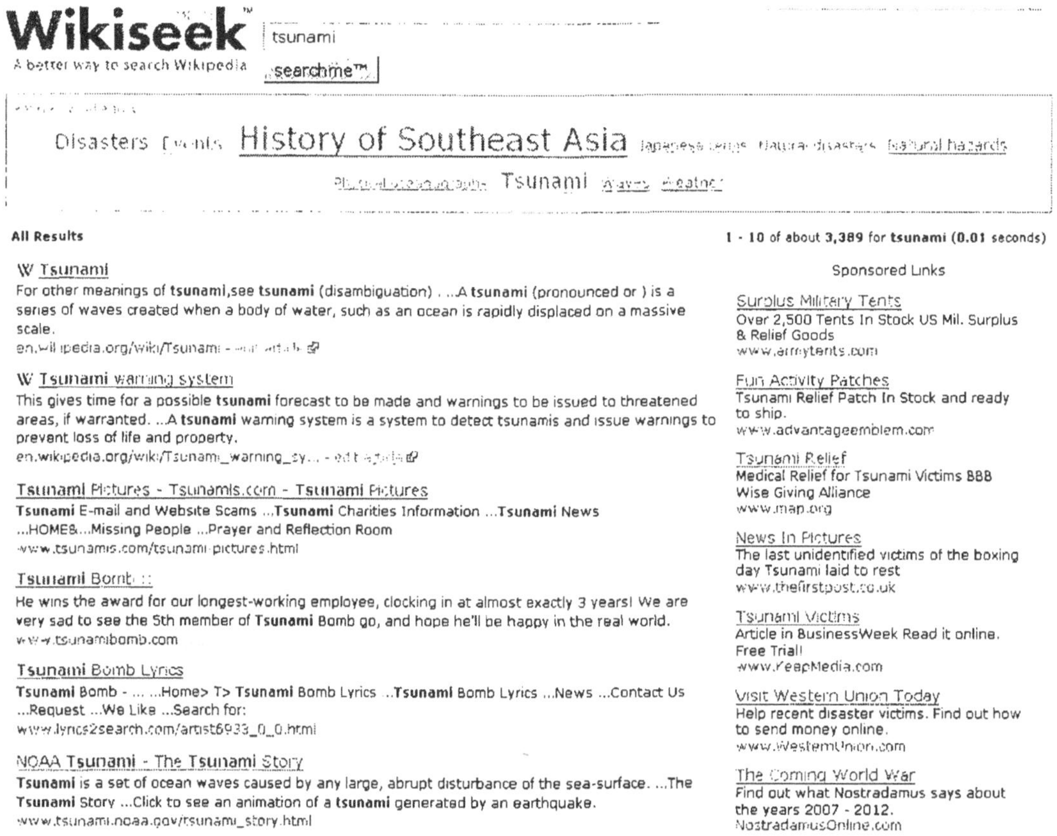
Wikiseek
tsunami
searchme
Disasters
History of Southeast Asia
Tsunami
All Results
1 - 10 of about 3,389 for tsunami (0.01 seconds)
Tsunami
For other meanings of tsunami,see tsunami (disambiguation) . ...A tsunami (pronounced or) is a series of waves created when a body of water, such as an ocean is rapidly displaced on a massive scale.
Tsunami warning system
This gives time for a possible tsunami forecast to be made and warnings to be issued to threatened areas, if warranted. ...A tsunami warning system is a system to detect tsunamis and issue warnings to prevent loss of life and property.
Tsunami Pictures - Tsunamis.com - Tsunami Pictures
Tsunami E-mail and Website Scams ...Tsunami Charities Information ...Tsunami News ...HOME&...Missing People ...Prayer and Reflection Room
Tsunami Bomb
He wins the award for our longest-working employee, clocking in at almost exactly 3 years! We are very sad to see the 5th member of Tsunami Bomb go, and hope he'll be happy in the real world.
Tsunami Bomb Lyrics
NOAA Tsunami - The Tsunami Story
Tsunami is a set of ocean waves caused by any large, abrupt disturbance of the sea-surface. ...The Tsunami Story ...Click to see an animation of a tsunami generated by an earthquake.
Sponsored Links
Over 2,500 Tents In Stock US Mil. Surplus & Relief Goods
Tsunami Relief Patch In Stock and ready to ship.
Tsunami Relief
Medical Relief for Tsunami Victims BBB Wise Giving Alliance
News In Pictures
The last unidentified victims of the boxing day Tsunami laid to rest
Tsunami Victims
Article in BusinessWeek Read it online. Free Trial!
Visit Western Union Today
Help recent disaster victims. Find out how to send money online.
The Coming World War
Find out what Nostradamus says about the years 2007 - 2012.

The drawbacks to Wikiseek are that it only searches the English language version of Wikipedia and it ***cannot parse non-Latin languages***. It touts itself as an "authoritative source of information less subject to spam and SEO schemes," but a search for [viagra] will quickly prove it is no better (in fact worse) than the major search engines in filtering spam. There are no preferences to change the number of results, for example, or to limit the search only to Wikipedia or only to links, but since Wikiseek is still in Beta, these features may appear later. Wikiseek also offers a Firefox plug-in to add Wikiseek to the Wikipedia search form on all Wikipedia pages.

WikiWax http://www.wikiwax.com/

WikiWax also uses "Look Ahead" AJAX technology to show very extensive lists of dynamically generated related terms. However, WikiWax cannot parse non-Latin search terms, e.g., Σμύρνη.

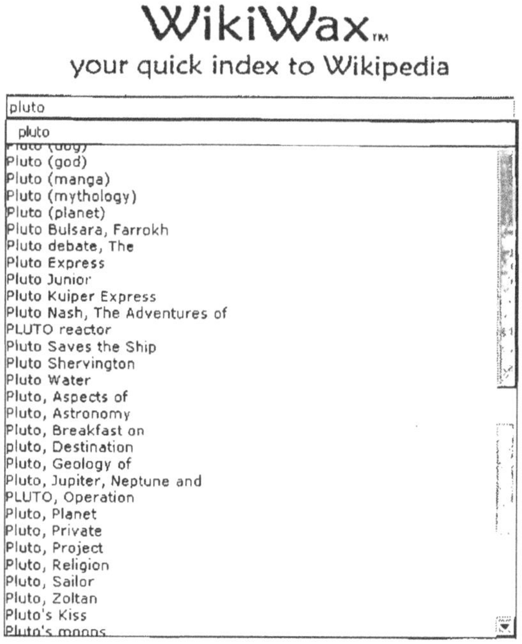
WikiWax™
your quick index to Wikipedia
pluto
pluto
Pluto (god)
Pluto (manga)
Pluto (mythology)
Pluto (planet)
Pluto Bulsara, Farrokh
Pluto debate, The
Pluto Express
Pluto Junior
Pluto Kuiper Express
Pluto Nash, The Adventures of
PLUTO reactor
Pluto Saves the Ship
Pluto Shervington
Pluto Water
Pluto, Aspects of
Pluto, Astronomy
Pluto, Breakfast on
pluto, Destination
Pluto, Geology of
Pluto, Jupiter, Neptune and
PLUTO, Operation
Pluto, Planet
Pluto, Private
Pluto, Project
Pluto, Religion
Pluto, Sailor
Pluto, Zoltan
Pluto's Kiss

Using Search Engines to Search Wikipedia

Yahoo now includes Quick Links for any Wikipedia results. For example, a Yahoo search for [internet] will return Wikipedia as result number eight and will include "Quick Links" to specific Wikipedia articles on this topic:

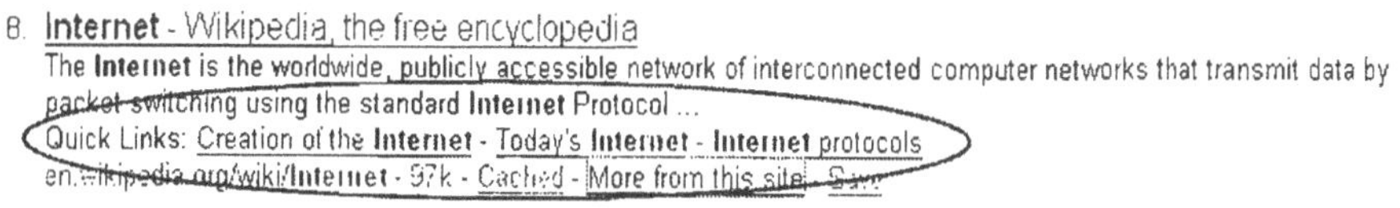
8. Internet - Wikipedia, the free encyclopedia
The Internet is the worldwide, publicly accessible network of interconnected computer networks that transmit data by packet switching using the standard Internet Protocol ...
Quick Links: Creation of the Internet - Today's Internet - Internet protocols
en.wikipedia.org/wiki/Internet - 97k - Cached - More from this site

You can further restrict the search to Wikipedia by clicking on "More from this site," which is an excellent way to search Wikipedia using Yahoo:

YAHOO! SEARCH
Web | Images | Video | Audio | Directory | Local | News | Shopping | More »
internet
Search
Answers My Web Search Services | Advanced Search Preferences
Search Results
1 - 10 of about 232,000 for internet - 0.05 sec. (About this page)
Also try: internet explorer, free internet, internet radio, internet tv More...
Your search was restricted to "en.wikipedia.org". For more matches, try searching all sources instead.
1. Internet - Wikipedia, the free encyclopedia
The Internet (also known simply as the Net) can be briefly understood as "a network of networks". Specifically, it is the worldwide, publicly accessible network of ...
Quick Links: Creation of the Internet - Today's Internet - Internet protocols
en.wikipedia.org/wiki/Internet - 97k - Cached - Save
2. Internet Explorer - Wikipedia
Encylopedia article about Internet Explorer, Microsoft's web browser. Covers its history, features, component architecture, common criticisms, and market share.
en.wikipedia.org/wiki/Internet_Explorer - 97k - Cached - Save
3. Internet troll - Wikipedia, the free encyclopedia
In Internet terminology, a troll is someone who comes into an established community such as an online discussion forum, and posts inflammatory, rude, repetitive or ...
Quick Links: Etymology - Vicious circles - Troll culture
en.wikipedia.org/wiki/Internet_troll - 87k - Cached - Save
4. History of the Internet - Wikipedia, the free encyclopedia
The history of the Internet dates back to the early development of communication networks. The idea of a computer network intended to allow general communication among ...
Quick Links: Before the Internet - A lack of inter-networking - Three terminals and an ARPA
en.wikipedia.org/wiki/History_of_the_Internet - Cached - Save

You can also use the **site:** syntax to search just the Wikipedia (or Wikipedias, if you like) in:

- Yahoo http://search.yahoo.com/
- Google http://www.google.com/
- Ask http://www.ask.com/
- Windows Live Search http://www.live.com/?searchonly=true
- A9 http://a9.com/
- Gigablast http://www.gigablast.com/
- Exalead http://www.exalead.com/search
- Clusty (site: and host: are interchangeable, but Clusty has a special Wikipedia search option) http://clusty.com/

This is an especially useful option for non-Latin searches, such as [site:wikipedia.org Çeşme], which returns results not only from the English and Turkish Wikipedias but from the German and Serbian Wikipedias as well:

Web | Images | Video | Audio | Directory | Local | News | Shopping | More »

YAHOO! SEARCH site:wikipedia.org Çeşme Search

Answers Search Services | Advanced Se

Search Results

1 - 100 of about 311 from wikipedia.org for Çeşme - 1.63 sec

1. Çeşme - Wikipedia, the free encyclopedia
Çeşme is a town on the west coast of Turkey and one of the districts of İzmir Province. It is a prominent center of international tourism in Turkey and is famous for ...
Quick Links: See also - External links
en.wikipedia.org/wiki/Çeşme - 13k - Cached - More from this site

2. Ilıca, Çeşme - Wikipedia, the free encyclopedia
Ilica is a small village near Çeşme (pronounced Tcheshme), which is a district of İzmir Province in Çeşme Peninsula in the extreme western tip of Turkey.
Quick Links: Aegean region of Turkey geography stubs - Districts of İzmir - İzmir
en.wikipedia.org/wiki/Il%C4%B1ca,_%C3%87e%C5%9Fme - 13k - Cached - More from this site

3. Çeşme - Wikipedia - Translate this page
Çeşme [] ist ein Ferienort etwa 100 Kilometer westlich von Izmir. Der Name "Çeşme", zu Deutsch "Brunnen", leitet sich von der großen Zahl dieser ab.
Quick Links: Ort in der Türkei
de.wikipedia.org/wiki/%C3%87e%C5%9Fme - 12k - Cached - More from this site

4. Cezayirli Gazi Hasan Pasha - Wikipedia, the free encyclopedia
Cezayirli Gazi Hasan Pasha (1713-1790), (Hasan Pasha of Algiers) was an Ottoman grand vizier and a navy and army commander of the late 18th century.
Quick Links: References
en.wikipedia.org/wiki/Cezayirli_Gazi_Hasan_Pasha - 16k - Cached - More from this site

5. Çeşme - Vikipedi
... anlam ayrım sayfası, Çeşme kavramının farklı kullanımlarını ... Retrieved from "http://tr.wikipedia.org/wiki/%C3%87e%C5%9Fme" Sayfa kategorisi: Anlam ayrım ...
tr.wikipedia.org/wiki/Çeşme - 10k - Cached - More from this site

6. Mehmet Culum - Wikipedia, the free encyclopedia
Mehmet Culum is a contemporary Turkish novelist who was born in a western town of Turkey called Çeşme in 1948. He studied political sciences at the University of Ankara.
Quick Links: European writer stubs - Turkish people stubs
en.wikipedia.org/wiki/Mehmet_Culum - 13k - Cached - More from this site

7. Marinas in Turkey - Wikipedia, the free encyclopedia
www.northbodrum

Search Tip:

To search all Wikipedias:

[site:wikipedia.org]

To search language-specific Wikipedias:

site:DIGRAPH.wikipedia.org, e.g.,
[site:**de**.wikipedia.org nordafrika]

Maps and Mapping

Maps and mapping technology continued to expand and improve during 2006, in large part because of competition with Google Maps. Virtually all the online maps get their data from one of two companies: Navteq or Tele Atlas. It is what each map site does with that data—how it implements its user interface and what features it offers—that makes it distinctive. Google Maps and Google's downloadable property Google Earth, helped inspire a cottage industry of what are termed "mashups," a music industry term transformed into computer slang meaning to "mix at least two different services from disparate, and even competing, websites." The best known mashups involve overlaying data such as crime statistics onto Google, Yahoo, Live Search's Virtual Earth, and other maps.

Some of these innovations are detailed in *Untangling the Web*; others, while no doubt useful in their own right, do not address specific research needs of this audience and are best left for other venues.

Google Maps & Google Earth

During 2005, Google and Microsoft both introduced a new dimension to their map sites with satellite maps. Google uses the Keyhole technology it purchased in 2004 and Microsoft uses its own TerraServer data to generate its satellite maps at its Virtual Earth page. Google went one step further, combining its satellite imagery with its traditional maps to offer a "hybrid" view, which overlays a map onto the satellite image of the U.S., Canada, most of Europe, and Japan. Google is facing stiff competition not only from its US competitors but also from exceptionally fine European mapping companies. See below for a comparison between Google and Mappy in the European map market.

Google Maps allows users to toggle among map, satellite, or hybrid (labeled) views. Google Maps uses the now ubiquitous address bubbles that can be closed and the ability to get directions between locations. Google Maps is one of the best sites for navigating around the map smoothly using with the mouse. In mid-2006 Google Maps added the zoom in/out feature using the mouse scroll wheel.

Google Maps also has a preview window, a small window inside the larger map window that shows a small image of the larger map. The intent is to let users see a larger view of an area

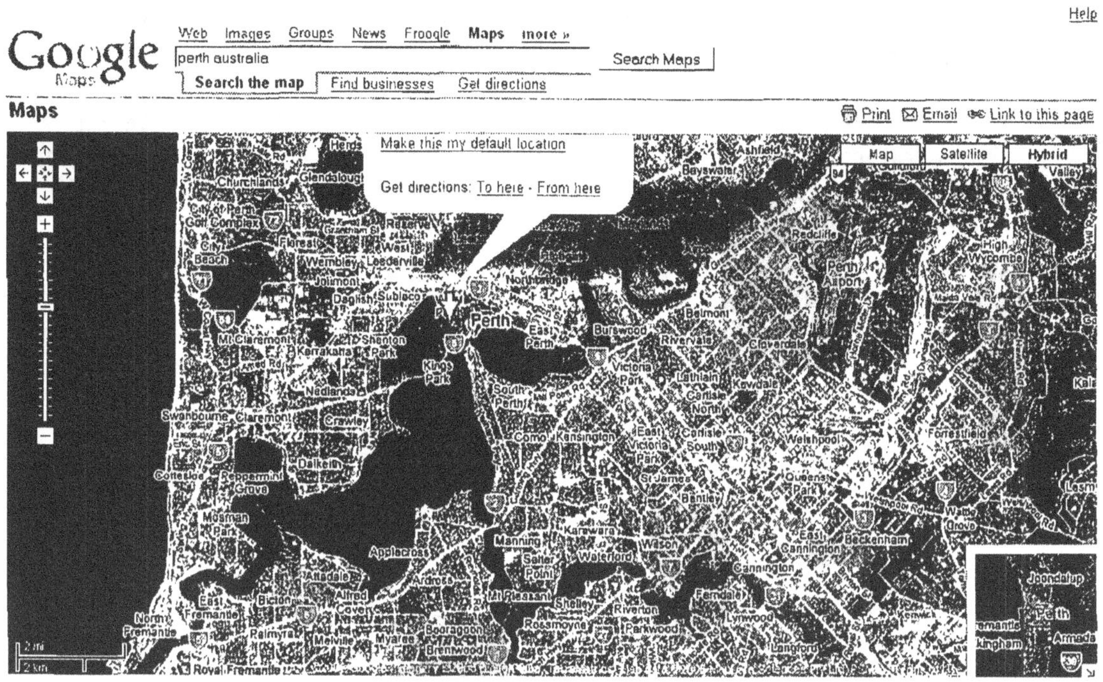
Help
Google Maps
Web Images Groups News Froogle Maps more »
perth australia
Search Maps
Search the map
Find businesses
Get directions
Maps
Print
Email
Link to this page
Make this my default location
Get directions: To here · From here
Map
Satellite
Hybrid
Perth

Google Maps Hybrid View with Preview Window http://maps.google.com/

During 2006 Google also "updated the satellite data used for Google Maps so it now has the many updates recently made to the Google Earth database. This means the new high resolution data for all of Germany, many places in Europe...and many other places are now available to Google Maps users."[83]

Google Earth represents a revolution in mapping technology. Google Earth is a geographic search and "fly" tool that combines the detailed three-dimensional satellite imagery from Google's Keyhole property with Google's local search and direction finding available at Google maps. Google Earth lets users:

- Fly from space to your neighborhood. Type in an address and zoom right in.
- Search for schools, parks, restaurants, and hotels. Get driving directions.
- Tilt and rotate the view to see 3D terrain and buildings.

---

[83] Frank Taylor, "Google Maps Gets Huge Satellite Update from Google Earth," Google Earth Blog, 24 April 2006, <http://www.gearthblog.com/blog/archives/2006/04/google_maps_and.html> (30 October 2006).

- Save and share your searches and favorites. Even add your own annotations.[84]

The Google Earth data is not limited to the US. "The whole world is covered with medium resolution imagery and terrain data. This resolution allows you to see major geographic features and man-made development such as towns, but not detail of individual buildings. Additional high-resolution imagery which reveals detail for individual buildings is available for most of the major cities in the US, Western Europe, Canada, and the UK. 3D buildings are represented in 38 US cities (the major urban areas). Detailed road maps are available for the US, Canada, the UK, and Western Europe. And Google Local search is available for the US, Canada, and the UK."[85] And all this is free for personal use and does not require registration.

All the news about Google Earth is not good, however.

- it's a download, which means many organizational policies will prohibit its use on the job.
- it will not run on Apples.
- it is only designed to work on Windows XP or 2000 (other MS operating systems are not supported).
- it's designed to run on broadband Internet connections, so I do not recommend running it on a slow connection.
- it requires a 3D graphics card: 3D-capable video card with 16MB VRAM, but many common graphics cards are not supported.
- while the basic software is free, there are two upgrades, a yearly subscription for Google Earth Plus with higher resolutions, GPS support, etc., and a more expensive business and professional version called Google Earth Pro (a 7-day free trial is available if you're curious).

However, if you are able to use Google Earth, it is amazing. I recommend using the extensive Google Earth help files available at Keyhole. One of the hottest trends on the web at the moment is map hacks or mash-ups. The Google Earth Hacks website brings you lots of free downloads designed to be used in conjunction with Google Earth.

Google Earth http://earth.google.com/

Google Earth Help http://www.keyhole.com/GoogleEarthHelp/GoogleEarth.htm

---

[84] Google Earth, <http://earth.google.com/> (30 October 2006).

[85] Google Earth FAQ, <http://earth.google.com/faq.html> (30 October 2006).

Google Earth Hacks http://www.googleearthhacks.com/

Google now has street-level mapping capabilities for Andorra, Australia, Austria, Belgium, Canada, Czech Republic, Denmark, Finland, France, Germany, Gibraltar, Greece, Hungary, Ireland, Italy, Japan, Liechtenstein, Luxembourg, Monaco, Netherlands, New Zealand, Norway, Poland, Portugal, Russia (Moscow only), San Marino, Slovakia, Spain, Sweden, Switzerland, Turkey (Istanbul only), the United Kingdom, and the United States. Although no separate websites exist for most of these country map sites, I expect they will eventually.

At present, Google has map websites for these countries:

- UK & Ireland — http://maps.google.co.uk/
- Canada — http://maps.google.ca/
- Australia (includes New Zealand) — http://maps.google.com.au/
- China — http://bendi.google.com/
- Germany — http://maps.google.de/
- Spain — http://maps.google.es/
- France — http://maps.google.fr/
- Italy — http://maps.google.it/
- The Netherlands — http://maps.google.nl/
- Japan — http://local.google.co.jp/

Google began offering local search and maps specifically for Japan during 2005. Both sites are in Japanese. However, Google maps Japan, which offers both street-level and satellite images for some of Japan, recognizes input in *romaji* (Latinized spellings of Japanese words). Here is a search on [Tokyo] showing the results in Google maps satellite images.

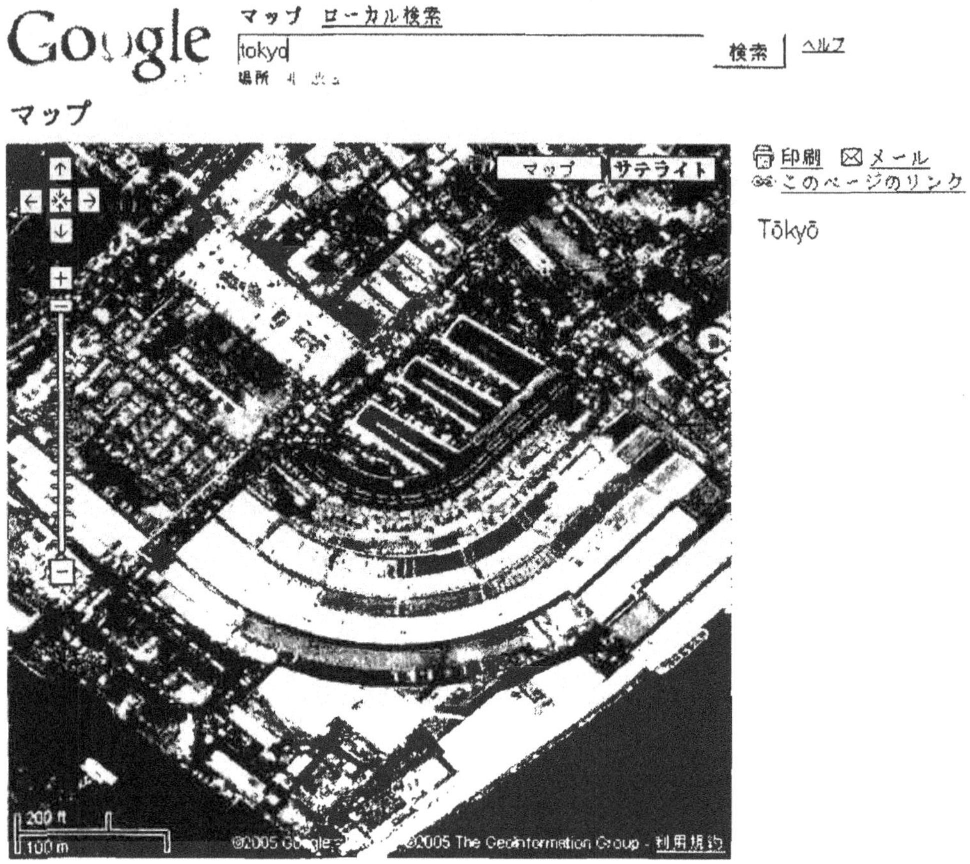
Google
マップ ローカル検索
tokyo
検索
ヘルプ
マップ
マップ
サテライト
印刷
メール
このページのリンク
Tōkyō
200 ft
100 m
©2005 Google
©2005 The GeoInformation Group - 利用規約

Google Maps Japan http://maps.google.co.jp/

Google Maps and Google Earth, also helped inspire a cottage industry of what are termed "mashups," a music industry term transformed into computer slang meaning to "mix at least two different services from disparate, and even competing, websites." The best known mashups involve overlaying data such as subway routes, and crime statistics onto Google, Yahoo, Live Local, or other maps. This is a Google Maps Mashup showing cellular towers in Boston using the Hybrid view, i.e., street and satellite maps combined.

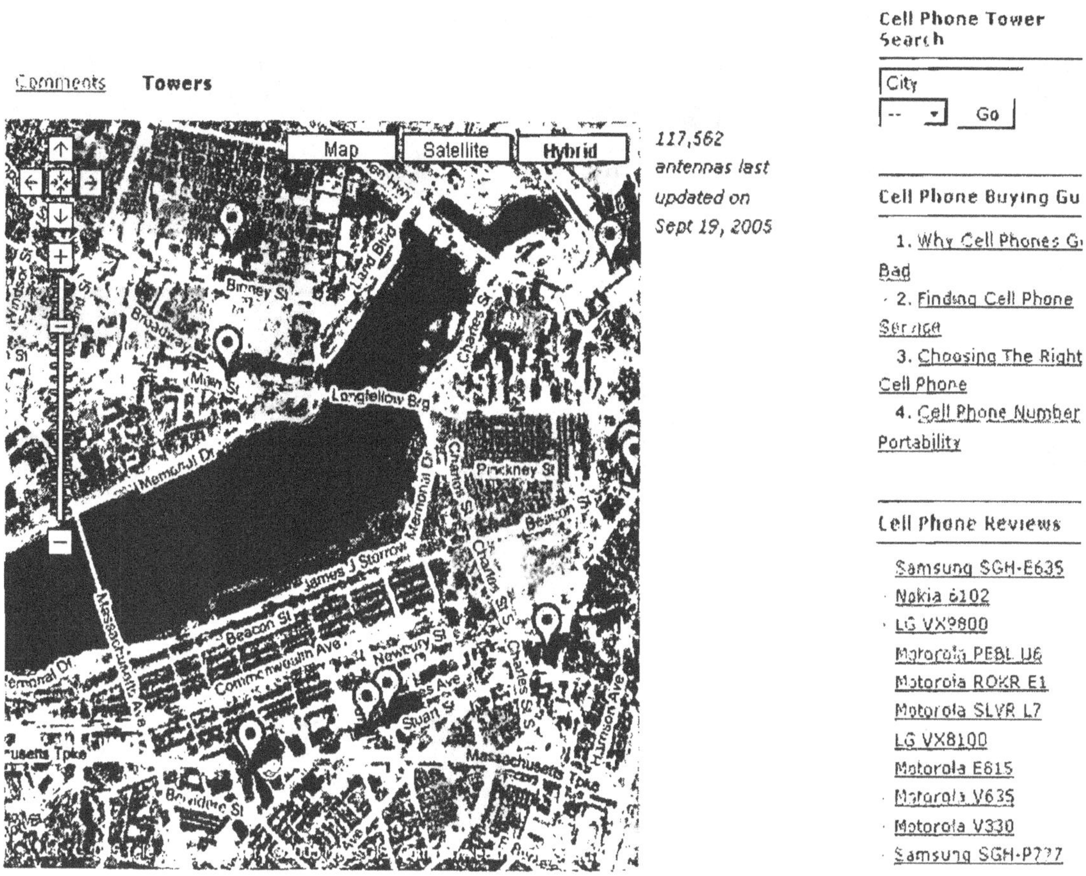
Comments
Towers
Map
Satellite
Hybrid
Binney St
Broadway
Main St
Longfellow Brg
Memorial Dr
Pinckney St
Charles St
Beacon St
James J Storrow
Commonwealth Ave
Newbury St
Massachusetts Ave
Massachusetts Tpke
117,562 antennas last updated on Sept 19, 2005
Cell Phone Tower Search
City
Go
Cell Phone Buying Gu
1. Why Cell Phones G Bad
2. Finding Cell Phone Service
3. Choosing The Right Cell Phone
4. Cell Phone Number Portability
Cell Phone Reviews
Samsung SGH-E635
Nokia 6102
LG VX9800
Motorola PEBL U6
Motorola ROKR E1
Motorola SLVR L7
LG VX8100
Motorola E815
Motorola V635
Motorola V330
Samsung SGH-P777

Mobiledia.com http://www.cellreception.com/

Microsoft's Live Local Powered by Virtual Earth

Microsoft fought back with some amazing technology of its own called Virtual Earth. With the appearance of Windows Live and Live Local, Virtual Earth became the power behind these searches and the former Virtual Earth site now takes users to Live Local.

Compare Google Maps' highest resolution hybrid view of the Natural History Museum in Washington, D.C., with the highest resolution image of Windows Live Local aerial view, with labels. You can see that Google Maps (using satellite imagery from Google Earth) now has the edge in better resolution and also offers a larger, easier to view image. Also, I find Google Maps not only much faster to load but easier to search and navigate.

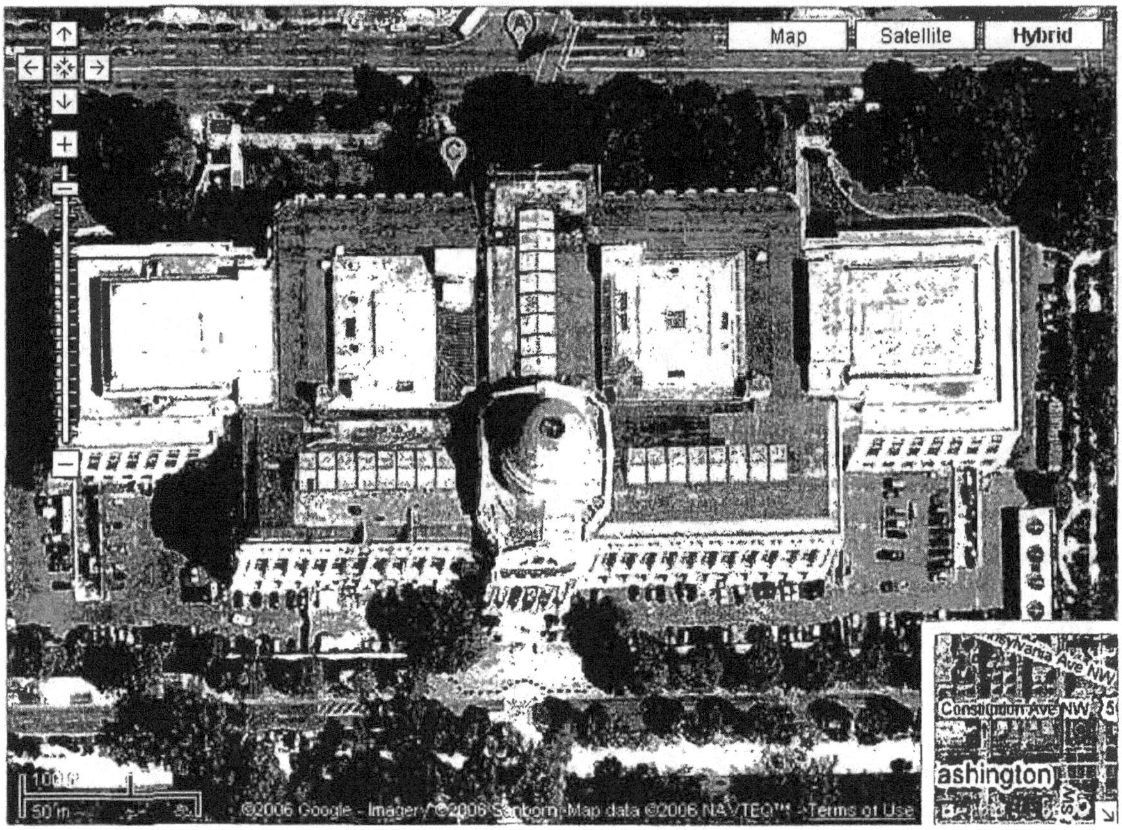
Map
Satellite
Hybrid
Constitution Ave NW
100 ft
50 m
©2006 Google - Imagery ©2006 Sanborn, Map data ©2006 NAVTEQ™ - Terms of Use

Google Maps Hybrid http://maps.google.com/

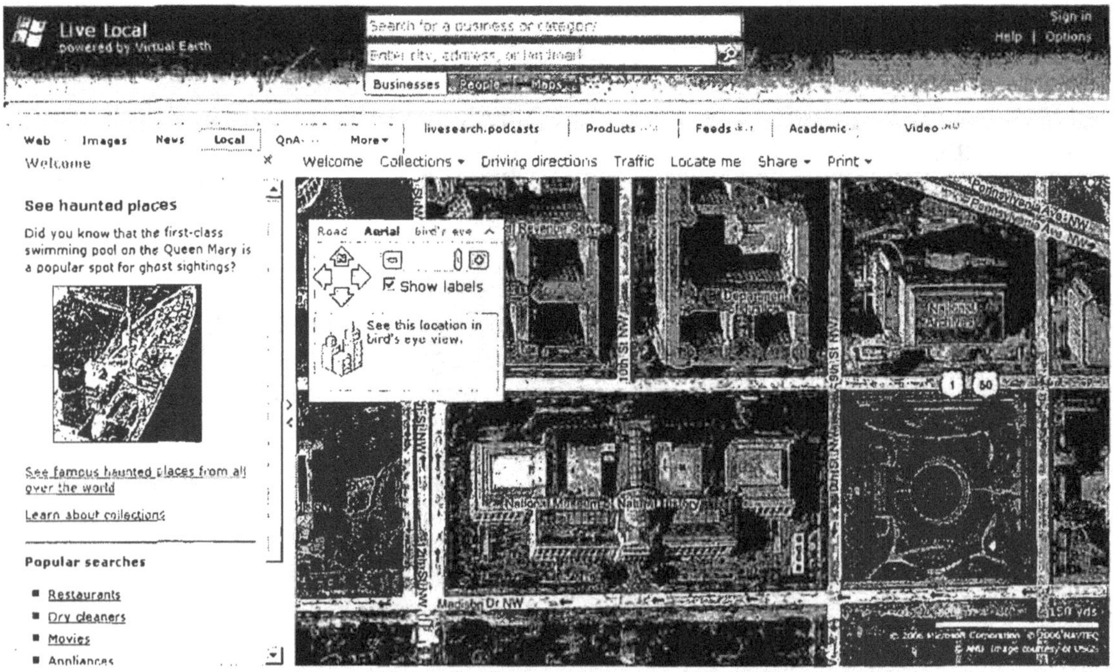
Live Local
powered by Virtual Earth
Search for a business or category
Enter city, address, or landmark
Businesses
People
Maps
Sign in
Help | Options
Web Images News Local QnA More
livesearch.podcasts
Products
Feeds
Academic
Video
Welcome
See haunted places
Did you know that the first-class swimming pool on the Queen Mary is a popular spot for ghost sightings?
See famous haunted places from all over the world
Learn about collections
Popular searches
Restaurants
Dry cleaners
Movies
Appliances
Welcome Collections Driving directions Traffic Locate me Share Print
Road Aerial Bird's eye
Show labels
See this location in bird's eye view.
10th St NW
Madison Dr NW
150 yds

Windows Live Local powered by Virtual Earth http://local.live.com/

However, Windows Live Local has one amazing feature that Google Maps does not. Windows Live Local offers "Bird's Eye View," which are images licensed from Pictometry International at a 45-degree angle instead of directly overhead. "To date, Pictometry information is available for geographic areas accounting for about 25 percent of the U.S. population, including the greater metropolitan areas of Manhattan, Seattle, Los Angeles and San Francisco. Microsoft will continue to work with Pictometry to shoot more landscapes, with a focus on highly populated areas and tourist destinations such as Las Vegas, which already shows up in bird's-eye view."[86] More locations are being added.

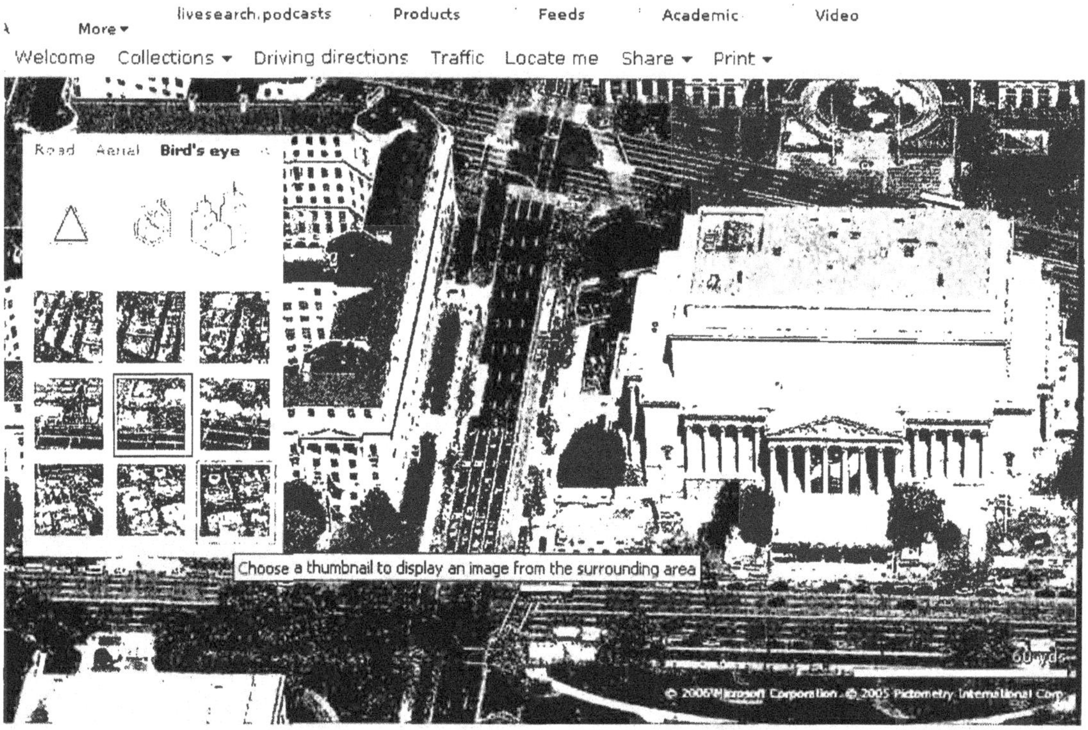
More
livesearch.podcasts
Products
Feeds
Academic
Video
Welcome
Collections
Driving directions
Traffic
Locate me
Share
Print
Road
Aerial
Bird's eye
Choose a thumbnail to display an image from the surrounding area

Clicking the rotation arrows (N S E W) in the navigation panel, displays a 360° panoramic view that is genuinely breathtaking. If you right-click anywhere on the map you can add your own pushpins. Live Local also offers a "Scratch Pad" to make notes about a location. For users who want to learn more about using Bird's Eye View and Live Local maps in general, there are several blogs devoted to Windows Live Local that I find far more useful than Microsoft's Live help.

Windows Live Spaces/Virtual Earth http://virtualearth.spaces.live.com/

---

[86] Susan Kuchinskas, "Windows Live Local Offers a New View," InternetNews.com, 8 December 2005, <http://www.internetnews.com/ent-news/article.php/3569386> (30 October 2006).

Virtual Earth Developers http://blogs.msdn.com/virtualearth/default.aspx

Bird's Eye Tourist http://www.birdseyetourist.com/

Early in 2006 the Virtual Earth Team introduced street views (something only A9 had previously tried to do in the US). Here's the announcement:

> The Virtual Earth team is pleased to launch a preview of a new feature we have been working on – interactive Street-side browsing. You can try it out at http://preview.local.live.com Street-side imagery allows you to drive around a city looking at the world around you as if you were in a car. But unlike the real world, you can stop your car anywhere you like and rotate your view around 360degrees. Currently we have street-side imagery for San Francisco and Seattle online.[87]

I love this technology because there is nothing like being able to "walk" or "drive" through an unfamiliar location to get your bearings. However, with the demise of A9 Maps, I am not sure this type of "you are there" technology is here to stay. The excellent European sites that offer similar options. France Telecom's Pages Jaunes and Spain's Callejero Fotographico offer similar "stroll" technologies.

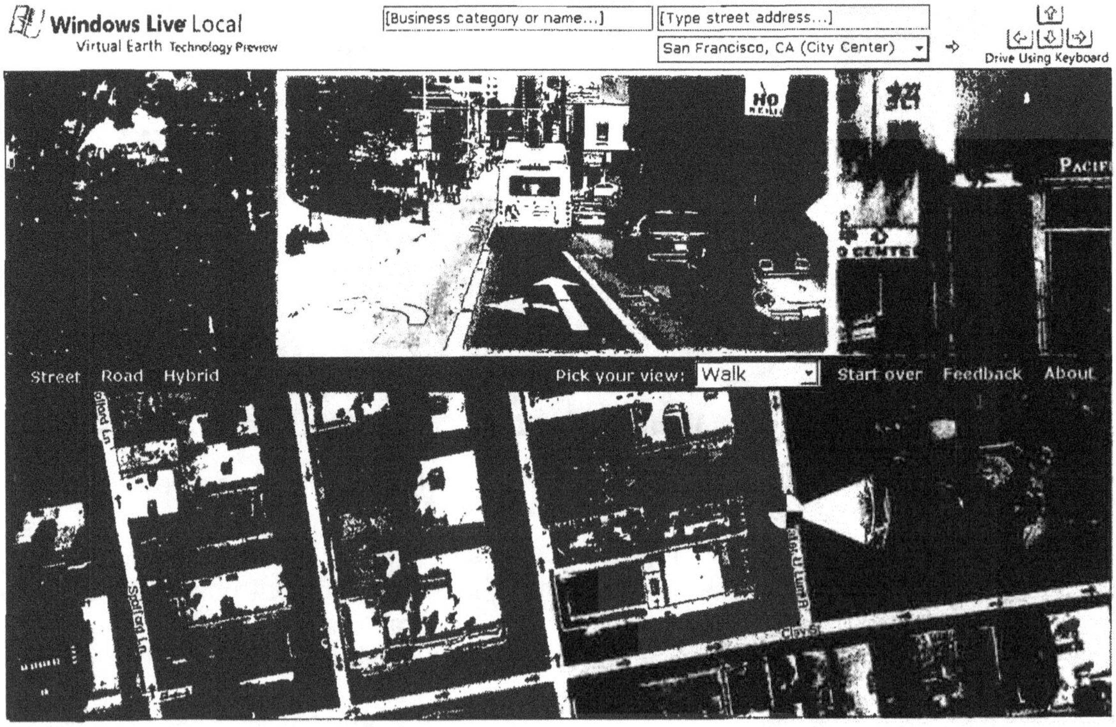
Windows Live Local
Virtual Earth Technology Preview
[Business category or name...]
[Type street address...]
San Francisco, CA (City Center)
Drive Using Keyboard
Street
Road
Hybrid
Pick your view:
Walk
Start over
Feedback
About

Virtual Earth Technology Preview http://preview.local.live.com/

---

[87] Sean Rowe, Program Manager, Virtual Earth, Live Search Weblog, 28 February 2006, <http://blogs.msdn.com/livesearch/archive/2006/02/28/540724.aspx > (31 October 2006).

At the end of 2006 Microsoft announced **Virtual Earth 3D**, which offers truly stunning views of many US cities. The downside of this technology is that it requires Internet Explorer and certain other Microsoft-specific software in order to run.

Microsoft Research, in partnership with the Government of India's Department of Science and Technology, is working on an interesting project powered by Microsoft Virtual Earth technology. According to the website, "The primary goal of this project is to explore novel and effective ways to collect and disseminate geospatial data and leverage multi-lingual technologies within maps. We currently have fairly limited data—1:1M India-wide data as well as 1:8000 Bangalore city data."[88]

Right now, the only street-level maps are for Bangalore; users cannot add pushpins (I tried unsuccessfully to view them but neither the annotation nor street search was working). However, the concept is a good one. Imagine having street-level and perhaps even aerial maps of an entire country in all that country's languages. Even now, you can view the maps in English, Hindi, Tamil, and Kannada, although Microsoft does caution there are inevitably errors in their rendering of the place names. There is no doubt this is an interesting experiment and I feel pretty confident MS didn't pick India randomly.

Here's a shot of a portion of Bangalore rendered in Hindi:

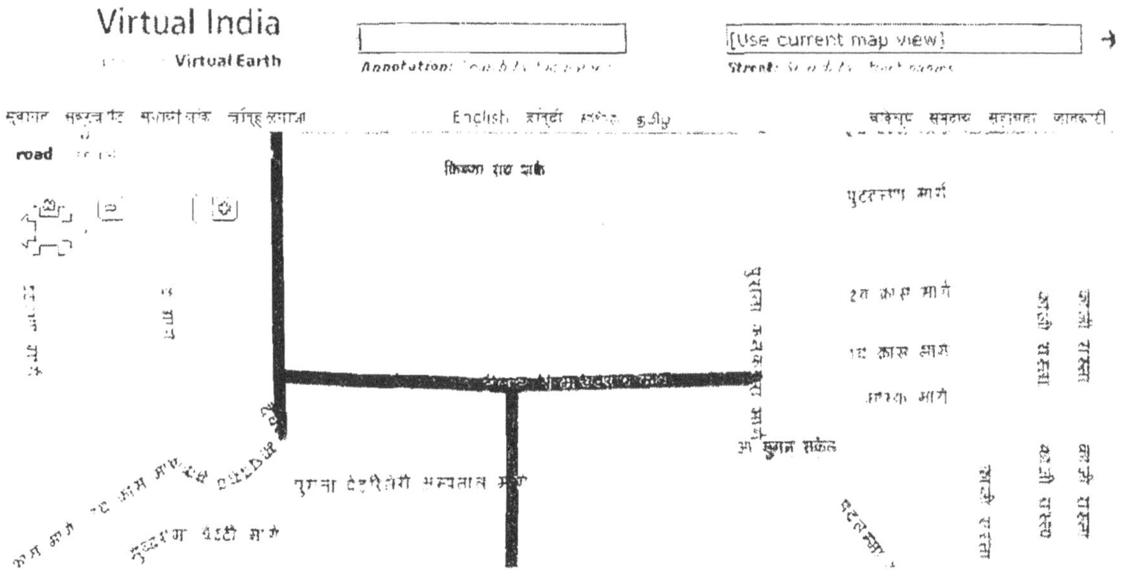
Virtual India
Virtual Earth
[Use current map view]
English
road

Microsoft Research Virtual India Project http://research.microsoft.com/virtualindia/

Virtual India Website http://virtualindia.msresearch.in/

[88] Welcome to the Virtual India Project! Microsoft Research-15, 2006, <http://research.microsoft.com/virtualindia/> (30 October 2006).

Yahoo Maps

In early 2006 Yahoo Maps introduced high-resolution satellite, aerial, and hybrid images of the US, medium resolution imagery of the rest of the world, and medium resolution global maps and overlays. Yahoo Maps imagery is provided by Aerials Express and icubed. Despite what you may read, Yahoo Maps is using ***both satellite imagery and aerial photography***, i.e., photographs taken from airplanes. Here's what Yahoo had to say about this development:

Here are the highlights:

> **Comprehensive Nationwide Satellite Imagery Coverage**
> Wall-to-wall coverage within the lower 48 states in the US. We are going for the best coverage nationwide, from the streets of New York to every inch of Redding, CA.
>
> **Global Satellite Imagery**
> The product features global images at 15 meters per pixel (zoom level 5, medium resolution), which basically lets you find and see every city, town, and major land feature in the world at medium resolution.
>
> **Global Maps**
> We're releasing maps and overlays at medium resolution for the whole world as well. This should help you view not only the suburbs of Bangkok, Thailand, but also help see the context of the imagery in hybrid mode.
>
> **APIs**
> The new imagery and global maps are available for API developers <http://developer.yahoo.com/maps/index.html> on the Yahoo! Developer Network. <http://developer.yahoo.com/> So whether you're new to the world of mashups or an experienced hacker, there is no better time to show off what you can do.
>
> **Better Views**
> In addition to getting all the data we can, we're processing the satellite imagery to make the visuals more aesthetically pleasing for users. We're blending away seam lines and normalizing the color pallet to create a continuous plane of imagery.[89]

---

[89] Michael Lawless and Vince Maniago, "Mo' Beta Maps," Yahoo! Search Blog, 11 April 2006, <http://www.ysearchblog.com/archives/000286.html> (30 October 2006).

Of course the first question is how do Yahoo's images stack up to Google's and Microsoft's? Here are three screenshots of downtown Washington, DC (Microsoft's Virtual Earth imagery slightly obfuscates the White House and surrounding areas, so I chose a clearer image of downtown DC).

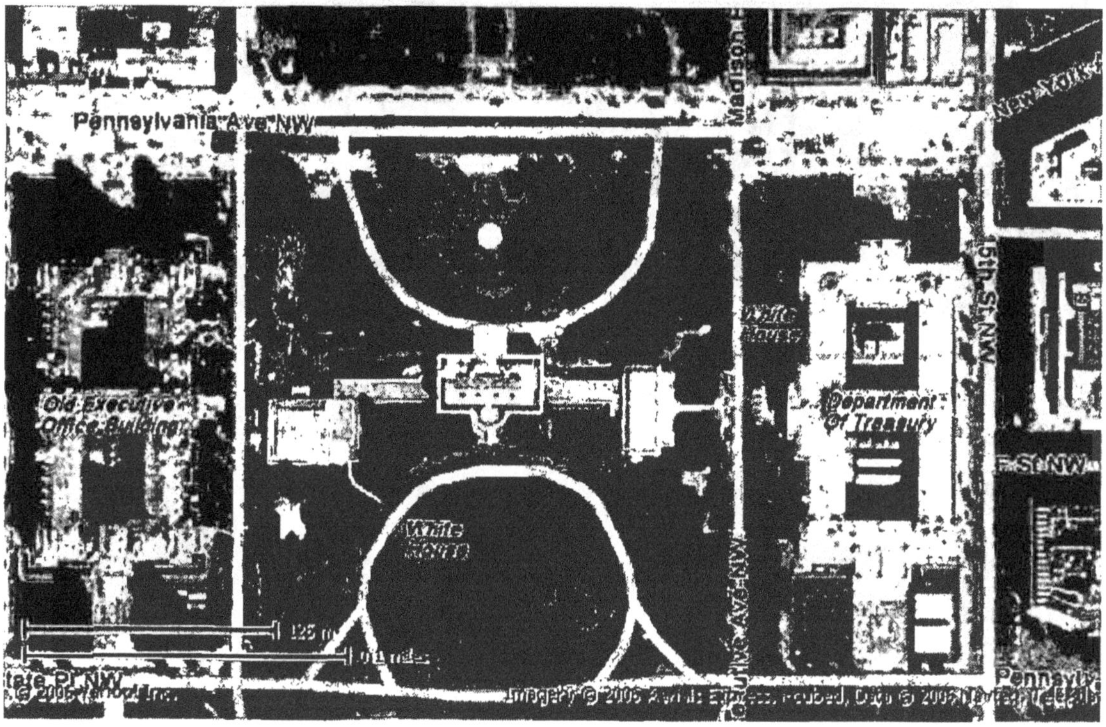
Pennsylvania Ave NW
Madison
15th St NW
White House
Old Executive Office Building
Department Of Treasury
F St NW
White House

Yahoo Maps' hybrid image of the White House and surrounding areas using Yahoo Maps Local (Beta). http://maps.yahoo.com/beta/

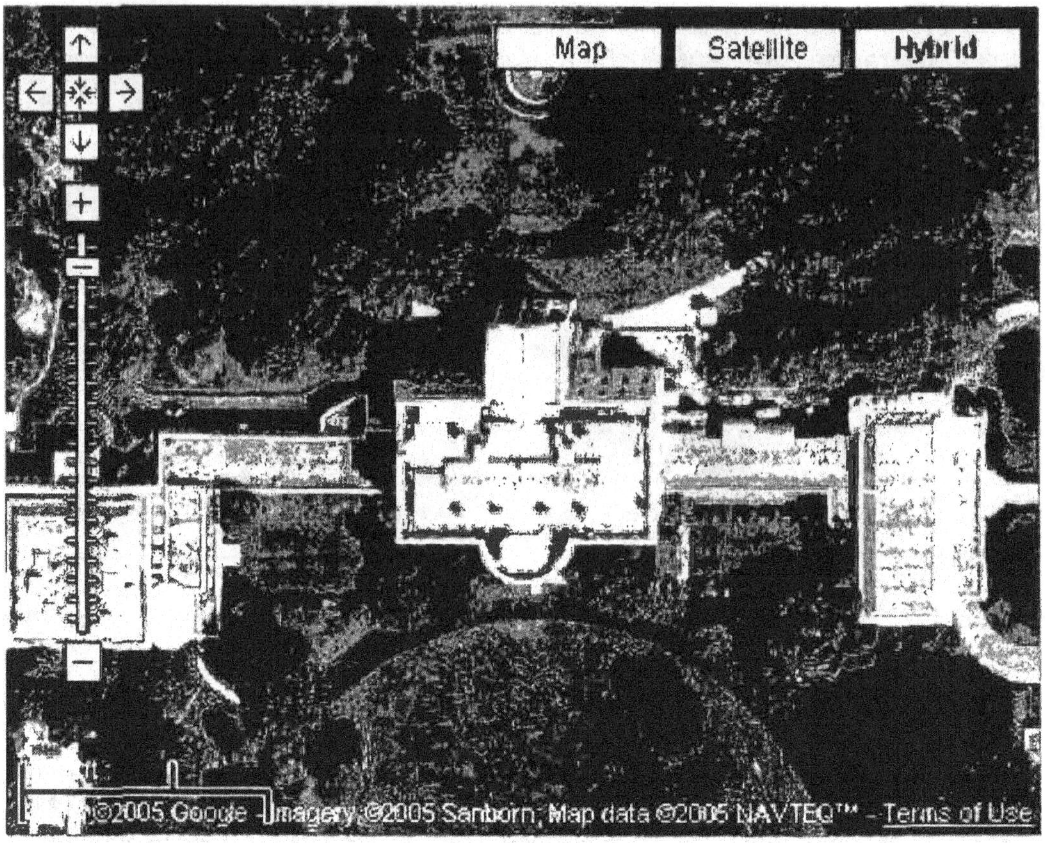
Map
Satellite
Hybrid
©2005 Google - Imagery ©2005 Sanborn, Map data ©2005 NAVTEQ™ - Terms of Use

Google Maps' hybrid image of the White House (at this resolution, the hybrid identifications disappear; there is actually one closer degree of resolution, but it is fuzzy).

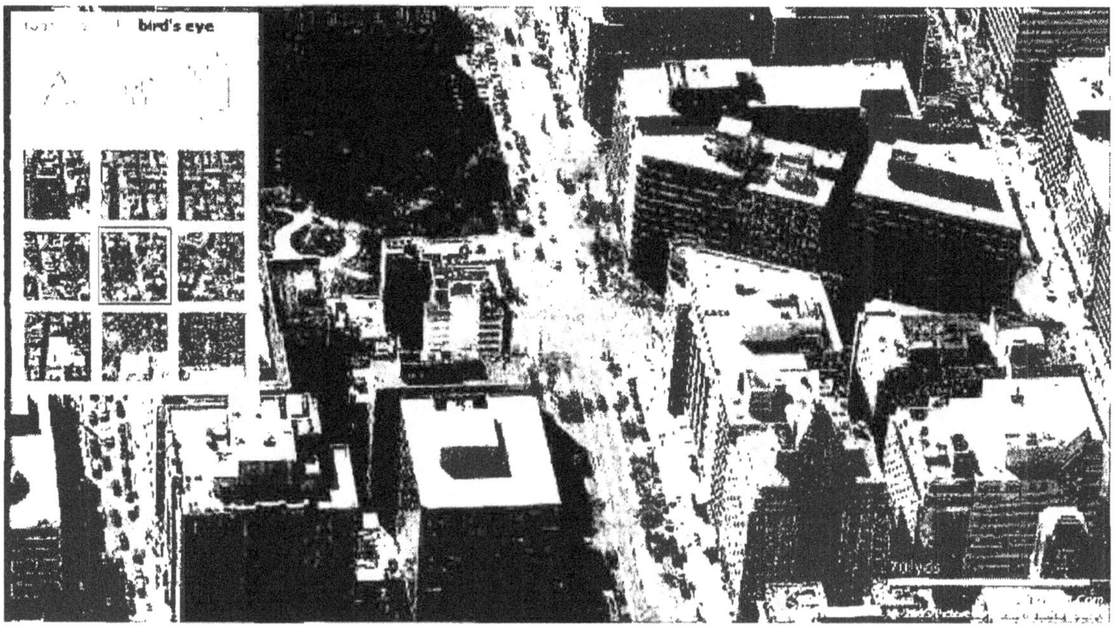
bird's eye

Microsoft's Virtual Earth "Bird's Eye View" of downtown Washington, DC.

Yahoo offers three APIs—their own Simple API, AJAX, and MacroMedia Flash—for use with Yahoo mapping technology. The Simple API lets users create customized Yahoo maps using just a text editor. By offering several different options for creating mashups, Yahoo appeals to differing levels of expertise, from "no programming" to fairly sophisticated embedded scripts. Yahoo maps also offers "building block APIs" including geocoding, Yahoo Local, traffic information, and map images.

Building Block Components

Several Yahoo! APIs help you create a powerful and useful Yahoo! Maps mashups. Use these together with the Yahoo! Maps APIs to enhance the user experience.

Geocoding API - Pass in location data by address and receive geocoded (encoded with latitude-longitude) responses.

Map Image API - Stitch map images together to build your own maps for usage in custom applications, including mobile and offline use.

Traffic APIs - Build applications that take dynamic traffic report data to help you plan optimal routes and keep on top of your commute using either our REST API or Dynamic RSS Feed.

Local Search APIs - Query against the Yahoo! Local service, which now returns longitude-latitude with every search result for easy plotting on a map. Also new is the inclusion of ratings from Yahoo! users for each establishment to give added context.

Yahoo offers a great deal of help to developers who want to use these APIs to create customized maps. Yahoo also permits third parties to host Yahoo map mashups on their own sites, which is something important for researchers to look for.

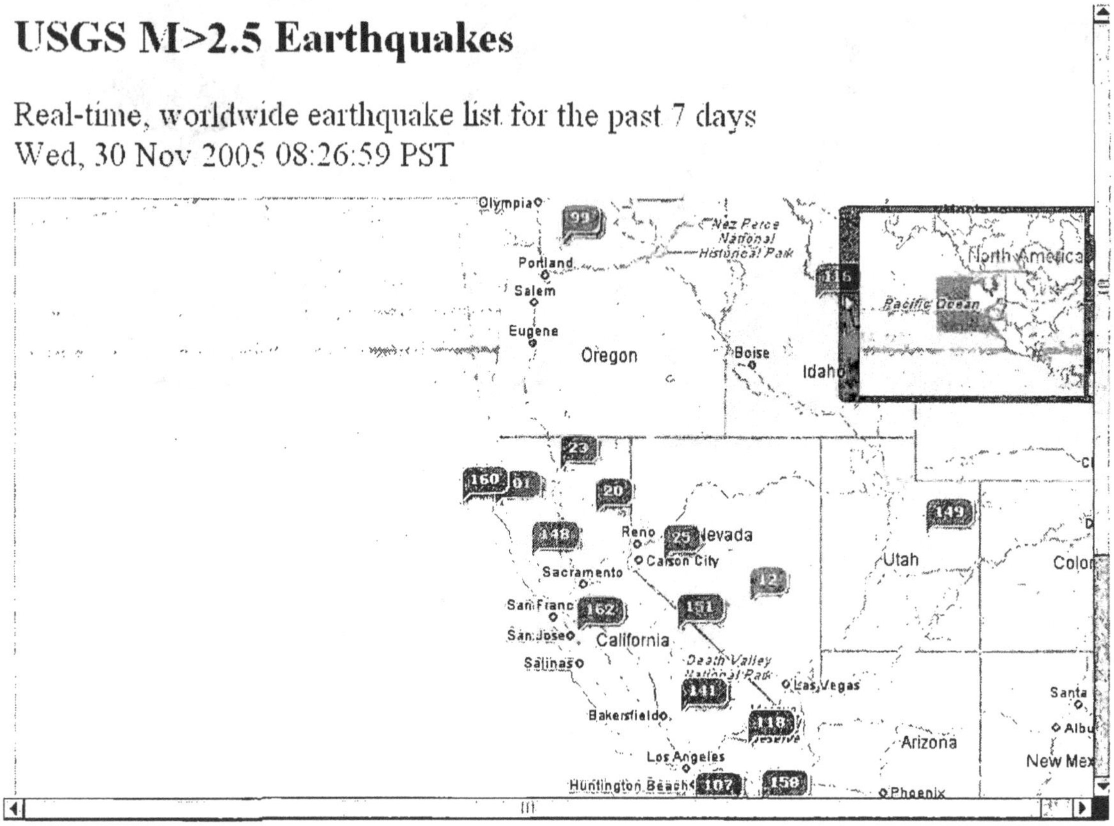
USGS M>2.5 Earthquakes
Real-time, worldwide earthquake list for the past 7 days
Wed, 30 Nov 2005 08:26:59 PST
Olympia
Portland
Salem
Eugene
Oregon
Boise
Idaho
North America
Pacific Ocean
Reno
Nevada
Carson City
Sacramento
Utah
California
San Jose
Salinas
Death Valley National Park
Las Vegas
Bakersfield
Arizona
Los Angeles
Huntington Beach
Phoenix
New Mex
Santa
Albu

Yahoo is adding non-US local maps, using mapping technology provided by other sources, such as Map24 for European locations. Here is a sample of what you get for Germany. Notice the choice between static and dynamic maps, which allow users to move around the map using a mouse. At this time, I recommend skipping Yahoo Local in Europe and going directly to Map24, which is much better.

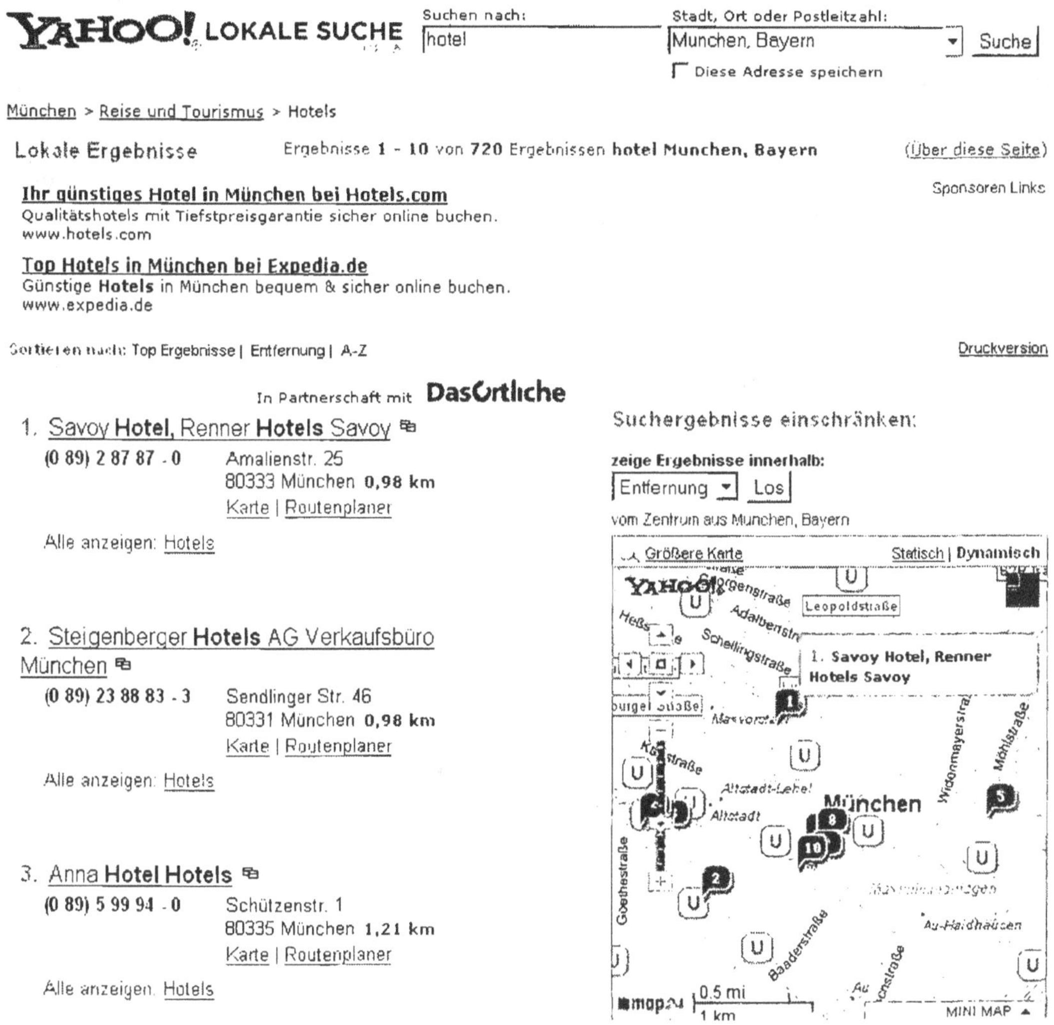
YAHOO! LOKALE SUCHE
Suchen nach:
hotel
Stadt, Ort oder Postleitzahl:
Munchen, Bayern
Suche
Diese Adresse speichern
München > Reise und Tourismus > Hotels
Lokale Ergebnisse
Ergebnisse 1 - 10 von 720 Ergebnissen hotel Munchen, Bayern
(Über diese Seite)
Ihr günstiges Hotel in München bei Hotels.com
Qualitätshotels mit Tiefstpreisgarantie sicher online buchen.
www.hotels.com
Sponsoren Links
Top Hotels in München bei Expedia.de
Günstige Hotels in München bequem & sicher online buchen.
www.expedia.de
Sortieren nach: Top Ergebnisse | Entfernung | A-Z
Druckversion
In Partnerschaft mit DasÖrtliche
1. Savoy Hotel, Renner Hotels Savoy
(0 89) 2 87 87 - 0
Amalienstr. 25
80333 München 0,98 km
Karte | Routenplaner
Alle anzeigen: Hotels
2. Steigenberger Hotels AG Verkaufsbüro München
(0 89) 23 88 83 - 3
Sendlinger Str. 46
80331 München 0,98 km
Karte | Routenplaner
Alle anzeigen: Hotels
3. Anna Hotel Hotels
(0 89) 5 99 94 - 0
Schützenstr. 1
80335 München 1,21 km
Karte | Routenplaner
Alle anzeigen: Hotels
Suchergebnisse einschränken:
zeige Ergebnisse innerhalb:
Entfernung
Los
vom Zentrum aus Munchen, Bayern
Größere Karte
Statisch | Dynamisch
Leopoldstraße
1. Savoy Hotel, Renner Hotels Savoy
München
Altstadt
0.5 mi
1 km
MINI MAP

Yahoo Maps http://maps.yahoo.com/

Yahoo Map APIs http://developer.yahoo.net/maps/

Ask Maps

The maps at Ask.com come from two familiar sources, Navteq and GlobeXplorer. However, what Ask has done with the source material makes them worth a special mention. As with other mapping sites, such as Google, Ask Maps allows users to choose between "aerial" and "physical" views with labels on or off (in Google, this means map, satellite, or hybrid (labeled) views). Ask Maps also uses address bubbles that can be closed and the ability to get directions between locations. As with Google Maps, Ask Maps offers the zoom in/out feature using the mouse scroll wheel.

Ask Maps adds the ability to:

- Show or hide directions (this is especially useful when you have very long, complicated directions).
- Choose between driving and walking directions; this is extremely useful when you are dealing with locations that have a lot of one-way streets, for example.
- Maximize the map with a simple click of the mouse, something Google Maps cannot do that annoys me to no end.
- The ability to add another location is much easier than with other mapping sites; right-click to add a push-pin location to the map.

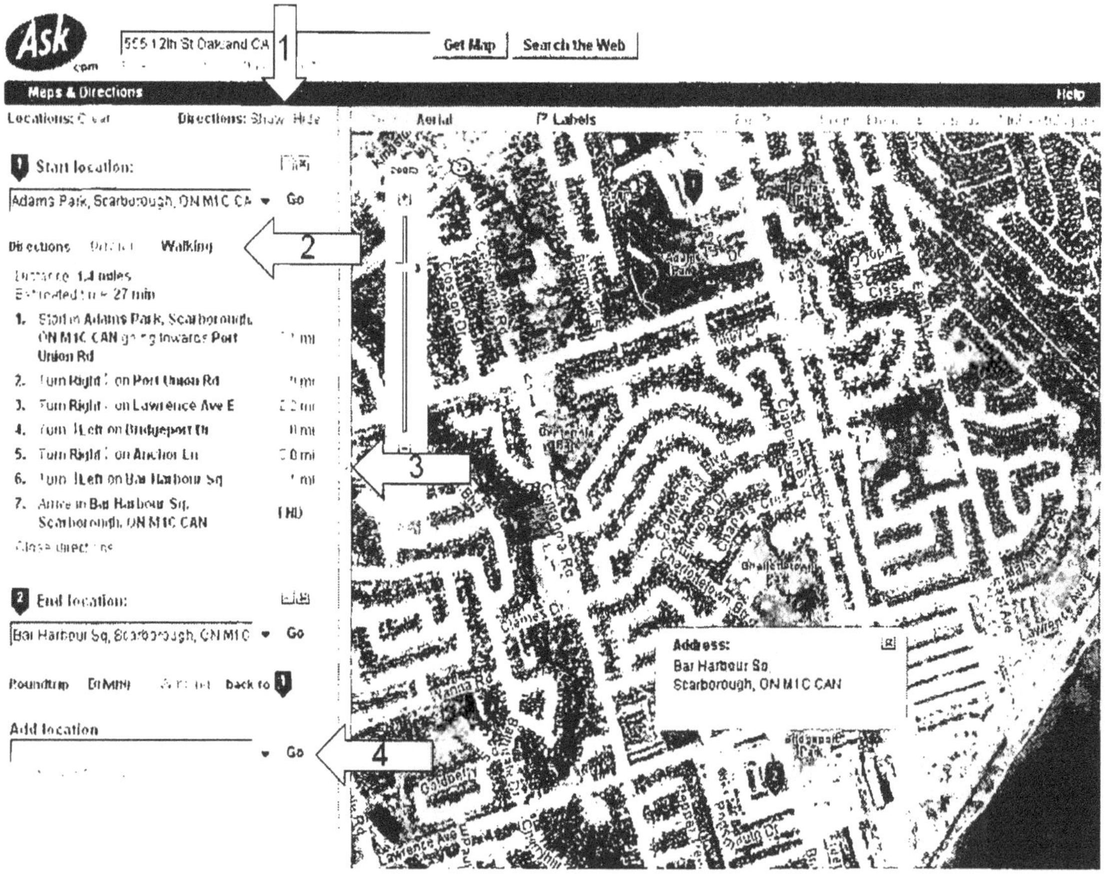
Ask
Get Map
Search the Web
Maps & Directions
Help
1
2
3
4
Aerial
Labels
Start location:
Adams Park, Scarborough, ON M1C CA
Go
Walking
End location:
Add location
Address:
Bar Harbour Sq,
Scarborough, ON M1C CAN

In addition to the US and Canada, Ask Maps also offers coverage of parts of Europe. To see the current coverage, I recommend you check GlobeXplorer's website.

Ask Maps http://maps.ask.com/maps

GlobeXplorer's Worldwide Satellite Imagery http://www.globexplorer.com/our-content/digital-globe.shtml

International Map Sites

Map24

I highly recommend **Map24**, which offers country and city maps for many international locations. Map24 is one of the most technically sophisticated free web-based map such as Google Maps or Virtual Earth. As with these sites, ***Map24 offers map, satellite, and hybrid views***. The site is set to redirect you to your country's map page, so if you want to see all the available international maps, you need to select ***Language: Change*** to see all the available languages and countries associated with them. For example, Portuguese is only associated with Brazil at Map24 even though a less detailed map of Portugal is offered.

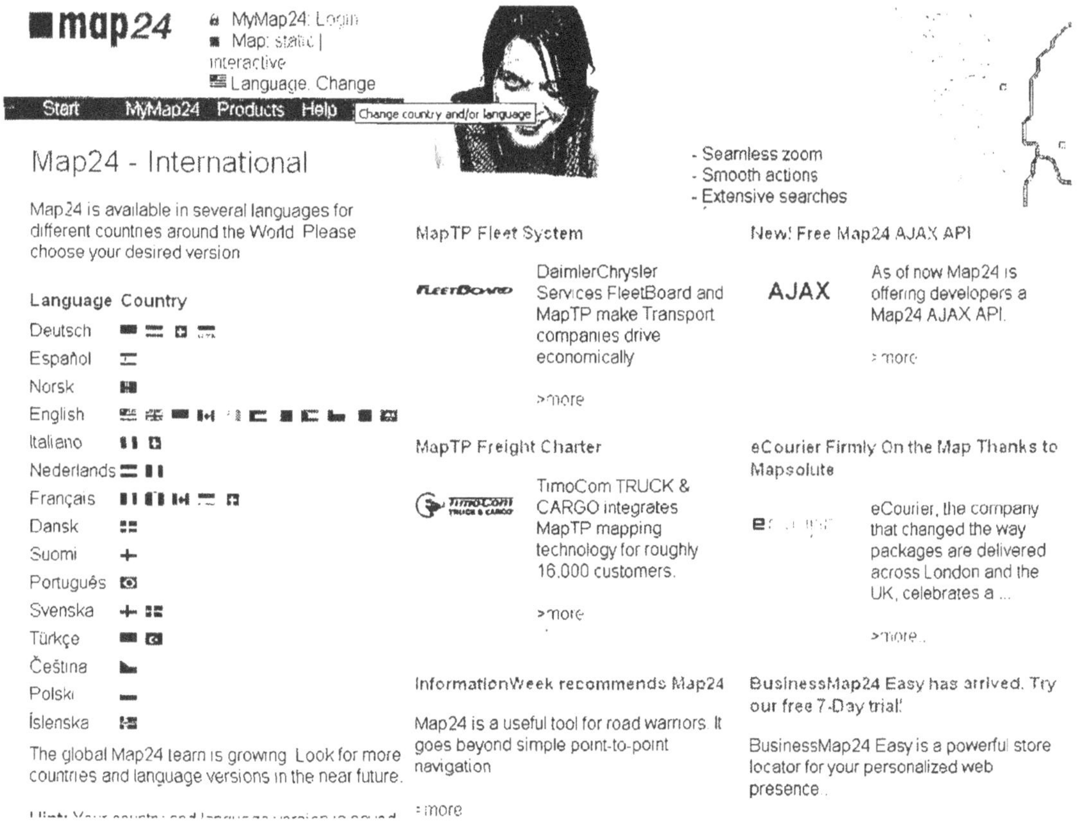
map24
MyMap24: Login
Map: static | interactive
Language: Change
Start MyMap24 Products Help
Change country and/or language
Map24 - International
Map24 is available in several languages for different countries around the World. Please choose your desired version
Language Country
Deutsch
Español
Norsk
English
Italiano
Nederlands
Français
Dansk
Suomi
Português
Svenska
Türkçe
Čeština
Polski
Íslenska
The global Map24 team is growing. Look for more countries and language versions in the near future.
- Seamless zoom
- Smooth actions
- Extensive searches
MapTP Fleet System
DaimlerChrysler Services FleetBoard and MapTP make Transport companies drive economically
>more
New! Free Map24 AJAX API
AJAX
As of now Map24 is offering developers a Map24 AJAX API.
>more
MapTP Freight Charter
TimoCom TRUCK & CARGO integrates MapTP mapping technology for roughly 16,000 customers.
>more
eCourier Firmly On the Map Thanks to Mapsolute
eCourier, the company that changed the way packages are delivered across London and the UK, celebrates a ...
>more
InformationWeek recommends Map24
Map24 is a useful tool for road warriors. It goes beyond simple point-to-point navigation
>more
BusinessMap24 Easy has arrived. Try our free 7-Day trial!
BusinessMap24 Easy is a powerful store locator for your personalized web presence.

Map24 http://www.map24.com/

Maporama

Maporama is especially useful because it has a number of detailed city and country maps. At present, Maporama offers more than 63 countries/locations. Where else are you going to find a street map of Riyadh, Saudi Arabia? Not every country or city is this detailed, but Maporama is an indispensable map and directions tool.

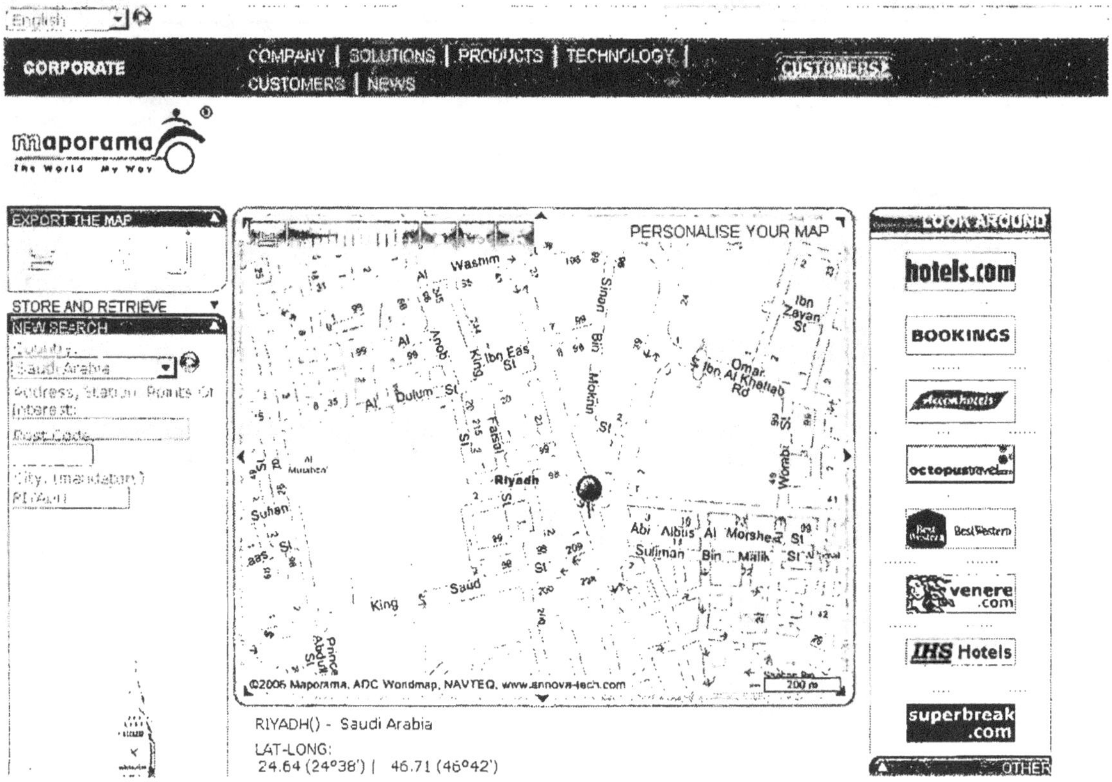
English
CORPORATE
COMPANY | SOLUTIONS | PRODUCTS | TECHNOLOGY |
CUSTOMERS | NEWS
CUSTOMERS
maporama
EXPORT THE MAP
STORE AND RETRIEVE
NEW SEARCH
Saudi Arabia
PERSONALISE YOUR MAP
Riyadh
©2006 Maporama, ADC Wordmap, NAVTEQ, www.annova-tech.com
200 m
RIYADH() - Saudi Arabia
LAT-LONG:
24.64 (24°38') | 46.71 (46°42')
LOOK AROUND
hotels.com
BOOKINGS
BestWestern
venere.com
IHS Hotels
superbreak.com
OTHER

Maporama http://world.maporama.com/

Mappy

Despite not having the most felicitous of names, Mappy is nonetheless a wonderful map site. Mappy's parent company is Pages Jaunes Groupe. Mappy recently added **aerial views of a number of European cities**—26 total as of now—in France, Spain, Belgium, Germany, and the Czech Republic (only Prague for now). The aerial photo maps require MacroMedia Flash.

Here is Mappy's aerial view of Flora Park in Amsterdam at the closest resolution; pay special attention to the roof of the large building where the arrow is pointing:

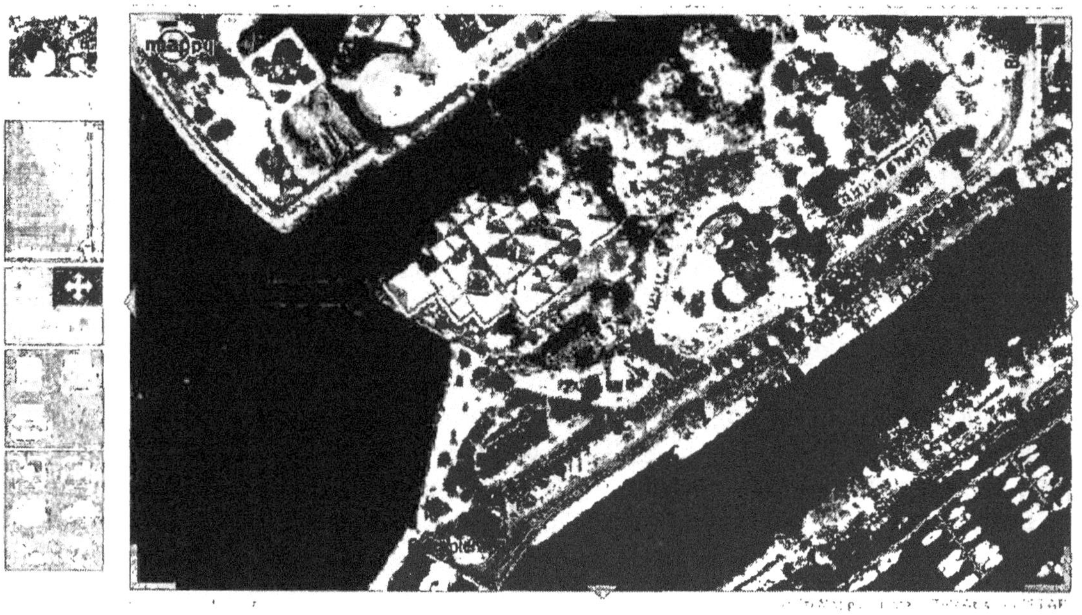

Not to be upstaged, in early 2006 Google "updated the satellite data used for Google Maps so it now has the many updates recently made to the Google Earth database. This means the new high resolution data for all of Germany, many places in Europe...and many other places are now available to Google Maps users."[90] Here is the Google Maps hybrid image of the same location at the closest resolution Google Maps offers:

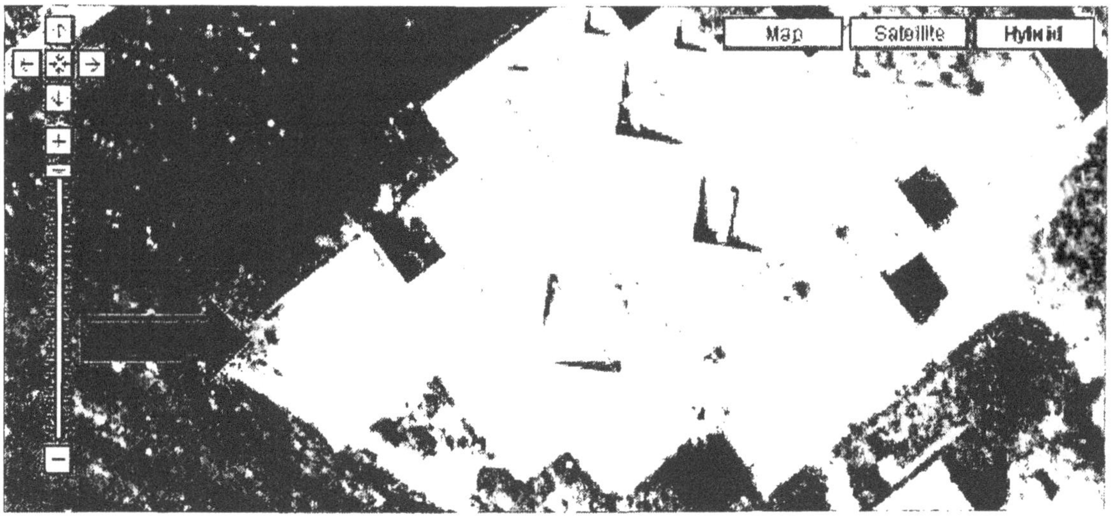
Map
Satellite
Hybrid

[90] Frank Taylor, "Google Maps Gets Huge Satellite Update from Google Earth," Google Earth Blog, 24 April 2006, <http://www.gearthblog.com/blog/archives/2006/04/google_maps_and.html> (30 October 2006).

As you can tell even from these snapshots, the Mappy image is not as clear as those provided by Google. Also, Google Maps continues to be the hands down winner in navigating around the map smoothly.

Mappy is very good not only for travel directions but also for identifying local landmarks, hotels, cash machines, parking lots, etc. None of the other mapping sites can match Mappy in these areas yet, but it is clear the competition is fierce and map sites and services are only going to get better and better.

Mappy's Aerial Photos http://www.mappy.com/ (select Maps | Aerial Photos)

Street-Level Map Views

Both France Telecom's **Pages Jaunes** and Spain's **Callejero Fotographico** (Photographic Street Guides) offer street-level views of cities using "stroll" technologies. These sites require the Macromedia Flash plug-in to use the visual maps. Both sites are worth a visit because they are both impressive and potentially very useful if you ever need to see a specific place in one of a number of French or Spanish cities. This is a snapshot from Pages Jaunes of the Place du Carrousel in downtown Paris. Note especially the icon below the photograph that lets you get a full 360-degree view of the location:

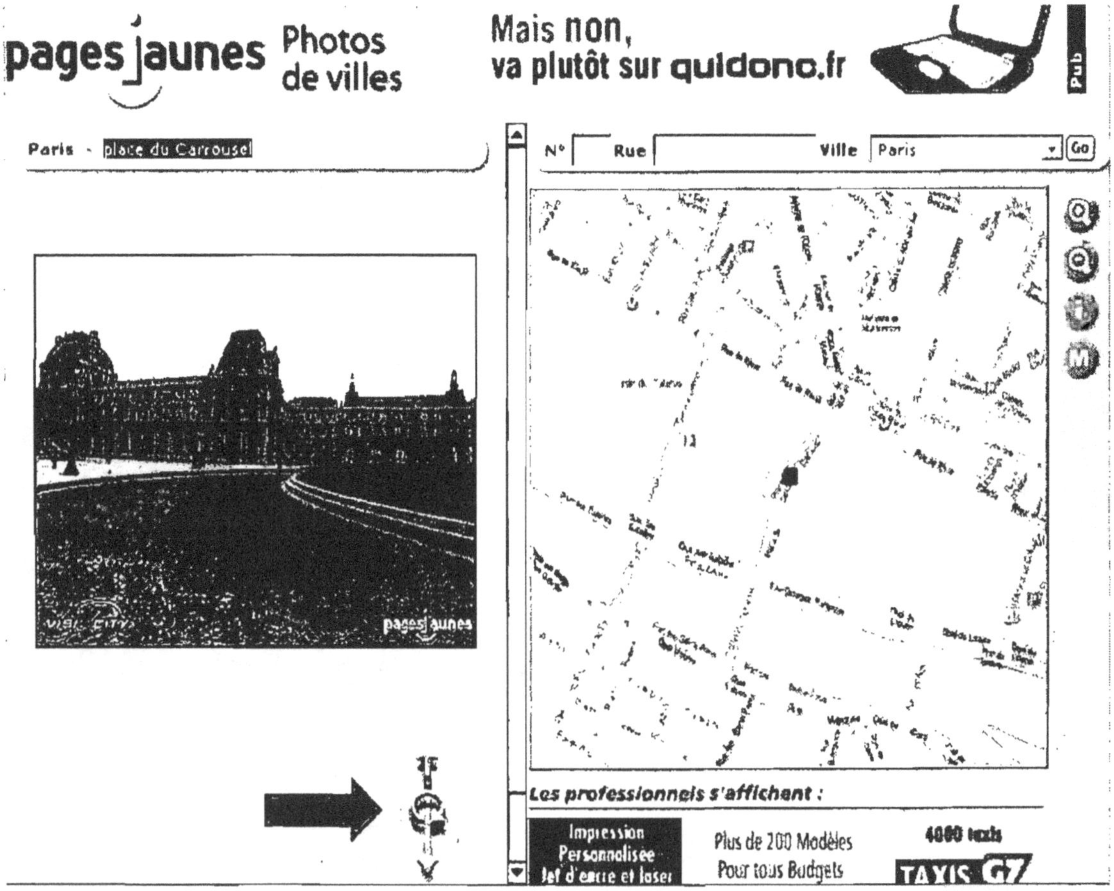
pagesjaunes
Photos de villes
Mais non, va plutôt sur quidonc.fr
Pub
Paris - place du Carrousel
N° Rue Ville Paris Go
Les professionnels s'affichent :
Impression Personnalisee Jet d'encre et laser
Plus de 200 Modèles Pour tous Budgets
4000 taxis
TAXIS G7

France Telecom's Pages Jaunes http://photos.pagesjaunes.fr/

Spain's Callejero Fotographico http://www.qdq.com/indexfotos.asp

Are they perfect? Of course not, but these sites do I think point to a future in which a variety of technologies—the Internet, GPS, digital photography—will be used together in increasingly creative ways to open the world in ways we are only beginning to imagine.

ViaMichelin

ViaMichelin maps cover the US and Canada as well as virtually all European countries. In fact, for Europe, ViaMichelin is hard to beat, especially for driving directions. Look at this street-level map of Lisbon (remember, as excellent as Map24 is, it doesn't cover Portugal). Users can even look for such things as speed cameras and roadwork.

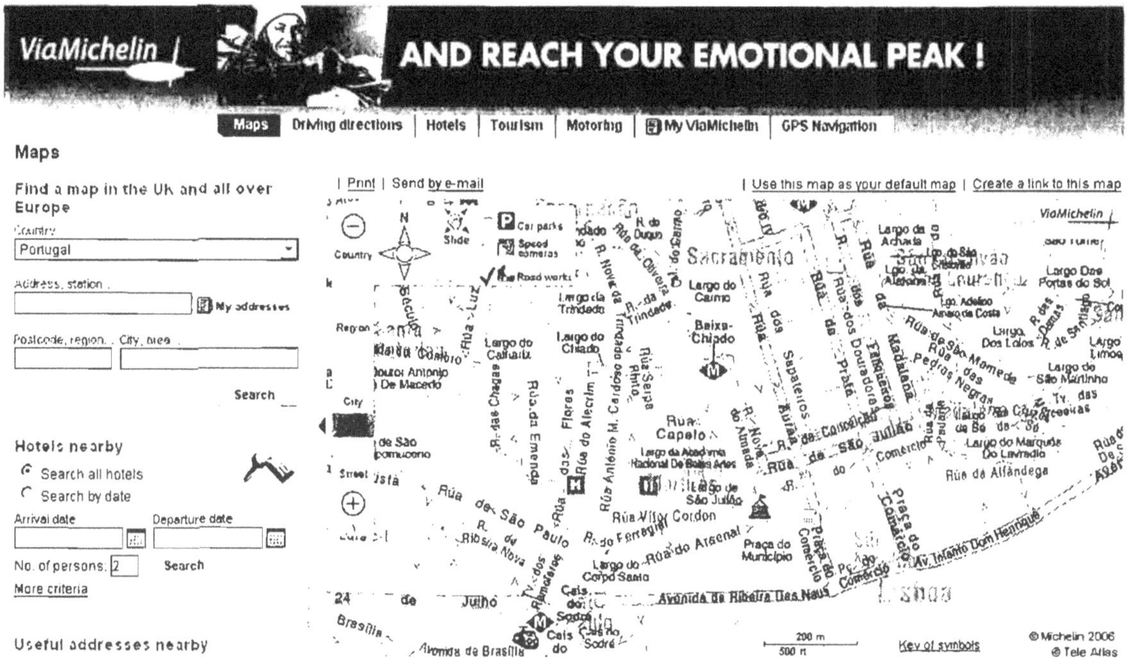
ViaMichelin
AND REACH YOUR EMOTIONAL PEAK !
Maps
Driving directions
Hotels
Tourism
Motoring
My ViaMichelin
GPS Navigation
Maps
Find a map in the UK and all over Europe
Portugal
My addresses
Search
Hotels nearby
Search all hotels
Search by date
Arrival date
Departure date
No. of persons
Search
More criteria
Useful addresses nearby
| Print | Send by e-mail
| Use this map as your default map | Create a link to this map
Car parks
Speed cameras
Road works
Key of symbols
© Michelin 2006
© Tele Atlas

ViaMichelin (Europe, US, Canada)
http://www.viamichelin.com/viamichelin/gbr/dyn/controller/Maps

Best Mapping Sites

Ask Maps http://maps.ask.com/maps

France Telecom's Pages Jaunes http://photos.pagesjaunes.fr/

Google Earth (must be downloaded) http://earth.google.com/

Google Maps http://maps.google.com/

Map24 http://www.map24.com/

MapQuest http://www.mapquest.com/

Maporama http://www.maporama.com/share/

Multimap (excellent source of maps worldwide) http://www.multimap.com/

Spain's Callejero Fotographico http://www.qdq.com/indexfotos.asp

Mappy (Europe) http://www.mappy.com/

ViaMichelin (Europe, US, Canada) http://www.viamichelin.com/viamichelin/gbr/dyn/controller/Maps

Windows Live Local/Virtual Earth http://local.live.com/

Yahoo Maps http://maps.yahoo.com/

Best Map MetaIndices

About's Maps http://geography.about.com/science/geography/msub1.htm

Martindale's "Virtual" Geoscience Center http://www.martindalecenter.com/GradGeoscience_5_GG.html

Odden's The Fascinating World of Maps and Map-Making http://oddens.geog.uu.nl/index.html

Perry-Castaneda Library Map Collection at the University of Texas Austin http://www.lib.utexas.edu/Libs/PCL/Map_collection/map_sites/map_sites.html

ReisWijs Route Planner Metasite http://www.reiswijs.co.uk/routeplanner/routeplanner.html

"Mystery Hits"

I followed all these suggestions but I'm still getting hits that seem completely unrelated to my query. Why? There are a number of possible explanations:

1) You've included "stop" words that may be read as wildcards.

2) You're the victim of misleading keywords, e.g., "lutefisk" was included as a metatag keyword by the St. Paul Star Tribune because of its popularity in Minnesota to drive traffic to the website.

3) You're the victim of "tiny words" (font so small as to be invisible) or "hidden text" (text and background are the same color) on a webpage, both of which are often treated as spam by search engines.

4) The webpage may have changed between the time it was indexed and today.

5) You're the victim of flawed search software.

Uncovering the "Invisible" Internet

One of the most frustrating things about Internet search tools is the fact that even the best index only a portion of the web, much less the entire Internet. The deep (aka hidden or invisible) web continues to elude most search services and users seeking to plumb its depths. We are still, for the most part, dependent upon specialty tools and sites to help us find and exploit deep web resources. The challenge is how to access that part of the web that remains invisible to search engines. It is important to understand that search engines are generally designed to index a certain subset of the Internet: web pages and, in some cases, certain types of files, e.g., video, audio, PDF[91]. Furthermore, ***most search engines limit their web page and document indexing***. For example, Google used to index approximately the first 100KB of HTML, and reportedly the first megabyte of PDF documents, but in October 2005, Google dramatically increased the size of its cache limit, although no one knows for sure what that limit is. Yahoo indexes at least the first 500KB of HTML and PDF documents. In any event, long documents usually are partially invisible to these and other search engines. You cannot rely upon a search engine spider to index long documents in their entirety.

A9 Search

At the end of September 2006, A9, the Amazon.com-owned search property, made sweeping changes, some good and some bad. Contrary to what some search bloggers said, A9 is not "dead" (at least not yet). But some of A9's best features are gone. As I feared, not enough people used the wonderful "street view" map resource and now it is gone. As of September 29, 2006, A9 "discontinued A9 Maps and the A9 Yellow Pages (including BlockView™) ... [and] discontinued the A9 Instant Reward program, and the A9 Toolbar and personalized services such as history, bookmarks, and diary." Other changes include "a new continuous scrolling feature, so you no longer have to bother with next and previous buttons to move from one page of results to the next. You can now also drag-and-drop the columns to change

[91] Google was the first major search engine to routinely index the *content* of many file types, including pdf, ps, xls, doc, ppt, and others. See "Google's Frequently Asked Questions — File Types," <http://www.google.com/help/faq_filetypes.html> (14 November 2006).

their order on the page."[92] A9 had earlier switched from Google to Windows Live Search for its web and news searches.

Another big change that is not perhaps obvious to the casual user is that A9 no longer offers (or requires) that users log in and have every single query ever made at A9 recorded and retrievable. I do not think that A9's eliminating the retention of personal search data is an accident given some infamous "leaks" of personal information by search services. In the past, I had recommended using the "generic" A9 interface if you did not want to log in to use A9; that requirement is now obviated.

The A9 homepage looks quite different. Notice that there are a number of options for searching that include dropdown menus to check the sources you wish to search.

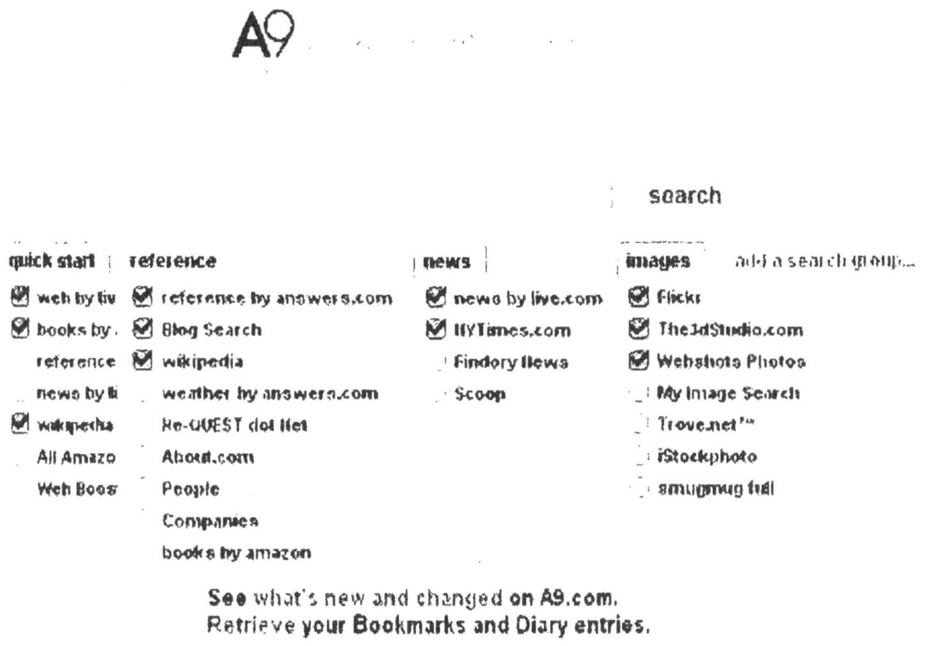
A9
search
quick start
reference
news
images
reference by answers.com
Blog Search
wikipedia
weather by answers.com
About.com
People
Companies
books by amazon
news by live.com
NYTimes.com
Findory News
Scoop
Flickr
Webshots Photos
My Image Search
iStockphoto
See what's new and changed on A9.com.
Retrieve your Bookmarks and Diary entries.

While this is nice, it is not much of an improvement over the old A9. Also note the "add a search group" option on the right side of the page. If you click on that link, you will see the other search categories you can add to your main A9 search page to tailor it to your own needs.

---

[92] "What's New at A9.com," A9, 29 September 2006, <http://a9.com/-/company/whatsNew.jsp> (5 October 2006).

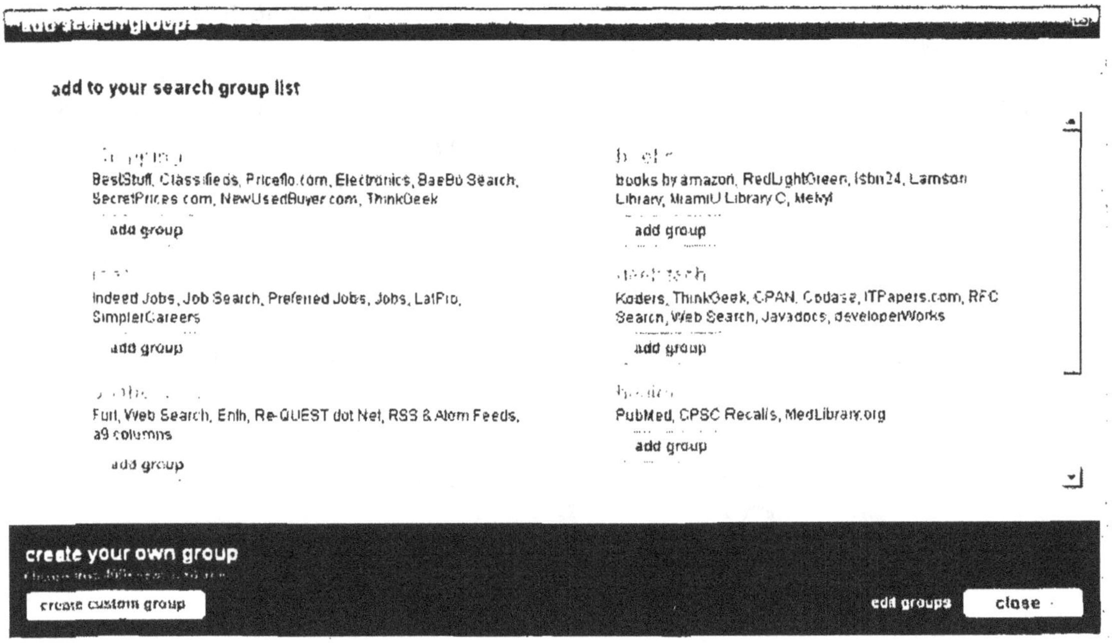
add to your search group list
BestStuff, Classifieds, Priceflo.com, Electronics, BaeBo Search, SecretPrices.com, NewUsedBuyer.com, ThinkGeek
add group
books by amazon, RedLightGreen, Isbn24, Lamson Library, MiamiU Library C, Melvyl
add group
Indeed Jobs, Job Search, Preferred Jobs, Jobs, LatPro, SimplerCareers
add group
Koders, ThinkGeek, CPAN, Codase, ITPapers.com, RFC Search, Web Search, Javadocs, developerWorks
add group
Furl, Web Search, Enth, Re-QUEST dot Net, RSS & Atom Feeds, a9 columns
add group
PubMed, CPSC Recalls, MedLibrary.org
add group
create your own group
create custom group
edit groups
close

There is a great deal of customization available at A9. By selecting "edit groups" you will be given the choice of nearly 500 sources either to add to the default search groups or, more likely, to create your own custom group and add sources to it. For example, you can easily create a Blog Search custom group to appear on the A9 homepage or you can edit the basic groups, adding and subtracting sources.

Another change at A9 that has not received the attention it deserves is the addition of "Web Booster" results. **Web Booster** is the Convera Excalibur web search service, which Clusty, Highbeam, and Govmine also use: "Web Booster makes standard search better by digging down into all the results pages you don't have time to read and pulling up information that you would otherwise miss. It is kind of like having a search helper who goes and gets information that might be of interest." In other word's Web Booster is a "deep web" resource, and it is a default search option on A9's quick start menu. A9 also retains what I think is one of its best features, i.e., results by column. Results from each of the search resources you choose to search appear in separate columns that are easily resizable. No other search service offers anything quite like this.

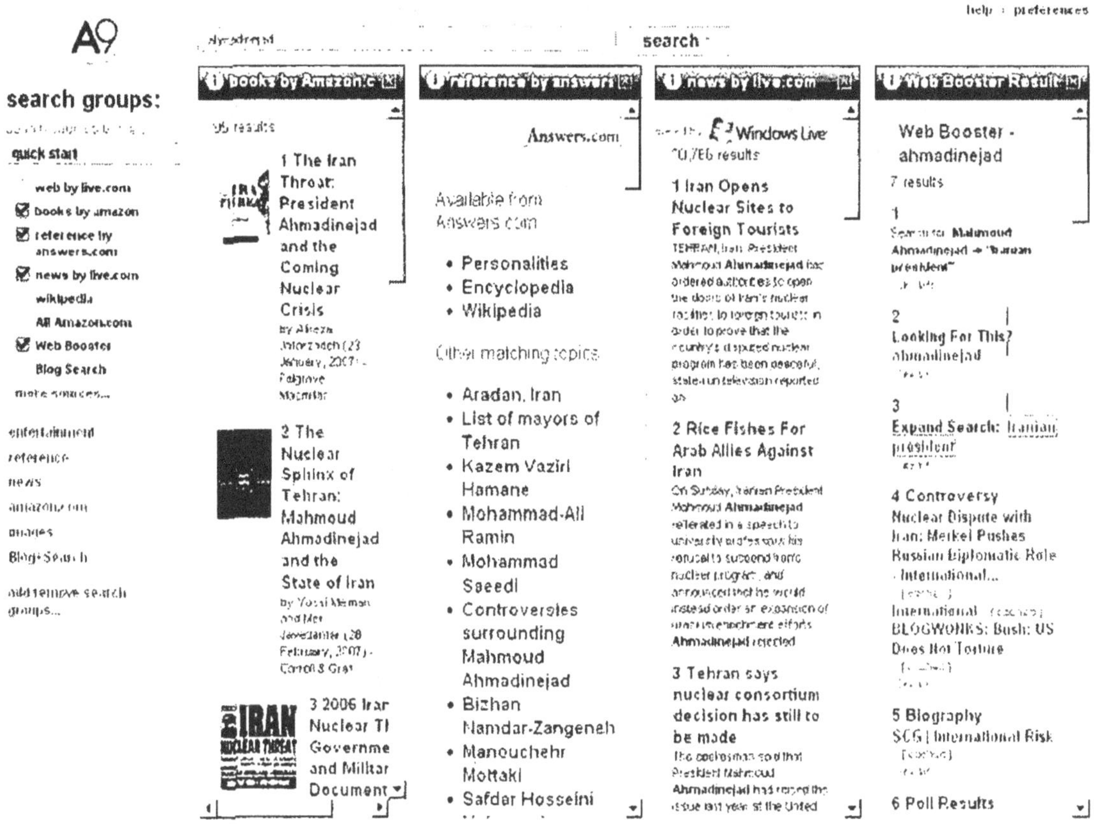
A9
search groups:
quick start
web by live.com
books by amazon
reference by answers.com
news by live.com
wikipedia
All Amazon.com
Web Booster
Blog Search
search
books by Amazon.c
reference by answers
news by live.com
Web Booster Result
1 The Iran Threat: President Ahmadinejad and the Coming Nuclear Crisis
2 The Nuclear Sphinx of Tehran: Mahmoud Ahmadinejad and the State of Iran
3 2006 Iran Nuclear Th Governme and Militar Document
Answers.com
Available from Answers.com
Personalities
Encyclopedia
Wikipedia
Other matching topics
Aradan, Iran
List of mayors of Tehran
Kazem Vaziri Hamane
Mohammad-Ali Ramin
Mohammad Saeedi
Controversies surrounding Mahmoud Ahmadinejad
Bizhan Namdar-Zangeneh
Manouchehr Mottaki
Safdar Hosseini
Windows Live
1 Iran Opens Nuclear Sites to Foreign Tourists
2 Rice Fishes For Arab Allies Against Iran
3 Tehran says nuclear consortium decision has still to be made
Web Booster - ahmadinejad
7 results
Looking For This? ahmadinejad
Expand Search: Iranian president
4 Controversy
Nuclear Dispute with Iran: Merkel Pushes Russian Diplomatic Role
5 Biography
6 Poll Results

A9's web and Web Booster results also include something called "Site Info." Simply moving the mouse over the "Site Info" box brings up this Alexa feature that shows various types of information Alexa has collected about the site.

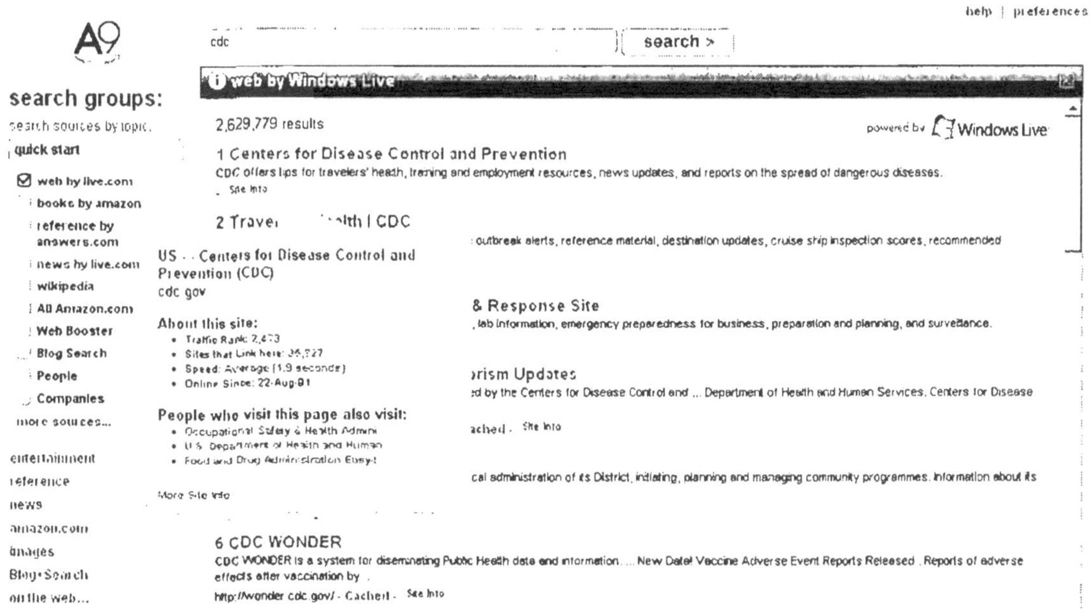
A9
help | preferences
cdc
search >
search groups:
search sources by topic
quick start
web by live.com
books by amazon
reference by answers.com
news by live.com
wikipedia
All Amazon.com
Web Booster
Blog Search
People
Companies
more sources...
entertainment
reference
news
amazon.com
images
Blog Search
on the web...
web by Windows Live
2,629,779 results
powered by Windows Live
1 Centers for Disease Control and Prevention
CDC offers tips for travelers' health, training and employment resources, news updates, and reports on the spread of dangerous diseases.
Site Info
US - Centers for Disease Control and Prevention (CDC)
cdc.gov
About this site:
People who visit this page also visit:
More Site Info
& Response Site
6 CDC WONDER
CDC WONDER is a system for disseminating Public Health data and information. ... New Data! Vaccine Adverse Event Reports Released. Reports of adverse effects after vaccination by ...
http://wonder.cdc.gov/ - Cached - Site Info

In early 2006, two great resources came together. A9 added people search using the excellent ZoomInfo search tool. ZoomInfo (formerly Eliyon) is a "web summarization" site that contains over 26 million summaries of people on the web. According to its FAQ, "ZoomInfo automatically and continuously grows its base of Web Summaries from corporate and personal websites, government filings, press releases and other public sources. All information found by ZoomInfo and used to create your Web Summary comes from public sources and can be found by anyone by using most major search engines like Yahoo and Google."[93] What ZoomInfo does for the user is to summarize that data and present it in easy to handle small packages. Basically, ZoomInfo does some of the "leg work" for you.

ZoomInfo is not one of the default settings on the A9 homepage, so you will need to add it by selecting "add a search group," "Edit groups," search for ZoomInfo, and add the ZoomInfo People and/or Company search to your A9 homepage. Here are the results for an A9 people and book search on Brewster Kahle of the Internet Archive fame:

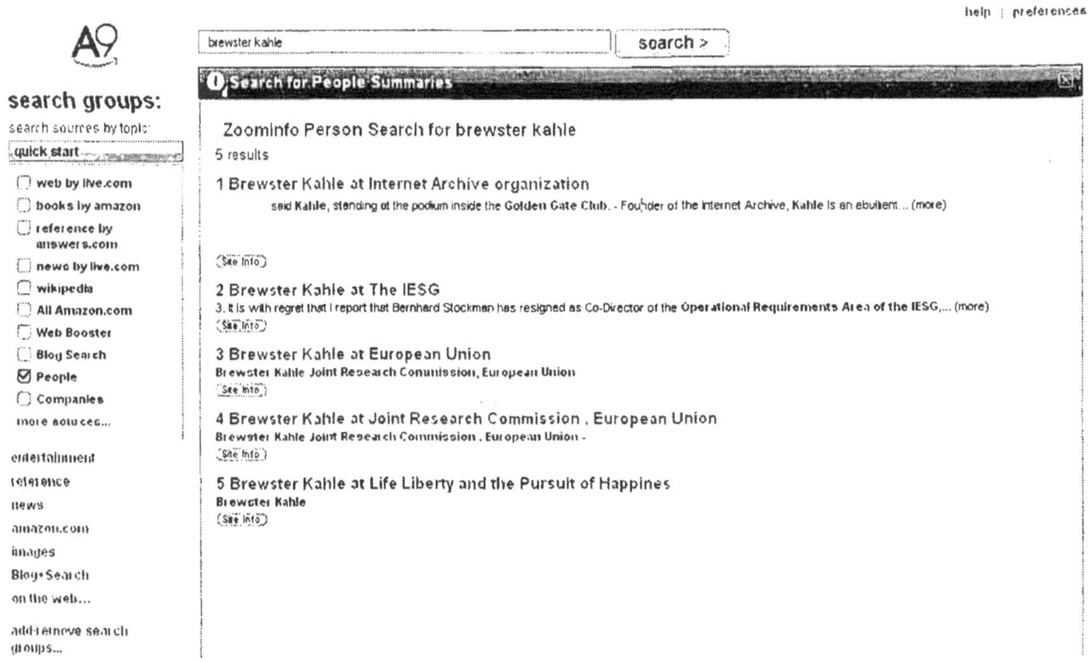
help | preferences
A9
brewster kahle
search >
Search for People Summaries
search groups:
search sources by topic:
quick start
web by live.com
books by amazon
reference by answers.com
news by live.com
wikipedia
All Amazon.com
Web Booster
Blog Search
People
Companies
more sources...
entertainment
reference
news
amazon.com
images
Blog+Search
on the web...
add/remove search groups...
ZoomInfo Person Search for brewster kahle
5 results
1 Brewster Kahle at Internet Archive organization
2 Brewster Kahle at The IESG
3 Brewster Kahle at European Union
Brewster Kahle Joint Research Commission, European Union
4 Brewster Kahle at Joint Research Commission , European Union
5 Brewster Kahle at Life Liberty and the Pursuit of Happines
Brewster Kahle

Now, when I click on the first hit under people summaries, I get a rather impressive dossier on Kahle. I can't show you nearly all of what is on this page—this is just the top part of an extensive entry that you will note was "automatically generated using 234 references found on the Internet."

---

[93] FAQ, ZoomInfo, <http://www.zoominfo.com/# > (14 November 2006).

OCID: 4046925

zoominfo
Person | Company
Find Person:
GO
Find
Be Found
PowerSearch
About Us
Mr. Brewster Kahle This is me
Board Member
Electronic Frontier Foundation
Add to Colleagues
Provide Feedback
This summary was automatically generated using 181 references found on the Internet. How did ZoomInfo get this information?
Employment History
Founder
Internet Archive - San Francisco, California
Founder
Alexa Internet
Founder
WAIS Inc. - California
Print View...
Keyword Search
Board Member
Electronic Frontier Foundation
Organizations
Charitable Organizations & Foundations
Founder
Internet Archive
Media & Internet
Search Engines & Internet Portals
Alexa Internet
Information Collection & Delivery
Show All
Upgrade to Premium Account
Colleagues & Associates
= Verified
= Not Verified

ZoomInfo also searches for "companies," but keep in mind that they index more than just companies or businesses—you will find all sorts of organizations, associations, academies, etc., in their database.

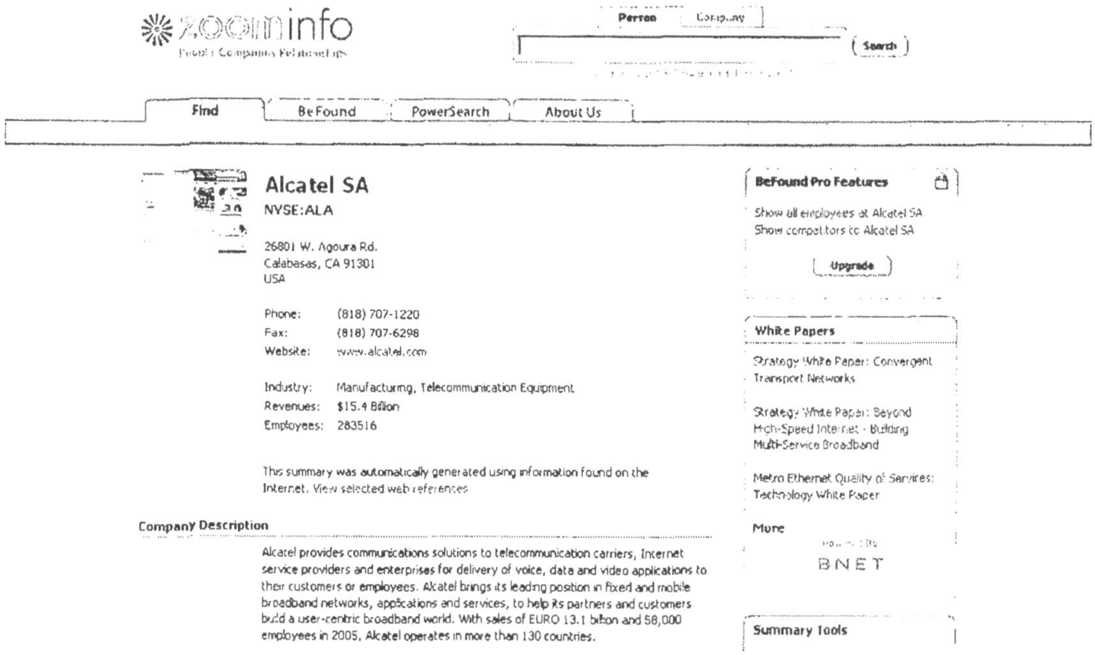
zoominfo
People Companies Relationships
Person
Company
Search
Find
Be Found
PowerSearch
About Us
Alcatel SA
NYSE:ALA
26801 W. Agoura Rd.
Calabasas, CA 91301
USA
Phone: (818) 707-1220
Fax: (818) 707-6298
Website: www.alcatel.com
Industry: Manufacturing, Telecommunication Equipment
Revenues: $15.4 Billion
Employees: 283516
This summary was automatically generated using information found on the Internet. View selected web references
Company Description
Alcatel provides communications solutions to telecommunication carriers, Internet service providers and enterprises for delivery of voice, data and video applications to their customers or employees. Alcatel brings its leading position in fixed and mobile broadband networks, applications and services, to help its partners and customers build a user-centric broadband world. With sales of EURO 13.1 billion and 58,000 employees in 2005, Alcatel operates in more than 130 countries.
BeFound Pro Features
Show all employees at Alcatel SA
Show competitors to Alcatel SA
Upgrade
White Papers
Strategy White Paper: Convergent Transport Networks
Strategy White Paper: Beyond High-Speed Internet - Building Multi-Service Broadband
Metro Ethernet Quality of Services: Technology White Paper
More
BNET
Summary Tools

You can also go directly to ZoomInfo to search companies and/or people.

ZoomInfo http://www.zoominfo.com/

Other features and facts about A9 you should know:

- "A9 Lite (lite.a9.com) is a version of A9.com that has been designed to work on older web browsers as well as many mobile devices. A9 Lite does not require JavaScript, so if JavaScript on your browser is turned off, or if you have an older browser, you will automatically be sent to A9 Lite. A9 Lite provides limited functionality, and does not have search groups, continuous scrolling, or the ability to add additional search sources."[94]
- advanced Live search options are available, and Live's special web syntax appears to work.
- ***continuous scroll*** eliminates need to click to see more results.
- ***cached*** option is available.
- no "similar pages" or "more results from..." options.

Clearly, Amazon is ambivalent about A9 and its future, but given Amazon's extraordinary accomplishments and the company's innovative approach to providing information (and selling stuff), I hope A9 will not merely survive but thrive.

A9 http://a9.com/
A9 Lite http://lite.a9.com/

Book Search

Google, Amazon/A9, Microsoft's Live Book Search, and Project Gutenberg <http://www.gutenberg.org/> provide an invaluable service for researchers that is not duplicated by any other type of search. This new approach to search is nothing short of a revolution in the way we are able to discover, access, and use information. Book search results are frequently better in terms of authoritativeness, utility, and thoroughness. However, I have found that the overlap among the three book search services appears to be even less than the overlap among web searches.

[94] A9 Help, <http://a9.com/-/company/help.jsp#lite> (10 October 2006).

If you are serious about in-depth research, you must use book search sites. The fact is, ***the information available through book search is for the most part entirely different from that provided by web search***. I urge you to use this deep web source, which *Newsweek* correctly described as "a lightning bolt from the future."[95]

Amazon's "Search Inside the Book"

A9 (select "books by Amazon") http://www.a9.com/

Amazon (search "Books") http://www.amazon.com/

Despite its many options, the main reason for using A9 is **Amazon's Search Inside the Book** feature, a major and unique tool for researching the invisible web. Until I started using it, I did not realize how important and valuable this feature is for researchers. Search Inside the Book lets users search through millions of pages from hundreds of thousands of books in the Amazon catalog. Unlike most book searches, ***most of the searchable books at Amazon are still under copyright***. To avoid copyright infringement problems, users can only access content of books for which Amazon has the publisher's permission to display copyrighted material. Also, while anyone can search inside the available books, only registered Amazon users can see the full text.

However, if you are able to register to use Amazon's service, Amazon's Search Inside the Book results are often better than web searches in terms of authoritativeness, utility, and thoroughness. Furthermore, ***Amazon's Search Inside the Book often lets users view the full content of books for which Google's and Live's book searches only provide limited views***.

Search Inside the Book is an option to search full text and full image content, but only as permitted by the publisher. This means that given two different editions of the same book—in this example, *Pride and Prejudice*—one is fully searchable while the other is not. How do you determine which books allow you to search their contents? ***Look for the "Search Inside" logo with any book at Amazon***; if it is there, you can search the contents of the book (including front and back covers, table of contents, index, and text). Only keyword searching works (no phrases, for example) and ALL terms are searched. Here is an example of a search for [pride prejudice] in Austen's classic:

[95] Steven Levy, "Welcome to History 2.0," *Newsweek*, 10 November 2003, <http://www.msnbc.msn.com/id/3339649> (14 November 2006).

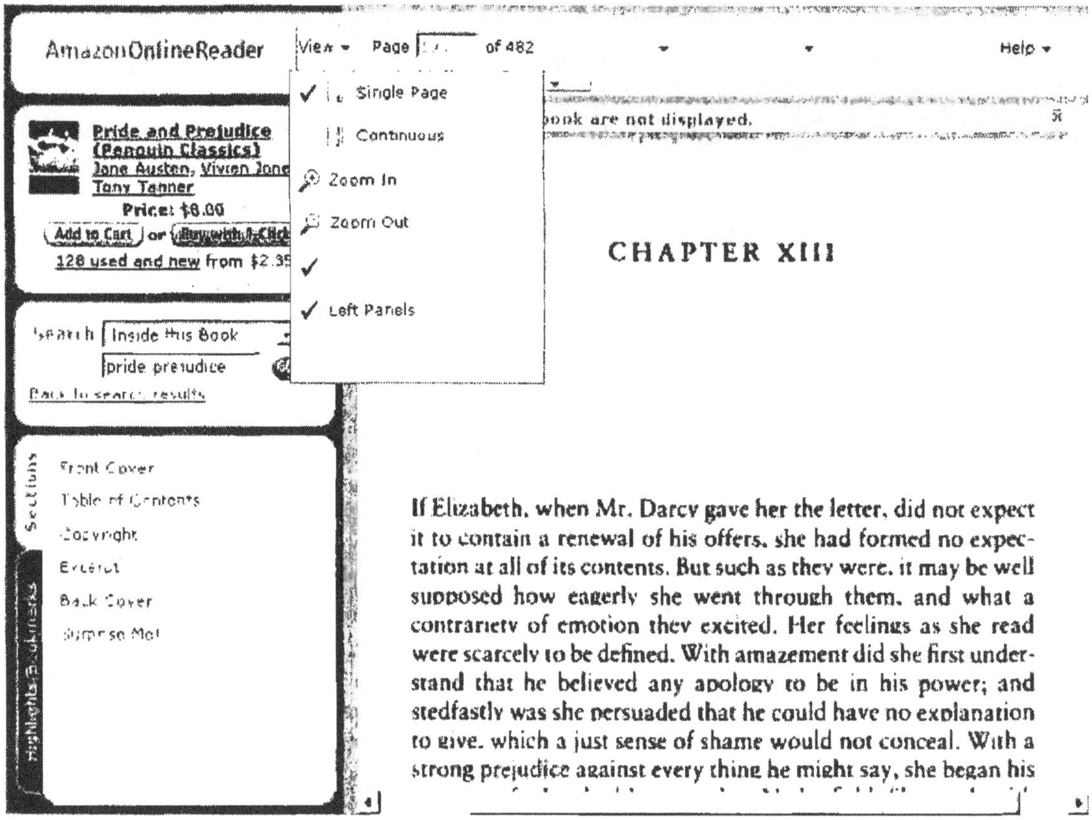
AmazonOnlineReader
View
Page
of 482
Help
Single Page
Continuous
Zoom In
Zoom Out
Left Panels
Pride and Prejudice (Penguin Classics)
Jane Austen, Vivien Jone
Tony Tanner
Add to Cart
128 used and new from $2.3
Search
Inside this Book
pride prejudice
Front Cover
Table of Contents
Copyright
Excerpt
Back Cover
Surprise Me!
Sections
Highlights/Bookmarks
CHAPTER XIII
If Elizabeth, when Mr. Darcy gave her the letter, did not expect it to contain a renewal of his offers, she had formed no expectation at all of its contents. But such as they were, it may be well supposed how eagerly she went through them, and what a contrariety of emotion they excited. Her feelings as she read were scarcely to be defined. With amazement did she first understand that he believed any apology to be in his power; and stedfastly was she persuaded that he could have no explanation to give, which a just sense of shame would not conceal. With a strong prejudice against every thing he might say, she began his

Notice you can view the pages continuously, view just the image of the page, and zoom in or out. What you cannot do, unless you buy an upgrade, is print the book (or any part of it), add bookmarks, highlight passages, or copy portions. The conditions for upgrading your search inside the book options require that you purchase or have purchased from Amazon.com in the past the specific book you wish to search. At present, upgrades are only available to US customers.

Here are a couple of important things to remember about Amazon's Search Inside the Book: you can actually search the entire Amazon book database or even the entire Amazon product line from any search inside the book query screen. The dropdown search menu includes options to search all Amazon books or products.

Search
Inside this Book
Inside this Book
Amazon.com Books
All Products
The Web (A9.com)

Furthermore, the A9 search engine will also search inside all books in the Amazon database and does so more efficiently. From my experience, I can tell you that Amazon may claim it cannot find any results that match your search; however, selecting "Click here to see additional results that may be relevant to your search" brings up the same results you would get from an A9 search for the same term, so always check those additional results just in case (or use A9 in the first place). As of now A9 remains the most efficient way to search inside the books at Amazon if you want to search across the Amazon database and not within a specific book. Look for the "**See more references to [keyword] in this book**" in the A9 book search.

The easiest way to understand the A9 search is to try it. Here are the results for the query [elliptic curve cryptography] limited to books by Amazon.

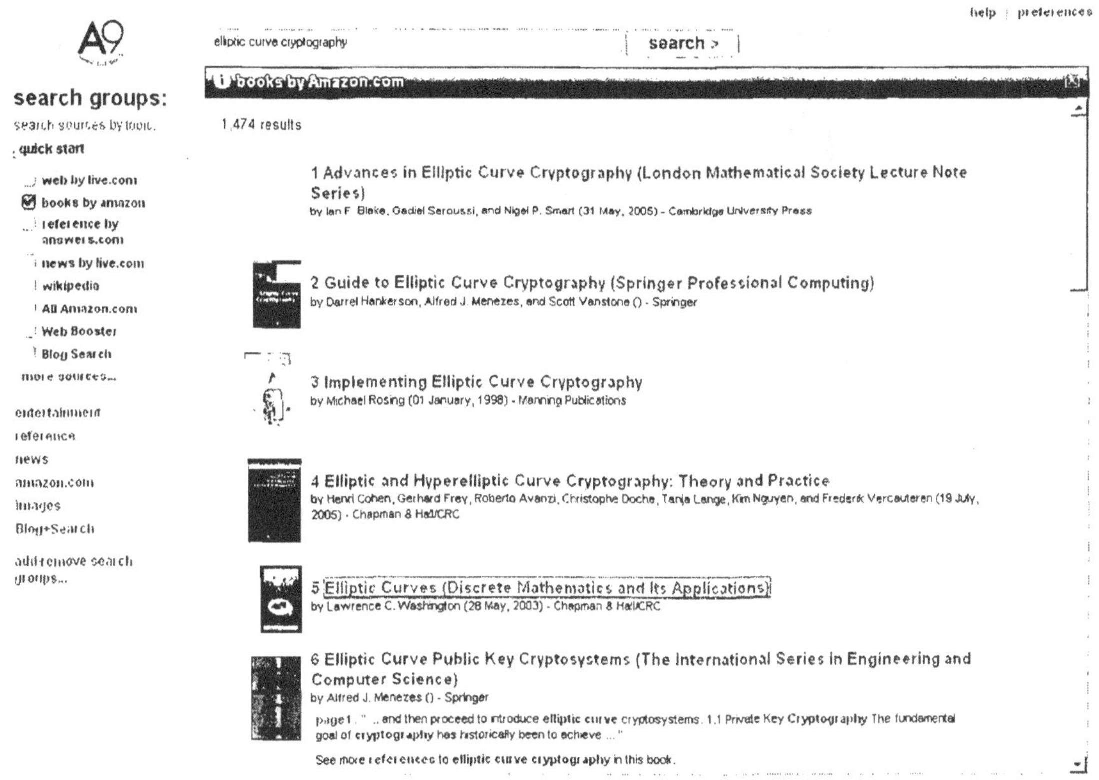
help | preferences
A9
elliptic curve cryptography
search >
search groups:
search sources by topic
quick start
web by live.com
books by amazon
reference by answers.com
news by live.com
wikipedia
All Amazon.com
Web Booster
Blog Search
more sources...
entertainment
reference
news
amazon.com
images
Blog+Search
add/remove search groups...
books by Amazon.com
1,474 results
1 Advances in Elliptic Curve Cryptography (London Mathematical Society Lecture Note Series)
by Ian F. Blake, Gadiel Seroussi, and Nigel P. Smart (31 May, 2005) - Cambridge University Press
2 Guide to Elliptic Curve Cryptography (Springer Professional Computing)
by Darrel Hankerson, Alfred J. Menezes, and Scott Vanstone () - Springer
3 Implementing Elliptic Curve Cryptography
by Michael Rosing (01 January, 1998) - Manning Publications
4 Elliptic and Hyperelliptic Curve Cryptography: Theory and Practice
by Henri Cohen, Gerhard Frey, Roberto Avanzi, Christophe Doche, Tanja Lange, Kim Nguyen, and Frederik Vercauteren (19 July, 2005) - Chapman & Hall/CRC
5 Elliptic Curves (Discrete Mathematics and Its Applications)
by Lawrence C. Washington (28 May, 2003) - Chapman & Hall/CRC
6 Elliptic Curve Public Key Cryptosystems (The International Series in Engineering and Computer Science)
by Alfred J. Menezes () - Springer
page 1: "...and then proceed to introduce elliptic curve cryptosystems. 1.1 Private Key Cryptography The fundamental goal of cryptography has historically been to achieve ..."
See more references to elliptic curve cryptography in this book.

On the right are results from a search across the vast Amazon database. If you click on a link to one of the books, that book's entry at Amazon appears, along with a new Search Inside the Book option. From here, you have to enter the search terms again (an extra step A9/Amazon would do well to eliminate). Here are the results for the query [fermat's last theorem] in this specific book:

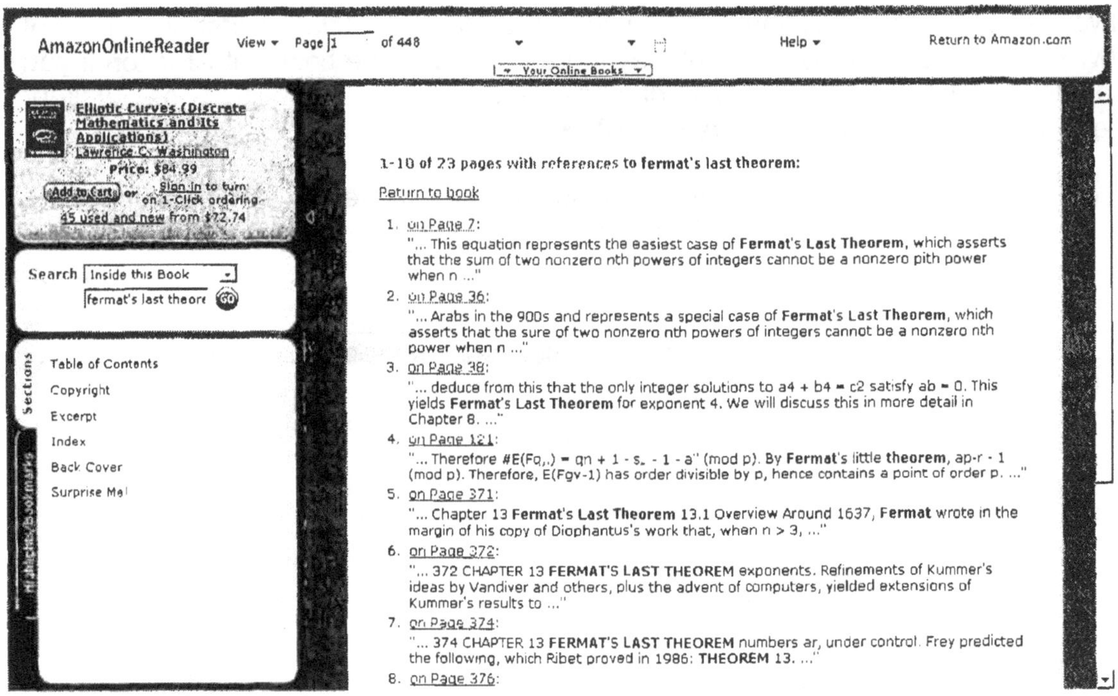
AmazonOnlineReader
View
Page 1 of 448
Help
Return to Amazon.com
Your Online Books
Elliptic Curves (Discrete Mathematics and Its Applications)
Lawrence C. Washington
Price: $84.99
Add to Cart or Sign in to turn on 1-Click ordering
45 used and new from $72.74
Search Inside this Book
fermat's last theore
Sections
Table of Contents
Copyright
Excerpt
Index
Back Cover
Surprise Me!
Highlights/Bookmarks
1-10 of 23 pages with references to fermat's last theorem:
Return to book
1. on Page 7:
"... This equation represents the easiest case of Fermat's Last Theorem, which asserts that the sum of two nonzero nth powers of integers cannot be a nonzero pith power when n ..."
2. on Page 36:
"... Arabs in the 900s and represents a special case of Fermat's Last Theorem, which asserts that the sure of two nonzero nth powers of integers cannot be a nonzero nth power when n ..."
3. on Page 38:
"... deduce from this that the only integer solutions to a4 + b4 = c2 satisfy ab = 0. This yields Fermat's Last Theorem for exponent 4. We will discuss this in more detail in Chapter 8. ..."
4. on Page 121:
"... Therefore #E(Fq,.) = qn + 1 - s. - 1 - a" (mod p). By Fermat's little theorem, ap-r - 1 (mod p). Therefore, E(Fgv-1) has order divisible by p, hence contains a point of order p. ..."
5. on Page 371:
"... Chapter 13 Fermat's Last Theorem 13.1 Overview Around 1637, Fermat wrote in the margin of his copy of Diophantus's work that, when n > 3, ..."
6. on Page 372:
"... 372 CHAPTER 13 FERMAT'S LAST THEOREM exponents. Refinements of Kummer's ideas by Vandiver and others, plus the advent of computers, yielded extensions of Kummer's results to ..."
7. on Page 374:
"... 374 CHAPTER 13 FERMAT'S LAST THEOREM numbers ar, under control. Frey predicted the following, which Ribet proved in 1986: THEOREM 13. ..."
8. on Page 376:

As you can see, each of the 23 pages with references to Fermat's Last Theorem in this book appear in order. From here, you can select a specific page and view it, ***but only if you are a registered Amazon user.***

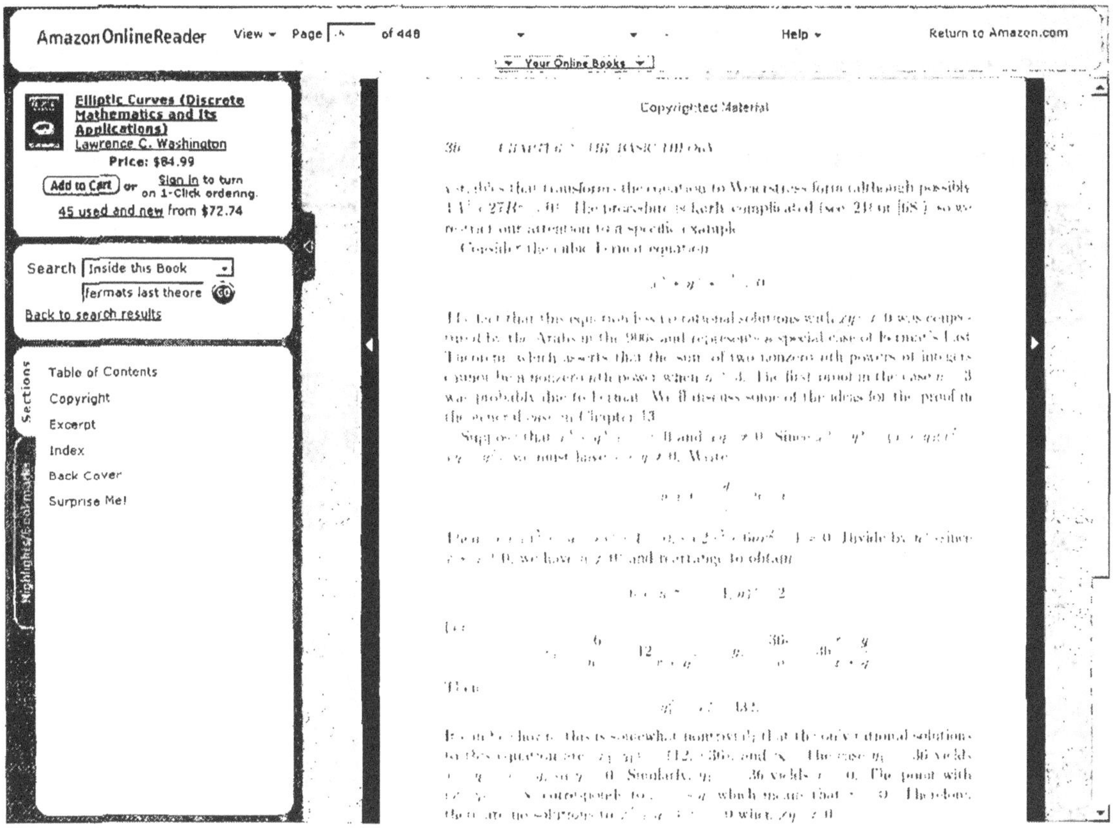
AmazonOnlineReader
View
Page
of 448
Help
Return to Amazon.com
Your Online Books
Elliptic Curves (Discrete Mathematics and Its Applications)
Lawrence C. Washington
Price: $84.99
Add to Cart
or
Sign in to turn on 1-Click ordering.
45 used and new from $72.74
Search
Inside this Book
fermats last theore
Back to search results
Sections
Table of Contents
Copyright
Excerpt
Index
Back Cover
Surprise Me!
Highlights/Bookmarks
Copyrighted Material

Registered Amazon users can view the full text, although Amazon prohibits downloading or printing[96] of these pages for copyright reasons.

Until recently, no one really came close to Amazon's book search option, but Google and Live Search are making great strides with their respective book searches.

Google Book Search

http://books.google.com/

The shift during 2005 from the old Google Print, which was designed to "provide links to some popular book sellers that may offer the full versions of these publications for sale," to Google Book Search transformed this service into a genuine competitor of Amazon's Search Inside the Book.

As a further sign of Google's commitment to digitizing print materials and making them searchable, Google added a distinct Google Book Search homepage during 2005. Unlike Google's web search, Google Book Search only searches an index of

---

[96] There is an inelegant work-around to enable printing of pages inside books at Amazon; put your browser into full screen and select "Print Screen"; then paste the image into either Word or PowerPoint and print from there.

books either from its publishers' collection or its library collection. It's important to understand the enormous significance of this search: Google Book Search searches the content of thousands of books. Here's what Google says about its service:

> "When we find a book whose content contains a match for your search terms, we'll link to it in your search results. Click a book title and you'll see the Snippet View which, like a card catalog, shows information about the book plus a few snippets - a few sentences of your search term in context. You may also see the Sample Pages View if the publisher or author has given us permission or the Full Book View if the book is out of copyright. In all cases, you'll also see 'Buy this Book' links that lead directly to online bookstores where you can buy the book."[97]

Each book indexed by Google includes an "About this book" page with basic bibliographic data like title, author, publication date, length and subject. There are three types of "views" for scanned books:

5. Full view: only for books that Google has publisher permission for or are in the public domain (out of copyright). If the book is in the public domain, users can download, save, and print a PDF version of the book.

6. Limited preview: books from Google's Partner Program permit users to view a few full pages. Even in limited preview users can run multiple searches within the book or browse the available pages.

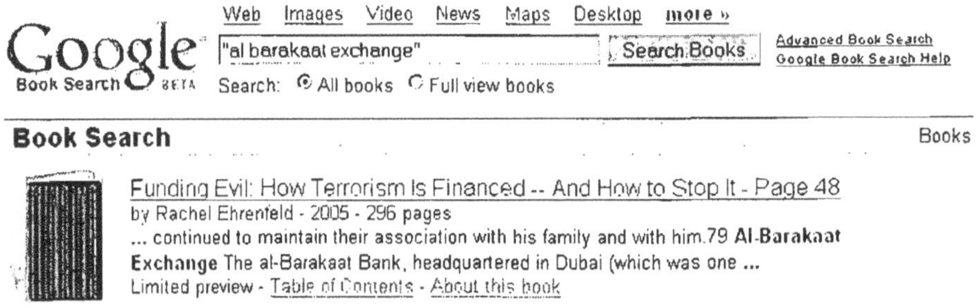
Web Images Video News Maps Desktop more »
Google Book Search BETA
"al barakaat exchange"
Search Books
Advanced Book Search
Google Book Search Help
Search: All books Full view books
Book Search
Books
Funding Evil: How Terrorism Is Financed -- And How to Stop It - Page 48
by Rachel Ehrenfeld - 2005 - 296 pages
... continued to maintain their association with his family and with him.79 Al-Barakaat Exchange The al-Barakaat Bank, headquartered in Dubai (which was one ...
Limited preview - Table of Contents - About this book

7. Snippet view: users can view the "About this book" page and search within the book. Search results will show a maximum of three snippets of text from the book with the query term highlighted.

8. No preview available: users will only see an "About this book" page with bibliographic information about the book and links to bookstores or libraries where the book may be available.

---

[97] About Google Book Search, <http://print.google.com/googleprint/about.html> (14 November 2006).

Even a limited preview can provide very useful information. As you can see, the query terms are highlighted in the text and there is a search box that permits additional searches inside the specific book. Furthermore, you can view the book's table of contents and index, read a summary, and find out more about the book.

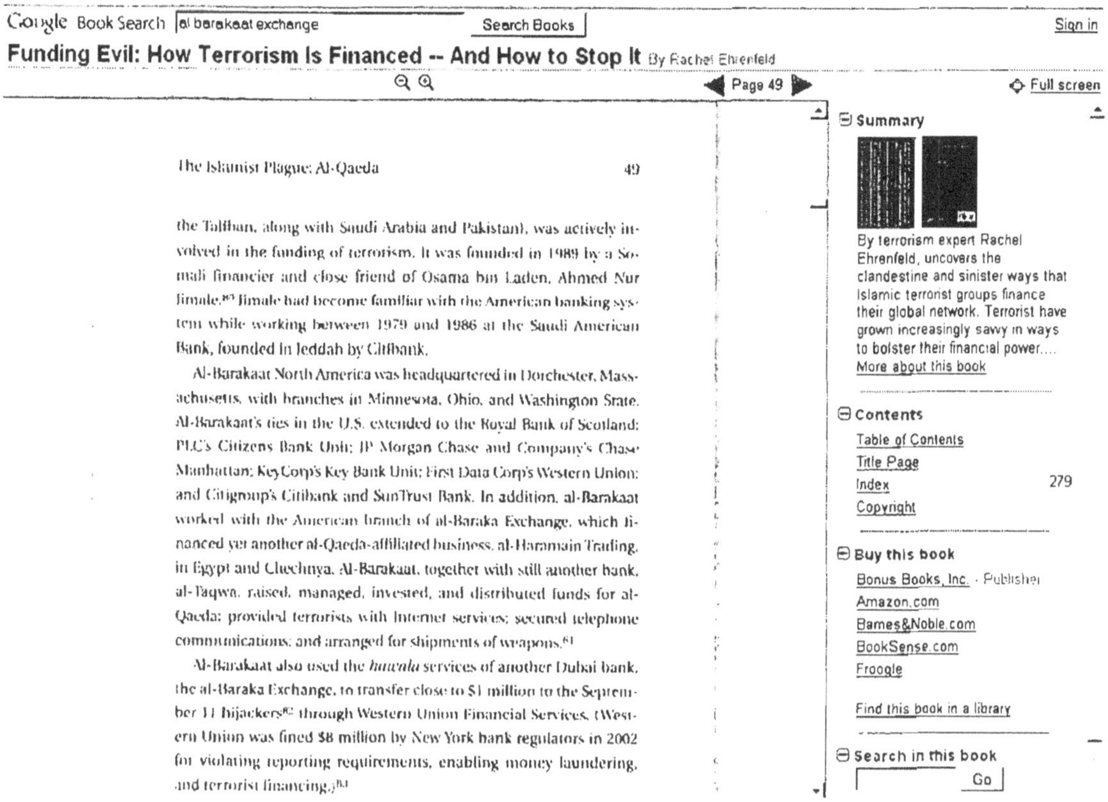
Google Book Search
al barakaat exchange
Search Books
Sign in
Funding Evil: How Terrorism Is Financed -- And How to Stop It By Rachel Ehrenfeld
Page 49
Full screen
The Islamist Plague: Al-Qaeda 49
the Taliban, along with Saudi Arabia and Pakistan), was actively involved in the funding of terrorism. It was founded in 1989 by a Somali financier and close friend of Osama bin Laden, Ahmed Nur Jimale. Jimale had become familiar with the American banking system while working between 1979 and 1986 at the Saudi American Bank, founded in Jeddah by Citibank.
Al-Barakaat North America was headquartered in Dorchester, Massachusetts, with branches in Minnesota, Ohio, and Washington State. Al-Barakaat's ties in the U.S. extended to the Royal Bank of Scotland: PLC's Citizens Bank Unit; JP Morgan Chase and Company's Chase Manhattan; KeyCorp's Key Bank Unit; First Data Corp's Western Union; and Citigroup's Citibank and SunTrust Bank. In addition, al-Barakaat worked with the American branch of al-Baraka Exchange, which financed yet another al-Qaeda-affiliated business, al-Haramain Trading, in Egypt and Chechnya. Al-Barakaat, together with still another bank, al-Taqwa, raised, managed, invested, and distributed funds for al-Qaeda; provided terrorists with Internet services; secured telephone communications; and arranged for shipments of weapons.
Al-Barakaat also used the hawala services of another Dubai bank, the al-Baraka Exchange, to transfer close to $1 million to the September 11 hijackers through Western Union Financial Services. (Western Union was fined $8 million by New York bank regulators in 2002 for violating reporting requirements, enabling money laundering, and terrorist financing.)
Summary
By terrorism expert Rachel Ehrenfeld, uncovers the clandestine and sinister ways that Islamic terrorist groups finance their global network. Terrorist have grown increasingly savvy in ways to bolster their financial power....
More about this book
Contents
Table of Contents
Title Page
Index 279
Copyright
Buy this book
Bonus Books, Inc. - Publisher
Amazon.com
Barnes&Noble.com
BookSense.com
Froogle
Find this book in a library
Search in this book
Go

Google Book Search offers a number of advanced search options either using the advanced search page or special syntax:

- **inpublisher:** Publishers such as W. W. Norton need only be searched as inpublisher:norton.
  Example of how to use the **inpublisher:** command: [inpublisher:o-reilly]

- **inauthor:** Unless the author is Shakespeare or Chaucer, a full name search is advisable.
  Example of how to use the **inauthor:** command: [inauthor:patrick-o-brian]

- **intitle:** Multiple word titles are best searched for as phrases, i.e., in double quotes.
  Example of how to use the **intitle:** command:
  [intitle:"nutmeg of consolation"]

- **isbn:** Very useful for finding a specific edition of a book because the International Standard Book Number uniquely identifies each edition,

variation, or format.
<u>Example of how to use the **isbn:** command</u>: [isbn:0393030326]

All of these special search operators—with the exception of ISBN, which identifies a unique entry—can be used in combination or alone. For example, the query [inpublisher:norton] returns almost 61,000 pages from books published by W. W. Norton. I can add to this query a keyword for a title, e.g., [inpublisher:o-reilly intitle:programming] to see how many books O'Reilly publishes with *programming* in the title. Or I can look for keywords occurring anywhere in a specific publisher's books: [inpublisher:o-reilly "network security"].

To find public domain books that can be viewed using Google Books' Viewer *and* downloaded as PDFs, first restrict the search to "full view books" because all public domain books can be downloaded. However, the converse is not true: all full view books are not downloadable. How can you tell if a book can be downloaded? As of now, the only way to do so is to select the book and see if the ***Download PDF option*** appears, as it does here:

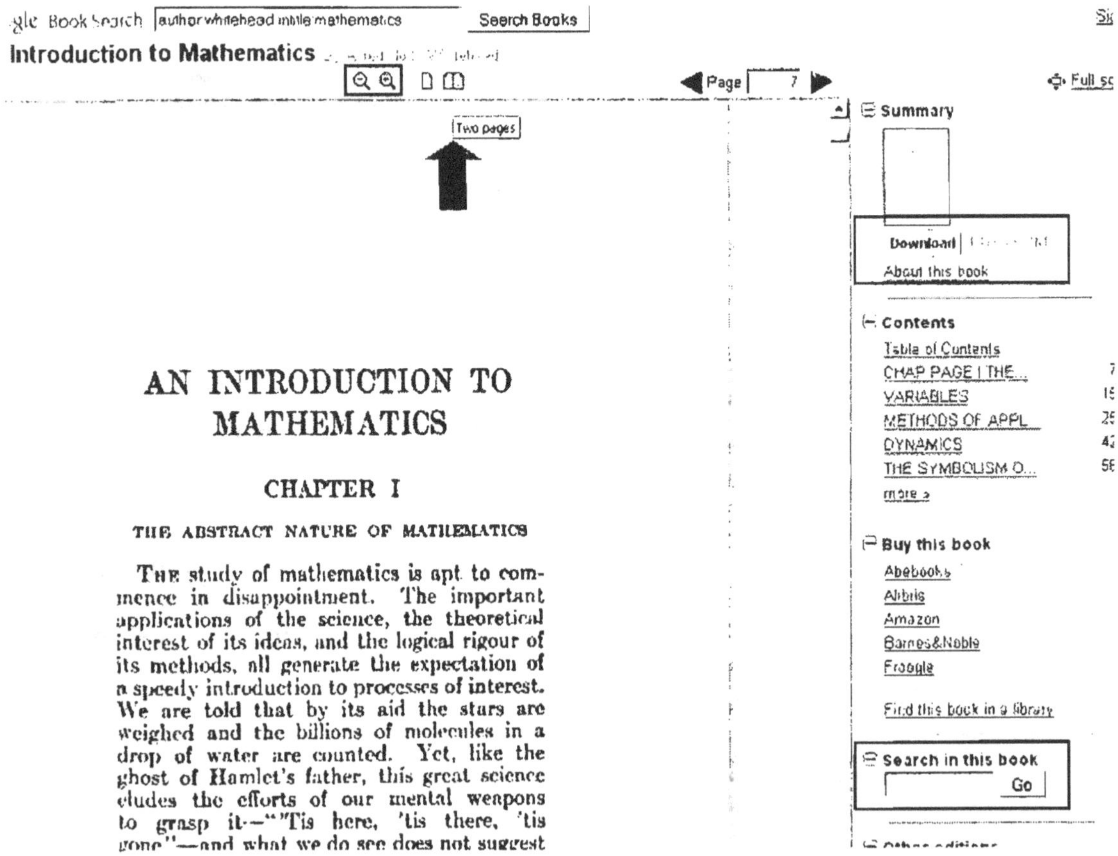
Book Search
author:whitehead intitle:mathematics
Search Books
Introduction to Mathematics
Page 7
Two pages
Summary
Download
About this book
Contents
Table of Contents
CHAP PAGE I THE...
VARIABLES
METHODS OF APPL
DYNAMICS
THE SYMBOLISM O...
more »
Buy this book
Abebooks
Alibris
Amazon
Barnes&Noble
Froogle
Find this book in a library
Search in this book
Go
AN INTRODUCTION TO MATHEMATICS
CHAPTER I
THE ABSTRACT NATURE OF MATHEMATICS
THE study of mathematics is apt to commence in disappointment. The important applications of the science, the theoretical interest of its ideas, and the logical rigour of its methods, all generate the expectation of a speedy introduction to processes of interest. We are told that by its aid the stars are weighed and the billions of molecules in a drop of water are counted. Yet, like the ghost of Hamlet's father, this great science eludes the efforts of our mental weapons to grasp it—"'Tis here, 'tis there, 'tis gone"—and what we do see does not suggest

Please note that you can also search inside the book from this page. Google Book Viewer includes a zoom option and the option to view one or two pages at a time, making the page-to-page scroll much smoother. I cannot recommend Google's "About This Book" option highly enough: it provides the book's publication information, a summary, links to find the book in a library or to buy it, the table of contents, selected pages, related books or references from the book, and key terms.

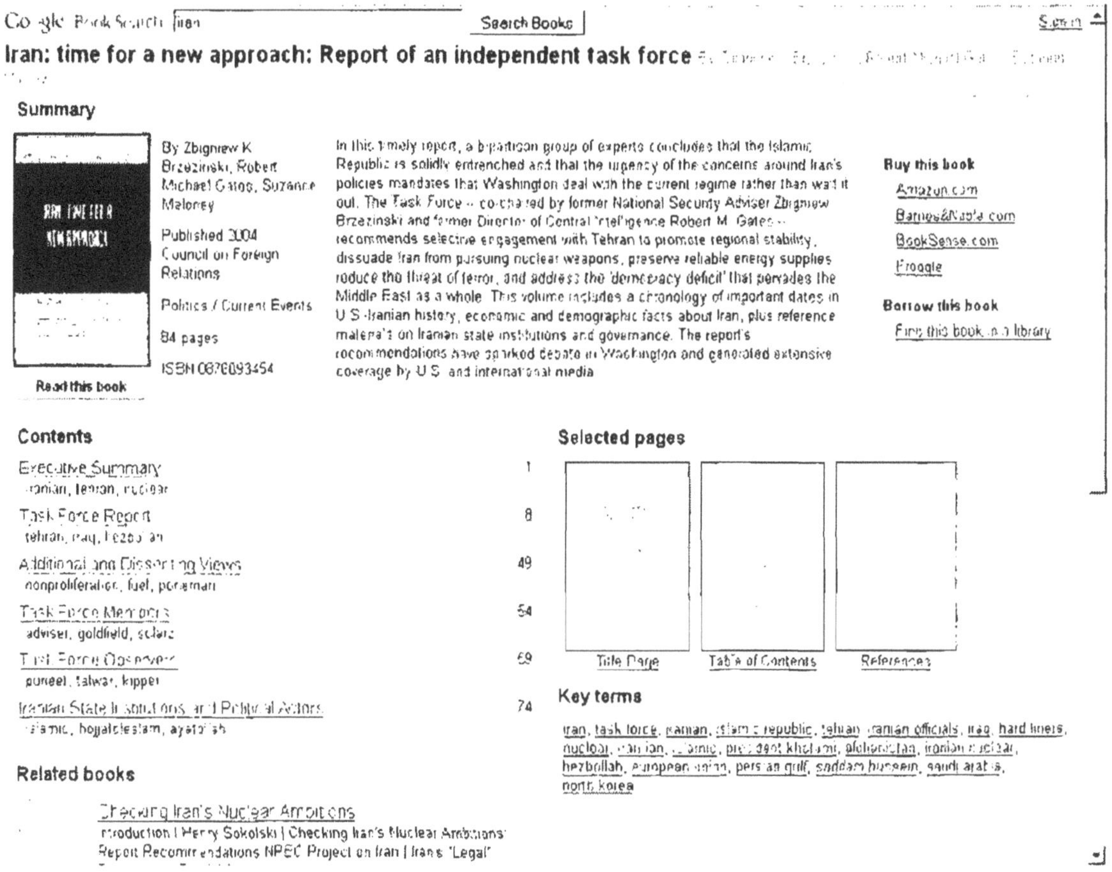
Search Books
Iran: time for a new approach: Report of an independent task force
Summary
Published 2004
Council on Foreign Relations
Politics / Current Events
84 pages
Read this book
Buy this book
Borrow this book
Contents
Executive Summary
Task Force Report
Task Force Members
Selected pages
Title Page
Table of Contents
References
Key terms
Related books
Checking Iran's Nuclear Ambitions

You will discover when you start looking at a PDF version of a book that these volumes were scanned by hand. You may even see images of people's hands as they place the books on the scanner, so don't be surprised if you see strange images. At this time, there are some problems with the downloaded files: "The PDF download seems to be have problems, as all the downloads stop at 1.3 MB, even though the files are much larger, but the bugs should be fixed."[98]

[98] Ionut Alex. Chitu, "Download Public Domain Books from Google," Google Operating System Blog, 30 August 2006, <http://googlesystem.blogspot.com/2006/08/download-public-domain-books-from.html> (12 September 2006).

While there is no option to limit your search by language, you can also search Google Books in languages other than English simply by entering a query in another language, e.g., ["quedó el moro"] returns only books written in Spanish. However, many public domain books in other languages have not been scanned because the Google Book project thus far has focused mainly on US libraries. There are many Spanish, French, German books scanned into Google Books, for example, but very few in Arabic.

At the same time many US publishers were fighting Google Book Search, several European nations complained that Google was not offering to add non-English language materials to the database. In September 2005, Google opened Google Book Search to European publishers and added a number of new discrete webpages for Google Book Search in several non-English speaking countries. They are:

- Austria
- Belgium
- France
- Germany
- The Netherlands/Holland
- Italy
- Switzerland
- Spain
- Brazil

A number of European publishers expressed interest in participating in the digitization project, but of course similar copyright concerns raised by US publishers will have to be addressed for European and, eventually, other non-US publishers. [99]

Even though users may use a specific country interface, they will probably see results from other countries' indices. Thus far, the European publishers that have signed pacts with Google are Grupo Planeta and Grupo Anaya of Spain, De Boeck and Editions De L'Eclat of France and Springer Science & Business Media of Holland. Interestingly, all this comes at a time when Google is embroiled in a lawsuit brought by the US book industry over alleged copyright violations. Google is trying to avoid a similar fight in Europe by limiting its book scanning to books that are either pre-1900 or in the public domain or from publishers with whom Google has signed agreements.

---

[99] Michael Liedtke, "Google Opens Digital Library to European Book Publishers," *The Detroit News*, 2 September 2005, <http://www.detnews.com/2005/technology/0509/05/0tech-300866.htm> (14 November 2006).

During 2006 Google also added a **Library Catalog Search** feature to Google Book Search. This feature will help searchers find books in libraries around the world. According to the Google Blog, "Queries on Google Book Search will automatically include results from library catalogs when appropriate. Each result includes a 'Find Libraries' link to help readers find libraries that hold the book—ideally a library nearby, or if need be, a library far away...we have worked with more than 15 library union catalogs that have information about libraries from more than 30 countries, as well as with our colleagues working on Google Scholar (which includes a similar feature just for scholarly books)."[100]

Google Books advanced search has an option to limit a search to library catalogs, so if you know the book and want to see where it is available, this search makes it easy to find a library. Keep in mind that at present Google Books and WorldCat are searching a limited number of libraries and catalogs.

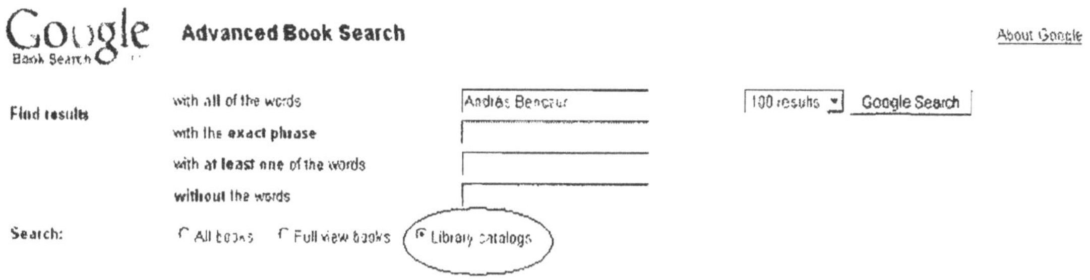
Google
Book Search
Advanced Book Search
About Google
Find results
with all of the words
András Benczúr
100 results
Google Search
with the exact phrase
with at least one of the words
without the words
Search:
All books
Full view books
Library catalogs

Use Google Books Advanced Search to limit query to library catalogs

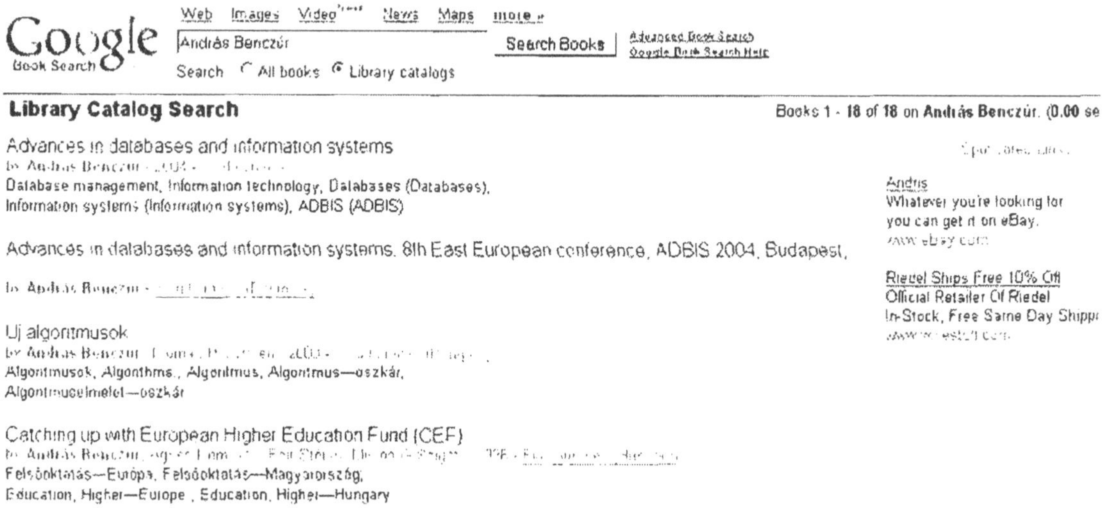
Google
Book Search
Web
Images
Video
News
Maps
more »
András Benczúr
Search Books
Search
All books
Library catalogs
Library Catalog Search
Books 1 - 18 of 18 on András Benczúr.
Advances in databases and information systems
Database management, Information technology, Databases (Databases), Information systems (Information systems), ADBIS (ADBIS)
Advances in databases and information systems. 8th East European conference, ADBIS 2004, Budapest,
Uj algoritmusok
Algoritmusok, Algorithms., Algoritmus, Algoritmus—oszkár, Algoritmuselmélet—oszkár
Catching up with European Higher Education Fund (CEF)
Felsőoktatás—Európa, Felsőoktatás—Magyarország, Education, Higher—Europe, Education, Higher—Hungary
Andris
Whatever you're looking for you can get it on eBay.
Riedel Ships Free 10% Off
Official Retailer Of Riedel
In-Stock, Free Same Day Shipp

---

[100] Bruno Fonseca, "Finding the wealth in your library (and everyone else's)," Google Blog, 24 August 24 2006, <http://googleblog.blogspot.com/2006/08/finding-wealth-in-your-library-and.html> (12 September 2006).

The other way to use the library search options is to click on the link for a book title and look for the "Find this book in a library" link. At this point, you enter a geographic location and the WorldCat search will try to locate the book as close to you as possible. In some cases, you may even be able to click through and reserve a book at a local library.

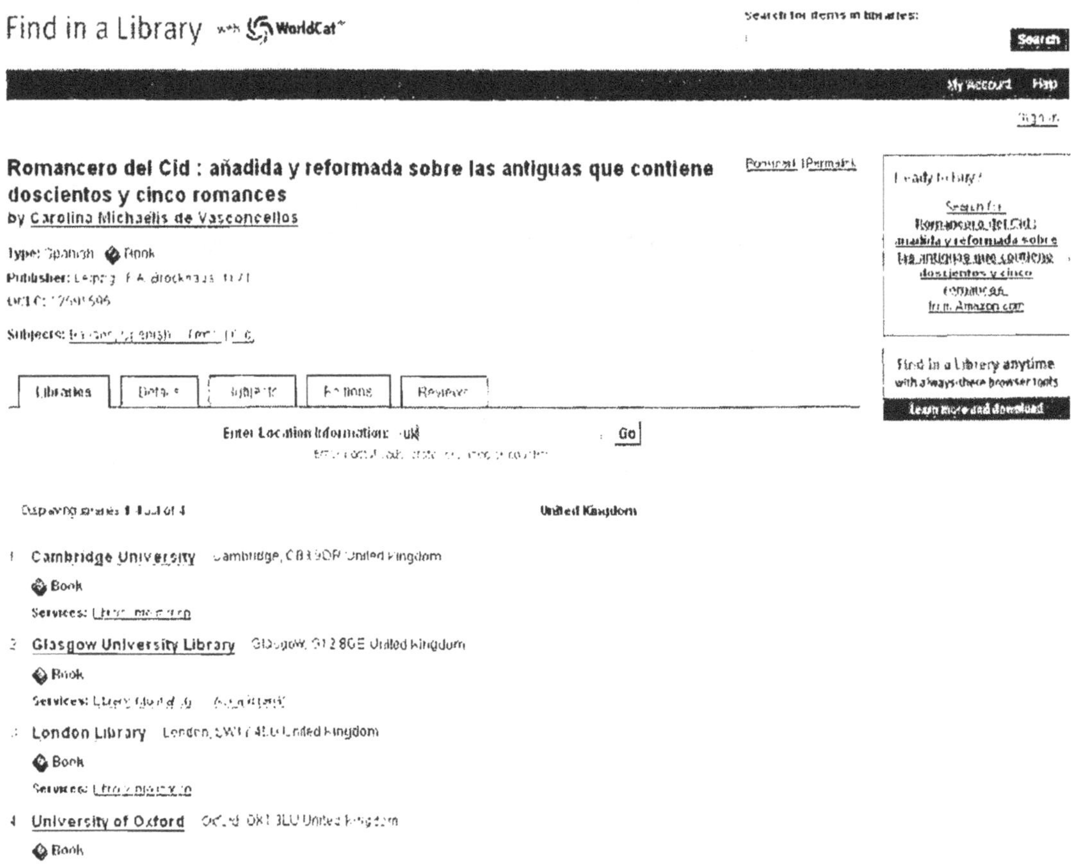
Find in a Library
WorldCat
Search
My Account
Romancero del Cid : añadida y reformada sobre las antiguas que contiene doscientos y cinco romances
by Carolina Michaëlis de Vasconcellos
Type:
Book
Publisher:
Subjects:
Libraries
Reviews
Enter Location Information:
Go
United Kingdom
Cambridge University
Book
Services:
Glasgow University Library
Book
Services:
London Library
Book
Services:
University of Oxford
Book
Find in a Library anytime

This is an excellent service that builds upon library catalog search and makes it more user friendly and accessible through a single interface.

Google is clearly trying to do for book content search what it did for web search, that is, make as much content freely available on line as possible. This may ultimately involve some sort of payment by users who wish to search, print, and download full text content for in copyright books, but for now Google is relying upon advertising to pay for Google Book Search. Google Book Search's future in large measure depends upon the outcome of the current battles between Google, Inc., and copyright holders and publishers, either through some mutual agreement or through litigation.

Live Book Search (Beta)

http://books.live.com/

In December 2006 Microsoft introduced full text searches of public domain (out of copyright) books only. Libraries participating in the digitization project include the University of California system, Trinity College, and the University of Toronto (and Microsoft is actively trying to enlist other libraries). Lest you think that there is no need for yet another book search tool, let me assure you there is. As with search engines, there is a startling lack of overlap in book searches. It did not take me two minutes to find an example: Google did not return a single copy of John McTaggart Ellis McTaggart's *The Nature of Existence*, while Live Book Search found the book in the University of Toronto collection. As you can see, the public domain books may be viewed and searched in their entirety online or downloaded as a PDF.

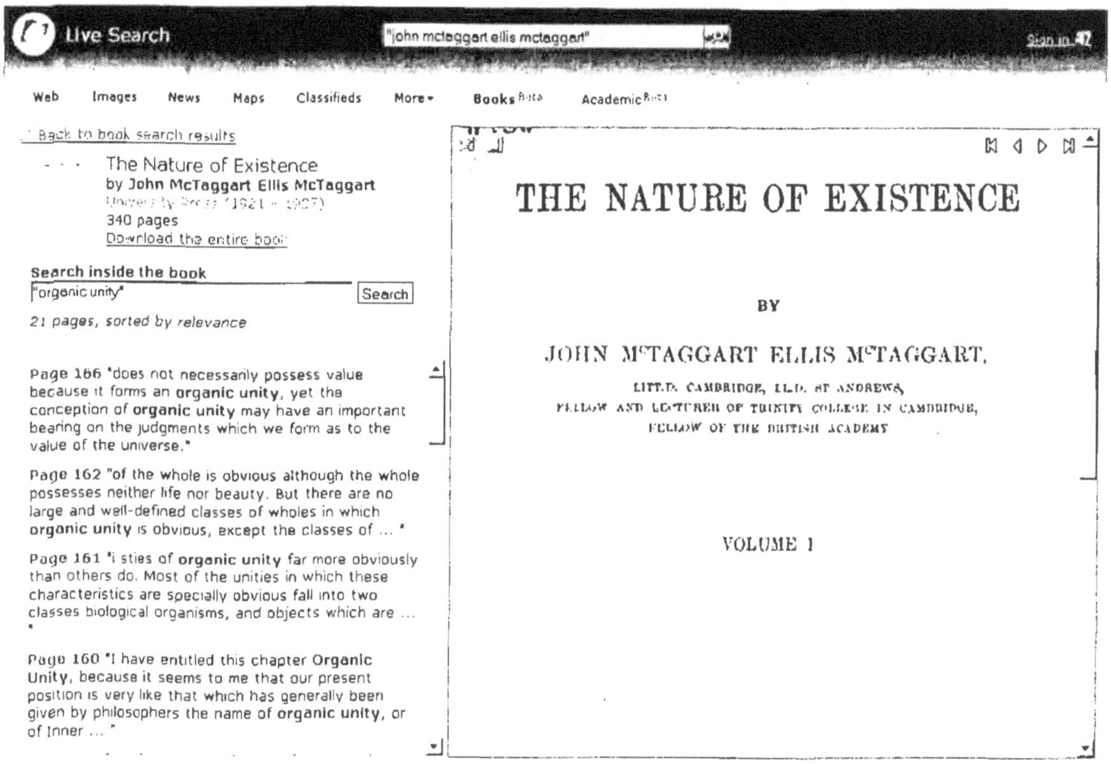
Live Search
"john mctaggart ellis mctaggart"
Web
Images
News
Maps
Classifieds
More
Books
Academic
Back to book search results
The Nature of Existence
by John McTaggart Ellis McTaggart
340 pages
Download the entire book
Search inside the book
"organic unity"
Search
21 pages, sorted by relevance
Page 166 "does not necessarily possess value because it forms an organic unity, yet the conception of organic unity may have an important bearing on the judgments which we form as to the value of the universe."
Page 162 "of the whole is obvious although the whole possesses neither life nor beauty. But there are no large and well-defined classes of wholes in which organic unity is obvious, except the classes of ... "
Page 161 "i sties of organic unity far more obviously than others do. Most of the unities in which these characteristics are specially obvious fall into two classes biological organisms, and objects which are ... "
Page 160 "I have entitled this chapter Organic Unity, because it seems to me that our present position is very like that which has generally been given by philosophers the name of organic unity, or of Inner ... "
THE NATURE OF EXISTENCE
BY
JOHN McTAGGART ELLIS McTAGGART,
VOLUME I

The most serious limitation of Live Book Search is the lack of advanced search features. At present the only one that seems to work is **intitle:**. Live Book Search needs a way to limit searches by author, at the very least, and would do well to emulate Google Book's advanced search page. Still, it is a thrill to be able to search the full text of so many of the myriad books the world has to offer.

Live Book Search is accessible either under the "More" tab on the Live homepage or directly at <http://books.live.com/>.

Metasearch Tool for Book Search

Alan Taylor, a professional Web Developer, created an ingenious tool that searches A9/Amazon Search Inside the Book, Google Books, and Live Books at once. Taylor recognized that, as wonderful as these book search engines are, "they require the searcher to either guess which website is most likely to house the results they want, or to try them one after the other." <http://kokogiak.com/>

This is an invaluable tool because there seems to be even less overlap among book searches than web searches. As you can see from this search, the three search engines returned a number of different books for one author.

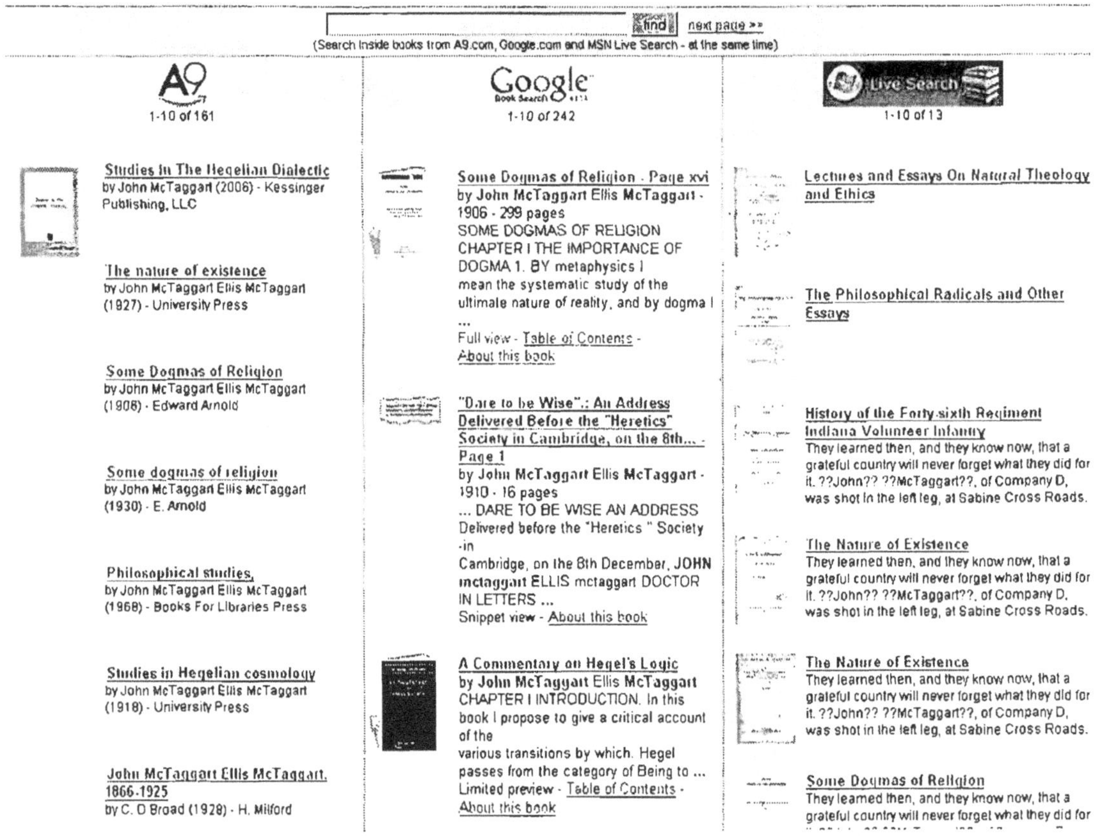
find
next page >>
(Search inside books from A9.com, Google.com and MSN Live Search - at the same time)
A9
1-10 of 161
Google Book Search
1-10 of 242
Live Search
1-10 of 13
Studies In The Hegelian Dialectic
by John McTaggart (2006) - Kessinger Publishing, LLC
The nature of existence
by John McTaggart Ellis McTaggart (1927) - University Press
Some Dogmas of Religion
by John McTaggart Ellis McTaggart (1906) - Edward Arnold
Some dogmas of religion
by John McTaggart Ellis McTaggart (1930) - E. Arnold
Philosophical studies,
by John McTaggart Ellis McTaggart (1968) - Books For Libraries Press
Studies in Hegelian cosmology
by John McTaggart Ellis McTaggart (1918) - University Press
John McTaggart Ellis McTaggart, 1866-1925
by C. D Broad (1928) - H. Milford
Some Dogmas of Religion - Page xvi
by John McTaggart Ellis McTaggart - 1906 - 299 pages
SOME DOGMAS OF RELIGION CHAPTER I THE IMPORTANCE OF DOGMA 1. BY metaphysics I mean the systematic study of the ultimate nature of reality, and by dogma I ...
Full view - Table of Contents - About this book
"Dare to be Wise".: An Address Delivered Before the "Heretics" Society in Cambridge, on the 8th... - Page 1
by John McTaggart Ellis McTaggart - 1910 - 16 pages
... DARE TO BE WISE AN ADDRESS Delivered before the "Heretics" Society -in Cambridge, on the 8th December, JOHN mctaggart ELLIS mctaggart DOCTOR IN LETTERS ...
Snippet view - About this book
A Commentary on Hegel's Logic
by John McTaggart Ellis McTaggart
CHAPTER I INTRODUCTION. In this book I propose to give a critical account of the various transitions by which. Hegel passes from the category of Being to ...
Limited preview - Table of Contents - About this book
Lectures and Essays On Natural Theology and Ethics
The Philosophical Radicals and Other Essays
History of the Forty-sixth Regiment Indiana Volunteer Infantry
They learned then, and they know now, that a grateful country will never forget what they did for it. ??John?? ??McTaggart??, of Company D, was shot in the left leg, at Sabine Cross Roads.
The Nature of Existence
They learned then, and they know now, that a grateful country will never forget what they did for it. ??John?? ??McTaggart??, of Company D, was shot in the left leg, at Sabine Cross Roads.
The Nature of Existence
They learned then, and they know now, that a grateful country will never forget what they did for it. ??John?? ??McTaggart??, of Company D, was shot in the left leg, at Sabine Cross Roads.
Some Dogmas of Religion
They learned then, and they know now, that a grateful country will never forget what they did for

Metasearch for Books http://kokogiak.com/booksearch/

Answers.com

I am usually skeptical about "ask a question" tools, but this one is definitely a cut above most. The reason it is better is simple: Answers.com is the "new" interface for **GuruNet**, which was subscription based and required users to download and install its software. In fact, before the Answers.com site came online, GuruNet was already an integral part of Amazon's A9 search service, powering A9's "Reference" results. If you aren't familiar with GuruNet, it provides information on over 750,000 topics. GuruNet has an extensive reference database, including dozens of encyclopedias, glossaries, and dictionaries. GuruNet will provide reference information from dictionaries (including definitions, pronunciation keys, and language translations), encyclopedias, thesauruses, a geography dictionary, key American historical documents, US Presidents, US Congress, US Supreme Court cases, conversions, abbreviations, Bible dictionary, Old & New Testament, English idioms, wine glossary, music glossary, planetary & astronomy glossary, Marine lingo, and many more. Answers also draws upon Wikipedia for some of its responses, a trend I believe is overdone.

Now users have many options for accessing the Answers.com/GuruNet resources:

1. **Answers.com** website, where users can enter questions or queries that will draw from GuruNet's huge resource repository.

2. **A9**, which will simultaneously query Answers.com/GuruNet, Google, Amazon, and other sources.

3. **Google**, which switched from Dictionary.com to Answers.com for definitions and other reference information.

Answers.com also earns its place as the premier "answer machine" on the Internet by adding a very handy "cite" option. At the bottom of the results page where all the copyrights for each source are listed is a new "Cite" button.

Copyrights

HOUGHTON MIFFLIN COMPANY

Dictionary definition of **François Mitterrand**

The American Heritage® Dictionary of the English Language, Fourth Edition Copyright © 2004, 2000 by Houghton Mifflin Company. Pu... ny. All rights reserved. More from Dictionary "Cite"

Click on "Cite"

Columbia University Press

Encyclopedia information about **François Mitterrand**

The Columbia Electronic Encyclopedia, Sixth Edition Copyright © 2005, Columbia University Press. Licensed from Columbia University Press. All rights reserved. www.cc.columbia.edu/cu/cup/ More from Encyclopedia "Cite"

Wikipedia information about **François Mitterrand**

This article is licensed under the GNU Free Documentation License. It uses material from the Wikipedia article "François Mitterrand". More from Wikipedia "Cite"

When you click on "Cite" you will then see a page that offers proper citations in MLA, Chicago, or APA formats. To use one of these formats for the citation, simply copy and paste the text from this page into your bibliography. This is a very handy tool for researchers who need to cite their sources correctly.

Cite this source from this AnswerPage

The URL of this AnswerPage is:

http://www.answers.com/topic/fran-ois-mitterrand   Select URL

Citation style

- (•) MLA
- ( ) Chicago
- ( ) APA
- ( ) List reference source

Add this to your Bibliography

Select the text below and then 'copy & paste' it to your document:

"François Mitterrand." The American Heritage® Dictionary of the English Language, Fourth Edition. Houghton Mifflin Company, 2004. *Answers.com* GuruNet Corp. 06 Jul. 2005. http://www.answers.com/topic/fran-ois-mitterrand

During 2006 Answers.com continued to update its service. One of the most noticeable changes was the "find as you type" option, which uses AJAX technology to anticipate your query. As you type, possible matching topics appear in a drop-down box as shown below:

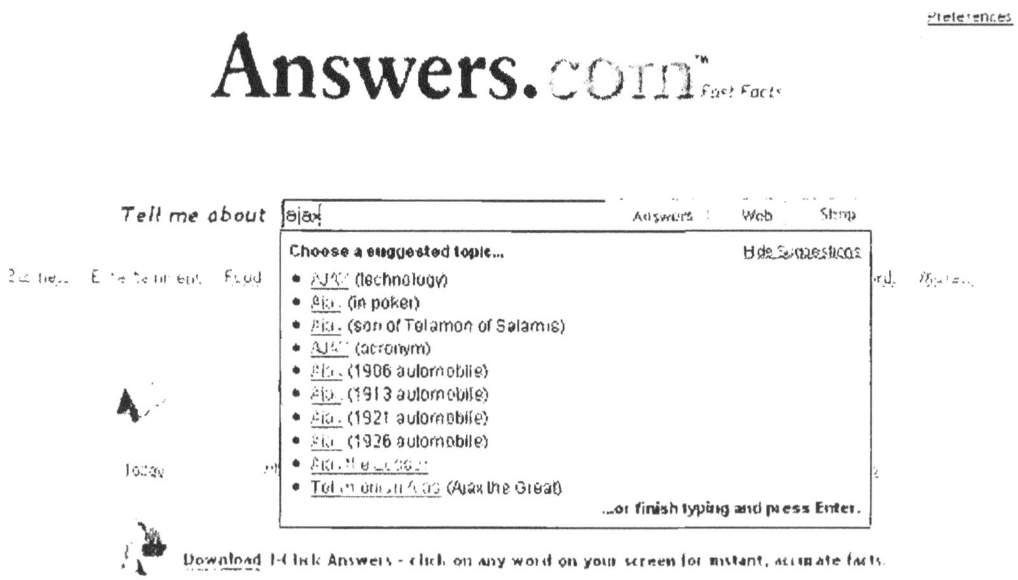
Preferences
Answers.com
Fast Facts
Tell me about
ajax
Answers
Web
Choose a suggested topic...
Hide Suggestions
(technology)
(in poker)
(son of Telamon of Salamis)
(acronym)
(1906 automobile)
(1913 automobile)
(1921 automobile)
(1926 automobile)
(Ajax the Great)
...or finish typing and press Enter.
Download 1-Click Answers - click on any word on your screen for instant, accurate facts.

Surfwax is aggressively marketing this technology through its company **LookAhead** <http://lookahead.surfwax.com/>. Answers.com is only one site among a growing number using AJAX technology; for example, Yahoo's Instant Search, A9 and A9 Maps, Ask, and many others all use AJAX. At the Answers.com site, if you do not like the Find as You Type feature, you can disable it on the preferences page.

The preferences also include a long list of languages to which Answers.com will translate from English, but not the other way around. I found this feature very confusing, so here is how to use it. If you have a term you want to translate from English, on the Answers.com home page enter the term and then scroll down the page until you see "Translations," and you will see the terms translated into all the available languages or only those languages you have selected on your preferences page. The translation also includes common idioms, which is very useful. You can also go to the translations' page <http://www.answers.com/library/Translations> and select the term from an alphabetic list:.

Here is a partial list of translations of the English word "abode"; ***this could be a very useful tool for copy/pasting search terms without needing non-Latin keyboards to input the terms***:

Deutsch (German)
n. - Wohnort, Wohnsitz

idioms:

- **no fixed abode** ohne festen Wohnsitz

Ελληνική (Greek)
n. - κατοικία, ενδιαίτημα, διαμονή

idioms:

- **no fixed abode** χωρίς μόνιμη κατοικία

Italiano (Italian)
residenza, domicilio

idioms:

- **no fixed abode** senza fissa dimora
- **right of abode** diritto di soggiorno

Português (Portuguese)
n. - lar (m), permanência (f) curta, residência (f)

idioms:

- **no fixed abode** sem residência fixa
- **right of abode** direito (m) de residência

Русский (Russian)
жилище, обиталище

idioms:

- **no fixed abode** без постоянного местожительства
- **right of abode** право на жилье

Svenska (Swedish)
n. - boning, bostad

中文(简体) (Chinese (Simplified))
住处, 住所

idioms:

- **no fixed abode** 无固定住处

中文(繁體) (Chinese (Traditional))
n. - 住處, 住所

idioms:

- **no fixed abode** 無固定住處

한국어 (Korean)
n. - 거주, 주소

日本語 (Japanese)
n. - 住所, 居住

idioms:

- **no fixed abode** 住所不定
- **right of abode** 居住権

العربية (Arabic)
(الاسم) مقر, إقامة

עברית (Hebrew)
n. - דירה, מגורים, בית, מעון

Answers.com continues to expand and improve. It is often my first stop for search and research.

As with any search, the best tool to use depends on what you need to find.

- If you want links to relevant websites about a topic, Google is probably the way to go.
- If you only need a fast answer in the form of a dictionary definition or encyclopedia entry, Answers.com is the best choice.
- But if you want thoroughness, A9 wins this one hands down because it quickly offers a choice among web, image, book, and a huge number and variety of reference results that are only a mouse click away.

Any way you look at it, Answers.com is a great addition to any researcher's resource set.

Answers.com http://www.answers.com/

OAIster

Add to your deep web bookmarks OAIster (pronounced "oyster") to help you "find the pearls." This information retrieval resource now contains a very impressive **9,950,256** records from **729 institutions** (as of 10 January 2007). OAIster states, "The service encompasses as broad a collection of resources as possible (i.e., with no subject parameters). It is accessible to the entire Internet community, without bounds." However, as you might have guessed, OAIster is not for run-of-the-mill searches.

So just what sort of information does OAIster index? The project seeks to provide easy access to *actual digital records*, not just lists, links, or bibliographies. What are these digital resources? They "can range from a 1959 photograph of an A&P bakery (from the Library of Congress American Memory project) to poems by Emily Dickinson (from the University of Michigan Digital Library Production Services American Verse project). Digital resources include items such as:

- electronic books
- online journals
- audio files (e.g., wav, mp3)
- images (e.g., tiff, gif)
- movies (e.g., mpeg, QuickTime)
- reference texts (e.g., dictionaries, directories)"[101]

Let's take a sample query and see what we come up with in OAIster. Here's a simple phrase query—["turing machine"]—asking for results to be listed in descending order by date (most recent first). Since many of the resources in OAIster have reliable dates (e.g., journal articles), using the date sort option should work fairly well.

---

[101] "What Are Digital Resources?" Wheeler Library, Otero Junior College, Colorado, <http://www.ojc.edu/library/digital.htm> (18 January 2007).

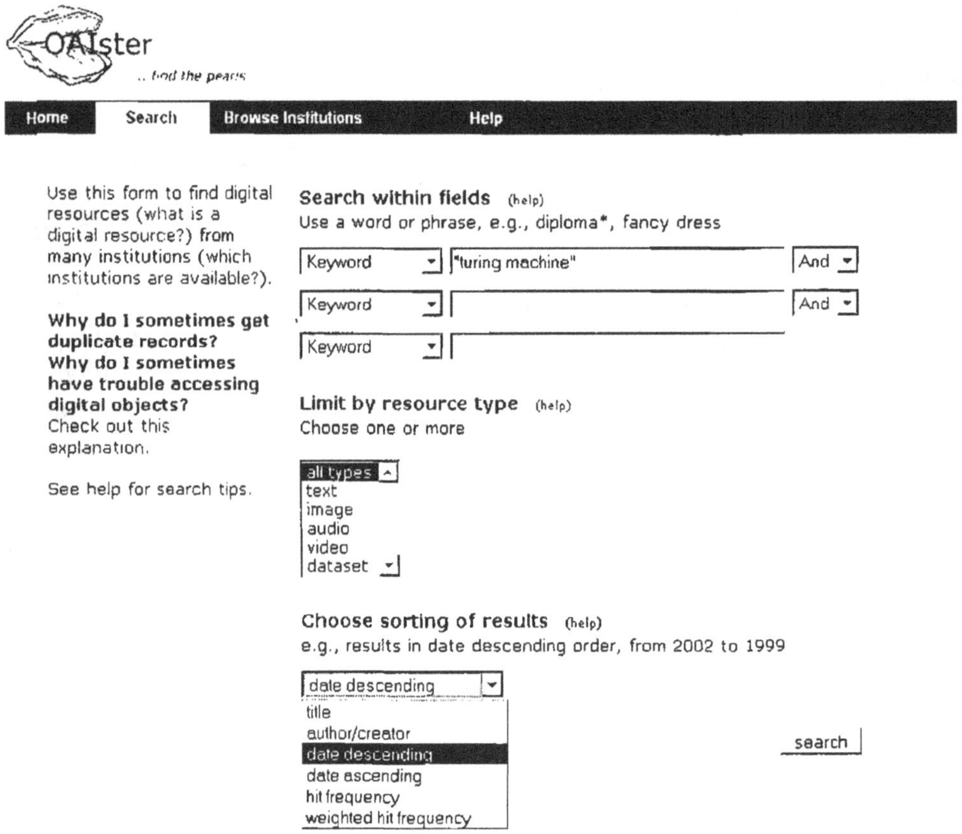
OAIster
.. find the pearls

Home | Search | Browse Institutions | Help

Use this form to find digital resources (what is a digital resource?) from many institutions (which institutions are available?).

Why do I sometimes get duplicate records?
Why do I sometimes have trouble accessing digital objects?
Check out this explanation.

See help for search tips.

Search within fields (help)
Use a word or phrase, e.g., diploma*, fancy dress

Keyword | "turing machine" | And
Keyword | | And
Keyword |

Limit by resource type (help)
Choose one or more

all types
text
image
audio
video
dataset

Choose sorting of results (help)
e.g., results in date descending order, from 2002 to 1999

date descending
title
author/creator
date descending
date ascending
hit frequency
weighted hit frequency

search

You can see from the "results by institution" on the left below that these 916 results are not your typical Google hits.

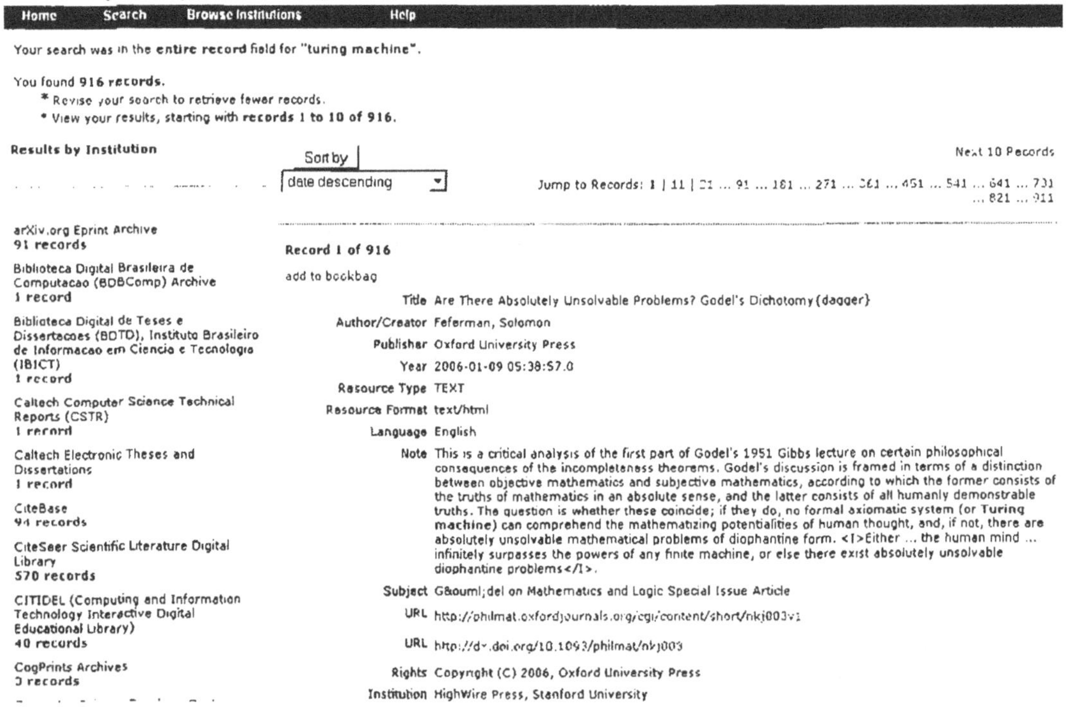
Home | Search | Browse Institutions | Help

Your search was in the **entire record** field for "**turing machine**".

You found **916 records.**
* Revise your search to retrieve fewer records.
* View your results, starting with **records 1 to 10 of 916.**

Results by Institution

arXiv.org Eprint Archive
91 records

Biblioteca Digital Brasileira de Computacao (BDBComp) Archive
1 record

Biblioteca Digital de Teses e Dissertacoes (BDTD), Instituto Brasileiro de Informacao em Ciencia e Tecnologia (IBICT)
1 record

Caltech Computer Science Technical Reports (CSTR)
1 record

Caltech Electronic Theses and Dissertations
1 record

CiteBase
94 records

CiteSeer Scientific Literature Digital Library
570 records

CITIDEL (Computing and Information Technology Interactive Digital Educational Library)
40 records

CogPrints Archives
3 records

Sort by
date descending

Next 10 Records

Jump to Records: 1 | 11 | 21 ... 91 ... 181 ... 271 ... 361 ... 451 ... 541 ... 641 ... 731 ... 821 ... 911

Record 1 of 916

add to bookbag

Title	Are There Absolutely Unsolvable Problems? Godel's Dichotomy{dagger}
Author/Creator	Feferman, Solomon
Publisher	Oxford University Press
Year	2006-01-09 05:38:57.0
Resource Type	TEXT
Resource Format	text/html
Language	English
Note	This is a critical analysis of the first part of Godel's 1951 Gibbs lecture on certain philosophical consequences of the incompleteness theorems. Godel's discussion is framed in terms of a distinction between objective mathematics and subjective mathematics, according to which the former consists of the truths of mathematics in an absolute sense, and the latter consists of all humanly demonstrable truths. The question is whether these coincide; if they do, no formal axiomatic system (or **Turing machine**) can comprehend the mathematizing potentialities of human thought, and, if not, there are absolutely unsolvable mathematical problems of diophantine form. <I>Either ... the human mind ... infinitely surpasses the powers of any finite machine, or else there exist absolutely unsolvable diophantine problems</I>.
Subject	Gödel on Mathematics and Logic Special Issue Article
URL	http://philmat.oxfordjournals.org/cgi/content/short/nkj003v1
URL	http://dx.doi.org/10.1093/philmat/nkj003
Rights	Copyright (C) 2006, Oxford University Press
Institution	HighWire Press, Stanford University

OAIster works best in finding scholarly work from such sources as CiteSeer, arcXiv Eprint Archive, a number of research institutes, and numerous digital libraries. And some of the data is truly impressive. For example, I searched in all fields for [neutrino*] (note the ability to use a **wildcard**) looking for the resource type *image*. Here is one of the 32 images I found:

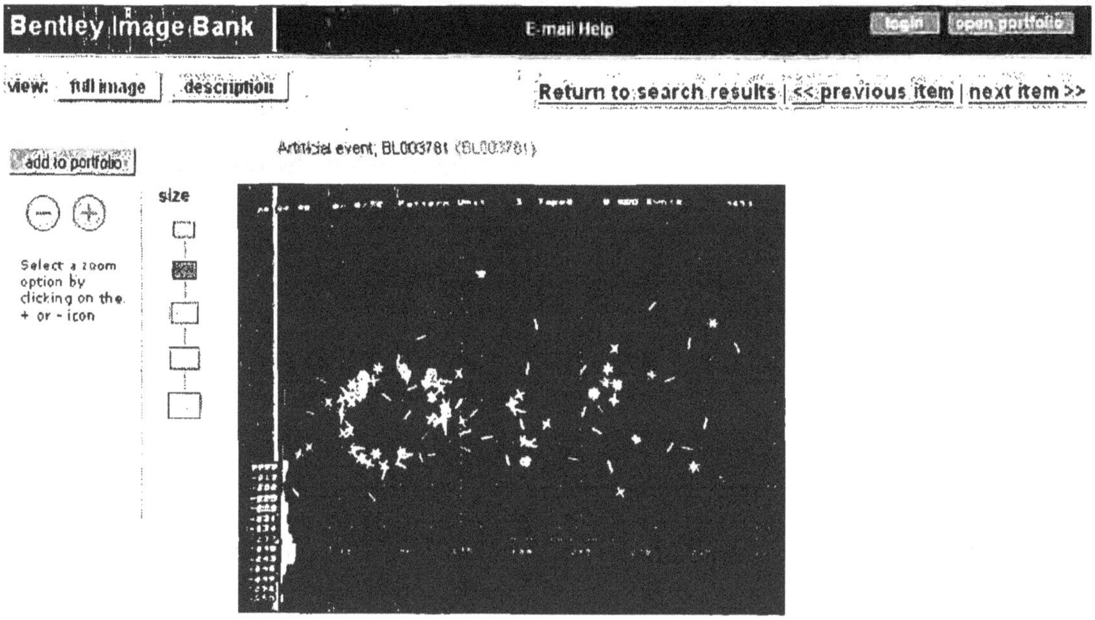
Bentley Image Bank
E-mail Help
login
open portfolio
view: full image | description
Return to search results | << previous item | next item >>
Artificial event; BL003781 (BL003781)
add to portfolio
size
Select a zoom option by clicking on the + or - icon

You are not going to find images like this with a Google, Live, and Yahoo image search.

It is important to keep in mind that ***OAIster is not a search engine but a repository of information that is structured in a very specific manner***, and retrieving the data you want has to be done in conformity to the way the data was entered into the record.

OAIster http://www.oaister.org/

More Scholarly Search Sites

Other academic search engines that may yield better results than Google Scholar and Windows Live Academic search at this point.

Citeseer ☆ http://citeseer.ist.psu.edu/

Scirus http://www.scirus.com/srsapp/

Cornell University's arXiv.org http://arxiv.org/

Research Now http://researchnow.bepress.com/

CiteULike http://www.citeulike.org/

Foreign Doctoral Dissertations http://www.crl.edu/content.asp?l1=5&l2=23&l3=44&l4=25

ISI Highly Cited http://isihighlycited.com/

Scholar Universe http://www.scholaruniverse.com/index.jsp

Ingenta Connect http://www.ingentaconnect.com/

Infomine's Electronic Journals Search http://infomine.ucr.edu/cgi-bin/search?ejournal

Science Direct (select Abstract Databases tab) http://www.sciencedirect.com/

Wiley InterScience Journal Search http://www3.interscience.wiley.com/

The Internet Archive & the Wayback Machine

You have to give Brewster Kahle credit for thinking big. The founder of the Internet Archive has a clear, if not easy, mission: to make all human knowledge universally accessible. And, who knows, he might just succeed. What has made Kahle's dream seem possible is extremely inexpensive storage technology. As of now, the Internet Archive houses "approximately 1 petabyte of data and is currently growing at a rate of 20 terabytes per month. This eclipses the amount of text contained in the world's largest libraries, including the Library of Congress. If you tried to place the entire contents of the archive onto floppy disks (we don't recommend this!) and laid them end to end, it would stretch from New York, past Los Angeles, and halfway to Hawaii."[102] In December 2006 the Archive announced it had indexed over 85 billion "web objects" and that its database contained over 1.5 petabytes of information.[103]

But that's not all that Kahle and company have archived. The Archive also now contains about 2 million audio works; over 10,000 music concerts; thousands of "moving images," including 300 feature films; its own and links to others' digitized texts, including printable and downloadable books; and 3 million hours of television shows (enough to satisfy even the most sedulous couch potato!). Kahle's long term dream includes scanning and digitizing the entire Library of Congress collection of about 28 million books (something that is technically within reach), but there are

---

[46] Internet Archive FAQs, <http://www.archive.org/about/faqs.php#The_Internet_Archive> (14 November 2006).

[103] Brewster Kahle, "Wayback Machine has 85 Billion Archived Webpages," Internet Archive Forum, 5 December 2006, <http://www.archive.org/iathreads/post-view.php?id=84843> (16 January 2007).

some nasty impediments such as copyrights and, of course, money. None of this deters Kahle, whose commitment to the preservation of the digital artifacts of our time drives the Internet Archive. As Kahle puts it, "If you don't have access to the past, you live in a very Orwellian world." Ironically, between the Internet Archive's voracious appetite for websites, which many view as an invasion of privacy, and the Alexa technology, which "monitors" web browsing through the "related links" feature in browsers, many people think of Kahle as the Internet's "Big Brother."

In addition to trying to capture every publicly available webpage and making them searchable via the Wayback Machine, Kahle is pressuring Google to give him a copy of its database with something like a six-month delay to avoid competition with "live" Google. So far, the search engine innovator has not yet come around to Kahle's way of thinking, but Kahle has a way of persuading people and institutions to make a "positive contribution to mankind" by contributing to the Archive.

Announced just 24 hours after Google went public with its own effort to digitize several major library collections, the Internet Archive's plan to digitize the collections of ten major libraries cannot be a coincidence. Among the libraries agreeing to participate in the Internet Archive's project are:

- Carnegie Mellon University and the Library of Congress Million Book Project
- University of Toronto, Canada
- Library of Congress American Memory Project
- McMaster University, Canada
- University of Ottawa, Canada
- Bibliotheca Alexandrina, Egypt
- Indian Institute of Science, India
- International Institute of Information Technology, India
- Zhejiang University, China
- European Archive, Netherlands

The goal? In Brewster Kahle's own words, "anyone with an Internet connection will have access to these collections and the growing set of tools to make use of them. In this way we are getting closer to the goal of Universal Access to All Knowledge."[104] Once again Internet Archive founder Kahle reinforces his reputation

[104] Brewster Kahle, "Announcement: Open-Access Text Archives," 15 December 2004 <http://www.archive.org/iathreads/post-view.php?id=25361> (14 November 2006).

as a true visionary: "Imagine being able to analyze the changes to the English language over time. Imagine being able to use the hand translated versions of past books as a way to train automatic translation technologies so we can more effectively translate any book into any language. Imagine being able to analyze the interrelation of papers through their footnotes and links to find new patterns of thought. Each of these projects is already proceeding using the digital holdings of the Internet Archive by researchers."[105] You have to love this guy.

Microsoft and Yahoo both threw in with Kahle and the Open Content Alliance (OCA) during 2005, Microsoft in advance of its new Live Book Search. This occurred as Google was embroiled in not one but two lawsuits to stop its book digitization project. The OCA has thus far avoided any such suits because it is only indexing books and other content in the public domain. But Microsoft has made it known it will not be content to stick with public domain content, which will put Microsoft on the horns of the same dilemma as Google. It will be interesting to see how OCA and its members handle copyright and other infringement issues.

Open Content Alliance http://www.opencontentalliance.org/

While some of its members may view the OCA project as a way to take on Google, Kahle is not at all unhappy about competition from other digitization projects. Quite the contrary, he sees his efforts as augmenting more commercial ventures while he openly seeks to emulate in the public domain Amazon's approach to full-text search. Any way you look at it, this is great news. Sometimes with all the petty annoyances in our everyday lives, it is hard to remember we really are witnessing and even participating in a revolution in human knowledge.

And just how does Kahle envision storing all these treasures? He worked with Capricorn Technologies to design what is called the PetaBox, basically a very large, affordable data repository that can store a million gigabytes of data. Capricorn shipped the first of its PetaBox products to the Internet Archive in June 2005.

All that data is accessible to users in a variety of ways, none more interesting or useful than the **Wayback Machine**. Using the Wayback Machine, you may very well be able to retrieve a page or an entire site even if it disappeared from the web years ago. Also keep in mind that Yahoo also offers an excellent way to search the Internet Archive to its fullest.

---

105 Kahle.

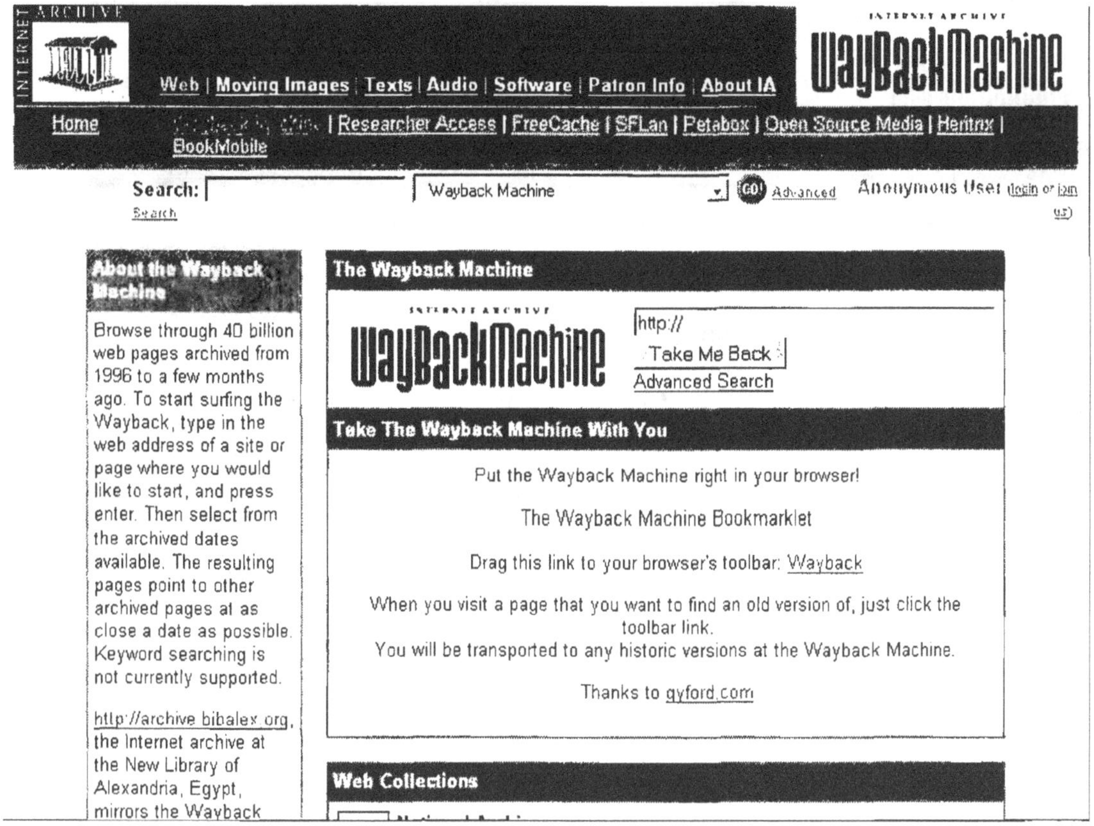
INTERNET ARCHIVE
Web | Moving Images | Texts | Audio | Software | Patron Info | About IA
INTERNET ARCHIVE
WaybackMachine
Home
| Researcher Access | FreeCache | SFLan | Petabox | Open Source Media | Heritrix |
BookMobile
Search:
Wayback Machine
GO!
Advanced
Anonymous User
About the Wayback Machine
Browse through 40 billion web pages archived from 1996 to a few months ago. To start surfing the Wayback, type in the web address of a site or page where you would like to start, and press enter. Then select from the archived dates available. The resulting pages point to other archived pages at as close a date as possible. Keyword searching is not currently supported.
http://archive.bibalex.org, the Internet archive at the New Library of Alexandria, Egypt, mirrors the Wayback
The Wayback Machine
INTERNET ARCHIVE
WaybackMachine
http://
Take Me Back
Advanced Search
Take The Wayback Machine With You
Put the Wayback Machine right in your browser!
The Wayback Machine Bookmarklet
Drag this link to your browser's toolbar: Wayback
When you visit a page that you want to find an old version of, just click the toolbar link.
You will be transported to any historic versions at the Wayback Machine.
Thanks to gyford.com
Web Collections

The Wayback Machine http://web.archive.org/

To use the Wayback Machine to bring back the past, simply search on a url or use the new bookmarklet that can be dragged and dropped onto your browser's toolbar. Whenever you are visiting a webpage, clicking on that bookmarklet searches the Internet Archive for earlier versions of that web address. What you will see first is a list of all previous versions of a website stored in the Archive.

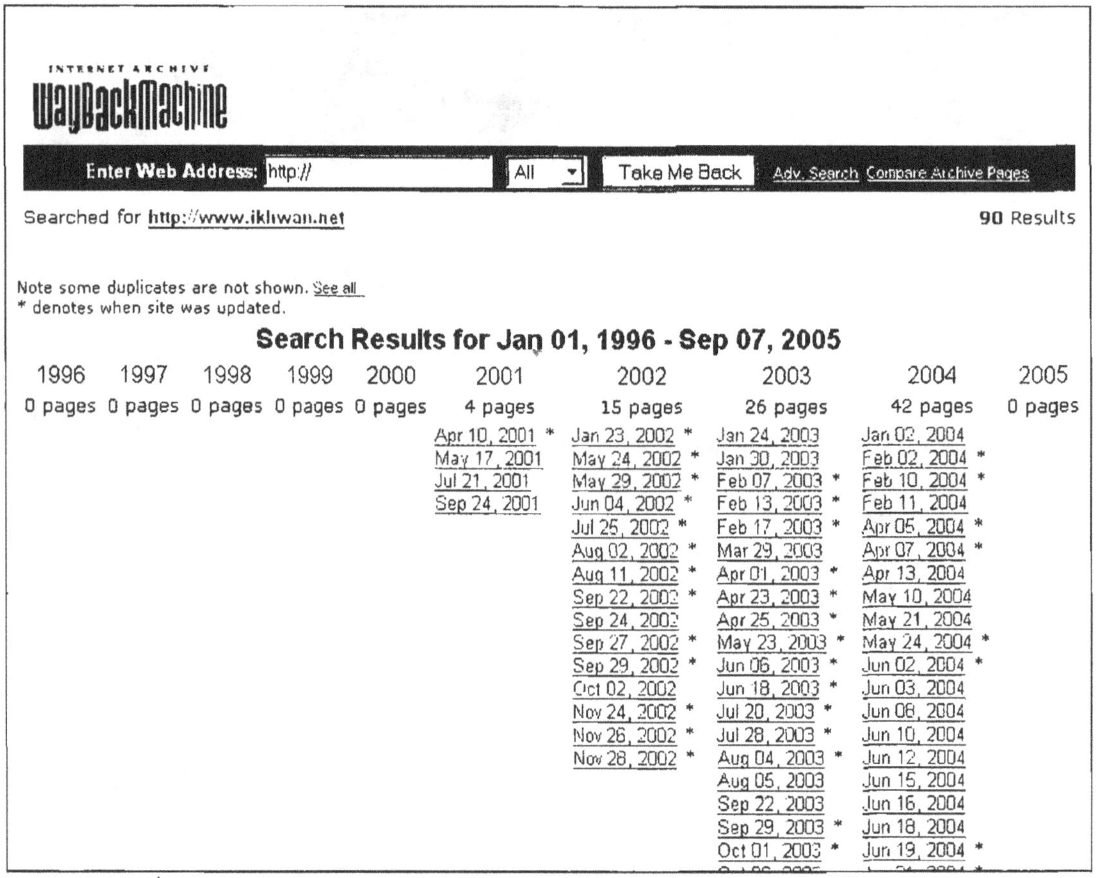
INTERNET ARCHIVE
WaybackMachine
Enter Web Address: http://
All
Take Me Back
Adv. Search
Compare Archive Pages
Searched for http://www.ikhwan.net
90 Results
Note some duplicates are not shown. See all
* denotes when site was updated.
Search Results for Jan 01, 1996 - Sep 07, 2005
1996 0 pages
1997 0 pages
1998 0 pages
1999 0 pages
2000 0 pages
2001 4 pages
Apr 10, 2001 * May 17, 2001 Jul 21, 2001 Sep 24, 2001
2002 15 pages
Jan 23, 2002 * May 24, 2002 * May 29, 2002 * Jun 04, 2002 * Jul 25, 2002 * Aug 02, 2002 * Aug 11, 2002 * Sep 22, 2002 * Sep 24, 2002 Sep 27, 2002 * Sep 29, 2002 * Oct 02, 2002 Nov 24, 2002 * Nov 26, 2002 * Nov 28, 2002 *
2003 26 pages
Jan 24, 2003 Jan 30, 2003 Feb 07, 2003 * Feb 13, 2003 * Feb 17, 2003 * Mar 29, 2003 Apr 01, 2003 * Apr 23, 2003 * Apr 25, 2003 * May 23, 2003 * Jun 06, 2003 * Jun 18, 2003 * Jul 20, 2003 * Jul 28, 2003 * Aug 04, 2003 * Aug 05, 2003 Sep 22, 2003 Sep 29, 2003 * Oct 01, 2003 *
2004 42 pages
Jan 02, 2004 Feb 02, 2004 * Feb 10, 2004 * Feb 11, 2004 Apr 05, 2004 * Apr 07, 2004 * Apr 13, 2004 May 10, 2004 May 21, 2004 May 24, 2004 * Jun 02, 2004 * Jun 03, 2004 Jun 08, 2004 Jun 10, 2004 Jun 12, 2004 Jun 15, 2004 Jun 16, 2004 Jun 18, 2004 Jun 19, 2004 *
2005 0 pages

Selecting any date will show you an archived copy of the site as it appeared on that date. Here is the (now defunct) Jihadist website Ikhwan.net as it appeared on 23 January 2002:

مرحب بكم في رحاب موقع الشبكة الإخوانية
الشبكة الإخوانية
الصفحة الرئيسية
أهداف الموقع
خطة الموقع
منهج الحركة
الحركة في سطور
الجديد
قيد الإضافة
كلمة الشهر
خدمات الموقع
ملتقى الإخوان
من أعلام الحركة
كتب للحركة
في ظلال القرآن
مقالات مختارة
رسالة الإخوان
مواقع مختارة
المكتبة الحركية الإلكترونية
أشرطة مضافة حديثا
مذكرات الدعوة والداعية
رسالة الإخوان الأسبوعية
موقع الشيخ فيصل مولوي

The Internet Archive doesn't just archive the homepage of each website. In most cases, you can dig down deeply into a site to find many additional pages from a website that might have disappeared or dramatically changed over time. A special feature in the Wayback Machine's advanced search lets users compare two versions of a webpage using technology from Docucomp. This option will identify any changes—inserted, deleted, replaced and moved text and code—in webpages and documents.[106]

The Wayback Machine is, very simply, one of the greatest deep web tools ever created.

---

[106] In case you are curious about what happened to the Internet Archive's foray into search, aka Recall, the programmer who wrote Recall was hired by Google in 2004 and that appears to be the end of Recall.

Other Invisible Web Resources

A great deal of what is on the Internet is going to remain inaccessible to search engines, either because the information is password-protected, is behind a firewall, requires registration and/or payment to view, etc. In short, this information either is not intended for public viewing or there is a price to do so. There is, however, a substantial amount of data that *is* meant for public consumption but is not indexed by search engine spiders: public databases. These run the gamut from very technical medical or scientific databases (e.g., MEDLINE, NASA EOSDIS) to frivolous (e.g., Jokes.com). How do you find these online databases and other "hidden" websites?

BUBL http://www.bubl.ac.uk/

There are a number of very good websites devoted entirely or mostly to tracking down web databases and other "hidden" resources. One of the most impressive catalogs of hard to find sites is **BUBL**. First established in 1990, the name stood for **BU**lletin **B**oard for **L**ibraries. Today, BUBL offers a huge index of resources primarily for academic researchers. BUBL's index can be sorted by topic, alphabetically, by country, or by Dewey Decimal System. **Deep Web Research**

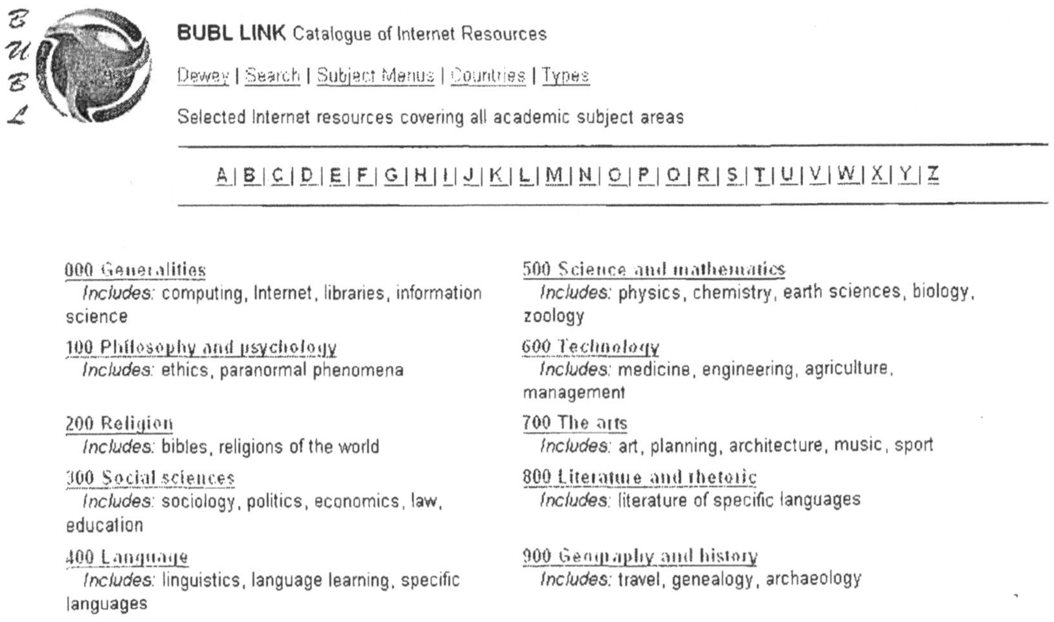
BUBL LINK Catalogue of Internet Resources
Dewey | Search | Subject Menus | Countries | Types
Selected Internet resources covering all academic subject areas
A | B | C | D | E | F | G | H | I | J | K | L | M | N | O | P | Q | R | S | T | U | V | W | X | Y | Z
000 Generalities
Includes: computing, Internet, libraries, information science
100 Philosophy and psychology
Includes: ethics, paranormal phenomena
200 Religion
Includes: bibles, religions of the world
300 Social sciences
Includes: sociology, politics, economics, law, education
400 Language
Includes: linguistics, language learning, specific languages
500 Science and mathematics
Includes: physics, chemistry, earth sciences, biology, zoology
600 Technology
Includes: medicine, engineering, agriculture, management
700 The arts
Includes: art, planning, architecture, music, sport
800 Literature and rhetoric
Includes: literature of specific languages
900 Geography and history
Includes: travel, genealogy, archaeology

Infomine http://infomine.ucr.edu/

Another scholarly resource is **Infomine**, from the University of California. Infomine is an excellent way of tracking down not only online databases, particularly in all the sciences, but also finding technical websites. For example, Infomine has a superb Maps and GIS webpage, with a large number of links to databases devoted to mapping and/or GIS.

Aardvark: Asian Resources for Librarians
http://www.aardvarknet.info/user/aardvarkwelcome/

Aardvark: Asian Resources for Librarians, owned and managed by iGroup, a database and eJournal distributor, keeps a low profile, which is too bad because it is a very good resource. Aardvark has two sections: Literature and Recommended resources and sites. I especially like the Asian Databases section, which currently lists over 650 databases. Thompson Gale published an excellent overview of Aardvark by Professor Péter Jacsó.[107]

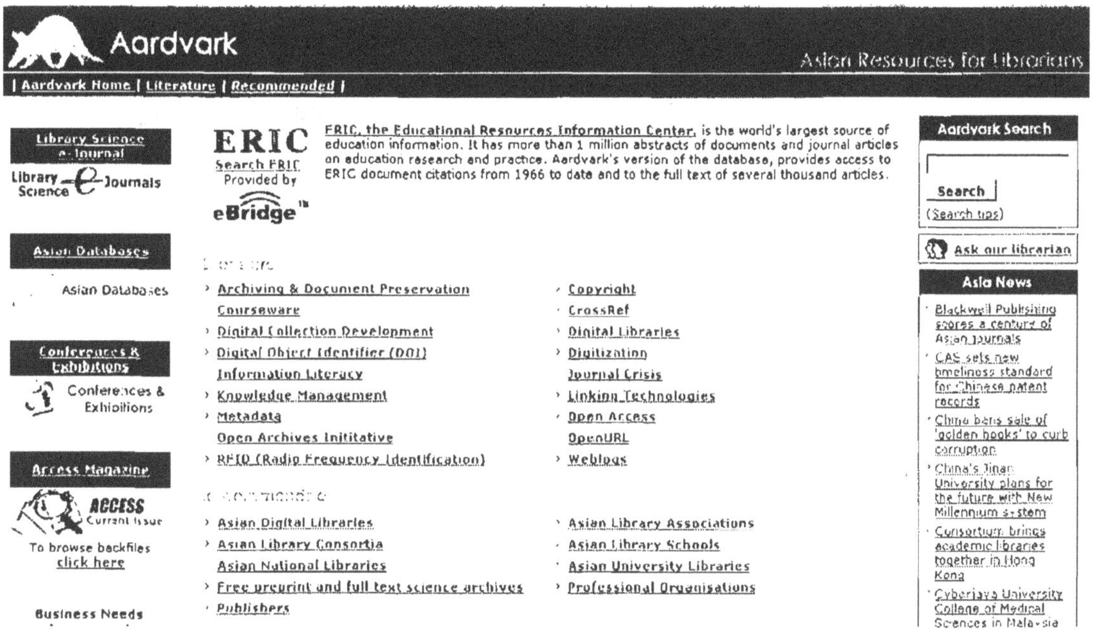
Aardvark
Asian Resources for Librarians
| Aardvark Home | Literature | Recommended |
ERIC
ERIC, the Educational Resources Information Center, is the world's largest source of education information. It has more than 1 million abstracts of documents and journal articles on education research and practice. Aardvark's version of the database, provides access to ERIC document citations from 1966 to date and to the full text of several thousand articles.
Search ERIC
Provided by
eBridge
Aardvark Search
Search
(Search tips)
Ask our librarian
Asia News
Library Science e-Journal
Asian Databases
Conferences & Exhibitions
Access Magazine
To browse backfiles click here
Business Needs
Archiving & Document Preservation
Courseware
Digital Collection Development
Digital Object Identifier (DOI)
Information Literacy
Knowledge Management
Metadata
Open Archives Inititative
RFID (Radio Frequency Identification)
Copyright
CrossRef
Digital Libraries
Digitization
Journal Crisis
Linking Technologies
Open Access
OpenURL
Weblogs
Asian Digital Libraries
Asian Library Consortia
Asian National Libraries
Free preprint and full text science archives
Publishers
Asian Library Associations
Asian Library Schools
Asian University Libraries
Professional Organisations

---

[107] Péter Jacsó, "Aardvark," Péter's Digital Reference Shelf, Thompson Gale, October 2006, <http://www.galegroup.com/reference/peter/aardvark.htm> (26 October 2006).

Deep Web Research http://www.deepwebresearch.com/

Deep Web Research is an information weblog created by the Virtual Private Library. As a blog, the site is designed to stay current with many links to a wide variety and number of sources related to deep web search and research. Topics covered include current and historical articles, papers, and videos; cross database articles, search services, and tools; presentations; and resources. Deep Web Research presents an impressive set of links that includes everything from articles on what constitutes the deep web and how to search it to online search sites such as SearchLight, the University of California's interface to publicly available databases in either the sciences and engineering or social sciences and humanities. The biggest drawback of Deep Web Research is the lack of annotation of the links. Most are simply listed by title and there is no way to know what services they offer without looking at each site individually. Still, Deep Web Research is an impressive collection of deep web resources.

A9 http://a9.com/

Amazon http://www.amazon.com/

BUBL Catalog http://www.bubl.ac.uk/

The Complete Planet http://www.completeplanet.com/

Deep Web Research http://www.deepwebresearch.com/

Infomine http://infomine.ucr.edu/

Intute (formerly Resource Discovery Network) http://www.intute.ac.uk/

Pinakes Subject Launchpad http://www.hw.ac.uk/libwww/irn/pinakes/pinakes.html

Research Beyond Google: 119 Authoritative, Invisible, and Comprehensive Resources http://oedb.org/library/college-basics/research-beyond-google

Ultimate Guide to the Invisible Web
http://oedb.org/library/college-basics/invisible-web

There is a lot more to the "hidden web" than just that information stored in databases, however. Many of the most interesting sites are simply not indexed by anyone, in part because search services have made a conscious decision not to try to index the entire web. How do you find them? Unfortunately, there is no magic, but here are a few tips that might help (and often really do work):

Tips for Navigating the "Hidden Web"

1. Always ***examine and follow the links*** from "links pages" at an interesting site (it's quite possible that those links lead to webpages that are not indexed).

2. Try ***url guessing***, i.e., just try what you think might be the address of a site you're looking for and you might get lucky. Do pay attention to any ***domain naming conventions*** widely used in a particular country. The Russians often use cities as part of the name, e.g., *http://www.pager.nnov.ru*; many other countries use the conventions of *.co* for companies, *.ac* for educational institutions, e.g., *http://www.aid.co.at* or *http://www.ua.ac.be*, etc.).

3. Spend time ***browsing a country's domain name registry***; many of these registries list "all" the websites in their top-level domain (more on how to find a country registry later).

4. Visit a ***country-specific website*** devoted to listing "all" the sites for that country and browse or search their list (you will undoubtedly find a number of websites not indexed by US-based and focused search engines).

Casting a Wider Net—International Search, Language Tools

International Search

I have tried to think of a better way to describe this topic, which I used to call "foreign search" until I realized that what is "foreign" to some readers is "home" to others. The concept I am trying to get across is simple: do not rely on your favorite search engine for research. Simple idea, difficult implementation, because users naturally and quite understandably have a very strong tendency to depend almost exclusively on certain search tools that focus on their location, whether a specific country, region, or city. That makes perfect sense until you try to find information about and/or from other locales. "Vanilla" Google, Yahoo, and Live Search are targeted at US users and locations, just as Baidu targets the Chinese audience. Researchers must first get out of the habit of using US-centric search engines, then look for and bookmark country and local search services to have them ready to hand.

US search engines have largely set the pace and the standards for search tools around the world and, in many cases, are the engines underlying international search sites. However, the target market for US search engines is the US user, not the rest of the world. This means that when you are "traveling" on the web, you must find and use search services that are appropriate to that region, country, or city. What are some of the advantages of using regional search engines?

- Focused search: regional or country-specific search engines often permit you to search in one region ("all European countries," for example) or one country.
- Focused data: most non-US search tools collect and store data primarily or exclusively from their region or country.
- Language selectivity: international search engines must offer the ability to search in the native language(s).
- Non-Latin code sets: the non-US search engines in countries that use non-Latin character sets often allow you to input your query in one or more character sets. Of course, your computer must already be configured to type in the non-Latin characters (unless you want to resort to the cut and paste approach).
- Translation function: you may come across an unusual translation option in an international search engine.

The major US-based search engines continue to expand into international markets, offering a myriad specialized interfaces by country and/or language. Search results may be much more precise using the international version of Google, for example, than the generic site. In fact, there is even a site that lets you compare the differences between Google searches using two different Google datacenters.

Look at the difference between the same search for the acronym ETA using Google.com compared to Google Spain:

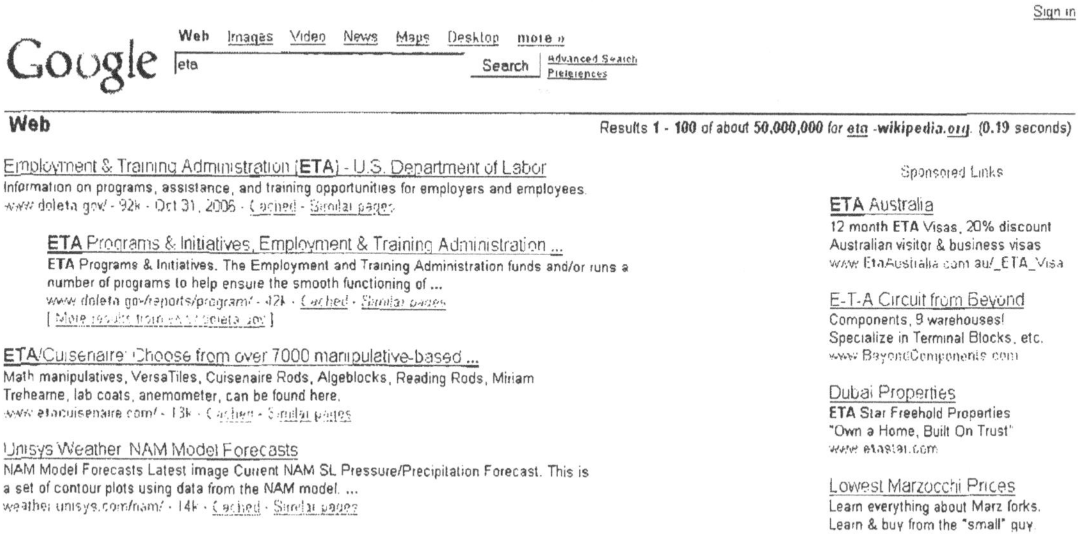
Sign in
Web Images Video News Maps Desktop more »
Google
eta
Search
Advanced Search
Preferences
Web
Results 1 - 100 of about 50,000,000 for eta -wikipedia.org. (0.19 seconds)
Employment & Training Administration (ETA) - U.S. Department of Labor
Information on programs, assistance, and training opportunities for employers and employees.
ETA Programs & Initiatives, Employment & Training Administration ...
ETA Programs & Initiatives. The Employment and Training Administration funds and/or runs a number of programs to help ensure the smooth functioning of ...
ETA/Cuisenaire: Choose from over 7000 manipulative-based ...
Math manipulatives, VersaTiles, Cuisenaire Rods, Algeblocks, Reading Rods, Miriam Trehearne, lab coats, anemometer, can be found here.
Unisys Weather: NAM Model Forecasts
NAM Model Forecasts Latest image Current NAM SL Pressure/Precipitation Forecast. This is a set of contour plots using data from the NAM model. ...
Sponsored Links
ETA Australia
12 month ETA Visas, 20% discount
Australian visitor & business visas
E-T-A Circuit from Beyond
Components, 9 warehouses!
Specialize in Terminal Blocks, etc.
Dubai Properties
ETA Star Freehold Properties
"Own a Home, Built On Trust"
Lowest Marzocchi Prices
Learn everything about Marz forks.
Learn & buy from the "small" guy.

Here is the identical search using Google Spain and restricting the search to Spanish sites. Clearly, Google Spain "understands" that to most Spanish users ETA means the Basque separatist group Euskadi Ta Askatasuna:

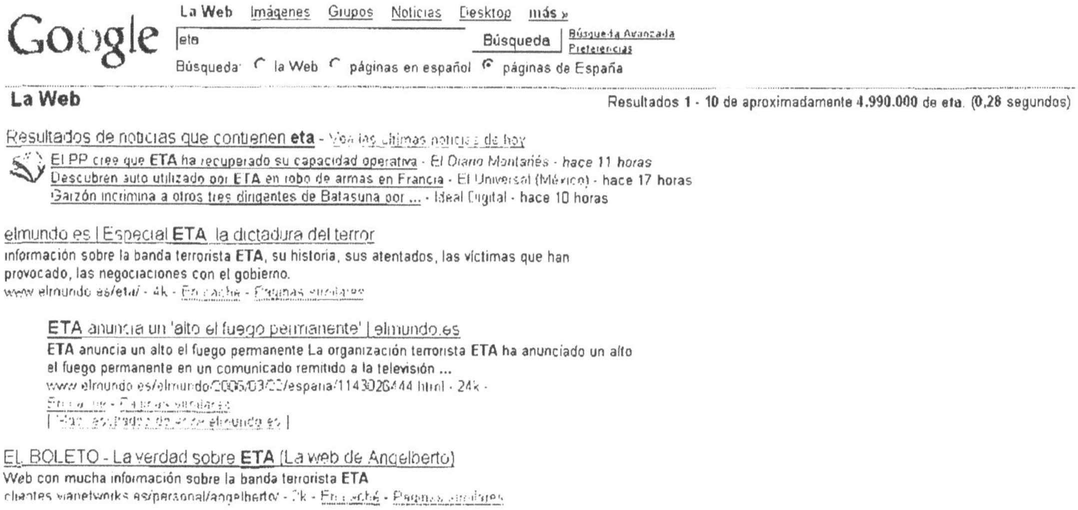
La Web Imágenes Grupos Noticias Desktop más »
Google
eta
Búsqueda
Búsqueda Avanzada
Preferencias
Búsqueda: la Web páginas en español páginas de España
La Web
Resultados 1 - 10 de aproximadamente 4.990.000 de eta. (0,28 segundos)
Resultados de noticias que contienen eta
El PP cree que ETA ha recuperado su capacidad operativa - El Diario Montañés - hace 11 horas
Descubren auto utilizado por ETA en robo de armas en Francia - El Universal (México) - hace 17 horas
Garzón incrimina a otros tres dirigentes de Batasuna por ... - Ideal Digital - hace 10 horas
elmundo.es | Especial ETA, la dictadura del terror
Información sobre la banda terrorista ETA, su historia, sus atentados, las víctimas que han provocado, las negociaciones con el gobierno.
ETA anuncia un 'alto el fuego permanente' | elmundo.es
ETA anuncia un alto el fuego permanente La organización terrorista ETA ha anunciado un alto el fuego permanente en un comunicado remitido a la televisión ...
EL BOLETO - La verdad sobre ETA (La web de Angelberto)
Web con mucha información sobre la banda terrorista ETA

Google provides the list of its "local" (meaning country) sites, identified by thumbnail flags, on its language tools page. There is also a very good list of the languages and

Google international interfaces that support them available at the French website Intelligence-Center.

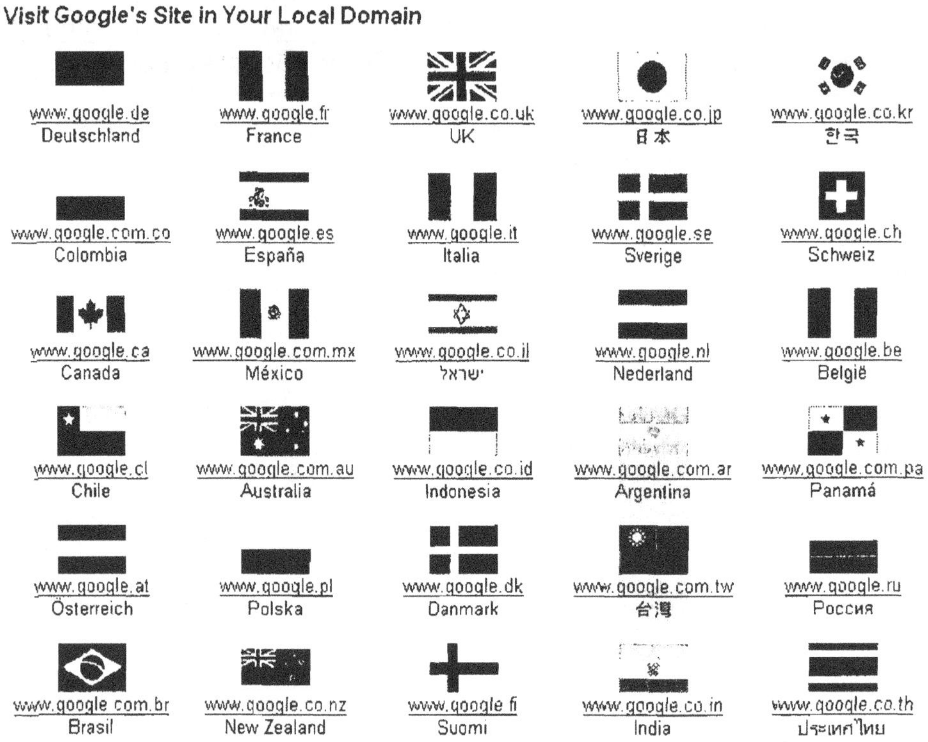
Visit Google's Site in Your Local Domain
www.google.de
Deutschland
www.google.fr
France
www.google.co.uk
UK
www.google.co.jp
日本
www.google.co.kr
한국
www.google.com.co
Colombia
www.google.es
España
www.google.it
Italia
www.google.se
Sverige
www.google.ch
Schweiz
www.google.ca
Canada
www.google.com.mx
México
www.google.co.il
ישראל
www.google.nl
Nederland
www.google.be
België
www.google.cl
Chile
www.google.com.au
Australia
www.google.co.id
Indonesia
www.google.com.ar
Argentina
www.google.com.pa
Panamá
www.google.at
Österreich
www.google.pl
Polska
www.google.dk
Danmark
www.google.com.tw
台灣
www.google.ru
Россия
www.google.com.br
Brasil
www.google.co.nz
New Zealand
www.google.fi
Suomi
www.google.co.in
India
www.google.co.th
ประเทศไทย

Google International Sites http://www.google.com/language_tools

Visit Google's Site in Your Local Domain
www.google.ae
الإمارات العربية المتحدة
www.google.com.af
افغانستان
www.google.com.ag
Antigua and Barbuda
www.google.off.ai
Anguilla
www.google.am
Հայաստան
www.google.com.ar
Argentina
www.google.as
American Samoa
www.google.at
Österreich
www.google.com.au
Australia
www.google.az
Azerbaycan
www.google.ba
Bosna i Hercegovina
www.google.com.bd
বাংলাদেশ
www.google.be
Belgie
www.google.bg
България
www.google.com.bh
البحرين
www.google.bi
Burundi
www.google.com.bo
Bolivia
www.google.com.br
Brasil
www.google.bs
The Bahamas
www.google.co.bw
Botswana
www.google.com.bz
Belize
www.google.ca
Canada
www.google.cd
Rep. Dem. du Congo
www.google.cg
Rep. du Congo
www.google.ch
Schweiz
www.google.ci
Cote D'Ivoire
www.google.co.ck
Cook Islands
www.google.cl
Chile
www.google.cn
中国
www.google.com.co
Colombia
www.google.co.cr
Costa Rica
www.google.com.cu
Cuba
www.google.cz
Česká republika
www.google.de
Deutschland
www.google.dj
Djibouti
www.google.dk
Danmark
www.google.dm
Dominica
www.google.com.do
Rep. Dominicana
www.google.com.ec
Ecuador
www.google.ee
Eesti

Intelligence Center: Google Worldwide
http://c.asselin.free.fr/french/googleworldwide.htm

While acronyms are a type of search that often renders ambiguous or misleading results, searching for the English-language name of a non-English-language entity can cause similar confusion. Here are three different searches for the Mexican group Fuerzas Armadas Revolucionarias del Pueblo, known in English as the Revolutionary Armed Forces of the People. The first search ["Revolutionary Armed Forces of the People"], using generic Google, generates the following results, which clearly illustrate the problem of searching for the translated name. This search comes up with groups of this same name in Guinea-Bissau, Colombia, and Mexico:

DOCID: 4046925

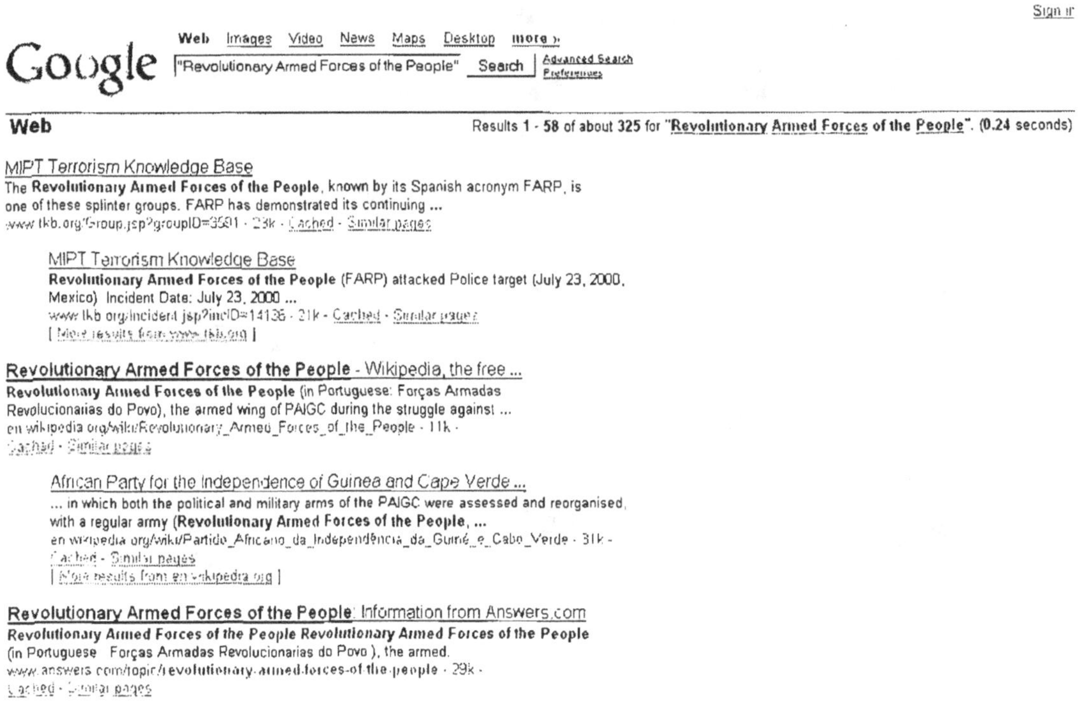

Sign in

Google

Web Images Video News Maps Desktop more »

"Revolutionary Armed Forces of the People" Search | Advanced Search Preferences

Web — Results 1 - 58 of about 325 for "Revolutionary Armed Forces of the People". (0.24 seconds)

MIPT Terrorism Knowledge Base
The **Revolutionary Armed Forces of the People**, known by its Spanish acronym FARP, is one of these splinter groups. FARP has demonstrated its continuing ...
www.tkb.org/Group.jsp?groupID=3591 - 23k - Cached - Similar pages

MIPT Terrorism Knowledge Base
Revolutionary Armed Forces of the People (FARP) attacked Police target (July 23, 2000, Mexico) Incident Date: July 23, 2000 ...
www.tkb.org/Incident.jsp?incID=14136 - 21k - Cached - Similar pages
[More results from www.tkb.org]

Revolutionary Armed Forces of the People - Wikipedia, the free ...
Revolutionary Armed Forces of the People (in Portuguese: Forças Armadas Revolucionarias do Povo), the armed wing of PAIGC during the struggle against ...
en.wikipedia.org/wiki/Revolutionary_Armed_Forces_of_the_People - 11k - Cached - Similar pages

African Party for the Independence of Guinea and Cape Verde ...
... in which both the political and military arms of the PAIGC were assessed and reorganised, with a regular army (**Revolutionary Armed Forces of the People**, ...
en.wikipedia.org/wiki/Partido_Africano_da_Independência_da_Guiné_e_Cabo_Verde - 31k - Cached - Similar pages
[More results from en.wikipedia.org]

Revolutionary Armed Forces of the People: Information from Answers.com
Revolutionary Armed Forces of the People Revolutionary Armed Forces of the People (in Portuguese Forças Armadas Revolucionarias do Povo), the armed.
www.answers.com/topic/revolutionary-armed-forces-of-the-people - 29k - Cached - Similar pages

Next, we search for the group by its Spanish name ["Fuerzas Armadas Revolucionarias del Pueblo"] using generic Google. The results are very different and most refer to the Mexican group:

Google

Web Images Video New! News Maps Desktop more »

"Fuerzas Armadas Revolucionarias del Puel Search | Advanced Search Preferences

Web — Results 1 - 100 of about 37,300 for "Fuerzas Armadas Revolucionarias del Pueblo". (0.29 seconds)

Tip: Search for **English** results only You can specify your search language in Preferences

APIA - Comunicado de las **Fuerzas Armadas Revolucionarias del** ...
Fuerzas Armadas Revolucionarias del Pueblo FARP Enviado el Friday, 01 September a las 17:34:35 por apia Login Nickname Password ...

APIA - Comunicado de las **Fuerzas Armadas Revolucionarias del** ...
Fuerzas Armadas Revolucionarias del Pueblo, FARP República Mexicana, a 7 de agosto de 2006 Enviado el Tuesday, 08 August a las 09:33:36 por apia ...

Cronología de Ejército Popular Revolucionario (EPR) y de las ...
Sujetos desconocidos, quienes estarían vinculados con el grupo subversivo **Fuerzas Armadas Revolucionarias del Pueblo** (FARP), que operaría en el estado de ...

CRONOLOGIA DEL EPR, el ERPI y las FARP (2000-2001)
La Coordinadora Guerrillera Nacional "Jose María Morelos", integrada por las **Fuerzas Armadas Revolucionarias del Pueblo** (FARP), el Ejército Villista ...

CEDEMA.ORG || Centro de Documentación de los Movimientos Armados
Documentación Fecha, País, Grupo, Categoría, Título 2006 08 26, Mexico, Partido Democrático Popular Revolucionario-Ejército Popular Revolucionario ...

Sponsored Links
Dave Solon Nissan
View our Huge Inventory of New and Used Nissan Cars, Trucks, and SUVs.

Notice how different the results are if we switch to Google Mexico <http://www.google.com.mx/> and limit the search to Mexican pages; virtually every result refers to the Mexican group FARP, the one we were seeking:

Accede

Google

La Web Imagenes Grupos Noticias Desktop más »

"Fuerzas Armadas Revolucionarias del Puel | Búsqueda | Búsqueda Avanzada | Preferencias

Búsqueda ○ la Web ○ páginas en español ◉ páginas de México

La Web Resultados 1 - 10 de aproximadamente 188 de "Fuerzas Armadas Revolucionarias del Pueblo". (0.36 segundos)

boletín
... SE TIENE CONOCIMIENTO QUE LAS **FUERZAS ARMADAS REVOLUCIONARIAS DEL PUEBLO** "FARP", SON UNA FRACCIÓN ESCINDIDA DEL AUTO DENOMINADO EJÉRCITO POPULAR ...

El Sur de Acapulco 26 de del 2006
... a pesar de todas las descalificaciones y que han usado a las FARP (**Fuerzas Armadas Revolucionarias del Pueblo**), para golpear al movimiento armado, ...

El pais, infestado de grupos guerrilleros
-**Fuerzas Armadas Revolucionarias del Pueblo**. Grupos con presencia regional:. Sonora. -Comité de Defensa del Pueblo Baja California Sur ...

Regreso de las bombas catalizadoras
... y acreditados a un grupo autodenominado **Fuerzas Armadas Revolucionarias del Pueblo** Nunca se supo más de esa organización supuestamente guerrillera. ...

Abrió PGR 15 averiguaciones en Oaxaca Cabeza de Vaca
... y de las **Fuerzas Armadas Revolucionarias del Pueblo** (FARP), escisión del PDPR-EPR, con una actividad de propaganda político-militar en Nazareno Etla, ...

La Jornada
... Nolasco buscaban información sobre actividades de las **Fuerzas Armadas Revolucionarias del Pueblo** (FARP), que hizo su aparición en Xochimilco en 1999. ...

Cuestiona la APPO a Rueda por el posible regreso a clases
Más aún, las **Fuerzas Armadas Revolucionarias del Pueblo** (FARP) convocaron a todos los actores políticos del movimiento popular y magisterial a llamar "a la

Google's local search usually offers the options to search all of Google, pages only from that country, or pages only in the language of the country (sometimes, as in the case of India, the Google local search has versions in multiple languages). As we have seen, you will get very different search results if you limit your search to a specific country and search in the language of that country; the results are especially dramatic when the native language uses non-Latin encoding. Here is a comparison of a search for [حسن نصر الله] (Nasrallah), first in generic Google:

Google
Web Images Video News Maps Desktop more »
حسن نصرالل
Search
Advanced Search
Preferences
Web
TunisiaGate Community > النص الكامل لكلمة ... - [Translate this page]
Full Version:
www.tunisiagate.net/Forum/lofiversion/index.php?t795.html - 27k - Cached - Similar pages
النص الكامل لكلمة السيد حسن نصر الله ... - [Translate this page]
www.tunisiagate.net/Forum/index.php?act=ST&f=24&t=795 - 108k - Supplemental Result - Cached - Similar pages
[More results from www.tunisiagate.net]
BBC Arabic.com | الشرق الأوسط | سؤال وجواب ... - [Translate this page]
news.bbc.co.uk/hi/arabic/middle_east_news/newsid_5205000/5205350.stm - 56k - Cached - Similar pages
Palestine Information Center - [Translate this page]
www.palestine-info-urdu.com/urdu/hamas/hamas statements/2004/1001.29.3004.shtml - 34k - Cached - Similar pages
Palestine Information Center - [Translate this page]
www.palestine-info-urdu.com/urdu/news/2004/april/11/index.shtml - 21k - Cached - Similar pages
[More results from www.palestine-info-urdu.com]
Iranian Bar Associations
www.iranbar.org/na21.php - 38k - Cached - Similar pages

Here is the same search at the *Google* Saudi Arabia site limited to Saudi sites:

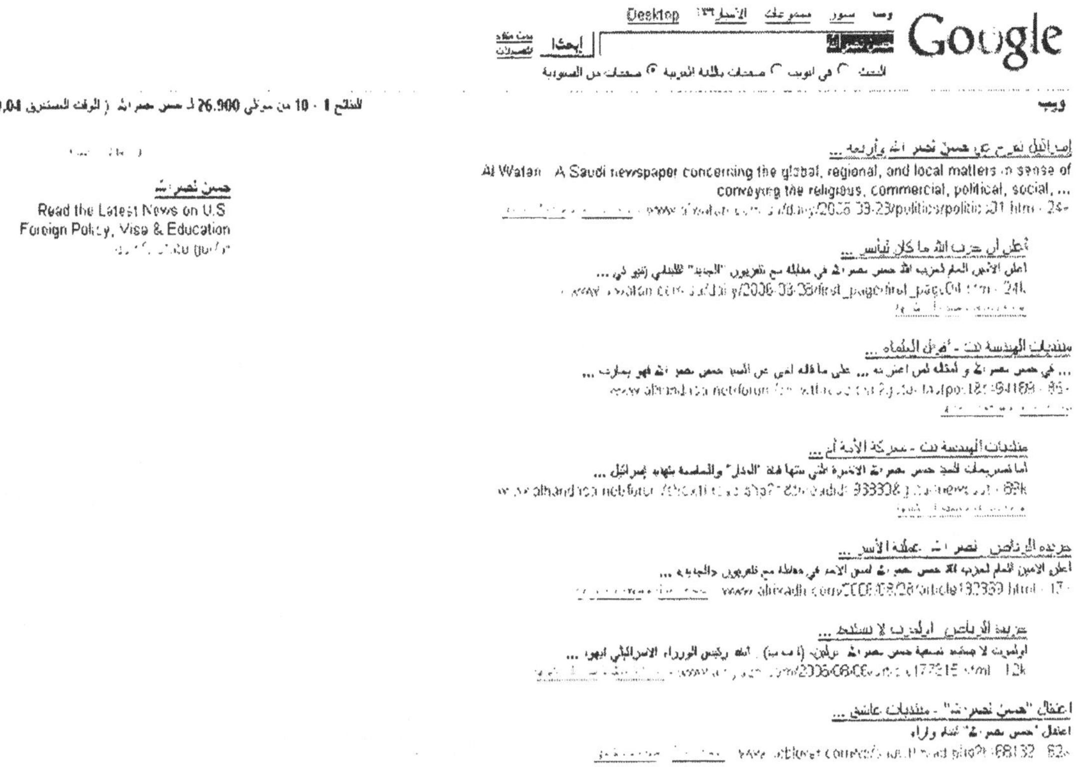
Google
Desktop
ويب
حسن نصر الله
Read the Latest News on U.S Foreign Policy, Visa & Education
Al Watan - A Saudi newspaper concerning the global, regional, and local matters in sense of conveying the religious, commercial, political, social, ...

You can, of course, search in Arabic at the generic Google website, or you can go to various specific Google locations to search. Your results will be different depending on where you search, how you limit your search (by domain, country), and in what language you search. While basic Google may be fine for a "quick and dirty" search, detailed research requires us to explore and exercise many search options.

I suspect we are all guilty of relying too heavily upon Google and thinking that if it is not in Google, it doesn't exist (well, at least not on the web). That, of course, is wrong. Here is an example. First are the results—rather, the absence of results—of a Google query for a Russian website [forummurata.h15.ru]:

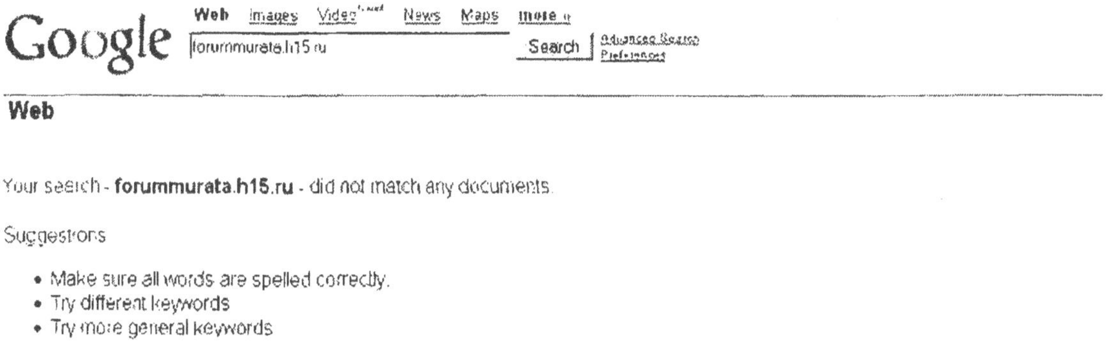
Google
Web Images Video News Maps more »
forummurata.h15.ru
Search
Web
Your search - forummurata.h15.ru - did not match any documents.
Suggestions:
Make sure all words are spelled correctly.
Try different keywords.
Try more general keywords.

Even Google Russia does not find this particular url. However, look at the results for the identical query from the Russian search engine Rambler, which returns 5054 documents from 15 sites:

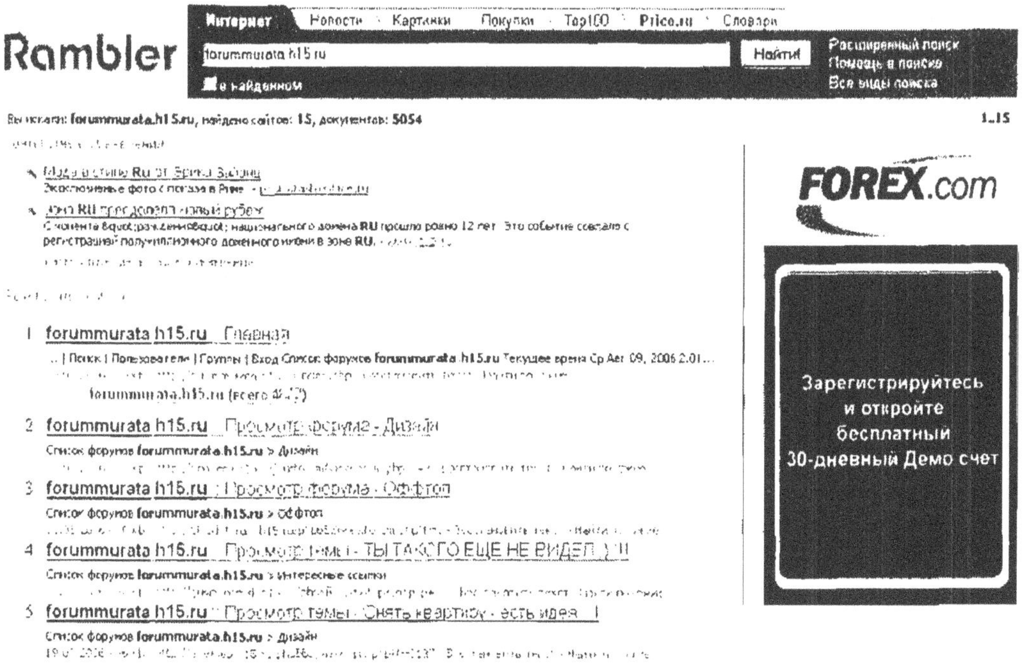
Rambler
Интернет Новости Картинки Покупки Top100 Price.ru Словари
forummurata.h15.ru
Найти
Расширенный поиск
Помощь в поиске
Все виды поиска
в найденном
forummurata.h15.ru
FOREX.com
Зарегистрируйтесь и откройте бесплатный 30-дневный Демо счет

Please believe me when I tell you that there is a remarkable lack of search engine overlap in general, and those differences become even more dramatic when we are focusing on international sites, terms, and non-English/non-Latin encoding.

Interestingly, in the example above, Yahoo performs well, returning 37 hits for the same query. The results for Yahoo and Yahoo Russia are virtually identical, which means they are searching essentially the same database, while in general Google's local searches produce very different results. Yahoo offers its own version of an international page.

Yahoo! International http://world.yahoo.com/

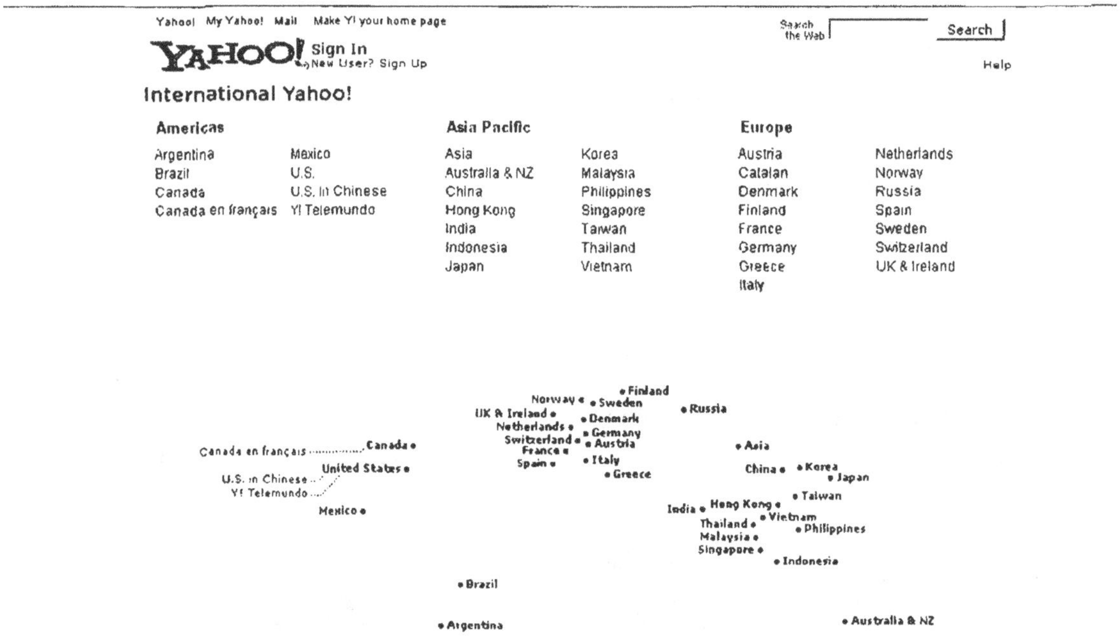
Yahoo! My Yahoo! Mail Make Y! your home page
Search the Web
Search
YAHOO!
Sign In
New User? Sign Up
Help
International Yahoo!
Americas
Argentina
Brazil
Canada
Canada en français
Mexico
U.S.
U.S. in Chinese
Y! Telemundo
Asia Pacific
Asia
Australia & NZ
China
Hong Kong
India
Indonesia
Japan
Korea
Malaysia
Philippines
Singapore
Taiwan
Thailand
Vietnam
Europe
Austria
Catalan
Denmark
Finland
France
Germany
Greece
Italy
Netherlands
Norway
Russia
Spain
Sweden
Switzerland
UK & Ireland

While non-US search sites are often the best places to perform country-specific searches, finding these resources can be a research project in itself. Your best bet for locating these search tools is to start with a directory of search engines that offers a section devoted to international search tools, such as the sites listed here. Also, remember to look at the megasearch sites for sections on international, regional, or country search engines.

All Search Engines.com http://www.allsearchengines.com/foreign.html

Beaucoup! http://www.beaucoup.com/

European Search Engines http://www.netmasters.co.uk/european_search_engines/

FetchFido European Search Engines
http://homepage.ntlworld.com/fetchfido2/interface/search_engines_european.htm

FetchFido World Search Engines
http://homepage.ntlworld.com/fetchfido2/interface/search_engines_worldwide.htm

FinderSeeker http://www.finderseeker.com/

Infisource Foreign Language Search Engines
http://www.infinisource.com/search-engines.html#foreign

International Search Engines http://www.arnoldit.com/lists/intlsearch.asp

ISEDB Local and Regional Search Engines
http://www.isedb.com/html/Internet_Search_Engines/Local_and_Regional_Search_Engines/

ISEDB Local and Regional Directories
http://www.isedb.com/html/Web_Directories/Local_and_Regional_Directories/

Phil Bradley's Country Based Search Engines http://www.philb.com/countryse.htm

Regional and Special Search Engines
http://www.ntu.edu.sg/lib/search/specialframe.htm

Search Engine Colossus☆ http://www.searchenginecolossus.com/

Search Engine Guide http://www.searchengineguide.com/pages/Regional/

Search Engine Index http://www.search-engine-index.co.uk/Regional_Search/

Search Engines 2 http://www.search-engines-2.com/

Search Engines Worldwide (2003) http://home.inter.net/takakuwa/search/

Ultimate Search Engines Links Page http://www.searchenginelinks.co.uk/

In summary, the basic rules for international search are:

- Search first using generic search engines such as Google, Yahoo, Live Search, Gigablast, Exalead for a quick overview and to size the query.
- Locate and bookmark multiple international search engines and directories, including country-specific search sites of major search engines such as Google and Yahoo.
- Search on the English-language term [Kurdistan Workers Party], the transliterated term [Partiya Karkeran Kurdistan], the native language term in its proper encoding, e.g., [Partiya Karkerên Kurdistan] in Kurdish, and the term in any related language, e.g., [Kürdistan İşçi Partisi] in Turkish.
- Search for variations on a name, e.g., PKK is also known as KADEK, Kongra-Gel, the Freedom and Democracy Congress of Kurdistan, Halu Mesru

Savunma Kuvveti (HSK), or the Kurdistan People's Congress (KHK)...you get the idea. Some queries demand a lot of attention and effort.

- Search for acronyms in their native language version [PHE], a transliterated version [RNE], or an English-language version [RNU], all of which stand for the group Russian National Unity, aka Russkoe Natsionalnoe Edinstvo (RNE) or РУССКОЕ НАЦИОНАЛЬНОЕ ЕДИНСТВО (PHE).
- Use sites such as Wikipedia to find native language spellings and encodings; these can be copied and pasted into search engines that can properly handle non-Latin encoding and/or diacritics.
- Limit queries to country-level domains (e.g., site: plus country digraph) or to a specific language (e.g., language: plus country digraph) or use the search engine's advanced search feature.

Rule Six

Look for specialized and/or unique functions in foreign search engines.

Online Dictionaries and Translators

Finding online translation tools is becoming easier as the quality of these machine translators steadily improves as well. There are three basic types of translation tools available via the Internet. At the simplest level are **online dictionaries**, which translate one word at a time, usually from one language to another. One step up from online dictionaries are text translators, which translate words, phrases or maybe a paragraph entered either by typing or by copy-and-paste. **Text translators** are most useful for translating an odd or unusual word or phrase to English. The crown jewels of Internet translation are the **web page translators**, which automatically translate a web page or even an entire website. The quality of the machine translation varies enormously from product to product (some are surprisingly good; others are dreadful). However, most are usually good enough to let you know if the page you're viewing is about soccer or cellular technology. And certainly if you are dealing with languages that do not use the Latin alphabet (Russian, Chinese, Hebrew, for example), almost any machine translation is a blessing.

yourDictionary http://www.yourdictionary.com/index.shtml

There are innumerable language dictionaries now on line, most of which offer some sort of automated look-up feature. I believe the best of the online dictionary metadirectories is the one created at Bucknell University that has become yourDictionary.com. This site has the most comprehensive, impressive set of links to dictionaries I have seen.

yourDictionary.com
Home
A Web of Online Dictionaries is now a part of yourDictionary.com
Find out more
Language | Specialty | Multi-lingual | Translation | Other Indexes
'nyms & more | Grammars | Language Identifiers | Research | Library | Fun & Games | More
Ad Info
Click here for translations by humans - not machines!
Professional Translation Service
Language Dictionaries
Top Ten Language Dictionaries:
English
Spanish
French
German
Japanese
Chinese
Hindi
Arabic
Russian
Thai
230 more!
Multi-lingual
Quick Look-up (English)
Dictionary:
TYPE IN YOUR WORD & CLICK GO!
GO
Thesaurus:
TYPE IN YOUR WORD & CLICK GO!
GO
Help with the Dictionary & Thesaurus
Merriam-Webster's Collegiate® Dictionary & Thesaurus
Word of the Day
YDC SELECT
PROFESSIONAL TRANSLATION
NAME A NEW PRODUCT
eLIBRARY FREE TRIAL
COPYRIGHT INTELLECTUAL PROPERTY
March Madness Glossary
EXCLUSIVE ANALYSIS
Inaugural Address

Foreignword http://www.foreignword.com/

Foreignword is an excellent language and translation resource that automatically links users with more than 265 online dictionaries for 73 languages and single point access to 28 text and url translators for over 38 different languages. Be sure to check the **Tools** page for **Translate Now!**. This page provides access to machine translation tools, including Systran, Arcnet, Worldlingo, InterTran, and Cybertrans, among others.

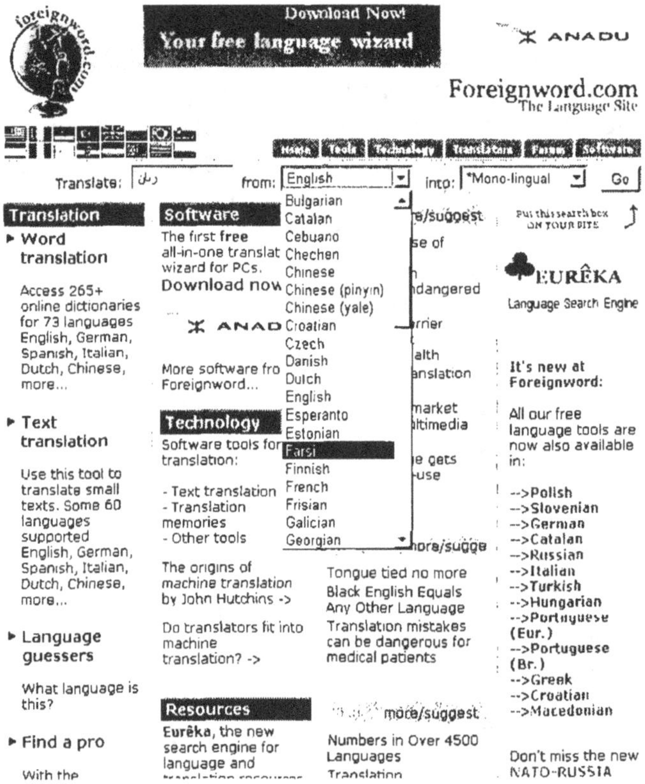
Download Now!
Your free language wizard
ANADU
Foreignword.com
Translate:
from:
English
into:
*Mono-lingual
Go
Translation
Word translation
Access 265+ online dictionaries for 73 languages English, German, Spanish, Italian, Dutch, Chinese, more...
Text translation
Use this tool to translate small texts. Some 60 languages supported English, German, Spanish, Italian, Dutch, Chinese, more...
Language guessers
What language is this?
Find a pro
Software
Technology
Resources
Bulgarian
Catalan
Cebuano
Chechen
Chinese
Chinese (pinyin)
Chinese (yale)
Croatian
Czech
Danish
Dutch
English
Esperanto
Estonian
Farsi
Finnish
French
Frisian
Galician
Georgian
EURÊKA
Language Search Engine
It's new at Foreignword:
All our free language tools are now also available in:
-->Polish
-->Slovenian
-->German
-->Catalan
-->Russian
-->Italian
-->Turkish
-->Hungarian
-->Portuguese (Eur.)
-->Portuguese (Br.)
-->Greek
-->Croatian
-->Macedonian
Tongue tied no more
Black English Equals Any Other Language
Translation mistakes can be dangerous for medical patients
Numbers in Over 4500 Languages

PROMT http://www.translate.ru/eng/srvurl.asp

PROMT maintains the best overall web page translation service. PROMT offers both text and url translations to and from English, Russian, French, German, and Spanish. Keep in mind there are many more single language machine translation services available. Use Word2Word and Foreignword to locate these sites.

The most widely used, though not the best, translation system on the Internet remains the **Systran** machine translation software. Systran's translation is available through AltaVista's Babelfish, Google, AOL, and the French search engine Voila.

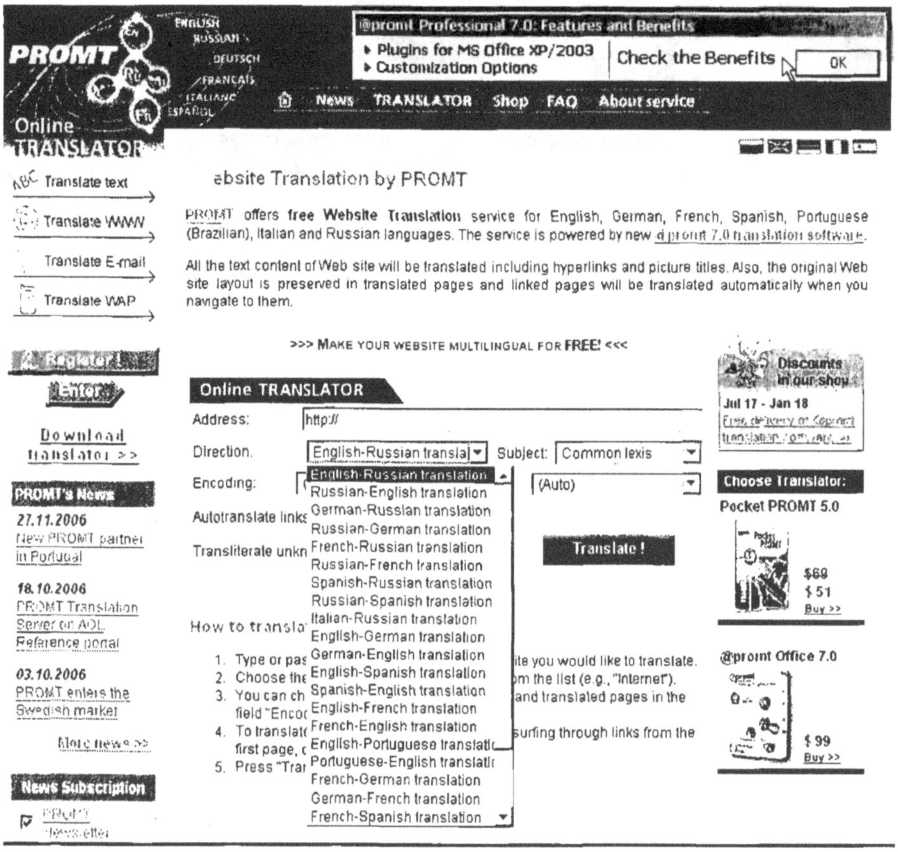
PROMT
Online TRANSLATOR
@promt Professional 7.0: Features and Benefits
▸ Plugins for MS Office XP/2003
▸ Customization Options
Check the Benefits
OK
News
TRANSLATOR
Shop
FAQ
About service
Translate text
Translate WWW
Translate E-mail
Translate WAP
ebsite Translation by PROMT
PROMT offers free Website Translation service for English, German, French, Spanish, Portuguese (Brazilian), Italian and Russian languages. The service is powered by new @promt 7.0 translation software.
All the text content of Web site will be translated including hyperlinks and picture titles. Also, the original Web site layout is preserved in translated pages and linked pages will be translated automatically when you navigate to them.
>>> MAKE YOUR WEBSITE MULTILINGUAL FOR FREE! <<<
Enter
Download translator >>
PROMT's News
27.11.2006
New PROMT partner in Portugal
18.10.2006
PROMT Translation Server on AOL Reference portal
03.10.2006
PROMT enters the Swedish market
More news >>
News Subscription
Online TRANSLATOR
Address:
http://
Direction:
English-Russian transla
Subject:
Common lexis
Encoding:
(Auto)
Autotranslate links
Transliterate unkn
Translate !
English-Russian translation
Russian-English translation
German-Russian translation
Russian-German translation
French-Russian translation
Russian-French translation
Spanish-Russian translation
Russian-Spanish translation
Italian-Russian translation
English-German translation
German-English translation
English-Spanish translation
Spanish-English translation
English-French translation
French-English translation
English-Portuguese translati
Portuguese-English translati
French-German translation
German-French translation
French-Spanish translation
ite you would like to translate.
om the list (e.g., "Internet").
and translated pages in the
urfing through links from the
Discounts in our shop
Jul 17 - Jan 18
Choose Translator:
Pocket PROMT 5.0
$69
$51
Buy >>
@promt Office 7.0
$99
Buy >>

Logos Multilingual Portal

Another site that has been around for years but was vastly improved is the **Logos Multilingual Portal**. Logos includes a dictionary, a universal conjugator, glossaries, a translation course, and links to other language and translation sites. The Logos dictionary contains over 7 million entries in 150 languages. In addition to translating terms, the Logos dictionary gives complete definitions, grammar, context, pronunciation, and, in some cases, associated pictures. Logos also offers access to over 1000 glossaries in different languages.

One of Logos' best features its ability to search in all languages at once, which is very helpful in **language identification**, i.e., identifying non-English terms without needing to know the language of origin. The easiest way to find terms in multiple languages is to go to the *Logos Dictionary main page*, select *Advanced Search*, then type in the term. Logos will present the term in all the languages in which it occurs:

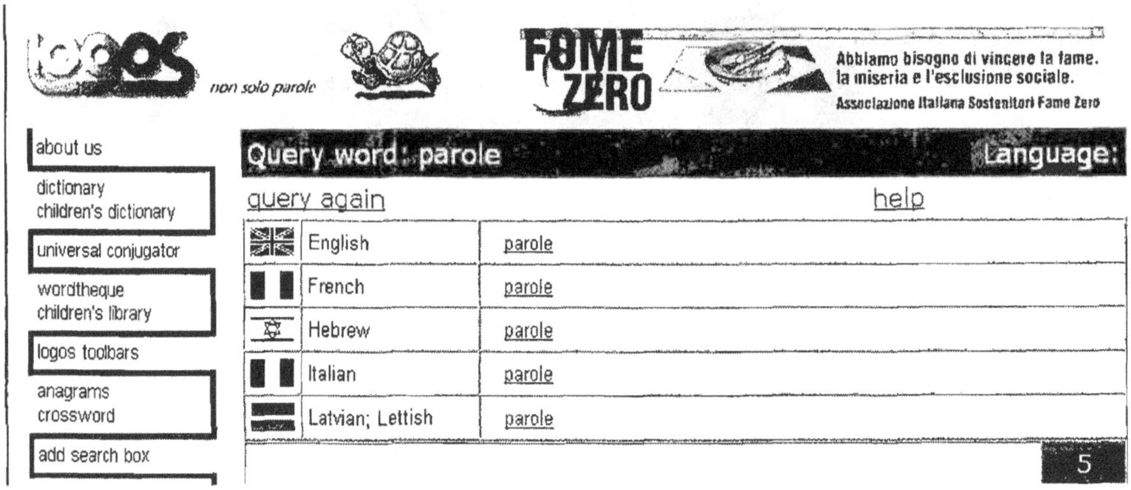
logos
non solo parole
FOME ZERO
Abbiamo bisogno di vincere la fame. la miseria e l'esclusione sociale.
Associazione Italiana Sostenitori Fame Zero
about us
dictionary
children's dictionary
universal conjugator
wordtheque
children's library
logos toolbars
anagrams
crossword
add search box
Query word: parole
Language:
query again
help
English parole
French parole
Hebrew parole
Italian parole
Latvian; Lettish parole
5

If I then select *parole* in Italian, I am presented with its definition (in Italian) and the option to translate *parole*, with the following results:

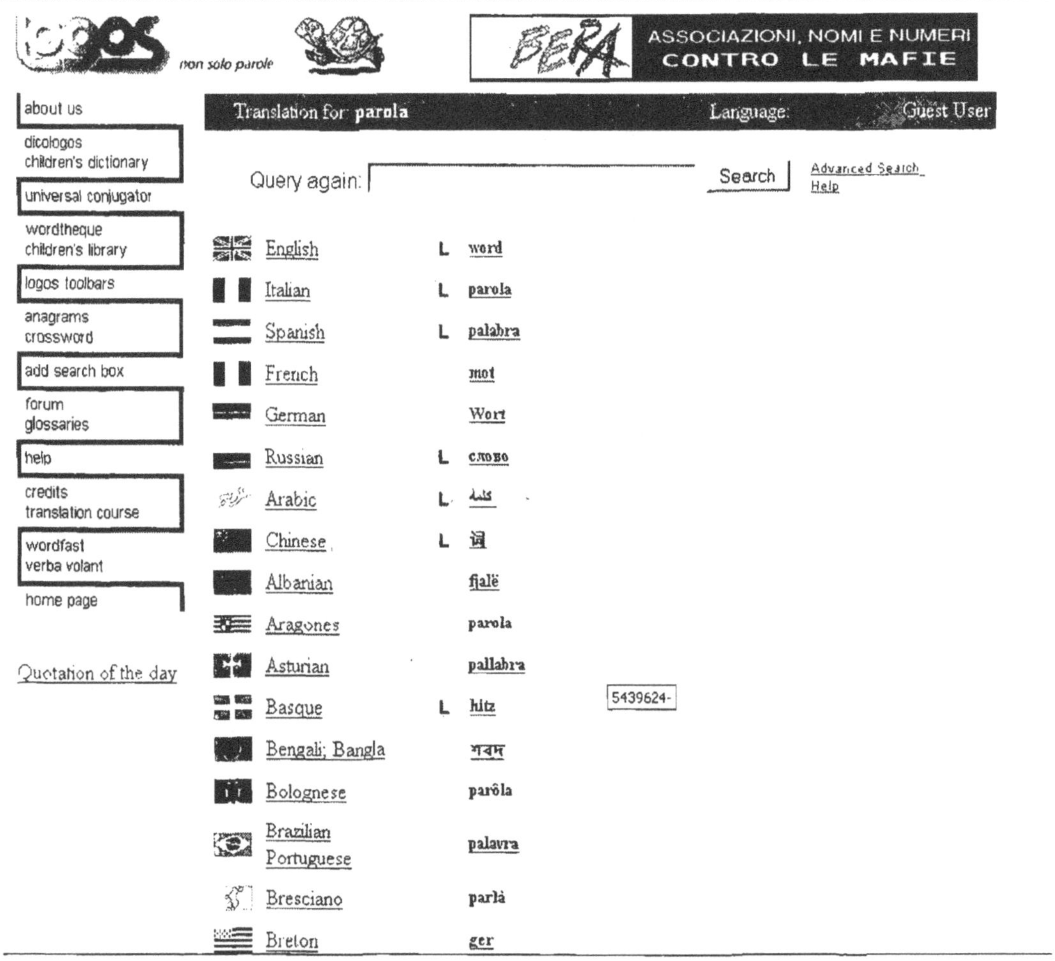
logos
non solo parole
BERA
ASSOCIAZIONI, NOMI E NUMERI
CONTRO LE MAFIE
about us
dicologos
children's dictionary
universal conjugator
wordtheque
children's library
logos toolbars
anagrams
crossword
add search box
forum
glossaries
help
credits
translation course
wordfast
verba volant
home page
Quotation of the day
Translation for: parola
Language:
Guest User
Query again:
Search
Advanced Search
Help
English L word
Italian L parola
Spanish L palabra
French mot
German Wort
Russian L слово
Arabic L
Chinese L
Albanian fjalë
Aragones parola
Asturian pallabra
Basque L hitz
5439624-
Bengali; Bangla
Bolognese paròla
Brazilian Portuguese palavra
Bresciano parlà
Breton ger

Logos also offers a feature called **Logos Library** (formerly Wordtheque), which is an interface to a vast database containing nearly 700 million words in almost 32,000 texts of novels, technical literature, and other texts in many languages. Logos is certainly one of the most valuable linguistic resources on the Internet. For example, here is a search on the keyword [web] and [shakespeare]:

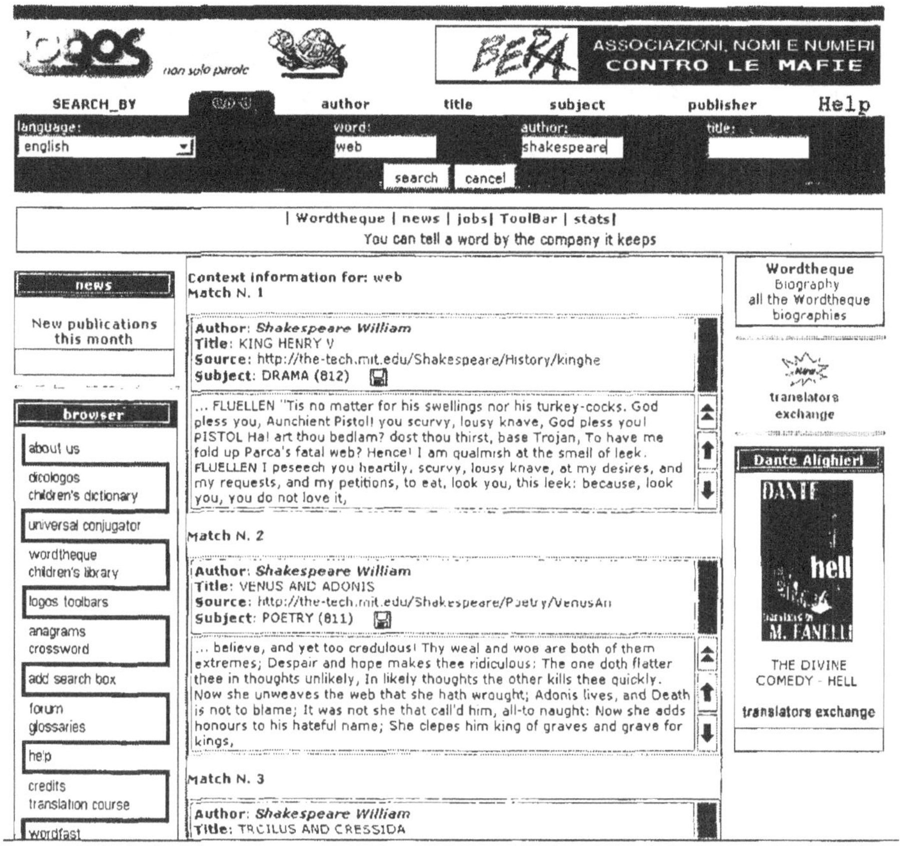
logos non solo parole
ASSOCIAZIONI, NOMI E NUMERI CONTRO LE MAFIE
SEARCH_BY
author
title
subject
publisher
Help
language:
english
word:
web
author:
shakespeare
title:
search
cancel
| Wordtheque | news | jobs| ToolBar | stats|
You can tell a word by the company it keeps
news
New publications this month
browser
about us
dicologos children's dictionary
universal conjugator
wordtheque children's library
logos toolbars
anagrams crossword
add search box
forum glossaries
help
credits translation course
wordfast
Context information for: web
Match N. 1
Author: Shakespeare William
Title: KING HENRY V
Source: http://the-tech.mit.edu/Shakespeare/History/kinghe
Subject: DRAMA (812)
... FLUELLEN "Tis no matter for his swellings nor his turkey-cocks. God pless you, Aunchient Pistol! you scurvy, lousy knave, God pless you! PISTOL Ha! art thou bedlam? dost thou thirst, base Trojan, To have me fold up Parca's fatal web? Hence! I am qualmish at the smell of leek. FLUELLEN I peseech you heartily, scurvy, lousy knave, at my desires, and my requests, and my petitions, to eat, look you, this leek: because, look you, you do not love it,
Match N. 2
Author: Shakespeare William
Title: VENUS AND ADONIS
Source: http://the-tech.mit.edu/Shakespeare/Poetry/VenusAn
Subject: POETRY (811)
... believe, and yet too credulous! Thy weal and woe are both of them extremes; Despair and hope makes thee ridiculous: The one doth flatter thee in thoughts unlikely, In likely thoughts the other kills thee quickly. Now she unweaves the web that she hath wrought; Adonis lives, and Death is not to blame; It was not she that call'd him, all-to naught: Now she adds honours to his hateful name; She clepes him king of graves and grave for kings,
Match N. 3
Author: Shakespeare William
Title: TROILUS AND CRESSIDA
Wordtheque
Biography
all the Wordtheque biographies
translators exchange
Dante Alighieri
THE DIVINE COMEDY - HELL
translators exchange

Logos http://www.logos.it/lang/transl_en.html

Logos Dictionary http://www.logosdictionary.com/pls/dictionary/new_dictionary.index_p

Fagan Finder Translation Wizard

The main page of Michael Fagan's excellent search site, **Fagan Finder** is designed to "help people find what they are looking for" by providing a variety of web, reference, media, news, and other search options. The home page also includes a number of links to "informational pages" (e.g., on RSS, popularity rankings) and useful tools. Probably the best of Fagan's useful tools is the **translation wizard**. The translation wizard brings together over two-dozen online translators from other websites. Many of the translators Fagan Finder accesses will already be familiar to users—Systran, PROMT, InterTran. But there are many others that users may not know about and Fagan Finder can help you locate and use them from one interface.

Keep several things in mind: first, Fagan Finder is simply an ***interface to translation tools located at other websites***. Second, some of the tools it accesses are only word-for-word ("dictionary") translations while other will translate blocks of text and still others are true url or webpage translators, translating entire webpages or even sites. Finally, some of the sites Fagan Finder accesses require registration to use (e.g., Ajeeb's Arabic to English webpage translation tool). All this being said, the Fagan Finder translation wizard is a wonderful resource not only because of all the translation tools it brings together in one place, but perhaps even more for the ***virtual international keyboards*** it offers. At present, Fagan Finder provides the following virtual keyboards:

- Arabic
- Cyrillic
- Greek
- Hebrew
- Thai

Virtual keyboards make it much easier to translate text from one of these languages to English because they mean users do not need international input locales (i.e., software for non-Latin keyboards) loaded on their computers. Fagan Finder's virtual keyboard option even supports right-to-left languages. Second, the international keyboards can be used not only to translate terms but also to search on non-English terms. How? It's simple. Select the international keyboard of your choice at Fagan Finder, type a word or phrase in, say, Hebrew.

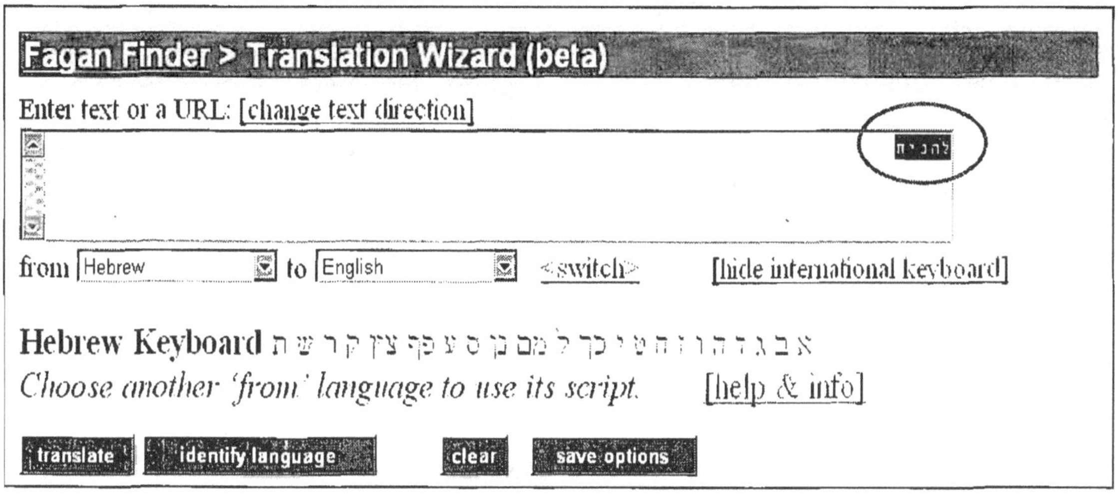
Fagan Finder > Translation Wizard (beta)
Enter text or a URL: [change text direction]
from Hebrew to English <switch> [hide international keyboard]
Hebrew Keyboard
Choose another 'from' language to use its script. [help & info]
translate
identify language
clear
save options

Fagan Finder Translation Wizard http://www.faganfinder.com/translate/

Then copy and paste the Hebrew term or phrase into a search engine such as Google and Yahoo (and many international search engines) that handle multiple character encodings and you can search on the term without having to load a Hebrew keyboard.

Here are some other interesting and important facts and features about the Fagan Finder translation wizard. The wizard gives users the option to *list all translation matches*, to *choose multiple translators* (when available), to *select dialects*, to list matches if there are several translation tools so users can pick one or more. Also, in addition to HTML, **PDF** and **SWF (Flash) file formats** are supported for webpage translations.

Another useful tool that is part of the Fagan Finder translation wizard is the **language identification interface**. Here users can ***enter either text or a url***, and Fagan Finder will query a set of language identification tools to try to determine the language and encoding of the text. However, keep in mind that it often easy to determine the language/encoding of a webpage either by looking at the **Page Info** (in Mozilla only; look at *charset*) or **Page Source** in both Mozilla and Internet Explorer. Look for *charset*, which will indicate a specific character set, which you can then use to identify the language. For example, Windows-1256, ISO 8859-6, ISO 8859-1 all are Arabic language encodings.

However, you cannot always find this information using Page Source/Info and, for individual words or phrases, Fagan Finder's language identification interface is especially valuable. My tests of the language identification tool indicate it usually does a good job of identifying either the language alone or the language and the encoding. Fagan Finder's language identification is not a single tool but a single interface to many language identification tools. Users have the option to choose among 10 language identification tools. Given the complexity and difficulty of language identification, it's probably a good idea to compare the results of several different tools and not rely on one.

For each supported language, Fagan Finder also provides a separate page with useful information about the language, translations, and links to other related sites. Fagan Finder does offer a complete list of supported languages, though there is no link to it from either the Fagan Finder home page or translation wizard page (the link is below). There is also a separate page listing all the translators invoked by Fagan Finder and a page devoted to the translation wizard's language identification tool.

Fagan Finder Translation Wizard http://www.faganfinder.com/translate/

Fagan Finder Translation Wizard Languages Page
http://www.faganfinder.com/translate/language.php

Fagan Finder Translation Wizard List of Translators
http://www.faganfinder.com/translate/tool.php

Fagan Finder Translation Wizard Language Identification
http://www.faganfinder.com/translate/identify.php

Google's Arabic ↔ English Webpage and Text
http://www.google.com/language_tools

Until recently, the only free online Arabic/English webpage translation tool was the Tarjim Site for Online Translation using the Sakhr translation software, where registration was required. Late last week Google introduced Arabic ↔ English text and—more importantly—webpage or url translations to its language tools page. Here's what Google had to say about its new tool:

> "Because we want to provide everyone with access to all the world's information, including information written in every language, one of the exciting projects at Google Research is machine translation. Most state-of-the-art commercial machine translation systems in use today have been developed using a rules-based approach and require a lot of work by linguists to define vocabularies and grammars.
>
> Several research systems, including ours, take a different approach: we feed the computer with billions of words of text, both monolingual text in the target language, and aligned text consisting of examples of human translations between the languages. We then apply statistical learning techniques to build a translation model. We have achieved very good results in research evaluations.
>
> Now you can see the results for yourself. We recently launched an online version of our system for Arabic-English and English-Arabic. Try it out! Arabic is a very challenging language to translate to and from: it requires long-distance reordering of words and has a very rich morphology. Our system works better for some types of text (e.g. news) than for others (e.g. novels) — and you probably

should not try to translate poetry ... but do stay tuned for more exciting developments."[108]

The big question in everyone's mind is: how good is this new tool? Take a look at this translation of the main Arabic Wikipedia page, first in Arabic…

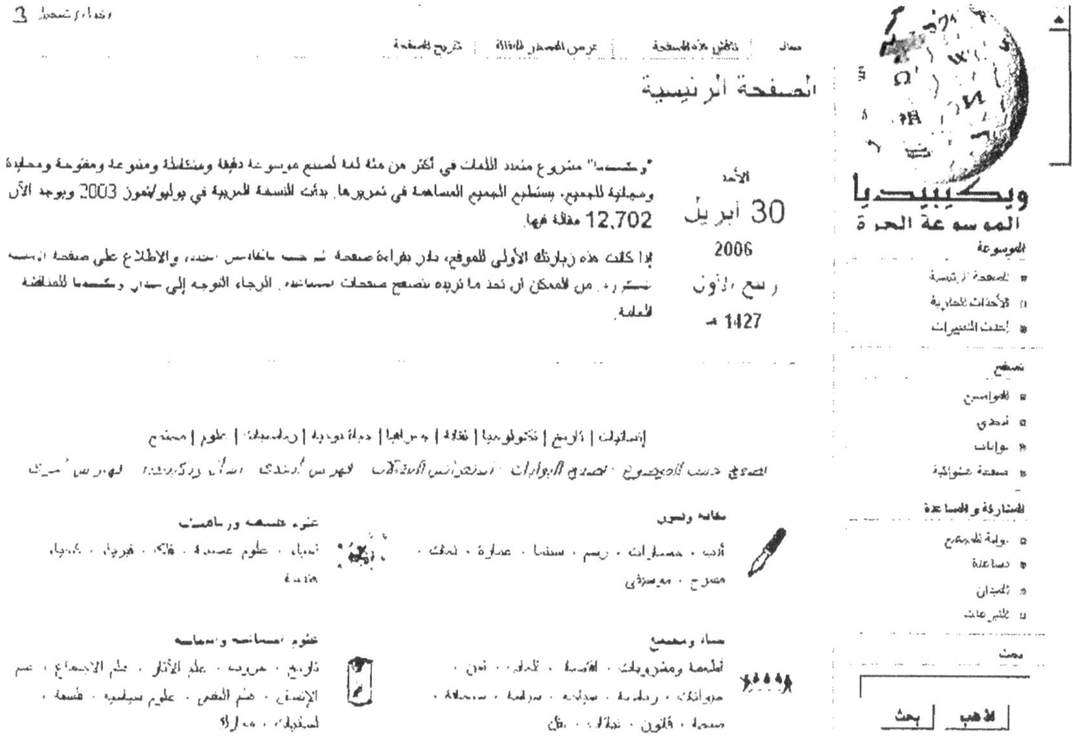
الصفحة الرئيسية
ويكيبيديا
الموسوعة الحرة
30 أبريل
2006
12,702

…then in the English translation provided by Google:

[108] Franz Och, Research Scientist, Google Research Blog, April 28, 2006, <http://googleresearch.blogspot.com/2006/04/statistical-machine-translation-live.html> (November 28, 2006).

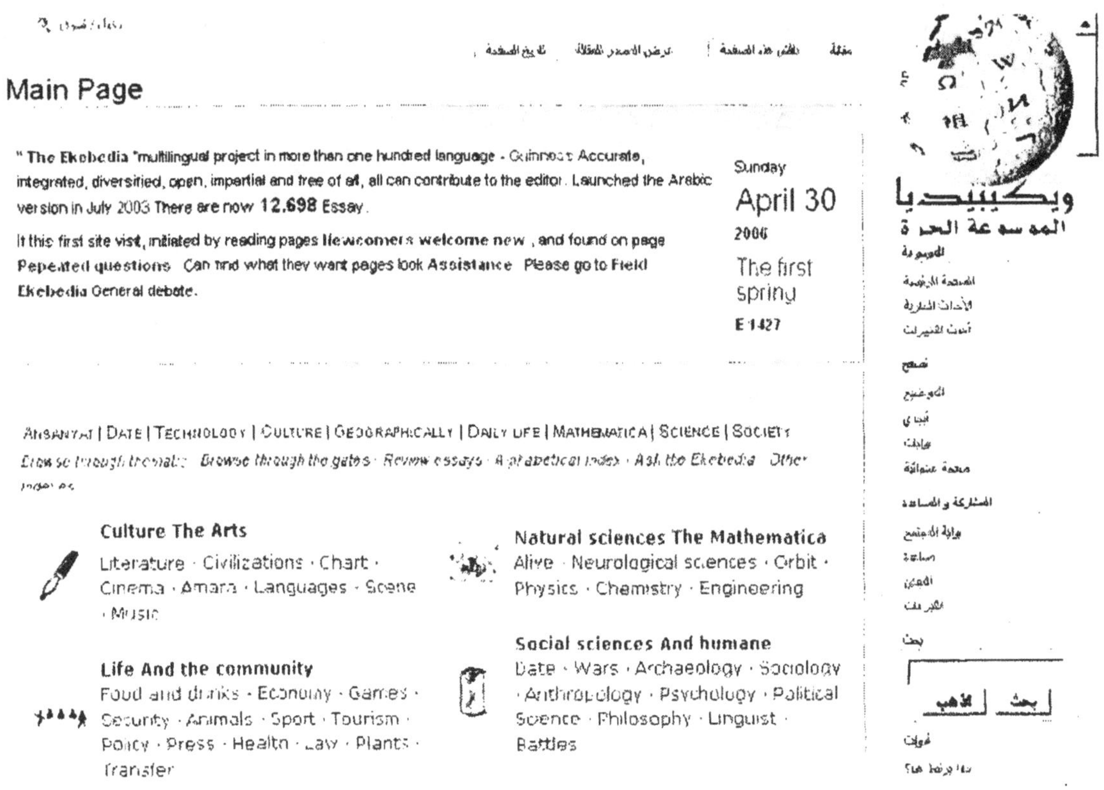
Main Page
" The Ekebedia "multilingual project in more than one hundred language - Guinness Accurate, integrated, diversified, open, impartial and free of all, all can contribute to the editor. Launched the Arabic version in July 2003 There are now 12.698 Essay.
It this first site visit, initiated by reading pages Newcomers welcome new , and found on page Repeated questions Can find what they want pages look Assistance Please go to Field Ekebedia General debate.
Sunday
April 30
2006
The first spring
E 1427
Ansanyat | Date | Technology | Culture | Geographically | Daily life | Mathematica | Science | Society
Culture The Arts
Literature · Civilizations · Chart · Cinema · Amara · Languages · Scene · Music
Natural sciences The Mathematica
Alive · Neurological sciences · Orbit · Physics · Chemistry · Engineering
Social sciences And humane
Date · Wars · Archaeology · Sociology · Anthropology · Psychology · Political Science · Philosophy · Linguist · Battles
Life And the community
Food and drinks · Economy · Games · Security · Animals · Sport · Tourism · Policy · Press · Health · Law · Plants · Transfer
ويكيبيديا
الموسوعة الحرة

Google's Arabic language tool is based on the company's own research and development and is not a product of Systran. Is it perfect? Hardly, and no one would expect it to be. But is it a whole lot better than nothing? Absolutely. I have always found machine translation tools to be most useful in translating between two very different languages, for example between English and languages, such as Arabic, that do not use the Latin alphabet. Interestingly, this tool will probably not be met with enthusiasm in some Arabic speaking countries. For example, the UAE bans online translation tools, which creates an enormous barrier for non-English speaking Arabs because so little of the web is in Arabic.[109] However, all in all, this is extremely welcome news.

Babelplex Bilingual Search http://www.babelplex.com/

Babelplex is a "bilingual search service that searches the web in one language and in another language via a cross-language information retrieval system."[110] Babelplex is the brainchild of HK Tang. As with many creations, Babelplex is the product both of Tang's frustration and his need for something that didn't exist. The frustration is

---

[109] United Press International, "The Web: Arabic Language Internet," 29 March 2006, <http://www.physorg.com/news62862307.html> (14 November 2006).

[110] About Babelplex, <http://www.babelplex.com/about.html> (14 November 2006).

one many of us have experienced: searching in non-Latin character sets when the keyboard is not configured for those languages. The need was driven by the fact that Tang and his family are bilingual, meaning they often want to search in more than just English. Tang also observed that "if you simply search down to the simplest equation, there are two sides. Output, which Google has solidly nailed down, and input, which is very relevant when searching in foreign text."[111] The input side is harder because computers are initially designed for the user to input text in one language, and anything else requires the user to make changes.

Babelplex does the work for you by using AltaVista's Babelfish Translation, Google Translate, or Yahoo Language Search tools and the AltaVista, Google, or Yahoo search tools to run parallel queries in two languages. Take a look at the Babelplex homepage and you'll get the idea:

Babelplex
Search
and search your
English query in Simplified Chinese
Choose a search engine to Babelplex:
AltaVista
Google
Google Deutsch
Google español
Google Français
Google Italiano
Google Nederlands
Google Português
Yahoo!
Babelplex
Help

What you don't see above is the list of *to* and *from* language pairs:

[111] John Battelle, "Babelplex: Search in Two Languages," *John Battelle's Searchblog*, 26 November 2004, <http://battellemedia.com/archives/001065.php> (14 November 2006).

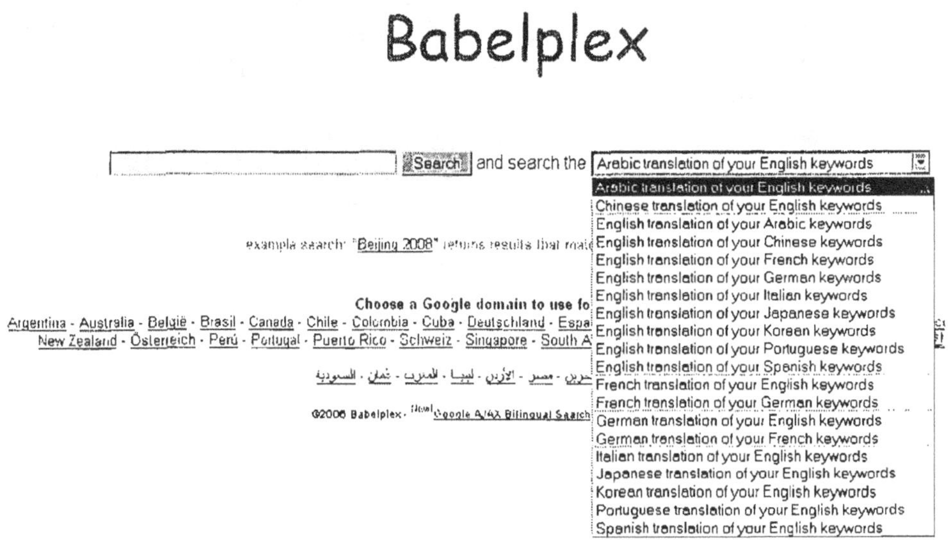
Babelplex
Search
and search the
Arabic translation of your English keywords
Chinese translation of your English keywords
English translation of your Arabic keywords
English translation of your Chinese keywords
English translation of your French keywords
English translation of your German keywords
English translation of your Italian keywords
English translation of your Japanese keywords
English translation of your Korean keywords
English translation of your Portuguese keywords
English translation of your Spanish keywords
French translation of your English keywords
French translation of your German keywords
German translation of your English keywords
German translation of your French keywords
Italian translation of your English keywords
Japanese translation of your English keywords
Korean translation of your English keywords
Portuguese translation of your English keywords
Spanish translation of your English keywords
Choose a Google domain to use fo
Argentina - Australia - België - Brasil - Canada - Chile - Colombia - Cuba - Deutschland - Espa
New Zealand - Österreich - Perú - Portugal - Puerto Rico - Schweiz - Singapore - South A

How do you use Babelplex? Enter a query in the SEARCH box in the language of your choice from the language pairs supported, then select a language to translate to/from. Babelplex now uses only Google to search. Here is a simple example of a search from English to Arabic:

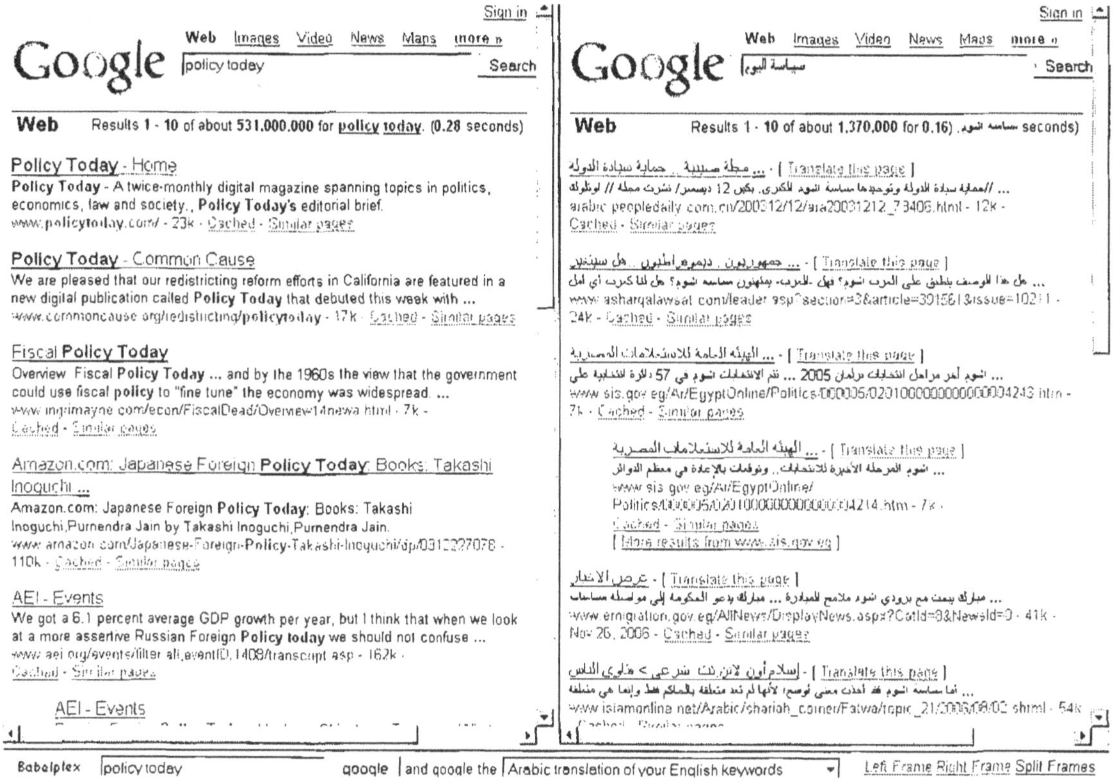
Sign in
Google
Web Images Video News Maps more »
policy today
Search
Web Results 1 - 10 of about 531,000,000 for policy today. (0.28 seconds)
Policy Today - Home
Policy Today - A twice-monthly digital magazine spanning topics in politics, economics, law and society., Policy Today's editorial brief.
www.policytoday.com/ - 23k - Cached - Similar pages
Policy Today - Common Cause
We are pleased that our redistricting reform efforts in California are featured in a new digital publication called Policy Today that debuted this week with ...
www.commoncause.org/redistricting/policytoday - 17k - Cached - Similar pages
Fiscal Policy Today
Overview. Fiscal Policy Today ... and by the 1960s the view that the government could use fiscal policy to "fine tune" the economy was widespread. ...
www.ingrimayne.com/econ/FiscalDead/Overview14newa.html - 7k - Cached - Similar pages
Amazon.com: Japanese Foreign Policy Today: Books: Takashi Inoguchi ...
Amazon.com: Japanese Foreign Policy Today: Books: Takashi Inoguchi,Purnendra Jain by Takashi Inoguchi,Purnendra Jain.
www.amazon.com/Japanese-Foreign-Policy-Takashi-Inoguchi/dp/0312227078 - 110k - Cached - Similar pages
AEI - Events
We got a 6.1 percent average GDP growth per year, but I think that when we look at a more assertive Russian Foreign Policy today we should not confuse ...
www.aei.org/events/filter.all,eventID.1408/transcript.asp - 162k - Cached - Similar pages
AEI - Events
Sign in
Google
Web Images Video News Maps more »
سياسة اليوم
Search
Web Results 1 - 10 of about 1,370,000 for سياسة اليوم. (0.16 seconds)
[Translate this page]
arabic.peopledaily.com.cn/200312/12/ara20031212_73406.html - 12k - Cached - Similar pages
[Translate this page]
www.asharqalawsat.com/leader.asp?section=3&article=391561&issue=10211 - 24k - Cached - Similar pages
[Translate this page]
www.sis.gov.eg/Ar/EgyptOnline/Politics/000005/0201000000000000004243.htm - 7k - Cached - Similar pages
[Translate this page]
www.sis.gov.eg/Ar/EgyptOnline/Politics/000005/0201000000000000004214.htm - 7k - Cached - Similar pages
[More results from www.sis.gov.eg]
[Translate this page]
www.emigration.gov.eg/AllNews/DisplayNews.aspx?CatId=8&NewsId=0 - 41k - Nov 26, 2006 - Cached - Similar pages
[Translate this page]
www.islamonline.net/Arabic/shariah_corner/Fatwa/topic_21/2006/08/02.shtml - 54k
Babelplex policy today google and google the Arabic translation of your English keywords Left Frame Right Frame Split Frames

And here is the same query translated from Arabic to English:

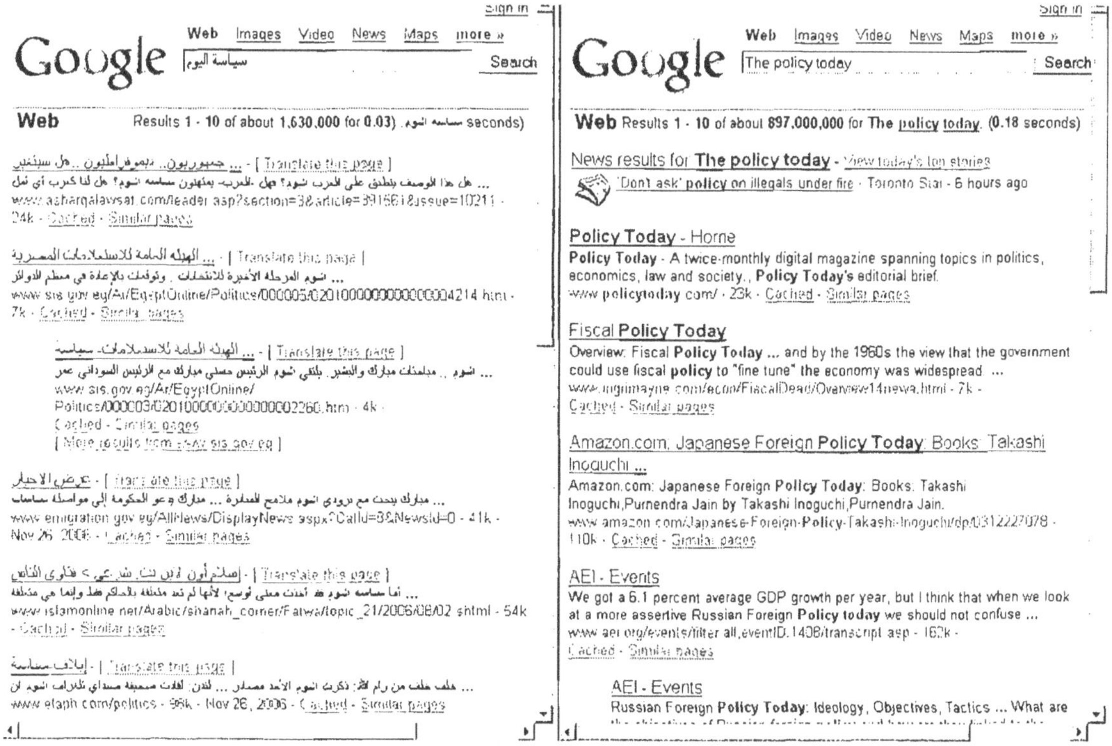
Sign in
Google
Web Images Video News Maps more »
سياسة اليوم
Search
Web Results 1 - 10 of about 1,630,000 for سياسة اليوم. (0.03 seconds)
[Translate this page]
www.asharqalawsat.com/leader.asp?section=3&article=391561&issue=10211 - 24k - Cached - Similar pages
[Translate this page]
www.sis.gov.eg/Ar/EgyptOnline/Politics/000005/0201000000000000004214.htm - 7k - Cached - Similar pages
[Translate this page]
www.sis.gov.eg/Ar/EgyptOnline/Politics/000003/0201000000000000002260.htm - 4k - Cached - Similar pages
[More results from www.sis.gov.eg]
[Translate this page]
www.emigration.gov.eg/AllNews/DisplayNews.aspx?CatId=8&NewsId=0 - 41k - Nov 26, 2006 - Cached - Similar pages
[Translate this page]
www.islamonline.net/Arabic/shariah_corner/Fatwa/topic_21/2006/08/02.shtml - 54k - Cached - Similar pages
[Translate this page]
www.elaph.com/politics - 98k - Nov 26, 2006 - Cached - Similar pages
Sign in
Google
Web Images Video News Maps more »
The policy today
Search
Web Results 1 - 10 of about 897,000,000 for The policy today. (0.18 seconds)
News results for The policy today - View today's top stories
'Don't ask' policy on illegals under fire - Toronto Star - 6 hours ago
Policy Today - Home
Policy Today - A twice-monthly digital magazine spanning topics in politics, economics, law and society., Policy Today's editorial brief.
www.policytoday.com/ - 23k - Cached - Similar pages
Fiscal Policy Today
Overview: Fiscal Policy Today ... and by the 1960s the view that the government could use fiscal policy to "fine tune" the economy was widespread. ...
www.ingrimayne.com/econ/FiscalDead/Overview14newa.html - 7k - Cached - Similar pages
Amazon.com: Japanese Foreign Policy Today: Books: Takashi Inoguchi ...
Amazon.com: Japanese Foreign Policy Today: Books: Takashi Inoguchi,Purnendra Jain by Takashi Inoguchi,Purnendra Jain.
www.amazon.com/Japanese-Foreign-Policy-Takashi-Inoguchi/dp/0312227078 - 110k - Cached - Similar pages
AEI - Events
We got a 6.1 percent average GDP growth per year, but I think that when we look at a more assertive Russian Foreign Policy today we should not confuse ...
www.aei.org/events/filter.all,eventID.1408/transcript.asp - 162k - Cached - Similar pages
AEI - Events
Russian Foreign Policy Today: Ideology, Objectives, Tactics ... What are

Not bad. However, the quality of results varies greatly, in part depending upon the complexity of queries and in part on how good the translation software works. Anyone who has used machine translation tools knows they are limited. Here are some tips that might help make your Babelplex queries more successful.

- Keep it simple: the query [syllogism] worked very well going from English to most languages, but [windows operating system] stumped most of the translation services in pretty much every language.

- Avoid proper names: proper names probably will not translate; a search for [quicken] finds *acelere* in Spanish; a search for [bill gates] from English to simplified Chinese and back to English produced interesting but unhelpful results.

- Syntax matters: you will get different results for the translation to Spanish of the English query [spanish military] versus [military spanish].

- Use synonyms judiciously: while it might seem intuitively obvious that using synonyms or related terms might produce better results, it may in fact confuse matters. If you are searching for information about cows, [cow] produces better results than [cow cattle bovine steer]. Also, remember that the default operator for all these search engines is AND, so they will only return webpages containing all the search terms.

Babelplex has nice user features for viewing results. The split screen shows both the original query and the query translated into a second language. The left frame displays the results of the original query and the right frame shows the results in the language chosen for translation. If no translation of a query is possible, you will see the translation tool's webpage and will have the option to try the translation and subsequent search from that page.

Babelplex enters the pantheon of search engine augmentation sites—including FindForward, Soople, FaganFinder's Ultimate Google Interface—as a welcome and useful addition.

Finding Online Dictionaries

Foreignword http://www.foreignword.com/Tools/dictsrch.htm

Language Automation's Glossaries http://www.rahul.net/lai/glossaries.html

Martindale's Language and Translation Center
http://www.martindalecenter.com/Language.html

Paderborn University List of Dictionaries
http://www-math.uni-paderborn.de/dictionaries/Dictionaries.html

Word2Word http://www.word2word.com/dictionary.html

yourDictionary http://www.yourdictionary.com/

Online Multilingual Dictionaries

Digital Dictionaries of South Asia http://dsal.uchicago.edu/dictionaries/

Eurodicautom* http://europa.eu.int/eurodicautom/Controller

Foreignword http://www.foreignword.com/Tools/dictsrch.htm

Language to Language http://www.langtolang.com/

Logos * http://www.logos.it/lang/transl_en.html

OneLook Dictionaries http://www.onelook.com/

Online Dictionary http://www.online-dictionary.biz/
English↔French, German, Spanish, Italian, Japanese, Chinese, Russian

Papillon Project http://www.papillon-dictionary.org/Home.po
English↔Estonian, German, French, Japanese, Vietnamese, Korean, Malay, Chinese

FreeDict http://www.freedict.com/

Travlang's Translating Dictionaries http://dictionaries.travlang.com/

UltraLingua http://www.ultralingua.net/
English↔German, French, Spanish, Italian, Portuguese, Esperanto, Latin

Word Reference http://www.wordreference.com/

Online Text Translators

AjaxTrans http://ajax.parish.ath.cx/translator/

Babelfish from Yahoo http://babelfish.yahoo.com/

FreeTranslation** http://www.FreeTranslation.com/

Foreignword http://foreignword.com/Tools/transnow.htm

InterTran** http://www.tranexp.com/win/itserver.htm

Mezzofanti Translations http://www.mezzofanti.org/translation/

PhraseBase http://www.phrasebase.com/translations/index.php?action=language

PopJisyo (Asian languages) http://www.online-dictionary.biz/

PROMT** http://www.translate.ru/eng/text.asp

Reverso** http://www.reverso.net/text_translation.asp

VoyCabulary http://www.voycabulary.com/

WorldLingo** http://www.worldlingo.com/products_services/worldlingo_translator.html

yourDictionary http://www.yourdictionary.com/diction1.html#translate

---

Online Web Page Translators

Ajeeb! Arabic ↔ English+ http://tarjim.ajeeb.com/ajeeb/default.asp?lang=1

Babelfish from Yahoo http://babelfish.yahoo.com/

InterTran** http://www.tranexp.com/win/itserver.htm

PROMT http://www.translate.ru/eng/srvurl.asp

Reverso** http://www.reverso.net/url_translation.asp

Systran http://www.systransoft.com/

VoyCabulary http://www.voycabulary.com/

WorldLingo** http://www.worldlingo.com/en/websites/url_translator.html

Other Language Sites & Tools

Computing with Accents, Symbols, & Foreign Scripts
http://tlt.its.psu.edu/suggestions/international/index.html

Detailed instructions from Penn State University for working on computers and the Internet in non-English characters and encodings.

+ Requires free registration

** Site offers virtual keyboard or special characters for non-English translations

You Gotta Know When to Fold 'Em

A last thought before bringing this section to a close. With the plethora of information available today via the Internet and the fact that more data is added every day, it is easy to fall prey to the erroneous belief that if we just know where to look on the web, the information we seek—*all* the information we seek—is "out there" somewhere. The danger is that this misguided idea will give rise to the ***never-ending search***, that is, the conviction that "if I just go to one more website, run one more query, or search one more database I'll find what I'm looking for."

Sometimes the information you need simply is not to be found on the Internet and no amount of searching, no amount of creativity or ingenuity can make it appear. I have the sneaking suspicion that most Internet researchers have fallen victim to the never-ending search on occasion because it is so tempting to believe that the answer is there, somewhere, only a few more clicks away. If the data is not available, you are not going to find it, no matter how clever or persistent you are, so try to keep this last in mind when you feel yourself becoming obsessed with a particular research topic:

Rule Seven

Know when to stop searching.

The problem with the Internet is that it is open-ended. Think of it in these terms. Let's say your job is to go into a large room piled high with papers and determine if a particular document is in that room. If you are careful and methodical, you will eventually find the document or be able to say with confidence that the document is not in the room. Not so with the Internet. ***You can never be sure something is not "out there"*** because there are always more possible places the information could be located than you can search and examine. While the number of documents on the Internet may not in fact be infinite, in practical terms the Internet is limitless. So give yourself a break if you can't find everything you're looking for!

The Rules of the Road:

1. Use the right tool for the job.
2. Let other people do as much work for you as possible.
3. Develop, maintain, and backup bookmarks.
4. Use more than one search engine.
5. Read the instructions.
6. Use the specialized and/or unique functions of foreign search engines.
7. Know when to stop searching.

Beyond Search Engines—Specialized Research Tools

Search engines are a good and natural starting place for performing research on the Internet, but they represent only a small portion of the data available on the web and only one way of tapping into that data. It is also important to understand that ***search engine spiders do not access*** (and therefore search engines do not index) ***most data contained in many databases or websites that require registration or payment*** to enter. For example, search engines do not normally index the data in PeopleData.com (a database) or any information beyond the first page or so of the *Chicago Tribune* (which requires registration). These types of sites require users to access them directly. The information at these sites is part of the invisible, hidden, or deep web.

The types of sites and information that are not generally accessible to search engines include:

- information in databases: phone and email directories, Whois registration & DNS data, dictionaries, encyclopedia articles, statistics, legal and medical data, financial information.
- rapidly changing information: news, airline flight information, stock, bond, currency market data, auctions.
- for-fee and subscription services.
- information behind a firewall (corporate, government, educational).

To give you a better idea just how vast the deep web is, consider these points from "The Deep Web: Surfacing Hidden Value"[112] by Michael K. Bergman.

[112] Michael K. Bergman, "The Deep Web: Surfacing Hidden Value," *BrightPlanet*, August 2001, <http://www.brightplanet.com/technology/deepweb.asp> (14 November 2006).

- Public information on the deep web is at least 400 to 550 times larger than the commonly defined World Wide Web.
- The deep web contains 7,500 terabytes of information compared to nineteen terabytes of information in the surface web.
- The deep web contains nearly 550 billion individual documents compared to the one billion of the surface web.
- More than 200,000 deep websites presently exist.
- Sixty of the largest deep-websites collectively contain about 750 terabytes of information—sufficient by themselves to exceed the size of the surface web forty times.

Therefore, it is vital to maintain a good set of bookmarks for a wide variety of research tools beyond search engines. Specialized search tools—database finders, email lookup tools, and online telephone and fax directories—are good first additions to a robust set of research tools.

Web Tip

If you don't want to "be found," never post to Usenet newsgroups. Once you do, expect to be spammed and to appear in directories, such as email lookup databases. Your only real solution at this point is to get a new Internet account.

Email Lookups

Email lookup tools vary widely in quality and the features they offer. Most, quite frankly, are not very good because their sources of data are poor. Most email lookup tools gather their information from two sources:

1. trolling Usenet news postings and
2. users who add themselves to the listing.

Also, don't expect to get a lot of information about someone even if you do manage to match a name and an email address. The most you usually get is a full name, a complete email address, and maybe an affiliation. This does not mean all email lookup tools are useless.[113] Sometimes it is a coup just to match an email address with a name.

Email lookup tools sometimes offer some type of domain or reverse lookup. **Infospace** permits reverse email lookups in the format [@eunet.yu]. This option is not under email search but under ***reverse lookups at:*** <http://www.infospace.com/home/white-pages/reverse-email>.
Furthermore, Infospace lets users enter ***partial addresses and will return all listings starting with those characters***, e.g., [@eunet].

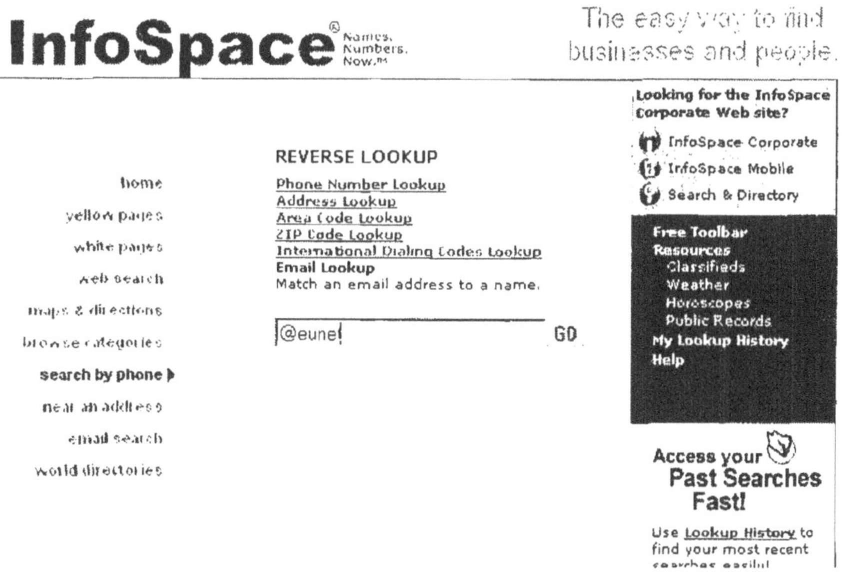
InfoSpace
Names. Numbers. Now.
The easy way to find businesses and people.
Looking for the InfoSpace Corporate Web site?
InfoSpace Corporate
InfoSpace Mobile
Search & Directory
REVERSE LOOKUP
Phone Number Lookup
Address Lookup
Area Code Lookup
ZIP Code Lookup
International Dialing Codes Lookup
Email Lookup
Match an email address to a name.
@eunet
GO
home
yellow pages
white pages
web search
maps & directions
browse categories
search by phone
near an address
email search
world directories
Free Toolbar
Resources
Classifieds
Weather
Horoscopes
Public Records
My Lookup History
Help
Access your Past Searches Fast!
Use Lookup History to find your most recent

[113] In fact, *Wired* raised quite a few eyebrows with its article on Iraqi leader Saddam Hussein's public email account that was provided at the Iraqi President's website. Brian McWilliams, "Dear Saddam. How Can I Help?" *Wired*, 28 October 2002, <http://www.wired.com/news/conflict/0,2100,55967,00.html> (14 November 2006).

Infospace http://infospace.com/

Infospace Email Lookup http://www.infospace.com/home/white-pages/email-search

Infospace Reverse Email Lookup
http://www.infospace.com/home/white-pages/reverse-email

Look4U claims to maintain a database of over 2 million Chinese names and email addresses not only from Taiwan and Hong Kong but also from China, Malaysia, Singapore, the US, and many more countries. The real advantage of Look4U is that it is designed to be searched in either GB or Big5 Chinese encoding but also permits the use of pinyin pronunciation of the name.

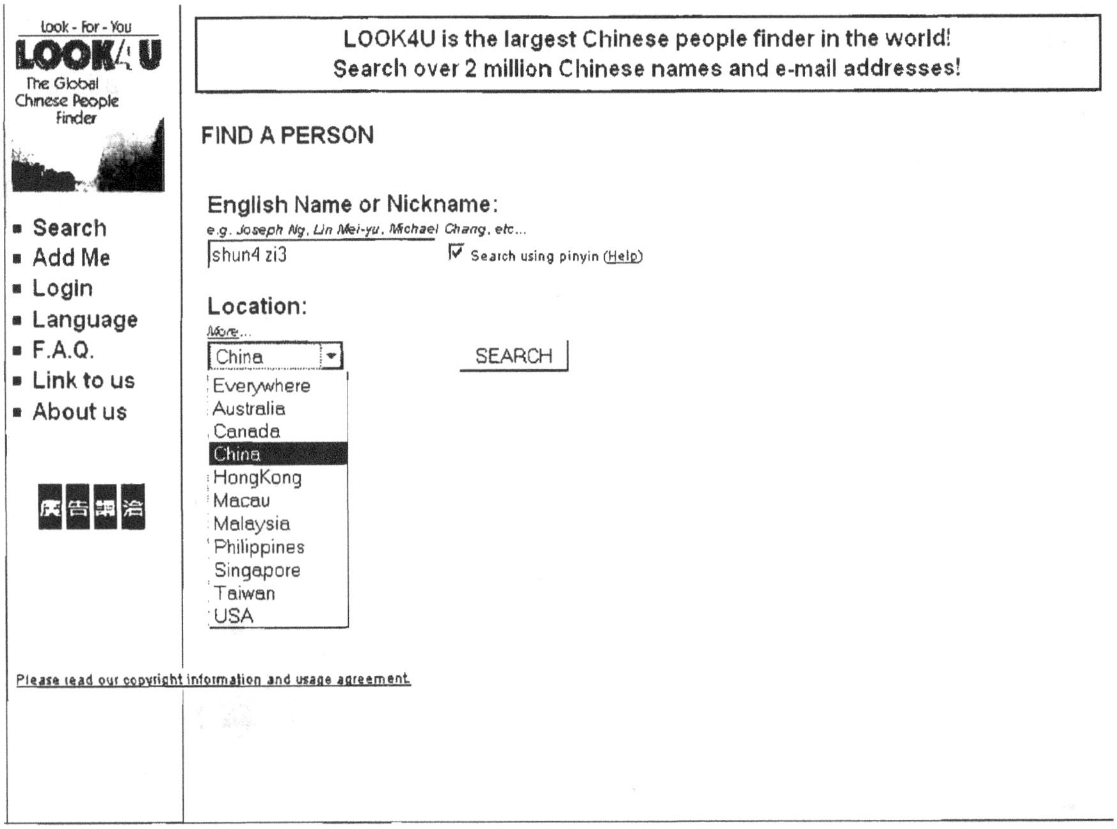
Look - For - You
LOOK4U
The Global Chinese People Finder
Search
Add Me
Login
Language
F.A.Q.
Link to us
About us
LOOK4U is the largest Chinese people finder in the world!
Search over 2 million Chinese names and e-mail addresses!
FIND A PERSON
English Name or Nickname:
e.g. Joseph Ng, Lin Mei-yu, Michael Chang, etc...
shun4 zi3
Search using pinyin (Help)
Location:
More...
China
Everywhere
Australia
Canada
China
HongKong
Macau
Malaysia
Philippines
Singapore
Taiwan
USA
SEARCH
Please read our copyright information and usage agreement.

Look4U http://www.look4u.com/english/

Finding Email Directories

Email-Directory.com http://www.email-directory.com/

Nedsite http://www.nedsite.nl/search/people.htm#email

International Email Lookup Tools

Addresses.com http://www.allemailaddresses.com/

Infospace Email Lookup http://www.infospace.com/home/white-pages/email-search

Infospace Reverse Email Lookup http://www.infospace.com/home/white-pages/reverse-email

Look4U http://www.look4u.com/english/

MESA MetaEmailSearchAgent http://mesa.rrzn.uni-hannover.de/

Peoplesearch Reverse Email Search http://peoplesearch.net/peoplesearch/peoplesearch_reverse_email_address.html

World Email Directory http://www.worldemail.com/freemail.htm

Mega/MetaDirectories

If these email look-up sites aren't sufficient (or you would just like to see what other email search tools are available), the following websites either link to many email lookup web pages (megadirectories) or actually run parallel email searches from their site (metadirectories). These are also the fastest and easiest way to see if a region or country has an email search tool specific to it. Some do, but I have not had a great deal of luck finding useful information using these email lookups. However, as the Internet expands, expect these services to improve.

Email Megadirectories

Freeality Email Lookup http://www.freeality.com/findet.htm

Infospace International Directories http://www.infospace.com/intl/int.html

MESA MetaEmailSearchAgent http://mesa.rrzn.uni-hannover.de/

Nedsite http://www.nedsite.nl/search/people.htm#email

Peoplesearch http://peoplesearch.net/peoplesearch/peoplesearch_reverse_email_address.html

Infobel's Email Addresses http://www.infobel.com/teldir/teldir.asp?page=/eng/more/email

Telephone and FAX Directories

Most of the email lookup tools above also have a telephone lookup service as well, so I am not going to discuss these sites again. Be aware, however, that no single source has every directory listed and that new directories come on line all the time. Also, many telephone directories are limited to or strongly emphasize North American telephone and fax numbers. Your best bet by far for international phone and fax numbers are ***country-specific directories***. Unlike the very limited international email directories, some international telephone and fax directories are outstanding.

There are two basic types of telephone directory sites: the first are specific telephone lookup sites where you can go and look up a number or name. The second are metasites, sites with many links to telephone directories on the web. These will help you find an online directory in a specific region, country, or city.

Infobel, the website of the Kapitol directory publishing company, is the best single source for online directories at present. It provides many types of directories for a number of countries, including Belgium, France, Luxembourg, Spain, Italy, Denmark, the Netherlands, UK, and more. In some cases, for example Belgium, there is even a GSM directory. Infobel allows users to pick any country, see which directories are available (white pages, yellow pages, GSM, fax), and go directly to that directory.

As of January 2003, **Infobel** and **Telephone Directories on the Web (Teldir)** merged, providing one huge interface to hundreds of links to white pages, yellow pages, fax listings, email addresses, and business directories by world regions. Teldir does not provide a search interface but is instead a directory of links to online directories for most of the countries that have them.

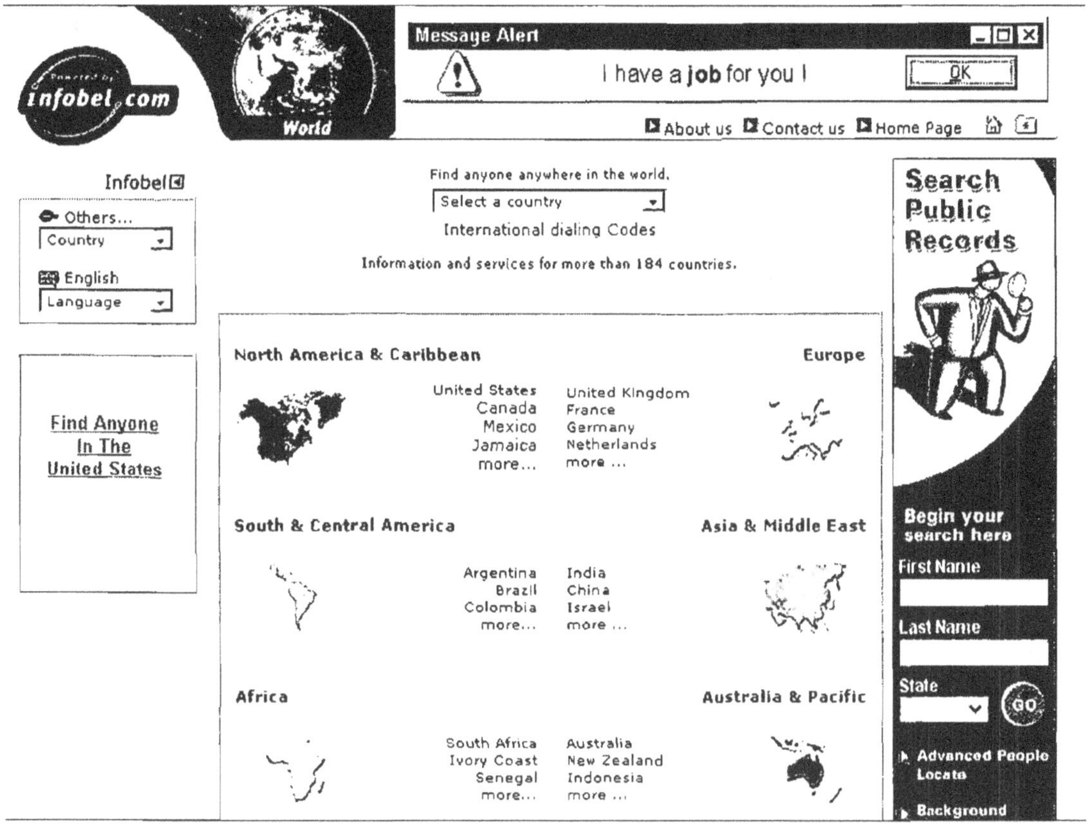
infobel.com
World
Message Alert
I have a job for you !
OK
About us
Contact us
Home Page
Infobel
Others...
Country
English
Language
Find Anyone In The United States
Find anyone anywhere in the world.
Select a country
International dialing Codes
Information and services for more than 184 countries.
North America & Caribbean
United States
Canada
Mexico
Jamaica
more...
Europe
United Kingdom
France
Germany
Netherlands
more ...
South & Central America
Argentina
Brazil
Colombia
more...
Asia & Middle East
India
China
Israel
more ...
Africa
South Africa
Ivory Coast
Senegal
more...
Australia & Pacific
Australia
New Zealand
Indonesia
more ...
Search Public Records
Begin your search here
First Name
Last Name
State
GO
Advanced People Locate
Background

Infobel's Telephone Directories on the Web http://www.infobel.com/teldir/

Phonebook of the World has emerged as one of the most comprehensive telephone directory sites. Click on the world map for your country of interest and Phonebook will tell you if there are white and/or yellow pages for that country online, then link you directly to the pages. Many online directories are, naturally, only in the native language. For example, each Brazilian state has an online directory and all are in Portuguese.

Phonebook of the World http://www.phonebookoftheworld.com/

Infospace International Directories is another good metaguide to international online phone directories. Selecting a specific country will show the types of online resources available—white pages, yellow pages, email directories—and, in a few cases, permit searches from the Infospace website. However, most of the phone directories are accessible only as links directly to their own webpages.

Infospace International Directories http://www.infospace.com/intl/int.html

International White and Yellow Pages, run by a Norwegian company, is another good starting place for finding online yellow pages, white pages, and fax directories.

The site lists countries alphabetically by continent, indicating the types of online directories—white, yellow, fax—available for each.

International White & Yellow Pages http://www.wayp.com/

Remember that city directories often produce better results than country directories, even though most phone directory lists lean towards country directories. Also keep in mind that it is important to search in the target language. Here is the Moscow City Telephone Directory, which has a very good search tool:

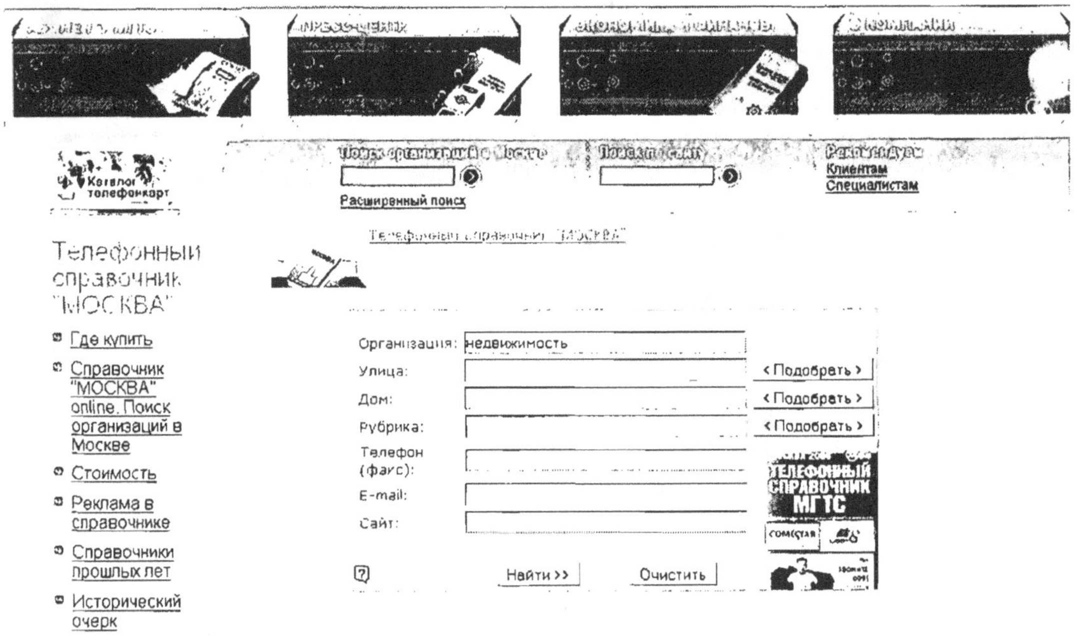
Расширенный поиск
Клиентам
Специалистам
Телефонный справочник "МОСКВА"
Где купить
Справочник "МОСКВА" online. Поиск организаций в Москве
Стоимость
Реклама в справочнике
Справочники прошлых лет
Исторический очерк
Организация: недвижимость
Улица:
Дом:
Рубрика:
Телефон (факс):
E-mail:
Сайт:
< Подобрать >
Найти >>
Очистить
ТЕЛЕФОННЫЙ СПРАВОЧНИК МГТС

Users can search by type of business, street name and/or number, telephone number, email address, or even web address. But searching in Russian is a precondition of using the search tool to the fullest.

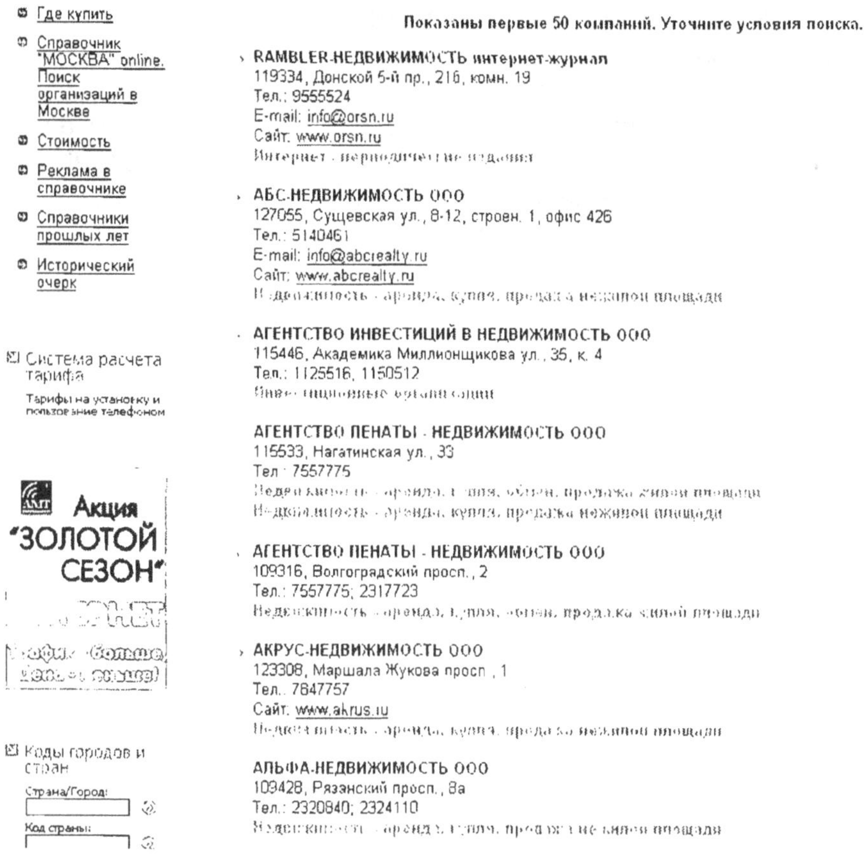
Где купить
Справочник "МОСКВА" online. Поиск организаций в Москве
Стоимость
Реклама в справочнике
Справочники прошлых лет
Исторический очерк
Система расчета тарифа
Тарифы на установку и пользование телефоном
Акция "ЗОЛОТОЙ СЕЗОН"
Коды городов и стран
Страна/Город:
Код страны:
Показаны первые 50 компаний. Уточните условия поиска.
RAMBLER-НЕДВИЖИМОСТЬ интернет-журнал
119334, Донской 5-й пр., 21б, комн. 19
Тел.: 9555524
E-mail: info@orsn.ru
Сайт: www.orsn.ru
АБС-НЕДВИЖИМОСТЬ ООО
127055, Сущевская ул., 8-12, строен. 1, офис 426
Тел.: 5140461
E-mail: info@abcrealty.ru
Сайт: www.abcrealty.ru
АГЕНТСТВО ИНВЕСТИЦИЙ В НЕДВИЖИМОСТЬ ООО
115446, Академика Миллионщикова ул., 35, к. 4
Тел.: 1125516, 1150512
АГЕНТСТВО ПЕНАТЫ - НЕДВИЖИМОСТЬ ООО
115533, Нагатинская ул., 33
Тел.: 7557775
АГЕНТСТВО ПЕНАТЫ - НЕДВИЖИМОСТЬ ООО
109316, Волгоградский просп., 2
Тел.: 7557775; 2317723
АКРУС-НЕДВИЖИМОСТЬ ООО
123308, Маршала Жукова просп., 1
Тел.: 7647757
Сайт: www.akrus.ru
АЛЬФА-НЕДВИЖИМОСТЬ ООО
109428, Рязанский просп., 8а
Тел.: 2320840; 2324110

Moscow City Telephone Network http://www.mgts.ru/menu.html

Tools for International Telephone Lookups

AnyWho International http://www.anywho.com/international.html

AOL International Directories http://www.aol.com/netfind/international.html

EscapeArtist Telephone Search Engine http://www.escapeartist.com/global/telephone.htm

Infobel's Telephone Directories on the Web http://www.infobel.com/teldir/

Infospace International Directories http://www.infospace.com/intl/int.html

International White & Yellow Pages http://www.wayp.com/

Nedsite http://www.nedsite.nl/search/people.htm#telephone

Phonebook of the World http://www.phonebookoftheworld.com/

SBN International Yellow Pages http://www.sbn.com/international/international.asp

Many specialty directories also exist. The **World Telephone Numbering Guide** "provides information on the world's telephone numbering formats. This includes various website links regarding telephone numbering. Area code lists, text articles, news of phone number changes, number-finding forms are included as much as feasible."[114] Unlike many sources, the WTNG includes information about mobile telephone systems. This is an exceptionally good and current resource not only for finding telephone numbers but also for garnering details about telephone systems, numbering schemes, and regulations around the world.

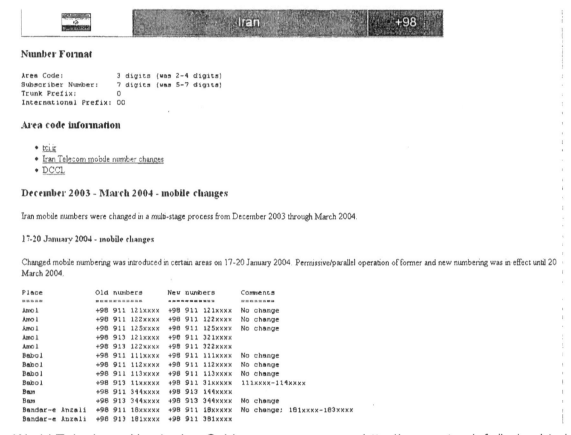

Iran | +98

Number Format

```
Area Code:              3 digits (was 2-4 digits)
Subscriber Number:      7 digits (was 5-7 digits)
Trunk Prefix:           0
International Prefix:   00
```

Area code information

- tci.ir
- Iran Telecom mobile number changes
- DCCL

December 2003 - March 2004 - mobile changes

Iran mobile numbers were changed *in a multi-stage* process from December 2003 through March 2004.

17-20 January 2004 - mobile changes

Changed mobile numbering was introduced in certain areas on 17-20 January 2004. Permissive/parallel operation of former and new numbering was in effect until 20 March 2004.

Place	Old numbers	New numbers	Comments
Amol	+98 911 121xxxx	+98 911 121xxxx	No change
Amol	+98 911 122xxxx	+98 911 122xxxx	No change
Amol	+98 911 125xxxx	+98 911 125xxxx	No change
Amol	+98 913 121xxxx	+98 911 321xxxx	
Amol	+98 913 122xxxx	+98 911 322xxxx	
Babol	+98 911 111xxxx	+98 911 111xxxx	No change
Babol	+98 911 112xxxx	+98 911 112xxxx	No change
Babol	+98 911 113xxxx	+98 911 113xxxx	No change
Babol	+98 913 11xxxxx	+98 911 31xxxxx	111xxxx-114xxxx
Bam	+98 911 344xxxx	+98 913 144xxxx	
Bam	+98 913 344xxxx	+98 913 344xxxx	No change
Bandar-e Anzali	+98 911 18xxxxx	+98 911 18xxxxx	No change; 181xxxx-183xxxx
Bandar-e Anzali	+98 913 181xxxx	+98 911 381xxxx	

World Telephone Numbering Guide http://www.wtng.info/index.html

ACR's International Calling Codes website provides country and city codes and adds the ability to list codes numerically. **International City Codes** also offers an excellent database of city calling codes. My favorite Internet telephone database is **Americom's International Decoder**, which permits you to look up either calling codes from city names or find out which city anywhere in the world matches a specific dialing code. Not only does Americom tell you the country and city codes,

[114] WTNG Help and Information Page, <http://www.wtng.info/wtng-hlp.html - HowTo>.

but often provides other useful information, even a telecommunications assessment, in some cases.

Americom's International Decoder http://decoder.americom.com/

Specialty Telephone Lookups

ACR's International Calling Codes by country http://www.the-acr.com/codes/cntrycd.htm

ACR's International Calling Codes listed numerically http://www.the-acr.com/codes/cntryno.htm

Americom's International Decoder http://decoder.americom.com/

International Dialing Codes http://kropla.com/dialcode.htm

International City Codes http://www.numberingplans.com/kropla/

World Telephone Numbering Guide http://www.wtng.info/index.html

Web Tip

Email lookup tools are not the only way to search for email addresses, phone/fax numbers, and street addresses. Search engines may actually be better at finding what you're looking for because most of them index entire webpages, and whatever data is on that page, including email addresses and phone numbers, is indexed.

Online Videos and Video Search

Online video of all types accelerated greatly over the past year. During the first half of 2006 it became apparent that online video/video search was going to be one of the year's biggest topics and developments, with new video sharing and video search sites gearing up and established companies such as Google, MICROSOFT, and Yahoo jumping on this particular bandwagon. "The market share of Internet visits to the 10 leading online video sites has increased by 164 percent in the past three months (week ending May 20, 2006 versus week ending February 25, 2006)," according to Hitwise, a competitive intelligence company.[115]

Furthermore, some in the "old media," e.g., ABC, CBS, BBC, AP, and Fox, saw this as a wave of the future they wanted to ride and began or expanded their video offerings via the Internet. To put the video revolution in perspective, "Apple Computer Inc. sold 12 million video clips at $1.99 each from its popular iTunes Music Store in just a few months"[116] and more than 40 million videos were viewed per day at the YouTube website.

Because we usually need to be able to find and get our hands on video quickly, there is a lot of interest in online video and video search. As with any new venture or technology, there are problems. In the case of Internet video, the main problems at the moment are:

- Quality—lots and lots of really awful, silly stuff.
- Quantity—still only a small number of videos are available compared to webpages, images, music.
- Format—there is no standard format in which videos are offered; some sites require you to download and install their proprietary software while some use MacroMedia Flash, QuickTime, RealPlayer, Windows Media Player, or something else.

---

[115] "Hitwise Data Shows Overall Visits to Video Search Sites Up 164%," Hitwise News Release, May 24, 2006, <http://www.hitwise.com/press-center/hitwiseHS2004/videosearch.php> (November 28, 2006).

[116] Walter Mossberg and Katherine Boehret, "Searching the Web for Video Clips," The Mossberg Solution, The Wall Street Journal, <http://ptech.wsj.com/archive/solution-20060301.html> (November 28, 2006).

What blogs are to traditional news, online videos are to television and, to a lesser extent, movies. The spread of broadband connections, cheap data storage, free or inexpensive video technology have all contributed to this boom. In a perverse way, al Qaeda in Iraq was an early adapter of Internet video, making effective if horrendous use of videos to spread its terror message, recruit new members, communicate across its terror network, and even offer training and support.

And, of course, online video has given rise to its own neologisms: **vlogs** (weblogs containing video), vloggers, vlogcasts, vlogcasting, and vlogcasters, as well as vodcasts/vidcasts. I suppose there are even vlogmasters, though that word conjures up images of a character in a Wagnerian opera.

Given the relative ease with which Internet videos can be produced and spread by anyone anywhere, we must take this technology very seriously and learn how to find videos of interest quickly and efficiently. Here are the some of the major players in online video and video search, excluding companies that are offering solely entertainment-related video uploads and downloads. Also, most news sites do not make video clips available for very long, usually no more than a week, before they are archived. Once archived, the clips may require registration and/or payment to view them.

The two basic types of video on line today are downloadable and streaming video. Some of the video sites I discuss below offer downloadable video, which may be saved and played later; this type of video may have a format that requires a specific type of video player. Virtually all the video sites below offer some form of streaming video, that is, one-way video transmissions over the Internet to a compatible media player. These do not require the user to download the video, nor can the video be saved for later viewing. You've seen this "buffering" message before, I'm sure, when requesting video be streamed to your computer:

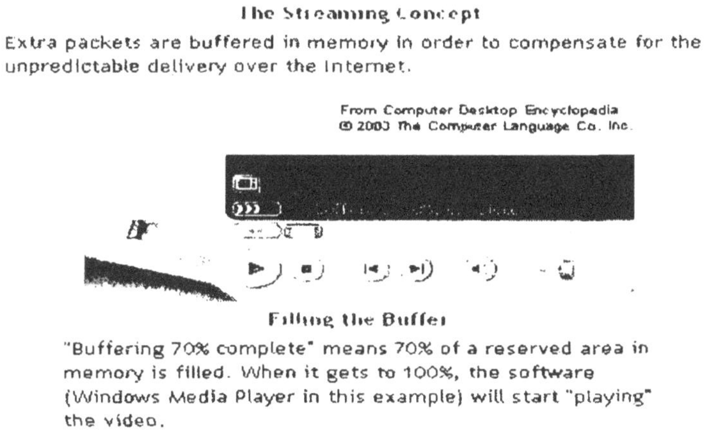
The Streaming Concept
Extra packets are buffered in memory in order to compensate for the unpredictable delivery over the Internet.
From Computer Desktop Encyclopedia
© 2003 The Computer Language Co. Inc.
Filling the Buffer
"Buffering 70% complete" means 70% of a reserved area in memory is filled. When it gets to 100%, the software (Windows Media Player in this example) will start "playing" the video.
[117]

[117] "Streaming video," Answers.com, Computer Desktop Encyclopedia, Computer Language Company Inc., 2005. <http://www.answers.com/topic/streaming-video> (15 November 2006).

What you are witnessing is streaming media. While not the smoothest, cleanest technology, it is cheap (the most popular software used for streaming media is free), and with broadband connections, it is relatively fast if sometimes choppy. The major free media players (all handle audio and video) are the following:

- Windows Media Player—Microsoft's proprietary software that plays many audio, video, and streaming formats, including MP3, WMA, CD audio, and MIDI.

- MacroMedia Flash Player—a browser plug-in that, according to the company "is installed on 98% of Internet-enabled desktops worldwide and on a wide range of popular devices."[118] The Flash Player displays content created with MacroMedia Flash and plays files with the .SWF extension.

- MacroMedia Shockwave Player—The Shockwave Player displays content created with MacroMedia Director and plays files with the .DCR extension.

Because there is so much confusion about these two products, MacroMedia created a webpage that compares these two products: <http://www.adobe.com/products/director/resources/integration/>

- QuickTime—the only one of these free media players not originally developed for the Windows environment, QuickTime is an Apple Computers' product capable of handling various formats of digital video, audio, text, animation, music, and more. The QuickTime media player is bundled with Apple computers but free downloads are available for Windows as well.

- RealPlayer—RealNetworks' media player that plays its own proprietary RealAudio and RealVideo as well as other formats including MP3, MPEG-4, and QuickTime. It was one of the first media players capable of handling streaming media over the Internet.

The Major Online Video and Video Search Sites

I decided to list these alphabetically because there is no good way to rank them: each has some advantages and usually quite a few drawbacks. Some things to keep in mind about the current state of video search:

- Most video search sites only search the metadata and/or text associated with a video. TVEyes is an exception; it uses voice recognition technology to index every word spoken in a video, but the technology is imperfect. Most video

[118] Macromedia Flash Player FAQ, <http://www.adobe.com/products/flashplayer/productinfo/faq/-item-1-1> (November 28, 2006).

search engines will index the text of television programs with closed captioning.

- Many video search sites only search their own video collection and do not search for videos across the web. I will note which sites do this below.

AOL Video Search http://search.aol.com/aolcom/videohome

AOL Video Search employs the video search engines of not one but two of the companies it owns—Singingfish and Truveo—for its own search site. AOL "Video Search results include multimedia streams and files (Real Media, Windows Media, QuickTime, MP3 and Flash) from AOL and the Web sorted by relevance." Note that you cannot sort the results by date. AOL Video search lets users limit the search to subfields including news, television, music, movies. Phrase searching works well. There is no way to limit your search by source, but if you include a source such as Reuters, AOL Video Search picks up on that keyword:

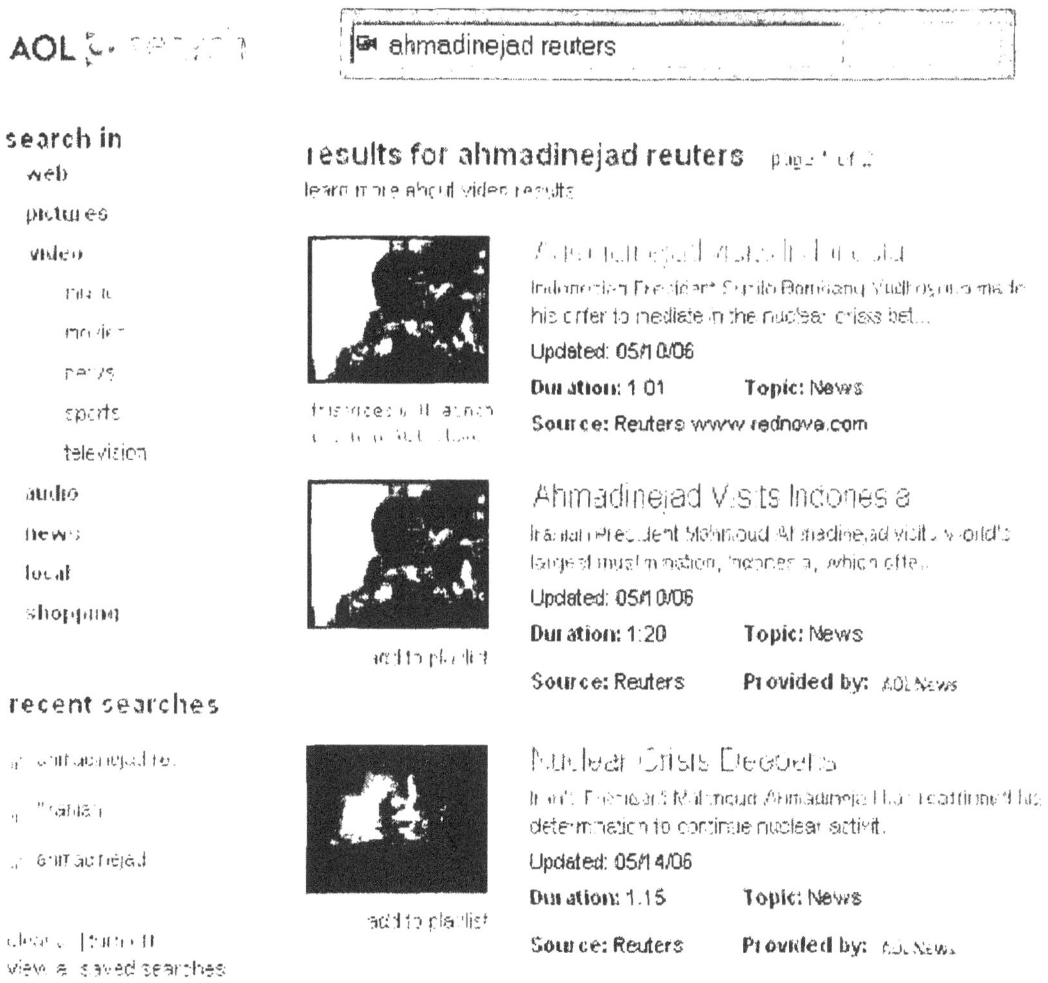
AOL
ahmadinejad reuters
search in
web
pictures
video
sports
television
audio
news
local
shopping
results for ahmadinejad reuters
Updated: 05/10/06
Duration: 1:01
Topic: News
Source: Reuters www.rednova.com
Ahmadinejad Visits Indonesia
Updated: 05/10/06
Duration: 1:20
Topic: News
Source: Reuters
Provided by:
add to playlist
recent searches
Updated: 05/14/06
Duration: 1:15
Topic: News
Source: Reuters
Provided by:
add to playlist

Summary:

- Sources of video data: multiple news and entertainment sources from AOL and across the web.
- Media types searched: Real Media, Windows Media, QuickTime, MP3 and Flash.
- Videos viewed: videos at AOL use AOL Video Player; videos from external sites use site's default media player.
- Search options: phrase searching; advanced search; no search by source option.
- Sort options: by relevance only.
- Upload video? Not directly; only via AOL's companies Singingfish or Truveo.

BBC Video http://news.bbc.co.uk/

BBC does not make it easy to find its news videos. For example, there is no separate BBC News video page. Instead, to view a BBC video, users must go to any News Online page and click the "Watch BBC news in video" or "Latest news in video and audio" button at the top of the page. This opens a new window where users can select from a list of current videos by news topic: headline, UK, business, politics, health, etc. The videos play in the same window using either RealPlayer (preferred) or Windows Media Player. Another option is BBC News 24, constantly updated national and international headlines. The best way to find BBC videos is using a third-party site such as Yahoo Video Search, which offers a search by source option.

In May the BBC announced it was opening its news archives, but only for users in the UK. If you are in the UK, "You can download nearly 80 news reports covering iconic events of the past 50 years including the fall of the Berlin Wall, crowds ejecting soldiers from Beijing's Tiananmen Square and behind-the-scenes footage of the England team prior to their victory over West Germany in 1966." <http://www.bbc.co.uk/calc/news/>

Summary:

- Sources of video data: BBC news video only.
- Videos viewed: Real Media (preferred) or Windows Media at BBC site.
- Search options: no separate video search.
- Sort options: by category (e.g., headlines or business news).
- Upload video? No.

Blinkx http://www.blinkx.tv/

Blinkx certainly has one of the snazziest sites, which is not necessarily a good thing because it can be distracting. However, it is a very good video search tool. Blinkx uses voice recognition software to transcribe the content of audio and video material, whether it is commercial television or personal vlogs. Users simply enter search terms in the box on the left of the screen and, once the results are returned, a slider appears that lets the user determine whether to rank the results by date or relevance or somewhere in between. Here is a snapshot on the Blinkx page showing results for [ahmadinejad] sorted by relevance.

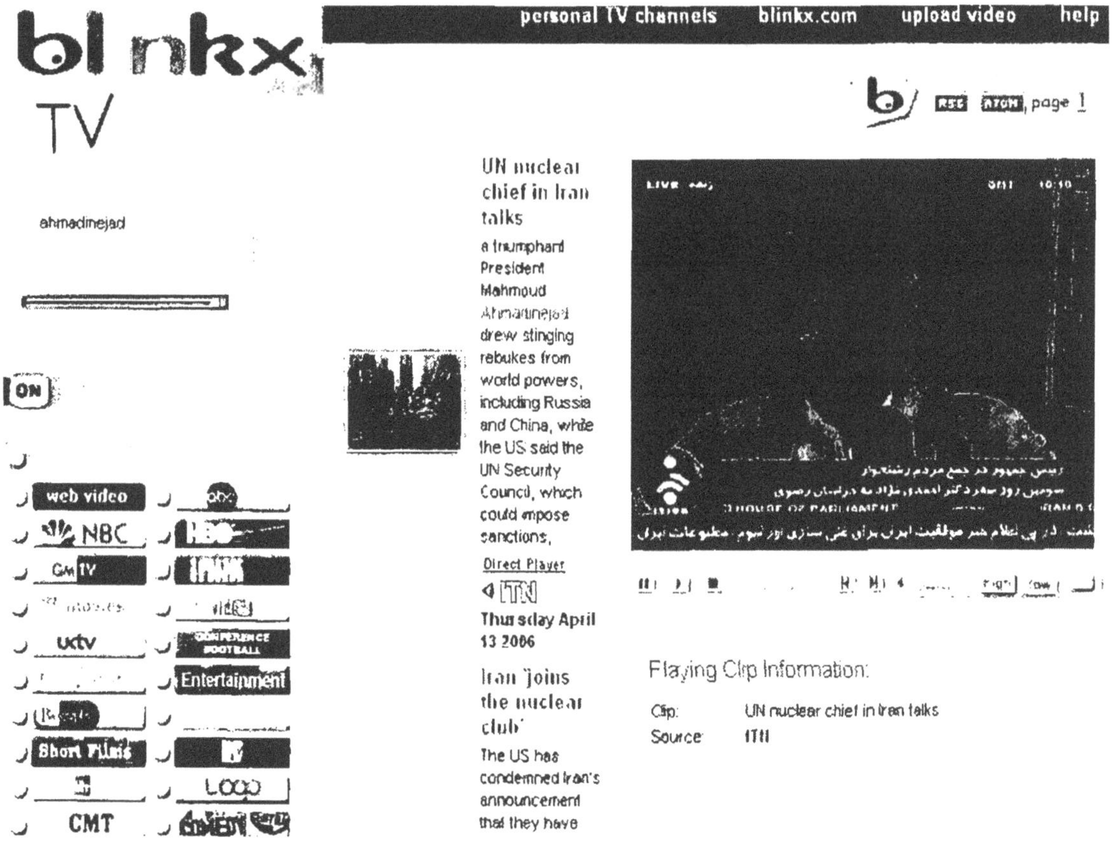
personal TV channels
blinkx.com
upload video
help
TV
page 1
ahmadinejad
UN nuclear chief in Iran talks
a triumphant President Mahmoud Ahmadinejad drew stinging rebukes from world powers, including Russia and China, while the US said the UN Security Council, which could impose sanctions,
Direct Player
Thursday April 13 2006
Iran 'joins the nuclear club'
The US has condemned Iran's announcement that they have
web video
NBC
uxtv
Entertainment
Short Films
CMT
Playing Clip Information:
Clip: UN nuclear chief in Iran talks

One of the best features in Blinkx is that it shows animated thumbnail images of each video. In some cases, Blinkx plays the full clip in the Blinkx window without users having to go to another website. Here is the key for determining the type of video and where it will play:

The blinkx b and camera icon located at the end of the result heading indicates that this result's video is hosted by blinkx.tv and can be viewed on the blinkx.tv Direct Play Screen located to the right of the result.

The eye and camera icon located at the end of the result heading indicates that this result's video can be viewed only from the original site.

The blinkx podcast icon located at the end of the result heading indicates that this result is a podcast.

Blinkx is much better than other video search sites at weeding out videos that have been archived and are thus no longer available at the originating site, although sometimes you want to know that a video was once available even if you cannot access it now.

One other thing you need to know about Blinkx is how to use the "channel" buttons on the left side of the homepage. Here is the "news" channel; users can select all, some, or none of these channels to search by clicking on the buttons. In this case, the news channels with the green dots have been selected to be searched:

CNN
REUTERS
BBC NEWS
Forbes
Bloomberg
EuroNews

Blinkx employs keyword and Boolean search options. Blinkx also has a special "Self Casting" channel for videos uploaded by users. Most of these are probably of dubious value, but it's worth keeping an eye on them.

Summary:

- Sources of video data: multiple news, entertainment, radio, podcasts from across the web.
- Video viewed: videos hosted on Blinkx servers use "Direct Player"; videos from external sites use site's default media player.
- Search options: keyword, Boolean, & phrase searching; "conceptual search"; option to limit search by source.
- Sort options: using slider, sort by date, relevance, or somewhere in between.
- Upload video? Yes.

CBS News Video Search

http://www.cbsnews.com/sections/i_video/main500251.shtml

CBS News has both an excellent video search, which appears to pull from the entire archive of CBS News videos. For example, I searched for ["hurricane isabel"] and found the very first CBS News video titled "Isabel Starting to Scare," September 12, 2003. Unfortunately, there is no way to force the search to look for a specific date, although there is the option to limit the search to specific topics (e.g., US news), to a specific CBS News show (e.g., "60 Minutes"), or to search in the Video Library by topic. However, there are almost no special search options, which makes it hard to find the nuggets in this treasure trove of videos. Despite this drawback, I believe the CBS News video search and archive is one of the great secret resources on the web. It is free to view the full videos, which open in the CBS News.com window using either RealPlayer or Windows Media Player (both free). All videos before November 20, 2003, play only in Real.

Summary:

- Sources of video data: CBS News video only.
- Video viewed: RealPlayer or Windows Media Player.
- Search options: keyword searching; search by ANY, ALL, EXACT PHRASE; no date limit.
- Sort options: sort by relevance or date.
- Upload video? No.

CNN Video Homepage http://www.cnn.com/video/

CNN Video Almanac http://www.cnn.com/resources/video.almanac/

CNN offers only seven days of free access to its vast video archive of news and features. After a week, users must sign up for CNN Pipeline, which requires both payment and downloading and installing a CNN video player.

Before CNN archives its video, you can search, browse, and view the free videos at the CNN website using Windows Media Player. All free videos are preceded by commercial advertisements. I find the CNN video set up very frustrating to use. If you go to the main CNN Video page <http://www.cnn.com/video/>, there is no way to limit your search to videos. However, if you go to the CNN homepage <http://www.cnn.com/>, scroll down to the middle of the page to **WATCH VIDEO**, and click on any video, the popup window includes a video only search box. Enter your search terms and CNN Video will return a list of video matches by date and time. You can then resort the results by Section or Most Popular. Note in this example that only the first two videos are still free. The rest have been archived to CNN Pipeline.

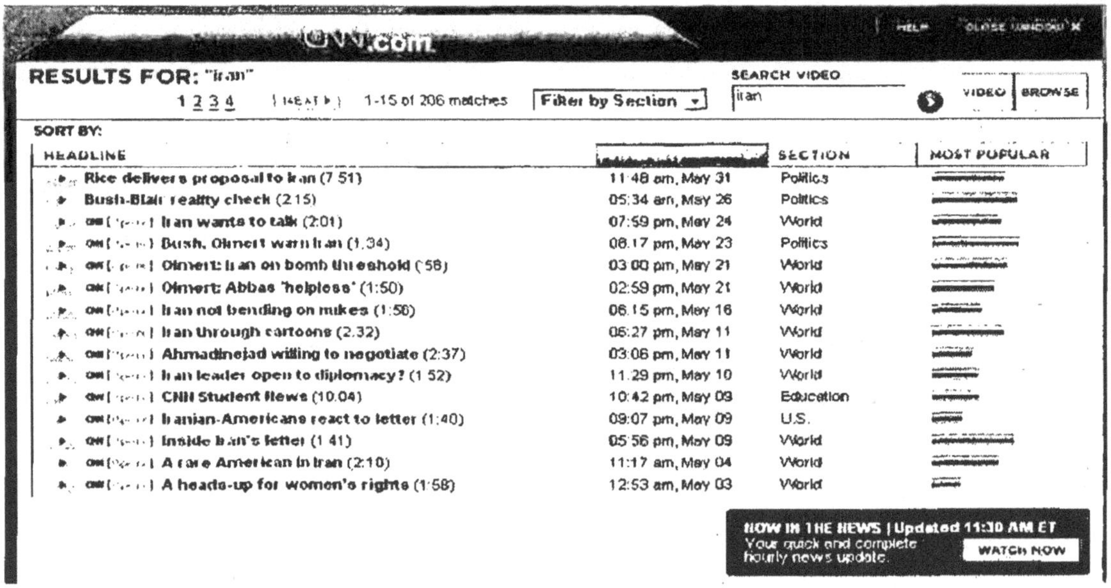
RESULTS FOR: "iran"
1 2 3 4
1-15 of 206 matches
Filter by Section
SEARCH VIDEO
iran
VIDEO
BROWSE
SORT BY:
HEADLINE
SECTION
MOST POPULAR
Rice delivers proposal to Iran (7 51) 11:48 am, May 31 Politics
Bush-Blair reality check (2 15) 05:34 am, May 26 Politics
Iran wants to talk (2:01) 07:59 pm, May 24 World
Bush, Olmert warn Iran (1:34) 08:17 pm, May 23 Politics
Olmert: Iran on bomb threshold (:58) 03:00 pm, May 21 World
Olmert: Abbas 'helpless' (1:50) 02:59 pm, May 21 World
Iran not bending on nukes (1:58) 06:15 pm, May 16 World
Iran through cartoons (2:32) 06:27 pm, May 11 World
Ahmadinejad willing to negotiate (2:37) 03:06 pm, May 11 World
Iran leader open to diplomacy? (1 52) 11:29 pm, May 10 World
CNN Student News (10:04) 10:42 pm, May 09 Education
Iranian-Americans react to letter (1:40) 09:07 pm, May 09 U.S.
Inside Iran's letter (1 41) 05:56 pm, May 09 World
A rare American in Iran (2:10) 11:17 am, May 04 World
A heads-up for women's rights (1:58) 12:53 am, May 03 World
NOW IN THE NEWS | Updated 11:30 AM ET
Your quick and complete hourly news update.
WATCH NOW

CNN has one other video feature you need to know about: its **Video Almanac**, a collection of the best video since the birth of CNN in 1980 through 1997. Users can select a year or topic, e.g., the video of the failed attempt to rescue the hostages in Iran in April 1980. The Video Almanac provides a very limited number of videos, most of which are very high profile events, though there are notable exceptions ("George Burns dies"). The videos play using QuickTime.

Summary:

- Sources of video data: CNN video only.
- Video viewed: Windows Media Player version 9 and above.
- Search options: search for one or more keywords or exact phrase in double quotes; filter by section; no date limit.
- Sort options: sort by date/time, section, or popularity.
- Upload video? No.

C-SPAN http://www.c-span.org/

C-SPAN Store http://www.c-spanstore.org/shop/

C-SPAN is the only one of the online video and video search sites that is a non-profit entity. In case anyone is unclear who and what C-SPAN is, here's what the network says about itself: "C-SPAN is a private, non-profit company, created in 1979 by the cable television industry as a public service. Our mission is to provide public access to the political process. C-SPAN receives no government funding; operations are funded by fees paid by cable and satellite affiliates who carry C-SPAN programming." <http://www.c-span.org/about/index.asp?code=About>

Until a video is archived, it is available for free at the C-SPAN site. C-SPAN offers both stored and streaming (live) media (video and audio, because there is also a C-SPAN radio). It is very important to remember that free C-SPAN video searches only index program titles, event descriptions, and air dates. "Searches are not based on text within the video files or closed-caption transcripts and do not include video available at the C-SPAN Archive <http://www.c-spanstore.org/shop/>, where you can buy almost any C-SPAN program." Because the C-SPAN search is not based upon text or closed-captioning, you may have to work harder to find a specific video.

Although the C-SPAN site says that "most events will remain in the archive for 15 days or less," I found many videos going back years that are still available, so it appears to be hit or miss as to what is or is not available for free. Keep in mind you can find and view many but not all Congressional hearings at C-SPAN. The site has an advanced search option that lets users limit a search by date ranges as well as by C-SPAN series and programs. The advanced search also offers the option to search by phrase and even a fuzzy word search ("spelled like" or "sounds like").

The C-SPAN video collection includes videos from these and other sources:

- Congressional hearings.
- US Government Executive branch (e.g., DoD, State Department) press conferences.
- Special commissions.
- Some State and Local videos.
- C-SPAN TV series, such as "Washington Journal," "Booknotes," etc.

Summary:

- Sources of video data: C-SPAN video (and audio) only.
- Video viewed: Stored clips require RealPlayer; streaming media offers a choice between RealPlayer and Windows Media Player.
- Search options: search for one or more keywords; advanced search offers date, program filters and fuzzy keyword search.
- Sort options: sort by relevance or date/time.
- Upload video? No.

Google Video http://video.google.com/

Google Video underwent a huge metamorphosis during December 2005. Google Video moved from being primarily a video search engine to being a store front to preview and buy video. This time Google's target isn't Microsoft but (primarily) Apple's iTunes and to a lesser degree cable television companies. Google Video is more focused now on selling videos, including television shows, than on searching current videos. However, there are many free videos still available on pretty much any subject you can imagine. Very importantly, Google Video only searches for videos hosted at its own website. Google Video has a link to AOL Video on its homepage, but the link takes you to the AOL Video website and Google Video does not search on AOL's videos from its site.

In May 2006 Google Video (finally) wised up and let users upload their videos directly online without having to install any software and without the previous submission time lag of up to 24 hours. ("Google Inc. Tweaks Its Video Service," AP report in Forbes.com,

http://www.forbes.com/technology/ebusiness/feeds/ap/2006/05/17/ap2751923.html) YouTube had always permitted direct video uploads and was basically cleaning Google Video's clock.

Google indexes the closed captioning and text descriptions of all the videos in its archive to facilitate search. Google Video search supports keyword and phrase searches. While Google Video does not support search by source, it does offer the option of searching by title. The correct syntax is [**title:keyword**]. Note that phrase searching does not work with the **title:** syntax.

Google Video has several options for viewing or limiting results. Users can choose to view the results as a list or as a grid (the list provides more information while the grid shows a larger image of the video clip). There is also an option to limit the results by the length of the video—long (20+ minutes), medium (4-20 minutes), or short (up to 4 minutes). Results can be sorted relevance, date, or title.

Google explains how it ranks videos in terms of popularity: "We use algorithms to identify videos that are suddenly becoming popular, and then rank them based on how popular they are—and how suddenly they became popular. We've been using this list internally, and now it's ready to share with you, so check it out. Right now this feature highlights videos from close to 40 countries, including Argentina, Australia, Belgium, Brazil, Canada, Estonia, Finland, Greece, Hong Kong, India, Israel, Japan, South Korea, Mexico, and New Zealand, to name a few." (Jon Steinback, "Movers, Shakers, and Hoops on Video," Google Blog, June 9, 2006, http://googleblog.blogspot.com/2006/06/movers-shakers-and-hoops-on-video.html)

The software required to view streaming video at Google Video is Macromedia Flash Player 7.0+. To view purchased and downloaded videos, users still must install Google's own Google Video Player, which has frustrated many users and infuriated others. It will be interesting to watch Google Video's evolution as the site tries to move higher in online video popularity. closed captioning and text descriptions of all the videos in our archive for relevant results.

Summary:

- Sources of video data: only news, entertainment, radio, podcasts hosted on Google servers.
- Video viewed: free streaming media uses Flash Player; purchased videos require Google Video Player.
- Search options: keyword & phrase searching; title: search; no search by source option.
- Sort options: sort by relevance, date, or title; view as List or Grid; limit results by length of video.
- Upload video? Yes.

IFILM http://www.ifilm.com/

IFILM "is a leading online video network, serving user-uploaded and professional content to over ten million viewers monthly. IFILM's extensive library includes movie clips, music videos, short films, TV clips, video game trailers, action sports and its popular 'viral videos' collection. IFILM is one of the leading streaming media networks on the internet." <http://www.ifilm.com/about/> Viacom purchased IFILM and made it a part of its MTV Network, so most of the commercial videos are from some segment of that network. However, what distinguishes IFILM is its vast supply of user created and uploaded videos.

IFILM specializes in viral videos, that is, videos that gain widespread popularity across the Internet through blogs, email, IM, websites, and old-fashioned word of mouth. Not all viral videos are humorous. One example of a viral video from IFILM was "Mercenary Sniper in Iraq," described in Defense Review.com as an "interesting 'Viral Video' clip that puts the viewer inside a helicopter in Iraq (urban environment), and then puts you right there with a sniper team engaging hostile enemy targets (i.e. insurgents/terrorists) from a rooftop (again, inside Iraq)." <http://www.defensereview.com/article826.html>

Summary:

- Sources of video data: multiple news, entertainment, and user-created videos.
- Video viewed: QuickTime, Windows Media Player, Macromedia Flash.
- Search options: keyword & phrase searching; limit search to specific collections, such as "User Video."
- Sort options: no.
- Upload video? Yes.

MSN Video[119] http://video.msn.com/

Don't bother trying to access this site from any browser other than Internet Explorer 6. Once you get past that extremely annoying requirement, MSN Video is not a bad site. Users have access to all NBC news, sports, and entertainment, as well as other sources such as the National Geographic Channel and the Discovery Networks. MSN Video also partners with IFILM, so you can access IFILM videos from this site.

---

119 As of this writing, MSN Video has not been renamed Live Video even though there is a video search incorporated into Live Search.

I cannot determine exactly how long videos are kept; some videos date back years, but most of the news videos appear to be from within the past 30 days. What I like best about the site are the many options for handling results. Users can choose between thumbnails or details; view all results; and sort by title, source, or date. Here is a screenshot of the results for the query ["hurricane katrina"] in detail view sorted by date:

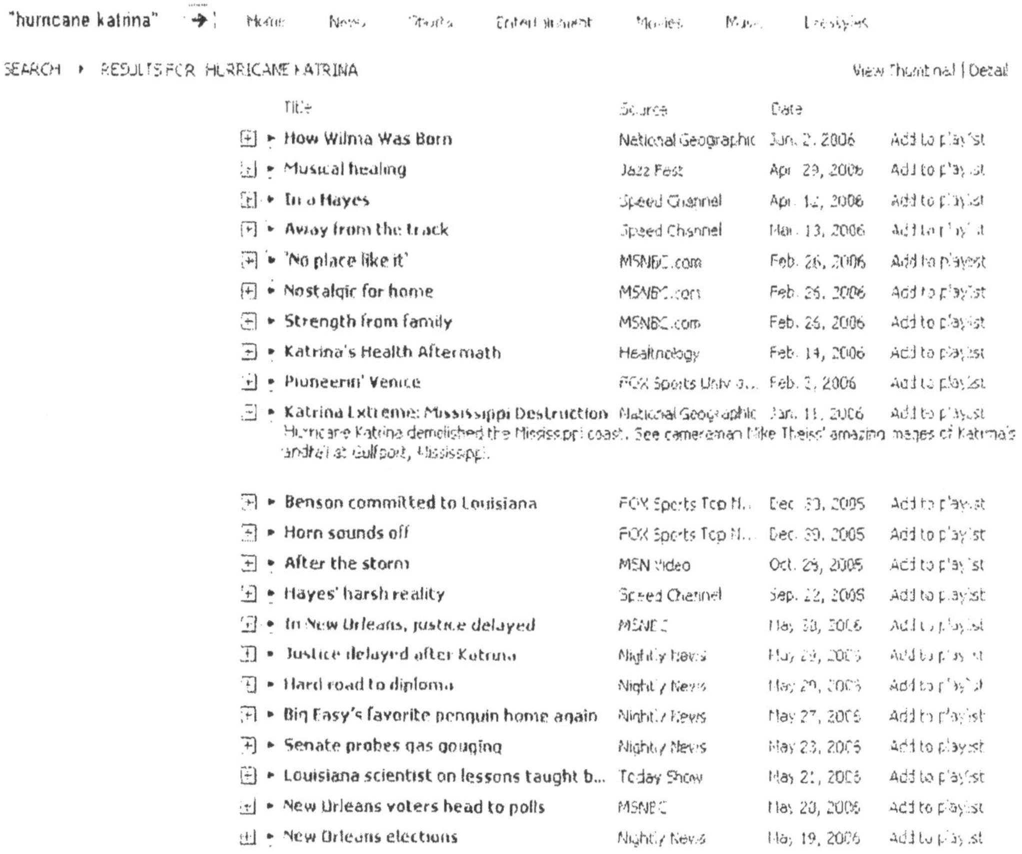
"hurricane katrina" Home News Sports Entertainment Movies Music Lifestyles

SEARCH › RESULTS FOR HURRICANE KATRINA View Thumbnail | Detail

Title	Source	Date	
How Wilma Was Born	National Geographic	Jun. 2, 2006	Add to playlist
Musical healing	Jazz Fest	Apr. 29, 2006	Add to playlist
In a Hayes	Speed Channel	Apr. 12, 2006	Add to playlist
Away from the track	Speed Channel	Mar. 13, 2006	Add to playlist
'No place like it'	MSNBC.com	Feb. 26, 2006	Add to playlist
Nostalgic for home	MSNBC.com	Feb. 26, 2006	Add to playlist
Strength from family	MSNBC.com	Feb. 26, 2006	Add to playlist
Katrina's Health Aftermath	Healthology	Feb. 14, 2006	Add to playlist
Pioneerin' Venice	FOX Sports Univ...	Feb. 3, 2006	Add to playlist
Katrina Extreme: Mississippi Destruction — Hurricane Katrina demolished the Mississippi coast. See cameraman Mike Theiss' amazing images of Katrina's landfall at Gulfport, Mississippi.	National Geographic	Jan. 11, 2006	Add to playlist
Benson committed to Louisiana	FOX Sports Top N...	Dec. 30, 2005	Add to playlist
Horn sounds off	FOX Sports Top N...	Dec. 30, 2005	Add to playlist
After the storm	MSN Video	Oct. 29, 2005	Add to playlist
Hayes' harsh reality	Speed Channel	Sep. 22, 2005	Add to playlist
In New Orleans, justice delayed	MSNBC	May 30, 2006	Add to playlist
Justice delayed after Katrina	Nightly News	May 29, 2006	Add to playlist
Hard road to diploma	Nightly News	May 29, 2006	Add to playlist
Big Easy's favorite penguin home again	Nightly News	May 27, 2006	Add to playlist
Senate probes gas gouging	Nightly News	May 23, 2006	Add to playlist
Louisiana scientist on lessons taught b...	Today Show	May 21, 2006	Add to playlist
New Orleans voters head to polls	MSNBC	May 20, 2006	Add to playlist
New Orleans elections	Nightly News	May 19, 2006	Add to playlist

<u>Summary</u>:

- Sources of video data: selected news, entertainment, etc., from NBC and MSN's partners as well as user-created videos from IFILM.
- Video viewed: Windows Media Player 7+, Macromedia Flash 7+ (site only works in Internet Explorer 6).
- Search options: keyword & phrase searching; limit search to specific collections, such as News.
- Sort options: by title, source, or date.
- Upload video? No.

Pixsy http://pixsy.com/

Pixsy is a metasearch engine for both still images and video that searches a large numbers of video content providers. "As a meta-aggregator of image and video thumbnails images from RSS providers, Pixsy provides a visual search alternative for consumers and a source of free, high quality traffic for RSS syndicating image & video content providers. New RSS providers include YouTube, Revver, SmugMug, RollingStone, StumbleUpon, Defamer, People Magazine, Pictopia, Metacafe, TheOnion, Rotten Tomatoes, Buzznet, CNN, NPR, PBS, and many more." (Loren Baker, "Pixsy Adds New Video Search Content from YouTube, CNN, Defamer & Others," Searchenginejournal.com, http://www.searchenginejournal.com/?p=3500)

The thumbnails are small enough to review quickly but still of good quality, and a mouseover of each thumbnail image reveals the title of the video, a summary of its contents, the date of the video, its source, and the date and location where it was indexed. The only sort options at Pixsy are by category. The Pixsy homepage also links to the latest and to featured videos.

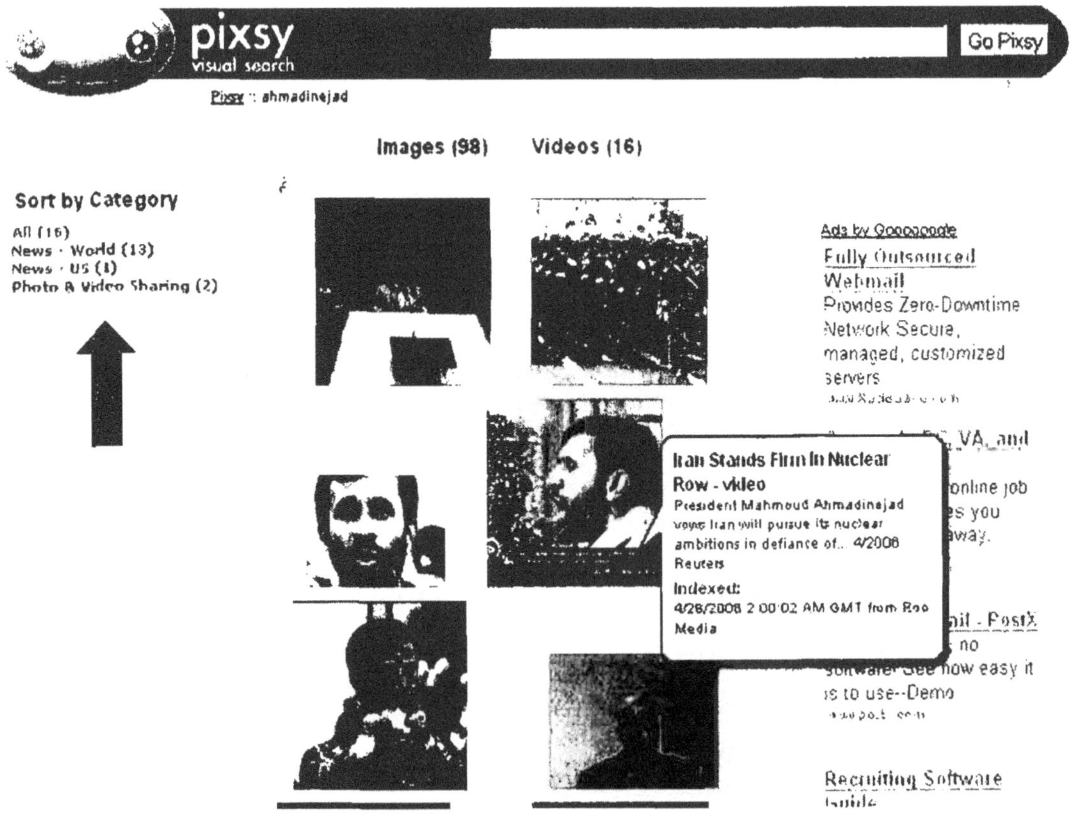
pixsy
visual search
Go Pixsy
Pixsy :: ahmadinejad
Images (98)
Videos (16)
Sort by Category
All (16)
News · World (13)
News · US (1)
Photo & Video Sharing (2)
Fully Outsourced Webmail
Provides Zero-Downtime Network Secure, managed, customized servers
Iran Stands Firm In Nuclear Row - video
President Mahmoud Ahmadinejad vows Iran will pursue its nuclear ambitions in defiance of... 4/2006 Reuters
Indexed:
4/26/2006 2:00:02 AM GMT from Roo Media
Recruiting Software

<u>Summary</u>:

- Sources of video data: multiple news, entertainment, radio, podcasts from across the web.
- Video viewed: videos play using external site's default media player inside a Frame at Pixsy with option to remove Frame.
- Search options: keyword and phrase searching; no search by source option; no advanced search.
- Sort options: searches for both images and videos; tab to video; only sort option is by category.
- Upload video? No.

Reuters Video http://today.reuters.com/tv

I have to mention Reuters because of its importance and reach, but the site is frustrating to use because there is no separate video search at the site. All you get is a list of available videos sorted by topic (news, entertainment, business, etc.). If you need Reuters' videos, I recommend using one of the video search sites—RocketInfo, SingingFish, RooTV, Blinkx, Yahoo Video Search, or Pixsy.

RocketInfo http://www.rocketnews.com/ [select the VIDEO tab]

RocketInfo offers some of the most extensive sources for news and video on the web, drawing from over 16,000 news sources on a continuous basis. The downside is that RocketInfo only searches for the past four days at most, and users can limit a search to today's videos. The search has no advanced features, not even phrase searching. It looks for all the keywords, so be careful not to search on too many terms.

RocketInfo is a metasearch engine that searches many video sources, including BBC, RedOrbit videos, Reuters, CBS, CNN, MSNBC, local news, etc. For current news searches, RocketInfo has to be ranked as one of the best and most comprehensive.

<u>Summary</u>:

- Sources of video data: news from over 16,000 sources across the web (not all or even most of these sources offer video)
- Video viewed: all videos play at originating site using that site's default video player
- Search options: keyword search only (searches for ALL terms)

- Sort options: by date or relevance
- Upload video? No

RooTV http://www.rootv.com/

RooTV or Roo is a powerful video search tool, but I find it annoying because it automatically plays a video when you visit the homepage (and it really doesn't want to stop!). There is a simple keyword search but no way to limit the search to just news, for example. The videos play at the RooTV site, using Windows Media Player or RealPlayer. RooTV requires ActiveX; if you are going to use RooTV, I recommend opening the site in Internet Explorer (you will have to add it to your Trusted Zone) so you do not need to install an ActiveX plug-in in Firefox or Netscape, something I would discourage. Sources include Reuters and AP. Supported by advertising.

Summary:

- Sources of video data: news from sources across the web, including Reuters and AP.
- Video viewed: all videos play at the RooTV site using either Windows Media Player or RealPlayer; requires ActiveX.
- Search options: keyword search only (searches for ALL terms).
- Sort options: none.
- Upload video? No.

Searchforvideo http://www.searchforvideo.com/home/index.html

Searchforvideo IM Service http://www.searchforvideo.com/misc/im.jsp

Searchforvideo Reel Time Lens http://www.searchforvideo.com/misc/reel.jsp

Searchforvideo is one of the best video search engines available. Searchforvideo searches a vast number of sources, including BBC, ABC, CBS, CNN, AP via iVillage, Reuters via RooTV, RedOrbit, the Discovery Networks, and many others. The site also has separate pages by topic—news, sports, entertainment, business, technology, health, and viral—where users can browse the top videos of the day.

Interestingly, even though there is no search by source, I found that if you use the name of a source and a keyword, that seems to work very well because you are searching on all metadata associated with a video clip. For example, [reuters iran] returns only videos from Reuters sources containing the keyword “iran”:

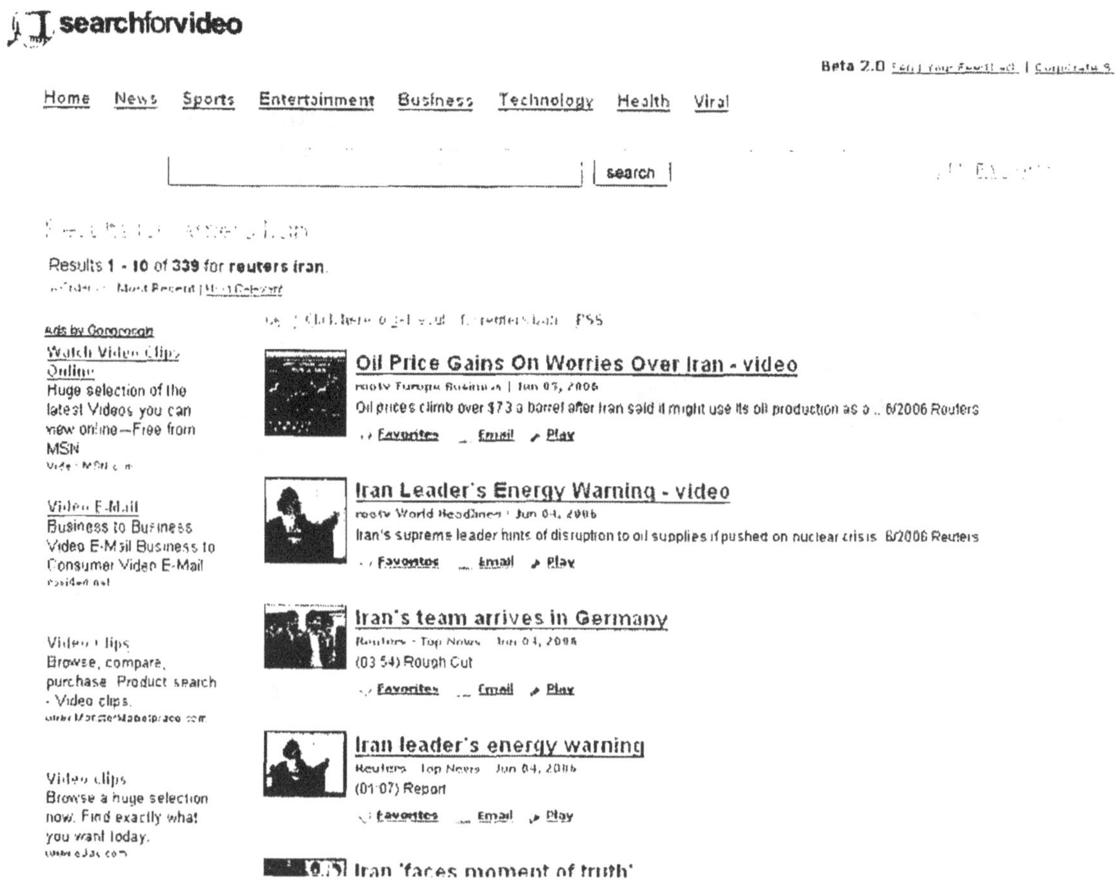
searchforvideo
Beta 2.0
Home News Sports Entertainment Business Technology Health Viral
search
Results 1 - 10 of 339 for reuters iran.
Ads by Goooooogle
Watch Video Clips Online
Huge selection of the latest Videos you can view online—Free from MSN
Video E-Mail
Business to Business Video E-Mail Business to Consumer Video E-Mail
Video Clips
Browse, compare, purchase. Product search - Video clips.
Video clips
Browse a huge selection now. Find exactly what you want today.
Oil Price Gains On Worries Over Iran - video
Oil prices climb over $73 a barrel after Iran said it might use its oil production as a .. 6/2006 Reuters
Favorites Email Play
Iran Leader's Energy Warning - video
Iran's supreme leader hints of disruption to oil supplies if pushed on nuclear crisis. 6/2006 Reuters
Favorites Email Play
Iran's team arrives in Germany
(03:54) Rough Cut
Favorites Email Play
Iran leader's energy warning
(01:07) Report
Favorites Email Play
Iran 'faces moment of truth'

Searchforvideo also offers IM search that works with AOL, Yahoo, and MSN. There is also a very interesting option to view videos that are being added or watched in real time at the Searchforvideo Real Time Lens page.

Summary:

- Sources of video data: searches wide variety of video sources across the web.
- Video viewed: at originating site using that site's default video player, except…
- Offers video podcasts that can be downloaded to video podcast software at Searchforvideo Publishers <http://www.searchforvideo.com/pub/>.
- Search options: keyword searching; add source keyword, e.g., [bbc], to limit by source.
- Sort options: sort by relevance or date.
- Upload video? No.

Sky News Video http://www.sky.com/skynews/video

The UK's Sky News, billed as Europe's first 24-hour news channel, offers a "video channel" and advanced search to limit queries to videos only. Keep in mind that the simple search on the main video page does not restrict your search to videos: you must use the advanced search page. Here users have the option to limit the search by "item type," and Video is one of the types. Notice you can also limit your search by section and date range.

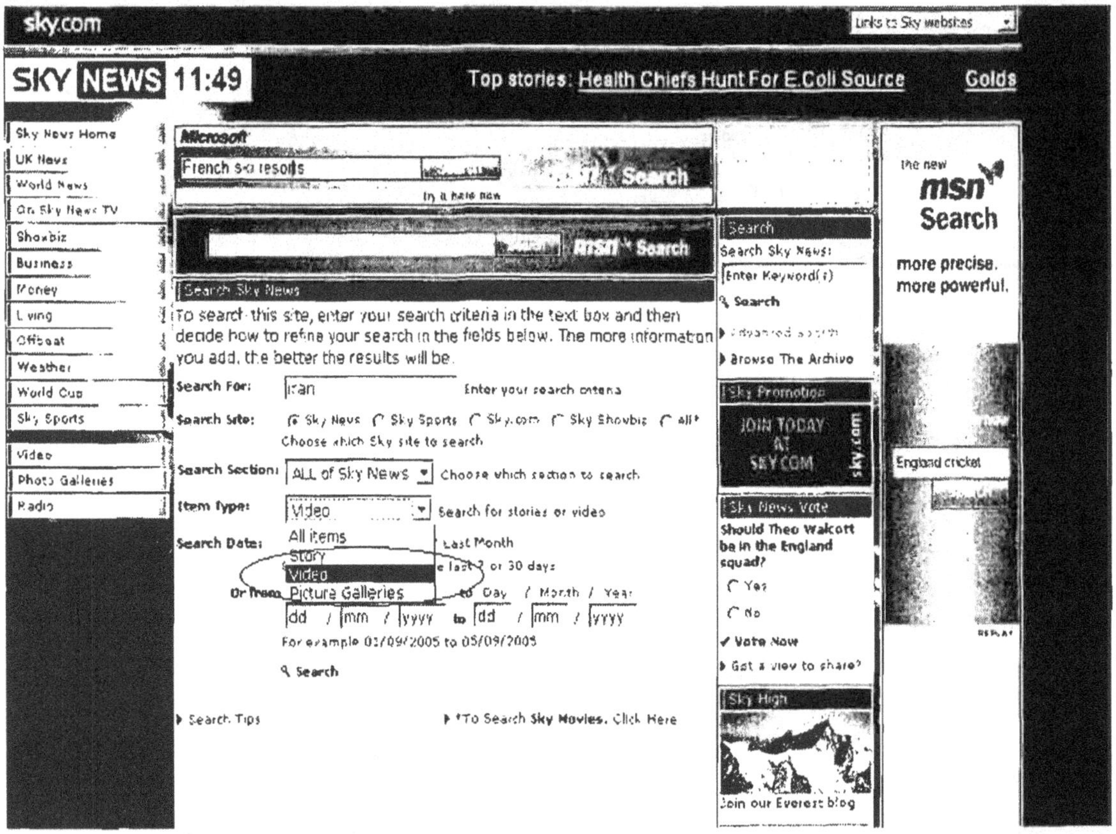
sky.com
SKY NEWS 11:49
Top stories: Health Chiefs Hunt For E.Coli Source
Sky News Home
UK News
World News
On Sky News TV
Showbiz
Business
Money
Living
Offbeat
Weather
World Cup
Sky Sports
Video
Photo Galleries
Radio
Search Sky News
To search this site, enter your search criteria in the text box and then decide how to refine your search in the fields below. The more information you add, the better the results will be.
Search For:
Enter your search criteria
Search Site:
Sky News
Sky Sports
Sky.com
Sky Showbiz
All
Choose which Sky site to search
Search Section:
ALL of Sky News
Choose which section to search
Item Type:
Video
Search for stories or video
Search Date:
All items
Story
Video
Picture Galleries
Last Month
For example 01/09/2005 to 05/09/2005
Search
Search Tips
To Search Sky Movies, Click Here
Search
Search Sky News:
Enter Keyword(s)
Search
Browse The Archive
Sky Promotion
JOIN TODAY AT SKY.COM
Sky News Vote
Should Theo Walcott be in the England squad?
Yes
No
Vote Now
Got a view to share?
Sky High
Join our Everest blog
the new msn Search
more precise.
more powerful.
England cricket

In the advanced search, the only characters recognized are letters, numbers, and hyphens. You can use a hyphen to create a phrase: [tony-blair]. Results are listed in date order with the most recent items listed first. Each result shows the relevance to your search. The videos play using Windows Media Player.

Summary:

- Sources of video data: Sky News video only.
- Video viewed: at Sky News site using Windows Media Player.
- Search options: keyword searching; query terms limited to alphanumerics and hyphens; search by section (e.g., news), type (e.g., video), date range.

- Sort options: None; listed by relevance.

- Upload video? No.

TVEyes http://tveyes.com/

If you want to search the content of TV and radio news broadcasts, I recommend the TVEyes search engine. TVEyes captures, indexes, analyzes, archives, and distributes in real-time content captured from television and radio broadcasts. "Currently live content alerts and searchable archives are available from:

- US

- UK

- Canada

- Australia

and the Al Jazeera network. Coverage is being extended to include France, Mexico, Guatemala, South Africa with these countries on stream by Q4." <http://www.tveyes.com/coverage/index.htm>

TVEyes uses voice recognition technology to create something the company calls its Spoken Word Index(TM), so that users can search every word in a broadcast. The search only seems to work if you use one keyword or a phrase because it interprets more than one word as a phrase. It works very well finding the right broadcast; unfortunately, once you go to the originating site of the news clip, you may well discover it is no longer available. However, TVEyes can help you locate a specific clip, especially if you did not know the source. You may have to pay for that video, but at least you know where to go to get it. TVEyes offers thumbnail versions of the clips, including some video no longer available at the originating site, which play from the TVEyes site. Full clips must be viewed at the originating site.

Summary:

- Sources of video data: television and radio broadcasts in US, UK, Canada, Australia, Al Jazeera network (to be extended in 4th Q 2006 to many other international sources).

- Video viewed: thumbnail versions of the clips, which play from the TVEyes site. Full clips must be viewed at the originating site and may no longer be available.

- Search options: keyword; more than one word is interpreted as a phrase.

- Sort options: sort by date or relevance.

➢ Upload video? No, but TVEyes "can respond immediately to any request to add TV channel or Radio station coverage."

Yahoo Video Search http://video.yahoo.com/

Yahoo News Video http://news.yahoo.com/video

Yahoo video search is one of the most popular on the web and for good reason. It draws from many sources across the web and around the world, and it has a powerful advanced search. Yahoo video search is one of the few that has an explicit site delimiter. If, for example, I only want to search for videos from the BBC, on the advanced search page, I would select "only search in this domain/site:" and enter [bbc.co.uk]. The syntax for this search (with or without keywords) is [fromsite:bbc.co.uk] if you prefer to use the simple search interface. The other advanced search options are to limit your search by format, size, or duration. Yahoo video advanced and simple search recognize + (must include), - (must exclude), and double quotes for a phrase.

Yahoo video search has another excellent option in the form of channels. Channels are collections of videos created by a common source; clicking on a "channel" displays all video results from that source. Unfortunately, there does not appear to be any central list of all channels. The best way to find the channels is to do the following: on the Yahoo video homepage, click the "Categories" tab, then select "News." Now you will see the channels listed, e.g., ABC News, CBS News, CBS.ca, AP, Reuters News, etc. Selecting a specific channel will show you all the available videos from that channel. To save channels to your list of favorites, you must be a registered member of Yahoo

In late 2006, Yahoo and CBS "announced an exclusive video syndication agreement in which local news video from 16 of CBS's owned stations will be made available on Yahoo! to the Internet's largest news audience. The relationship...marks the first video agreement between a network-owned television station group and an Internet news provider."[120] Yahoo's video focus has always been on commercial and not homemade video, the opposite of the YouTube/Google approach, even though Yahoo does have a video site that tries to compete with YouTube. Yahoo already had deals with CBS's "60 Minutes," as well as with ABC, Disney, and CNN. This deal is with local news CBS affiliates. While some are criticizing Yahoo's approach to offering video from commercial/professional sources, it is an approach that continues to work for them. When users want to find and view news videos, they do not think YouTube, they think Yahoo News Video. I use it heavily not only because of the number and quality of videos available but also because it is very easy to use,

---

[120] Michael Liedtkey, "Yahoo adds CBS news to video lineup," AP/San Francisco in Businessweek.com, 16 October 2006, <http://www.businessweek.com/ap/financialnews/D8KPGBUO1.htm> (24 October 2006).

the videos open at the Yahoo site, and you do not need to register to find and view the videos.

Summary:

- Sources of video data: searches wide variety of video sources across the web, including user-submitted videos.
- Video viewed: at originating site using that site's default video player.
- Search options: keyword searching; + (must include), - (must exclude), and double quotes for a phrase; limit search to specific site [fromsite:domain]; search by format, size, or length of video.
- Sort options: None; listed by relevance.
- Upload video? Yes.

YouTube http://www.youtube.com/

If you have heard of only one online video and video search site, it's probably YouTube, the most popular such site by far, garnering almost 43 percent of all visits to video websites in mid-2006.[121] YouTube is currently serving 100 million videos per day, with more than 65,000 videos being uploaded daily.[122] YouTube started as a personal video sharing site and opened to the public in February 2005. While other video sites try to mix commercial and homemade videos, individual users create almost everything at YouTube. However, because of YouTube's phenomenal success, some commercial enterprises have latched onto the site's popularity to promote their own products via videos. Most notably, YouTube now offers official movie trailers and promotion videos from established media companies.

It came as a surprise to no one when Google bought YouTube in late 2006 (when Video replaced Froogle on the Google homepage, that was a pretty clear indication of Google's direction). But this is a gamble, possibly a big gamble. Not everyone has forgotten what happened to Napster; Mark Cuban, chairman of HDNet (among other things) says in his blog, "It will be interesting to see what happens next and what happens in the copyright world. I still think Google Lawyers will be a busy, busy bunch. I dont think you can sue Google into oblivion, but as others have mentioned, if Google gets nailed one single time for copyright violation, there are going to be

[121] "Hitwise Data Shows Overall Visits to Video Search Sites Up 164%," Hitwise Competitive Intelligence, 24 May 2006, <http://www.hitwise.com/press-center/hitwiseHS2004/videosearch.php> (24 October 2006).

[122] YouTube Fact Sheet, YouTube.com, <http://www.youtube.com/t/fact_sheet> (24 October 2006).

more shareholder lawsuits than doans has pills to go with the pile on copyright suits that follow."[123]

However, YouTube was in the process of cleaning up its copyright act before the acquisition, and Google is certain to ensure that happens. Both Google and YouTube have recently made deals with major video partners, so Google may dodge most of the lawsuits, especially if video producers realize there is serious money to be made with this partnership. Nonetheless, big, rich Google makes a much more tempting target for potential litigants than YouTube ever did. For now Google says that YouTube will continue "to operate independently to preserve its successful brand and passionate community."[124]

Nonetheless, as of the first of this year, Google Video began to include results from YouTube. For now, when users click on the YouTube results, they are taken to the YouTube website. As of this writing, YouTube videos do not appear on the Google video homepage, only in search results where they are recognizable from their address.

With YouTube, people can:

- Upload, tag and share videos worldwide.
- Browse millions of original videos uploaded by community members.
- Find, join and create video groups to connect with people who have similar interests.
- Customize the experience by subscribing to member videos, saving favorites, and creating playlists.
- Integrate YouTube videos on websites using video embeds or APIs.
- Make videos public or private—users can elect to broadcast their videos publicly or share them privately with friends and family upon upload.

YouTube is building a community that is highly motivated to watch and share videos. The service is free for everyone." <http://www.youtube.com/t/about>

In May 2006 YouTube launched a new service that allows users to upload videos directly from their mobile phones and PDAs to the YouTube website. Clearly, two big concerns for a site such as YouTube are copyright infringement and pornography,

---

[123] Mark Cuban, "I Still Think Google is Crazy," : Blog Maverick: The Mark Cuban Weblog, 9 October 2006, <http://www.blogmaverick.com/2006/10/09/i-still-think-google-is-crazy/> (24 October 2006).

[124] "Google To Acquire YouTube for $1.65 Billion in Stock," Google Press Center, 9 October 2006, <http://www.google.com/intl/en/press/pressrel/google_youtube.html> (14 November 2006).

both of which they constantly try to avoid, in part by employing a "community policing" policy where users "turn in" offenders. When found, copyrighted material and pornography are removed.

Lest you think every video at YouTube is of the "Snakes on a Plane" variety, there is so much user-created video available that there is at least a little bit of everything as well as a lot of some things. Take a look at this snapshot of the News & Blogs category, which shows the most popular tags in this category and thumbnail images of the most popular videos. Video tags created and added to uploaded videos is crucial to the success of any video being found using search, which motivates video creators to properly and adequately tag their videos. However, the tags are solely the discretion of the video's creator (i.e., there is no standard taxonomy).

News & Blogs
News, Blogs, Local Issues
Popular Tags in News & Blogs
osama bush funeral techno 911 rave ultrabeat dance watch funny
america schwarzenegger american bigot ban fun apbt september melody born
terror Videos in News & Blogs 1-15 of 559
IRAQ WAR
04:57
18 hours ago
95
1 rating
Kurds against Turkish TERRORISM Part 2
07:39
1 day ago
48
3 ratings
The Truth
00:59
1 day ago
13
the real terrorists
02:49
1 day ago
182
1 rating
05:22
1 day ago
22

We cannot afford to ignore sites such as YouTube because this is where users go every day to share and view their own homespun videos. It is an important new form of communication via the Internet.

<u>Summary</u>:

- Sources of video data: only searches for user-submitted videos hosted at its own website.
- Video viewed: at YouTube using MacroMedia Flash.

- Search options: keyword searching for tags from uploaded videos; double quotes for a phrase will search for a video title.
- Sort options: Relevance, date added, title, view count, users' rating (using star system).
- Upload video? Yes.

Sites requiring registration, payment, and/or software downloads

AP Archive

The Associated Press video archive is different from the other video search services in several important ways. First, it only searches the AP archive, but that is hardly a small thing. With over half a million stories in the archive, the AP probably has what you want. There is a catch: this is not a free service. The process for ordering video clips is complicated, but if you really need a specific video, you should read the AP Archive's "How We Work" page to learn the particulars. <http://www.aparchive.com/APArchive/pages/admin/how_we_work.html>

However, you do get some very useful data from the site for free. Look at the information provided about a May 2006 video entitled "Iran Nuclear":

Title: Iran Nuclear

Tape Number: EF06/0422

Duration: 00:01:53

Intime: 10:51;09

Date: 2006-05-17

Source: IRINN

SHOTLIST:

1. Zoom-out from banners to wide of rally
2. Close-up of people holding picture of Ahmadinejad
3. SOUNDBITE: (Farsi) Mahmoud Ahmadinejad, Iranian President
"They say they want to give us incentives! Do you think you are dealing with a four-year old child to whom you can give some walnuts and chocolates and get gold from him?"
4. Cutaway crowd waving and chanting
5. SOUNDBITE: (Farsi) Mahmoud Ahmadinejad, Iranian President
"I tell you that we do not want anything beyond our legitimate rights. We want our rights within the NPT (Non Proliferation Treaty) and we will not accept one iota less or more than our rights."
6. Pan of crowd
7. SOUNDBITE: (Farsi) Mahmoud Ahmadinejad, Iranian President
"Don't force governments and nations to renounce their membership of the Nuclear Nonproliferation Treaty."
8. Crowd chanting and waving flags
9. SOUNDBITE: (Farsi) Mahmoud Ahmadinejad, Iranian President
"The Iranian nation won't accept any suspension or end to its nuclear activities."
10. More of crowd at rally

STORYLINE:

Iran's President Mahmoud Ahmadinejad on Wednesday rejected a European plan to offer his country incentives, including a light-water nuclear reactor, in return for giving up uranium enrichment.

"Do you think you are dealing with a four-year old child to whom you can give some walnuts and chocolates and get gold from him?" Ahmadinejad told thousands of people in central Iran, in a speech broadcast live on state television.

European nations have weighed up adding a light-water reactor to a package of incentives meant to persuade Tehran to permanently give up uranium enrichment - or face the threat of UN Security Council sanctions.

On Tuesday senior diplomats and EU government officials said tentative plans were being discussed by France, Britain and Germany as part of a possible package to be presented to senior representatives of the five permanent UN Security Council members.

Ahmadinejad also repeated his threat to pull out of nuclear Non Proliferation Treaty (NPT) if international pressure to give up uranium enrichment continued.

"Don't force governments and nations to renounce their membership of the Nuclear Nonproliferation Treaty," he said, asserting that Iran had the right to a civilian nuclear power programme.

"The Iranian nation won't accept any suspension or end to its uranium enrichment activities," Ahmadinejad said.

He also said Iran trusted the European Union in 2003 and suspended its nuclear activities as a gesture to boost negotiations over its nuclear programme - only to have the Europeans eventually demand Iran permanently halt the programme.

The 2003 deal called for guarantees that Iran's nuclear programme wouldn't diverge from civilian ends toward producing weapons.

Iran agreed to the request, but negotiations collapsed in August 2005 when the Europeans said the best guarantee was for Iran to permanently give up its uranium enrichment programme.

Iran responded by resuming uranium reprocessing activities at its uranium conversion facility in Isfahan, central Iran.

Earlier this year it resumed research and uranium enrichment after the International Atomic Energy Agency, the UN nuclear watchdog, referred Tehran to the UN Security Council.

As you can see, the "storyline" provides extensive information about this clip and is free at the AP Archive website. Users would be hard-pressed to find more or better news video and text than in the AP Video Archive.

AP Video Archive http://www.aparchive.com/

CNN Pipeline

CNN offers only seven days of free access to its vast video archive of news and features. After a week, users must sign up for CNN Pipeline, which requires both payment and downloading and installing a CNN video player. Here is what CNN Pipeline offers:

"CNN Pipeline empowers you to watch up to four live news feeds at once, changing feeds at any time with a single click. Our free video player only allows you to view one video at a time. Additionally, while our free video contains commercial advertisements, Pipeline offers commercial free access to video content on-demand, including free video, with multiple features for ease in searching and browsing. Our extensive archive, not available through the Free Video Player, gives you unlimited access to search and browse CNN.com's online video library." <http://www.cnn.com/help/pipeline/#28>

CNN Pipeline http://www.cnn.com/pipeline/index.html

Vanderbilt Television News Archive

If you must have a video of a news broadcast and you are willing to register and pay for it, the Vanderbilt Television News Archive is the place to go. "The Television News Archive collection at Vanderbilt University is the world's most extensive and complete archive of television news. The collection holds more than 30,000 individual network evening news broadcasts from the major U.S. national broadcast networks: ABC, CBS, NBC, and CNN, and more than 9,000 hours of special news-related programming including ABC's Nightline since 1989...The archive makes two kinds of video tape loans to clients: duplications and compilations. We charge fees for loans to offset the costs in providing this service." <http://tvnews.vanderbilt.edu/>

Vanderbilt Television News Archive http://tvnews.vanderbilt.edu/

Conclusion

There seems to be no end to the number of sites that provide some sort of online video access, whether it is news or homemade videos, for-fee sites or free ones, archives or live feeds. Every time I look, I find more sites, but I had to decide where to draw the line and when to stop compiling sites. If you believe I missed a critically important source, please let me know (and tell me what features make it special).

Online Audio, Podcasts, and Audio Search

Last year I predicted that podcasting would be the "next big thing" for the Internet. At first, I thought I was wrong. While podcasting initially garnered a lot of attention, it did not take off until the second half of 2006. Fueled in large part by the immense popularity of digital audio devices such as iPod and the spread of broadband connections, podcasts and online audio have revolutionized the way people share and get news, entertainment, and information. While music downloads continue to dominate online audio, podcasting is coming on strong. And podcasting is not just for audio: video podcasting—know as **vodcasting** or **vidcasting**—is increasing in popularity, too. Podcasting is recording and broadcasting any non-musical information—be it news, radio shows, sporting events, audio tours, or personal opinions—for playback on a computer or a mobile device. "Though podcasters' web sites may also offer direct download or streaming of their content, a podcast is distinguished from other digital audio formats by its ability to be downloaded automatically using software capable of reading feed formats such as RSS or Atom."[125]

Think of a podcast as analogous to a radio or television broadcast. The podcaster first makes a file available on the Internet, either an audio file (usually in MP3 format) or video file[126] (these can be in a number of formats). Then the podcaster announces the availability of the file using a feed (RSS or Atom) that lists the available podcasts (very much like individual radio or TV shows) with the title, date, and a short text description of each episode. Finally, the user either plays the file on a computer or downloads the file to play it on a portable device such as an iPod capable of playing MP3 and/or video files.

Podcasting has caught on because it is easy, inexpensive, mobile, flexible, and powerful. Many websites now serve as directories to help users find podcasts of every variety anywhere in the world. Podcast search engines generally index podcast metadata such as title, description, and length, which usually makes searching for podcasts fairly accurate. To date, audio podcasts remain more common and popular than video podcasts, but that is changing.

---

[125] "Podcast," Wikipedia, <http://en.wikipedia.org/wiki/Podcast> (19 October 2006).

[126] While the terminology is changing, I would say the term "video podcast" subsumes vlogs, which are a special type of video podcast, i.e., a weblog containing video.

Yahoo Podcast Search

Yahoo got out in front of the podcasting trend with its new Podcasts Search (Beta) site after a study the search giant published with Ipsos Insight, which disclosed that most of the people who are using do so without even knowing it.[127] Yahoo's audio search option also finds podcasts, but if you are looking specifically for podcasts, the Podcast Search site is better. Yahoo Podcast Search indexes metadata such as keywords, categories, or user-generated topic tags to match queries to podcasts. Users do not need to register at Yahoo to listen to podcasts, but registration is required to subscribe to podcasts. Yahoo Podcast Search includes its own player that launches at the site so users can listen to podcasts.

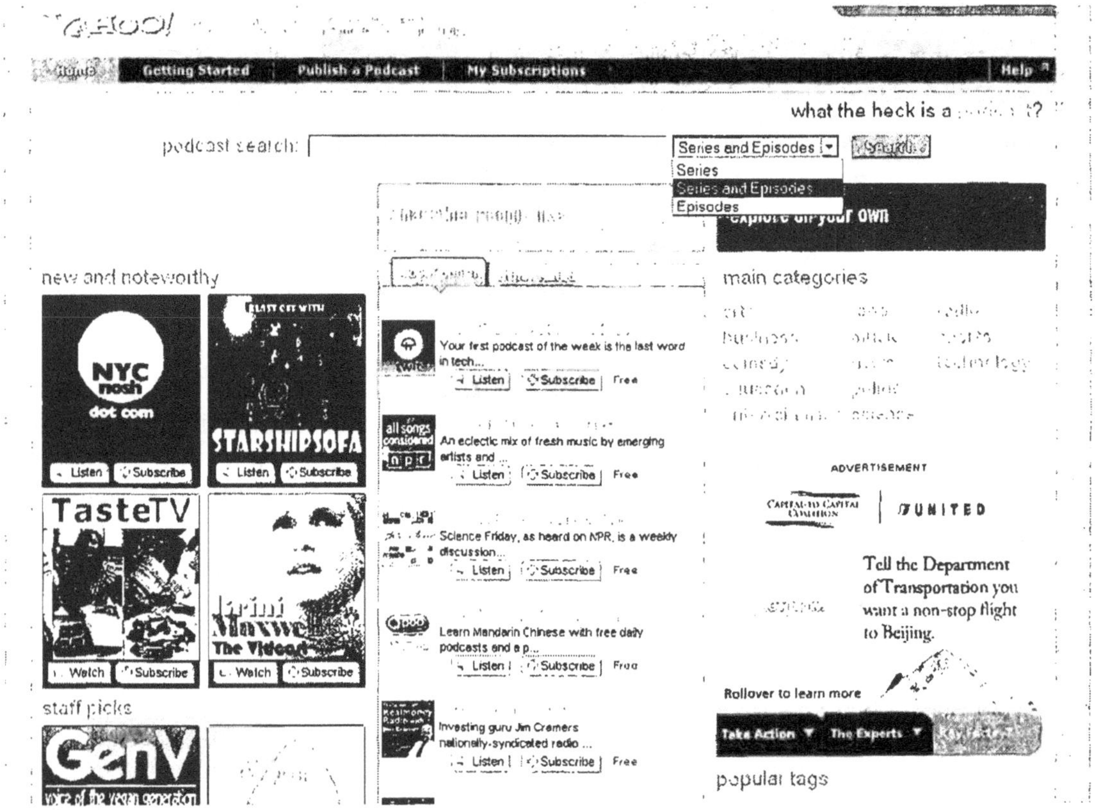
Getting Started
Publish a Podcast
My Subscriptions
Help
podcast search:
Series and Episodes
Series
Episodes
new and noteworthy
main categories
NYC nosh dot com
STARSHIPSOFA
TasteTV
Listen
Subscribe
Watch
Your first podcast of the week is the last word in tech...
An eclectic mix of fresh music by emerging artists and ...
Science Friday, as heard on NPR, is a weekly discussion...
Learn Mandarin Chinese with free daily podcasts and a p...
Investing guru Jim Cramers nationally-syndicated radio ...
Free
ADVERTISEMENT
UNITED
Tell the Department of Transportation you want a non-stop flight to Beijing.
Rollover to learn more
Take Action
The Experts
staff picks
GenV
popular tags

Yahoo Podcast Search (Beta) http://podcasts.yahoo.com/

Podzinger

Podzinger takes a completely different approach to indexing podcasts. "Podzinger is a podcast search engine that lets you search the full audio of podcasts just like you search for any other information on the web...Podzinger takes search a step further

---

[127] Joshua Grossnickle, et al., Yahoo! and Ipsos Insight, "RSS: Crossing into the Mainstream," [PDF] October 2005 , <http://publisher.yahoo.com/rss/RSS_whitePaper1004.pdf> (24 October 2006).

by searching the spoken words inside the podcast in order to find more specific and relevant results. The text-based search results include snippets from the audio to help you figure out if the result is relevant. You can even click on the words to listen to the audio from that point."[128] While this speech recognition technology works well, it only indexes English and Spanish podcasts at this time and the Podzinger index is much smaller than others. Podzinger searches audio, video, or both types of podcasts.

Podzinger also introduced a special tab on its homepage that allows user to "search inside" YouTube videos, that is, to search the spoken text of YouTube videos. A search of YouTube using Podzinger will return the results with the keywords highlighted within the transcribed text at the exact time at which they occurred:

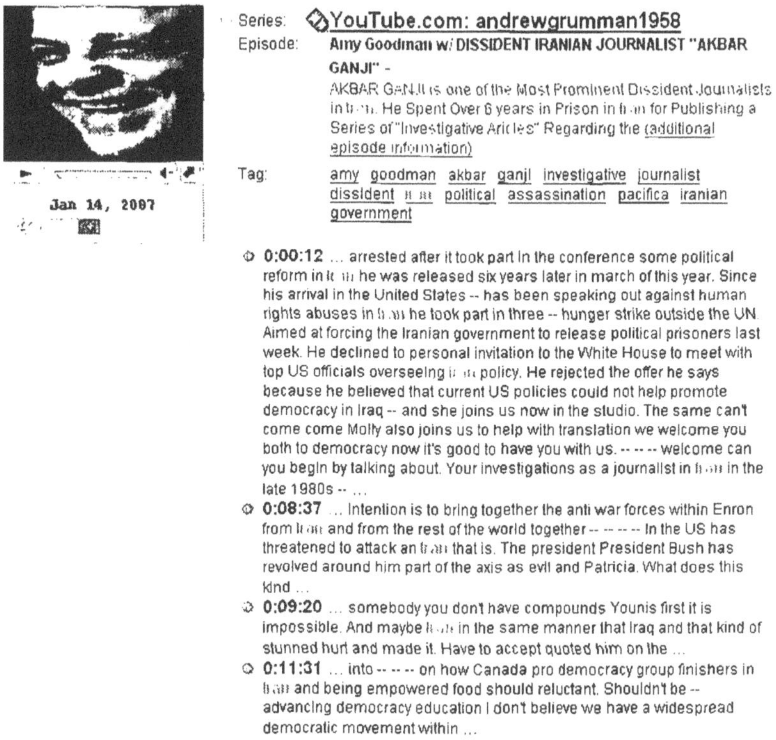
Jan 14, 2007
Series: YouTube.com: andrewgrumman1958
Episode: Amy Goodman w/ DISSIDENT IRANIAN JOURNALIST "AKBAR GANJI" -
AKBAR GANJI is one of the Most Prominent Dissident Journalists in Iran. He Spent Over 6 years in Prison in Iran for Publishing a Series of "Investigative Articles" Regarding the (additional episode information)
Tag: amy goodman akbar ganji investigative journalist dissident Iran political assassination pacifica iranian government
0:00:12 ... arrested after it took part in the conference some political reform in Iran he was released six years later in march of this year. Since his arrival in the United States -- has been speaking out against human rights abuses in Iran he took part in three -- hunger strike outside the UN. Aimed at forcing the Iranian government to release political prisoners last week. He declined to personal invitation to the White House to meet with top US officials overseeing Iran policy. He rejected the offer he says because he believed that current US policies could not help promote democracy in Iraq -- and she joins us now in the studio. The same can't come come Molly also joins us to help with translation we welcome you both to democracy now it's good to have you with us. -- -- -- welcome can you begin by talking about. Your investigations as a journalist in Iran in the late 1980s -- ...
0:08:37 ... Intention is to bring together the anti war forces within Enron from Iran and from the rest of the world together -- -- -- -- In the US has threatened to attack an Iran that is. The president President Bush has revolved around him part of the axis as evil and Patricia. What does this kind ...
0:09:20 ... somebody you don't have compounds Younis first it is impossible. And maybe Iran in the same manner that Iraq and that kind of stunned hurt and made it. Have to accept quoted him on the ...
0:11:31 ... into -- -- -- on how Canada pro democracy group finishers in Iran and being empowered food should reluctant. Shouldn't be -- advancing democracy education I don't believe we have a widespread democratic movement within ...

Requirements for Podzinger are, for Windows Systems

---

[128] Podzinger FAQ, Podzinger.com,<http://www.podzinger.com/about.jsp?section=faq> (19 October 2006).

- Internet Explorer 6.0 or higher and RealPlayer 9.0 or higher
- Internet Explorer 6.0 or higher and Windows Media Player 9.0 or higher
- Firefox 1.0 or higher and Quicktime 6.0 or higher

For Macintosh Systems

- Safari 1.0 or higher and Quicktime 6.0 or higher

If your system does not meet these requirements, you can find podcasts and read the text transcript but you cannot play them. Video podcasts can be downloaded and played later or opened and played using a multimedia player such as Windows Media Player or Apple Quicktime. Podzinger supports the following video formats: mp4, mov, m4v, flv, mpg, or mpeg.

Podzinger http://www.podzinger.com/

Podzinger Spanish Search http://www.podzinger.com/index.jsp?il=es

Odeo

Odeo has been around since 2004 and now boasts over a million audio files, mostly podcasts from all over the world. It offers both a search and browse option. Users can either download the file or listen to it in a neat little player at the Odeo site. There is an option to create an account so that you can subscribe to feeds and save them at Odeo, but an account is not necessary. The search and categorization scheme is based upon tags, simple keywords or categories.

Odeo http://odeo.com/

Podcast.net

Podcast.net has more search options than most podcast search engines. Users can search by title & description, keyword, location, host, or episodes. I think the location search is the most useful because you can locate podcasts around the world.

Podcast.net http://www.podcast.net/

Podscope

Like Podzinger, Podscope transcribes audio from podcasts into searchable text. This means you can search on words and phrases that occur in the podcast; however, I found that Podscope's search is not entirely reliable (the search misses some words that have been transcribed). Nonetheless, it is a valuable podcast search site. Podcasts play at the site using Flash 8+: simply click on the green play button 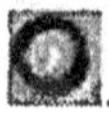.

OCID: 4046925

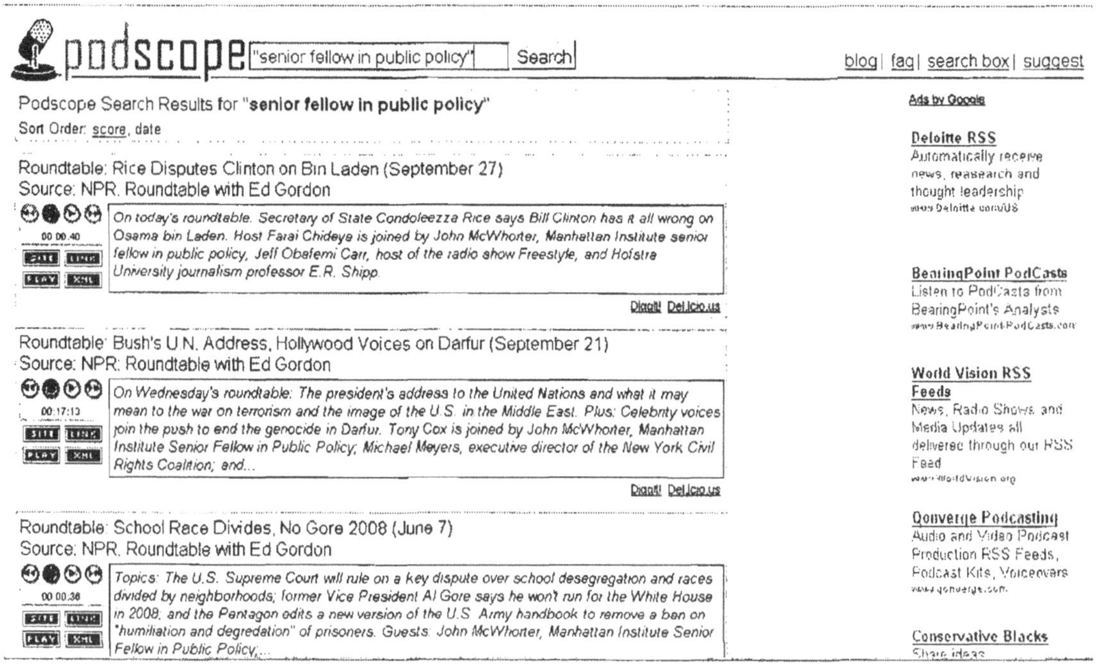
podscope "senior fellow in public policy" Search
blog | faq | search box | suggest
Podscope Search Results for "senior fellow in public policy"
Sort Order: score, date
Roundtable: Rice Disputes Clinton on Bin Laden (September 27)
Source: NPR: Roundtable with Ed Gordon
On today's roundtable: Secretary of State Condoleezza Rice says Bill Clinton has it all wrong on Osama bin Laden. Host Farai Chideya is joined by John McWhorter, Manhattan Institute senior fellow in public policy, Jeff Obafemi Carr, host of the radio show Freestyle, and Hofstra University journalism professor E.R. Shipp.
Roundtable: Bush's U.N. Address, Hollywood Voices on Darfur (September 21)
Source: NPR: Roundtable with Ed Gordon
On Wednesday's roundtable: The president's address to the United Nations and what it may mean to the war on terrorism and the image of the U.S. in the Middle East. Plus: Celebrity voices join the push to end the genocide in Darfur. Tony Cox is joined by John McWhorter, Manhattan Institute Senior Fellow in Public Policy; Michael Meyers, executive director of the New York Civil Rights Coalition; and...
Roundtable: School Race Divides, No Gore 2008 (June 7)
Source: NPR: Roundtable with Ed Gordon
Topics: The U.S. Supreme Court will rule on a key dispute over school desegregation and races divided by neighborhoods; former Vice President Al Gore says he won't run for the White House in 2008; and the Pentagon edits a new version of the U.S. Army handbook to remove a ban on "humiliation and degredation" of prisoners. Guests: John McWhorter, Manhattan Institute Senior Fellow in Public Policy,...
Ads by Google
Deloitte RSS
Automatically receive news, reasearch and thought leadership
BearingPoint PodCasts
Listen to PodCasts from BearingPoint's Analysts
World Vision RSS Feeds
News, Radio Shows and Media Updates all delivered through our RSS Feed
Qonverge Podcasting
Audio and Video Podcast Production RSS Feeds, Podcast Kits, Voiceovers
Conservative Blacks

Podscope http://www.podscope.com/

Other Podcast Search Sites & Directories

A number of sites offer podcast search or function as directories that index podcasts usually by topic.

Podcast Directories

iPodder	http://www.ipodder.org/directory/4/podcasts
Podcast Directory	http://www.podcastdirectory.com/
Podfeed	http://www.podfeed.net/
Podcasting Station	http://www.podcasting-station.com/
Podcast Shuffle	http://www.podcastshuffle.com/

Other Podcast Search Sites

Blinkx	http://www.blinkx.tv/
Podcast Alley	http://www.podcastalley.com/

Special Topics—News, Blogs, & Technology Search

Newsgroups, Forums, & Mailing Lists

News is one of the types of information most amenable to the Internet: both are fast moving, worldwide, and in high demand. Along with the proliferation of news sources on the Internet have come an even greater number of ways to share opinions on every subject, limited only by the scope of the human imagination. In this article, I am going to discuss several different ways of accessing both news and opinions on the Internet: newsgroups, message boards or forums, and mailing lists. I discuss weblogs in the next section because they require special software and thus demand a separate discussion.

Newsgroups in general and Usenet in particular have lost a great deal of their prominence in recent years as more outlets for sharing information and communicating on the Internet have appeared. Usenet is a system that allows individuals to post messages and have them read around the world within about three hours. One of the drawbacks of Usenet is that it requires special software (a newsreader) configured to send and receive data via a user's Internet Service Provider's news server. Microsoft embeds its newsreader into software such as Outlook and Outlook Express.

Usenet Warning

This is a place on the Internet I recommend exercising great caution and skepticism because anyone with Internet access can literally post anything to a newsgroup, and often do.

Usenet newsgroups are noted for being chaotic, notoriously unreliable (lots of gossip), and confusing. So why would anyone look at them? Because occasionally there is valuable and unique information posted to newsgroups.

By August 2000, there was only one Usenet search engine remaining: Deja (formerly Dejanews). Fortunately, Google acquired Deja in 2001 and made available the entire Deja archive consisting of over 700 million posts from 1981 to today.

Google Groups indexes more than 1 billion Usenet postings, and new postings normally appear in Google Groups within 10 minutes.

Google Groups put forward a new look in 2004 (out of Beta in early 2007) while retaining all the old postings in its index.

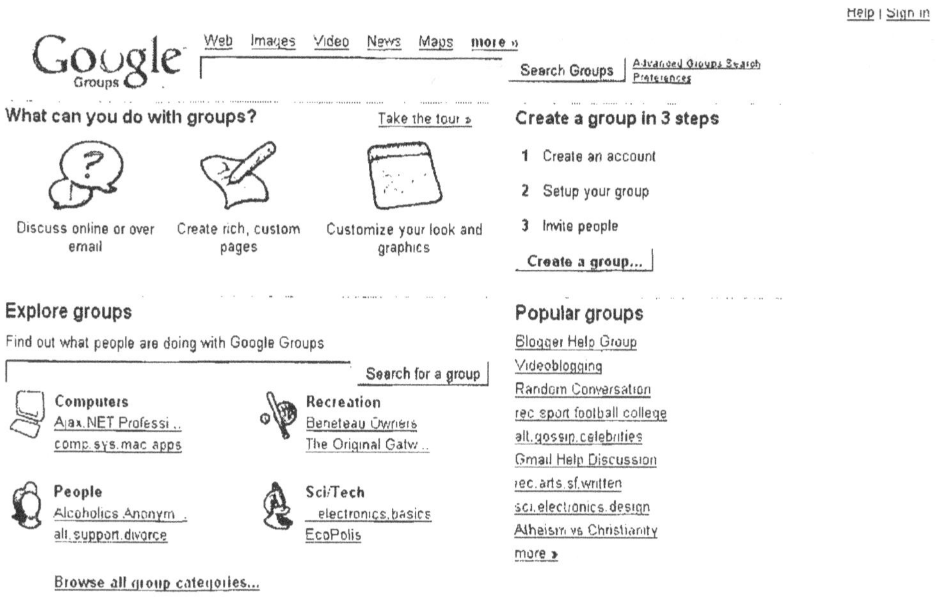
Help | Sign in
Google Groups
Web Images Video News Maps more »
Search Groups
Advanced Groups Search
Preferences
What can you do with groups?
Take the tour »
Discuss online or over email
Create rich, custom pages
Customize your look and graphics
Create a group in 3 steps
1 Create an account
2 Setup your group
3 Invite people
Create a group...
Explore groups
Find out what people are doing with Google Groups
Search for a group
Computers
Ajax.NET Professi...
comp.sys.mac.apps
Recreation
Beneteau Owners
The Original Gatw...
People
Alcoholics Anonym...
alt.support.divorce
Sci/Tech
electronics.basics
EcoPolis
Browse all group categories...
Popular groups
Blogger Help Group
Videoblogging
Random Conversation
rec.sport.football.college
alt.gossip.celebrities
Gmail Help Discussion
rec.arts.sf.written
sci.electronics.design
Atheism vs Christianity
more »

While Google Groups has many options and features that require an account, many of the most valuable features do not:

ACTIVITIES THAT DON'T REQUIRE A GOOGLE ACCOUNT:

- Reading posts in public groups
- Searching for groups, posts, or authors
- Posting to groups via email if they are unrestricted or you're already a member
- Joining a public Google Group via email

ACTIVITIES THAT DO REQUIRE A GOOGLE ACCOUNT:

- Creating and managing your own Google Group
- Posting to groups via our web interface
- Creating pages and uploading files
- Subscribing to a Usenet newsgroup and receiving posts via email
- Joining a Google Group via our web interface
- Changing your subscription type (No Email, Abridged Email...)
- Reading a restricted group's posts online

Google Groups has an **advanced search** that lets you limit your search by language, subject, forum, author, message ID, and/or date. It also has a feature known as "author profile." When you are viewing a posting, click on "More options" (next to the date); then select "View profile" next the author's From address. By clicking on a "Show options" next to an author's name, Google will present the option to "Find messages by this author" and automatically search for every posting from a particular address. ***Google Groups no longer shows the author's entire email address for privacy reasons, but many addresses are guessable.***

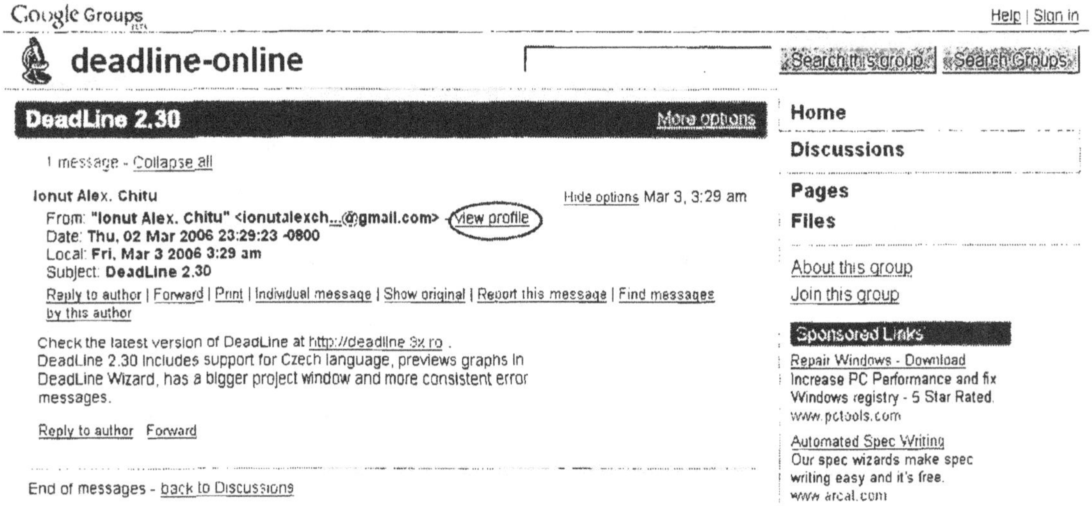
Google Groups
Help | Sign in
deadline-online
Search this group
Search Groups
DeadLine 2.30
More options
Home
Discussions
Pages
Files
About this group
Join this group
1 message - Collapse all
Ionut Alex. Chitu
Hide options Mar 3, 3:29 am
From: "Ionut Alex. Chitu" <ionutalexch...@gmail.com> View profile
Date: Thu, 02 Mar 2006 23:29:23 -0800
Local: Fri, Mar 3 2006 3:29 am
Subject: DeadLine 2.30
Reply to author | Forward | Print | Individual message | Show original | Report this message | Find messages by this author
Check the latest version of DeadLine at http://deadline.3x.ro .
DeadLine 2.30 includes support for Czech language, previews graphs in DeadLine Wizard, has a bigger project window and more consistent error messages.
Reply to author Forward
End of messages - back to Discussions
Sponsored Links
Repair Windows - Download
Increase PC Performance and fix Windows registry - 5 Star Rated.
www.pctools.com
Automated Spec Writing
Our spec wizards make spec writing easy and it's free.
www.arcat.com

Google Groups' Option to "Find messages by this author"

Google Groups is also one of the most valuable sources of information on technical and computer-related problems. Someone has probably encountered that same complex and mysterious error message that has you flummoxed. Simply by copying and pasting the error message into Google Groups, you have a very good change of finding out what it means and, even better, how to fix the problem. ***Despite what the site says, Google Groups no longer limits the query to 10 words, but to 32*** as Google Search does. Also, the Google wildcard (*) does not work in Google Groups.

Google Groups http://groups.google.com/

Yahoo Groups is also a huge source of newsgroups on virtually every topic. Yahoo Groups is very easy to use because it is arranged like the Yahoo general subject guide, i.e., it is browsable by topic and keyword searchable. Some groups require membership (registration), but many are publicly accessible.

Yahoo Groups does not offer a search by "author's profile," but it does link directly to its **Member Directory**. You may be able to view the member profile, which may provide the user's real name, location, age, occupation, and email address.

Yahoo Member Directory http://members.yahoo.com/

However, the popularity of newsgroups has declined with the rise of forums, also known as message boards. One of the main advantages of forums as a means of communicating via the Internet is that they do not require software other than a web browser for users to read and post to them. Forums usually are focused on a specific topic, anything from computer games to politics. "In terms of countable posts, Japan is far in the lead with over two million posts per day on their largest forum, 2ch. The United States does not have any one large forum, but instead several hundred thousand smaller forums...China, the Netherlands, and France are also home to hundreds of independent forums."[129]

There are several major search engines specifically designed to index and search forums, message boards, and discussion groups. BoardTracker, Lycos Discussions and BoardReader make it possible for users to search many more online discussion communities than ever before.

BoardReader, which was offline for most of 2005, is back and improved. The site has several software partners, including Vivisimo, the clustering metasearch engine behind Clusty, and Thunderstone to search and group information from Internet forums and message boards. There is an option to limit searches by date, site, and language. The sources indexed by BoardReader still tend to be very different from those accessed by Lycos Discussion, Google Groups, or Yahoo Groups, making it a valuable addition to "hidden web" search. **BoardReader does recognize non-Latin character sets**. The biggest change to BoardReader is the addition of a Domain Profile option, which uses data BoardReader gathers by indexing forums to create a detailed picture of a domain, including number and sources of inbound links, the pages in a domain with the most active inbound links.

BoardReader http://www.boardreader.com/

BoardTracker searches only for the content of message boards/forums and can be used as a traditional keyword search engine or a browsable directory of message boards by topic. BoardTracker has an "instant alerts system" whereby it "will notify you in a number of ways (email, Jabber, ICQ) as soon as a thread matching your search term is posted on any of the thousands of forums we track." While BroadTracker does not require registration to use, its alert service obviously must require registration, which in turn requires a valid email address. Nonetheless, BoardTracker's registration does not ask for your name or address, so users can employ a free email service such as Hotmail or Yahoo Mail for BoardTracker alerts.

Notice the search options BoardTracker offers, including the search for Threads or by Author, to sort by Relevancy or by Date, and to search in All Categories, your own selected threads, or in a specific category, which often has more subcategories within it, e.g., World & Regional includes specific continents, General, Middle East, and Warfare. I am also happy to report that ***BoardTracker searches equally well in non-Latin character sets as it does in English***. I am also intrigued that it will search for results from the "past 6142 years."

---

[129] "Internet Forum," *Wikipedia*, <http://en.wikipedia.org/wiki/Internet_forum> (14 November 2006).

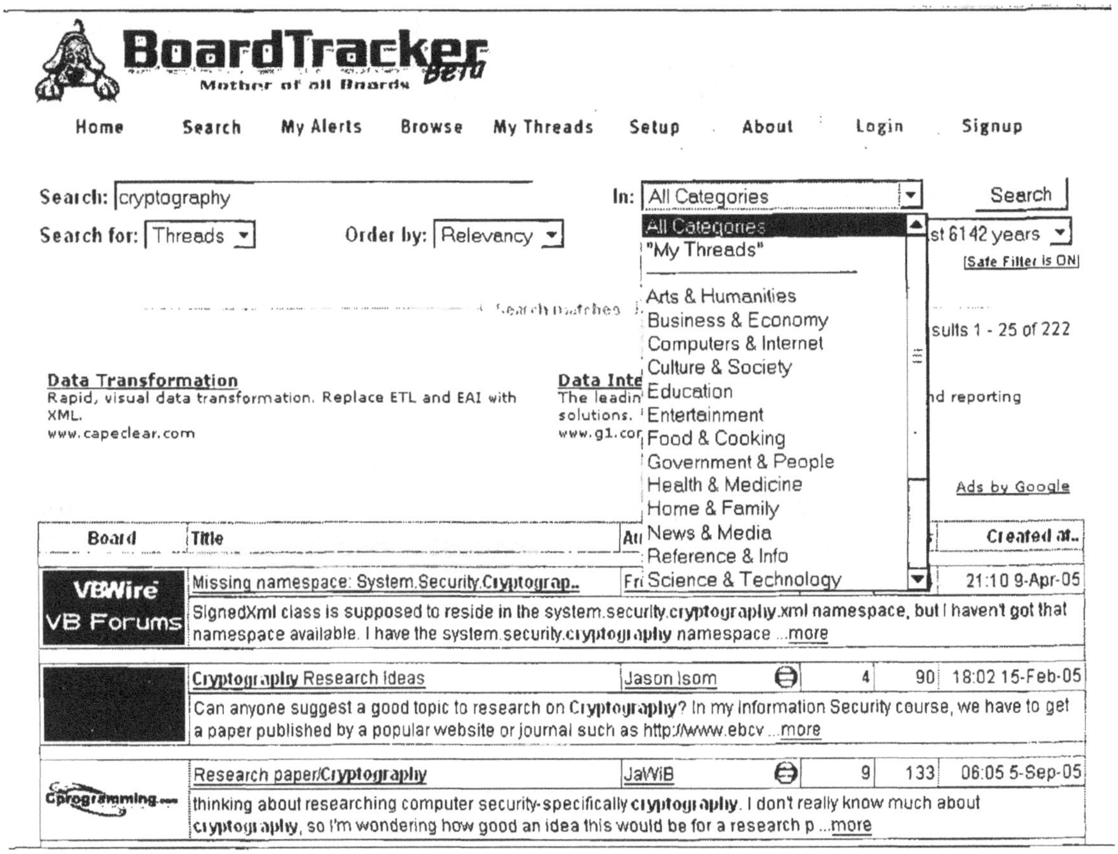
BoardTracker
Mother of all Boards
Home Search My Alerts Browse My Threads Setup About Login Signup
Search: cryptography
In: All Categories
Search
Search for: Threads
Order by: Relevancy
All Categories
"My Threads"
Arts & Humanities
Business & Economy
Computers & Internet
Culture & Society
Education
Entertainment
Food & Cooking
Government & People
Health & Medicine
Home & Family
News & Media
Reference & Info
Science & Technology
Safe Filter is ON
sults 1 - 25 of 222
Data Transformation
Rapid, visual data transformation. Replace ETL and EAI with XML.
www.capeclear.com
Ads by Google
Board
Title
Created at
VBWire VB Forums
Missing namespace: System.Security.Cryptograp..
21:10 9-Apr-05
SignedXml class is supposed to reside in the system.security.cryptography.xml namespace, but I haven't got that namespace available. I have the system.security.cryptography namespace ...more
Cryptography Research Ideas
Jason Isom
4
90
18:02 15-Feb-05
Can anyone suggest a good topic to research on Cryptography? In my Information Security course, we have to get a paper published by a popular website or journal such as http://www.ebcv ...more
Research paper/Cryptography
JaWiB
9
133
06:05 5-Sep-05
thinking about researching computer security-specifically cryptography. I don't really know much about cryptography, so I'm wondering how good an idea this would be for a research p ...more

BoardTracker http://www.boardtracker.com/

OMGILI is a new vertical search engine that only searches the content of web forums. Why would anyone want to limit a search to discussion forums? "The information contained in online forums is typically presented in a 'question and answer' or debate style format. How is this significant? Many times you will have a question that has already been answered. Using Omgili, you can avoid posting already asked questions and quickly find your answer...Unlike ordinary search engines that prioritize articles and edited web pages, Omgili only indexes discussion forums. Using Omgili's advanced search capabilities you can choose to independently search titles, topics or just the replies of a discussion."[130]

Think how many times you have needed to find the answer to a question—for example, about trojan horse removal—and had to struggle through typical search engine results that were littered with useless advertisements. Now look at the results for that query using Omgili:

[130] "About Omgili," Omgili.com, <http://www.omgili.com/about.html> (31 October 2006).

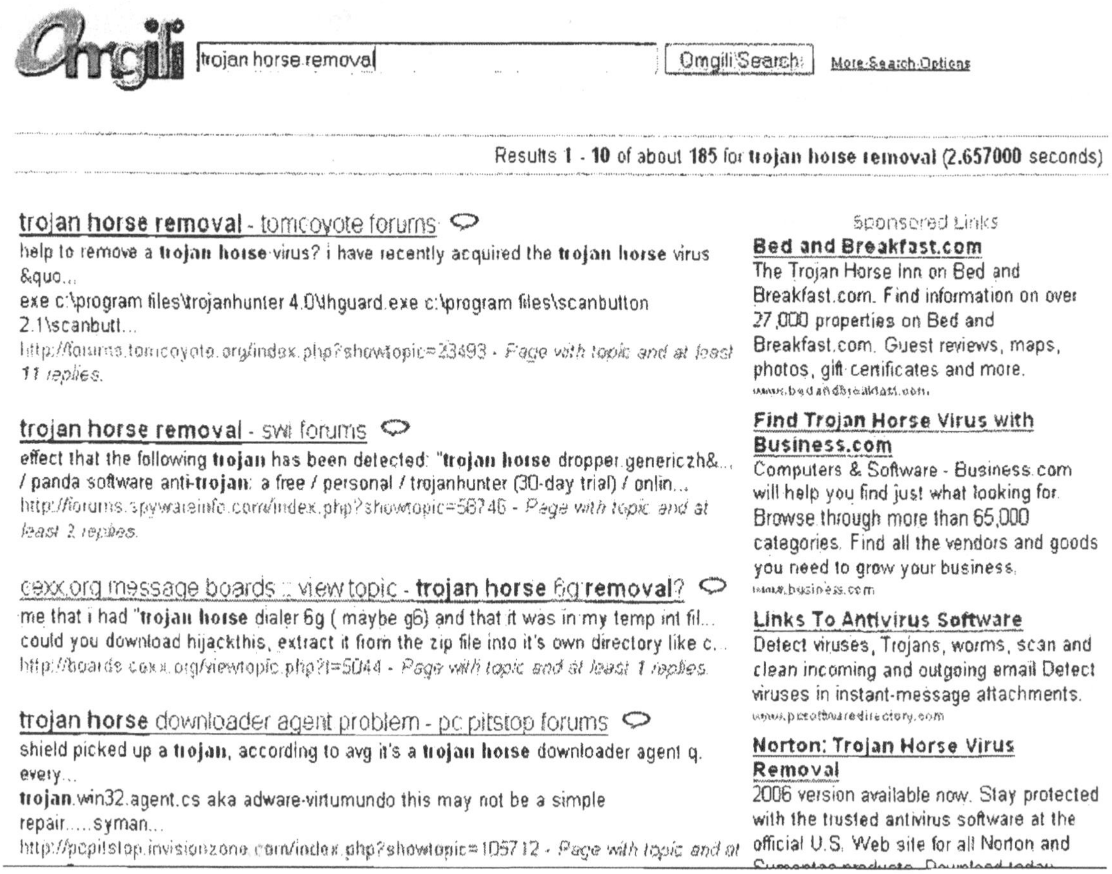
Omgili
trojan horse removal
Omgili Search
More Search Options
Results 1 - 10 of about 185 for trojan horse removal (2.657000 seconds)
trojan horse removal - tomcoyote forums
help to remove a trojan horse virus? i have recently acquired the trojan horse virus &quo...
exe c:\program files\trojanhunter 4.0\thguard.exe c:\program files\scanbutton 2.1\scanbutt...
http://forums.tomcoyote.org/index.php?showtopic=23493 - Page with topic and at least 11 replies.
trojan horse removal - swi forums
effect that the following trojan has been detected: "trojan horse dropper.genericzh&...
/ panda software anti-trojan: a free / personal / trojanhunter (30-day trial) / onlin...
http://forums.spywareinfo.com/index.php?showtopic=58746 - Page with topic and at least 2 replies.
cexx.org message boards :: view topic - trojan horse 6g removal?
me that i had "trojan horse dialer 6g (maybe g6) and that it was in my temp int fil...
could you download hijackthis, extract it from the zip file into it's own directory like c...
http://boards.cexx.org/viewtopic.php?t=5044 - Page with topic and at least 1 replies.
trojan horse downloader agent problem - pc pitstop forums
shield picked up a trojan, according to avg it's a trojan horse downloader agent q. every...
trojan.win32.agent.cs aka adware-virtumundo this may not be a simple repair.....syman...
http://pcpitstop.invisionzone.com/index.php?showtopic=105712 - Page with topic and at
Sponsored Links
Bed and Breakfast.com
The Trojan Horse Inn on Bed and Breakfast.com. Find information on over 27,000 properties on Bed and Breakfast.com. Guest reviews, maps, photos, gift certificates and more.
Find Trojan Horse Virus with Business.com
Computers & Software - Business.com will help you find just what looking for. Browse through more than 65,000 categories. Find all the vendors and goods you need to grow your business.
www.business.com
Links To Antivirus Software
Detect viruses, Trojans, worms, scan and clean incoming and outgoing email Detect viruses in instant-message attachments.
Norton: Trojan Horse Virus Removal
2006 version available now. Stay protected with the trusted antivirus software at the official U.S. Web site for all Norton and

The Omgili crawler analyzes a forum differently from a webpage, recognizing and assigning different weights to the topic, title, and replies. One thing to keep in mind is that you will have to register at forums where registration is required, i.e., Omgili does not offer a cached option. I would recommend adding Omgili to your forum search bookmarks. Oh, in case you're interested, Omgili stands for "oh my God I love it!"

Omgili http://www.omgili.com/

Finding News Groups and Mailing Lists

In contrast to searching newsgroup *content*, **Tile.net** will help you find newsgroups or mailing lists of interest. Tile.net offers an alphabetical listing of all Usenet newsgroups by description or newsgroup hierarchy. Don't worry if you don't understand the hierarchy because Tile.net has a search engine that will find the appropriate newsgroups to match your keywords. *Tile.net also provides* a searchable listing of mailing lists and discussion groups.

CataList is the official source to "browse any of the 58,638 public LISTSERV lists on the Internet, search for mailing lists of interest, and get information about LISTSERV host sites." LISTSERV is one of the most useful and now venerable Internet programs (dating from 1986), scanning email messages for the words "subscribe" and "unsubscribe" to automatically update mailing list subscriptions. At the CataList site, users can get the following public List and Site information:

List information

- Search for a mailing list of interest
- View lists by host country
- View lists with 10,000 subscribers or more
- View lists with 1,000 subscribers or more

Site information

- Search for a LISTSERV site of interest
- View sites by country

Finally, both Google and Yahoo Groups allow users to search or browse groups by topic.

Google Groups	http://groups.google.com/
Yahoo Groups	http://groups.yahoo.com/
Tile.net	http://www.tile.net/
CataList	http://www.lsoft.com/lists/listref.html

Weblogs & RSS Feeds

Weblogs (more often known simply as **blogs**) are web pages that are usually defined as online journals or diaries but can be many things. "Blog posts are like instant messages to the web. Many blogs are personal, 'what's on my mind' type musings. Others are collaborative efforts based on a specific topic or area of mutual interest. Some blogs are for play. Some are for work. Some are both."[131] Blogs can be wonderful sources of news and information, or they can be absolute schlock. I find blogs most useful as sources of rumors and opinions on particular topics or breaking news stories. Making blogs useful is tricky, but some of the best ways to access them are via specialty blog search tools because many search engines do not offer blog-specific searches. The first major blog search tool was Daypop, an excellent news search engine that lets users limit a search to blogs or RSS feeds. Google has a separate blog search engine, Yahoo subsumes blog searches under its news search, and Live has a "feed" search.

Many—in fact probably most—blogs make their content available in **RSS**. What is RSS? RSS stands either for Rich Site Summary or RDF Site Summary (there is some dispute about this). "RSS is an XML format for syndicating web content. A website that wants to allow other sites to publish some of its content creates an RSS document and registers the document with an RSS publisher. A user that can read RSS-distributed content can use the content on a different site. Syndicated content includes such data as news feeds, events listings...excerpts from discussion forums or even corporate information."[132] Look for the RSS/Atom feed icon at websites to subscribe to a feed.

To add to the confusion, RSS is not the only format used for blogging and newsfeeds. The other major format is Atom. Here's a good explanation of the difference:

> RSS/XML/Atom are technologies, but syndication is a process. RSS and Atom are two flavors of what is more or less the same thing: a 'feed' which is a wrapper for pieces of regularly and sequentially-updated content, be they news articles, weblog posts, a series of photographs, and more. For the purposes of this article, ***consider the terms interchangeable***. XML is the base technology both are built on, but that's almost totally irrelevant; the

[131] "About Blogger," Blogger.com, <http://www.blogger.com/about.pyra> (14 November 2006).

[132] "RSS," Webopedia, <http://www.webopedia.com/TERM/R/RSS.html> (14 November 2006).

orange buttons are mislabeled, and should read 'RSS' or 'Atom' instead. Strange, but true.[133]

News aggregators (sometimes referred to as RSS aggregators) are programs designed to read XML formatted content, which is very popular in the blogging community. News aggregators retrieve RSS/Atom feeds and present these feeds in an easy to read format. Bloggers and many news websites use RSS/Atom feeds in XML format to publish information. Most news aggregators are downloadable programs that need to be installed on your computer, but some are implemented on websites.

There are now a number of RSS and blog search engines. Unfortunately, the relationship between RSS feeds and blogs is not as clean and clear as one might hope. While RSS search engines get their content from RSS feeds and not from crawling the web, any type of website can distribute content using RSS. This means RSS search engines are searching more than just blogs. On the flip side, not every blog uses RSS to distribute its content, so some blogs are not searched by RSS search engines.

So what exactly do blog search engines provide that traditional news search engines do not? What users often get from blogs are biased insights and opinions. For good or for ill, blogs are somewhere between newsgroups/chat rooms and true journalistic sites. Why use blog search engines? First, remember that traditional search engines are the least useful for news or date-sensitive information. News search tools are best for timely objective reporting. What blog search adds is diversified opinion (sometimes useful, sometimes not) on virtually every topic imaginable. Also, even good news search sites may index a limited number of sources whereas a good blog site may get news out faster and more efficiently. However, for a number of reasons, ***blog search engines' algorithms are not as good as general search engines' algorithms at weeding out spam***, so you will probably have to wade through a lot of inappropriate and useless sites when using a blog search tool.

The list of blog search engines seems to be growing weekly, but I expect some will fall by the wayside as others become more popular. I recommend trying several of these tools because you will probably get very different results. In late 2006 Google Blogsearch overtook **Technorati** as the most popular blog search site. **Technorati** offers a somewhat different approach. The concept behind Technorati is that it "watches" over **63 million weblogs**, analyzing who is linking to a blog, website, or news article. By entering the url of any blog, website, or news article, users can see how many blogs link to it, which bloggers are linking to that page, and what they are saying about it. In essence, Technorati is a very simple website analysis tool that

---

[133] "What is RSS/XML/Atom Syndication?" Mezzoblue.com, <http://www.mezzoblue.com/archives/2004/05/19/what_is_rssx/> (14 November 2006).

lets users see what others are saying about almost any site of interest. Technorati also offers a keyword search of all the weblogs it tracks.

Technorati and the Associated Press "initiated a service to connect bloggers to more than 440 AP member newspapers nationwide...Increasingly, what the blogosphere says about a news story becomes part of a more complete story, lending diverse perspectives and often expert commentary. The new service will bring blogger commentary about AP news stories to communities large and small throughout the USA, giving bloggers a voice in trusted local papers throughout the nation...When readers visit an AP member Web site that uses AP Hosted Custom News, they will see a module featuring the 'Top Five Most Blogged About' AP articles right next to the article text, dynamically powered by Technorati. Additionally, when readers click on an AP article, Technorati will deliver 'Who's Blogging About' that article."[134]

At the participating sites you will see the "Blog Roundup" from Technorati:

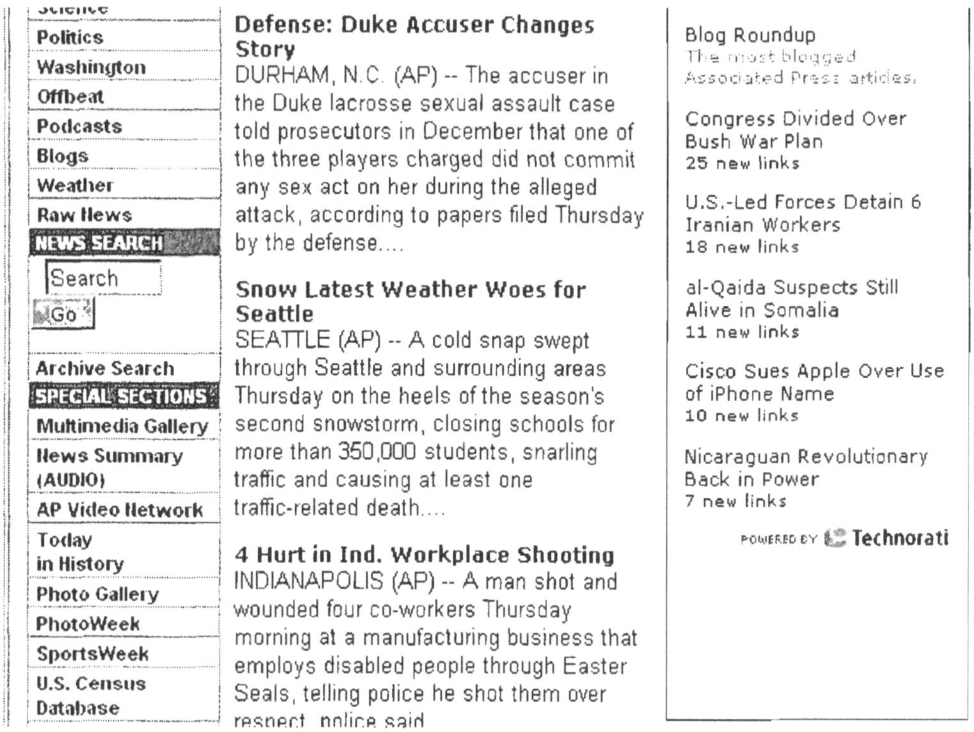

Politics
Washington
Offbeat
Podcasts
Blogs
Weather
Raw News
NEWS SEARCH
Search
Go
Archive Search
SPECIAL SECTIONS
Multimedia Gallery
News Summary (AUDIO)
AP Video Network
Today in History
Photo Gallery
PhotoWeek
SportsWeek
U.S. Census Database

Defense: Duke Accuser Changes Story
DURHAM, N.C. (AP) -- The accuser in the Duke lacrosse sexual assault case told prosecutors in December that one of the three players charged did not commit any sex act on her during the alleged attack, according to papers filed Thursday by the defense....

Snow Latest Weather Woes for Seattle
SEATTLE (AP) -- A cold snap swept through Seattle and surrounding areas Thursday on the heels of the season's second snowstorm, closing schools for more than 350,000 students, snarling traffic and causing at least one traffic-related death....

4 Hurt in Ind. Workplace Shooting
INDIANAPOLIS (AP) -- A man shot and wounded four co-workers Thursday morning at a manufacturing business that employs disabled people through Easter Seals, telling police he shot them over respect, police said

Blog Roundup
The most blogged Associated Press articles.

Congress Divided Over Bush War Plan
25 new links

U.S.-Led Forces Detain 6 Iranian Workers
18 new links

al-Qaida Suspects Still Alive in Somalia
11 new links

Cisco Sues Apple Over Use of iPhone Name
10 new links

Nicaraguan Revolutionary Back in Power
7 new links

POWERED BY Technorati

---

[134] Peter Hirshberg, "Technorati Teams With The Associated Press to Connect Bloggers to More Than 440 Newspapers Nationwide," Technorati Weblog, 23 May 2006, <http://technorati.com/weblog/2006/05/107.html> (31 October 2006).

Clicking on the headline links you to the article; clicking on the "Links to this article" takes you to Technorati to read blog entries about the story. I think this development is further evidence that the definition of "journalism" is fuzzier than ever.

The Diarist Registry indexes more than 7100 blog sites, is fully searchable, and lists blogs by country. **Bloogz** permits searches in four European languages plus English and **Blogwise** offers an option to list blogs by country. **IceRocket** has a "trend tool" that shows keyword trends for the last one to three months as well as other information to allow users to get a broad sense of a topic's popularity.

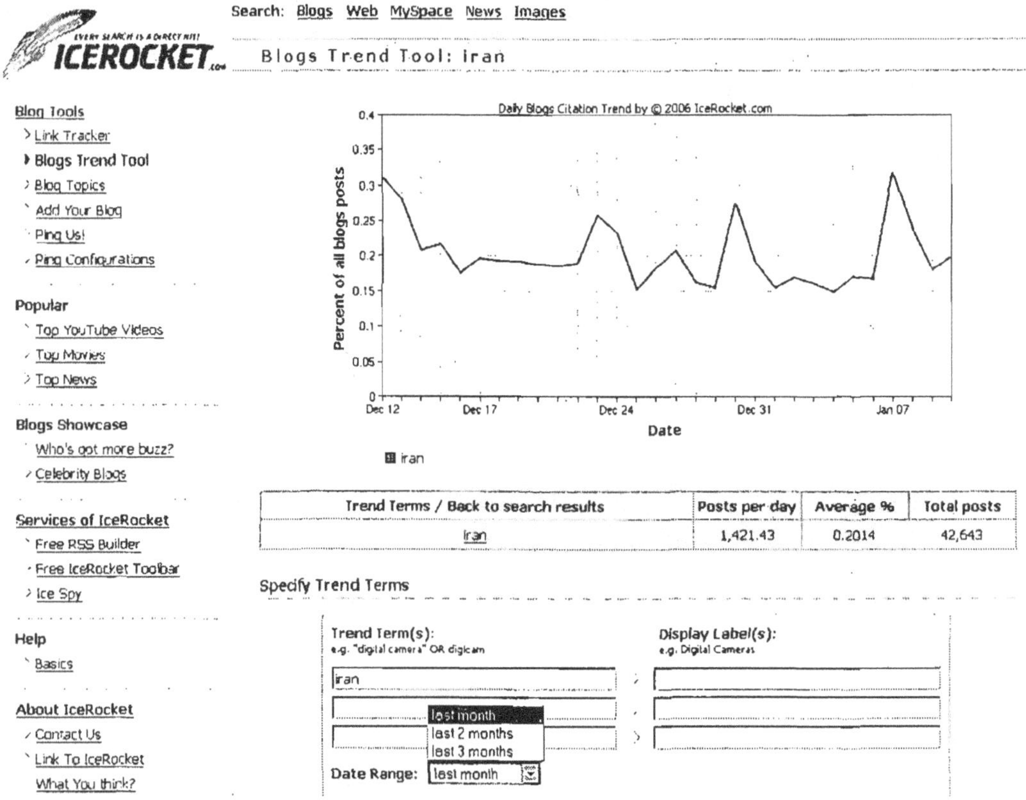
Search: Blogs Web MySpace News Images
ICEROCKET
Blogs Trend Tool: iran
Blog Tools
Link Tracker
Blogs Trend Tool
Blog Topics
Add Your Blog
Ping Us!
Ping Configurations
Popular
Top YouTube Videos
Top Movies
Top News
Blogs Showcase
Who's got more buzz?
Celebrity Blogs
Services of IceRocket
Free RSS Builder
Free IceRocket Toolbar
Ice Spy
Help
Basics
About IceRocket
Contact Us
Link To IceRocket
What You think?
Daily Blogs Citation Trend by © 2006 IceRocket.com
Percent of all blogs posts
Date
Dec 12
Dec 17
Dec 24
Dec 31
Jan 07
iran
Trend Terms / Back to search results
Posts per day
Average %
Total posts
iran
1,421.43
0.2014
42,643
Specify Trend Terms
Trend Term(s):
e.g. "digital camera" OR digicam
Display Label(s):
e.g. Digital Cameras
last month
last 2 months
last 3 months
Date Range:

Also, don't forget Waypath, which lets you see blogs that link to specific websites almost in real time. Waypath also searches the full text of blogs as well as by keyword.

Of course, we now have Google Blogsearch, which offers many search features and options but contains only fairly recent content. Also, keep in mind that ***Google Blog Search only indexes the site feed, not the full content at the website*** that originated the feed.

All the blog search engines permit ***sort by date*** in addition to sort by relevance. Date sorting works perfectly well with weblogs, which (unlike webpages) have date/time tags.

Blogdigger	http://www.blogdigger.com/
Blog Search Engine	http://www.blogsearchengine.com/
Blogwise	http://www.blogwise.com/
Bloogz	http://www.bloogz.com/
Clusty Blog Metasearch	http://blogs.clusty.com/
Daypop	http://www.daypop.com/
Feedster	http://www.feedster.com/
Google Blogsearch	http://blogsearch.google.com/
IceRocket	http://blogs.icerocket.com/
Sphere	http://www.sphere.com/
The Diarist Registry	http://www.diarist.net/registry/
Technorati	http://www.technorati.com/

There are as many potential blogs as there are people willing to create and maintain them. Quite honestly, most are as dull as ditchwater. However, there is a type of blog—or perhaps it's more accurate to say, a type of blogger—that is inherently interesting: the insider. For example, there are several hundred Microsoft employees who are blogging, and a lot of what they are talking about is directly related to Microsoft products that are still in the planning stages. "Prolific Microsoft blogger Robert Scoble mentioned Windows XP Tablet PC Edition 2005 (and complained that the real name was far less interesting than the codename Lone Star)."[135] A few days after his blog entry appeared, Microsoft officially announced that the service pack Tablet PC is tied to wouldn't be out as planned by June (the last month of the Microsoft fiscal year) but was being delayed until at the earliest July 2005, when Microsoft's 2005 fiscal year begins. Other Microsoft bloggers have discussed such things as technical details of work with the latest developer tools, Longhorn (which became Vista, the current Windows OS for desktop PCs) architecture, and how features are developed for Microsoft products like Word.

Microsoft has definitely taken a "if you can't beat 'em, join 'em" attitude towards blogs (so far). In 2004 the company went so far as to launch Channel 9, a website that combines blogs, discussion forums and other technology to improve communications with developers. Channel 9 refers to the audio channel many

---

[135] "Inside Track," *The Guardian*, 20 May 2004, <http://technology.guardian.co.uk/online/story/0,3605,1220085,00.html> (14 November 2006).

airlines use to let passengers listen in on crew conversations, and, according to the welcome message at the site, "We think developers need their own channel 9, a way to listen in to the cockpit at Microsoft, an opportunity to learn how we fly, a chance to get to know our pilots...Five of us in Redmond are crazy enough to think we just might learn something from getting to know each other...Join in, and have a look inside our cockpit and help us fly the plane."

Channel 9 http://channel9.msdn.com/

Companies walk a fine line between encouraging people to share information and protecting their proprietary "crown jewels." In fact, Microsoft fired a long-time temporary employee for posting photographs of Apple G5 Macs being delivered to the Microsoft campus. The employee originally posted the photo on his own site and intended it mainly for friends and family (he even cropped the photo so that the Microsoft campus wouldn't be visible). Nonetheless, the photo was quickly discovered and widely discussed across the web, leading Microsoft to dismiss him for "violating company security policy." Given the wide-open nature of the Internet, the ease of blogging, and the fact that people love to share information for whatever reason, blogging will undoubtedly become a bigger problem for companies and organizations. Employees may innocently or inadvertently post inside information that could be harmful to their organization. And it's not hard to imagine what damage a disgruntled employee could do.

General News Sources

Finding news sources around the globe has become fairly easy. All of the following are very good sources of different types of news sources: newspapers, magazines, wire services, etc. Many have links to online publications and news sources, and many of these publications are available free over the Internet.

ABYZ Newslinks http://www.abyznewslinks.com/
Guardian's World News Guide http://www.guardian.co.uk/worldnewsguide/
HeadlineSpot http://www.headlinespot.com/
Kiosken http://www.esperanto.se/kiosk/engindex.html
Metagrid (newspapers & magazines) http://www.metagrid.com/
NewsCentral (online newspaper links) http://www.all-links.com/newscentral/
NewsDirectory http://newsdirectory.com/
Newslink http://newslink.org/
Online Newspapers http://www.onlinenewspapers.com//index.htm
RefDesk (My Virtual Newspaper) http://www.refdesk.com/papmain.html

News Sites & Search Engines

In addition to the sites above that help you locate news sources, there are also a growing number of excellent search services dedicated to delivering and searching the news. In fact, so many people all over the world are relying on the Internet for their news that "old media" sources continue to lose revenue as they compete for readers and viewers. It turns out that some of the "new media" Internet news services such as Google and Yahoo now compensate old media sources such as AP on a pay-per-click basis. "Major Internet portals such as Yahoo and America Online have been paying for content since their creation in the mid-1990s… Earlier this year [2006], Google signed a deal with the Associated Press, one of more than 50 agreements AP Chief Executive Tom Curley has obtained from Internet players" since 2003.[136]

Google News and **Yahoo News** are among the very best news search tools. Both have superb advanced search options that include the ability to limit searches to specific publications, dates, etc. Both let you sort the results by relevance or by date and time. ***News searching is one area in which it makes sense to sort results by date because news stories, unlike webpages, have meaningful date/time "stamps" or tags.***

Because of partnerships between many news websites and news providers, readers often have access to subscription news stories without having to register (e.g., *The New York Times*) or give personal information at a news website. Occasionally, you may even discover you can access an article requiring payment because of a partnership arrangement. For example, this article is only available by paid subscription at its website, but by using Google News to find it, a reader can access the full article for free and without registering.

The **Patent** Office: Getting Wiki With It
Law.com (subscription), CA - Jan 12, 2007
In August, when the **Patent** and Trademark Office acknowledged that it had taken **Wikipedia** off its list of acceptable research sources, the surprise was not ...

[136]Elise Ackerman, "New media making deals with 'old' news providers," *Mercury News*, 30 July 2006, <http://www.mercurynews.com/mld/mercurynews/news/15157800.htm> (14 November 2006).

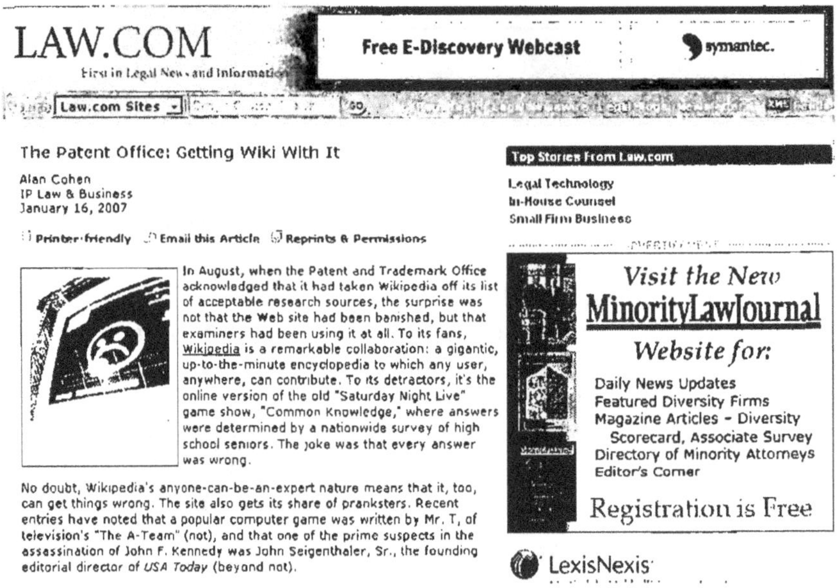

LAW.COM
First in Legal News and Information

Free E-Discovery Webcast — symantec.

Law.com Sites

The Patent Office: Getting Wiki With It

Alan Cohen
IP Law & Business
January 16, 2007

Printer-friendly | Email this Article | Reprints & Permissions

In August, when the Patent and Trademark Office acknowledged that it had taken Wikipedia off its list of acceptable research sources, the surprise was not that the Web site had been banished, but that examiners had been using it at all. To its fans, Wikipedia is a remarkable collaboration: a gigantic, up-to-the-minute encyclopedia to which any user, anywhere, can contribute. To its detractors, it's the online version of the old "Saturday Night Live" game show, "Common Knowledge," where answers were determined by a nationwide survey of high school seniors. The joke was that every answer was wrong.

No doubt, Wikipedia's anyone-can-be-an-expert nature means that it, too, can get things wrong. The site also gets its share of pranksters. Recent entries have noted that a popular computer game was written by Mr. T, of television's "The A-Team" (not), and that one of the prime suspects in the assassination of John F. Kennedy was John Seigenthaler, Sr., the founding editorial director of *USA Today* (beyond not).

Top Stories From Law.com

Legal Technology
In-House Counsel
Small Firm Business

Visit the New
MinorityLawJournal
Website for:
Daily News Updates
Featured Diversity Firms
Magazine Articles - Diversity Scorecard, Associate Survey
Directory of Minority Attorneys
Editor's Corner
Registration is Free

LexisNexis

However, finding and understanding how to use these services can be confusing because most sites that permit unimpeded access to subscription news do not do a good job of advertising how to do it.

Some of the major news sources that require registration or personal information to access some or all of their news, but which are accessible via news search sites include:

The New York Times
The Washington Post
The LA Times
BusinessWeek
Forbes
The Guardian (UK)
The Times of London

I have included some ways to access subscription news without providing personal information in the discussion of news search engines below.

Google News http://news.google.com/

Google News headlines are entirely generated using a computer algorithm that scours over 10,000 worldwide news sources.[137] Google News also offers **international editions** for France, Germany, India, Italy, Spain, and several other countries.

Google News lets users:

- Sort results by ***date*** or ***relevance*** .
- Search on a ***title*** or headline using the syntax intitle: [intitle:iraq].
- Use both the *intitle:* and *inurl:* or *source:* commands together to find a ***specific topic in a specific publication***.
- Limit a search to a ***specific U.S. state*** or a ***country*** [location:ny] or [location:germany].
- Limit a search to a ***specific news site*** using the syntax *inurl:* or *source:* [inurl:washingtonpost.com].

Google also permits users to limit their search by news source in both simple and advanced news search. The ***source:*** command must be used with a keyword. The syntax in simple news search is ***source:*** and the query is:

[source:news_source_name keyword]

For example: [source:new_york_times iraq]

The ***source:*** syntax can be tricky to use. Here's a partial list I've come up with which works (as of now) in simple search:

source:bbc_news
source:international_herald_tribune
source:united_press_international
source:guardian
source:christian_science_monitor
source:cbs_news
source:abc_news

---

[137] Philipp Lenssen's research during 2006 indicated at least 10.584 unique sources for Google News. "Which sources does Google News index?" Google Blogoscoped, <http://blog.outer-court.com/googlenews/> (14 November 2006).

source:washington_post
source:reuters
source:cnn
source:msnbc
source:forbes

Another option is to use the ***Advanced Google News Search*** that includes these options:

- Find results—all words, exact phrase, at least one word, without the word(s).
- Sort by relevance or date.
- Location by country or state.
- Occurrences of terms—anywhere or in headline, body, or url of article.
- Date—anytime, last hour, last day, past week, past month, or range of dates within past month.
- Source by simple name, e.g., CNN, New York Times.

The *source* option in advanced search is an improvement over having to use the ***source:*** syntax that required such entries as [source:united_press_international]. However, there are still two problems with the ***source:*** option. Google doesn't list its sources anywhere that I can find (it merely says there are about 4500 of them), and you still have to write out the full name of the source, so UPI or Fox will not work but United Press International or Fox News will. Google News now offers a Google Blogsearch link from the News homepage.

Google News Advanced Search http://news.google.com/advanced_news_search

Google News also offers **customization features** that let you add or remove news categories, decrease or increase the number of headlines you see, select from numerous regional editions of Google News from around the world, and more. For example, you could customize your Google News page to show only the World, Sci/Tech, Business, and U.S. pages, then add standard sections for German Business and another for France in general. Google News also permits adding a custom section in which you enter keywords (all must be in the story for it to appear in your section, so choose carefully). You can even use the "Advanced" option under custom section to limit your keyword search to a specific topic, e.g., "internet security" only in the Sci/Tech section. You can change the language of the stories in your custom section by clicking on "Advanced" and selecting a language in the pulldown menu. While these new customization options may sound a little complicated, they are fairly intuitive, and Google News is very forgiving. If you make a mess of your customization, there is an option to "Reset page to default" link to return your page to its default settings (i.e., the standard version without any customizations). This will, however, obliterate any customization you ever made, so you might want to delete unwanted sections one at a time instead.

Google News now permits users to get any Google News page, section, search, or customized page via RSS or Atom feeds. Google News provides detailed information on how to set up and access these feeds.

Google News Feeds http://news.google.com/intl/en_us/news_feed_terms.html

Google News keeps news stories in its database for about 30 days. This might be a bit misleading since much of the freely available news on the Internet is only available for less than 30 days (usually somewhere between 1 and 2 weeks). Unlike the Google search engine, Google News does not offer cached (stored) copies of these news stories. However, some news articles that cannot be accessed via the original site's archives because they require registration were indeed accessible using Google. So if you cannot get to a news article that is less than 30 days old via the original source, try searching for it in Google News and linking directly to the page.

Google News Archive http://news.google.com/archivesearch

Google News unveiled a new service that, from the looks of it, will be very popular and useful—and potentially highly addictive. It's Google News Archive Search, which links from the Google News homepage. Here are the results for the search [ronald reagan]. On the right are the results listed by relevance, though they are a bit odd, considering that the third result is from CNN but dated 1952 (the article actually dates from Reagan's death in 2004 and is about his life in the 1950's). On the left you can limit the search by date range or by publication.

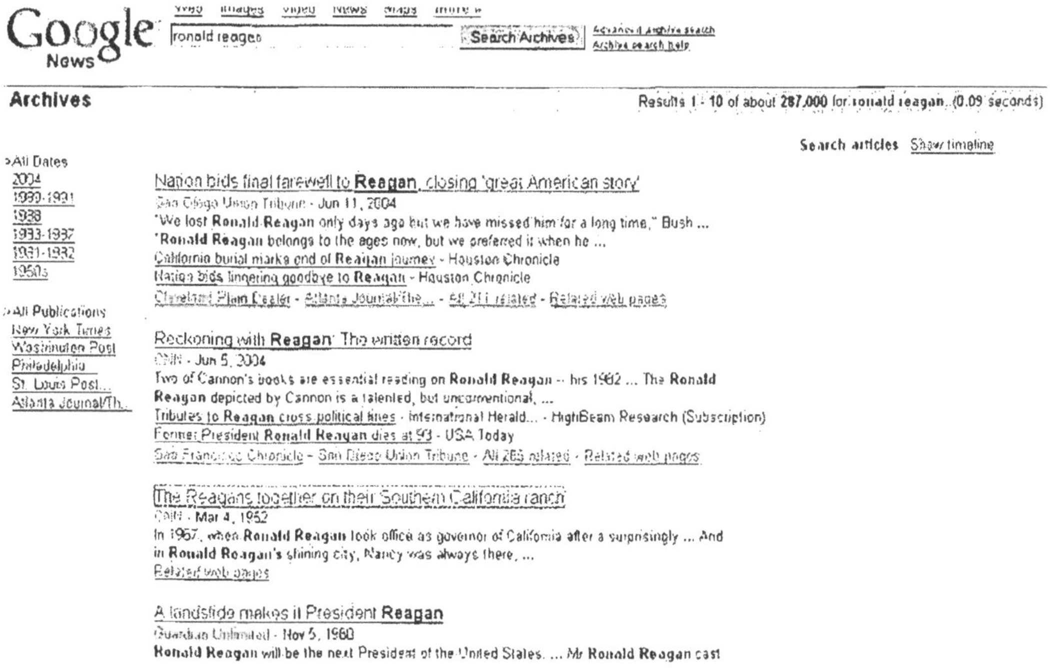
Google News
ronald reagan
Search Archives
Archives
Results 1 - 10 of about 287,000 for ronald reagan. (0.09 seconds)
Search articles Show timeline
All Dates
2004
1989-1991
1988
1983-1987
1981-1982
1950s
All Publications
New York Times
Washington Post
Philadelphia
St. Louis Post...
Atlanta Journal/Th...
Nation bids final farewell to Reagan, closing 'great American story'
Reckoning with Reagan: The written record
The Reagans together on their Southern California ranch
A landslide makes it President Reagan

This is where the Google News Archive search is irresistible to researchers (not to mention to history and news buffs). Take a look at the results from the 1950s:

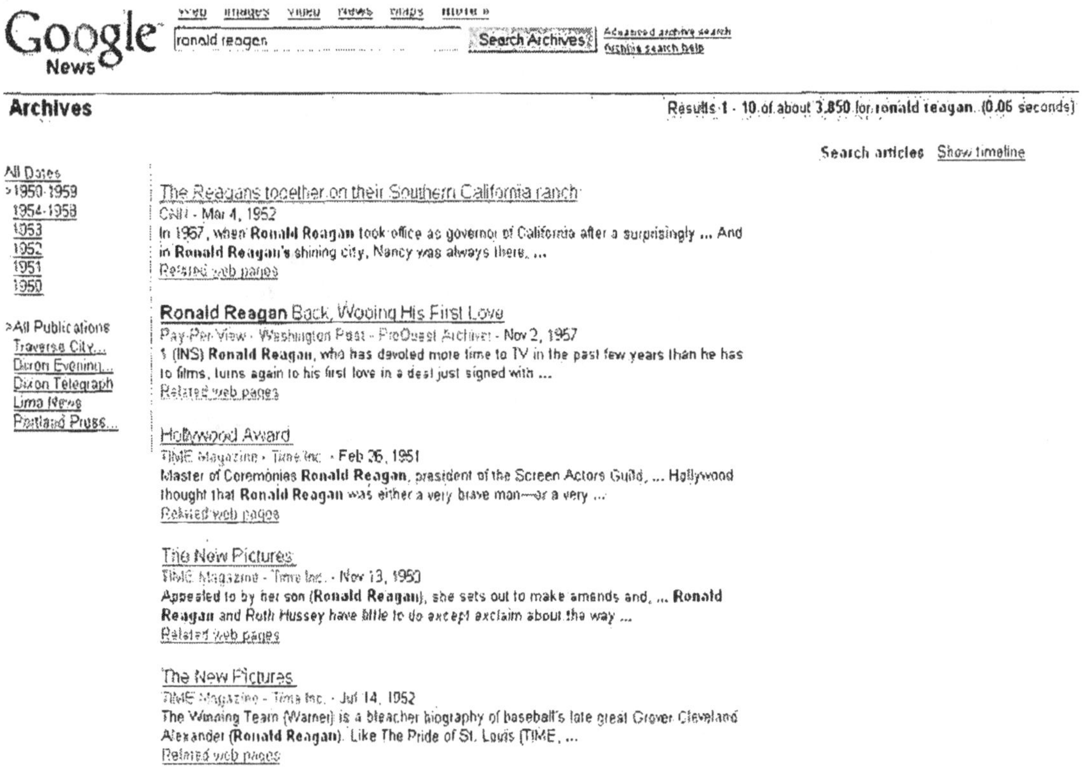
Google News
ronald reagan
Search Archives
Archives
Results 1 - 10 of about 3,850 for ronald reagan. (0.06 seconds)
Search articles Show timeline
All Dates
»1950-1959
1954-1958
1953
1952
1951
1950
»All Publications
Traverse City...
Dixon Evening...
Dixon Telegraph
Lima News
Portland Press...
The Reagans together on their Southern California ranch
CNN - Mar 4, 1952
In 1967, when Ronald Reagan took office as governor of California after a surprisingly ... And in Ronald Reagan's shining city, Nancy was always there. ...
Related web pages
Ronald Reagan Back, Wooing His First Love
Pay-Per-View - Washington Post - ProQuest Archiver - Nov 2, 1957
1 (INS) Ronald Reagan, who has devoted more time to TV in the past few years than he has to films, turns again to his first love in a deal just signed with ...
Related web pages
Hollywood Award
TIME Magazine - Time Inc. - Feb 26, 1951
Master of Ceremonies Ronald Reagan, president of the Screen Actors Guild, ... Hollywood thought that Ronald Reagan was either a very brave man—or a very ...
Related web pages
The New Pictures
TIME Magazine - Time Inc. - Nov 13, 1950
Appealed to by her son (Ronald Reagan), she sets out to make amends and, ... Ronald Reagan and Ruth Hussey have little to do except exclaim about the way ...
Related web pages
The New Pictures
TIME Magazine - Time Inc. - Jul 14, 1952
The Winning Team (Warner) is a bleacher biography of baseball's late great Grover Cleveland Alexander (Ronald Reagan). Like The Pride of St. Louis (TIME, ...
Related web pages

Ignore the first result, which we know isn't really from 1952. Most of the oldest news appears to come either from Time (if it's free) or from the Newspaper Archive. The Newspaper Archive is "the single largest historical newspaper database online, containing more newspaper pages from 1759 to present than any other service." It is subscription based, but free to public libraries and K-12 schools, so if you need a copy of the full sized newspaper image, you can get it at your public library.

The good news is that Google News Archive searches the Newspaper Archive and returns the results in a window. For example:

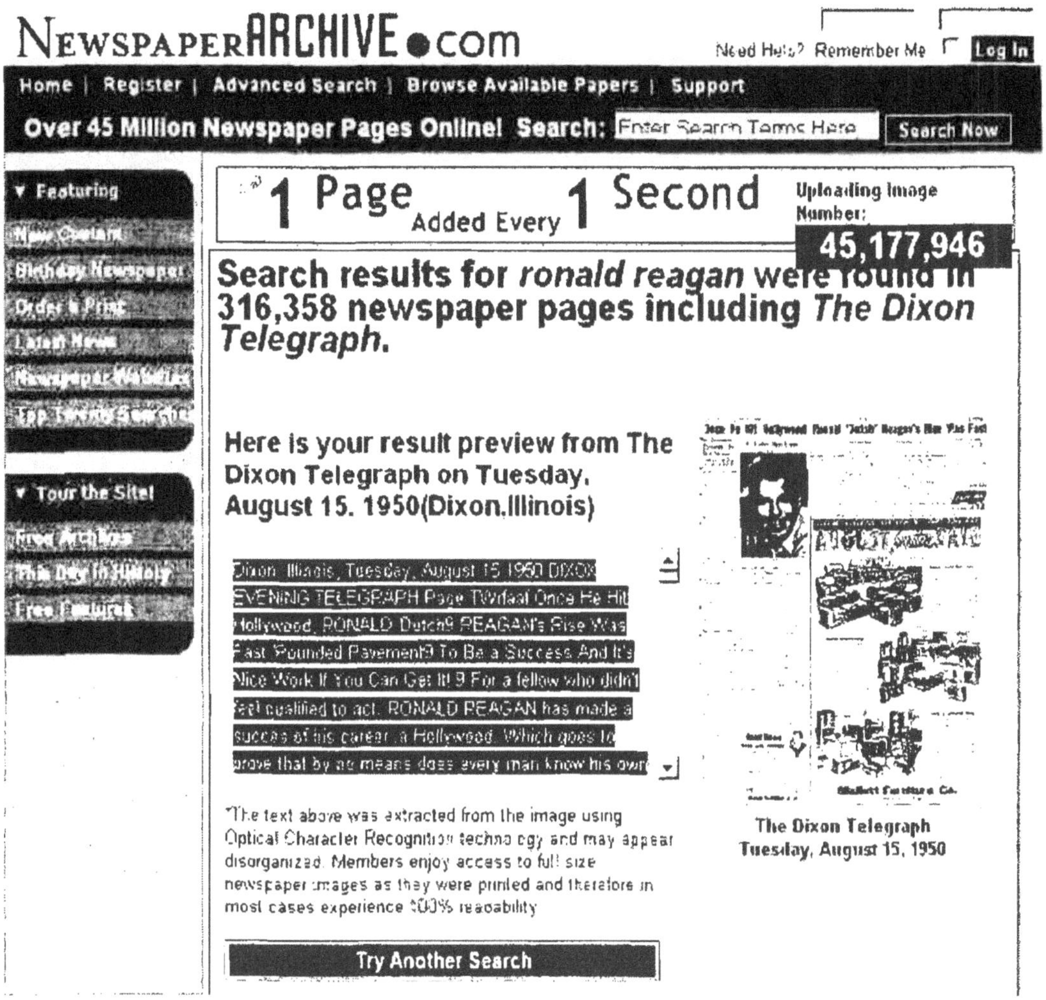
NewspaperARCHIVE.com
Need Help? Remember Me Log In
Home | Register | Advanced Search | Browse Available Papers | Support
Over 45 Million Newspaper Pages Online! Search: Enter Search Terms Here Search Now
Featuring
1 Page Added Every 1 Second
Uploading Image Number: 45,177,946
Search results for ronald reagan were found in 316,358 newspaper pages including The Dixon Telegraph.
Here is your result preview from The Dixon Telegraph on Tuesday, August 15. 1950(Dixon,Illinois)
Tour the Site!
The text above was extracted from the image using Optical Character Recognition technology and may appear disorganized. Members enjoy access to full size newspaper images as they were printed and therefore in most cases experience 100% readability
The Dixon Telegraph Tuesday, August 15, 1950
Try Another Search

Although you cannot get the full size image using Google New Archive search, the complete article is here. You can scroll down, read, and even copy the entire newspaper article from the August 15, 1950, Dixon Telegraph using Google News Archive search, whereas even viewing results at the Newspaper Archive is unavailable to users unless they are paid members of the Newspaper Archive. You can limit your Google News Archive searches to the Newspaper Archive by adding [source:"newspaperarchive"]. Many of the results in the Google News Archive search come from subscription services, and these are clearly marked as such. The advanced search contains an option to "Return articles with the following price," including articles with "no price," meaning "free," not "priceless."

You can also limit your search to a specific news source, e.g., [source:"daily herald"], and Google News Archive will show the possible publications if more than one fits that query:

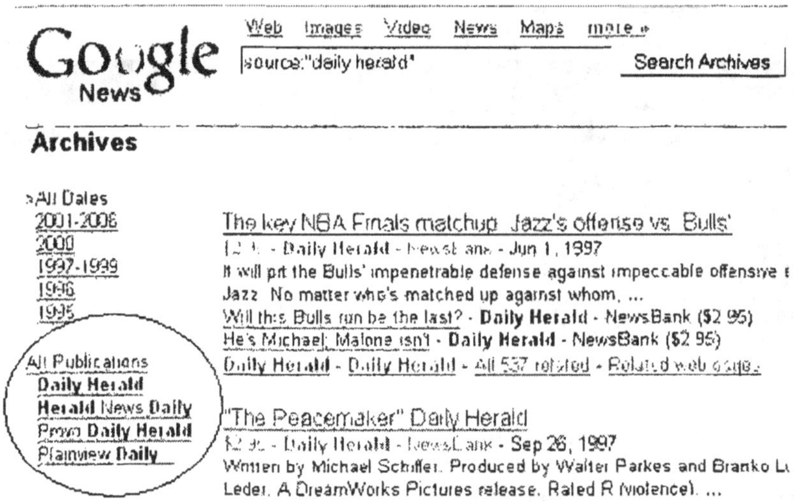
Web Images Video News Maps more »
Google News
source:"daily herald"
Search Archives
Archives
All Dates
2001-2006
2000
1997-1999
1996
1995
All Publications
Daily Herald
Herald News Daily
Provo Daily Herald
Plainview Daily
The key NBA Finals matchup Jazz's offense vs. Bulls'
Daily Herald - Jun 1, 1997
It will pit the Bulls' impenetrable defense against impeccable offensive
Jazz. No matter who's matched up against whom, ...
Will this Bulls run be the last? - Daily Herald - NewsBank ($2.95)
He's Michael; Malone isn't - Daily Herald - NewsBank ($2.95)
Daily Herald - Daily Herald - All 537 related - Related web pages
"The Peacemaker" Daily Herald
Daily Herald - Sep 26, 1997
Written by Michael Schiffer. Produced by Walter Parkes and Branko Lu
Leder. A DreamWorks Pictures release. Rated R (violence). ...

You may have noticed the option to "view timeline" above. When you select this, Google News Archive puts all the search results into a date ordered timeline that may make complex historical events and news stories easier to research:

Web Images Video News Maps more »
Google News
ronald reagan
Search Archives
Advanced archive search
Archive search help
Archives > Timeline: All dates > Before 1981
Timeline for ronald reagan (0.09 seconds)
Search articles Show timeline
1966-67» TIME Magazine Archive Article -- Oct. 7, 1966
Oct 7, 1966
Marion Pike's impression of Ronald Reagan may strike some people as a bit youthful
looking — like that fellow in the 1940s movies ...
Related web pages
1968-73» Poor Ronald Reagan
May 20, 1973
Ronald Reagan's personal financial situation. As everyone knows by now, Gov. Reagan had
such bad financial reverses last year that he didn't have to pay ...
Related web pages
Ronald Reagan, Laying the Groundwork for '76
Jun 22, 1973
Ronald Reagan has publicly downgraded the Watergate scandal and unreservedly defended
President Nixon, his top advisers are privately urging cool detachment ...
Related web pages
1974-75» Reagan 13, Ford 0
Apr 20, 1975
"Under today's conditions, Ronald Reagan would sweep delegates from the 13 southern and
border states against Gerald R. Ford." ...
Unconventional Wisdom TIME Magazine - Time Inc
Related web pages
1976» TIME Magazine Archive Article -- How Ford Won and Reagan Lost ...
Mar 8, 1976
At Ronald Reagan's headquarters in Concord's dingy New Hampshire Highway Hotel,
confident aides had chilled several bottles of Almadén blanc de blancs ...
TIME Magazine Archive Article -- Never underestima - TIME Magazine - Time Inc
Related web pages
The Rousing Numbers Game

The oldest newspaper articles I saw are from the Public Register or Freeman's Journal (dating back to 1763), Edinburgh Advertiser (to 1772), the Times of London (to 1788), and the Daily Universal Register of London (to 1786). You can read articles in British newspapers about the upstart American colonials and follow the war news up to a report on November 30, 1781, in the Edinburgh Advertiser that news "arrived off Cape Charles [Virginia] on the 24th [of October], when we had the mortification to hear that Lord Cornwallis had proposed terms of capitulation to the enemy on the 19th. This intelligence was brought us by the pilot of the Charon, and some other persons Who came off from the shore." If that doesn't make you salivate, nothing will.

Yahoo News http://news.yahoo.com/

Yahoo Advanced News Search http://news.search.yahoo.com/news/advanced

Yahoo News is one of the most impressive news sites on the Internet. In fact, it is my favorite news search site. Yahoo News crawls over **7,000** news sources in **35** languages. In each topical section, Yahoo News offers the option of viewing all articles from a specific source and access to the full-text articles. The news sources include major news sources that require registration or personal information to access some or all of their news, but which are accessible via Yahoo News, including but not limited to:

The New York Times
The Washington Post
The Los Angeles Times
Baltimore Sun
Chicago Tribune
Agence France-Presse (AFP)
The Guardian (UK)
Canadian Press (CP)
BusinessWeek
Forbes
Motley Fool
The Deal

News from the wire services, Reuters and the Associated Press, is stored for two weeks. News from other sources is usually stored for one month, although for some providers it could be as little as seven days.

Yahoo News search now includes results from blogs as well as from traditional news sources. When you search using Yahoo News you get results from blogs, the Flickr photograph sharing site, and links from My Web 2.0 Beta, a Yahoo "social search" property that encourages grassroots "reporting." Traditional news source results appear on the left side of the screen and the new blog results on the right.

YAHOO! NEWS
Web | Images | Video | Directory | Local | News | Shopping
iraq
Search
Subscriptions Advanced Search Preferences
News Results
Got feedback? | Suggest a Site
Also try: iraq constitution, iraq election, iraq war More...
Full Coverage: • World > Iraq Archive • World > Iraq • World > Antiwar Movement
Sort Results by: Relevance | Date
BLOGS BETA
1. Rice defends policy on Iraq
Reuters via Yahoo! News - 42 minutes ago
Secretary of State Condoleezza Rice sought to reassure critical lawmakers on Wednesday that the Bush administration had a strategy in Iraq that would "assure victory."
Save to My Web
2. Rice won't say if US troops will quit Iraq in 10 years
AFP via Yahoo! News - 29 minutes ago
US Secretary of State Condoleezza Rice refused to speculate on whether US troops would be out of Iraq in five or even 10 years.
Save to My Web
3. Rice Wants to Follow Afghan Model in Iraq
AP via Yahoo! News - 2 hours, 14 minutes ago
Secretary of State Condoleezza Rice on Wednesday outlined a strategy for helping Iraqis clear out insurgents and build durable, national institutions as she sought to reassure jittery members of Congress about the path to peace in Iraq.
Save to My Web
10/19/2005: Soldier Killed: Iraq Operati...
American Forces Information Service - News Articles
1 hour, 18 minutes ago
Washington, Oct 19, 2005 - A U.S. soldie...
Quarter of Iraq Vets Have Health Problem:...
Outside The Beltway
1 hour, 34 minutes ago
A new survey reveals that more than a qu...
Add to My Yahoo! - Save
9:04 am: Rice outlines strategy of "clea...
Free New Mexican
2 hours, 47 minutes ago
WASHINGTON - Secretary of State Condolee...

Yahoo Advanced News Search offers many options, including **date**, **location**, **language**, and **source** limits. To limit your search to a specific source using Advanced News Search, scroll down to **Source** and enter the source, e.g., [Los Angeles Times]. Users can also search by source" or site in the simple search screen as follows:

- Limit a search to a **specific news source** using the syntax url: inurl:, or source:
  [url:nytimes.com] or [inurl:nytimes.com] or [source:new_york_times]

Yahoo News also lets users:

- Sort results by date or relevance.
- Search on headline: or title word to see exact title (headline) matches [headline:iraq].

- Search on headline: before each of multiple keywords to search for titles that must contain *all* selected words.
  [headline:mideast headline:peace]
- Use parentheses to search for headlines that contain at least one of your selected words.
  [(headline:mideast headline:peace)]
- Enter phrases to search for titles that contain a specific terms.
  [headline:"mideast peace"]
- Limit a search to a specific U.S. state or a country: [location:ny] or [location:germany]

These queries can be used in combination as well as alone or with keywords.

Pandia Newsfinder http://www.pandia.com/news/

Pandia's newsfeed is powered by **Moreover**, which means the number of sources queried is huge and includes the usual—*The New York Times*, AP, Reuters, BBC, the *Washington Post*, ABC, MSNBC—as well as the uncommon—*Financial Times*, Stratfor, Wikinews, and more. Use the *Pandia News Extended Search* on the pull-down menu for a greater number of news sources. Pandia Newsfinder is like everything else at this excellent Internet search site: easy to use, well organized, and high quality.

In addition to news search engines, there are some excellent sites that offer one interface to many news search sources. These are some of the best.

HavenWorks http://havenworks.com/news/search/

Don't give up on this resource just because HavenWorks has one of the ugliest, most cluttered webpages you'll ever see. HavenWorks lets you use some of the best news search engines from one page: AltaVista and AlltheWeb News (both use Yahoo News sources), Google News, Daypop (news and blog search), and Rocketinfo (only five days worth). Unfortunately, Moreover no longer permits individual users to search from its home page, so the HavenWorks' Moreover search is no longer operational. Havenworks also has some excellent news search tips and specialized news searches if you can find your way through the maze of clashing colors and general untidiness.

JournalismNet http://www.journalismnet.com/

As its name implies, JournalismNet is designed primarily as a tool for journalists.

Journalism Net's international newsfeed is powered by Moreover, but it's not designed for searching. Jnet provides links by country and region—North America, Europe, and Africa—to the respective newspapers, news archives, television, radio, etc. There are also links to search tools, people finders, fact finders, newspapers, magazines, TV, radio, etc. The site is very heavily oriented towards tutorials—lots of "how to" pages, which are very useful.

NewsNow http://www.newsnow.co.uk/

NewsNow offers a newsfeed that automatically updates itself every five minutes from news sources around the world. NewsNow is a UK resource that provides constant news monitoring of nearly **22000** worldwide sources on a large variety of topics. From the home page users can select a specific newsfeed from top-level subjects that include hot topics, business & finance, industry sectors, IT, Internet, current affairs, and regions. The newsfeeds are listed by "new in the last 5 minutes," "10 to 15 minutes old," "15 to 30 minutes old," and earlier articles. Although earlier articles from the same or previous days are available, **NewsNow is primarily a current news source**, not an archive. NewsNow also offers a free simple search that is quite good.

All sources are clearly indicated both by the name of the media outlet and by a national flag of the source's home country. Almost all the articles are in English, though I have seen a few non-English language articles on occasion. NewsNow is a great resource and is one of the best "breaking news" sources because of the automatic updates, the number and quality of news sources, the fact it does not require registration/payment, and because it has no ads. NewsNow will not display properly in older versions of Netscape.

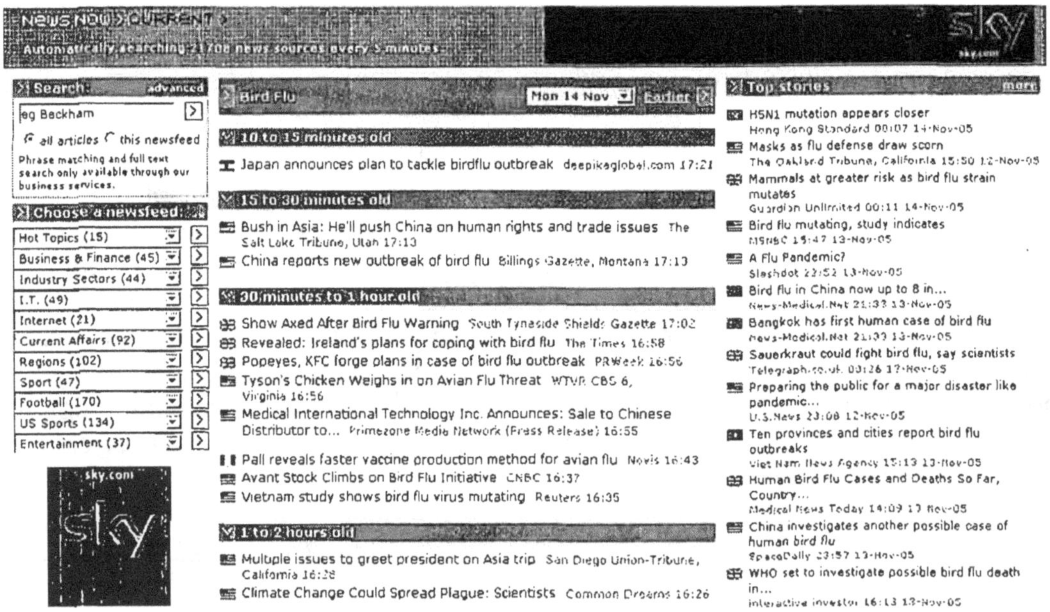
Search: advanced
eg Beckham
all articles this newsfeed
Phrase matching and full text search only available through our business services.
Choose a newsfeed:
Hot Topics (15)
Business & Finance (45)
Industry Sectors (44)
I.T. (49)
Internet (21)
Current Affairs (92)
Regions (102)
Sport (47)
Football (170)
US Sports (134)
Entertainment (37)
sky.com
Bird Flu
Mon 14 Nov
Earlier
10 to 15 minutes old
Japan announces plan to tackle birdflu outbreak deepikaglobal.com 17:21
15 to 30 minutes old
Bush in Asia: He'll push China on human rights and trade issues The Salt Lake Tribune, Utah 17:13
China reports new outbreak of bird flu Billings Gazette, Montana 17:13
30 minutes to 1 hour old
Show Axed After Bird Flu Warning South Tyneside Shields Gazette 17:02
Revealed: Ireland's plans for coping with bird flu The Times 16:58
Popeyes, KFC forge plans in case of bird flu outbreak PRWeek 16:56
Tyson's Chicken Weighs in on Avian Flu Threat WTVR CBS 6, Virginia 16:56
Medical International Technology Inc. Announces: Sale to Chinese Distributor to... Primezone Media Network (Press Release) 16:55
Pall reveals faster vaccine production method for avian flu Novis 16:43
Avant Stock Climbs on Bird Flu Initiative CNBC 16:37
Vietnam study shows bird flu virus mutating Reuters 16:35
1 to 2 hours old
Multiple issues to greet president on Asia trip San Diego Union-Tribune, California 16:28
Climate Change Could Spread Plague: Scientists Common Dreams 16:26
Top stories
more
H5N1 mutation appears closer
Hong Kong Standard 00:07 14-Nov-05
Masks as flu defense draw scorn
The Oakland Tribune, California 15:50 12-Nov-05
Mammals at greater risk as bird flu strain mutates
Guardian Unlimited 00:11 14-Nov-05
Bird flu mutating, study indicates
MSNBC 15:47 13-Nov-05
A Flu Pandemic?
Slashdot 22:52 13-Nov-05
Bird flu in China now up to 8 in...
News-Medical.Net 21:33 13-Nov-05
Bangkok has first human case of bird flu
News-Medical.Net 21:33 13-Nov-05
Sauerkraut could fight bird flu, say scientists
Telegraph.co.uk 03:26 13-Nov-05
Preparing the public for a major disaster like pandemic...
U.S.News 23:08 12-Nov-05
Ten provinces and cities report bird flu outbreaks
Viet Nam News Agency 15:13 13-Nov-05
Human Bird Flu Cases and Deaths So Far, Country...
Medical News Today 14:09 13-Nov-05
China investigates another possible case of human bird flu
SpaceDaily 23:57 13-Nov-05
WHO set to investigate possible bird flu death in...
interactive investor 16:13 13-Nov-05

NewsNow's Bird Flu News Page

Topix.net http://www.topix.net/

Topix gained a lot of attention and new users during the past year. Here are some of its features:

- It indexes more than 10,000 news sources (mostly US, but not all).
- It has more than 360,000 topically based pages (including world news pages on virtually every country).
- It offers a "live" newsfeed of the news stories Topix is indexing at the moment.
- It has a large searchable database.
- It offers perhaps the best local US news aggregation and search available (I've not seen better).

Topix tweaked its news ranking algorithm during 2006, producing much better results for top stories than in the past. Topix also added two new features during 2006 that have elevated it at least in one regard. You are supposed to be able to search using capitalization, which has sadly become a rare search feature. The only problem is that Topix's new feature is not working properly or consistently. I tried a number of variations and the results were inconsistent. For example, a search on [ICon] contrasted with [icon] or [Icon] worked, but searches on [ZIP] [zip] [Zip] were unpredictable. This feature needs more work.

However, the **archive function** works just fine. Here is a search for Hurricane Katrina, and you can clearly see by the blue graph the peaks and valleys of news stories on Katrina. In order to see the news from earlier times, users simply need to click anywhere along the timeline:

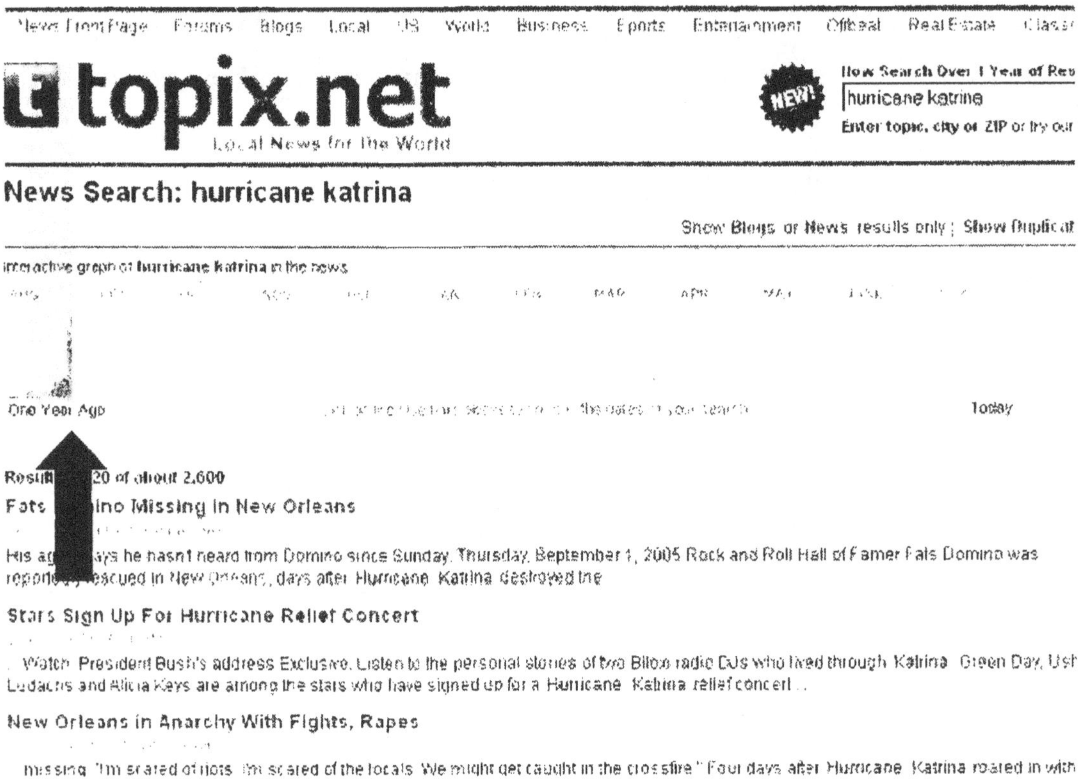
topix.net
Local News for the World
hurricane katrina
News Search: hurricane katrina
Show Blogs or News results only | Show Duplicat
One Year Ago
Today
Fats ino Missing In New Orleans
His ag ays he hasn't heard from Domino since Sunday. Thursday, September 1, 2005 Rock and Roll Hall of Famer Fats Domino was
Stars Sign Up For Hurricane Relief Concert
Watch: President Bush's address Exclusive: Listen to the personal stories of two Biloxi radio DJs who lived through Katrina. Green Day, Usher, Ludacris and Alicia Keys are among the stars who have signed up for a Hurricane Katrina relief concert ...
New Orleans in Anarchy With Fights, Rapes
missing. "I'm scared of riots. I'm scared of the locals. We might get caught in the crossfire." Four days after Hurricane Katrina roared in with

The archive feature is very significant because finding and accessing news older than the past 30 days is usually both hard and/or costly (until Google News Archive Search showed up). Most news sources require at least registration and often payment to get their archived news. Topix will find news from up to one year ago, but what you can actually read for free and without registration is hit or miss. For example, I had very good luck with *Washington Post* stories, but not so much with the *New York Times*. CNN was perfect: I was able to find and read all the CNN stories I tried from last year, something you cannot do at CNN itself. However, I found that the "Restrict to Source" [cnn] option did not work well whereas the "Restrict to Url" [cnn.com] option worked perfectly. Also, I recommend selecting "No Blogs" if you only want to search news sources.

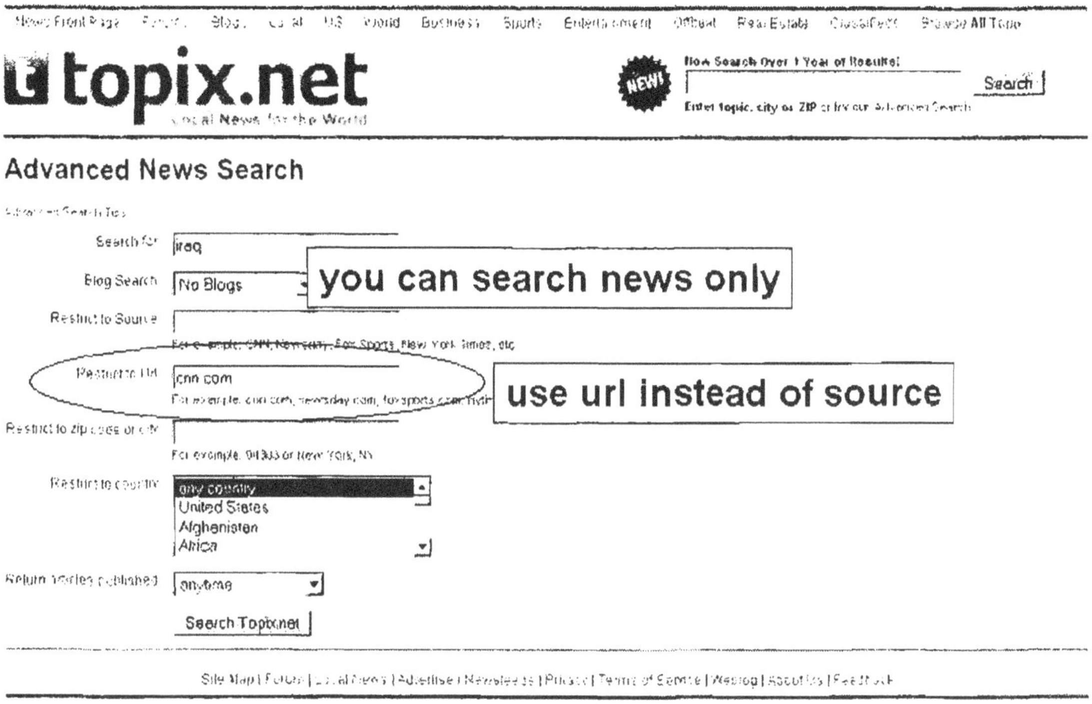
topix.net
Local News for the World
Search
Advanced News Search
Search for
iraq
Blog Search
No Blogs
you can search news only
Restrict to Source
Restrict to Url
cnn com
use url instead of source
United States
Afghanistan
Africa
anytime
Search Topix.net

Despite the fact that the capitalization search is not (yet) working properly, the fact that the archive search is working makes these Topix upgrades important and valuable to researchers.

In short, Topix is a good entry into the news aggregation, archive search, and email alerts arena. For those who can use email news alerts, Topix makes it very easy. Just click the "email alert" link on any topical page to get related results delivered either daily or weekly. In case you're interested, Google News has an "as it happens" option for news delivery <http://www.google.com/alerts>, something Topic needs to add.

Daypop http://www.daypop.com/

Don't forget that Daypop is also a news search site and draws from a large number of international news sources. Daypop's Advanced Search lets you limit your search by time (from three hours to two weeks), by language (all or one of a dozen), and by country. Daypop is one of the few news search sites that caches news stories, so you have the option of viewing the stored copy.

Worldnews http://www.wn.com/

Worldnews is one of the best international news sources, offering the option of searching in English or one of 20+ other languages. Worldnews also presents separate home pages for many world regions, e.g., "Arab World News" as well as specialty news searches by topics such as business, politics, and science.

Technology News Sources

There are many fine telecommunication news sources on the web, so I will limit my discussion to a few of the best. ***Users can now subscribe to the all of the following publications' technology feeds and blogs*** without registering at the various websites. Look for the RSS/Atom feed icon at the news websites.

TechNews.com http://www.washingtonpost.com/wp-dyn/technology/

The *Washington Post* absorbed Newsbytes to create TechNews, which offers current headlines, technology news by topic (biotech, telecom, software, Internet, etc.), and breaking technology news off the wire from Reuters. ***The site now requires free registration, but both Yahoo News and Google News let you read full-text TechNews articles without registering.*** For both Yahoo News and Google News, the search syntax is:

[source:washington_post keyword]

Because of the way the *Post* names its links, I have not been able to come up with a unique keyword to limit the query to TechNews articles only. The best way to find them is to use the author's name or a word from the title.

TechWeb http://www.techweb.com/

TechWeb is part of the CMP Technology Network, a huge enterprise that publishes InformationWeek, *EE Times*, and *Windows Developer* magazines. TechWeb does not do original reporting; instead, it compiles the latest technology news from around the world and updates the website continuously. TechWeb is one of the single best sources of technology news on the web; it and all other parts of the CMP network are fully searchable.

Wired News http://www.wired.com/

Wired News, part of the Terra Lycos family, has always been known as a slightly "edgy" site, and there is no doubt their original tech news features are among the most popular and sometimes controversial on the web. *Wired* doesn't only cover technology but brings in all aspects of our "wired" world: culture, business, and politics as well as technology. Wired also receives breaking news from AP and Reuters.

New York Times Technology News
http://www.nytimes.com/pages/technology/index.html

Consistently a leader in technology news, *The New York Times*' feature writers include John Markoff and David Pogue. The Times also publishes the best tech stories from the wire services. I consider the Times technology section a "must read" to stay abreast of significant developments and to find in-depth articles on topics others deal with superficially.

The Register http://theregister.co.uk/

If you like irascible journalism, check out The Register from the UK. Their slogan–"biting the hand that feeds IT"–says a great deal about the tone of the site. However, *The Register* is known for gutsy original reporting and is unafraid to take on controversial topics in a straightforward way. They are often cited and quoted by other sources. One of my favorites.

ZDNet News http://news.zdnet.com/

One of *Wired's* major competitors is ZDNet News, part of the vast **CNET** network. In addition to original reporting and features, ZDNet News offers news by section: wired and wireless, security, IT management web technology, personal tech, etc.

Newsfactor Network http://www.newsfactor.com/

The Newsfactor Network does both original reporting and compilations of technology news from many sources. Newsfactor has many special sections, such as a superb Enterprise Security section, data management, tech trends, Internet life, among others.

Telecommunications on the Web

PTTs and Private Telecommunication Providers

Researching Public Telephone and Telegraph organizations (PTTs) and major private telecommunication providers, both national and international, is at the heart of much telecommunications research, and most providers, as well as the growing field of private telecom companies, are coming on line with websites. The hard part can be finding them. Fortunately, several fine websites have compiled lists of links to both public and private telecommunication companies.

Bandwidth Market Telecom Links
http://www.bandwidthmarket.com/component/option,com_weblinks/Itemid,4/

BWM's links pages offer an exceptionally fine collection of telecommunications-related information. The site includes a hyperlinked list of **telecom operators** alphabetically by country. This page makes it extremely easy to see which telecom companies operate in which country and go directly to the company pages. The site also has an impressive hyperlinked list of over 200 **telecom manufacturers** from around the world listed alphabetically. These are the categories and number of links at BWM's site:

- Analysts and Consultants (57)
- Associations and Organizations (32)
- General (6)
- Government Sites (36)
- Internet Exchanges (11)
- Internet Telephony (15)
- Maps (165)
- Satellite Telecom (26)
- Telecom Manufacturers (203)
- Telecom Operators (164)
- Telecom Publications (153)
- Telecom Standards (16)

ITU's Global Directory of Regulators (select ***Regulators*** for PTTs)
http://www.itu.int/cgi-bin/htsh/mm/scripts/mm.search

The ITU's Global Directory may be the most authoritative source of PTT's available. The site lists all members and can be searched by country, company, or organization. The site may load slowly.

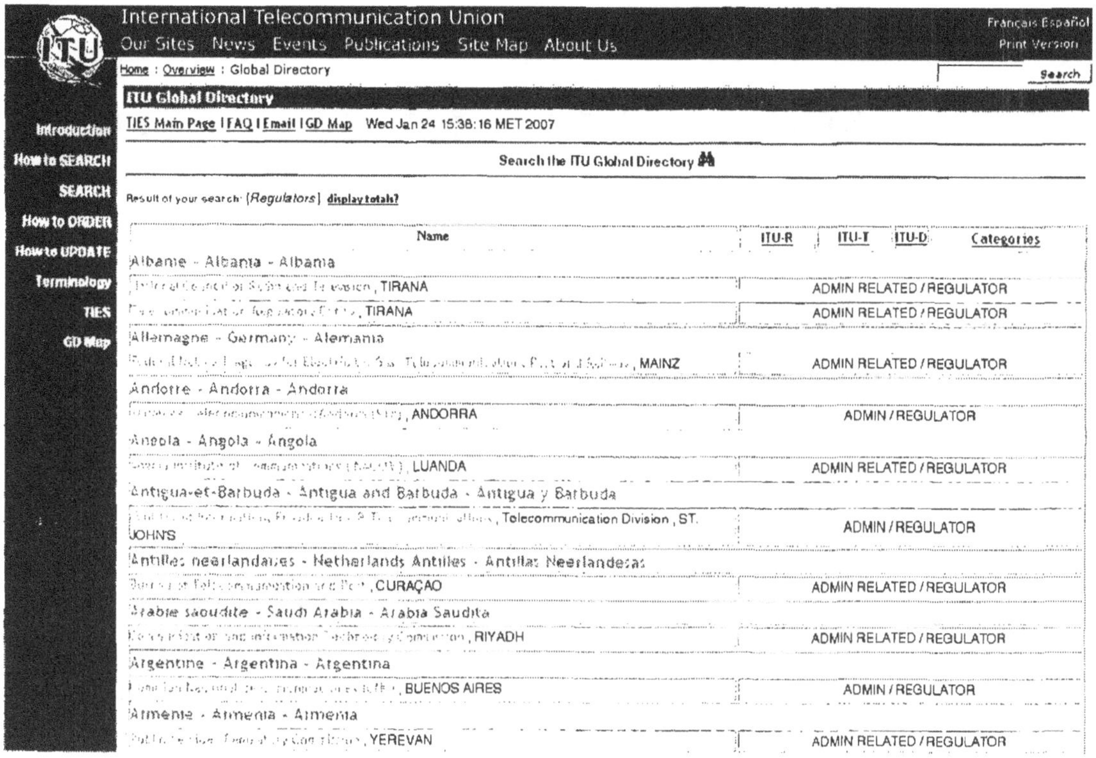
International Telecommunication Union
Our Sites News Events Publications Site Map About Us
Français Español
Print Version
Home : Overview : Global Directory
Search
ITU Global Directory
Introduction
How to SEARCH
SEARCH
How to ORDER
How to UPDATE
Terminology
TIES
GD Map
TIES Main Page | FAQ | Email | GD Map Wed Jan 24 15:38:16 MET 2007
Search the ITU Global Directory
Result of your search: [Regulators] display totals?
Name
ITU-R
ITU-T
ITU-D
Categories
Albanie - Albania - Albania
TIRANA
ADMIN RELATED / REGULATOR
TIRANA
ADMIN RELATED / REGULATOR
Allemagne - Germany - Alemania
MAINZ
ADMIN RELATED / REGULATOR
Andorre - Andorra - Andorra
ANDORRA
ADMIN / REGULATOR
Angola - Angola - Angola
LUANDA
ADMIN RELATED / REGULATOR
Antigua-et-Barbuda - Antigua and Barbuda - Antigua y Barbuda
Telecommunication Division, ST. JOHN'S
ADMIN / REGULATOR
Antilles neerlandaises - Netherlands Antilles - Antillas Neerlandesas
CURAÇAO
ADMIN RELATED / REGULATOR
Arabie saoudite - Saudi Arabia - Arabia Saudita
RIYADH
ADMIN RELATED / REGULATOR
Argentine - Argentina - Argentina
BUENOS AIRES
ADMIN / REGULATOR
Arménie - Armenia - Armenia
YEREVAN
ADMIN RELATED / REGULATOR

World Wide Web Telecommunication Resource Center

http://home.planet.nl/~wvhwvh/teletop.htm

Country Index for Major PTTs, PTOs, and Major Service Providers

http://home.planet.nl/~wvhwvh/countidx.htm

The World Wide Web Telecommunication Resource Center has a separate listing of over 300 links to PTTs and major service providers in 120 countries. This site was last updated in 2004, but much of this information has not changed.

Goodman's International Telecom Companies

http://www.gbmarks.com/html/international.html

Another excellent source is Goodman's International Telecom Companies, which lists and links to telecom and telephone providers and authorities for each country alphabetically.

Finally, I suggest you stop by a website maintained by the American University's Management of Global Information Technology program. Each semester students prepare descriptive and analytical reports on all aspects of one country's information technology. Be aware the reports vary in quality, but every report is worth reading if only for the bibliography and links to other websites. The best reports (as judged by the instructor) on **IT Landscape in Nations Around the World** are indicated with stars.

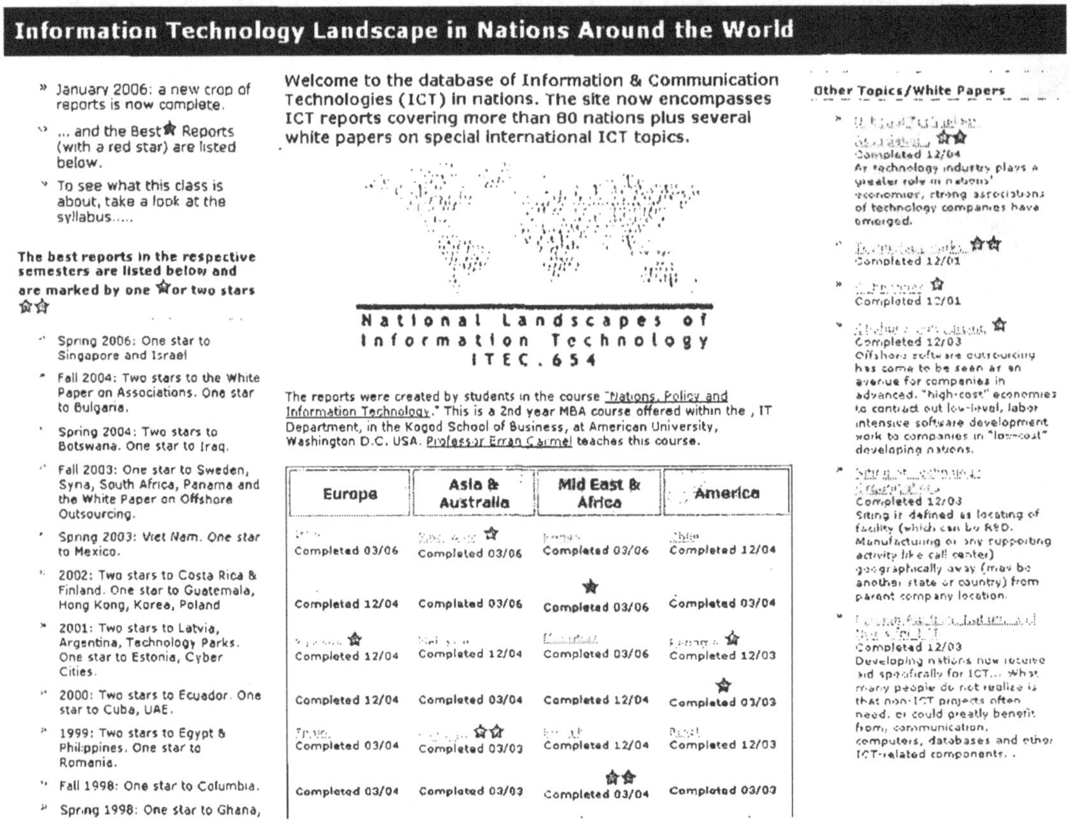
Information Technology Landscape in Nations Around the World
January 2006: a new crop of reports is now complete.
... and the Best Reports (with a red star) are listed below.
To see what this class is about, take a look at the syllabus.....
The best reports in the respective semesters are listed below and are marked by one or two stars
Spring 2006: One star to Singapore and Israel
Fall 2004: Two stars to the White Paper on Associations. One star to Bulgaria.
Spring 2004: Two stars to Botswana. One star to Iraq.
Fall 2003: One star to Sweden, Syria, South Africa, Panama and the White Paper on Offshore Outsourcing.
Spring 2003: Viet Nam. One star to Mexico.
2002: Two stars to Costa Rica & Finland. One star to Guatemala, Hong Kong, Korea, Poland
2001: Two stars to Latvia, Argentina, Technology Parks. One star to Estonia, Cyber Cities.
2000: Two stars to Ecuador. One star to Cuba, UAE.
1999: Two stars to Egypt & Philippines. One star to Romania.
Fall 1998: One star to Columbia.
Spring 1998: One star to Ghana,
Welcome to the database of Information & Communication Technologies (ICT) in nations. The site now encompasses ICT reports covering more than 80 nations plus several white papers on special international ICT topics.
National Landscapes of Information Technology ITEC.654
The reports were created by students in the course "Nations, Policy and Information Technology." This is a 2nd year MBA course offered within the , IT Department, in the Kogod School of Business, at American University, Washington D.C. USA. Professor Erran Carmel teaches this course.
Europe
Asia & Australia
Mid East & Africa
America
Other Topics/White Papers
Completed 12/04
As technology industry plays a greater role in nations' economies, strong associations of technology companies have emerged.
Offshore software outsourcing has come to be seen as an avenue for companies in advanced, "high-cost" economies to contract out low-level, labor intensive software development work to companies in "low-cost" developing nations.
Siting is defined as locating of facility (which can be R&D, Manufacturing or any supporting activity like call center) geographically away (may be another state or country) from parent company location.
Developing nations now receive aid specifically for ICT... What many people do not realize is that non-ICT projects often need, or could greatly benefit from, communication, computers, databases and other ICT-related components.

IT Landscape in Nations Around the World
http://www.american.edu/academic.depts/ksb/mogit/country.html

---

Telecommunications Directories

Analysys Telecoms Virtual Library http://www.analysys.com/vlib

Analysys Telecom Virtual Library has over 8000 links to international telecom sites. The links can be browsed by topic or searched by keyword. This site is one of the best for finding international telecom providers of all types.

LIDO Telecom Web Central http://www.telecomwebcentral.com/secure/links/

One of the best directories of telecommunications information on the web, LIDO offers an easy to use directory or internal search for full range of telecom topics. - LIDO Telecom Web Central is a searchable web directory covering such tele-communications topics as applications, IT technology, companies, regulations, and research resources. The impressive directory is kept fairly current and contains thousands of links.

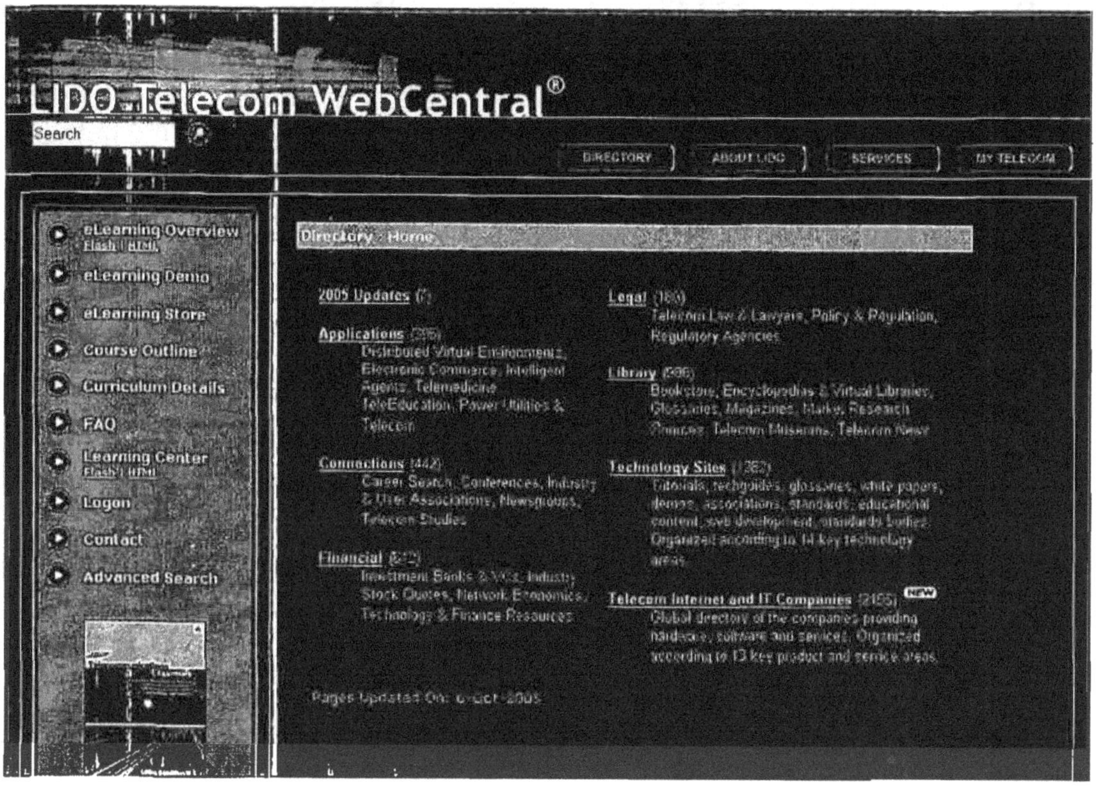
LIDO Telecom WebCentral®
Search
DIRECTORY
ABOUT LIDO
SERVICES
MY TELECOM
eLearning Overview
eLearning Demo
eLearning Store
Course Outline
Curriculum Details
FAQ
Learning Center
Logon
Contact
Advanced Search
Directory : Home
2005 Updates
Applications
Connections
Financial
Legal
Telecom Law & Lawyers, Policy & Regulation, Regulatory Agencies
Library
Technology Sites
Telecom Internet and IT Companies
NEW

Computer and Communication Entry Page http://www.cmpcmm.com/cc

The Computer and Communications Entry Page has great links to companies, mostly in the US but some international. It's also an excellent source for standards as well as for programs and projects. The site is internally searchable and once again appears to be up to date.

Goodman's Bookmarks http://www.gbmarks.com/

Goodman's Bookmarks have also been around for years and has undergone a major makeover. Goodman covers such topics as wireless, IP telephony, vendors, and international telecommunications providers. In fact, his list of international telecom operators is one of the best on the web, providing links to the websites of almost all international telephone and telecom companies and organizations.

Radio, Television, and Satellite Broadcast Information

Other useful sources of telecommunications information include sites devoted to broadcast radio and TV stations and to satellites. Many radio stations around the world now have audio feeds to permit users to listen to them over the Internet, and television is also making more and more of its content accessible over the Internet. There are a number of excellent sites that will help you locate these stations and find out which ones have Internet feeds. Communications satellites are often the primary source in many parts of the world that lack the infrastructure for terrestrial communications.

I have included a link to the Union of Concerned Scientists' satellite database, which includes information not only on communications but also other types of satellites: "the UCS Satellite Database is a listing of operational satellites currently in orbit around Earth. It...contains only the official name of the satellite in the case of government and military satellites, and the most commonly used name in the case of commercial and civil satellites."

Radio Locator	http://www.radio-locator.com/
Radio, TV, and Satellite Links	http://www.liensutiles.org/sat.htm
Live Radio	http://www.live-radio.net/info.shtml
Radio Station World	http://radiostationworld.com/default.asp
Mike's Radio World	http://www.mikesradioworld.com/
vTuner	http://www.vtuner.com/
USC Satellite Database	http://www.ucsusa.org/global_security/space_weapons/satellite_database.html
Heaven's Above Satellite Database	http://www.heavens-above.com/selectsat.asp
SatcoDX Satellite Chart	http://www.satcodx.com/eng/
NASA's J-Track Satellite Tracking	http://science.nasa.gov/RealTime/JTrack/
Small Satellites Home Page	http://centaur.sstl.co.uk/SSHP/

Research How-Tos

Finding People

While searching for people remains one of the most popular and important Internet research projects, it also continues to be one of the most frustrating. It is also one of the most lucrative for businesses—legitimate, shady, or simply incompetent. Sites that claim "you can find anyone, anywhere!" are all over the web. Try running a search for the phrase "search for people" and you are likely to find these categories of results, each of which has its drawbacks and most of which are so limited in some way as to be useless except under special circumstances:

- pay services provided by individuals or companies (everything from sleazy to legitimate PI services)
- for-fee software ("net detective"-type software)
- free software (this probably is a case of "you get what you pay for")
- links to a gazillion "people finder" websites

It all boils down to this: there are three basic types of information on the Internet useful in searching for people:

1. information you must pay for
2. information you must register to get
3. information that is free

While it is true you may have to pay for help and/or information either to find a person or to retrieve information about a person, it isn't always a case of "you get what you pay for" with people searches. People are actually selling books with bloated claims and largely useless tips on looking for people. Most of this so-called "expert" advice is worthless. If you have searched for people on the Internet, you've probably encountered some or all of the suggestions, claims, offers, and outright scams from people claiming to help you find someone using the Internet. Many sites will gladly take your money to search for people. Some of these are legitimate companies with access to huge databases containing a great deal of personal information. Some of these for-profit companies give you nothing more than a list of web links that in turn may only lead you to a 404 error. In many, many cases, ***you end up paying for information that was freely available elsewhere on the web!***

Internet researchers agree: searching for information about a person is one of the hardest, most frustrating types of research to perform, especially when you're limited to using free resources that do not require registration. The major reasons this is so difficult are:

1. The decentralized nature of the Internet. Despite what some people may fear, there is no giant database "out there" where everyone's names and addresses are stored. There are many, many databases all independent of each other spread around the world. Also, *there are as yet no reliable publicly available mobile telephone directories.*
2. How quickly the Internet changes. Web pages come and go, individuals frequently change email addresses and/or service providers, networks go up and down, etc.
3. Data about people is often found in the deep web. In most cases, the information is in databases that search engine spiders cannot crawl and index. This means that, in addition to traditional search engines, we have to use other resources to search for people.
4. People have become much more conscious of and careful about their personal information. In the early days of the Internet, people thought nothing of putting their contact information on their webpages, using de-obfuscated email addresses in newsgroups, the finger utility was still in use, etc. Today, people are much more likely to hide personal information to protect their privacy and their computers from malicious hackers.

The Usual Suspects Round-up: Debunking People Finder Lists

I get really irritated at sites and people who proclaim how fast and easy it is to find people using the Internet. Because you are going to encounter lots of bogus claims about how simple it is to find people via the Internet, I thought it would be a good idea to go down the list of the usual types of sites recommended for people searches and discuss their advantages (few) and drawbacks (many). What follows is an overview of the usual sources cited as ways to find people and the problems with and benefits of each.

- Sites that claim "you can find anyone, anywhere!"

  These are all over the web. The bigger the hype, the smaller the likelihood they are worth anything. Approach with care and skepticism or, better yet, avoid altogether.

- Big Directories

  Number one on virtually all "search for people" lists are large directories (telephone and/or email). They are usually disappointing. What's the problem? The big residential telephone directories on the web often come from printed directories that have been scanned in by companies who then make them searchable via the Internet. Email directories are almost entirely made up of addressed culled from USENET newsgroups, voluntary registration, ICQ users, and (sometimes) web pages. Other limitations to big phone and email directories are:

- they generally focus on the US (maybe Canada, too)
- their sources of information are very limited (people registering themselves or from trolling newsgroups)
- they require you already know details about the person you're seeking
- they are woefully out of date, and/or
- their data is inaccurate

➢ Public Databases

Public databases are similarly frustrating. Almost invariably, the public databases listed on "people finder" sites are US-based and many of the best ones require at least registration and even a justification of use if not payment. I recommend you look for public databases in the country you're researching, but expect to encounter similar problems. In general, public databases suffer from many of the same problems as big directories as well as having drawbacks of their own. Public databases are wonderful sources of information if you are seeking such public records as real estate transactions, lawsuits, liens, judgments, bankruptcies, fire arm transactions, sex offenders, motor vehicle registrations, corporate records, etc. However, this kind of research simply is beyond the purview of this book. Other drawbacks of public databases are:

- they generally focus on the US (maybe Canada, too)
- they often require registration and/or payment
- they tend to be restricted to specific types of data (e.g., sex offenders or property records in a specific state)
- they require you already know details about the person you're seeking
- they are out of date, and/or
- their data is inaccurate

➢ Private or Internal Directories

These are not inherently bad, but they are by their very nature extremely limited. Many private directories belong to universities, government agencies, corporations, and organizations. If, however, you are looking for someone whom you know works for an organization or attends (or is even an alumnus of) a specific university, these directories can be gold! See <u>below</u> for details.

- Genealogy Sites

Genealogy sites always pop up on "search for people" lists—nice for generating your family tree, but not usually pertinent for current research.

- Biographical Directories

The big biographical directories are fine for one thing: finding information about well-known and/or successful people living and dead. Otherwise, they are useless. People search lists always include links to "celebrity searches." However, a few specialized biographical directories might prove useful.

- The US Census

Census data for a number of different years is now online and searchable. Again, a great genealogy tool, but probably not very useful for your research.

- Fee for Service Sites

There are a myriad for-profit people finder websites on the Internet. They may or may not be good, but I am only discussing free products, services, and sites.

- Finger

Finger is a UNIX utility that uses an email address to return information about the owner of the address. Even today finger always seems to appear on people finder webpages, but the fact is that the vast majority of sites disabled the finger command years ago for privacy and security reasons, so it's basically useless.

- The MIT interface to the USENET Address Database

USENET is still accessible via the old MIT interface. I mention the MIT interface (which contains over 4 million email addresses collected from USENET newsgroup postings between July 1991 and February 1996 only) because it continues to show up on "people finder" lists. However, it's almost a decade old, which may as well be a century in Internet time. Instead of wasting time on the MIT USENET interface, why not go directly to Google Groups? One reason: *MIT does not obfuscate email addresses*; Google now does, so you might find an older but full email address in the MIT USENET database that is obscured in Google Groups.

MIT USENET Address Database (1991-1996) http://usenet-addresses.mit.edu/

Recommended People Search Techniques

Now that you have a pretty good idea of sites to avoid or at least not waste a lot of time on, what kinds of resources are available to help you find people? How do you find information about a person? Here are some recommended steps to use in searching for people, keeping in mind that there is no one way to do this and certainly no guarantee of success.

1. Start a search for a person by name, address, email address, phone number or any other uniquely identifying information using not *one* but several good search engines (Google, Yahoo, and several foreign search engines as appropriate). Common names will return too many hits to be useful, so *you must find some limiting query term to narrow down the search.*
2. Try to find the person in Google and Yahoo Groups, weblogs, and news stories using several high-quality news search engines.
3. Try telephone, email, address, and other lookups at a variety of online directories. The quality of these directories varies greatly. Be sure to look for directories specific to a city or country.
4. If you know the person's profession, you might find additional information about him in a database that contains such things as licensing information. The US is very good about licensing all types of professions; check other countries for similar information.
5. Property ownership and transactions are carefully recorded in the US and many such records are publicly accessible. This may also be true in other countries. Look for such public databases of these records and transactions.
6. If a person has other interests, such as hobbies, you may find more information about him at a site devoted to that sport or hobby, especially if it's an unusual one.
7. If you know where a person works, that organization (be it government, academic, or corporate) may have a publicly accessible directory of its employees, faculty, students, alumni, or members.
8. Check the ICQ directory. Instant Messaging (IM) is popular worldwide.
9. Whois databases contain information about thousands of people associated with the Internet. It's worth a look. The Whois databases maintained by ARIN, APNIC, AfriNIC, LACNIC, and RIPE are all searchable by name using their advanced search forms.
10. Famous dead people are not the only ones listed in biographical directories. If the person you're seeking is at all well known, he might have an entry.

Recommended People Search Resources

There is no Internet equivalent to the Rosetta Stone for finding people. Searchers need a good set of bookmarks that include at least the following sites.

ICQ User Directory http://people.icq.com/whitepages/

Unlike AOL's AIM instant messaging system directory, ICQ's phone and email directory is publicly available, listing more than 120 million users in over 245 countries. The ICQ User Directory's Advanced People Search is impressive. You can search on as few or as many fields—including first/last name, nickname, email address, language, phone number, country, occupation, etc.—as you wish. While there are several interfaces to ICQ search (depending on whether you want to search by email address or phone number, for example), I recommend starting with the main People Search page and then going to the Advanced People Search for more options.

Google Groups http://groups.google.com/

Not only is Google Groups much more user friendly than the MIT USENET interface, it is also current, including postings in almost real time. Google Groups can be searched in many ways, but for the purpose of locating people, the *Author* search is exceptionally valuable. However, keep in mind that anyone can make up an alias, change addresses, and use other people's email, so USENET postings are of real but limited value. Also remember that USENET postings are one of the places where a sort by date works and makes sense because all newsgroup postings are date/time stamped. There are three basic ways to run an *Author* search (important note: ***Google Groups now obscures the userid in email addresses by inserting an ellipsis (...) for privacy reasons***; however, many obscured addresses can be easily guessed). However, ***if you know the author's full email address***, you can use search syntax in the main Groups & Groups Beta query screen—**[author:email address]**. This will return all the postings associated with that email address.

Even if you do not know the full email address, you can still see all the posting associated with one email address (these instructions apply to Google Groups *Beta*):

- While viewing a specific posting, click on "More options" (next to the date);
- Select "View profile" next the author's From address;
- Google will automatically search for and display every posting from a particular address.

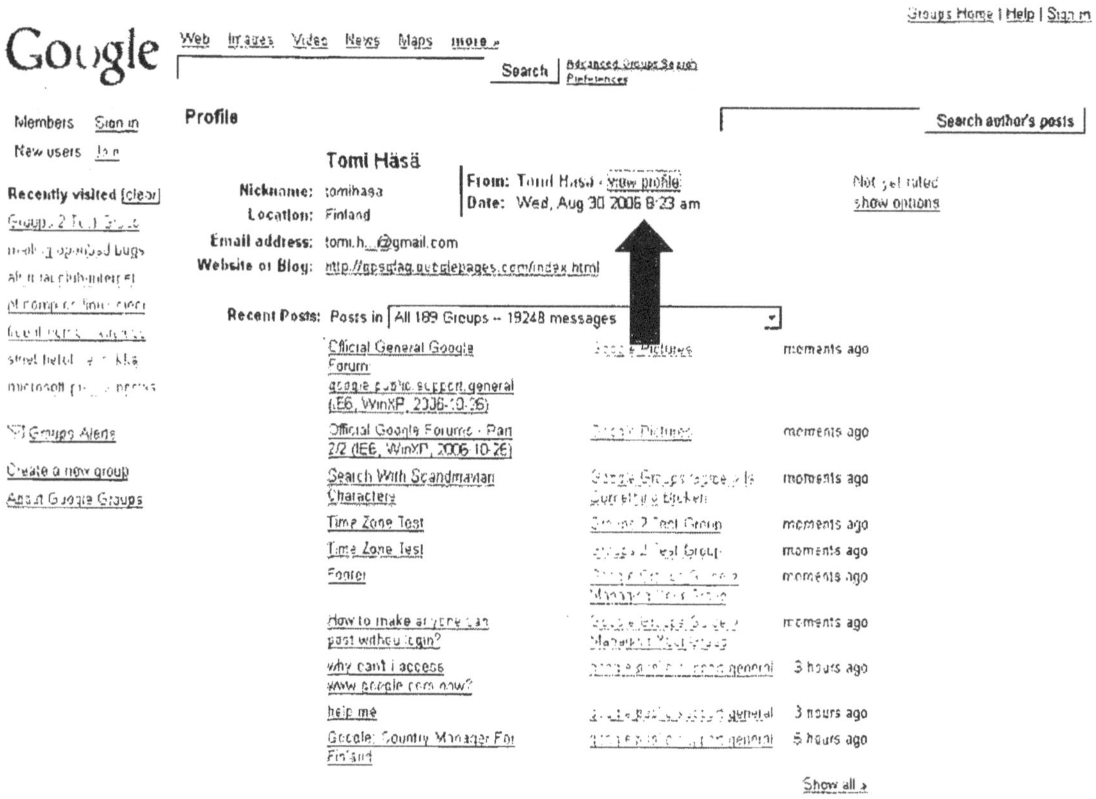
Groups Home | Help | Sign in
Google
Web Images Video News Maps more »
Search
Members Sign in
New users
Profile
Search author's posts
Tomi Häsä
Nickname: tomihasa
Location: Finland
Email address: tomi.h...@gmail.com
Website or Blog:
From: Tomi Häsä - view profile
Date: Wed, Aug 30 2006 8:23 am
Not yet rated
show options
Recently visited [clear]
Groups Alerts
Create a new group
About Google Groups
Recent Posts: Posts in All 189 Groups -- 19248 messages
Official General Google Forum
moments ago
Search With Scandinavian Characters
Time Zone Test
Time Zone Test
Footer
How to make anyone can post without login?
why can't i access www.google.com now?
help me
Google: Country Manager For Finland
3 hours ago
5 hours ago
Show all »

Yahoo Member Directory

http://members.yahoo.com/

Similar to Google Groups, Yahoo's Member Directory lets you view information about the registered members of its groups. Some people provide a lot of personal data, including their real name, location, age, occupation, and email address (and sometimes even a photograph). I am skeptical that you will find anyone of great interest unwise enough to provide this kind of data, but it is worth a look.

Specialized Biographical Directories

While the big biographical directories are virtually useless, there are a few specialty directories that include the less than famous that you might want to know more about. The Forbes People Lists include: 400 Richest Americans, Best Paid CEOs, World's Richest People, and the Midas List of the top investors in information technology and life sciences. The Biography Center lists nearly 25,000 biographies total, with almost 11,000 in English, 6,000 in French, 7,000 in German, and 2,000 in Spanish. The search form requires you to search in one of these languages; there is no way to search in all at one time and the results are different for each. There is also a Chinese Biographical Database containing 3500 biographies, all in English. Wolfram's Science World Biography includes famous and not so famous scientists living and dead.

Forbes People Lists http://www.forbes.com/lists/

Biography Center http://www.biography-center.com/

Biography Reference Center from MacGill University http://www.library.mcgill.ca/refshelf/biograph.htm

Chinese Biographical Database http://www.lcsc.edu/cbiouser/

ISI Highly Cited Researchers http://www.isihighlycited.com/

Wolfram's Science World Biography http://scienceworld.wolfram.com/biography/

Edgar Online Database

There are several free search tools of SEC filings that allow you to search by a person's name or by a specific company name. As of mid-November 2006, the users can "search the contents of the disclosure documents filed electronically with the SEC using a new full-text search tool on the Commission's website. The newly searchable information includes registration statements, annual and quarterly reports, and other filings by companies and mutual funds filed during the past four years on the Commission's EDGAR database."[138] This means users can search on names of people as well as companies.

SurfWax offers a wonderful tool using its "LookAhead" (AJAX) technology to search the SEC EDGAR filings for 2004-2006. This is a very flexible and powerful search tool, and it permits searches on the names of individuals.

The Price Waterhouse Coopers' website now offers free full-text search of EDGAR filings, which means it can search on names. Even though others have said this query tool is slow, my trial searches ran fairly quickly. My impression is this is a remarkably powerful tool, especially given it is free and does not require registration. Make sure you use the link below to access the advanced search of the free service.

Search the SEC's Edgar Database http://searchwww.sec.gov/EDGARFSClient/jsp/EDGAR_MainAccess.jsp

SurfWax SEC Search http://lookahead.surfwax.com/edgar/

EdgarScan Advanced Search http://edgarscan.pwcglobal.com/servlets/advancedsearch

[138] "SEC Enhances Online Search Capabilities for Investors," SEC Press Release, US Security and Exchange Commission, 14 November 2006, <http://www.sec.gov/news/press/2006/2006-190.htm> (20 November 2006).

University and other organizational phonebook directories

Such directories are usually very good sources of current information about personnel at a specific university, in an entire university system, employed by a company, or working for a government agency. Many universities use the PH system or WebPH, which is a web interface PH. PH is a UNIX-based Internet utility that lets you search for someone's email address if their email provider has a Ph server program. The problem with these PH systems is that there is no reliable way to locate them even if you know the domain name of the institution. However, as with most things on the Internet there are lists of publicly accessible PH servers available in several places. There are about 330 PH servers around the world currently on the PH server list, which is accessible at various locations such as:

Queen's University (Canada) http://www.queensu.ca/cgi-bin/ph/lookup?Query=.

Drake University http://alpha9.drake.edu/cgi-bin/WebPh?other_ph_servers

University of Medicine and Dentistry, New Jersey
http://www2.umdnj.edu/~webph/cgi-bin/WebPh?do=show_others

Your best bet for using some sort of internal directory successfully is to identify the university, company, or other organization (including government agencies) and locate the directory at that website. Directory services are generally very popular, so most sites will have a link to theirs prominently displayed on their home page. If you don't see a link on the home page, try browsing the site map.

Whois Lookups

These should never be overlooked as a source of information about people. All five of the Regional Internet Registries (ARIN, APNIC, AfriNIC, LACNIC, and RIPE) have Whois access that permits users to search on any of the database fields, including the person object, i.e., the technical or administrative contact responsible for the registration. However, keep in mind people lie all the time when registering in Whois databases, so all information obtained from Whois entries must be validated using other sources. To learn how to search the three identically formatted Whois databases, see the instructions (RIPE's documentation is the most thorough, APNIC's the easiest to follow).

APNIC Whois Database Query Option
http://www.apnicnet/db/search/all-options.html

ARIN Whois Database Search Help http://www.arin.net/tools/whois_help.html

AfriNIC Database User Manual
http://www.afrinic.net/docs/db/afsup-dbgs200501.htm

RIPE Database Reference Manual
http://wwwripe.net/ripe/docs/databaseref-manual.html

Weblogs

Blogs may indeed mention people that traditional sources miss. Blogs are better than traditional search engines for finding people whose names just surfaced in the news. See the blogsearch section for the best blog search engines.

News Search Engines

These are great for timely searching, which means they may be the best source of information on someone who has recently gained his/her fifteen minutes of fame. As with USENET groups, news searching is amenable to date/time order because news stories, unlike web pages, have dates and times associated with them. They are not generally good for historical research, i.e., more than one to three months old. There are numerous excellent news search engines detailed in the News Search Section.

Online Telephone Directories

Telephone directories may be useful or merely frustrating. Despite the drawbacks of many telephone directories, some are quite good. More and more cities and even whole countries are putting their white and yellow pages on line, often with flexible lookups, including reverse lookup. Check a country's search engines and directories for listings of telephone directories. Information on finding and using telephone directories on the web is available in the Telephone and Fax Directories section.

Email Directories

While these remain spotty and often unreliable, do not overlook this potential source of information that is growing and improving all the time. Information on finding and using email directories on the web is available in the Email Lookups section.

Recommended People Finder Websites

I have tried to give you some idea of the types of information and links you will encounter at people finder websites so you won't be surprised or become discouraged if they do not turn out to be all they claim. Still, you will have to sift through a lot of silt and sand to find the nuggets, but they are there. I think these are the best general sites for getting started looking for people.

Pandia People Search http://www.pandia.com/people/

I would start here. One of the best web search sites for locating people, Pandia Search has a terrific section on people search that includes both metasearch interfaces as well as many links to the best people finder resources.

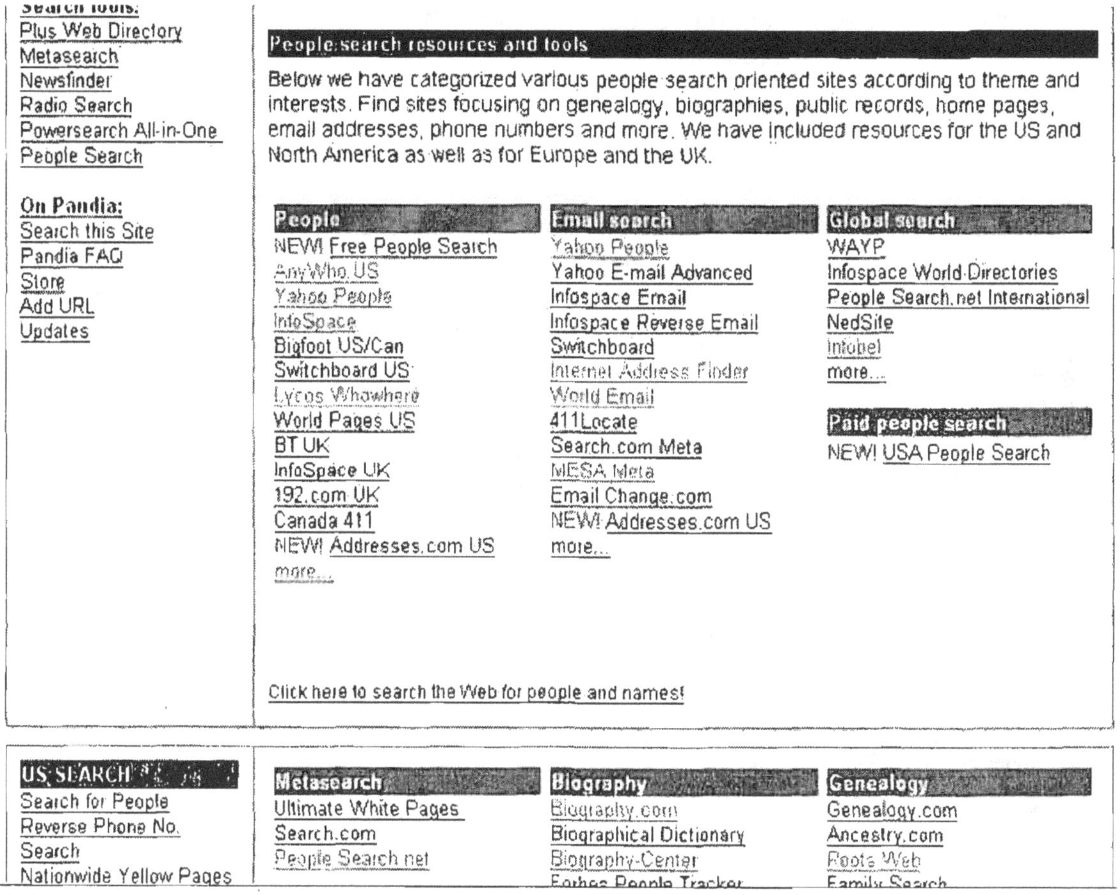
Search tools:
Plus Web Directory
Metasearch
Newsfinder
Radio Search
Powersearch All-in-One
People Search
On Pandia:
Search this Site
Pandia FAQ
Store
Add URL
Updates
People search resources and tools
Below we have categorized various people search oriented sites according to theme and interests. Find sites focusing on genealogy, biographies, public records, home pages, email addresses, phone numbers and more. We have included resources for the US and North America as well as for Europe and the UK.
People
NEW! Free People Search
AnyWho US
Yahoo People
InfoSpace
Bigfoot US/Can
Switchboard US
Lycos Whowhere
World Pages US
BT UK
InfoSpace UK
192.com UK
Canada 411
NEW! Addresses.com US
more...
Email search
Yahoo People
Yahoo E-mail Advanced
Infospace Email
Infospace Reverse Email
Switchboard
Internet Address Finder
World Email
411Locate
Search.com Meta
MESA Meta
Email Change.com
NEW! Addresses.com US
more...
Global search
WAYP
Infospace World Directories
People Search.net International
NedSite
Infobel
more...
Paid people search
NEW! USA People Search
Click here to search the Web for people and names!
US SEARCH
Search for People
Reverse Phone No.
Search
Nationwide Yellow Pages
Metasearch
Ultimate White Pages
Search.com
People Search.net
Biography
Biography.com
Biographical Dictionary
Biography-Center
Genealogy
Genealogy.com
Ancestry.com
Roots Web

The Virtual Chase People Finder Guides

http://www.virtualchase.com/topics/people_finder_index.shtml

Another excellent web search site, the Virtual Chase is devoted to legal research and, as such, its people finder guide focuses on US public records but includes reverse lookups, white pages, email directories, and non-US phone books. The first link will help you find online sources of public records and other people search tools. The second link is to an "annotated research guide to find sources of information about people" as well as articles on people research.

Searchbug People Finder http://www.searchbug.com/peoplefinder/

Searchbug is a metasearch site with links to many resources, but as with most such resources, the most potentially lucrative are not free or at least require registration. You are probably going to be disappointed when you discover that most of the links are to for-fee services. Also, the focus is distinctly US.

People Search Engines http://www.people-search-engines.com/

This is a good metasearch site that searches some or all of multiple sources by name, email, phone, or address. This site typifies a respectable people metasearch site. Unlike most people search sites, this one also has an "international" page where you can query resources that are not limited to the US and Canada:

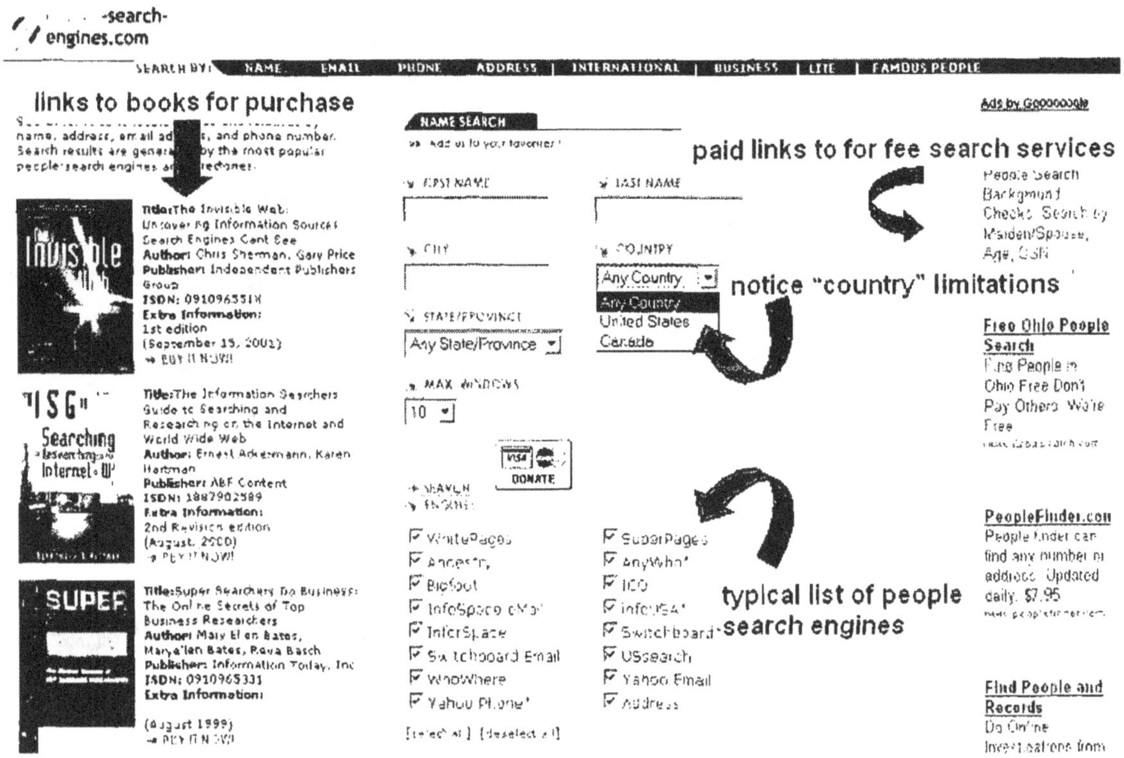
links to books for purchase
paid links to for fee search services
notice "country" limitations
typical list of people search engines

PeopleSearch.net http://peoplesearch.net/peoplesearch/peoplesearch_deluxe.html

This site is almost identical in the search engines it queries to People Search Engines, but without the international option.

ZoomInfo http://www.zoominfo.com/

ZoomInfo is a "web summarization" site that contains over 27 million summaries of people on the web. According to its FAQ, "ZoomInfo automatically and continuously grows its base of Web Summaries from corporate and personal websites, government filings, press releases and other public sources. All information found by ZoomInfo and used to create your Web Summary comes from public sources and can be found by anyone by using most major search engines like Yahoo and Google." ZoomInfo does summarizes that data and presents it in easy to handle small packages. ZoomInfo is integrated into A9 search; see that section for details.

The Virtual Gumshoe http://www.virtualgumshoe.com/

The Virtual Gumshoe provides lots of links to all sorts of investigative resources. The site contains many interesting categories, including gangs, terrorists, cults, criminal histories, etc. According to the site, "We provide links to every information source known to us that can be obtained on the Internet for free. Some categories may list a professional company as a source for information, which is not free. This is merely an option when there is no legitimate source online (known to us) that provides the information for free." And there are plenty of categories of information you are not going to get (legitimately) for free.

Search Systems Free Public Records Database http://www.searchsystems.net

With links to nearly 30,000 mostly free databases, this is one of the most impressive of the people finder megasites. What makes this site especially valuable is the large Worldwide section, which is further divided into continents and regions. Here is a snapshot of the Africa page and, as you can see, the vast majority of resources really are free:

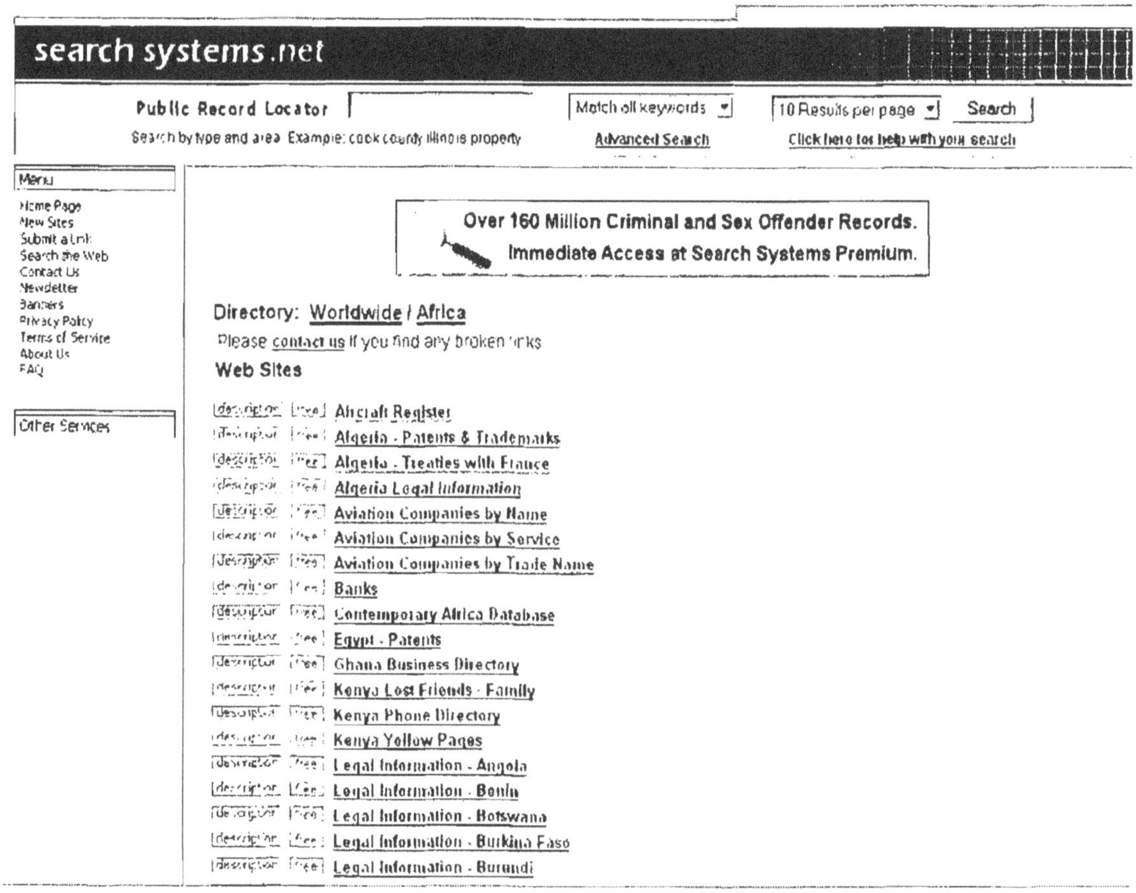
search systems.net
Public Record Locator
Match all keywords
10 Results per page
Search
Advanced Search
Click here for help with your search
Menu
Other Services
Over 160 Million Criminal and Sex Offender Records.
Immediate Access at Search Systems Premium.
Directory: Worldwide / Africa
Web Sites
Aircraft Register
Algeria - Patents & Trademarks
Algeria - Treaties with France
Algeria Legal Information
Aviation Companies by Name
Aviation Companies by Service
Aviation Companies by Trade Name
Banks
Contemporary Africa Database
Egypt - Patents
Ghana Business Directory
Kenya Lost Friends - Family
Kenya Phone Directory
Kenya Yellow Pages
Legal Information - Angola
Legal Information - Benin
Legal Information - Botswana
Legal Information - Burkina Faso
Legal Information - Burundi

Langenberg.com Person Finder http://person.langenberg.com/

This is an easy to use megasearch site that lets you query individual search tools from one page. Resources include the Lycos' WhoWhere phone book, Yahoo members search, Google Groups search, ICQ, and Live Spaces (still labeled as MSN Member Directory), among others.

Public Record Finder Outside the US
http://www.publicrecordfinder.com/outside_usa.html

Notable for its unusual nature, this site lists a moderate number of international public records' resources.

Power Reporting People Finders
http://powerreporting.com/category/People_finders

A site devoted to helping journalists, this is a no-nonsense guide to realistic people searching. No ads!

People Search Sites http://www.nettrace.com.au/resource/search/people.html

This Australian site will try to sell you people search products but has an impressive list of free resources as well. There are also many more non-US resources at this site for obvious reasons.

Deadline Online's People Finders
http://www.deadlineonline.com/peoplefinders.html

Another good megasite with no ads! A very clean interface and many useful people finder tools neatly organized by topic.

People Search Links http://www.peoplesearchlinks.com/

This site is actually a directory with categories that include international search, criminal searches, phone & address search, public records searches, vocations, etc.

By this time you are going to start seeing the same resources again and again, which should tell you that you're exhausting the free resources.

Miscellaneous People Finder Websites

These are sources for finding people that do not easily fit into any category.

The Virtual Chase Criminal Records
http://www.virtualchase.com/topics/criminal_records.shtml

This site has many links to Federal, State, and international online criminal resources, including US government lists of blocked, debarred, or denied persons. The Federal Bureau of Prisons inmate locator, which contains records about federal inmates dating back to 1982, is now online and free. Many state prison systems now have online searches for current and former inmates.

CrimeNet http://www.crimenet.com.au/

This Australian site provides information on criminals, missing persons, wanted persons, and unsolved crimes in Australia.

The Black Book Online http://www.blackbookonline.info/

Designed for use in legal investigations, this resource has links to some potentially lucrative US databases. From Crimetime.com, the Black Book includes unusual links such as a mail drop search form (to find out if a legitimate looking suite number is in fact a PO box). Many of the resources listed require payment.

Landings Certified Pilots Database
http://www.landings.com/_landings/pages/search/certs-pilot.html

A database of about 600,000 pilots registered with the FAA who have a current medical certificate. Updated monthly.

NameBase http://www.namebase.org/

A review describes NameBase as "an index of people influential in politics, the military, intelligence, crime, business, and the media since WWII. It started in the late 1960's when New Left activist Daniel Brandt began clipping magazine and newspaper articles and collecting investigative books about the power structure...NameBase includes close to 100,000 names from approximately 260,000 citations. The names are drawn from over 700 books and serials...most books and articles come from a leftist perspective." Many of the resources require payment, but a basic search does not. The site is worth a look because of the extensive cross-referencing of names from various sources and the "Social Network Diagram" available for each name mentioned.

Final Thoughts on Finding People

Do not overlook search engines. They may still be your best way of finding people on the Internet because they crawl the web almost constantly, index vast amounts of data, index everything on a website (including names, addresses, phone numbers, email addresses, etc.), and they are fast and easy to use. Also keep in mind there is a remarkable lack of overlap among search engine results, so use a number of different ones or several metasearch engines in your search.

Web Tip
Look for information in unconventional sources.

Sometimes calling codes and dialing information can come from unusual sources. For example, one of the best explanations of the arcane dialing rules and list of up-to-date calling codes for the CIS countries is located, of all places, at the *"Russian Bride"* website. And why not? After all, these folks are running a business and must provide accurate information about how to contact their clients.

Using the Internet to Research Companies

One of the main issues researchers face when looking for information about companies is there either seems to be too much or too little available data. How is it possible to have both a superfluity and a paucity of information at the same time? When it comes to businesses, there is too much data that is not useful (advertisements, sale's pitches, etc.) and too little of the kinds of data most researchers need (details about ownership, personnel, products, contracts, etc.). However, there are some good starting points for researching companies on the Internet, and I have put together a set of tips and techniques to address this challenging topic.

But what about using free Internet resources to research companies? Yes, it can be done with success and sometimes there is information on the Internet that is not available anywhere else. What follows are some of the steps I recommend for researching companies on the Internet using only free and non-subscription resources. Those that require payment and/or registration are not discussed here.

Company Research Guides & Tutorials

Several sites have excellent online tutorials and guides for how to research businesses.

Researching Businesses and Non-Profits on the Web
http://www.ojr.org/ojr/technology/1028068074.php

From the University of Southern California's Online Journalism Review; heavily US-oriented.

Researching Companies Online http://www.learnwebskills.com/company/

Excellent tutorial that includes separate section on identifying international business resources.

Virtual Business Information Center http://www.vbic.umd.edu/

From the University of Maryland School of Business; offers a segment devoted to international business and specific countries.

Virtual International Business and Economic Sources
http://library.uncc.edu/display/?dept=reference&format=open&page=68

From the University of North Carolina, Charlotte; the site's emphasis is entirely international.

Free Company Research Websites

Two well-known company research tools have been available on the Internet since 2004, but both have severe limitations. In the past the excellent Thomas Register's search tool ThomasNet was restricted to US and Canadian companies. Thankfully, in 2006 Thomas Register combined its search products into ThomasGlobal, which still does not require either registration or payment to use. ThomasGlobal is a search tool for international products/services or company names and includes an option to limit a search by location and to browse by product/service category.

ThomasGlobal http://www.thomasglobal.com/

ThomasGlobal provides a brief company profile (including links to any website), contact and location information, other headings under which a company is listed, and a link to a product catalog where available.

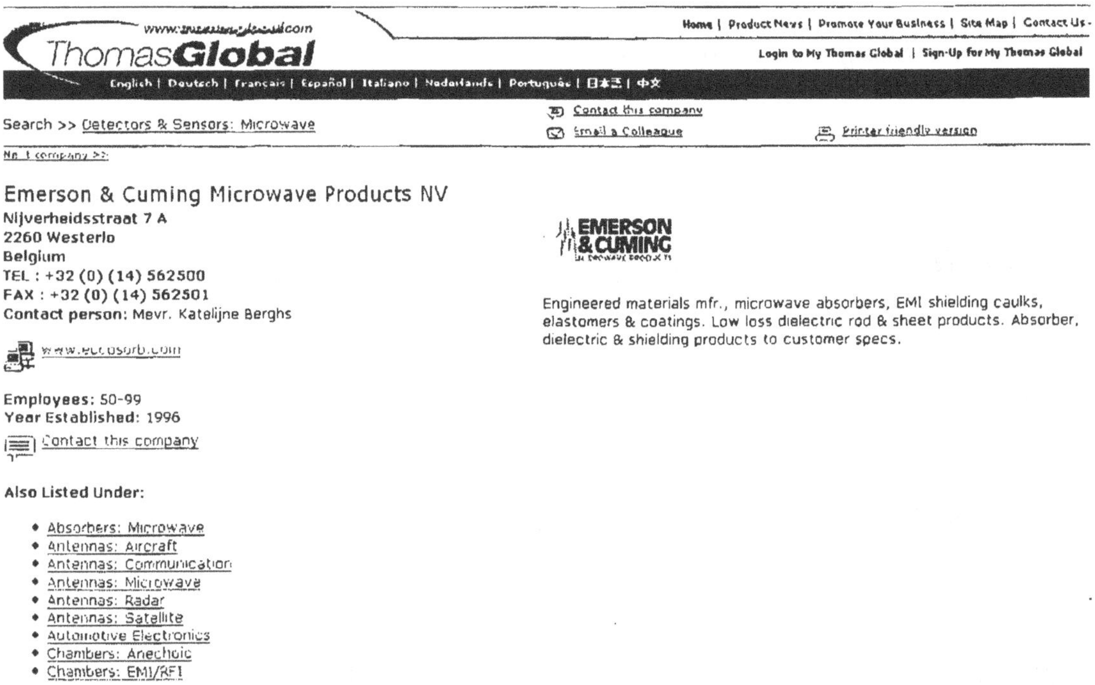
ThomasGlobal
Home | Product News | Promote Your Business | Site Map | Contact Us
Login to My Thomas Global | Sign-Up for My Thomas Global
English | Deutsch | Français | Español | Italiano | Nederlands | Português | 日本語 | 中文
Search >> Detectors & Sensors: Microwave
Contact this company
Email a Colleague
Printer friendly version
Emerson & Cuming Microwave Products NV
Nijverheidsstraat 7 A
2260 Westerlo
Belgium
TEL : +32 (0) (14) 562500
FAX : +32 (0) (14) 562501
Contact person: Mevr. Katelijne Berghs
Employees: 50-99
Year Established: 1996
Contact this company
Also Listed Under:
Absorbers: Microwave
Antennas: Aircraft
Antennas: Communication
Antennas: Microwave
Antennas: Radar
Antennas: Satellite
Automotive Electronics
Chambers: Anechoic
Chambers: EMI/RFI
EMERSON & CUMING
Engineered materials mfr., microwave absorbers, EMI shielding caulks, elastomers & coatings. Low loss dielectric rod & sheet products. Absorber, dielectric & shielding products to customer specs.

This is a very well designed tool and its global reach makes it extremely powerful and, in fact, invaluable for company research.

Virtual Chase Company Information Guide
http://www.virtualchase.com/topics/company_information_index.shtml

Virtual Chase provides an annotated research guide to sources of information about companies, executives, court records, SEC filings, news and even public opinion about companies. The site also includes a number of articles on how to research companies.

Kompass

http://www.kompass.com/

Kompass business-to-business search is among the best international company search sites. Users can search the huge Kompass database of over 2 million companies in 70 countries by products/services or company name, and limit the search to any combination of countries or world regions. While users must pay for full company profiles, searches are free and provide full company names, addresses, telephone/FAX numbers, website addresses, and a full list of companies' products and services. Here are the results of a search for "integrated circuit" in the Russian Federation, which produced 97 hits:

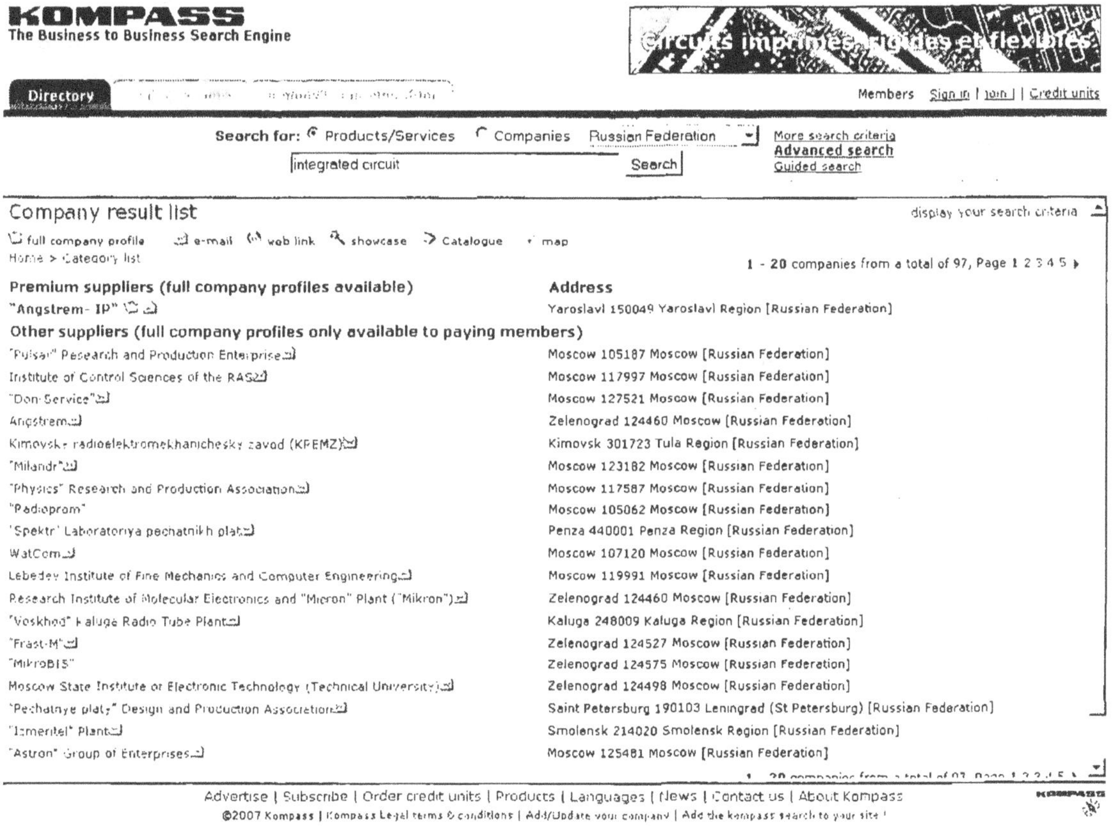
KOMPASS
The Business to Business Search Engine
Directory
Members Sign in | Join | | Credit units
Search for: Products/Services Companies Russian Federation
integrated circuit
Search
More search criteria
Advanced search
Guided search
Company result list
display your search criteria
full company profile e-mail web link showcase Catalogue map
Home > Category list
1 - 20 companies from a total of 97, Page 1 2 3 4 5
Premium suppliers (full company profiles available)
Address
"Angstrem- IP"
Yaroslavl 150049 Yaroslavl Region [Russian Federation]
Other suppliers (full company profiles only available to paying members)
"Pulsar" Research and Production Enterprise
Moscow 105187 Moscow [Russian Federation]
Institute of Control Sciences of the RAS
Moscow 117997 Moscow [Russian Federation]
"Don-Service"
Moscow 127521 Moscow [Russian Federation]
Angstrem
Zelenograd 124460 Moscow [Russian Federation]
Kimovsky radioelektromekhanichesky zavod (KREMZ)
Kimovsk 301723 Tula Region [Russian Federation]
"Milandr"
Moscow 123182 Moscow [Russian Federation]
"Physics" Research and Production Association
Moscow 117587 Moscow [Russian Federation]
"Radioprom"
Moscow 105062 Moscow [Russian Federation]
'Spektr' Laboratoriya pechatnikh plat
Penza 440001 Penza Region [Russian Federation]
WatCom
Moscow 107120 Moscow [Russian Federation]
Lebedev Institute of Fine Mechanics and Computer Engineering
Moscow 119991 Moscow [Russian Federation]
Research Institute of Molecular Electronics and "Micron" Plant ("Mikron")
Zelenograd 124460 Moscow [Russian Federation]
"Voskhod" Kaluga Radio Tube Plant
Kaluga 248009 Kaluga Region [Russian Federation]
"Frast-M"
Zelenograd 124527 Moscow [Russian Federation]
"MikroBIS"
Zelenograd 124575 Moscow [Russian Federation]
Moscow State Institute of Electronic Technology (Technical University)
Zelenograd 124498 Moscow [Russian Federation]
"Pechatnye platy" Design and Production Association
Saint Petersburg 190103 Leningrad (St Petersburg) [Russian Federation]
"Izmeritel" Plant
Smolensk 214020 Smolensk Region [Russian Federation]
"Astron" Group of Enterprises
Moscow 125481 Moscow [Russian Federation]
Advertise | Subscribe | Order credit units | Products | Languages | News | Contact us | About Kompass
©2007 Kompass | Kompass Legal terms & conditions | Add/Update your company | Add the kompass search to your site !

The Scannery

http://www.thescannery.com/

The second tool is much harder to use, or at least that is my experience. Called "The Scannery," it is a database of more than 12,000 public companies around the world organized by country and sector. The database contains many types of documents, including HTML, PDF, PowerPoint, Word, and more. The Scannery also offers many search options, including <u>boolean</u> operators, wildcards, "fuzzy" matches, synonym searching, etc. Sounds great, doesn't it? The only problem is, it's royally confusing and the results are often completely puzzling. For example, an unlimited search on IBM brings up as the top three hits, QLOGIC, Cable & Wireless, and Synopsys, but not IBM. Why? Because according to Joseph Pols, the developer of The Scannery, "not all companies allow their websites to be indexed...Therefore, the only results would be those documents on other websites which contain the search phrase."

However, even given this limitation, The Scannery is a very useful tool. Most companies do permit indexing, so chances are the ones you are researching will be accessible via The Scannery. The problem usually isn't too little information but too much. For example, an unrestricted search on [alcatel] brings back more than 3100 hits! When I limit my Alcatel search to France, I get a more manageable 74 hits. Keep in mind that what you are seeing is any mention of your query term in any document indexed by The Scannery, so my search for Alcatel will return not only documents from the Alcatel sites but also documents at other company websites that mention Alcatel.

The Scannery has a related site with the oh-so-clever name TimBuckOne. TimBuckOne will permit users to choose a country, then choose a specific company in that country, and then search that company's website. However, TimBuckOne *requires free registration* in order to use.

SEC's Edgar Database

Search the SEC's Edgar Database
http://searchwww.sec.gov/EDGARFSClient/jsp/EDGAR_MainAccess.jsp

SurfWax SEC Search http://lookahead.surfwax.com/edgar/

EdgarScan Advanced Search
http://edgarscan.pwcglobal.com/servlets/advancedsearch

There are several free search tools of SEC filings that allow you to search by a person's name or by a specific company name. As of mid-November 2006, the users can "search the contents of the disclosure documents filed electronically with the SEC using a new full-text search tool on the Commission's website. The newly searchable information includes registration statements, annual and quarterly reports, and other filings by companies and mutual funds filed during the past four years on the Commission's EDGAR database."[139] This means users can search on names of people and companies.

SurfWax now offers a wonderful tool using its "LookAhead" (AJAX) technology to search the SEC EDGAR filings for 2004-2006. This is a very flexible and powerful search tool, and it permits searches on the names of individuals.

---

[139] "SEC Enhances Online Search Capabilities for Investors," SEC Press Release, US Security and Exchange Commission, 14 November 2006, <http://www.sec.gov/news/press/2006/2006-190.htm> (20 November 2006).

The Price Waterhouse Coopers' website now offers free full-text search of EDGAR filings. Even though others have said this query tool is slow, my trial searches ran fairly quickly. My impression is this is a remarkably powerful tool, especially given it is free and does not require registration. Make sure you use the link above to access the advanced search of the free service.

More Company Research Sites

Keep in mind that even those sites that do require payment and/or registration for full access often provide some information for non-subscribers. Try these sites:

Annual Reports from Report Gallery http://www.reportgallery.com/

Free annual reports searchable by company name, stock ticker, or sector. Reports may be in PDF or HTML formats.

Arab Data Net http://www.arabdatanet.com/

Large amounts of information about business in the Arab world; be sure to check the Directory for data on companies.

Business.com http://www.business.com/

Excellent business-related search engine/directory; not limited to US companies

Business Information on the Internet http://www.rba.co.uk/sources/index.htm

Great UK site with huge number of resources for researching businesses not only in the UK but around the world. Working through this site is like taking a research tutorial.

Corporate Information http://www.corporateinformation.com/

Reports by Wright Investors' Service; full access requires registration; the available free information is good.

Free Reports for Top 20 European Companies
http://amadeus.bvdep.com/amadeus/top20/_top20.htm

Only 20 companies, but if you are researching of them, you have hit the jackpot with this site. These are full financial reports for free.

Global Edge International Business Research (Michigan State University)
http://globaledge.msu.edu/ibrd/ibrd.asp

The Center for International Business Education and Research at Michigan State University maintains Global Edge, a site of international country studies and business information with links to over 5000 resources. The site includes current information on the business climate, economic landscape, and relevant statistical data for 197 countries. "Powerful features such as comparing countries using multiple statistical indicators and ranking countries based on a selected statistical indicator are available. A rich collection of country-specific international business links adds to the vast collection of information."

Hoovers http://www.hoovers.com/

Hoovers is one of the best and most comprehensive company, industry, and market research resources. However, full access requires subscription. Even so, the information at Hoovers' free site is fairly extensive.

Irasia Investor Relations Asia http://www.irasia.com/listco/

Asian Company Reports; excellent data on Asian companies by name or sector.

MacRae's Blue Book http://www.macraesbluebook.com/
MacRae's EuroPages Search
http://www.europages.net/co_branding/macraesbluebook/home-en.html

MacRae's Blue Book is a guide to more than a half a million US and Canadian industries and over 2 million product listings; search by product or company. MacRae's now offers a EuroPages search of European companies at a separate link.

Market Access and Compliance http://www.mac.doc.gov/

US Department of Commerce site that is packed with information about trade and development around the world.

MSN Money's Key Developments
http://news.moneycentral.msn.com/ticker/sigdev.asp

Major milestones for publicly traded US companies; search by name or ticker symbol and see all developments, earnings, or product announcements.

PRNewswire http://www.prnewswire.com/

A great source for press releases; search by company, keyword, or limit your search by country.

SEDAR http://www.sedar.com/

Canadian securities-related information.

Yahoo Finance Press Releases http://biz.yahoo.com/prnews/

Easily searchable source of PRNewswire press releases.

Company Websites

Look at a company's own websites, but don't limit your search to the official home page. Be sure to check out country-specific sites (most international companies have sites specific to a country or region) and alternative domain names. Many companies have more than one domain name associated with them. For example, the Russian firm Kaspersky Labs has sites in multiple countries. In addition to the *.com* and *.net* domains, I counted 20 country top-level domains associated with the domain name *kaspersky*. I explored these names using Yahoo's Site Explorer and Google's *info:* command.

Google's Info: Command

This is a valuable tool for getting a fast overview of a website. Simply by inserting an address—*kaspersky.com*—into the Google search box, you will see the following options:

- view Google's cache of *kaspersky.com*
- find other pages similar *kaspersky.com*
- find external pages linking to *kaspersky.com*
- find pages containing *kaspersky.com*

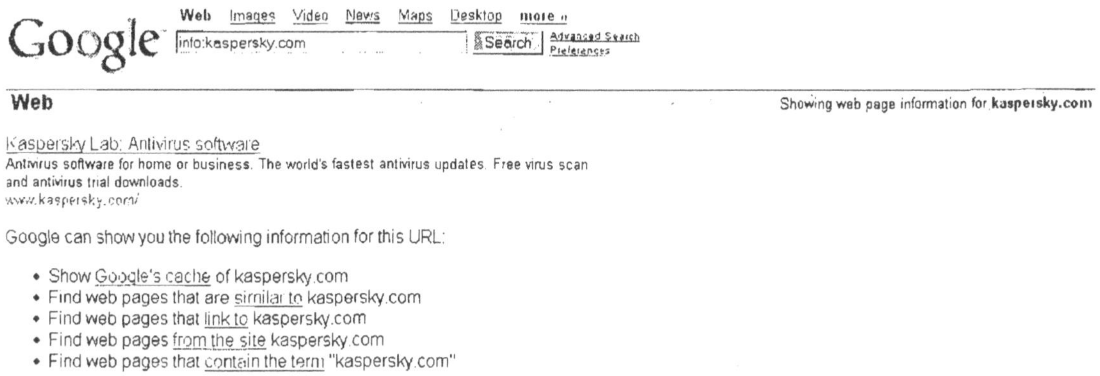
Google
Web Images Video News Maps Desktop more »
info:kaspersky.com
Search
Advanced Search
Preferences
Web
Showing web page information for kaspersky.com
Kaspersky Lab: Antivirus software
Antivirus software for home or business. The world's fastest antivirus updates. Free virus scan and antivirus trial downloads.
www.kaspersky.com/
Google can show you the following information for this URL:
• Show Google's cache of kaspersky.com
• Find web pages that are similar to kaspersky.com
• Find web pages that link to kaspersky.com
• Find web pages from the site kaspersky.com
• Find web pages that contain the term "kaspersky.com"

Whois Lookups

While people can and do lie about the information they enter into domain registration forms, the Whois databases remain an excellent source of information about companies and people (addresses, phone/fax numbers, email addresses, ISPs, mail servers, etc.). However, I strongly caution you to double-check these data with other sources to validate them.

ARIN (North America and some Caribbean)	http://www.arin.net/whois/index.html
RIPE (Europe, Middle East, North Africa)	http://www.ripe.net/perl/whois
APNIC (Asian-Pacific)	http://www.apnic.net/apnic-bin/whois.pl
AfriNIC (Africa)	http://www.afrinic.net/cgi-bin/whois
LACNIC (Latin America and some Caribbean)	http://lacnic.net/cgi-bin/lacnic/whois

Look for Additional Domains in Whois Database

As an example, Kaspersky Labs owns and operates websites under the domain names *kaspersky-labs* and *avp* (for Anti-Virus Protection) for many country top-level domains as well as the *.com* and *.net* domains. How did I find these alternate domain names and websites? By investigating Whois data on Kaspersky Labs and noting email addresses *@avp.ru* and other server names (*kaspersky.com, kasperskylabs.net, kaspersky-labs.com*).

Correlate Whois Data

For example, when you find a person's name (the *person* object is the technical or administrative contact for a Whois entry), I suggest you search on that name in all the major Whois databases to see if his/her name shows up anywhere else, thus providing possible leads to relationships between seemingly unrelated companies or organizations. To see how this works, search on the name [vladimir] in the RIPE, ARIN, and APNIC databases at IP-Plus:

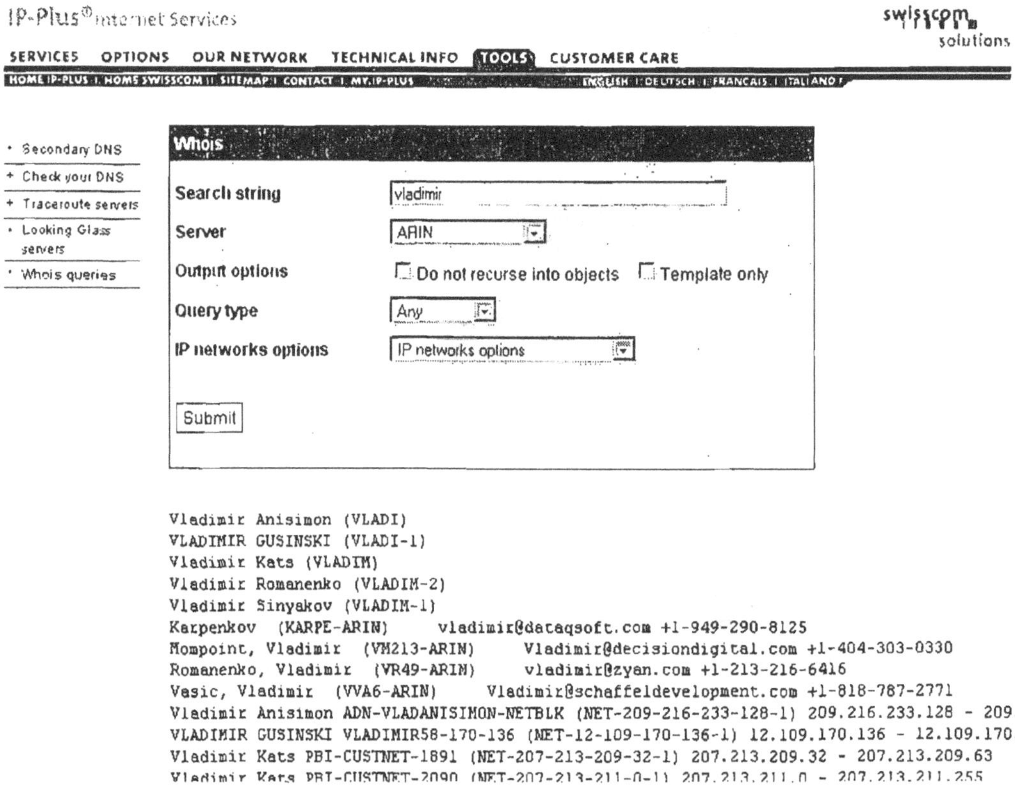
IP-Plus Internet Services
swisscom solutions
SERVICES OPTIONS OUR NETWORK TECHNICAL INFO TOOLS CUSTOMER CARE
Secondary DNS
Check your DNS
Traceroute servers
Looking Glass servers
Whois queries
Whois
Search string
vladimir
Server
ARIN
Output options
Do not recurse into objects
Template only
Query type
Any
IP networks options
IP networks options
Submit
Vladimir Anisimon (VLADI)
VLADIMIR GUSINSKI (VLADI-1)
Vladimir Kats (VLADIM)
Vladimir Romanenko (VLADIM-2)
Vladimir Sinyakov (VLADIM-1)
Karpenkov (KARPE-ARIN) vladimir@dataqsoft.com +1-949-290-8125
Mompoint, Vladimir (VM213-ARIN) Vladimir@decisiondigital.com +1-404-303-0330
Romanenko, Vladimir (VR49-ARIN) vladimir@zyan.com +1-213-216-6416
Vasic, Vladimir (VVA6-ARIN) Vladimir@schaffeldevelopment.com +1-818-787-2771
Vladimir Anisimon ADN-VLADANISIMON-NETBLK (NET-209-216-233-128-1) 209.216.233.128 - 209.216.
VLADIMIR GUSINSKI VLADIMIR58-170-136 (NET-12-109-170-136-1) 12.109.170.136 - 12.109.170.143
Vladimir Kats PBI-CUSTNET-1891 (NET-207-213-209-32-1) 207.213.209.32 - 207.213.209.63

IP-Plus http://www.ip-plus.ch/tools/whois_set.en.html

Also, search on the name using search engines, phone directories, email lookup tools, etc.

Wildcard Whois Search Tools

One of the best ways to find other domains registered to a company is to use a wildcard Whois lookup tool. There are a number of these available on the Internet and they are detailed in a later section of this book. Wildcard searches work best with unusual or distinctive domain names.

Domain Surfer	http://www.domainsurfer.com/
Domain Tools Whois Source	http://whois.domaintools.com/
Namedroppers.com	http://www.namedroppers.com/
Netcraft	http://www.netcraft.com/
Whois.net	http://www.whois.net/
Whoix?	http://www.whoix.com/
Whoix? Advanced Search	http://www.whoix.com/advdomsearch.html

Explore the Site

There are any number of things you can learn from looking at the various pages at a company website. Fortunately, it is easy to display all the pages at a website (at least all the pages indexed by a specific search engine); unfortunately, for a large website, finding useful information within all these can be a daunting task. The best tools for site exploration are Yahoo's Site Explorer and the various *site:* and *link:* commands available at major search engines.

In addition to Yahoo's Site Explorer, here is a summary of the various "site exploring" syntax available in Yahoo, Google, and Live Search:

Yahoo Search:

- **site/domain:** restricts results to a specific website or domain, including a specific top-level domain. May be used with or without keywords.
  [site:fr] returns all the pages Yahoo has indexed in the French top-level domain
  [site:who.int sars] finds all the pages at the WHO website that contain the keyword SARS; will find pages at wpro.who.int as well as www.who.int.

Google Search:

- **site:** restricts results to websites in a given domain. May be used with or without keywords.
  [site:in] returns all the pages Google has indexed in the Indian top-level domain

[site:who.int "avian flu"] finds all the pages at the WHO website that contain the phrase "avian flu"
[site:who.int "avian flu" -inurl:speeches] finds all the pages at the WHO website that contain the phrase "avian flu" and exclude any page with a url containing the keyword "speeches."
[cirrus -site:mastercard.com] finds pages containing the keyword cirrus that are not on the Mastercard website

Live Search:

- **site/domain:** restricts results to a specific website or domain, including a specific top-level domain. May be used with or without keywords.
  [site:gov.it] returns all the pages Live Search has indexed in the Italian Government's domain
  [site:gov.it language:en] finds all the pages in the Italian Government's domain that are in English
  [site:gov.it -filetype:html] finds all the pages in the Italian Government's domain that are not in HTML format

Link: Command at Major Search Engines

It is always worthwhile to see who is linking to a site. Best bets for running the link command are the biggest search engines because, you guessed it, they have the biggest databases! If you get too many hits, try limiting the search by restricting the results to a specific domain or adding a keyword when permitted (Google does not permit the use of keywords with its link: command). All these search engines use the same syntax for link searching, i.e., *link*:, but remember, ***Yahoo has two link commands.***

Also, the Yahoo Site Explorer is extremely good at showing "inlinks" to an entire domain or just a specific url. Unlike Google, Yahoo Site Explorer provides all the links to a site (which Yahoo calls "inlinks") instead of a limited subset of links.

Yahoo Search:

- **link:** restricts results to pages containing links to a specific url. Yahoo ***requires users to enter the full url including the http:// for the link: command.*** May be used with or without additional keywords and syntax.

  [link:http://www.who.int] finds all the incoming links to *this specific page*

  [link:http://www.who.int "avian flu"] finds all the incoming links to this specific page that also contain the phrase "avian flu"

  [link:http://www.who.int "avian flu" inurl:h5n1] finds all the incoming links to this specific page that also contain the phrase "avian flu" and have H5N1 anywhere in the linking page's url.

- **linkdomain:** restricts results to all incoming links to an *entire domain*, not a specific page or url.

  [linkdomain:who.int] finds all the incoming links to *this entire domain*

  [linkdomain:who.int "avian flu"] finds all the incoming links to *this entire domain* that contain the phrase "avian flu"

  [linkdomain:who.int "avian flu" inurl:h5n1] finds all the incoming links to *this entire domain* that contain the phrase "avian flu" and have H5N1 anywhere in the linking page's url.

Google Search:

- **link:** restricts results to documents that link to a specific url, but excludes many inlinks as a Google policy to try to reduce webpage or keyword spamming. *No other syntax or keywords can be used with the Google link: command*

Live Search:

- **link:** restricts results to documents that link to a specific url. May be used with or without additional keywords and/or additional syntax.

  [link:who.int loc:de geflügelpest] finds all the incoming links to this domain that are in the German (de) domain and contain the keyword Geflügelpest.

Link Harvester: http://www.linkhounds.com/link-harvester/

- Link search on steroids. This incredibly powerful tool shows where all the links come from in groups and by common Class C IP Address blocks. Link Harvester also links to the Internet Archives, two types of WhoIs Source data, and several variations of Google cache. Data can be exported as a CSV file and imported into a spreadsheet or database for additional analysis.

Google	http://www.google.com/
Live Search (beta)	http://www.live.com/
Yahoo	http://search.yahoo.com/

Also remember that you can use Fagan Finder's URLinfo page to run the link queries from all these search engines one at a time.

Fagan Finder's URLinfo	http://www.faganfinder.com/urlinfo/

Newspapers, Magazines, Journals, Newsgroups, Blogs

There is a lot of hard news, opinion, and old-fashioned gossip on the Internet about companies and their goings-on...new products, new partnerships, contracts, lawsuits, you name it. Disgruntled employees may post to newsgroups or publish blogs. In fact, people have been fired for leaking company secrets in blogs. Obviously, the validity of this information ranges from very high to scandalously poor, but in many cases researchers are piecing together a monochrome jigsaw puzzle, so any bit of information may help.

Press Releases

These can be a great source of cutting edge information about companies. They want to make money and, if they are public, sell stock, so they love to tell the world of their successes. The two best ways to find press releases is using PRNewswire or going to the horse's mouth (the company website—look for the magic words *press releases* on the home page). PRNewswire has improved its search and I especially recommend using the advanced search option. The PRNewswire site also lets you select a specific region or country, so it's easier to find international information here than at Yahoo.

PRNewswire	http://www.prnewswire.com/
Yahoo Press Releases via PR Newswire	http://biz.yahoo.com/prnews/

How to Research a Specific Country

In order to perform in-depth research on a particular country, it is a good idea to have numerous sites bookmarked as general starting places for country-specific Internet research. Some countries have a large web presence while others, for all intents and purposes, are absent or have a very limited presence. For the latter category (countries such as North Korea or Burma) you will have to rely largely on "third party" sites, that is, webpages *about* the country. Be aware that expatriates, exiles, or dissidents with large axes to grind created many of these. Third party country pages are usually most useful as sources for links to other webpages, and some are impressive in their size and complexity.

My best advice in beginning to research a specific country is to try to find the most comprehensive country metasites, and this is not always easy. Good starting places are ***university area studies departments***. Many of these have country or regional webpages with links to other metasites. For example, the University of Texas's Russia and Eastern Europe Network Information Center (REENIC) has compiled an impressive set of links for this part of the world and makes a good starting place, as do Bucknell University's Russian Studies Department, Columbia University's Middle East and Jewish Studies.

Russia and Eastern Europe Network Information Center (REENIC) http://reenic.utexas.edu/

Bucknell University's Russian Studies http://www.departments.bucknell.edu/russian/

Middle East and Jewish Studies http://www.columbia.edu/cu/lweb/indiv/mideast/cuvlm/

Occasionally ***for-profit enterprises*** will provide excellent free resources, such as *Aardvark: Asian Resources for Librarians.* Aardvark has two sections: Literature and Recommended resources and sites. I especially like the Asian Databases section, which currently lists over 650 databases. Thompson Gale published an excellent overview of Aardvark by Professor Péter Jacsó.[140]

Aardvark: Asian Resources for Librarians http://www.aardvarknet.info/user/aardvarkwelcome/

Also, do not overlook ***non-profit organizations*** as a resource. Friends and Partners is probably the most famous non-profit organization on the Internet devoted to improving Russian-American relations and, as a result has become a tremendous source of information about Russia—everything from current news to telecommunications to downloading Cyrillic character sets.

Friends and Partners http://www.friends-partners.org/friends/

Try to find webpages that claim to index ***"all" the sites in a country***. Even though that claim is undoubtedly exaggerated, these sites are still good starting places. Another type of excellent country resource is a commercially operated country website that attempts to create an overarching ***subject directory of a particular country or region***. Every world region and most countries have such guides or directories, but the quality can range from poor to excellent. A good test is to look at categories such as "governments" or "technology." If these are sparsely populated, chances are the directory is oriented more to entertainment, travel, shopping, etc., than to research.

---

[140] Péter Jacsó, "Aardvark," Péter's Digital Reference Shelf, Thompson Gale, October 2006, <http://www.galegroup.com/reference/peter/aardvark.htm> (26 October 2006).

Al Bawaba
البوابة
Music
Email
E-Cards
Blogs
Forums
Webguide
عربي
Algeria Bahrain Cyprus Egypt Iran Iraq Jordan Kuwait Lebanon Libya Mauritania
Morocco Oman Palestine Qatar Saudi Arabia Sudan Syria Tunisia Turkey UAE Yemen
Search Albawaba
Search
Main
News
Entertainment
Business
Kids
Travel
Forums
E-cards
Albawaba's Email
Blogs
Music
Webguide
Amman suicide bomber was former US detainee released from custody
A US military spokesman revealed on Sunday that one of the suspects in Wednesday's attack on three of Amman's luxury hotels was a detainee who was released last year by the US after determining that he was not a security threat. The attack, for which Al Qaeda claimed responsibility, killed 57 people and injured at least 100.
Iraq: Roadside bomb kills three as 37 resistance fighters die in clashes
Israeli troops kill two Palestinians, including senior Hamas activist
Progress made on Gaza Strip matters at Rice-Abbas meeting
Iranian FM arrives in Damascus: Regional action needed to face US, Israel cooperation
More
Actor Bruce Willis offers $1 ...
Palestinian boy's organs donated ...
"The Simpsons" debuts in the ...
Tell Us What You Think
Do you think that the recent terror attacks in Amman were a result of the US invasion of Iraq?
Yes
No
Don't know
Downtown Living at Burj Dubai
EMAAR
Albawaba.com launches the first Arabic/English language 'blogging' service
Report: Egypt's mobile connections outstrip their fixed-line counterparts
Palestine: Inflation up 1.93% during October
Asteco welcomes ceiling on Dubai rents
National Bank of Dubai establishes US$ 1 billion Euro Medium Term Note Program

Al Bawaba: The Middle East Gateway http://www.albawaba.com/

Yahoo Countries http://dir.yahoo.com/regional/countries/index.html

If I am looking for information about a country with whose Internet presence I am unfamiliar, I often start with Yahoo's Country index. A simple alphabetic list of the nations of the world leads you to individual pages within Yahoo for each country, ranging in quality and quantity from the vast (the UK with nearly 222,000 entries) to the tiny (Kiribati with 20). Most Yahoo country pages, however, are fairly robust, with subject headings such as "business and finance," "computers and Internet," "government," "education," "news and media," etc. However, never assume the Yahoo country lists and links are thorough or even adequate representations of a country's Internet presence because they are not. Yahoo is nonetheless a very fine starting place for researching a country on the web.

Library of Congress Country Studies http://lcweb2.loc.gov/frd/cs/cshome.html

The Library of Congress website "contains online versions of books previously published in hard copy by the Federal Research Division of the Library of Congress under the Country Studies/Area Handbook Program sponsored by the US Department of the Army. Because the original intent of the Series' sponsor was to focus primarily on lesser known areas of the world or regions in which US forces might be deployed, the series is not all-inclusive. At present, 101 countries and regions are covered. Notable omissions include Canada, France, the United Kingdom, and other Western nations, as well as a number of African nations. The date of information for each country appears on the title page of each country and at the end of each section of text." Each country study presents historical, political, economic, social, and military analyses and descriptions. Users may search across all countries or any combination of countries or browse an alphabetical list of countries.

The Economist Country Briefings http://www.economist.com/countries/

The Economist is a well-respected publication that has made a huge contribution to the Internet with its website. Specifically, the Economist Intelligence Unit has a resource known as "Country Briefings" that looks at about 60 countries and provides superb profiles, forecasts, statistics, and more about each. All the content at the site is original with the Economist and not merely a compilation of information from other sources. Some of the material is "premium content," meaning users must pay to view it. However, there is more than enough free information to make the site very valuable.

BBC Country Profiles http://news.bbc.co.uk/1/hi/country_profiles/

Similar to its counterparts at the Library of Congress and The Economist, the BBC offers Country Profiles at its website. The profiles not only include information on the history, politics, and economy of most of the world's nations, they also offer audio and video clips from the BBC's extensive archives. The main Country Profiles' page presents six drop-down menus for world regions with countries in each region listed alphabetically and one menu for International Organizations. Once users select a specific country to profile, they are offered an overview, a set of basic facts, information on current leadership, and details about the country's media. The BBC profiles are not quite as detailed as those from the Library of Congress or The Economist, but the BBC does provide a very good and user-friendly overview of most nations.

Northwestern University Library Foreign Governments
http://www.library.northwestern.edu/govpub/resource/internat/foreign.html

Northwestern University Library International Governmental Organizations
http://www.library.northwestern.edu/govpub/resource/internat/igo.html

Unrepresented Nations and Peoples Organization (UNPO) http://www.unpo.org/

The Northwestern University Library has a superb collection of documents related to governments, international organizations, and unrepresented nations and peoples. The governments' page has links to websites of countries from Afghanistan to Zambia. Even better, the page is kept up to date. Obviously, every website for every government is not listed, but the resource focuses on the major sites, such as president/prime minister, MFA, central bank, parliament, etc.

The international governmental organizations page lists intergovernmental organizations (IGOs) that have webpages. Most are in English and generally the list links only to the main page, though there are exceptions. Organizations cover a wide range of topics, from the African Development Bank to the Chemical Weapons Convention to the International Criminal Court.

Finally, the Unrepresented Nations and Peoples Organization (UNPO) page, which resides at a separate website, lists both current and former members. UNPO members comprise nations and peoples inadequately represented at the UN. Among the UNPO current members are Australian Aboriginals, Assyrians, Chechens, Iraqi Turkoman, etc. For each group, the site links to a UNPO page containing information about the people and their homeland, including geographical, economic, and historical information as well as links to UNPO documents relating to the nation/people.

Academic Info http://www.academicinfo.net/

I am somewhat hesitant to recommend this site only because I can't find out much about who is running it (Mark Madin and a group of "volunteer subject matter experts"). However, many reputable sources link to and recommend Academic Info, including the superb University of California, Berkeley, library, which describes it as a "Rich selection of about 25,000 pages, selected as 'college and research level Internet resources' aimed at "at the undergraduate level or above." And it does contain a lot of useful information. The information appears to be fairly high-level and most of it is collected elsewhere. However, this is a good general starting place and has the advantage of having been updated frequently. Because of the currency, quality, and amount of information throughout the website, I would recommend bookmarking and using the site as a good starting place for research on a number of topics.

Google Directory Country Index
http://directory.google.com/Top/Regional/Countries/

Google Directory uses the Open Directory Project listings as a basis but adds to them and includes more links.

Admi.net http://admi.net/world/

Admi.net is a French website that attempts to be "the cyber-documentation center delivering general information about government authorities and public services." Admi.net's scope, therefore, is more limited than some others, but may thereby be more useful under some circumstances. The site is very well organized: *General Information/Regions/*Countries. Within each subheading are the same subject headings, from "political organizations" to "law" to "companies" to "telecom" and much more in between. Obviously, much more information is indexed about some countries than others and not every heading has an entry for each country.

BUBL Country List http://bubl.ac.uk/link/world/index.html

Carefully researched and annotated web directory by information professionals.

WWW Virtual Library Regional Studies http://vlib.org/Regional

Links to African, Asian, Indigenous, Latin American, Middle Eastern, Russian, Eastern European, Western European, and Pacific studies. The Virtual Library was established in 1993 and is still kept up to date.

The Organization for Economic Co-operation and Development (OECD)
http://www.oecd.org/

The Organization for Economic Co-operation and Development (OECD) has launched webpages for each member country. The main features for each country website are:

- *What's new*: News, recent and forthcoming events or conferences, latest publications.
- *Statistics*: OECD statistical profile, OECD free access databases by country, and more.
- *Country and the OECD:* ratification of the convention, role and activities of the delegation, financial contribution to the budget, etc.
- *Publications & documents:* documentation listed in chronological order, by topic or document categories.
- *Information by topic:* every topic from the ageing society to transportation.
- *Don't Miss:* Country profile including over 100 indicators, how to obtain OECD publications, list of translations, useful links and contact us.

What countries are included in the OECD project? All thirty member nations have country webpages and most of those country's also have a separate statistical profile page. OECD also has active relationships with 70 non-member countries. The OECD site includes numerous publications and documents, such as an Economic Survey of Russia (July 2004), Main Economic Indicators: Non-Member Countries, April 2005 **(PDF)**, as well as information about co-operation operation between OECD member and non-member nations by country. The OECD site also permits users to browse the site by topics, including biotechnology, corruption, energy, finance, money laundering—32 topics in all.

NationMaster http://www.nationmaster.com/

NationMaster takes the data from the CIA World Factbook, the UN, and OECD and mines it for statistics, then makes all that information easily accessible. Also, users can easily generate maps and graphs on all kinds of statistics. One of NationMaster's nicest features is the easy ability to view profiles of individual countries, which includes many details about a country beyond just statistical data. There is no original data at NationMaster, but it is the single best source for pulling together, organizing, searching, and displaying data about every country on earth.

Finally, look for ***directories and country-specific search engines within the country itself***. In my experience, the quality of such sites varies greatly, so do not give up if the first few you look at are not impressive. Once you have identified the major ISPs in a country, look to see if they have subject directories; many ISPs are creating both national and local guides for Internet users. The best will provide internal search engines and maybe even a translation feature. For example, one of Greece's major telecommunications and Internet companies, FORTHNet, offers a directory of more than 32,000 Greek websites.

FORTHNet Directory http://dir.forthnet.gr/index-0-en.html

How do you find these sites? There is no one way of detecting the best websites in or about a country but try this: once you find a good website in or about a country, ***run a "link" command*** in Yahoo, Google, Live Search, and Gigablast to see webpages with links to the page of interest. You can quickly scan these for interesting-looking new websites.

How To Research a Country

There is no sure-fire methodology for researching a specific country on the Internet. Try these sources:

1. University area studies departments
2. Non-profit organizations or foundations
3. Country-specific subject directories
4. In-country guides, portals, directories (check big ISPs)
5. Search engines' "link" command

Finding Political Sites on the Web

If a country has a parliamentary form of government (and most do by the broad definition used by this organization), be sure to check out the **InterParliamentary Union**'s web page. Established in 1889, the IPU is an organization of parliaments of sovereign states. The IPE site has its own superb PARLINE database of country parliaments, which includes all relevant details about the form and structure of governments. In addition, the IPU provides links to official parliament web pages. Of the more than 200 countries listed, not all have official parliamentary web pages, but a surprising number do.

InterParliamentary Union http://www.ipu.org/english/home.htm

Northwestern University's Foreign Governments
http://www.library.northwestern.edu/govpub/resource/internat/foreign.html

Northwestern University's International Governmental Organizations
http://www.library.northwestern.edu/govpub/resource/internat/igo.html

Also be sure to look at Northwestern University's Foreign Governments list as well as its International Governmental Organizations. Currently up to date.

Global Edge http://globaledge.msu.edu/

The Center for International Business Education and Research at Michigan State University maintains Global Edge, a site of international country studies and business information with links to over 5000 resources. Much of the information comes from the CIA's Country Studies and the Department of State's Country Background Notes, but the site adds a tremendous amount of value to this data.

Foreign Government Resources on the Web
http://www.lib.umich.edu/govdocs/foreign.html

The University of Michigan also maintains Foreign Government Resources on the Web, a site with links to many types of foreign government sites, including official sites, embassies, constitutions, laws, etc. Last updated in July 2006.

Political Resources on the Net http://www.politicalresources.net/

Political Resources on the Net: Unrepresented People
http://www.politicalresources.net/int6.htm

Political Database of the Americas http://www.georgetown.edu/pdba

One of the best political websites is Political Resources on the Net. This is a superb site for locating websites for governments, political parties (mainstream and fringe), NGOs, institutes, and "unrepresented" peoples or areas, such as Kosovo, East Timor, Hmong, etc. The topics are arranged from global to regional to country-specific. For *anything* political, this is a very good first stop. However, the internal search does not work. For political information on the Americas, try Georgetown University's Political Database of the Americas; the site contains both original information as well as links to reference sources.

European Countries http://europa.eu/abc/european_countries/index_en.htm

At the EU's Europa site, each of the European Union member states has a link from this page. The newly redesigned site has far less information about member nations' government site and now only provides a link to each country's official website. Europa still provides information about and links to EU candidate countries and to other European nations.

Council of the Baltic Sea States http://www.cbss.st/

The Council of Baltic Sea States also offers a website with information including news, history, the structure of the Council, lots of documents, contact lists, etc. The site is completely in English, though external links to member states are usually in the appropriate national language. The homepage includes a map with links to the foreign ministries of each country.

East & Southeast Asia: An Annotated Directory of Internet Resources
http://newton.uor.edu/Departments&Programs/AsianStudiesDept/index.html

The East & Southeast Asia annotated directory of Internet resources from the University of Redlands (in California) is an excellent portal for information specific to Asian nations as well as Asian resources in general as well as "hot topics" such as the avian flu, the North Korean nuclear crisis, and the tsunami of 2004.

Current Rulers Worldwide http://www.terra.es/personal2/monolith/

Rulers of the World http://rulers.org/

For information on world leaders, be sure to see Zarate's Political Collections, which includes Current Rulers Worldwide. The site is updated regularly. Rulers of the World is also kept current but has an accent on history: "This site contains lists of heads of state and heads of government (and, in certain cases, de facto leaders not occupying either of those formal positions) of all countries and territories, going back to about 1700 in most cases." Current rulers are also included.

Finding Ministries of Foreign Affairs

To track down ministries of foreign affairs, you can try directories such as Google's, Yahoo's, and the Open Directory Project, or you can go directly to one of these resources that provide links to MFAs. Please keep in mind that no resource has everything. ***Just because the MFA you seek isn't listed by any of these sources does not mean it's not on the Web.*** Sometimes url guessing will work, e.g., *www.mfa.gov.yu* for the former Yugoslavia; sometimes it won't, e.g., *www.ud.se* for Sweden.

US Institute of Peace Library Foreign Affairs Ministries on the Web
http://www.usip.org/library/formin.html

The site has about 140 links to most MFAs and a few other international agencies. Current information.

Library of Congress: Portals to the World
http://www.loc.gov/rr/international/portals.html

First-rate alphabetical list of countries, with links to government sites. For most, finding the MFA is a simple click to the Government, Politics, and Law link.

Stefano Baldi's Ministries of Foreign Affairs Online
http://hostings.diplomacy.edu/baldi/mofa.htm

Hosted by Diplo Directory site, this is a superb resource, with 110 links listed.

Ministries of Foreign Affairs from Lawresearch
http://www.lawresearch.com/v10/global/ciministries.htm

Links to about 100 countries' MFA websites.

Finding Embassies

One task the Internet has certainly made much easier is locating information about embassies, including their addresses, phone/fax numbers, names of individuals associated with them, and even email addresses and, sometimes, websites. Even more fortunately, there are several websites that specialize in providing information about and links to foreign ministries on the web.

Also keep in mind that ministries of foreign affairs (MFAs) are usually in the business of providing information rather than protecting it, so they are often great resources for learning a wide range of details about a country. ***I strongly recommend that you always check out a country's MFA Web site for the most current information about its missions.*** For example, a country such as Afghanistan has gone through many changes in the past few years both in terms of representation from and within the country. Never assume any site is absolutely current; it's best to go directly to the source when possible. Here is the Afghani MFA's list of its embassies and consulates from its website.

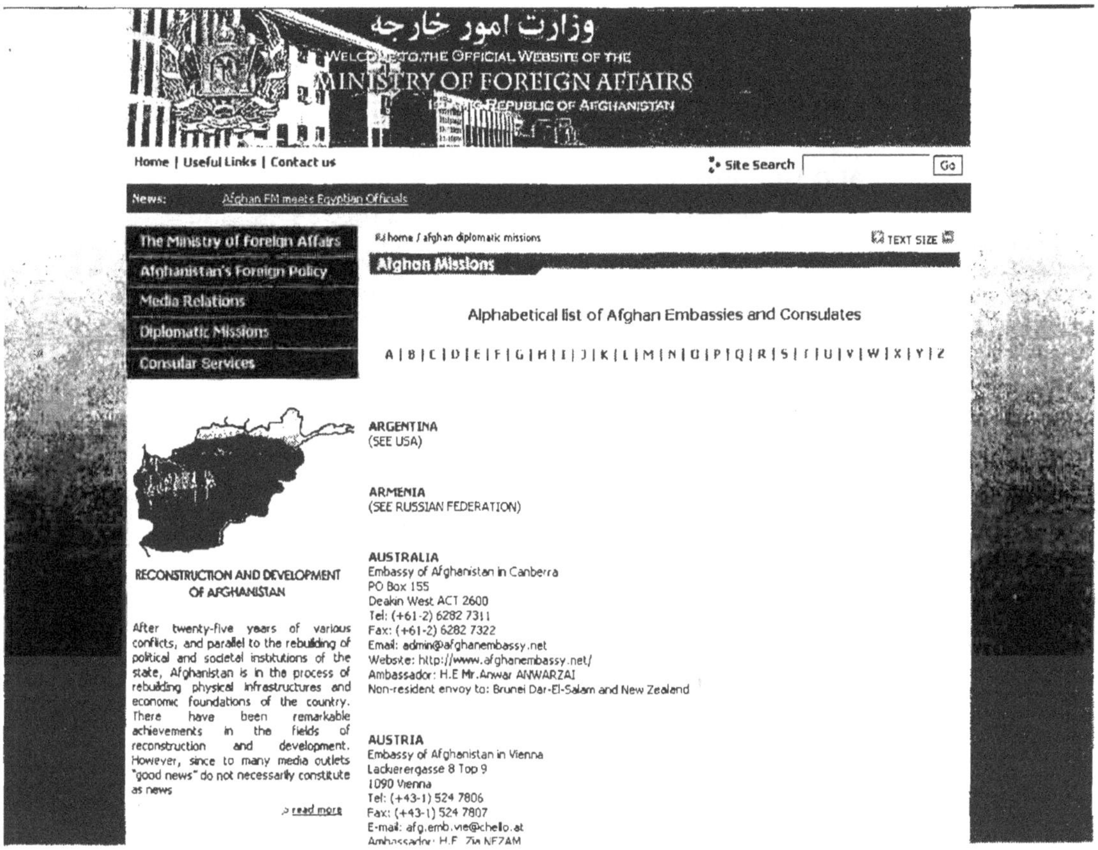
وزارت امور خارجه
MINISTRY OF FOREIGN AFFAIRS
Home | Useful Links | Contact us
Site Search
Go
News:
Afghan FM meets Egyptian Officials
The Ministry of Foreign Affairs
Afghanistan's Foreign Policy
Media Relations
Diplomatic Missions
Consular Services
TEXT SIZE
Afghan Missions
Alphabetical list of Afghan Embassies and Consulates
A | B | C | D | E | F | G | H | I | J | K | L | M | N | O | P | Q | R | S | T | U | V | W | X | Y | Z
ARGENTINA
(SEE USA)
ARMENIA
(SEE RUSSIAN FEDERATION)
AUSTRALIA
Embassy of Afghanistan in Canberra
PO Box 155
Deakin West ACT 2600
Tel: (+61-2) 6282 7311
Fax: (+61-2) 6282 7322
Email: admin@afghanembassy.net
Website: http://www.afghanembassy.net/
Ambassador: H.E Mr.Anwar ANWARZAI
Non-resident envoy to: Brunei Dar-El-Salam and New Zealand
AUSTRIA
Embassy of Afghanistan in Vienna
Lackierergasse 8 Top 9
1090 Vienna
Tel: (+43-1) 524 7806
Fax: (+43-1) 524 7807
E-mail: afg.emb.vie@chello.at
RECONSTRUCTION AND DEVELOPMENT OF AFGHANISTAN
After twenty-five years of various conflicts, and parallel to the rebuilding of political and societal institutions of the state, Afghanistan is in the process of rebuilding physical infrastructures and economic foundations of the country. There have been remarkable achievements in the fields of reconstruction and development. However, since to many media outlets "good news" do not necessarily constitute as news
read more

The following are sites helpful in finding embassies:

Embassy World http://www.embassyworld.com/

Embassy World is a commercial site that has improved dramatically over the last few years. Today, it is probably the single best resource for finding embassies and consulates. The home page offers users the option to search for a specific country's mission in any or a single location. You can also use the alphabetical list by country to find a country's missions in other countries or other countries' missions in a specific country. I recommend searching both ways because the results are not always identical. Embassy World also includes UN Missions and an international telephone directory that varies in quality depending on the country of interest (poor for Palau, great for France...you get the idea).

Library of Congress: Portals to the World
http://www.loc.gov/rr/international/portals.html

First-class alphabetical list of countries, with links to government sites. Virtually all include a link to Embassies.

Embassies & Consulates Worldwide http://www.mypage.bluewin.ch/caccia/

Search by country in which embassy or consulate is located. This site is current and contains other useful information about countries, generally geared toward visitors.

Yahoo Embassies and Consulates
http://dir.yahoo.com/Government/Embassies_and_Consulates/

An always-reliable if incomplete source; countries listed alphabetically with both embassies of and embassies in each country.

Tagish Worldwide Embassies http://www2.tagish.co.uk/Links/embassy1b.nsf/

Tagish has hosted the European Union's Ethos site for some years and has a separate page devoted to diplomatic missions derived from the Ethos data, which now appears to be a couple of years out of date. Specifically, the Tagish site also includes a separate ***list of embassy Web sites and embassy email addresses*** (both sorted by either host country or country of origin).

Embassy.org http://www.embassy.org/

Embassy.org is *the* source for information about any embassy in Washington, D.C.

Research Round-up: The Best Research Tips & Techniques

Tip 1: Use the Right Tool

I am repeating Rule #1 because it is so important. The single biggest mistake researchers make is using the wrong search tool. For example, search engines are generally not useful for finding current news (use a specialized news search service). Wikis, custom search engines, and directories are generally better when researching a broad topic because they have a select group of sites (but watch out for paid placement; for a directory that contains *no* paid placement, use Open Directory). There are also many specialized or vertical search services that cover a huge variety of topics, everything from chemistry to message boards.

Tip 2: Search for the Most Obscure Term

I needed to find the url for an article at CNET's *News.com*. Sounds easy. I knew the date and title, but the internal CNET search tool wouldn't let me search by date and the title words returned nothing. Solution? I picked an obscure word from the article—in this case, the last name of the Microsoft executive interviewed for the article, ("wallent") and searched on that term. Bingo. Only one hit and it was the article I wanted.

Tip 3: Put the Most Important Search Term First

While it's not always true, search engines usually give more weight to the first term you list because the search software assumes it's the most important term (otherwise, why would you list it first)? Try these two queries in Google one after the other: [gardening roses] then [roses gardening]. The results are similar but not identical.

Tip 4: Search on the Singular Form First

While it is not always the case that search engines automatically search for plural forms of search terms, many (including Yahoo and Google) do. The converse, however, is not true, i.e., a search on [rose] will find *roses*, but a search on [roses] will not find *rose*. Therefore, it makes sense to search first on the singular form of a term.

Tip 5: Use Regional Search Services, Directories, and Databases

It's so easy to fall into the trap of using US-based search engines (and, it's true, their databases are the biggest in the world). But I believe it is critical to use search engines, directories, and databases that are country- and region-specific, or language-specific. These international sites focus on collecting data from a particular part of the world and also offer more language options than US-based services. Don't just think in terms of search engines—local phone and email directories can be invaluable.

Tip 6: Search in the Native Language

If you really want to exploit the power of an international search engine, directory, or database, you need to search in the native language. Doing so will vastly improve your chances for finding what you seek. Also, remember that many if not most international databases—for example, phone directories—only list information in the native language, so searching the online Moscow phone book for a name in English will produce zero results because all the listings are in Russian Cyrillic.

Look at the difference between searching using Google in English and in Russian [Emercom Russia] with 32K hits vs. [МЧС России] with 1.25M hits and, more importantly, the Emercom site as the top result:

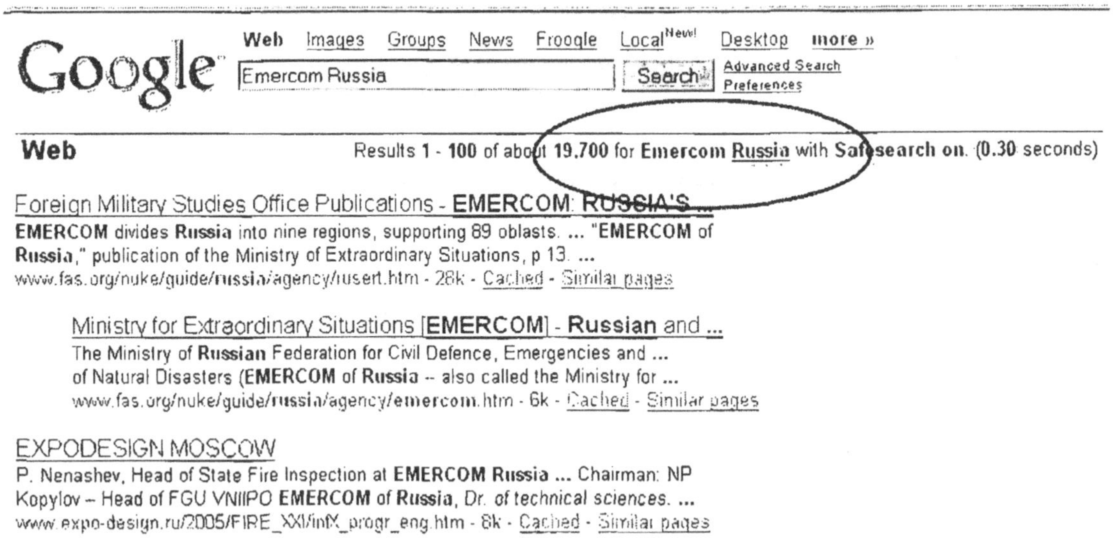
Google
Web Images Groups News Froogle Local New! Desktop more »
Emercom Russia
Search
Advanced Search
Preferences
Web
Results 1 - 100 of about 19,700 for Emercom Russia with Safesearch on. (0.30 seconds)
Foreign Military Studies Office Publications - EMERCOM: RUSSIA'S ...
EMERCOM divides Russia into nine regions, supporting 89 oblasts. ... "EMERCOM of Russia," publication of the Ministry of Extraordinary Situations, p 13. ...
www.fas.org/nuke/guide/russia/agency/rusert.htm - 28k - Cached - Similar pages
Ministry for Extraordinary Situations [EMERCOM] - Russian and ...
The Ministry of Russian Federation for Civil Defence, Emergencies and ...
of Natural Disasters (EMERCOM of Russia -- also called the Ministry for ...
www.fas.org/nuke/guide/russia/agency/emercom.htm - 6k - Cached - Similar pages
EXPODESIGN MOSCOW
P. Nenashev, Head of State Fire Inspection at EMERCOM Russia ... Chairman: NP
Kopylov – Head of FGU VNIIPO EMERCOM of Russia, Dr. of technical sciences. ...
www.expo-design.ru/2005/FIRE_XXI/infX_progr_eng.htm - 8k - Cached - Similar pages

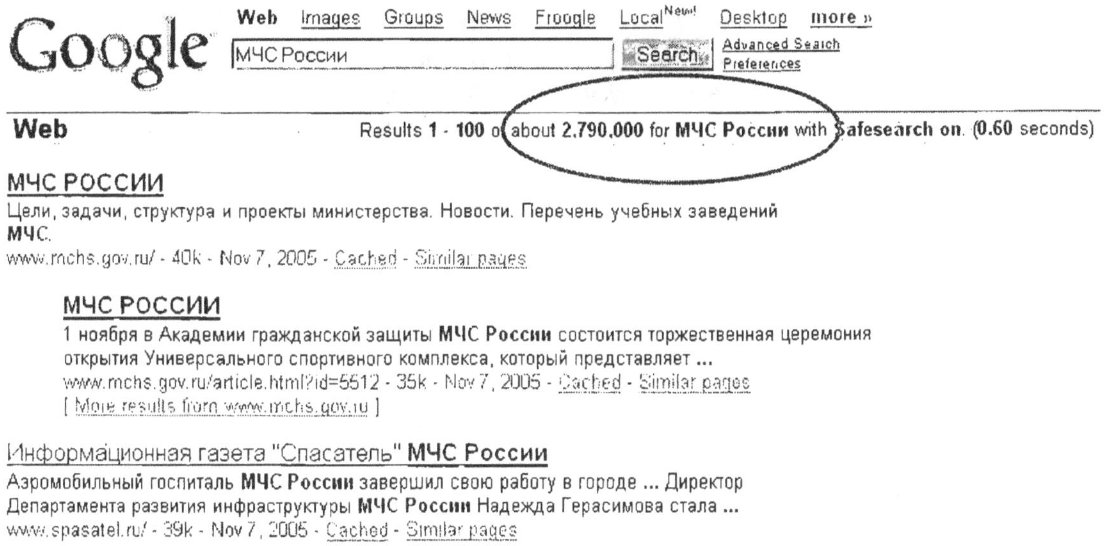
Google
Web Images Groups News Froogle Local New! Desktop more »
МЧС России
Search
Advanced Search
Preferences
Web
Results 1 - 100 of about 2,790,000 for МЧС России with Safesearch on. (0.60 seconds)
МЧС РОССИИ
Цели, задачи, структура и проекты министерства. Новости. Перечень учебных заведений МЧС.
www.mchs.gov.ru/ - 40k - Nov 7, 2005 - Cached - Similar pages
МЧС РОССИИ
1 ноября в Академии гражданской защиты МЧС России состоится торжественная церемония открытия Универсального спортивного комплекса, который представляет ...
www.mchs.gov.ru/article.html?id=5512 - 35k - Nov 7, 2005 - Cached - Similar pages
[More results from www.mchs.gov.ru]
Информационная газета "Спасатель" МЧС России
Аэромобильный госпиталь МЧС России завершил свою работу в городе ... Директор Департамента развития инфраструктуры МЧС России Надежда Герасимова стала ...
www.spasatel.ru/ - 39k - Nov 7, 2005 - Cached - Similar pages

Tip 7: Follow Those Links

Whenever you find a good website, always check its links. While in theory links at a web page that is indexed by a search engine should also have been indexed, the reality is often different. "Links" pages are often a gold mine of sites with similar information.

Tip 8: Learn Two Words in Any Non-English Language in Which You are Searching

Those two words are *search* and *links*. You need to be able to push the *search* or *find* button on a non-English web page, and you need to be able to find the *links* page.

Tip 9: Search on the LINK Field

Every time you find a good website, go to a search engine, such as Yahoo[141], Gigablast, the new Live Search, or Google[142] that permits link: command searches and find the pages that link to your newly discovered web page. ***Any page that is really interesting to you is also interesting to others interested in the same subject.*** Check the search syntax of each site: some search engines, such as Yahoo, require a full address, including the http://.

---

[141] Yahoo and Live Search have two different link commands (link: and linkdomain:) that serve different purposes. Linkdomain: is a broader, more comprehensive search option.

[142] The Google *link:* command is not showing all links as it once did because of Google's efforts to try to limit webspam.

Here's an interesting twist on link searching, that is, finding sites that link to a specific address. This search, which works with Yahoo and (to a lesser extent) Windows Live Search, ***finds pages that link to a specific domain or domains but not to another specific domain or domains***. An example would help. Let's say I start by finding the sites that link to the Iranian Ministry of Defense. Here is the query I would use:

[linkdomain:mod.ir]

In Yahoo this query returns 545 hits. Now, suppose I want to see which sites link to both the Iranian MOD and the Iranian Electronics Industries. I can do that easily with this query:

[linkdomain:mod.ir linkdomain:ieimil.com]

However, I see lots of sites that also link to the ever-present CIA World Factbook, which, while a wonderful resource, isn't want I want. I would really like to see the sites that link to both the Iranian MOD and IEI sites but not to the CIA Factbook. Can I do this? Sure:

[linkdomain:mod.ir linkdomain:ieimil.com -linkdomain:cia.gov]

While this technique has obvious applicability for search engine optimization ("who is linking to my competitors but not linking to me?"), I think it is worth knowing about because you may come up with some creative ways to use it. Just as an interesting example, try these two queries in both Live Search and Yahoo. It's interesting to see what drops from the results' list on the second query.

[linkdomain:cia.gov linkdomain:nsa.gov]

[linkdomain:cia.gov linkdomain:nsa.gov -linkdomain:fbi.gov]

I believe you will consistently find that Yahoo! provides more results than Live Search for the linkdomain: searches. However, the results will vary, so it's worth using both search engines. Google does not offer a linkdomain: search, and its link: search has been hobbled.

Yahoo Search http://search.yahoo.com/

Windows Live Search http://www.live.com/

Tip 10: Look Beyond Search Engines and the Web

Search engines and directories index only a tiny portion of the Internet. With some notable exceptions, they are basically designed to index web pages. A vast amount of data is stored, for example, in online databases, many of which are free and open to the public. They often contain information useful to researchers—phone directories, Whois databases, NIH PubMed, SEC Edgar, Amazon's "Search Inside the Book" feature, digital library collections. The Domain Name System/Service (DNS) itself is the largest distributed database ever created and freely accessible to any user via a simple query (NSLookup). Also, don't overlook mailing lists, newsgroups, and other non-web segments of the Internet.

Tip 11: Configure and Use Two Browsers

If you spend much time on the Internet, especially viewing non-US sites, you will probably encounter certain webpages that will not display at all, will not display properly, and/or will not print in the browser you're using. However, if you open the page in the other browser, it may be fine. So don't despair if a page isn't displaying or printing as it should. Chances are there are simply problems with the way the page was created and it will look fine in the other browser.

Tip 12: Try URL Guessing

It works more frequently than you would imagine. For example, I found the Iranian Ministry of Foreign Affairs by guessing *www.mfa.gov.ir*. And guess what the address for the Russian Ministry of Internal Affairs (MVD) is? Yes, it's *www.mvd.ru*. No search engine indexed either of these sites at the time I first found them.

Tip 13: Change URLs to Find "Hidden" Webpages

Sometimes a simple change inside a long url will disclose interesting pages deep within a website. For example, look at these two pages from the Federal Trade Commission:

http://www.ftc.gov/opa/2006/02/

http://www.ftc.gov/opa/2007/02/

Simply by changing the portion of the url that indicates year and month, you can view the FTC News Releases for a specific date. This is a simple example of a technique that can be used to uncover "hidden" webpages. It's especially useful on sites that update on a regular schedule, e.g., sites for press releases or news.

Tip 14: Be on the Lookout for URL Errors

Not surprisingly, many urls listed on webpages are incorrect. Among the most common mistakes are misspellings, putting a backslash (\) where a slash (/) should be, including or excluding the L in HTML, e.g.:

http://www.examlpe.com/pathname\bigmistake.html

Reload
http://beyond-asso.com/enlgish/projects.htm
Print
Completed Projects
Beyond A. L. has extensive experience in working with multinational companies and a proven track record of successfully managing and completing projects on time and with in budget.
Project List
"English" is often misspelled as "Enlgish"
MaryKay (China) Co., Ltd. (美国玫琳凯化妆品公司)
Beyond designed and built new office for Marykay's in 10 cities in China, which total spaces is more than 12,000 sq. m2.
Siemens (西门子公司)
Beyond designed and built 1500 sq. m2 new office at 20/F, Majesty Tower.
Bellalcatel Business System Shanghai Co., Ltd (阿尔卡特通信公司)
Beyond designed and built 2000 sq. m2 new office at Waigaoqiao area.
Simmons &Simmons Solicitors(西蒙斯律师行)
Beyond designed and constructed 500 sq. m2 new office at 3/F, Dynasty Tower.
New York Bank Shanghai Office(美国纽约银行上海代表处)
Beyond designed and built 800 sq. m2 totally new office at 5/F, Dynasty Tower.

Correct the spelling or use the tried and true method of "backward hacking" (backspace deleting) file and path names and retrying the link (you'll be amazed how often this works!).

Tip 15: Take a Look at the "Site Map"

Most large, well-organized websites have maps of what is on their web server. These site maps can be extremely useful and revealing, often showing pathways to more topics than were revealed on the home page. Also, internal search tools are notoriously unreliable; a site map may be much more useful at digging down inside a website.

Tip 16: Try Using the "Mouseover"

For non-English sites where you don't know the language, try the "mouseover" trick, i.e., move your mouse over hyperlinks. Often, the link information is in English or, if it isn't, quite often the url that appears in the toolbar at the bottom of the browser is revealing because it is likely to be written in English.

Tip 17: Try Alternative Spellings, Especially of Non-English Names or Terms, or Even Purposeful Misspellings

This hint especially applies to searching for non-English websites. For example, the Arabic word *khalifah* (usually written as *Caliph* in English) generally refers to the person who is the successor to Muhammed for leadership of the Islamic community. This term is transliterated as *kalifa*, *kalifah*, *califah*, or some other variant. However, there is a related term *khilafah*, which is the office itself, i.e., *Caliphate* in English. This term is often associated with some of the most radical Islamic websites. I found I had to search on many variations of these terms or I would have missed a number of important sites. Even though many of the websites are in Arabic, the addresses for the sites are still written in the Latin alphabet, and searching for these key words using syntax as *inurl:*, *site:*, *domain:* revealed many useful links.

The same is true for commonly misspelled words, e.g., *http referrer* is usually misspelled *http referer* and *genealogy* is often misspelled *geneology.* A wildcard makes it possible to find both correctly and incorrectly spelled terms.

As you can see from the following example, a search on ["ministry of foregin affairs islamic republic of iran"] elicits 918 hits while a search with *foreign* spelled correctly returned only 38 results.

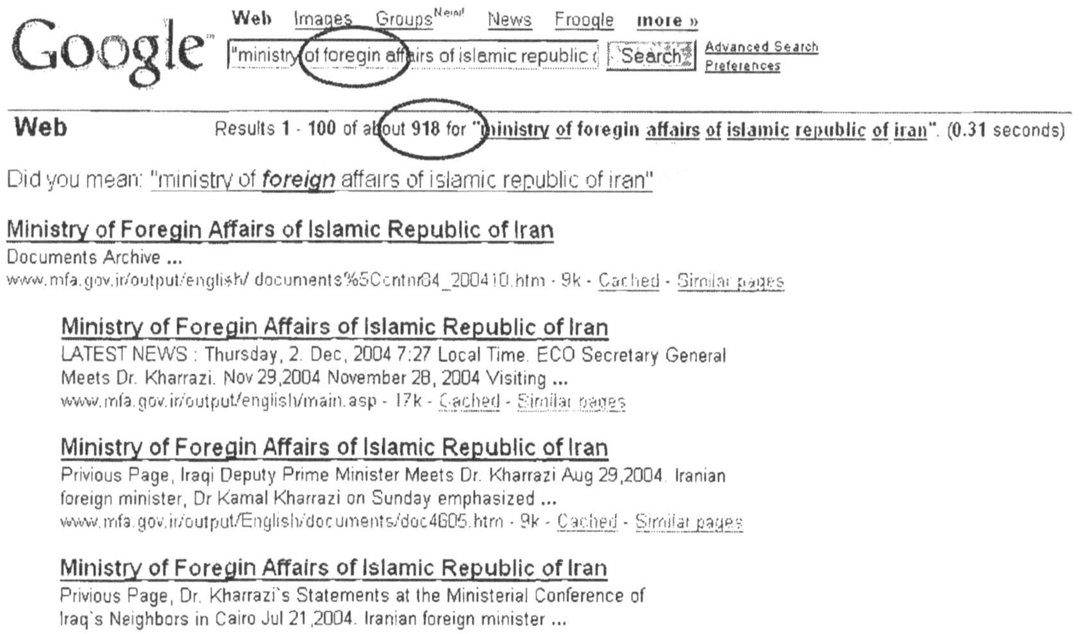
Google
Web Images Groups New! News Froogle more »
"ministry of foregin affairs of islamic republic (Search
Advanced Search
Preferences
Web
Results 1 - 100 of about 918 for "ministry of foregin affairs of islamic republic of iran". (0.31 seconds)
Did you mean: "ministry of *foreign* affairs of islamic republic of iran"
Ministry of Foregin Affairs of Islamic Republic of Iran
Documents Archive ...
www.mfa.gov.ir/output/english/ documents%5Cntnr34_200410.htm - 9k - Cached - Similar pages
Ministry of Foregin Affairs of Islamic Republic of Iran
LATEST NEWS : Thursday, 2. Dec, 2004 7:27 Local Time. ECO Secretary General Meets Dr. Kharrazi. Nov 29,2004 November 28, 2004 Visiting ...
www.mfa.gov.ir/output/english/main.asp - 17k - Cached - Similar pages
Ministry of Foregin Affairs of Islamic Republic of Iran
Privious Page, Iraqi Deputy Prime Minister Meets Dr. Kharrazi Aug 29,2004. Iranian foreign minister, Dr Kamal Kharrazi on Sunday emphasized ...
www.mfa.gov.ir/output/English/documents/doc4605.htm - 9k - Cached - Similar pages
Ministry of Foregin Affairs of Islamic Republic of Iran
Privious Page, Dr. Kharrazi's Statements at the Ministerial Conference of Iraq's Neighbors in Cairo Jul 21,2004. Iranian foreign minister ...

Tip 18: Always Look at a Website's Native Language Version

Usually, the native language version of a website will differ from the English version, sometimes a little, sometimes a great deal. To illustrate this point, check out the differences between the Al Jazeera Arabic and English home pages. You will see a different set of stories with different emphases on each.

DOCID: 4046925

ALJAZEERA.NET
Updated on: Monday 14 November 2005, 16:44 Makka Time, 13:44 GMT
TOP NEWS
SEARCH ALJAZEERA
Advanced Search
Homepage
News
Economy
Culture
Sci-Tech
Special Reports
Weather
Polls
Your feedback
Contact Us
About Aljazeera
Code of Ethics
Services
Frequencies
Gaza border deal near, Abbas says
Palestinian President Mahmoud Abbas said an Israeli-Palestinian deal on reopening the Gaza-Egypt border was imminent, after talks with US Secretary of State Condoleezza Rice. FULL STORY
• Rice pays surprise visit to Iraq
• Middle East reform talks collapse
Three killed in Baghdad bombing
French cabinet renews emergency law
Israeli troops kill Hamas leader
Palestinian polls in Hebron delayed
AUDIENCE PANEL
Click here to join
Market Watch
Interviews
"Peretz' win could be a new watershed moment for Israel"
"The bomb was designed to injure and kill as many people"
In pictures
Jordanians protest hotel blasts
Palestinians remember Arafat
Arab World
• Failed bomber confesses on Jordan TV
• Sudan rejects US peace initiative
• Another journalist assaulted in Egypt
• Syria official rejects Lebanon hearings
Global
• Four killed Kashmir separatist attack
• Three arrested over New Delhi blasts
• Azeri opposition calls for poll rerun
• Bomb kills two judges in Bangladesh
Economy
• S Africa mines attract black partners
• Airbus A380 completes Asia test flight
• Workers doubt Mubarak job pledge
• Boeing sets new flight record
Science & Technology
• UN summit to debate internet control
• Conservationists warn of new dams
• Australia headed for hottest year ever
• HIV man claims body cured itself
Culture
Special Reports
islamweb.net

Notice the completely different set of articles and even the way the pages are organized:

ابحث عن
تفجيرات فنادق عمان
تيسير علوني
سامي الحاج

Tip 19: Use Wildcards to Maximize Effectiveness

I love to use wildcards in searching. Unfortunately, few search engines permit wildcard searches, which means the researcher must enter the term in a variety of forms for a thorough search. Here is how the major search engines handle wildcards.

Google: one limited wildcard (*); can only replace any single term with white space on either side (e.g., ["what a * web"] will find "what a tangled web" and "what a coiled web"). Cannot be used within or at the end of a search term (for example, to pluralize a term).

Yahoo: Yahoo does not support wildcard searching. The old cheat of using a "small" word, such as *a*, no longer works in Yahoo.

Live Search: no truncation, no wildcard. A search for [cat] finds *cat*, not *cats*.

Exalead: a search on [child*] return pages with *children* highlighted as a search result. The wildcard also can be used inside a search term, e.g., [kazak*stan]. This search accurately finds *kazakhstan*, *kazakh*, and *kazak* as well as *kazakstan*.

Tip 20: Examine Page Source Code

In addition to often revealing the webpage's language encoding, page source can provide other helpful details, including names, dates, email addresses, type of software used to create the page, etc. However, experienced webpage designers have learned by now that putting these sorts of details into the source code is an open invitation to spammers to harvest them, so finding useful bits of information in source code is less likely now than in the past. Still, many people are not experienced Internet users and have not yet learned to keep this information out of the source code, so it is worth a look. Below is a very good example of how analyzing page source code helped one company track down the person responsible for a fabricated television interview that spread potentially libelous information about the company. The page source contained an email address that ultimately led to the person responsible for the false information.

Daniel Janal, "How to Deal with Lies About Your Company (and You) on the Internet," Scambusters.org http://www.scambusters.org/Scambusters29.html

Tip 21: Ask for Help

One of the hardest questions to answer is, "how do you know when you aren't finding something because you're not searching correctly or because it's not there to find?" The best rule of thumb I can think of is to ask a more experienced Internet researcher for advice and assistance if you are hitting a brick wall. Experienced researchers generally have developed a pretty good sense of where to look for different kinds of information and, most importantly, the types of information you are not likely to find on the Internet.

Researching & Understanding the Internet

A Plain English Guide to Internetworking

The Internet, were it a person, would be something of a narcissist. It probably has more information about itself than any other single topic, which means the best place to research the Internet is the Internet. However, understanding how the Internet actually works can be frustrating and most books, websites, and other resources that purport to explain the inner working of the Internet exacerbate rather than enhance knowledge because they make assumptions about their readers that are often wildly inaccurate.

There are generally two types of guides on how the Internet works: one is the technical guide for network engineers and system administrators; the other type is a users' guide, a how-to book for people who want to use the Internet but don't really need to know anything about what's under the hood. The first type of guide is fine if you are an engineer, but these manuals or websites are written by and for technical people, meaning you already have to know their jargon in order to understand what they are saying. The second type of guide merely avoids the problem all together by ignoring technical explanations. After all, you don't need to know how an internal combustion engine works to drive a car.

I have attempted to create a little niche in between the extremes by offering a non-technical explanation of some of the basic concepts behind how the Internet gets traffic from one point to another and does it with remarkable speed, accuracy, and efficiency. For many of us who are not technically inclined, the answer has often seemed to be "magic happens here." But more and more often in our jobs we find we need to know a little bit about the mysterious inner workings of the Internet. I have no pretensions that this overview is complete. I knowingly gloss over some subtleties that I think are simply too arcane for a high-level look at Internetworking.

The Internet

The Internet isn't a single network; it is a network of networks. However, all these many networks are not directly connected to each other and, in fact, do not "talk" or exchange data directly with each other. Rather, the Internet employs an elaborate system of rules (known as protocols), numbering schemes (e.g., autonomous systems, IP addresses), hardware (e.g., routers and servers), and arrangements (e.g., peering agreements) to ensure that your request to view a website in Pakistan or send an email to Seoul is fulfilled almost in the blink of an eye.

Packets

At the (almost) simplest level of the Internet are things called packets. For our purposes, a packet is the fundamental unit of data that is sent between an origin and a destination on the Internet. How does a packet get from Point A to Point B? First, it needs two addresses: one for its origin and one for its destination. The origin address is known as the source IP address and the end address is known as the destination IP address.

IP Addresses

Every device attached to the Internet must have an address. In IPv4,[143] these addresses are written as "dotted quads," i.e., four sets of numbers separated by periods, as in 204.180.95.2.[144] Today, IP addresses are much more likely to identify the public end point of a path than an individual computer. The last IP address you see is likely the point at which the packet enters a private network, and you can't normally see beyond that point.

IP addresses are similar to telephone numbers or house addresses, but that analogy only goes so far because many IP addresses are temporary. Temporary IP addresses are called dynamic IP addresses; the IP address your cable or DSL provider assigns to you is probably dynamic. On the other hand, certain types of devices on the Internet are usually assigned permanent or static IP addresses. One of these types of devices is a router.

This brings us back to our traveling packet. Once we know that the packet has its own address and a destination address, we then need to know how the packet gets from one place to another. It gets there because of something called a router.

Routers

A router is a device that forwards packets from one network to another. Routers used to be called gateways and you may still encounter that term. So, how does routing work?

Routing

Think of routing as the process in which your computer (host) tries to find a destination computer (host). Basically, your source IP address host with data to send asks your ISP's router two very important questions:

---

[143] IPv4 is Internet Protocol version 4, the current Internet Protocol or set of rules for exchanging data across the Internet. It is in the process of being replaced by IP version 6. IPv4 uses 32-bit addresses (allowing 4.3 billion addresses); IPv6 uses 128 bits (allowing 3.4x1038 addresses).

[144] This type of IP address is an IPv4 address; the next generation Internet will use IPv6.

A) how do I get to my desired destination address? and

B) would you please take me there?

On the Internet, it is highly likely that there is more than one path from computer A to computer B. So the router has the job not only of knowing where to send the packet but deciding which path or route the packet should take. But how do all these routers on the Internet know which path to use to send packets so they get to their destinations?

Routers use routing tables to determine where a packet is going and how to send it. The Internet in part consists of a huge grid of routers, each of which only knows a piece of the whole picture, not the entire Internet (it's too big for that!).

Look at this image. You can see that in this simple example, Router A does not know Router B's connections. It literally cannot see Router B. The hub or switch at 10.0.0.0 knows both Router A and Router B's destination paths and can, therefore, route packets from 20.0.0.0 to 40.0.0.0 and back. All Router A needs to know is to send those packets via its Port 1 to 10.0.0.0 and, from there, Router A has no more to do with the packets. The reverse is true of Router B.[145]

From Computer Desktop Encyclopedia
© 1998 The Computer Language Co. Inc.

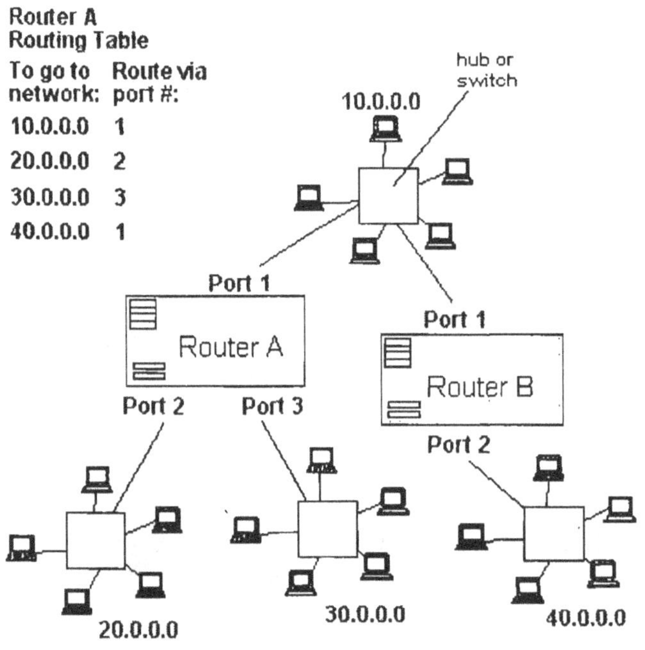
Router A
Routing Table
To go to network: Route via port #:
10.0.0.0 1
20.0.0.0 2
30.0.0.0 3
40.0.0.0 1
hub or switch
10.0.0.0
Port 1
Router A
Port 2
Port 3
Port 1
Router B
Port 2
20.0.0.0
30.0.0.0
40.0.0.0

---

[145] "Router." Answers.com. Computer Desktop Encyclopedia, Computer Language Company Inc., 2005. <http://www.answers.com/topic/router> (14 November 2006).

There is a further layer to the addressing and routing of Internet traffic that we need to know about. As the Internet grew, it quickly became apparent that it would be impossible for every network to know every other networks' "gateways," as they used to be known. This works only a small scale, and the Internet is anything but small. Thus, the idea of autonomous systems was introduced, by which only those devices within a specific AS would know all of that AS's routes.

Autonomous System

An autonomous system (AS) is a network or group of IP networks run by a single entity using common routing policies or protocols. Another way of looking at an AS is to think of it as a lot of networks being handled on the Internet as one logical domain. There are two types of Autonomous System numbers: Public AS numbers and Private AS numbers. A public AS has a unique number (ASN) associated with it; this number is used in both the exchange of exterior routing information (between neighboring Autonomous Systems), and as an identifier of the AS itself. The introduction of autonomous systems also meant that AS's could run their own internal network any way they pleased, which made sysadmins happy. We are only interested in Public ASNs.

In order to keep track of these new AS's, each AS was assigned an AS number (a 16 bit integer) by IANA, the Internet Authority for Assigned Names and Addresses. IANA assigns both IP addresses and ASNs to Regional Internet Registries (the three biggest are ARIN, APNIC, and RIPE), who then assign both types of numbers to customers.[146]

The rules for how traffic is handed off among and between AS's on the Internet are called protocols, and there are a number of them currently in use. The core external routing protocol for the Internet is the Border Gateway Protocol (BGP), which we'll look at next.

Border Gateway Protocol (BGP)

BGP works by maintaining a table of IP networks or 'prefixes' that designate network accessibility between autonomous systems (AS).

In this example, imagine Customer 1 in Baltimore wants to reach a website maintained by an ISP in Los Angeles, represented here by Customer 4 at Regional ISP (AS2). Customer 1's Regional ISP (AS1) routes the request through National

---

[146] Elquapo's Guide to Routing—Part 3, BPG, 20 August 2003, <http://www.kuro5hin.org/story/2003/8/19/18378/4228> (14 November 2006).

ISP (AS3) in Chicago because there is no direct connection between Regional ISP (AS1) in Baltimore and Regional ISP (AS2) in Los Angeles and going through multiple third party networks would be both slow and costly. Does this look familiar? It is the same principle that was at work in the way routers handle IP addresses.

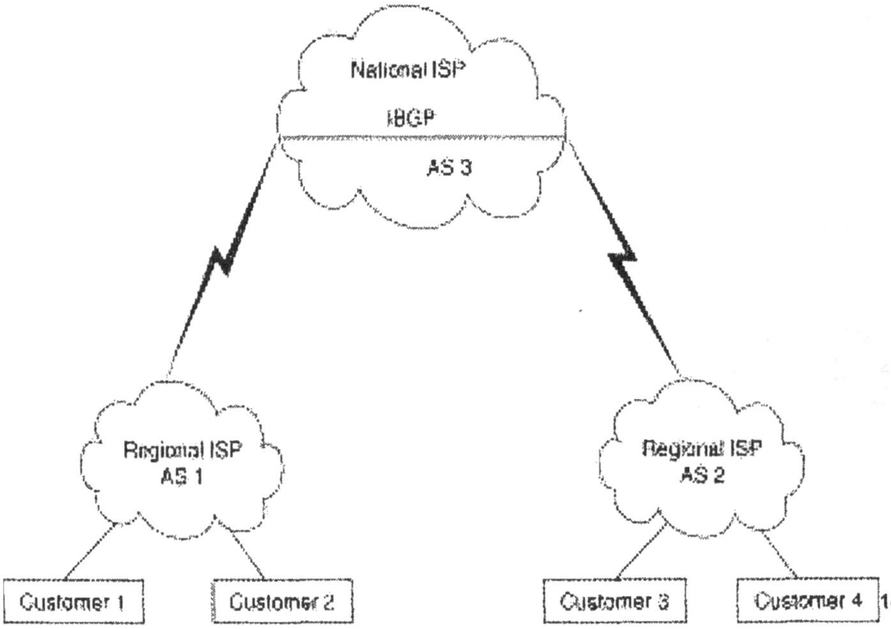
National ISP
IBGP
AS 3
Regional ISP
AS 1
Regional ISP
AS 2
Customer 1
Customer 2
Customer 3
Customer 4
[147]

Think of the routing of Internet packets as analogous to airline travel. Imagine you need to fly from Dallas to Damascus, Syria. There is no direct flight, so you must first fly to Washington Dulles Airport, change planes and carriers, then fly to Nicosia, Cyprus. Once there, you must again change planes and carriers to get to Damascus.

Internet traffic moves in much the same way. The airline is analogous to an Autonomous System. Each AS has a limited number of other AS's with which they exchange data, just as American Airlines does not have a direct connection with Syrian Airlines. Similarly, Dulles Airport is analogous to MAE-East, a major **Internet Exchange Point** along the East Coast of the US.

Internet Exchange Points

Internet Exchange Points are physical locations where different networks exchange traffic, much like large airport hubs such as Washington's Dulles or London's

---

[147] "Border Gateway Protocol," Cisco Documentation, Cisco.com, <http://www.cisco.com/univercd/cc/td/doc/cisintwk/ito_doc/bgp.htm> (14 November 2006).

Heathrow airports. Internet Exchanges (IX) are also known less frequently as Network Access Points or NAPs.

The IX system is extremely important because it greatly facilitates the movement of Internet traffic both within a country and around the world. IXs permit traffic to be handed off directly between two different Autonomous Systems without incurring a cost instead of having to travel through third party networks, which is usually slower and more expensive. Over the past decade, there has been a marked increase in the number of IXs in the world. It used to be the case that in order for Internet traffic to travel within a country, it had to be routed to an Internet Exchange on another continent because there was no way for a network in one city to get its traffic directly to a network in another city in the same country, at least without paying a rather steep price to use a third party network.

Internet Exchanges principally consist of a large number of **switches**, network devices that select a path for sending data to its next stop. In general, a switch is simpler than a router because it does not require as much information about the network and routing policies.

Some IXs are relatively small while others, such as MAE-East and MAE-West in the US, may have many providers as members and many different locations. At the heart of this system, in which traffic is voluntarily handed off or exchanged among Autonomous Systems, is something called peering.

Peering

"Peering is the practice of voluntarily interconnecting distinctly separate data networks on the Internet, for the purposes of exchanging traffic between the customers of the peered networks...Peering usually indicates that neither party pays the other for the traffic being exchanged. There are, of course, examples of paid peering.

The act of peering typically involves the following elements:

- The physical interconnection of the networks involved.
- The exchange of routing information, through the BGP routing protocol.
- Commercial and contractual peering contracts or agreements".[148]

Why is peering so important? One of the original concepts behind the creation of the Internet was the concept of global reachability, that is, every point on the Internet is

---

[148] "Peering," Wikipedia, <http://en.wikipedia.org/wiki/Internet_exchange#Exchange_Points_and_Colocation_Facilities> (14 November 2006).

in theory reachable by and from any other point on the Internet. But at the same time, every network most certainly is not connected directly to every other network on the Internet, so there had to be accommodations made to ensure that traffic could flow freely around the globe. This mutually beneficial arrangement to exchange Internet data freely between two networks is a peering agreement. I'll accept and forward your traffic if you'll accept and forward mine.

CIDR

One final concept needs to be mentioned. About a decade ago people realized that something had to be done about Internet addressing. There were real concerns that we would run out of IP addresses and run out of capacity in global routing tables, with new networks being connected to the Internet at the rate of about one every thirty minutes. The solution was something called CIDR (pronounced "cider").

CIDR "CIDR stands for Classless Inter-Domain Routing…CIDR is an effective method to stem the tide of IP address allocation as well as routing table overflow. Without CIDR having been implemented in 1994 & 1995, the Internet would not be functioning today. Basically, CIDR eliminates the concept of class A, B, and C networks and replaces this with a generalized 'IP prefix'."[149] CIDR permits blocks of addresses to be assigned to networks as small as 32 hosts or as large as those with over a half a million hosts. This type of address assignment more accurately reflects an organization's actual size and specific network requirements.

What is the relationship between CIDR and the Autonomous System? Perhaps the easiest way to understand their relationship is to think of "an Autonomous System as a collection of CIDR IP address prefixes under common technical management. For example, the CIDR block from 208.130.28.0 to 208.130.31.255, as well as network 201.64.75.0 might be AS2934."[150] In turn, AS2934 may represent one network made up of 50 computers or fifty networks made up of thousands of computers.

---

[149] The CIDR FAQ, version 7.0, May 2004, <http://www.interall.co.il/cidr.html> (14 November 2006).

[150] "Autonomous System," Connected: An Internet Encyclopedia, Third Edition, April 1997, <http://www.freesoft.org/CIE/Topics/4.htm> (14 November 2006).

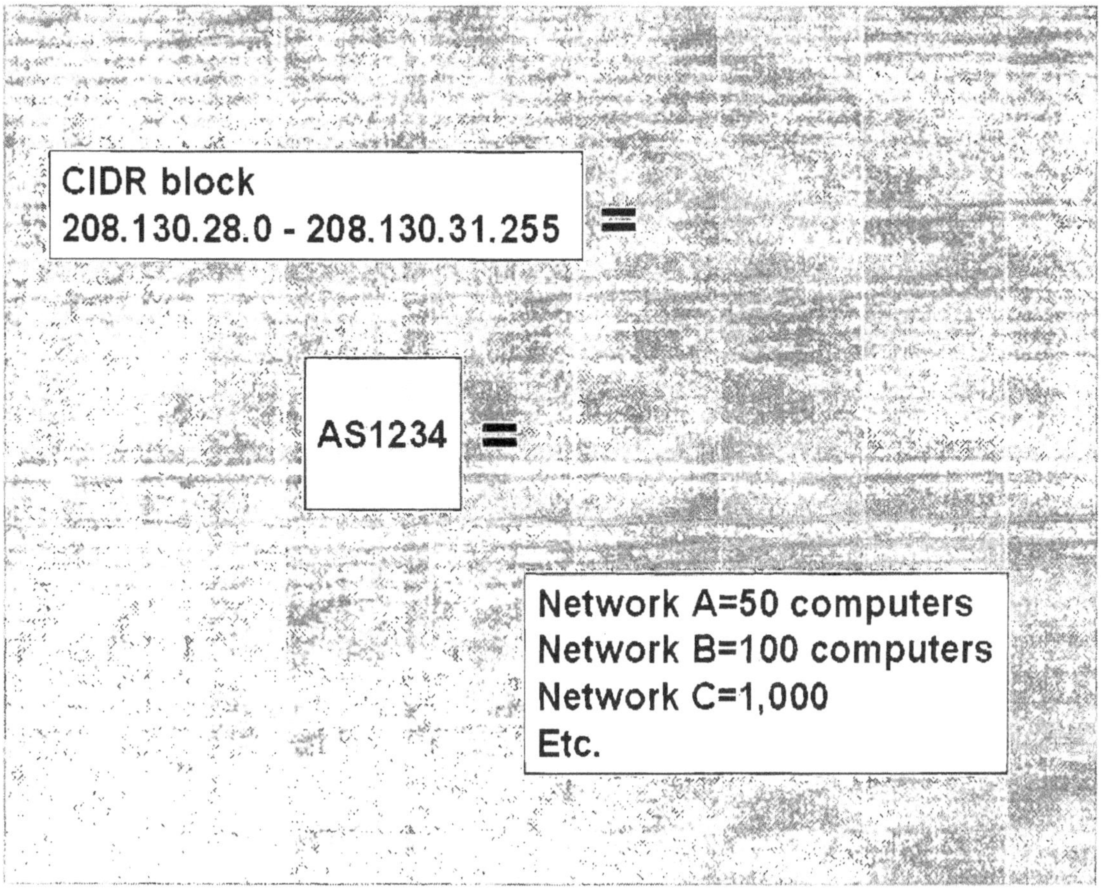
CIDR block
208.130.28.0 - 208.130.31.255
=
AS1234
=
Network A=50 computers
Network B=100 computers
Network C=1,000
Etc.

Researching Internet Statistics

Typical research questions about the Internet concern its size, numbers of websites, numbers of users, where they are, who they are, what ISPs serve a particular city, etc. This information is available on the Internet and knowing a few of the best basic sources of Internet-related data will give you a head start in finding these facts fast.

Network Wizards Domain Survey http://www.isc.org/ds

One of the best sites that tracks the growth of the Internet is the Network Wizards Internet Domain Survey. Network Wizards has been surveying the Internet since 1987 and provides very straightforward data about numbers of hosts on the Internet and their distribution by top-level domain. Thus, researchers can quickly find out the .INFO domain went from 2128 hosts in Jan 2002 to 76692 in the July 2006 survey. Network Wizards also provides a very good explanation of how they conduct their surveys. They make the important point that their numbers are "fairly good estimates of the ***minimum*** size of the Internet." *They have no way of telling how many hosts or domains they did not locate.*

Netcraft http://news.netcraft.com/

Another Internet statistics site of long-standing is Netcraft, best known for its "what's that site running" data. Netcraft keeps track of a wide range of data about the Internet, including "market share of web servers, operating systems, hosting providers, ISPs, encrypted transactions, electronic commerce, scripting languages and content technologies on the Internet." Netcraft is probably the top site for information about Internet technology usage.

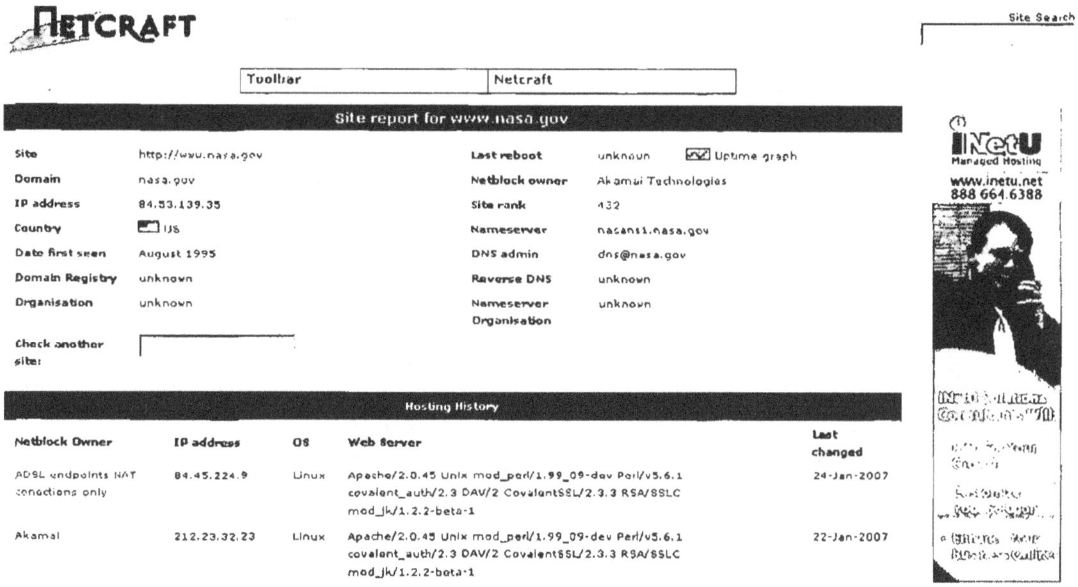

NETCRAFT

Site Search

Toolbar	Netcraft

Site report for www.nasa.gov

Site	http://www.nasa.gov	**Last reboot**	unknown Uptime graph
Domain	nasa.gov	**Netblock owner**	Akamai Technologies
IP address	84.53.139.35	**Site rank**	432
Country	US	**Nameserver**	nasans1.nasa.gov
Date first seen	August 1995	**DNS admin**	dns@nasa.gov
Domain Registry	unknown	**Reverse DNS**	unknown
Organisation	unknown	**Nameserver Organisation**	unknown
Check another site:			

Hosting History

Netblock Owner	IP address	OS	Web Server	Last changed
ADSL endpoints NAT conections only	84.45.224.9	Linux	Apache/2.0.45 Unix mod_perl/1.99_09-dev Perl/v5.6.1 covalent_auth/2.3 DAV/2 CovalentSSL/2.3.3 RSA/SSLC mod_jk/1.2.2-beta-1	24-Jan-2007
Akamai	212.23.32.23	Linux	Apache/2.0.45 Unix mod_perl/1.99_09-dev Perl/v5.6.1 covalent_auth/2.3 DAV/2 CovalentSSL/2.3.3 RSA/SSLC mod_jk/1.2.2-beta-1	22-Jan-2007

iNetU
Managed Hosting
www.inetu.net
888 664.6388

ClickZ Stats http://www.clickz.com/stats/

Jupiter Media bought NUA, the Network Users Association, a European Internet consultancy that tracked many types of Internet trends and surveys around the world, and combined it with its former competitor. The ClickZ Stats sites provides detailed information about Internet usage in many sectors and topics: demographics, geographics, wireless, government, finance, hardware, etc. The statistics tend to be consumer-oriented (e.g., DVR usage) but there are a number of technology surveys as well, e.g., the global VoIP market.

More good sources of statistical and usage information

Global Reach's Global Internet Statistics by Language
http://www.glreach.com/globstats/

Internet Traffic Report http://www.internettrafficreport.com/main.htm

Zooknic Internet Statistics http://www.zooknic.com/

Regional Registries and NICs

Despite the "wild west" metaphors applied to the Internet, in one respect it is quite orderly. Domain names and IP addresses are assigned, registered, and stored in repositories around the world that are publicly accessible from anywhere. This means it is usually a simple matter to find an IP address given a domain name and vice versa. But the registries store other valuable information as well, such as who has registered the IP address, associated domain names, the network name, the registering organization name and address, and points of contact (usually system administrators), along with their phone/fax numbers and email addresses.

The **Internet Corporation for Assigned Names and Numbers (ICANN)** is the non-profit corporation that was formed to take over responsibility for the IP address space allocation, protocol parameter assignment, domain name system management, and root server system management functions previously performed under US-government contract by IANA, InterNIC, and other entities. While there is virtually no limit to the number of possible domain names, there are very definite limits to the number of IP addresses available, which means distribution and allocation of IP addresses is strictly controlled. IP address space is distributed hierarchically by **ICANN**, which allocates blocks of IP address space to Regional Internet Registries (**RIRs**) (more often called **Network Information Centers—NICs**). These RIRs in turn allocate blocks of IP addresses to **Local Internet Registries**. It is these local registries that assign IP addresses to **local ISPs**, who in turn allocate addresses to end users.

There are five **Regional Internet Registries:**

RIPE NCC for Europe and the Middle East

APNIC for the Asian/Pacific countries and

ARIN, the American Registry for Internet Numbers

LACNIC, the Latin American and Caribbean Network Information Center

AfriNIC, the African Network Information Center

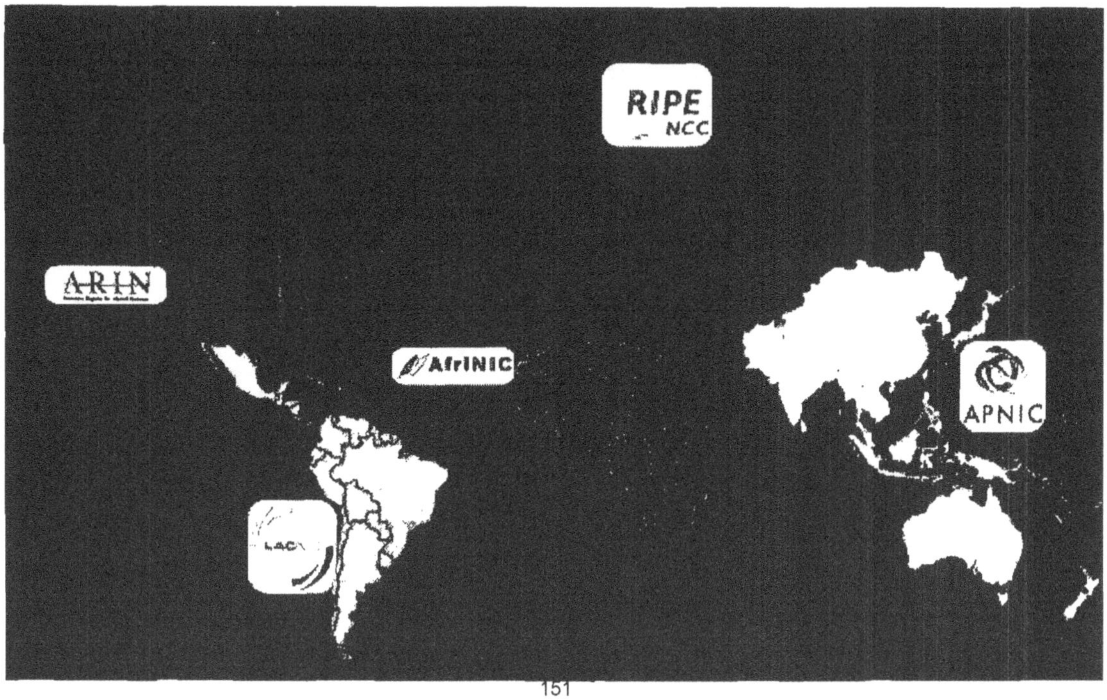
RIPE
NCC
ARIN
AfrINIC
APNIC

[151]

ARIN took over **InterNIC's** responsibility for managing IP number assignments, though ARIN does not have the responsibility InterNIC once did for registering domain names. ARIN is modeled after RIPE and APNIC, that is, it is an independent non-profit corporation responsible for managing IP addresses for North America.

In December 2005, the new EU top-level domain opened for business. According to the European Registry of Internet Domain Names (EURid) website, "anyone based in the European Union can register a .eu domain name as long as it is available...EURid has more than 1000 accredited registrars from all over the world." At the same time as registration began, EUint opened a Whois search site for .eu domains. One indication of the popularity of the EU domain is that the Google search [site:eu] returns nearly 39 million hits.

European Registry of Internet Domain Names (EURid)
http://www.eurid.eu/en/registrant/

Complete List of Accredited EURid Registrars
http://list.eurid.eu/registrars/ListRegistrars.htm?lang=en

EURid Whois http://www.whois.eu/whois/GetDomainStatus.htm

---

[151] Number Resources Organization, <http://www.nro.net/about/get_resources.html> (30 January 2007).

I believe the best site for finding top-level domains and their associated Regional Internet Registry is maintained by the Number Resource Organization (NRO), which was formed by the five RIRs in 2003 "to formalise their co-operative efforts." The site includes a great deal of information about the RIRs, including a chart of each country and its associated RIR. RIPE also provides such a list.

NRO's List of Country Codes and RIRs http://www.nro.net/about/rir-areas-rir.html

List of Country Codes and RIRs http://www.ripe.net/info/resource-admin/rir-areas.html

ICANN and the Regional Internet Registries (aka NICs):

ICANN	http://www.icann.org
AfriNIC	http://www.afrinic.net/
APNIC	http://www.apnic.net
ARIN	http://www.arin.net
European Registry of Internet Domain Names (EURid)	http://www.eurid.eu/
LACNIC	http://lacnic.net/en
RIPE	http://www.ripe.net

The Ipv6 Transition

The current Internet addressing system uses IP version 4, and has for more than 20 years. However, the transition to IP version 6 (IPv6) is underway, especially in Europe and Asia.

The Internet Corporation for Assigned Names and Numbers (ICANN), the independent body that coordinates the Internet's address system, announced in July 2004 that IP version 6 (commonly written IPv6 and referred to as the "next generation" IP addressing system), has been added to the Internet's root server system. Vincent Cerf, an Internet "founding father" and a member of ICANN's board of directors, made the announcement at ICANN's annual meeting in Kuala Lumpur, Malaysia.

In practical terms, what does the addition of IPv6 mean? Today ***there are still only 13 root domain name servers that contain the master records for all Internet address mapping***.

root-servers.org

VeriSign · USC-ISI · Cogent · UMD · NASA-ARC · ISC · DOD-NIC · ARL · Autonomica · RIPE · ICANN · WIDE

See also: AS112 · presentations

The root hints file (**named.cache, root.ca, root.hints**, ...) can be obtained via IANA's page for popular links.

NEWS		PRESENTATIONS	
Date	**Subject**	**Date**	**Occasion**
2004-01-29	New IP address for b.root-servers.net. (192.228.79.201)	2003-03-24	GAC meeting during ICANN meeting in Rio de Janeiro (PDF)
2004-01-26	New AS number for i.root-servers.net.	2003-12-09	WSIS meeting in Geneva (PDF)
		ALL	Complete list of presentations

Root Servers.org http://www.root-servers.org/

This current set of rules used for Internet addresses, IPv4, became the Internet standard in 1981 and at that time it seemed as though it would provide enough addresses to last forever. But forever turned out to be a lot shorter than anyone anticipated, and the range of numbers used under the IPv4 is slowly running out; about two thirds of the 4.3 billion numbers allocated have been used. So a new set of rules, called IPv6, was created. IPv6 will increase the number of numerical addresses to 340 billion, billion, billion, billion numbers. This should be enough for the life of the Internet, in any event, regardless of how many computers, devices, or imaginary beings need to connect to it. IPv6 is also supposed to add reliability and security enhancements because of such features as built-in encryption.

I would not panic about some overnight shift from IPv4 to IPv6. Cerf said that the plan is for IPv6 to run parallel to IPv4 for about 20 years to ensure that any bugs or system errors are discovered and corrected. For more information on IPv6, see the IPv6 Information Page and read the November 2005 interview with Latif Ladid, founder of the IPv6 Forum, in which he asserts, "Overall, I would say that the show is happening in Asia, and in five years time you can expect China to be the biggest IPv6 user base in the world. By 2010 they will have two or three hundred million people using IPv6. Today the Western world will be taken by surprise. We are staying in denial."[152] IPv6 migration will happen—it's only a matter of when, not if.

Ipv6 Information Page http://www.ipv6.org/

---

[152] Dahna McConnachie, "IPv6 Forum Chief: The New Internet is Ready for Consumption," Computerworld, 18 November 2005, <http://www.computerworld.com.au/index.php?id=75779762> (15 November 2006).

The Shared Registration System

InterNIC, which was the original US-government sponsored domain name registry, no longer allocates IP addresses or registers domain names. Just as its IP address allocation functions were turned over to ARIN, its authority for registering the Top Level Domains (**TLDs**) .COM, .NET, .ORG and .EDU and, even more importantly, managing the vast database of registered domain names, was first assumed by **Network Solutions**.

In 1999, the **Shared Registration System** for the .COM, .NET, and .ORG domains was opened on equal terms to all accredited registrars, meaning that any company that meets ICANN's standards for accreditation is able to enter the market as a registrar and offer customers competitive domain name registration services in the . .COM, .NET, and .ORG domains.

More ***new generic top-level domains*** are being approved by ICANN all the time. Among those now in use are .BIZ, .INFO, .COOP, .AERO, .NAME, .PRO, and .MUSEUM with more on the way. For more information on these new TLDs, how to and who may register them, see the InterNIC FAQ on new top-level domains. <http://www.internic.net/faqs/new-tlds.html>

Shared Registration System (SRS) Warning!

If you look up a .com, .org, or .net domain name at the website of any SRS member (NSI, etc.), be careful! Your "domain name search will contain only technical information about the registered domain name and referral information for the registrar of the domain name. In the Shared Registration System model, registrars are responsible for maintaining Whois domain name contact information. To obtain information on the Registrant of a domain name, go to the registrar's web site listed."

What does this mean? To see the complete Whois record, you must go to the registrar's website or use a Whois browser interface that queries multiple Whois databases (more on this below).

The accuracy of the data in the Whois databases, however, is problematic. ICANN issued its first report on the accuracy of Whois database information in March 2004. The report is the first in what will be a series of annual reports on the status of complaints about Whois entries. Since ICANN instituted the Whois Data

Problem Report System (WDPRS) in September 2002 to let individual users report incorrect or incomplete domain registration information, they received over 24,000 complaints. Of those, nearly 5,000 of the complaints concerned domains containing incorrect or incomplete contact information—telephone numbers, email addresses, street addresses—for known or suspected spammers.

One problem with correcting inaccurate Whois information is that there is no real enforcement mechanism in place. Although all registrars accredited by ICANN to register domain names are required to provide "accurate and reliable contact details and promptly correct and update them during the term of the registered name registration," there is no process for following up to make sure reported inaccuracies are corrected or domain names taken away from offenders. The ICANN report claimed that "ICANN's experience has been that accredited registrars by and large do conscientiously comply with their contractual obligations by acting promptly to correct incomplete or inaccurate data that is brought to their attention." However, without stronger means of enforcing compliance, there is no way of ensuring Whois data is accurate and complete.

Whois Data Problem Report System http://wdprs.internic.net/

Internet Myth:

You can tell a domain's country of origin by looking at its TLD.

Internet Fact:

MANY foreign websites use .com, .net, and (to a lesser extent) other top-level domains, so not every foreign site will accurately reflect its country of origin by digraph. Moreover, many country registries will sell their TLD to anyone willing to pay. You cannot reliably determine country of origin by looking at a TLD.

Domain Name Registries

Every country has its own assigned top-level domain (ccTLD) in the form of a digraph: .de for Germany, .bo for Bolivia, etc. Finding the ccTLD administrator that registered a particular domain name is vital for finding all the available data registered about that name and its assigned IP address. Because local registries have the added responsibility for assigning domain names, they usually maintain separate databases that often provide more information than the regional repositories. Most national registries offer some kind of a searchable database of their registered domain names. Country-level registries are some of the best places to search for information about specific domains.

A number of websites list links to **domain name registries around the world**. All these sites present complete, up-to-date lists of **Internet country codes** (aka ccTLDs). Domain Name Registries Around the World offers both a country and digraph sort; IANA only lists the codes alphabetically by digraph.

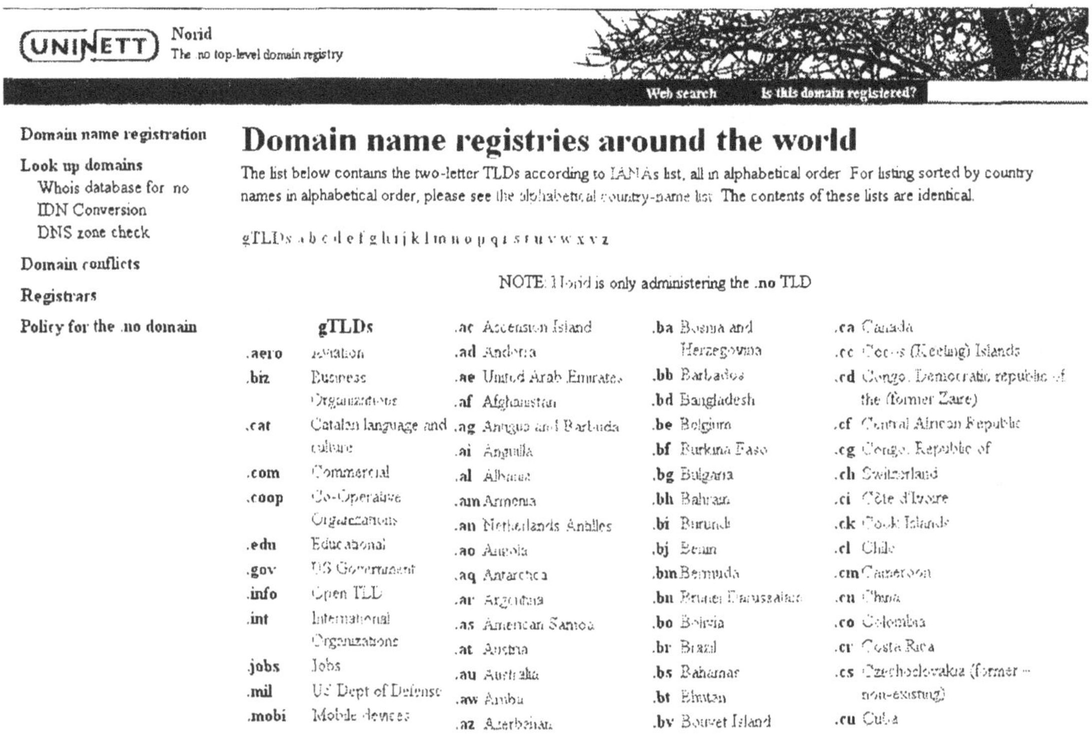
UNINETT
Norid
The .no top-level domain registry
Web search
Is this domain registered?
Domain name registration
Look up domains
Whois database for .no
IDN Conversion
DNS zone check
Domain conflicts
Registrars
Policy for the .no domain
Domain name registries around the world
The list below contains the two-letter TLDs according to IANA's list, all in alphabetical order. For listing sorted by country names in alphabetical order, please see the alphabetical country-name list. The contents of these lists are identical.
gTLDs a b c d e f g h i j k l m n o p q r s t u v w x y z
NOTE: Norid is only administering the .no TLD
gTLDs
.aero Aviation
.biz Business Organizations
.cat Catalan language and culture
.com Commercial
.coop Co-Operative Organizations
.edu Educational
.gov US Government
.info Open TLD
.int International Organizations
.jobs Jobs
.mil US Dept of Defense
.mobi Mobile devices
.ac Ascension Island
.ad Andorra
.ae United Arab Emirates
.af Afghanistan
.ag Antigua and Barbuda
.ai Anguilla
.al Albania
.am Armenia
.an Netherlands Antilles
.ao Angola
.aq Antarctica
.ar Argentina
.as American Samoa
.at Austria
.au Australia
.aw Aruba
.az Azerbaijan
.ba Bosnia and Herzegovina
.bb Barbados
.bd Bangladesh
.be Belgium
.bf Burkina Faso
.bg Bulgaria
.bh Bahrain
.bi Burundi
.bj Benin
.bm Bermuda
.bn Brunei Darussalam
.bo Bolivia
.br Brazil
.bs Bahamas
.bt Bhutan
.bv Bouvet Island
.ca Canada
.cc Cocos (Keeling) Islands
.cd Congo, Democratic republic of the (former Zaire)
.cf Central African Republic
.cg Congo, Republic of
.ch Switzerland
.ci Côte d'Ivoire
.ck Cook Islands
.cl Chile
.cm Cameroon
.cn China
.co Colombia
.cr Costa Rica
.cs Czechoslovakia (former – non-existing)
.cu Cuba

Domain Name Registries Around the World
http://www.norid.no/domenenavnbaser/domreg.html

IANA's Contact List for TLD Administrators http://www.iana.org/root-whois/index.html

Yahoo's Computers and Internet Domain Name Registration
http://dir.yahoo.com/Computers_and_Internet/Internet/Domain_Name_Registration/→
Top_Level_Domains__TLDs_/Registry_Operators/International_Country_Codes/

The Least You Need to Know about IPs and Domains

The Regional Internet Registries (aka NICs) are responsible solely for **IP address allocation and registration** (usually to Local Internet Registries or very large ISPs).

All five NICs (**ARIN**, **APNIC**, **AfriNIC**, **LACNIC**, and **RIPE**) maintain databases that contain formatted information about all IP addresses they have allocated or assigned, including those IP addresses its regional registries have authority to assign. **You can look up an IP number—but not a domain—at a NIC.**

"Local" registries (usually operating at country level, e.g., DENIC for Germany) register domain names and, in many cases, also allocate IP addresses.

Domain name registration for the **.com**, **.org.**, **.net**, **.biz**, **.info**, etc. TLDs has become a wide-open commercial enterprise. Anyone anywhere can register these TLDs.

Understanding Domain Name and Whois Lookup Tools

Domain names are nothing more than aliases for IP addresses; they are a convenience for human beings, who have more trouble remembering long numbers than letters, words, and names. But computers love numbers, so every time you enter a url (an address that contains the protocol, host, domain, directory, and file information), the computers that really *are* the Internet translate that name into its corresponding IP address. How do the computers know what domain name matches which IP number? **DNS**: the domain name system (or service). DNS is the Internet service that translates domain names, such as *www.amazon.com*, into IP addresses.

DNS is a huge *distributed database* residing on a network of servers. If a DNS server does not know how to translate a domain name, it will simply query another DNS server until it finds one that can translate the domain name to the correct IP address. Every server (a computer that offers services to other computers) connected to the Internet has some information about domain names; **DNS servers** keep huge lists of domain names matched with their IP addresses. Invisible to us the Internet users, the domain names are converted to their numerical addresses using the tables on DNS servers, and we get to where we want to be.

Web Tip

For a more detailed explanation of the Domain Name System/Service see:

The Domain Name Service
http://www.scit.wlv.ac.uk/~jphb/comms/dns.html

DNS and BIND, 3rd Edition, O'Reilly Online Catalog
http://www.oreilly.com/catalog/dns3/chapter/ch02.html

Fortunately, there are many freely accessible tools for accessing and analyzing the domain name system. Domain name and Whois lookup are among the most useful tools available on the Internet. Many different kinds of tools are used for gathering information about domain names and/or IP addresses. Before using these tools, it is helpful to review how domain names and IP addresses work together to make it possible for users to navigate the web and which Internet resources these tools access in order to provide information about domain names and IP addresses.

After the DNS, the Whois databases maintained by the Regional Internet Registries are the second major source of information about Internet addresses. The Whois databases contain records of IP address registrations. Searches that access the Whois databases are generally known as **network lookups** or **network Whois lookups**. In contrast to the DNS, the five Whois databases—ARIN, APNIC, AfriNIC, LACNIC, and RIPE—are *not distributed on servers across the Internet* but must be queried individually. In addition, these databases are designed to search on IP addresses, not domain names. A different type of tool should be used when searching on domain names. Unfortunately, to add to the confusion, this tool is also referred to as a Whois lookup!

In order to distinguish this third major resource from "true" Whois lookups, I will call it domain name (Whois) lookup. Domain name lookups search one or more **domain name registries**. Domain name registries are the places on the Internet where individuals and organizations go to register a name generally intended to be associated with a website, e.g., *amazon.com*. These registries include the **generic top-level domain (gTLD) registries** for .com, .net, .org, .biz, .info, and a few other gTLDs, as well as **country code top-level domain (ccTLD) registries**, e.g., .uk, .it, .ru, et al. Domain name (Whois) lookups differ from network Whois lookups in that they are primarily designed to search on a domain name instead of an IP address. All of the domain lookup tools described here can search more than one domain registry, and in some cases, will automatically search across all of them.

These are the various types of lookups that can be performed to learn more about Internet addresses.

- **NSLookup**—input format: either IP address or domain name

  Using the DNS, a domain name is converted to its IP address (forward DNS lookup) or an IP address is converted to its host name (reverse DNS lookup).

  NSLookup is a UNIX tool that allows anyone using the Internet to access the DNS and match domain names to IP addresses, or vice versa. Web interface query forms for NSLookup are very numerous. Although NSLookup doesn't do anything more than match domain names and IP addresses, this may be all you need or may give you the information you need to keep looking. For example, I have had domain names I could not match to any IP number using a Whois search. By running an NSLookup query I have found an IP number that I could then use to track down registration information about that domain.

  The advantage of NSLookup is that, unlike Whois requests, ***NSLookup queries can be run from anywhere*** on the Internet because the DNS data is not located in one database on one server but is distributed across a collection of intercommunicating computers, known as name servers, that translate domain names to IP addresses and vice versa.

NSLookup Tools

AnalogX	http://www.analogx.com/contents/dnsdig.htm
Check DNS	http://www.checkdns.net/quickcheck.aspx
DNS Stuff*	http://www.dnsstuff.com/
Eye-Net Consulting*	http://www.enc.com.au/itools/
Infobear	http://www.infobear.com/nslookup.shtml
Multiple NSLookup	http://www.bankes.com/nslookup.htm
SmartWhois NSLookup	http://swhois.net/
Squish DNS Lookup	http://www.squish.net/dnscheck/

A free service for DNS experts. "Given a record name, and a record type, you will receive a report detailing all possible answers. This is accomplished by traversing the DNS tree from the root examining all possible routes that a client could travel, calculating percentage probabilities on the way."

WebReference NsLookup Gateway
http://www.webreference.com/cgi-bin/nslookup.cgi

ZoneEdit NSLookup http://www.zoneedit.com/lookup.html?ad=goto&kw=nslookup

NSLookup

This DNS utility is provided by ZoneEdit.Com, the industry leader in DNS and domain mangement solutions.
Click here to sign up for a free, no obligation trial of our dns services.
Click here to use our SMTP test utility.
Click here to use our global whois utility (domain ownership info).

DNS Lookup

1. Enter a host name for Forward DNS Lookup:
   (IE: yahoo.com)
2. Select record type (optional):
   Ip Address (A)
3. Enter server name or IP (optional):
4. Look it up

Reverse Lookup

1. Enter an IP address for Reverse Lookup:
   (IE: 216.115.108.245)
2. Enter server name or IP (optional):
3. Look it up

*These sites provide Ipv6 lookups in addition to Ipv4.

- **Network Whois Lookup**—input format: IP address

  IP address blocks are maintained by ARIN, RIPE, AfriNIC, LACNIC, and APNIC in separate, non-distributed databases. Formatted Whois data provides a wealth of registration information. All five Whois databases allow for advanced searches on fields other than IP address.

- **Domain Name (Whois) Lookup**—input format: domain name

  Checks a domain name against registration records based upon that domain name's TLD (.com, .uk, .ru, etc.); some automated programs can search some or all domain name registries at once.

The confusion about domain name and Whois lookups is probably in part caused by the fact that domain name registration is separate from IP address assignment. Perhaps the easiest way to understand this is to consider the following fictitious, overly simplistic example.

An imaginary Russian company named Moscow Motors wants to register <u>two domain names</u>: *moscowmotors.ru* and *moscowmotors.com*. But Moscow Motors only wants to use *one IP address* through its ISP, RT Communications (RTComm) Network in Moscow. The European Registry, RIPE, maintains the block of IP addresses handled by RTComm. Next, Moscow Motors goes to Network Solutions, Inc., to register its *moscowmotors.com* domain name and to RU-Center, the Russian top-level domain name registration service, to register its *moscowmotors.ru* domain name. Both domain names resolve to the same IP address registered with RIPE.

Now let's say a user runs a domain name lookup against *moscowmotors.com*. He finds the *domain name* is registered with Network Solutions, but when he tries to look up the *corresponding IP address* belonging to *moscowmotors.com,* he finds there is no network Whois record in the American Registry (ARIN) Whois database. Why? Because the ARIN Whois database only contains IP addresses assigned to it, and *moscowmotors.com* resolves to an IP address in the European (RIPE) database.

If you would like to see a real-life example of what I've just described, try the following:

1. Go to Domain Dossier (<u>http://centralops.net/co/DomainDossier.vbs.asp</u>) and enter the following query:
   ripe.net
   search on *domain whois record* and *network whois record*

2. Look at the IP address for *ripe.net*: 193.0.0.203

3. Look at the Domain Whois record: *ripe.net* is registered with Network Solutions, Inc.

4. Look at the Network Whois records: 193.0.0.203 is not registered in the ARIN (American) Whois database; it is registered in the RIPE (European) Whois database.

One point of this example is to warn people against making assumptions about the relationship between domain names and IP addresses, especially the generic TLDs such as .com, .net, .org, .info, .and .biz. Do not assume that a domain name in any of these generic TLDs will have an IP address registered by ARIN, the North American Registry. However, ***Local Registries (usually country-level registries) often have restrictions on who can and cannot register their top-level domains***. In these cases, IP addresses and domain names are more likely to correspond (e.g., a .ru domain almost certainly has an IP address registered in the RIPE database), but even in these cases, there are exceptions.

There are many variations in the precise type, format, quantity, and quality of the information each tool described below provides. I will explain the features of each so you can get an idea of which tool to use in which circumstance. For each tool I have listed its special features so you can quickly determine which tool is right for your specific research need.

World Network Whois Databases: AfriNIC, APNIC, ARIN, LACNIC, & RIPE

Special features: official international registrars of IP addresses; advanced search on many other fields in Whois database.

All five international registrars permit simple and advanced queries of the Whois databases via web interfaces. ***Each database must be queried separately***; however, there are third-party sites that will query all these databases at one time. Advanced search pages at each site give instructions on which fields in the Whois database can be searched. For example, all the Whois advanced searches permit lookups of network name, person (may or may not be an individual), AS (autonomous system) number, and even an associated domain, although these Whois databases should not be used for domain name lookups.

For anything and everything you ever wanted to know about Whois databases, refer to the **RIPE Database Reference Manual**. The manual's instructions also apply to the ARIN and APNIC databases.

RIPE Database Reference Manual
http://www.ripe.net/ripe/docs/databaseref-manual.html

APNIC Whois lookups http://www.apnic.net/search/index.html

APNIC Whois help http://www.apnic.net/db/search/all-options.html

IP address registration for Asia-Pacific countries.

ARIN http://www.arin.net/whois/index.html

IP address registration for North America.

ARIN Whois help http://www.arin.net/tools/whois_help.html

AfriNIC Whois http://www.afrinic.net/cgi-bin/whois

AfriNIC Database User Manual
http://www.afrinic.net/docs/db/afsup-dbgs200501.htm

EURid Whois http://www.whois.eu/whois/GetDomainStatus.htm

LACNIC Whois http://lacnic.net/cgi-bin/lacnic/whois

RIPE http://www.ripe.net/perl/whois/

IP address registration for Europe, North Africa, and the Middle East .

RIPE Whois help http://www.ripe.net/ripencc/pub-services/db/whois/whoishelp.html

---

Global Network Whois Search Tools

The following sites offer web interfaces that will either search all five regional Whois databases at once or individually.

Whois at Webhosting.info http://whois.webhosting.info/

Special features: shows hosting company name with total domains and gives reverse IP list.

Describing itself as "power whois," Web Hosting's advanced Whois service gives some additional bits of information that don't come with traditional whois lookups:

- Web Hosting Company hosting the domain name.
- Current IP Address of the domain.
- Geographical location of the IP Address.
- Reverse IP Facility - Lists all domains hosted on an IP Address.
- Domain Owner Details - Registrant, Admin Contact, etc.

MSV.DK Network Whois http://msv.dk/ms593.aspx

Special feature: searches across all registries.

Will quickly search all three main Whois databases ARIN, APNIC, and RIPE.

IP-Plus http://www.ip-plus.ch/tools/whois_set.en.html

Special feature: in addition to the main Whois databases, option to search several other country-level Network Information Centers.

Does not search all Whois databases at once; each must be searched individually from pull-down menu.

DNS411 http://dns411.com/

Special feature: "smart Whois" searches across all Whois databases automatically.

DNS411 uses the "smart Whois" search to perform universal IP address and domain names lookups. Even though the search form says to enter a domain name, it searches on IP addresses as well. In addition, DNS411 will search on second-level domain names, e.g., [amazon.co.uk], NIC Handles (unique identifiers assigned to each domain name, contact, and network records in a registrar's database), IP addresses, Autonomous System (AS) numbers, and Netblock Handles automatically across all registries. See the "tips" page for details.

Network-Tools http://network-tools.com/

Special feature: in addition to network Whois lookups, offers full range of network lookup tools on one page.

Does not search all Whois databases at once; each must be searched individually from pull-down menu. Also offers other network tools: ping, NSLookup, traceroute, DNS records, etc.

Geektools http://www.geektools.com/whois.php

Special feature: automatic domain name/IP address search across all registries.

Geektools has long been a staple of many research toolkits by providing a fast and easy universal Whois lookup for both domain names and IP addresses. Must enter full name (domain plus TLD) or IP number (no partial searches or wildcard function).

Domain Dossier http://centralops.net/co/DomainDossier.vbs.asp

Special Features: searches across all the network Whois Registries, shows DNS records, and performs service scan from one query. Domain Dossier now supports internationalized domain names (IDNs).

Domain Dossier is one of the best lookup tools for a fast, thorough search of the three main network Whois Registries (ARIN, RIPE, APNIC). Select "network whois record" for automatic search of IP or domain name; results include canonical name, aliases, and IP addresses.

Domain Dossier
Investigate domains and IP addresses
domain or IP address
domain whois record
DNS records
traceroute
network whois record
service scan
go
user: 206.112.75.238 [anonymous] 49/50
log in | get account
CentralOps.net
-- end --
Domain Dossier now supports internationalized domain names (IDNs).
Try a few:
日本語.jp

Domain Name Whois Lookups

Each site described below either has the ability to search for domain names across all TLDs or has some sort of wildcard function when searching for domain names. **Wildcard functions** are highlighted in **red**. ***Some sites will search on second-level domain names*** (e.g., *mfa.co.jp*).

Allwhois http://www.allwhois.com/

Special features: search any TLD; full Whois output directly from appropriate database; alphabetical list of country NICs with links to websites; second-level domain name search.

Allwhois automatically queries the appropriate Whois database for registration information. The output, which appears in a small window below the query box, is directly from the Whois database without any changes. Allwhois's other useful feature is an alphabetical list of registries around the world (with the corresponding country code listed next to the country name). You can "jump to" any registry's Whois page directly from Allwhois. However, keep in mind that this list of Whois pages is not complete. For example. Saudi Arabia has a NIC with a Whois tool, but it is not on the list.

Checkdomains http://www.checkdomain.com/

Special features: universal search domain names across all TLDs at once; search on second-level domains.

Checkdomain.com will check for registrations of any domain name across all registries. Checkdomain will also search for second-level domain names, such as cars.co.uk, and provide registration information for them.

CoolWhois http://www.coolwhois.com/

Special Features: universal domain name search across all TLDs; search on second-level domains; bookmarklet can be added to the browser toolbar.

CoolWhois really is. It is a true cross-domain name search tool that automatically looks up a domain name in any TLD. An added feature of CoolWhois is a neat little bookmarklet, a piece of JavaScript that you add to your browser toolbar simply by dragging and dropping it there. Once it's there, all you have to do to perform a domain name lookup on any page you're viewing is to click on the CoolWhois bookmarklet (which looks exactly like a personal toolbar link).

DNS411 http://dns411.com

Special features: universal domain name search across all TLDs; search on second-level domains; many other unusual lookup options.

DNS411 uses the "smart Whois" search to perform universal domain names lookups. In addition, DNS411 will search on second-level domain names (e.g., *amazon.co.uk*), NIC Handles (unique identifiers assigned to each domain name, contact, and network records in a registrar's database), IP addresses, Autonomous System (AS) numbers, and Netblock Handles automatically across all registries. See the "tips" page for details.

DrWhois http://www.drwhois.com/

Special feature: universal domain name/IP address search across all TLDs.

DrWhois searches for domain names or IP addresses across all registries and will also search on second-level domain names such as [mfa.gov.ir].

Domain Dossier http://centralops.net/co/DomainDossier.vbs.asp

Special Features: searches across all the network Whois Registries, shows DNS records, and performs service scan from one query. Domain Dossier now supports internationalized domain names (IDNs).

Domain Dossier is one of the best lookup tools for a fast, thorough search of the three major network Whois Registries (ARIN, RIPE, APNIC). Domain Dossier also automatically searches the InterNIC and Open SRS domain Whois records, provides DNS records (name servers, mail exchanges, etc.), and performs a service scan, which shows the status of FTP, SMTP, HTTP, POP3, and NNTP Ports.

Domainsearch http://www.domainsearch.com/

Special feature: search domain name against many combinations of TLDs, including many second-level domains.

The advantage of Domainsearch.com is that it allows you to check a domain name's registration against many available extensions at once. You enter a domain name without any extension into the search box, then select the extensions you want to search using the CTRL key for multiple selections. Domainsearch includes not only TLDs (e.g., .au) but also many second-level domains, (e.g., com.au). If a name is registered, it will appear with a small "i" inside a blue circle; clicking on the "i" brings up the Whois data for that domain name. Not all TLDs are listed as extensions.

CheckDNS http://www.checkdns.net/quickcheckdomainf.aspx

Special feature: generates the most detailed domain report available for free on the web.

CheckDNS.NET can check DNS delegation, mail and web servers for any domain. The detailed report that is generated is, I believe, unique among the various Whois tools available for free:

CheckDNS.NET is testing mfa.gov.ir

CheckDNS.NET is asking root servers about authoritative NS for domain

Got DNS list for 'mfa.gov.ir' from ns.nic.ir or ns.nic.ir

Found NS record: web-srv.mfa.gov.ir[82.100.96.245], was resolved to IP address by ns.nic.ir or ns.nic.ir or ns.nic.ir

Found NS record: irserv1.iredco.com[194.165.0.10], was resolved to IP address by ns.nic.ir

Domain has 2 DNS server(s)

CheckDNS.NET is verifying if NS are alive

Error fetching SOA from web-srv.mfa.gov.ir [82.100.96.245], request timed out. Probably DNS server is offline.

DNS server irserv1.iredco.com[194.165.0.10] is alive and authoritative for domain mfa.gov.ir

1 server(s) are alive

DNS server web-srv.mfa.gov.ir failed and will be dropped from other tests

CheckDNS.NET checks if all NS have the same version

Your server has zone version 2000091808

CheckDNS.NET is verifying if NS are alive

NS list mismatch: registration authority reports that domain is hosted on the following servers: 'web-srv.mfa.gov.ir; irserv1.iredco.com', but DNS server irserv1.iredco.com reports domain to be hosted on 'irserv1.iredco.com; irserv2.iredco.com'.

CheckDNS.NET verifies www servers

DNS round-robing with multiple web servers detected

Checking HTTP server www.mfa.gov.ir [217.172.99.246]

HTTP server www.mfa.gov.ir[217.172.99.246] answers on port 80

Received: HTTP/1.1 200 OK (Server: Microsoft-IIS/6.0) . Welcome to Ministry of Foreign Affairs of I.R of Iran. H2 { .FONT-SIZE: 10pt; MARGIN: 0cm 0cm 0pt; DIRECTION: rtl; FONT-FAMILY: Times New Roman; unicode-bidi: embed; TEXT-ALIGN: right .} .A:hover {color: #FF0000} . . http://www.mfa.gov.ir. . . ((. . The Islamic Republic of Iran is an active partner in the global . co

Checking HTTP server www.mfa.gov.ir [217.172.99.12]

Error connecting to HTTP server www.mfa.gov.ir [217.172.99.12] port 80 : timed out waiting for connection

Checking HTTP server www.mfa.gov.ir [217.172.99.245]

Error connecting to HTTP server www.mfa.gov.ir [217.172.99.245] port 80 : timed out waiting for connection

CheckDNS.NET tests mail-servers

Domain mfa.gov.ir has only one mail-server

Checking mail server (PRI=10) email.mfa.gov.ir [194.165.0.14]

Mail server email.mfa.gov.ir[194.165.0.14] answers on port 25

```
<<< 220 email.mfa.gov.ir ESMTP
>>> HELO www.checkdns.net
<<< 250 email.mfa.gov.ir
>>> MAIL FROM: <dnscheck@uniplace.com>
<<< 250 ok
>>> RCPT TO: <postmaster@mfa.gov.ir>
<<< 250 ok
>>> QUIT
```

Mail server email.mfa.gov.ir [194.165.0.14] accepts mail for mfa.gov.ir

All MX are configured properly

Domainsurfer http://www.domainsurfer.com/

Special feature: offers some **wildcard** search capability for domain names registered in major gTLDs.

Domainsurfer offers some degree of wildcard searching. If you simply search for any domain name without a TLD extension, Domainsurfer will find all appearances of that search string regardless of where the string appears in the domain name. For example, if I search for [sonia], Domainsurfer will find *smith**sonian*** as well as ***sonia**-net*. In order to find domain names that begin with a string, prepend a carat (^) to the string, e.g., to find *sonia-net* but not *smithsonian*, the query is [^sonia]. Only queries the .com, .org, .net, .biz, .info gTLDs.

EasyWhois http://www.easywhois.com/

Special feature: automatic universal first- and second-level Whois lookups.

Automatically searches for most gTLD and country-level domains, including second-level domains, e.g., [amazon.co.uk] In March 2004, EasyWhois found it necessary to add a Turing number field to cut down the autobots and data mining activity. All this means is that you must enter the number in the green box before running a query.

Geektools http://www.geektools.com/whois.php

Special feature: universal domain name/IP address search across all TLDs.

Geektools has long been a staple of many analyst toolkits by providing a fast and easy universal Whois lookup for both domain names and IP addresses. Users must enter full name (domain plus TLD) or IP number (no partial searches or wildcard function). Geektools also requires users to enter the text shown below in the **Key** field before submitting a query because of autobots and data mining activity.

Namedroppers http://www.namedroppers.com/

Special feature: unusual **wildcard** option to run multiple keyword searches for domain names in the major gTLDs.

Namedroppers offers a different approach: you can enter a number of keywords to include or exclude in your search. So, a search for domain names that contain the keywords [national] and [security] and exclude the keyword association will find 153 domain names out of 29 million, including *nationalairportsecurity.com* and *internationalnetsecurity.com*. Unfortunately, Namedroppers only queries .com, .net, .org, and .edu TLDs.

Multiple DNS Lookup Engine http://www.bankes.com/nslookup.htm

Special feature: performs simple automatic network mapping.

Another invaluable web tool, the Multiple DNS Lookup Engine does something unique, as far as I know: it steps you through an entire IP address block (up or down). You either enter an IP address or a domain name and decide whether you want the lookup tool to step up or down from that address. The tool then lists the next address in the "neighborhood," associated IP or domain name, and in some cases, links to registration information. This tool is ideal for mapping a network.

Netcraft http://searchdns.netcraft.com/?host

Netcraft Search Help http://searchdns.netcraft.com/?help=yes

Special features: true **wildcard** function against all TLDs; provides information about what a site is running.

Netcraft is many researchers' favorite lookup tool because of its flexibility and power. Netcraft is designed to perform wildcard searches of domain names and it queries all TLD extensions. The simple search offers the option of searching for a domain name where the search string contains/starts with/ends with or matches a subdomain.[153] Netcraft also offers three wildcard options: * matches any number of characters; ? matches a single character; [ ] matches on specified characters. A search for [*.sc[aeio].com] returns all domains that contain *sca.com*, *sce.com*, *sci.com*, or *sco.com*. A search for sites that contain ".ir" returns sites such as www.irs.gov, whereas a search for [*.ir] finds only those sites in the Iranian domain. If you search for [*.gov.ir] Netcraft returns 11 sites, including "www.mfa.gov.ir" and "intranet.mim.gov.ir." What's more, for each hit, you have the option of seeing "what's that site running," which provides information about the operating system and web server at the site.

Whois.net http://www.whois.net/

Special feature: **wildcard** matches search string anywhere in domain name, or string anchored right or left.

Now that Whois.net searches most TLDs, it is one of the very best lookup tools on the web. As with Domainsurfer, the Whois.net keyword search will find *smith**sonian*** as well as ***sonia**-net* in its default mode. In order to find domain names that begin or end with a string, use the advanced search form, which is accessible from the ***search results' page***.

---

[153] As of this writing, the "site starts with" and "site ends with" options are all resolving to "site contains."

Domain Tools Whois Source http://whois.domaintools.com/

Domain Tools and Services http://www.domaintools.com/services/

Special features: option-rich search includes using **wildcards** to search for domain names; displays an image of the home page for domain records with websites.

One of the best Internet tool sites, Whois Source, was put under the umbrella of DomainTools.com. In most ways, it remains the same, but its interface is new and, I think, even nicer. The biggest and best change is the fact that ***the site now queries all top-level domains*** and not just the usual .com, .net, etc., it had queried in the past. Unfortunately, many of the tools offered require free registration (with an email address), but there are still many options available.

Another useful trick that still works is to ***type [whois.sc/sampledomain.com] into your browser's address bar to get the Whois records for that domain***. This is a very nice to know tip that doesn't require any special software or script.

Here is a snapshot of DomainTools.com's tools and services, and I have indicated which tools require registration. The remainder are fundamentally the same as those at the original site.

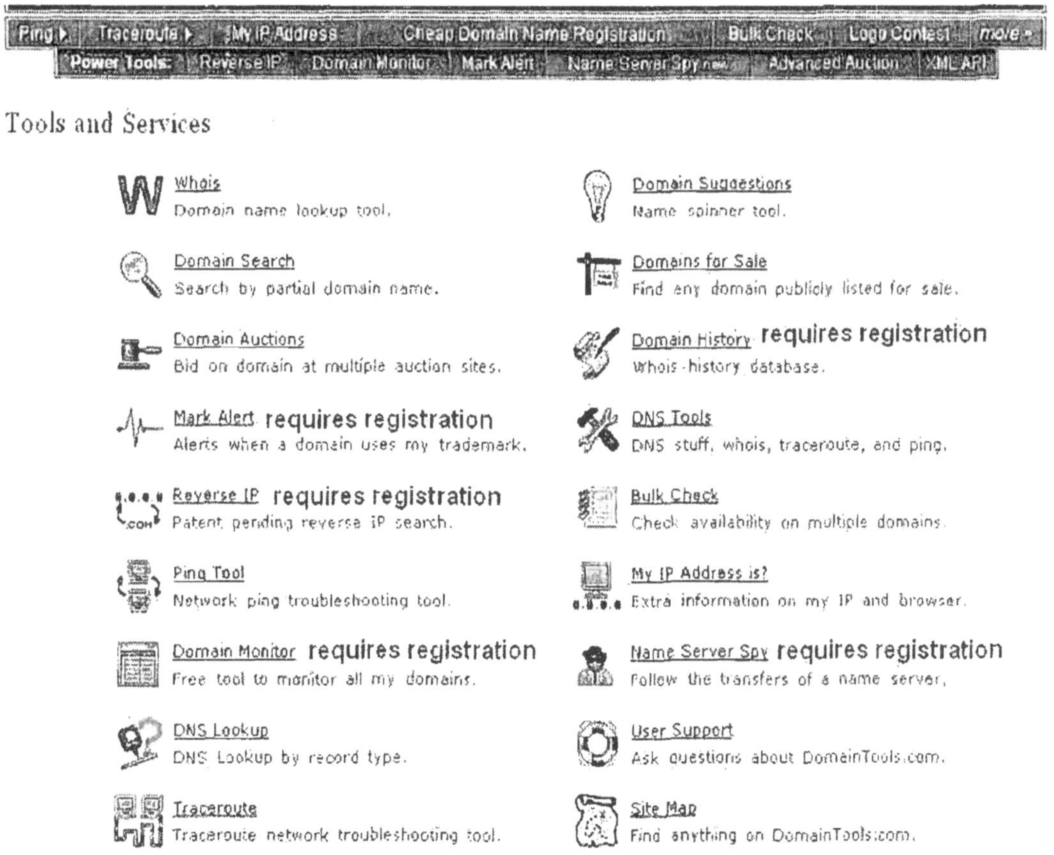
Ping | Traceroute | My IP Address | Cheap Domain Name Registration | Bulk Check | Logo Contest | more
Power Tools: | Reverse IP | Domain Monitor | Mark Alert | Name Server Spy new | Advanced Auction | XML API
Tools and Services
Whois
Domain name lookup tool.
Domain Suggestions
Name spinner tool.
Domain Search
Search by partial domain name.
Domains for Sale
Find any domain publicly listed for sale.
Domain Auctions
Bid on domain at multiple auction sites.
Domain History requires registration
Whois-history database.
Mark Alert requires registration
Alerts when a domain uses my trademark.
DNS Tools
DNS stuff, whois, traceroute, and ping.
Reverse IP requires registration
Patent pending reverse IP search.
Bulk Check
Check availability on multiple domains.
Ping Tool
Network ping troubleshooting tool.
My IP Address is?
Extra information on my IP and browser.
Domain Monitor requires registration
Free tool to monitor all my domains.
Name Server Spy requires registration
Follow the transfers of a name server.
DNS Lookup
DNS Lookup by record type.
User Support
Ask questions about DomainTools.com.
Traceroute
Traceroute network troubleshooting tool.
Site Map
Find anything on DomainTools.com.

Domain Tools offers more features than just about any other domain name search tool. Users can search on one or more terms, using only partial word(s). It also displays an image of the home page that corresponds to the domain registration record. When you search on a domain and view the Domain Tools information, the data provided includes:

➢ Page information

- website title with hyperlink to the homepage;
- metadata description from website's own HTML meta description tag;
- meta-keywords from website's own HTML meta keywords tag;
- About us: link to Wikipedia article (if applicable);
- related sites.

➢ Indexed Data

- entries at Open Directory, Alexa Ranking and Trend tracking, and Y! Directory;
- the number of listings on the site in both the Yahoo directory and DMOZ Open Directory with a link to all those listings; the directory listings show the category within the directory (e.g., History, News & Media), the title of the specific link within the domain (e.g., "History of Religions in Iran"), the description of the link, and the url itself.
- The Alexa data shows both the site's rank for the past one and three months in terms of traffic and whether the traffic trend is up or down.

➢ Server Data

- type of server on which the site runs;
- IP address plus options for **W** (run whois lookup), **P** (ping this address), **D** (run a DNS lookup on this address), and **T** (run a traceroute to this address);
- IP location, i.e., the physical location of the server hosting the domain (this may not always be accurate);
- Response code[154], which indicates status of server;

---

[52] Joe Burns, "Server Response Codes," HTML Goodies, <http://www.htmlgoodies.com/tutors/src.html> (14 November 2006).

- Blacklist status, which indicates if a site has been blacklisted for sending spam;
- SSL cert: whether or not site has a site security certificate and when it expires;
- Website status: active, parked/redirected, on-hold, deleted, etc.

➢ Registry Data

- ICANN Registrar: the registry with which the domain name registered (e.g., Network Solutions);
- Created: date domain name was originally created;
- Expires: date domain name will expire;
- Registrar Status: the registry generally sets domain name status codes. Domain Tools provides a detailed explanation of each code;
- Whois server for the domain (where to get a full whois record);
- Name server for the domain (host that that enables the domain name to be resolved to an IP address).

➢ Domaintools Exclusive: special information provided by Domain Tools, including:

- NS history: how often the nameserver for this domain has changed;
- IP history: how often the IP address for this domain has changed;
- Whois history: historical whois lookups done by users of Domain Tools (requires registration);
- Reverse IP or the number of hosts (computers) this webserver hosts (to see them requires free registration).

➢ Whois record is displayed at the bottom of the page.

DOCID: 4046925

Whois Record for Amazon.com

Page Information

Website Title: Amazon.com: Online Shopping for Electronics, Apparel, Computers, Books, DVDs & more

Record Type: Domain Name

Meta Description: Online shopping from the earth's biggest selection of books, magazines, music, DVDs, videos, electronics, computers, software, apparel & accessories, shoes, jewelry, tools & hardware, housewares, furniture, sporting goods, beauty & personal care

Meta Keywords: Amazon, Amazon.com, Books, Online Shopping, Book Store, Magazine, Subscription, Music, CDs, DVDs, Videos, Electronics, Video Games, Computers, Cell Phones, Toys, Games, Apparel, Accessories, Shoes, Jewelry, Watches, Office Products, Sports & Outdoors

About Us: Wiki article on Amazon.com

Related Sites: alibris.com, amazon.co.uk, barrypublications.com, bestsellersexchange.com, bsrmedia.com, buy.com, cduniverse.com, google.com, yahoo.com,

Indexed Data

DMOZ: 112 listings

Alexa Trend/Rank: 19 (1 Month) 18 (3 Month)

Yl Directory: 39 listings

Server Data

Server Type: Server

IP Address: 72.21.206.5 W P D T

IP Location: - Washington - Seattle - Amazon.com Inc

Response Code: 200

Blacklist Status: Clear

SSL Cert: www.amazon.com expires in 53 days

Website Status: Active

Registry Data

ICANN Registrar: NETWORK SOLUTIONS, LLC.

Created: 1994-11-01

Expires: 2013-10-31

Registrar Status: REGISTRAR-LOCK

Whois Server: whois.networksolutions.com

Name Server: UDNS1.ULTRADNS.NET

DomainTools Exclusive

NS History: 2 changes. Using 3 unique name servers in 2 years.

IP History: 93 changes. Using 7 unique IP addresses in 2 years.

Whois History: 953 records have been archived since 2000-04-03

Reverse IP: 120 other sites hosted on this server.

Whois Record

```
Registrant:
Amazon.com, Inc
  Legal Dept. P.O. Box 81226
  Seattle, WA 98108-1226
  US

  Domain Name: AMAZON.COM
```

Thumbnail: 2006-08-27

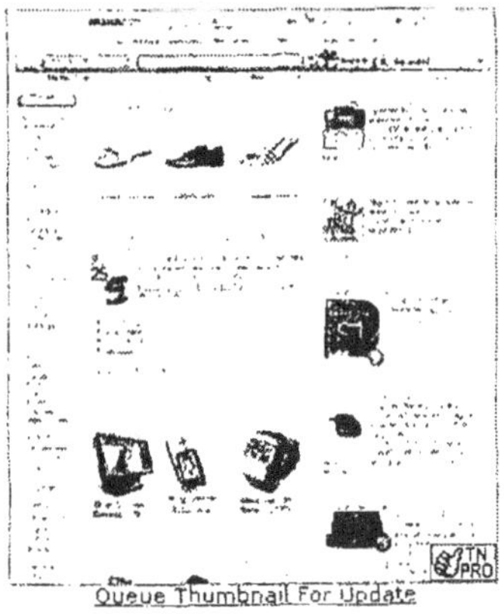

Queue Thumbnail For Update

Other TLDs

Show Key

.com .net .org .info .biz .us

Reverse IP

There are 120 other sites hosted on this webserver. View a sample with Reverse IP.

Domains for Sale

Domain	Price
LadyAmazon.com	$10.00
AmazonDoMmes.com	$100.00
AmazonBabes.com	$251.00
AmazonDiscounts.com	$500.00
AmazonMart.com	$650.00
LittleAmazon.com	$1,000.00
DomainAmazon.com	$1,000.00
AmazonCrafts.com	$1,188.00
AmazonAnt.com	$1,300.00
FemaleAmazon.com	$1,300.00
AmazonOnline.com	$1,688.00

Domains At Auction

Domain	Auction Date
Amazon-MarketPlaced.com	10-30-2006
Amazon-Protect.com	10-30-2006
Amazon-Servers1.com	10-30-2006
Amazon-Upgrades.com	10-30-2006
AmazonBeautyCo.com	10-30-2006
My-Store-Amazon.com	10-30-2006
AmazonUsa.net	11-01-2006
Amazon-Sy.com	11-01-2006
Amazon-Sy.net	11-01-2006
AmazonRePricer.com	11-01-2006
DomainsAmazon.com	11-01-2006

In addition, Domain Tools offers very sophisticated and somewhat confusing **wildcard** options. From the Domain Tools home page, go to More Tools and Services, then select Domain Search.

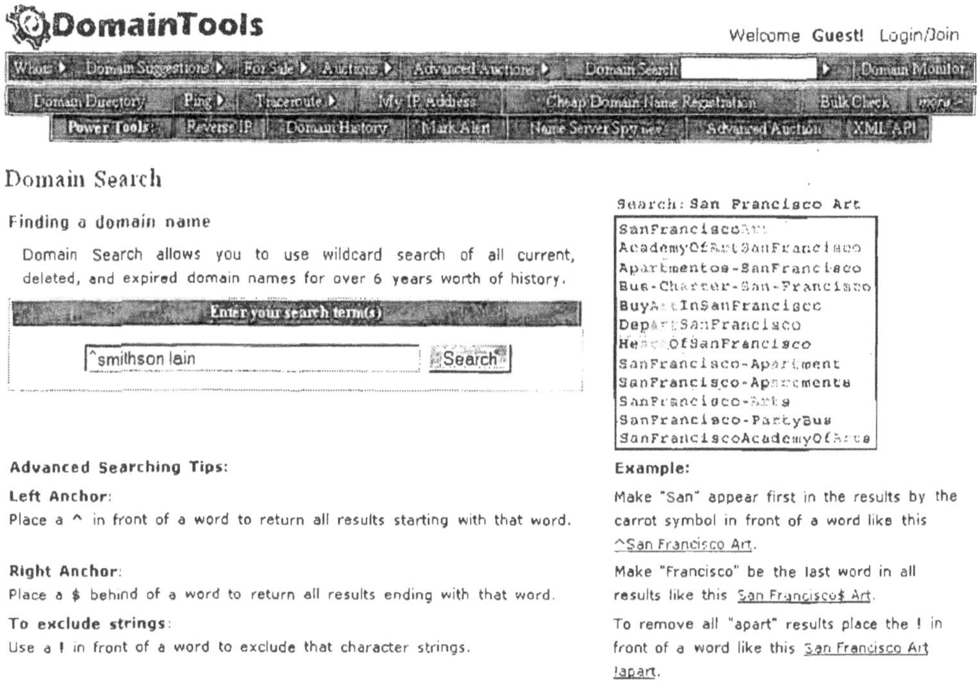
DomainTools
Welcome Guest! Login/Join
Domain Search
Finding a domain name
Domain Search allows you to use wildcard search of all current, deleted, and expired domain names for over 6 years worth of history.
Enter your search term(s)
^smithson lain
Search
Search: San Francisco Art
Advanced Searching Tips:
Left Anchor:
Place a ^ in front of a word to return all results starting with that word.
Right Anchor:
Place a $ behind of a word to return all results ending with that word.
To exclude strings:
Use a ! in front of a word to exclude that character strings.
Example:
Make "San" appear first in the results by the carrot symbol in front of a word like this ^San Francisco Art.
Make "Francisco" be the last word in all results like this San Francisco$ Art.
To remove all "apart" results place the ! in front of a word like this San Francisco Art !apart.

On this page, you can use three operators to perform wildcard searches:

- A **caret (^)** serves as a Left Anchor to find all domains that **start with a series of letters**. So ^smithson will find will find not only *smithsonian* but *smithsonbrothers*.

- A **dollar sign ($)** serves as a Right Anchor to find all domains that **end with a series of letters**. So $afghan will find *absolutelyafghan* as well as *yellowpageafghan* and everything in between.

- An **exclamation point (!)** serves to **exclude a string of characters**. So [^park !ing !ave] will find domain names that start with *park* and do not include the strings *ing* or *ave* anywhere in the name, thus eliminating *parking* or *parkave* or *parkavenue* from the search. In this search I have also chosen to limit the search to 25 character names, exclude hyphenated domains, show only active domain names, and turn on the adult content filter.

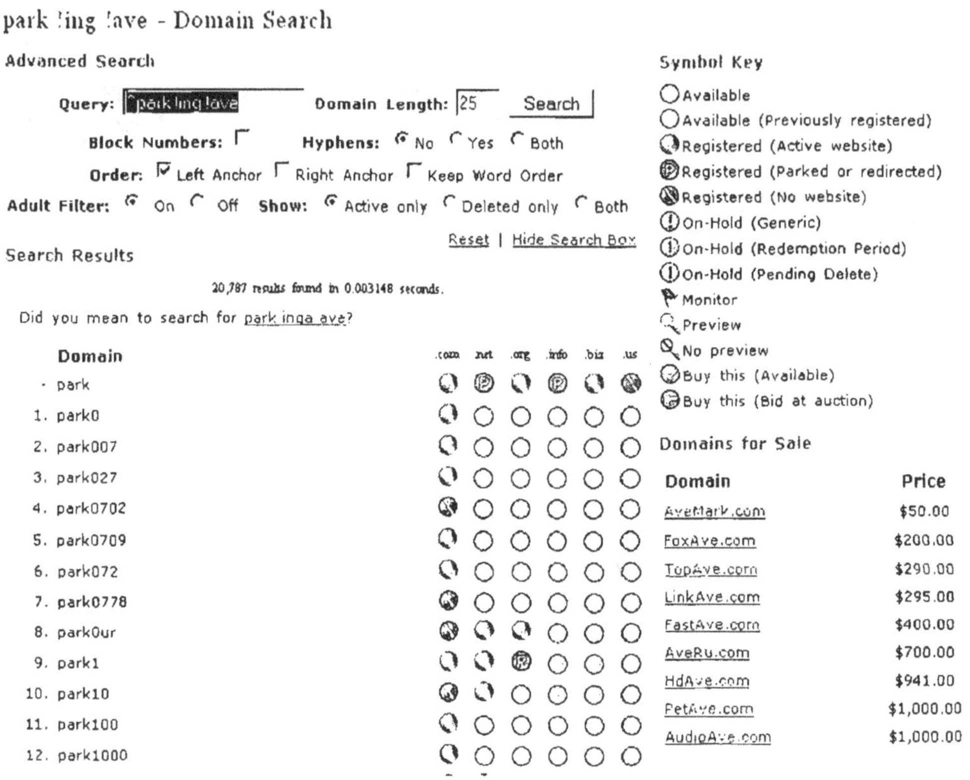

park ing ave - Domain Search

Advanced Search

Query: [park ing ave]   Domain Length: [25]   Search

Block Numbers: ☐   Hyphens: ⦿ No ○ Yes ○ Both
Order: ☑ Left Anchor ☐ Right Anchor ☐ Keep Word Order
Adult Filter: ⦿ On ○ Off   Show: ⦿ Active only ○ Deleted only ○ Both

Reset | Hide Search Box

Search Results

20,787 results found in 0.003148 seconds.

Did you mean to search for park inga ave?

Domain	.com	.net	.org	.info	.biz	.us
· park	◒	ⓑ	◒	ⓟ	◒	⊗
1. park0	◒	○	○	○	○	○
2. park007	◒	○	○	○	○	○
3. park027	◒	○	○	○	○	○
4. park0702	⊗	○	○	○	○	○
5. park0709	◒	○	○	○	○	○
6. park072	◒	○	○	○	○	○
7. park0778	⊗	○	○	○	○	○
8. park0ur	⊗	◒	◒	○	○	○
9. park1	◒	◒	ⓟ	○	○	○
10. park10	⊗	◒	○	○	○	○
11. park100	◒	○	○	○	○	○
12. park1000	◒	○	○	○	○	○

Symbol Key

○ Available
○ Available (Previously registered)
◒ Registered (Active website)
ⓑ Registered (Parked or redirected)
⊗ Registered (No website)
ⓘ On-Hold (Generic)
ⓘ On-Hold (Redemption Period)
ⓘ On-Hold (Pending Delete)
⚑ Monitor
🔍 Preview
🔍 No preview
◎ Buy this (Available)
ⓑ Buy this (Bid at auction)

Domains for Sale

Domain	Price
AveMark.com	$50.00
FoxAve.com	$200.00
TopAve.com	$290.00
LinkAve.com	$295.00
FastAve.com	$400.00
AveRu.com	$700.00
HdAve.com	$941.00
PetAve.com	$1,000.00
AudioAve.com	$1,000.00

In this tool, Domain Tools searches generic TLDs .com, .net, .org, .info, .biz, and .us with the status of domain names indicated by color images.

The old Whois.sc page has not actually gone away and DomainTools.com claims it will not go away. By the way, the Whois.sc page claims that the SC TLD (Seychelles) used to stand for 'source' and now stands for 'short cut.' I still think Seychelles every time I see it, though.

Whoix?   http://www.whoix.com/

Whoix? Advanced Search   http://www.whoix.com/advdomsearch.html

Special features: **wildcard** search on keyword across all domains; enter up to 25 domain names at one time.

From one page, users can search domains in gTLDs, domains in any country-level TLD, or second-level domain names (e.g., co.jp). In addition, Whoix has a keyword search that shows all domains that include the search term as part of their names, but only for the most popular domains. Be sure to click on the "Show who owns this domain" link for information about registered domains. Whoix's most unusual feature, however, is its option to search up to 25 domain names at a time. Results of this search show status (taken or free) and the option to view the Whois data. Home page is Yahoo-like directory devoted to domain names, trademarks, etc. Use the advanced search for options mentioned here.

Xwhois http://www.xwhois.com/

Special features: automatic universal first-, second-level Whois lookups; links to all country-level domain name registries.

Enter any domain name, including first- and second-level domain names—e.g., [bgu.ac.il]—for automatic search of correct database. Homepage includes list of all country code TLDs and links to country-level registries.

Whois the Oldest? http://www.whoisd.com/oldestcom.php

Finally, just for fun, take a look at the 100 oldest .com domains that are still registered! The very first "create date" was 15 March 1985. Try to guess the domain name ... you may be surprised.

Web Tip

To find an almost unlimited number of downloadable tools for your PC, visit:

http://www.tucows.com/
http://download.com.com/
http://downloads-zdnet.com.com/

Many very good utility programs that will allow you to run UNIX commands, such as Whois, ping, traceroute, etc., from a PC are available for a small fee or for free. Be sure to virus scan any executable file BEFORE you install it. Also, beware of ALL SHAREWARE...it could be SPYWARE! Don't install software promiscuously.

Internet Toolkits

Some of the most useful websites are those with many **Windows-based forms** that let you run a number of different queries from one page. These are often called **Internet Tool Kits, IP Tools**, or **Nettools,** like the page at Demon.net, which gives users a range of query options as well as the opportunity to query RIPE, "InterNIC," and the UKNIC databases from one screen.

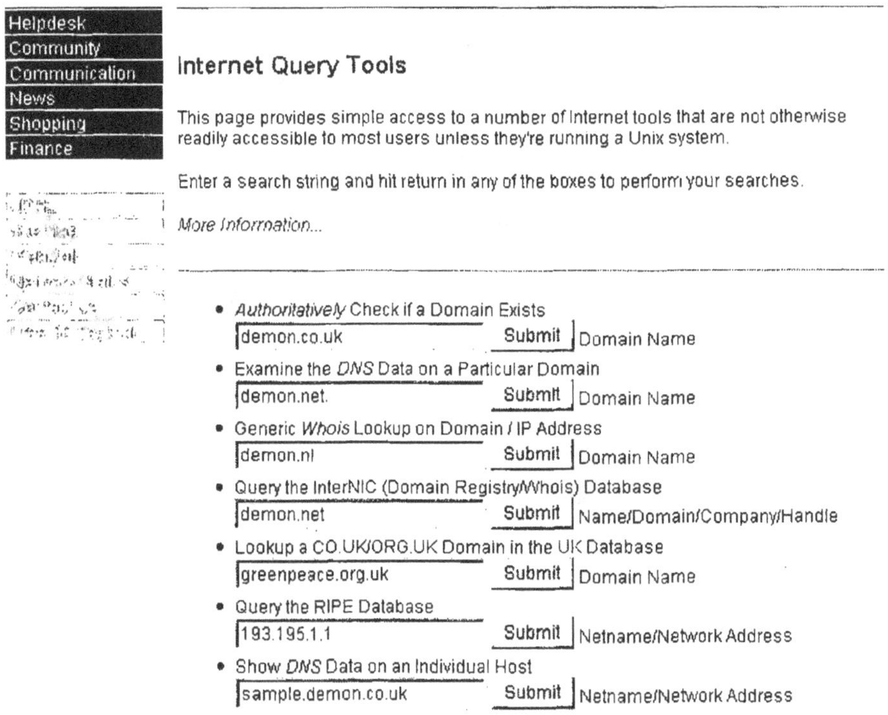
Helpdesk
Community
Communication
News
Shopping
Finance

Internet Query Tools

This page provides simple access to a number of Internet tools that are not otherwise readily accessible to most users unless they're running a Unix system.

Enter a search string and hit return in any of the boxes to perform your searches.

More Information...

- *Authoritatively* Check if a Domain Exists — demon.co.uk — Submit — Domain Name
- Examine the *DNS* Data on a Particular Domain — demon.net. — Submit — Domain Name
- Generic *Whois* Lookup on Domain / IP Address — demon.nl — Submit — Domain Name
- Query the InterNIC (Domain Registry/Whois) Database — demon.net — Submit — Name/Domain/Company/Handle
- Lookup a CO.UK/ORG.UK Domain in the UK Database — greenpeace.org.uk — Submit — Domain Name
- Query the RIPE Database — 193.195.1.1 — Submit — Netname/Network Address
- Show *DNS* Data on an Individual Host — sample.demon.co.uk — Submit — Netname/Network Address

Internet Query Tools http://www.demon.net/external/

RodentNet Ad Hoc IP Tools http://tatumweb.com/iptools.htm

RodentNet Ad Hoc IP Tools lets users run NSLookup, ping, traceroute, and an extremely powerful and fast IP range DNS query from one convenient menu. The IP range query will show all the IP addresses, any associated domain names, websites, and/or email addresses, as well as registration information for an entire block of addresses.

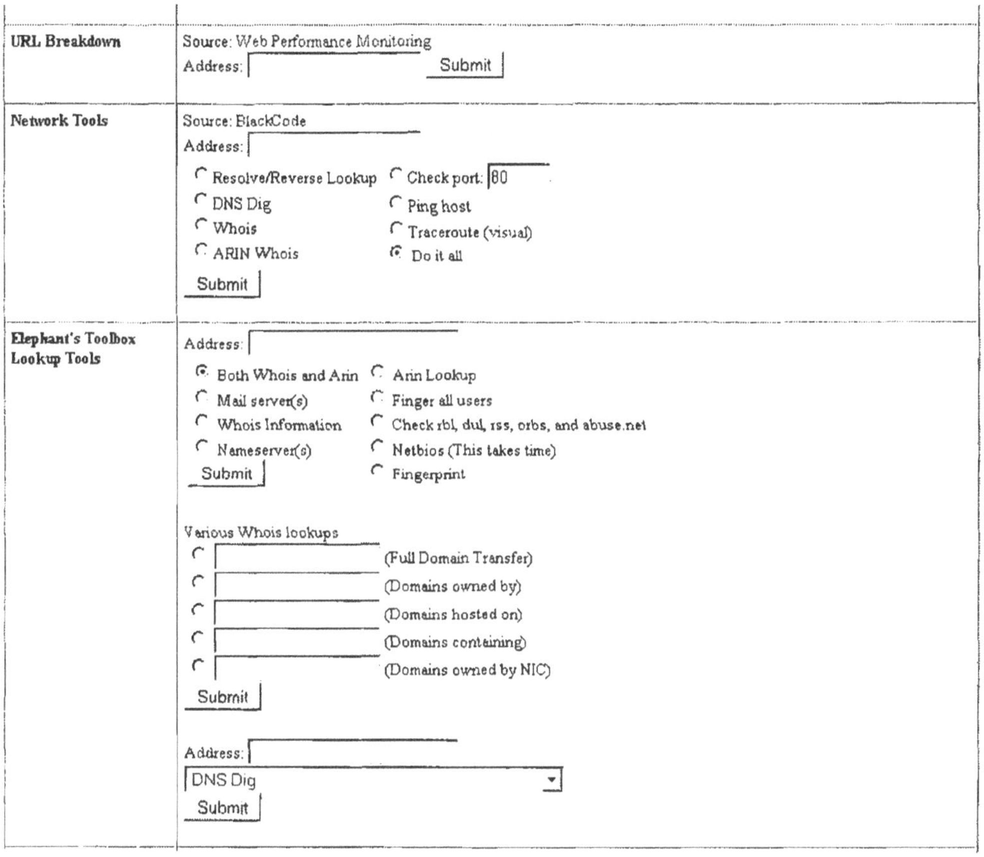

URL Breakdown	Source: Web Performance Monitoring Address: [ ] [Submit]
Network Tools	Source: BlackCode Address: [ ] ○ Resolve/Reverse Lookup ○ Check port: [80] ○ DNS Dig ○ Ping host ○ Whois ○ Traceroute (visual) ○ ARIN Whois ◉ Do it all [Submit]
Elephant's Toolbox Lookup Tools	Address: [ ] ◉ Both Whois and Arin ○ Arin Lookup ○ Mail server(s) ○ Finger all users ○ Whois Information ○ Check rbl, dul, rss, orbs, and abuse.net ○ Nameserver(s) ○ Netbios (This takes time) [Submit] ○ Fingerprint Various Whois lookups ○ [ ] (Full Domain Transfer) ○ [ ] (Domains owned by) ○ [ ] (Domains hosted on) ○ [ ] (Domains containing) ○ [ ] (Domains owned by NIC) [Submit] Address: [ ] [DNS Dig ▼] [Submit]

The other sites listed below offer web interfaces to all the basic network tools and, in some cases, add interesting new ones.

All-Nettools.com	http://www.all-nettools.com/tools1.htm
Centralops	http://centralops.net/co/body.asp
Internet Query Tools	http://www.demon.net/external/
iTools Internet Tools	http://www.itools.com/internet/
Network-Tools	http://www.network-tools.com/
RodentNet Ad Hoc IP Tools	http://tatumweb.com/iptools.htm

Centralops Hexillion Tools includes the AspTcpQuery, which will run an HTTP GET query to retrieve the page source code data about any webpage or IP address. Here's part of the result for [search.yahoo.com]:

AspTcpQuery sample

service ○ whois ○ finger ◉ HTTP ○ echo

server search.yahoo.com

query GET / HTTP/1.0 Go

powered by HexGadgets
view source | download

HexTcpQuery 1.0.18 1-processor license
webmaster, Hexillion Technologies

HexLookup 1.0.14 1-processor license
webmaster, Hexillion Technologies

Querying **search.yahoo.com [216.109.117.133]**...

[begin response]

```
HTTP/1.1 200 OK
Date: Wed, 17 Nov 2004 17:56:44 GMT
P3P: policyref="http://p3p.yahoo.com/w3c/p3p.xml", CP="CAO DSP COR CUR ADM DEV TAI PSA PSD IVAi I
Cache-Control: private
Connection: close
Content-Type: text/html; charset=ISO-8859-1

<!doctype html public "-//W3C//DTD HTML 4.01//EN" "http://www.w3.org/TR/html4/strict.dtd">
<html>
<head>
<meta http-equiv="content-type" content="text/html; charset=ISO-8859-1"><title>Yahoo! Search</tit
<link rel="stylesheet" href="http://us.i1.yimg.com/us.yimg.com/lib/s/ysch_hp_041018.css" type="te
<style>#yschtgln em {font:normal 100% arial;display:block;}</style>
<![if !IE]>
<script language="javaScript1.2">
if (document.layers&&!document.getElementById)
```

Central OPS TcpQuery http://www.hexillion.com/samples/AspTcpQuery.asp

How to Research a Domain Name or IP Address

In a previous section, I discussed the differences between domain name and whois lookups. Now I am going to walk through some basic research steps to learn how to analyze who registered a domain name and/or who owns an IP address. There are a number of ways to research domain names and IP addresses and many tools on the Internet that can help provide information about a domain or IP address.

Important Caveat: despite what you may hear or read elsewhere, you cannot ascertain the location or ownership of a domain name or IP address based solely on the fact it is in one of the most commonly used top-level domains, i.e., .com, .org, or .net. Names in these domains may be registered by anyone anywhere in the world. Likewise, domains registered in specific country top-level domains, e.g., .ru, .pk, .fr, are only *presumed* to be registered by non-US entities, but there is no guarantee based on the top-level domain alone this is the case. The point is simple: all domain names must be researched, with a few exceptions. The exceptions are .mil, .gov, and .edu, all of which are, at least theoretically, restricted to US-entities. There are even some exceptions in these cases. However, you may safely assume domain names or IP addresses associated with a .mil, .gov, or .edu top-level domain are US entities.

Steps for researching a domain name and IP address:

1. Does the domain name or IP address correspond to an Internet website?

If so, the first step is to use a good search tool such as Google to find out more about the site. The *info:* command at Google will show you links to Google's cache of the page, pages that are similar to the webpage, pages that link to this site, and pages that contain the search term. For this article, I will use a high-profile Russian anti-virus company Kaspersky Labs because it has registered domain names in both the .com and .ru top-level domains.

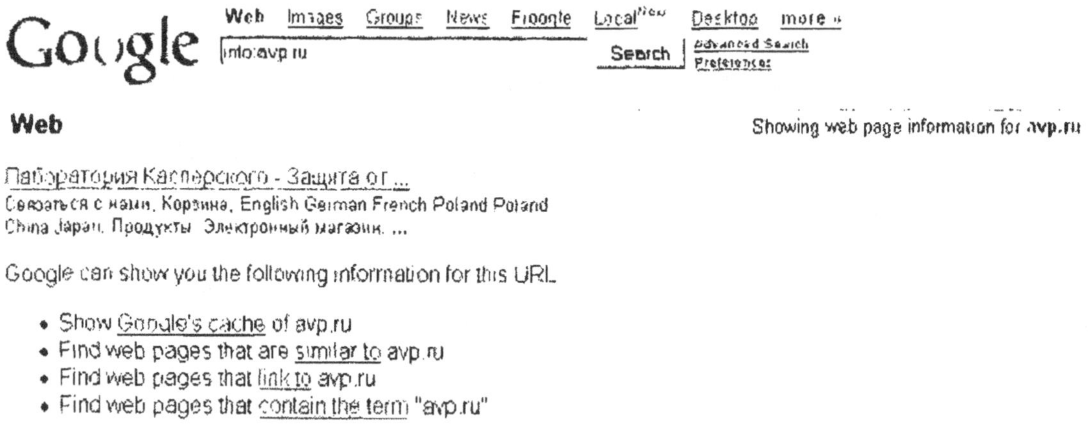
Google
Web Images Groups News Froogle Local New Desktop more »
info:avp.ru
Search
Advanced Search
Preferences
Web
Showing web page information for avp.ru
Лаборатория Касперского - Защита от ...
Связаться с нами, Корзина, English German French Poland Poland
China Japan, Продукты Электронный магазин. ...
Google can show you the following information for this URL
Show Google's cache of avp.ru
Find web pages that are similar to avp.ru
Find web pages that link to avp.ru
Find web pages that contain the term "avp.ru"

The next step would be to look at the website itself. The first two pages—About Us and Contact Us—are probably the most important, but there are other good pages to examine at the site itself:

- "About Us," which often tells you a great deal about the company (who owns it, where it's located, branch offices, subsidiaries).
- Other company websites in different languages, which may provide new domain names and IP addresses, as in the case of *avp.ru*: *kaspersky.ru*, *kasperksy.com*, *kaspersky.com.cn*, etc.
- "Contact Us," which may reveal numerous email addresses, physical locations, etc.
- "Site Map," which may show more pages on the site than are obvious from the homepage.
- "Press" or "News Releases," which often have the latest news about a company's activities, including company locations, ownership, ISPs, etc.
- "Page Source," which sometimes contains information about the person or organization that created or maintains the site. To view page source in Mozilla-based browsers, View | Page Source; in Internet Explorer, View | Source. Look or search for [author] in the HTML code:

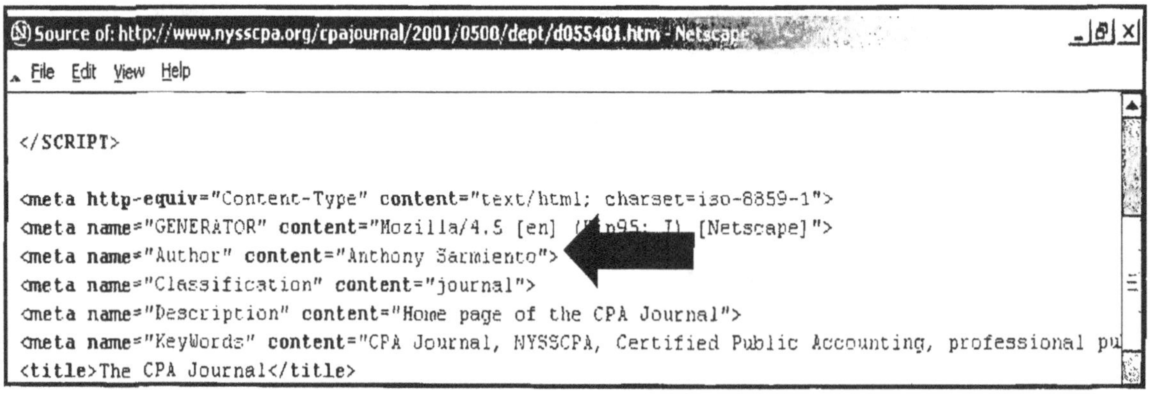
Source of: http://www.nysscpa.org/cpajournal/2001/0500/dept/d055401.htm - Netscape
File Edit View Help
</SCRIPT>
<meta http-equiv="Content-Type" content="text/html; charset=iso-8859-1">
<meta name="GENERATOR" content="Mozilla/4.5 [en] [Netscape]">
<meta name="Author" content="Anthony Sarmiento">
<meta name="Classification" content="journal">
<meta name="Description" content="Home page of the CPA Journal">
<meta name="KeyWords" content="CPA Journal, NYSSCPA, Certified Public Accounting, professional pu
<title>The CPA Journal</title>

2. Who registered the domain name?

Many, many domain name and Whois lookup tools exist on the Internet to permit you to find out information about a domain name. These tools will reveal such information as the servers associated with a specific domain name. Continuing with the example of *avp.ru*, using a lookup tool such as **Domain Dossier** (http://centralops.net/co/DomainDossier.aspx), I can quickly find the domain whois records, the network whois records, and the DNS records for *avp.ru*.

The domain whois record is formatted information about the domain name *avp.ru*, in this case pulled from RIPE's Whois Service database; notice the IP address, who registered the domain (Kaspersky Labs), and the domain's registrar (RIPN, the Russian Network Information Center).

Address lookup

canonical name avp.ru.

aliases

addresses 81.176.69.78

Domain Whois record

Queried **whois.ripn.net** with "avp.ru"...

```
% By submitting a query to RIPN's Whois Service
% you agree to abide by the following terms of use:
% http://www.ripn.net/about/servpol.html#3.2 (in Russian)
% http://www.ripn.net/about/en/servpol.html#3.2 (in English).

domain:     AVP.RU
type:       CORPORATE
nserver:    ns.kasperskylabs.net.
nserver:    ns.macomnet.ru.
nserver:    ns1.kasperskylabs.net.
nserver:    ns2.gldn.net.
nserver:    ns3.gldn.net.
state:      REGISTERED, DELEGATED
org:        Joint Stock Company
org:         "Kaspersky Lab"
phone:      +7 095 7978700
phone:      +7 095 9484331
fax-no:     +7 095 7978700
fax-no:     +7 095 9484331
e-mail:     webmaster@avp.ru
e-mail:     sales@kaspersky.com
e-mail:     rudomen@kaspersky.com
registrar:  RUCENTER-REG-RIPN
created:    1996.11.04
paid-till:  2005.04.01
source:     TC-RIPN
```

The network whois record for *avp.ru*. This is formatted information about the IP address from RIPE's database. Notice the block of IP addresses in which the IP address falls, which usually provides insight into the physical location of the servers and service providers. In this case, we can see the IP address owner (Kaspersky Labs), the Internet Service Provider (RT-COMM, a huge Russian ISP), street address, phone numbers, email addresses (interestingly, at *kasperksy.com*).

Network Whois record

Queried **whois.ripe.net** with "81.176.69.78"...

```
% This is the RIPE Whois query server #1.
% The objects are in RPSL format.
%
% Rights restricted by copyright.
% See http://www.ripe.net/db/copyright.html

inetnum:        81.176.69.64 - 81.176.69.127
netname:        KASPERSKY-RTCOMM
descr:          Kaspersky Labs
descr:          Geroyev Panfilovtcev St., 10 Moscow, 123363, RU
country:        RU
admin-c:        SF1624-RIPE
tech-c:         SF1624-RIPE
status:         ASSIGNED PA
notify:         registry@rt.ru
notify:         ism@kaspersky.com
mnt-by:         AS8342-MNT
changed:        luda@rt.ru 20030730
source:         RIPE

route:          81.176.0.0/15
descr:          RTCOMM-RU
origin:         AS8342
notify:         noc@rtcomm.ru
mnt-by:         AS8342-MNT
changed:        rus@rt.ru 20030120
changed:        rus@rt.ru 20031105
changed:        rus@rt.ru 20040809
source:         RIPE

person:         Sergey Fomin
address:        System Administrator /Kaspersky Lab Ltd
address:        10, Geroyev Panfilovtsev Str.,
address:        123363, Moscow, Russia
e-mail:         sergey@kaspersky.com
phone:          +7 095 797 87 00
```

The DNS record for *avp.ru* is the formatted information about this domain name used by the Domain Name Service/System to route data, including email, over the Internet to and from this host. Notice the different server names (*kasperskylabs.net*, macomnet.*ru*, kaspersky.*com*, *kaspersky-labs.com*, *gldn.net*) associated with *avp.ru*. For example, the preferred mail exchange server for *avp.ru* is *relay1.macomnet.ru* (a Russian telecom provider).

DNS records

name	class	type	data	time to live
avp.ru	IN	A	81.176.69.78	3600s (01:00:00)
avp.ru	IN	MX	preference: 10 exchange: sf.kaspersky.com	86400s (1.00:00:00)
avp.ru	IN	MX	preference: 400 exchange: relay1.macomnet.ru	86400s (1.00:00:00)
avp.ru	IN	MX	preference: 500 exchange: relay2.macomnet.ru	86400s (1.00:00:00)
avp.ru	IN	NS	ns.kasperskylabs.net	86400s (1.00:00:00)
avp.ru	IN	NS	ns1.kasperskylabs.net	86400s (1.00:00:00)
avp.ru	IN	NS	ns.macomnet.ru	86400s (1.00:00:00)
avp.ru	IN	NS	ns2.gldn.net	86400s (1.00:00:00)
avp.ru	IN	NS	ns3.gldn.net	86400s (1.00:00:00)
avp.ru	IN	SOA	server: ns1.kasperskylabs.net email: dnsadmin.kaspersky.com serial: 2004100701 refresh: 7200 retry: 3600 expire: 8640000 minimum ttl: 86400	86400s (1.00:00:00)
78.69.176.81.in-addr.arpa	IN	CNAME	78.64/26.69.176.81.in-addr.arpa	86399s (23:59:59)
78.64/26.69.176.81.in-addr.arpa	IN	PTR	proxy2-ru.kaspersky-labs.com	86400s (1.00:00:00)

3. Who registered the *.com* domain name? What can we find out about it?

Just because there are *.com* domain names associated with Kaspersky Labs does not mean they are US entities. First, let's look at the very thorough records available using CheckDNS.net .

CheckDNS.NET is testing kaspersky.com

CheckDNS.NET is asking root servers about authoritative NS for domain

Got DNS list for 'kaspersky.com' from a.gtld-servers.net

Found NS record: ns.kasperskylabs.net[195.170.248.13], was resolved to IP address by a.gtld-servers.net

Found NS record: ns1.kasperskylabs.net[212.5.80.3], was resolved to IP address by a.gtld-servers.net

Domain has 2 DNS server(s)

CheckDNS.NET is verifying if NS are alive

DNS server ns.kasperskylabs.net[195.170.248.13] is alive and authoritative for domain kaspersky.com

DNS server ns1.kasperskylabs.net[212.5.80.3] is alive and authoritative for domain kaspersky.com

2 server(s) are alive

CheckDNS.NET checks if all NS have the same version

All 2 your servers have the same zone version 2004092201

CheckDNS.NET is verifying if NS are alive

NS list mismatch: registration authority reports that domain is hosted on the following servers: 'ns.kasperskylabs.net; ns1.kasperskylabs.net', but DNS server ns1.kasperskylabs.net reports domain to be hosted on 'ns.macomnet.ru; ns2.gldn.net; ns3.gldn.net; ns.kasperskylabs.net; ns1.kasperskylabs.net'. Please make sure that you configure the same DNS servers in register database and on your DNS

CheckDNS.NET verifies www servers

Checking HTTP server www.kaspersky.com [81.176.69.79]

HTTP server www.kaspersky.com[81.176.69.79] answers on port 80

Received: HTTP/1.1 200 OK (Server: Apache) read .Kaspersky Lab - antivirus protection - protect your cyberspace...... Products E-Store Threats Viruses Hackers Spam Services Downloads Partners About Us . Buy in your country License renewal Buy online. code green.. virus activity is normal .... Viruslist.com. Virus Encycl

CheckDNS.NET tests mail-servers

Domain kaspersky.com has 3 mail-servers.

Checking mail server (PRI=10) sl.kaspersky.com [212.5.80.6]

Mail server sl.kaspersky.com[212.5.80.6] answers on port 25

<<< 220 sl.kaspersky.com ESMTP service ready
>>> HELO www.checkdns.net
<<< 250 sl.kaspersky.com
>>> MAIL FROM: <dnscheck@unplace.com>
<<< 250 Ok
>>> RCPT TO: <postmaster@kaspersky.com>
<<< 250 Ok
>>> QUIT

Mail server sl.kaspersky.com [212.5.80.6] accepts mail for kaspersky.com

Checking mail server (PRI=400) relay1.macomnet.ru [195.128.64.2]

Mail server relay1.macomnet.ru[195.128.64.2] answers on port 25

<<< 220 relay1.macomnet.ru ESMTP Sendmail 8.12.10/8.12.10; Tue, 8 Feb 2005 20:25:11 +0300 (MSK)
>>> HELO www.checkdns.net
<<< 250 relay1.macomnet.ru Hello [195.60.98.252], pleased to meet you
>>> MAIL FROM: <dnscheck@unplace.com>
<<< 250 2.1.0 <dnscheck@unplace.com>... Sender ok
>>> RCPT TO: <postmaster@kaspersky.com>
<<< 250 2.1.5 <postmaster@kaspersky.com>... Recipient ok
>>> QUIT

Of special note are the two IP addresses that appear in these records as mail servers: 195.128.64.9 and 212.5.80.6. The first resolves to Macomnet in Moscow and the second to *kaspersky.com*, a domain name registered with Tucows.com. Sounds like it might be in the US.

4. Where is *kaspersky.com* or 212.5.80.6 physically located?

This is usually harder to determine because truly accurate geolocation tools are not available for free on the Internet. However, we can get some pretty good clues from the network analysis tool traceroute. Below is the traceroute to *kaspersky*.com using Domain Dossier. Notice in particular the last three hops before reaching 212.5.80.6: they are Frankfurt, Germany (frankfurt1.de.alter.net), Moscow (msk.macomnet.net), and Macomnet's address 195.128.64.9 in Moscow. Traceroute shows the name of routers along the path the data is traveling, and these routers frequently (but certainly not always) use airport codes (e.g. LON, ATL). Below, for example, *dca4.alter.net* is almost certainly in the Washington, DC, area. There is much more that can be learned from traceroutes, a topic covered in the next section.

Traceroute

Tracing route to **sl.kaspersky.com [212.5.80.6]**...

hop	rtt	rtt	rtt	ip address	fully qualified domain name
1	0	0	1	216.46.228.229	port-216-3073253-es128.devices.datareturn.com
2	0	0	0	64.29.192.145	port-64-1949841-zzt0prespect.devices.datareturn.com
3	0	0	0	64.29.192.226	daa.g921.ispb.datareturn.com
4	0	0	0	168.215.241.133	hagg-01-ae0-1001.dlfw.twtelecom.net
5	0	0	0	66.192.253.124	core-02-ge-0-3-1-504.dlfw.twtelecom.net
6	0	0	0	168.215.54.74	tran-02-ge-0-3-0-0.dlfw.twtelecom.net
7	2	2	2	160.81.227.105	sl-gw40-fw-4-2.sprintlink.net
8	2	2	2	144.232.8.245	sl-bb21-fw-4-3.sprintlink.net
9	2	2	2	144.232.11.217	sl-bb20-fw-14-0.sprintlink.net
10	3	3	3	144.232.20.17	sl-st21-dal-1-0.sprintlink.net
11	3	3	3	144.232.9.134	
12	3	3	3	152.63.97.57	0.so-1-0-0.xl1.dfw9.alter.net
13	3	3	3	152.63.0.193	0.so-0-0-0.tl1.dfw9.alter.net
14	36	36	36	152.63.9.193	0.so-7-0-0.il1.dca6.alter.net
15	36	36	36	146.188.13.38	so-1-0-0.ir1.dca4.alter.net
16	123	123	123	146.188.8.162	so-6-0-0.tr1.fft1.alter.net
17	123	123	131	146.188.8.141	so-0-1-0.xr2.fft4.alter.net
18	123	123	123	149.227.48.30	pos6-0.gw9.fft4.alter.net
19	168	168	168	139.4.174.210	macomnet.frankfurt1.de.alter.net
20	168	168	169	195.128.64.80	ncc-3-eth-100.msk.macomnet.net
21	179	170	170	195.128.75.89	macom-i010301193-labkasper.macomnet.net

There are several geolocation tools available for free on the Internet, but no one should rely upon them alone because they may not be completely accurate. **HostIP.info** uses two sources of information to generate its geolocation tool: people identifying their city as associated with an IP address and automatic traceroutes. **Networldmap** also uses information entered by people visiting their site and has recently added links to its commercial site, **Geobytes**, which provides a more detailed report on IP address geolocation. For more details on using these tools, go to the section on Geolocating Internet Addresses.

5. What else can we learn?

Autonomous System Numbers (ASN) can help physically locate domain names and IP addresses. ASNs are unique numbers (written AS1234) associated with something called an Autonomous System (AS). An AS is an IP network with a single, clear external routing policy, which is used to exchange routing information between various Autonomous Systems. In short, each AS establishes a set of rules for how Internet traffic can travel between and among IP networks around the globe. In the case of *avp.ru/kaspersky.com*, the ASN is AS8342. Using the AS Whois Lookup tool available at Eye-Net Consulting, we see that AS8342 belongs to RT-RU, which is the Russian telecommunications giant, Rostelecom, one of the two principal shareholders of RT-COMM.

Autonomous System Whois Lookup

This page looks up Autonomous System Numbers found in the various registries

AS Number 8342

NIC (optional)

Search

Querying Network Information Centres whois servers about the AS Number ...

Results from Reseaux IP Europeans

Registrant	RTComm.RU Autonomous System Moscow, Russia
Admin Contact	RT-RU
Tech Contact	RT-RU
Maintainer Contact	AS8342-MNT
	from AS702 146.188.66.49 at 195.161.1.152 action pref=200; from AS702 146.188.68.149 at 195.161.1.149 action pref=200; accept ANY from AS1299 213.248.99.89 at 195.161.1.5 action pref=200; from AS1299 213.248.101.33 at 195.161.1.149 action pref=200; accept ANY

Eye-Net Consulting Autonomous System Number Whois Tool
http://www.enc.com.au/itools/aut-num.php

What did we ultimately learn about *avp.ru* and its associated domain names and IP addresses? They are most likely all located inside Russia, even those in the .com and .net domains. While not every domain name or IP address is as straightforward as the example of Kaspersky Labs, the techniques and on-line tools discussed here usually can provide sufficient information to draw a pretty clear picture of who owns an IP address, who registered a domain name, and where the host computers associated with those addresses are physically located.

Confused by all the acronyms and numbers in the various DNS and Whois records? These sites provide excellent information on understanding these important records.

Paul Adams, "Ins and Outs of DNS," *Webmonkey*
http://www.webmonkey.com/webmonkey/02/31/index3a.html

DNS for Rocket Scientists http://newweb.zytrax.com/books/dns/ch1/

IBM iSeries Information Center: DNS Resource Records
http://publib.boulder.ibm.com/infocenter/iseries/v5r3/index.jsp?topic=/rzakk/rzakkconceptresourcerec.htm

Traceroute

Traceroute is an Internet utility that shows in real time the complete logical connection path between a local host and the remote host it is contacting, i.e., it is a tool for mapping the path from one computer to another computer's IP address while showing all the IP addresses of the routers between these two points as well as the time between each step along the way.

> "A traceroute utility maps the path that data packets take between two points on the Internet, showing all of the intermediate nodes traversed, along with an indication of the speed of travel. Traceroute was invented in 1988 by Van Jacobson at Lawrence Berkeley National Laboratory in the US Today a traceroute utility often comes as part of the operating system. Windows, for example, has a small utility called tracert, which is used by typing, at the MS-DOS prompt, tracert."[155]

Traceroute, which shows all the intermediate routers that packets pass through to get to their destination, was written as a network troubleshooting utility to reveal problems with routers along a specific path. It is also very useful in showing how systems on the Internet are connected to each other. While traceroute only shows logical connections, it can nonetheless help you understand possible physical connections because the Internet is not only highly redundant but also very repetitive. In other words, Internet traffic will usually take the same routes over and over again unless something, such as equipment overload or failure, interferes with it.

Let's take a look at a simple traceroute run using a free visual traceroute tool available at IP Address Guide:[156]

---

[155] Martin Dodge, "Mapping Where the Data Goes," *Internet Society*, March 2000, <http://www.isoc.org/oti/articles/0200/dodge.html> (14 November 2006).

[156] The Internetipaddress.com website is currently unavailable.

Traceroute Enter host name (or IP/IPv6)
Traces the route packets take to this host
213.248.64.1 Tracert

```
Tracing route to 213.248.64.1 over a maximum of 20 hops:

   1    0 ms    2 ms    0 ms   67.18.29.185     UNITED STATES   -   -   THEPLANET.COM INTERNET SERVICES INC
   2    0 ms    0 ms    0 ms   69.41.250.73     UNITED STATES   -   -   THEPLANET.COM INTERNET SERVICES INC
   3    0 ms    0 ms    0 ms   12.96.160.10     UNITED STATES   TEXAS   FLOWER MOUND   THEPLANET.COM INTERNET
SERVICES
   4    0 ms    0 ms    1 ms   70.85.127.74     UNITED STATES   -   -   THEPLANET.COM INTERNET SERVICES INC
   5    0 ms    0 ms    0 ms   70.85.127.5      UNITED STATES   -   -   THEPLANET.COM INTERNET SERVICES INC
   6    1 ms    1 ms    0 ms   64.210.104.57    UNITED STATES   NEW YORK   BRONX   GLOBAL CROSSING
   7   32 ms   32 ms   32 ms   67.17.67.94      UNITED STATES   CALIFORNIA   LOS ANGELES   GLOBAL CROSSING
   8   32 ms   32 ms   33 ms   208.50.13.122    UNITED STATES   ILLINOIS   MACOMB   GLOBAL CROSSING
   9   45 ms   45 ms   45 ms   213.248.80.74    UNITED KINGDOM   -   -   TELIA INTERNATIONAL CARRIER
  10  141 ms  141 ms    *      213.248.64.33    GREECE   IRAKLION (CRETE)   IRAKLEION   TELIA INTERNATIONAL
CARRIER
  11  142 ms  142 ms  142 ms   213.248.64.1     GREECE   IRAKLION (CRETE)   IRAKLEION   TELIA INTERNATIONAL
CARRIER

Traceroute complete.
```

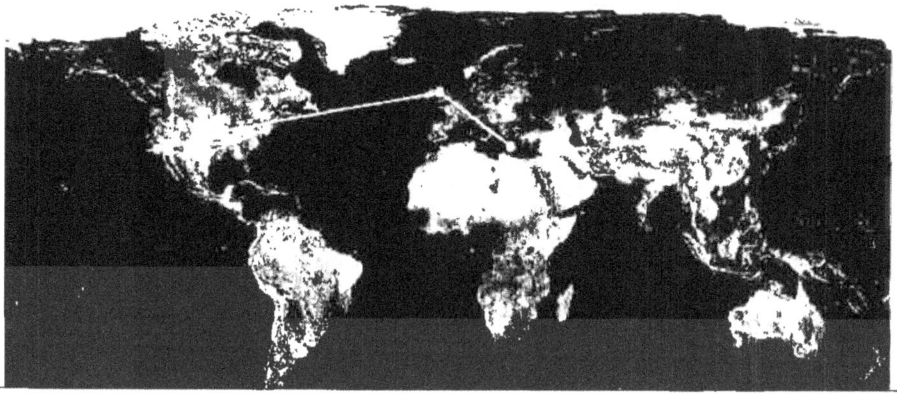

All this looks rather cryptic, but upon examination, the traceroute output is not usually very hard to read. Let's look at this traceroute step by step. We begin by entering either an IPv4 or IPv6 address or a host name, e.g., cnn.com, in the "Tracert" query box.

Traceroute Enter host name (or IP/IPv6)
Traces the route packets take to this host
213.248.64.1 Tracert

Enter the IP address or host name of the destination computer; in this case, I am using an IPv4 address.

The traceroute results should begin to appear almost instantaneously. What the readout below shows is a "road map" from the starting point of the trace, in this case, IP number 67.18.29.185, *theplanet.com* in Dallas, Texas, to the requested host 213.248.64.1, Telia International, Iraklion Crete. On the first line you see that this traceroute will allow a "maximum of 20 hops," which means the trace will show up to 20 stops of a packet as it moves from router to router before it reaches its destination. A packet is the fundamental data unit sent on the Internet or any other packet-switched network. Each line, numbered from one up to possibly 20 for this traceroute, represents a node. The node number is followed by three time measurements in milliseconds, such as line 8 with 32 ms 32 ms 33 ms. These numbers represent three different measurements of the time—known as the Round Trip Time (RTT)—it took the packet to travel from the origin IP address to that note (router) and back again. Next comes the IP address of the node, its geographical location (sometimes only a country, sometimes a country and city), and finally the domain name (THEPLANET.COM) or the name of the organization that owns that domain (e.g., TELIA INTERNATIONAL CARRIER).

```
Tracing route to 213.248.64.1 over a maximum of 20 hops:

   1     0 ms     2 ms     0 ms   67.18.29.185      UNITED STATES   -   -   THEPLANET.COM INTERNET SERVICES INC
   2     0 ms     0 ms     0 ms   69.41.250.73      UNITED STATES   -   -   THEPLANET.COM INTERNET SERVICES INC
   3     0 ms     0 ms     0 ms   12.96.160.10      UNITED STATES   TEXAS   FLOWER MOUND   THEPLANET.COM INTERNET
SERVICES
   4     0 ms     0 ms     1 ms   70.85.127.74      UNITED STATES   -   -   THEPLANET.COM INTERNET SERVICES INC
   5     0 ms     0 ms     0 ms   70.85.127.5       UNITED STATES   -   -   THEPLANET.COM INTERNET SERVICES INC
   6     1 ms     1 ms     0 ms   64.210.104.57     UNITED STATES   NEW YORK   BRONX   GLOBAL CROSSING
   7    32 ms    32 ms    32 ms   67.17.67.94       UNITED STATES   CALIFORNIA   LOS ANGELES   GLOBAL CROSSING
   8    32 ms    33 ms    33 ms   208.50.13.122     UNITED STATES   ILLINOIS   MACOMB   GLOBAL CROSSING
   9    45 ms    45 ms    45 ms   213.248.80.74     UNITED KINGDOM   -   -   TELIA INTERNATIONAL CARRIER
  10   141 ms   141 ms     *      213.248.64.33     GREECE   IRAKLION (CRETE)   IRAKLEION   TELIA INTERNATIONAL
CARRIER
  11   142 ms   142 ms   142 ms   213.248.64.1      GREECE   IRAKLION (CRETE)   IRAKLEION   TELIA INTERNATIONAL
CARRIER

Traceroute complete.
```

Next, we can look at the visualization of the traceroute on the world map. This makes it graphically clear that packets can take very strange routes to get from point A to point B.

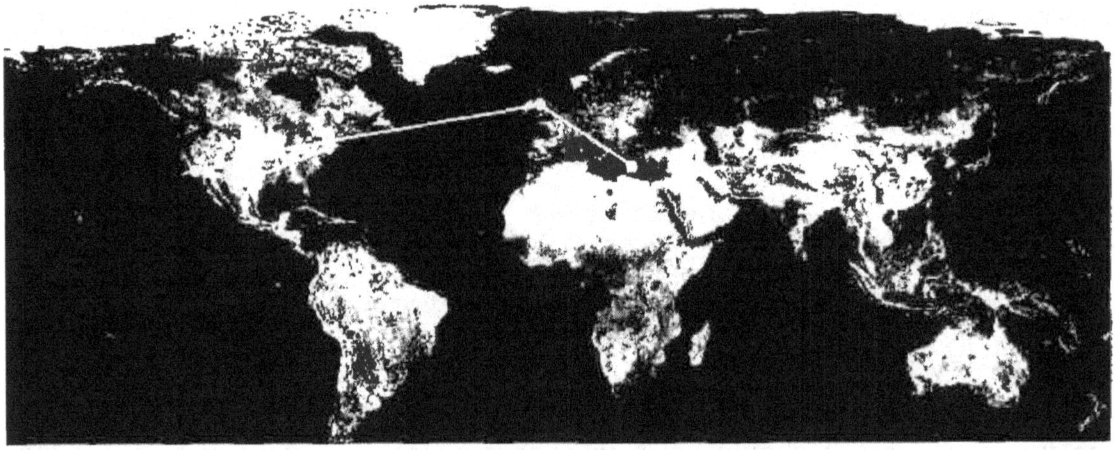

In our example, the packet travels from Texas to New York to California to Illinois before crossing the Atlantic to the UK via one of Global Crossing's three undersea cables.

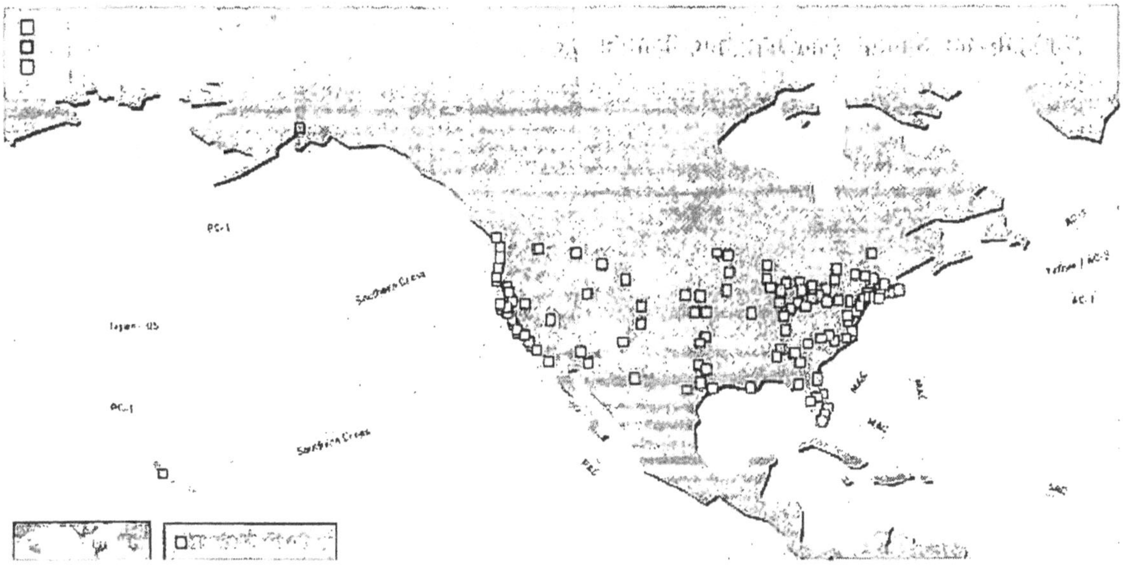

Global Crossing's Interactive Network Map
http://www.globalcrossing.com/html/map05_11_05.html

Not all traceroutes look like the one above. There are many variations on the original traceroute program now available and many of them are free to use at websites. Frequently you will see a traceroute that is somewhat more cryptic than the one we just examined, but it may reveal more information about the geographical route your packet has taken. Here is a typical traceroute run at All Net Tools (http://www.all-nettools.com/toolbox)

Hop	IP Address	Hostname	Average RTT[1]
1	192.168.1.8	baal.pair.net	2.82 ms
2	192.168.1.9	bodhi.pair.net	0.49 ms
3	64.214.174.177	so-2-1-0.ar2.de1.gblx.net	6.56 ms
4	67.17.67.57	so2-1-0-2488M.ar1.DCA3.gblx.net	17.13 ms
5	208.50.13.206	reach.ar1.DCA3.gblx.net	176.40 ms
6	202.84.143.78	i-12-0.mia-core01.net.reach.com	198.18 ms
7	202.84.143.73	i-12-0.dal-core01.net.reach.com	60.16 ms
8	202.84.143.65	i-2-0.wil-core02.net.reach.com	83.93 ms
9	202.84.144.102	i-0-0.syd-core02.net.reach.com	233.02 ms
10	203.50.13.41	10GigabitEthernet6-0.pad-core4.Sydney.telstra.net	240.21 ms
11	203.50.6.89	10GigabitEthernet9-0.chw-core2.Sydney.telstra.net	231.74 ms
12	203.50.6.226	Pos2-0.cha-core4.Brisbane.telstra.net	245.26 ms
13	203.50.51.33	GigabitEthernet5-1.cha23.Brisbane.telstra.net	245.59 ms
14	139.130.97.62	apnic1-new.lnk.telstra.net	247.08 ms
15	202.12.29.20	nori.apnic.net	245.87 ms

Looking at line 4 in this example, the designation DCA3 indicates this router is in or near Washington, D.C. How do we know this? Because routers often use airport trigraphs to show their location and DCA=Washington's Reagan National Airport. A traceroute frequently provides us insights into the physical path the packet travels by indicating the locations of the routers along the way.

Following hops 4 through 7 we can see the packet moves from Washington to Miami to Dallas and then on line 8 to the mysterious *wil-core02*. What is "*wil*"? It is not an airport code but rather refers to a building, the One Wilshire Building in Los Angeles. "One Wilshire is home to virtually all of the market leaders in the telecommunications industry. The property currently houses over 120 telecom related companies including: AT&T, Cable & Wireless, China Telecom, Global Crossing, Level 3

Communications, MCI Worldcom, MFS, PacBell, Qwest Communications, Sprint, Time Warner (AOL), Verizon (GTE) and XO Communications."[157]

At this point note the rather dramatic increase in the RTT from 83.93 ms to 233.02 ms. The packet leaves the US and travels to Sydney, Australia, (*syd-core02*) between hops 8 and 9. In this example, *wil-core02* is the gateway router to the trans-Pacific link of the Reach network. Reach is an Asia-Pacific focused backbone provider operating a "high-speed cable network in the Asia-Pacific region. It has significant interests in all major submarine cables consortia in the Asia-Pacific region." The other indication that the packet traveled via submarine cable instead of satellite is the fact that the RTT remains relatively low. A good rule of thumb (though not by any means a guarantee) is that satellite traffic usually has a >500 ms RTT.

The packet stays on the Reach network all the way to hop 10 in Sydney where it is handed off to Telstra, Australia's largest telecommunications company. Here we see another indication of the network infrastructure: the packet is now in Telstra's 10GigabitEthernet6-0. In February 2003, Telstra activated the first 10 Gigabit Ethernet link on its Internet Direct backbone, the network that delivers broadband Internet services across Australia. The final geographical hops in this traceroute occur between hops 11 and 12 where the packet travels from Sydney to Brisbane. In fact, APNIC (the Asia Pacific Network Information Center) is located in Milton, Brisbane, Australia, the same location as the server *nori.apnic.net.* However, be cautious about geolocating a final destination server with the actual location of the organization you're researching because that organization's site may well be hosted in a separate location.

Traceroute Anomalies and Failures

Watch out for traceroute anomalies. For example, look at line 6 in this traceroute at NYC-gw12.USA.net.DTAG.DE. Despite the fact the router has a German top-level domain (DE), this router is in the US (NYC). Note also the RTT increase from 37 ms to 141 ms.

---

[157] "Our Building," *One Wilshire*, 2002, <http://www.onewilshire.com/our_building/index.htm> (14 November 2006).

```
TraceRoute to host thing.net

Timeout 5
Start from hop 1
Maximum Hops 30

#    Address          Host Name                               Msg Type          Time
1    193.158.142.213  Unavailable                             TTL Exceeded      33 ms
2    217.5.112.254    Unavailable                             TTL Exceeded      302 ms
3    194.25.7.175     KN-ag1.KN.net.DTAG.DE                   TTL Exceeded      31 ms
4    212.185.11.129   KN-gw1.KN.net.DTAG.DE                   TTL Exceeded      37 ms
5    212.185.8.177    F-gw12.F.net.DTAG.DE                    TTL Exceeded      37 ms
6    194.25.6.110     NYC-gw12.USA.net.DTAG.DE                TTL Exceeded      141 ms
7    194.25.6.90      dc.nyc1.verio.net                       TTL Exceeded      141 ms
8    129.250.16.210   p1-0-0-0.r00.nycmny02.us.ra.verio.netTTL Exceeded           146 ms
9    129.250.126.137  d3-0-1-0.a03.nycmny05.us.ra.verio.netTTL Exceeded           147 ms
10   209.14.148.161   fa-5-0.a00.nycmny05.us.ra.verio.netTTL Exceeded           153 ms
11   209.227.40.182   thing-gw.spacelab.net                   TTL Exceeded      156 ms
12   209.14.134.3     thing.net                               Echo Reply        157 ms
```

You may also run into incomplete traceroutes, usually indicated by asterisks and, sometimes, by the warning "Request timed out." There are several reasons for a traceroute to fail.

- a network problem, e.g., a server or router on the network is down (you will probably see "Request timed out"). The router immediately after the last visible one is usually the culprit.
- a server or router along the path has rejected your packet. Again, the router immediately after the last visible one is usually the culprit.
- the target host does not exist on the network because it has been disconnected, turned off, or is otherwise unreachable. You may see a !H or !N message in the traceroute.
- the traceroute may have encountered a routing loop, in which case the packet will simply bounce between two routers and never reach its destination.
- a firewall is in the route path (you may or may not see "Request timed out").
- the traceroute encounters a private IP address.
- there is packet filtering occurring somewhere along the traceroute path.

Here is a typical incomplete traceroute due to a network problem. The traceroute timed out at hop 9 because of what turned out to be an ATM router problem on the Quest network. This was determined by checking the Quest network status at the time of the interruption.

```
Tracing route to www.gwww.aol.com [64.12.187.22]
over a maximum of 30 hops:

  1    13 ms    19 ms     1 ms  gw.ziv-127.brandeis.edu [129.64.165.1]
  2     1 ms    <1 ms    <1 ms  129.64.253.1
  3     2 ms     2 ms     1 ms  bos-edge-02.inet.qwest.net [65.115.97.217]
  4     2 ms     2 ms     1 ms  bos-core-01.inet.qwest.net [205.171.28.13]
  5     7 ms     7 ms     7 ms  ewr-core-02.inet.qwest.net [205.171.8.26]
  6     8 ms     7 ms     7 ms  ewr-core-03.inet.qwest.net [205.171.17.34]
  7    11 ms    11 ms    11 ms  dca-core-02.inet.qwest.net [205.171.8.181]
  8    11 ms    11 ms    11 ms  dca-edge-04.inet.qwest.net [205.171.9.66]
  9     *        *        *     Request timed out.
```

The next example shows what is probably a firewall that caused the traceroute to fail:

```
traceroute to gov.ru (194.226.80.160), 30 hops max, 40 byte packets
 1  gw-casablanca.logix.cz (217.11.251.254)  0.347 ms  0.000 ms  0.000 ms
 2  81.0.225.5  0.548 ms  0.405 ms  0.365 ms
 3  THP-NE40-ge4-0-3.cas.ip-anywhere.net (217.11.224.224)  1.569 ms  2.577 ms  1.613 ms
 4  nix.interoute.cz (194.50.100.127)  0.779 ms  5.696 ms  0.501 ms
 5  PO6-0.prg-001-access-1.interoute.net (212.23.50.77)  36.447 ms  37.910 ms  36.427 ms
 6  PO10-0.fra-006-core-2.interoute.net (312.23.50.70)  233.174 ms  233.089 ms  219.003 ms
 7  PO8-0.dus-001-access-1.interoute.net (84.233.146.13)  45.651 ms  36.736 ms  36.627 ms
 8  PO7-0.ham-001-access-1.interoute.net (84.233.146.10)  36.840 ms  36.687 ms  36.548 ms
 9  PO9-0.cph-002-access-1.interoute.net (84.233.168.145)  36.635 ms  36.677 ms  36.512 ms
10  PO9-0.Sto-002-access-2.interoute.net (212.23.43.73)  37.064 ms  36.947 ms  36.612 ms
11  PO9-0.Sto-002-access-1.interoute.net (212.23.43.69)  36.522 ms  36.576 ms  36.558 ms
12  84.233.135.46  38.608 ms  37.867 ms  38.036 ms
13  vlan102-g0-1.r2-sth2.se.ionip.net (195.7.95.116)  38.973 ms  38.901 ms  38.424 ms
14  194.88.115.226  63.348 ms  65.553 ms  62.652 ms
15  vlan1-r5-MSK-NIR.ionip.ru (213.152.128.79)  63.288 ms  63.313 ms  63.261 ms
16  rosniiros-gw.ionip.ru (213.152.129.94)  81.458 ms  82.137 ms  83.405 ms
17  * * *
18  * * *
19  * * *
20  * * *
21  * * *
22  * * *
23  * * *
24  * * *
25  * * *
26  * * *
27  * * *
28  * * *
29  * * *
30  * * *
```

You may also encounter a routing loop in a traceroute. This occurs when the packet is simply bounced between two routers until the traceroute reaches its maximum number of hops. In this example, the packet was bouncing between routers at 186.40.64.94 and 186.40.64.93:

```
Tracing route to lostserver.confusion.net [186.9.17.153]
over a maximum of 30 hops:

  1   <10 ms   <10 ms   <10 ms  186.217.33.1
  2    60 ms    70 ms    60 ms  rtr-2.confusion.net [186.40.64.94]
  3    70 ms    71 ms    70 ms  rtr-1.confusion.net [186.40.64.93]
  4    60 ms    70 ms    60 ms  rtr-2.confusion.net [186.40.64.94]
  5    70 ms    70 ms    70 ms  rtr-1.confusion.net [186.40.64.93]
  6    60 ms    70 ms    61 ms  rtr-2.confusion.net [186.40.64.94]
  7    70 ms    70 ms    70 ms  rtr-1.confusion.net [186.40.64.93]
  8    60 ms    70 ms    60 ms  rtr-2.confusion.net [186.40.64.94]
  9    70 ms    70 ms    70 ms  rtr-1.confusion.net [186.40.64.93]
...
...
...
Trace complete.
```

Here is what you might see in the case of an unreachable host. In this example, the traceroute attempted to reach a private IP address, but that host was not reachable on the network. Note the H! message, which is usually appears as !H.

```
traceroute to 10.1.2.5 (10.1.2.5), 30 hops max, 40 byte packets
 1  gw-casablanca.logix.cz (217.11.251.254)  0.000 ms   0.000 ms   0.000 ms
 2  81.0.225.5  0.608 ms   1.803 ms   0.855 ms
 3  vip.cas.ip-anywhere.net (217.11.224.240)  1.005 ms   2.648 ms   0.490 ms
 4  * * *
 5  * * *
 6  * * *
 7  * * *
 8  * * *
 9  vip.cas.ip-anywhere.net (217.11.224.240)(H!)  2728.031 ms (H!)  2708.043 ms (H!)  2688.135 ms
```

A private IP address will cause a traceroute to fail. Normally, you should not see these IP address blocks in a traceroute (the traceroute should time out before it reaches the private address). However, if you reach a private IP address, it is easy to spot because it belongs to a block of IPv4 addresses that are reserved for private use, meaning these address ranges are unassigned non-Internet addresses. Because they cannot be routed over the Internet, these private address are only for use on internal systems. In the following example, the traceroute should have timed out at hop 10:

- 10.*

- 172.[16-31].*
- 192.168.*

```
10 ebay-2-gw.customer.ALTER.NET (157.130.197.90) 114.204 ms 123.232 ms 120.957 ms
11 10.1.2.5 (10.1.2.5) 110.693 ms 114.475 ms 107.747 ms
12 * * *
13 * * *
```

The private address 10.1.2.5 within another network should not be visible to us. In this case, though, it is the last visible address before the trace ends in timeouts.

Traceroute Servers

As with other Internet utilities, there are many sites that let you run a traceroute from their site to the domain or IP of your choice. Multiple Traceroute Gateway deserves a special mention because it will run a traceroute to any host or IP address from multiple starting points. The starting points can be anywhere in the world. Unfortunately, many of the traceroute starting points listed at the site no longer work, so you may have to try several to find a good one for a particular region. I recommend you not try to run too many traceroutes at once because that can be a very slow process. There are also now numerous sites offering traceroute utilities for IPv6.

The Logbud Toolkit

Logbud's set of webmaster tools is definitely worth adding to your own set of Internet toolkits. What first attracted me to it was the visual traceroute that does the following:

> **Visual Traceroute** shows geo information of the gateways it traverses: Country, Region, (State for the US), City and Network organization. The way of the trace is shown on small and big geographical maps. Also Visual Traceroute displays network names (e.g., [GNTY-NETBLK-4]) and AS (autonomous system) numbers (e.g., [AS6846]).

As you can see from this traceroute to a Japanese domain, the traceroute information from Logbud is much easier to read and understand. There is also an accompanying map, but it is not always accurate (in this case, the trace ended in New Jersey). This is certainly one of the clearest, most useful traceroute tools I have seen.

IP or Domain response ip | Traceroute!

Traceroute Output:

1	21844	[illegible]	[illegible]	70.85.54.51	0.9ms
2	21844	[illegible]	[illegible]	12.96.160.42	0.4ms
3	21844	[illegible]	[illegible]	70.85.127.74	0.8ms
4	21844	[illegible]	[illegible]	70.85.127.1	1.4ms
5	7018	[illegible]		12.119.136.49	0.8ms
6	7018	[illegible]	[illegible]	12.122.82.38	28.5ms
7	7018	[illegible]	[illegible]	12.122.10.50	26.7ms
8	7018	[illegible]	[illegible]	12.123.199.185	27.0ms
9	7018	[illegible]		12.119.138.34	40.9ms
10	10026	[illegible]	[illegible]	202.147.0.214	147.8ms
11	10026	[illegible]	[illegible]	202.147.1.185	147.8ms
12	10026	[illegible]	[illegible]	203.192.149.198	147.6ms
13	9607	[illegible]	[illegible]	211.14.3.49	156.5ms
14	9607	[illegible]	[illegible]	203.141.63.39	156.7ms
15	9607	[illegible]	[illegible]	211.14.30.245	145.8ms
16	9607	[illegible]	[illegible]	211.14.31.66	159.6ms

LFT spent 0.65s tracing and 14.68s resolving names and ASNs.

United States	Texas	Dallas	THEPLANET.COM INTERNET SERVICES	-97	33
United States	Texas	Dallas	THEPLANET.COM INTERNET SERVICES	-97	33
United States	Texas	Dallas	THEPLANET.COM INTERNET SERVICES	-97	33
United States	Texas	Dallas	THEPLANET.COM INTERNET SERVICES	-97	33
United States	New Jersey	Middletown	AT&T WorldNet Services	-74	40
United States	New Jersey	Middletown	AT&T WorldNet Services	-74	40
United States	New Jersey	Middletown	AT&T WorldNet Services	-74	40
Japan			Asia Netcom Corporation	105	35
Japan			Asia Netcom Corporation	105	35
Japan			Asia Netcom NRT HUB	138	36
Japan			BroadBand Tower, Inc.	138	36
Japan			BroadBand Tower, Inc.	138	36
Japan			BroadBand Tower, Inc.	138	36
Japan			IRI Commerce&Technology, Inc.	138	36

Logbud also offers a traceroute manual that has some interesting scenarios of hard to understand hops that are sometimes seen in a traceroute. It's somehow comforting when the author looks at a particular traceroute hop and comments "God only knows what's going on with [hop] 12."

Visual traceroute is only one of a number of very useful tools Logbud offers. Of special interest are those tools that help provide information about domain names and IP addresses. Note, in particular, the **IP Range query**, which does a bulk query of IP addresses in a certain range and resolves them to their host names.

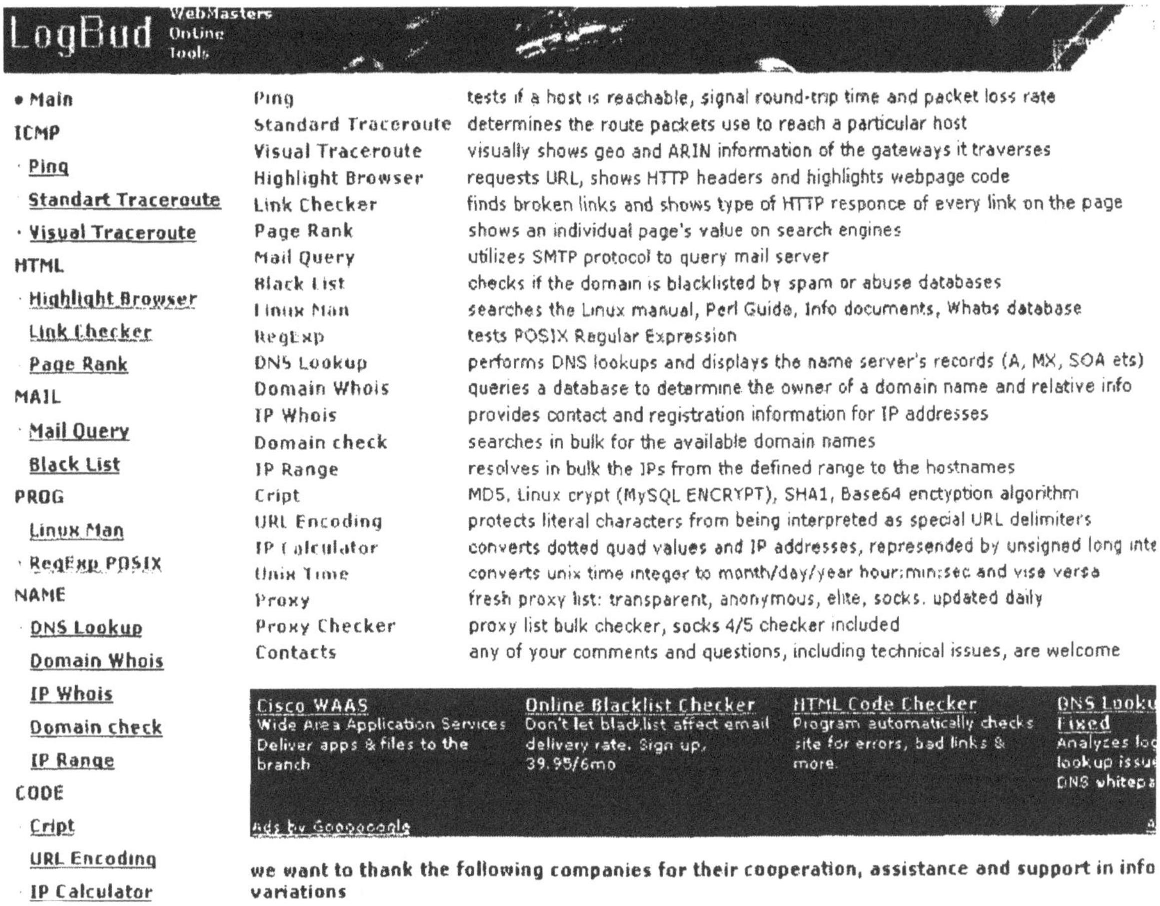
LogBud
WebMasters Online Tools
Main
ICMP
Ping
Standart Traceroute
Visual Traceroute
HTML
Highlight Browser
Link Checker
Page Rank
MAIL
Mail Query
Black List
PROG
Linux Man
RegExp POSIX
NAME
DNS Lookup
Domain Whois
IP Whois
Domain check
IP Range
CODE
Cript
URL Encoding
IP Calculator
Ping tests if a host is reachable, signal round-trip time and packet loss rate
Standard Traceroute determines the route packets use to reach a particular host
Visual Traceroute visually shows geo and ARIN information of the gateways it traverses
Highlight Browser requests URL, shows HTTP headers and highlights webpage code
Link Checker finds broken links and shows type of HTTP responce of every link on the page
Page Rank shows an individual page's value on search engines
Mail Query utilizes SMTP protocol to query mail server
Black List checks if the domain is blacklisted by spam or abuse databases
Linux Man searches the Linux manual, Perl Guide, Info documents, Whatis database
RegExp tests POSIX Regular Expression
DNS Lookup performs DNS lookups and displays the name server's records (A, MX, SOA ets)
Domain Whois queries a database to determine the owner of a domain name and relative info
IP Whois provides contact and registration information for IP addresses
Domain check searches in bulk for the available domain names
IP Range resolves in bulk the IPs from the defined range to the hostnames
Cript MD5, Linux crypt (MySQL ENCRYPT), SHA1, Base64 enctyption algorithm
URL Encoding protects literal characters from being interpreted as special URL delimiters
IP Calculator converts dotted quad values and IP addresses, represended by unsigned long inte
Unix Time converts unix time integer to month/day/year hour:min:sec and vise versa
Proxy fresh proxy list: transparent, anonymous, elite, socks. updated daily
Proxy Checker proxy list bulk checker, socks 4/5 checker included
Contacts any of your comments and questions, including technical issues, are welcome
Cisco WAAS
Wide Area Application Services Deliver apps & files to the branch
Online Blacklist Checker
Don't let blacklist affect email delivery rate. Sign up. 39.95/6mo
HTML Code Checker
Program automatically checks site for errors, bad links & more.
Ads by Goooooogle
we want to thank the following companies for their cooperation, assistance and support in info variations

The Logbud site packages a number of basic network analysis tools and provides some improved interfaces and displays to make them more useful. Highly recommended.

Logbud Online Tools http://www.logbud.com/

More Traceroute Sites and Toolkits

IP Address Guide http://www.internetipaddress.com/traceroute.aspx

All Nettools.com http://www.all-nettools.com/toolbox

Cogentco http://www.cogentco.com/htdocs/glass.php

Geektools Traceroute http://www.geektools.com/traceroute.php

IP-Plus Traceroute Servers http://www.ip-plus.ch/tools/traceroute.en.html

Multiple Traceroute Gateway http://www.tracert.com/cgi-bin/trace.pl

New York Internet Traceroute Links	http://www.nyi.net/traceroute.html
Opus One Traceroute	http://www.opus1.com/www/traceroute.html
SixXs IPv4 and IPv6 Traceroute	http://www.sixxs.net/tools/traceroute/
Traceroute.org	http://www.traceroute.org/
Tracerouters Around the World	http://tracerouters.nielssen.com/
BGPNet IPv4 Wiki	http://www.bgp4.net/tr
BGPNet IPv6 Wiki	http://www.bgp4.net/tr6

An excellent resource for finding IPv6 traceroute sites; an example of an IPv6 traceroute is shown below:

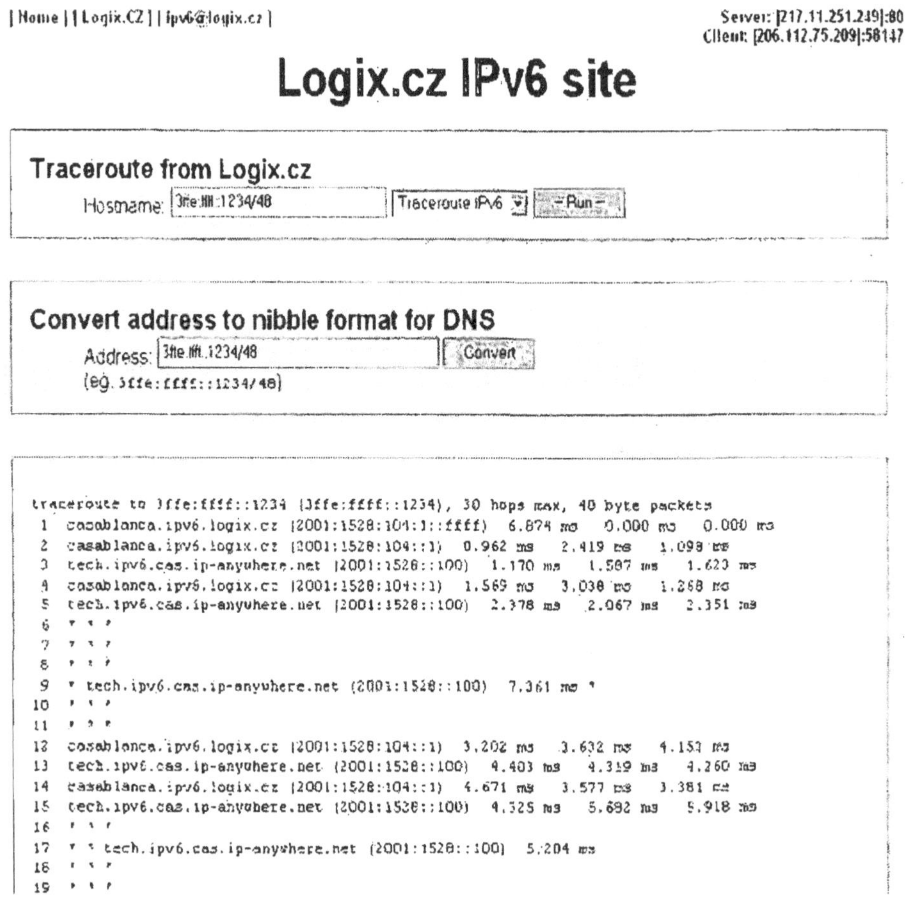
| Home | | Logix.CZ | | ipv6@logix.cz |
Server: [217.11.251.249]:80
Client: [206.112.75.209]:58147
Logix.cz IPv6 site
Traceroute from Logix.cz
Hostname: 3ffe:ffff::1234/48
Traceroute IPv6
Run
Convert address to nibble format for DNS
Address: 3ffe:ffff::1234/48
Convert
(eg. 3ffe:ffff::1234/48)
traceroute to 3ffe:ffff::1234 (3ffe:ffff::1234), 30 hops max, 40 byte packets
1 casablanca.ipv6.logix.cz (2001:1528:104:1::ffff) 6.874 ms 0.000 ms 0.000 ms
2 casablanca.ipv6.logix.cz (2001:1528:104::1) 0.962 ms 2.419 ms 1.098 ms
3 tech.ipv6.cas.ip-anywhere.net (2001:1528::100) 1.170 ms 1.587 ms 1.623 ms
4 casablanca.ipv6.logix.cz (2001:1528:104::1) 1.569 ms 3.038 ms 1.268 ms
5 tech.ipv6.cas.ip-anywhere.net (2001:1528::100) 2.378 ms 2.067 ms 2.351 ms
6 * * *
7 * * *
8 * * *
9 * tech.ipv6.cas.ip-anywhere.net (2001:1528::100) 7.361 ms *
10 * * *
11 * * *
12 casablanca.ipv6.logix.cz (2001:1528:104::1) 3.202 ms 3.632 ms 4.153 ms
13 tech.ipv6.cas.ip-anywhere.net (2001:1528::100) 4.403 ms 4.319 ms 4.260 ms
14 casablanca.ipv6.logix.cz (2001:1528:104::1) 4.671 ms 3.577 ms 3.381 ms
15 tech.ipv6.cas.ip-anywhere.net (2001:1528::100) 4.325 ms 5.692 ms 5.918 ms
16 * * *
17 * * tech.ipv6.cas.ip-anywhere.net (2001:1528::100) 5.204 ms
18 * * *
19 * * *

More Traceroute Tools

Learning to read and interpret traceroute data can be very frustrating and confusing in part because there is no standard way of naming routers and any number of different traceroute programs showing a variety of types of data. Here are some sites that are useful in helping to explain traceroutes.

Airport and City Code Database
http://www.airportcitycodes.com/aaa/CCDBFrame.html

World Airport Codes http://www.world-airport-codes.com/

Airlines of the Web Airport Codes http://flyaow.com/airportcode.htm

Sarangworld Traceroute Project Known Hostname Codes
http://www.sarangworld.com/TRACEROUTE/showdb-2.php3

A huge file of codes and IP addresses seen in traceroutes translated into city names with latitudes and longitudes

International Locations

Codes	Country	City	Latitude	Longitude
aep, bue, buenosaeres, buenosaires.ar, eze	Argentina	Buenos Aires	34°35'S	58°22'W
rdc.nsw.au	Australia	*	34°00'S	151°00'E
ade.au, adelaide, adl, adl.au	Australia	Adelaide	34°55'S	138°36'E
bal.au, ballarat, ballarat.au	Australia	Ballarat	37°34'S	143°52'E
bne, bne.au, bri.au, brisbane, brs.au	Australia	Brisbane	27°29'S	153°08'E
bundaberg.au	Australia	Bundaberg	24°52'S	152°21'E
cai.au	Australia	Cairns	16°55'S	145°46'E
campbelltown.au	Australia	Campbelltown	34°04'S	150°49'E
can.au, canberra, cbr, cbr.au	Australia	Canberra	35°17'S	149°08'E
darwin	Australia	Darwin	12°28'S	130°51'E
dubbo.au	Australia	Dubbo	32°15'S	148°36'E
eburwd.au, eburwd.vic.au	Australia	East Burwood	37°51'S	145°09'E
frank.au	Australia	Frankston	38°08'S	145°07'E
free.au	Australia	Freestone	28°08'S	152°08'E
gee.au, geelong.au, glg.au, gln.au	Australia	Geelong	38°08'S	144°21'E
gct.au	Australia	Gold Coast	27°58'S	153°25'E
gos.au, gosford	Australia	Gosford	33°26'S	151°21'E
hobart	Australia	Hobart	42°53'S	147°19'E
livrp.au	Australia	Liverpool	33°54'S	150°56'E
bur.au, mel, mel.au, melbourne	Australia	Melbourne	37°47'S	144°58'E

Traceroute Articles and Tutorials

It is important to understand the value and the pitfalls of using traceroute for network analysis. While much Internet traffic is symmetric, in some cases it is not, i.e., the traffic does not travel the same way in both directions. Although traceroute can tell you whether two networks communicate directly or indirectly, ***it cannot tell you anything with certainty about the nature of their relationship***, such as who is the provider and who is the customer. For a good explanation of how traceroute works (and the some of the drawbacks of using traceroute), look at these traceroute articles and tutorials:

Mapping Where the Data Flows http://www.isoc.org/oti/articles/0200/dodge.html

Traceroute Tutorial http://www.exit109.com/~jeremy/news/providers/traceroute.html

Russ Haynal's Traceroute Overview http://navigators.com/traceroute.html

---

Geolocating Internet Addresses

Geolocating Internet addresses using IP addresses has become big business and has applications for individuals as well. Why is knowing where someone is (actually, where the host computer is that he is using to connect to the Internet is physically located) important or useful? Some of the many uses of IP address geolocation include, but are not limited to:

- Tailored search results: some search engines will put local sites higher in the results' list, so if you search for "orthodontists" and you are in Boise, Idaho, local orthodontists may come up first.
- Online companies may use geolocation to tailor currency, sales tax, shipping rates, and even in some cases, prices by locality
- Automatic localization of configuration profiles (no need to reset a browser or chat software, it automatically detects where you are).
- Reducing network congestion by routing users to the closest servers that mirror the original content.
- Targeting advertising to a specific city, state, or country.
- Complying with local laws; this especially applies to online gambling, which is legal in some places and not in others.
- Controlling access; France uses geolocation to prohibit French Internet users from accessing pro-Nazi websites; China has a long, effective, and disturbing history of using geolocation to prevent access to many sites the Chinese government deems "unacceptable."

There are a number of companies, such as Quova and Digital Envoy, that sell IP geolocation/IP mapping products, but they do not give this technology away for free. However, a few websites do provide free online tools to geolocate IP addresses and/or domain names. All these tools have inherent limitations because of the ways in which they go about determining the physical location of an address. Knowing where your users are located has become increasingly important over time, whether you're a company that wants to know which products to target at a certain audience or a government seeking to control your Internet space. Therefore, because there is so much money to be made in this area, the free tools are simply not nearly as good as the ones you pay for.

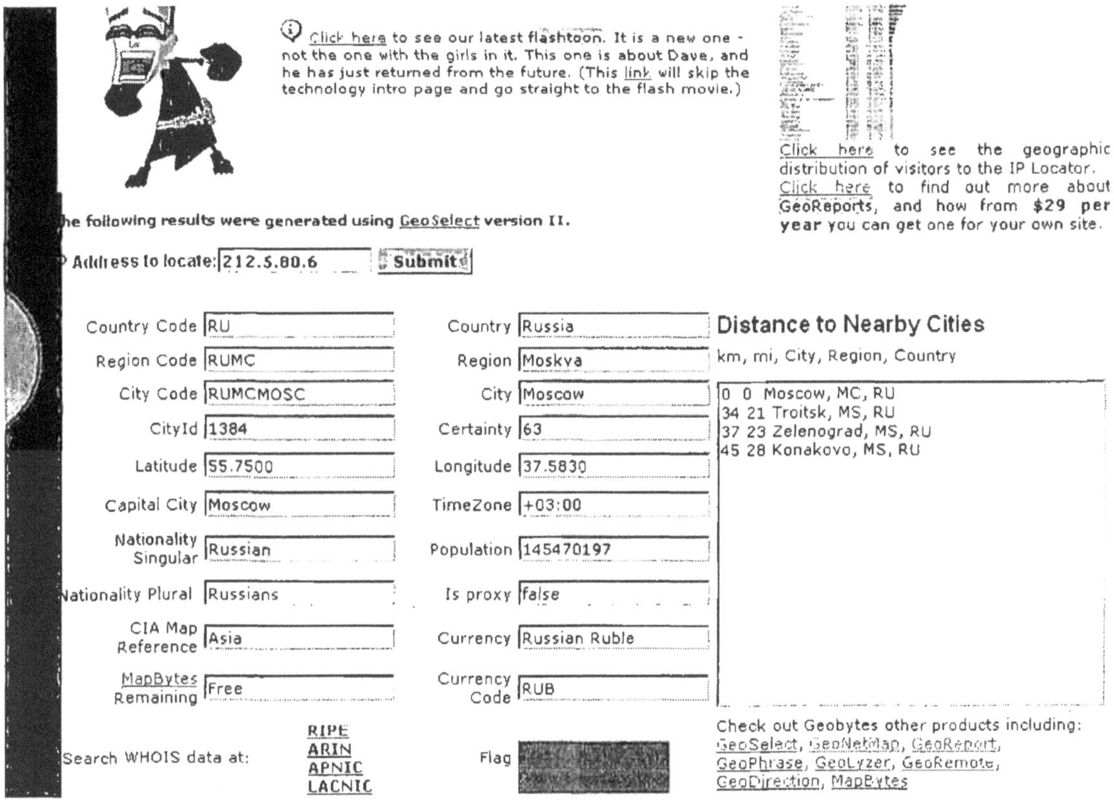
Click here to see our latest flashtoon. It is a new one - not the one with the girls in it. This one is about Dave, and he has just returned from the future. (This link will skip the technology intro page and go straight to the flash movie.)
Click here to see the geographic distribution of visitors to the IP Locator.
Click here to find out more about GeoReports, and how from $29 per year you can get one for your own site.
he following results were generated using GeoSelect version II.
Address to locate: 212.5.80.6
Submit
Country Code RU
Country Russia
Region Code RUMC
Region Moskva
City Code RUMCMOSC
City Moscow
CityId 1384
Certainty 63
Latitude 55.7500
Longitude 37.5830
Capital City Moscow
TimeZone +03:00
Nationality Singular Russian
Population 145470197
Nationality Plural Russians
Is proxy false
CIA Map Reference Asia
Currency Russian Ruble
MapBytes Remaining Free
Currency Code RUB
Search WHOIS data at:
RIPE
ARIN
APNIC
LACNIC
Flag
Distance to Nearby Cities
km, mi, City, Region, Country
0 0 Moscow, MC, RU
34 21 Troitsk, MS, RU
37 23 Zelenograd, MS, RU
45 28 Konakovo, MS, RU
Check out Geobytes other products including:
GeoSelect, GeoNetMap, GeoReport, GeoPhrase, GeoLyzer, GeoRemote, GeoDirection, MapBytes

Geobytes IP Locator http://www.geobytes.com/IpLocator.htm

The best of the free online geolocation tools, Geobytes is the commercial version of NetWorldMap that provides a more detailed report on IP address geolocation. GeoSelect does not use any DNS reverse lookups or Whois lookups to determine location. GeoSelect uses its proprietary GeoNetMap database to determine IP location. "The purpose of the Geobytes map is to map IP Addresses to geographical locations. To achieve this we acquire seed data from a number of sources. All of these sites ask the web surfer to provide their geographic location, and this location along with the user's IP Address is forwarded to us as seed data. We then run this data through a series of algorithms which identify and extract collaborating seed points."[158] If you want to see the original site where Geobytes began collecting its data, visit NetWorldMap.

NetWorldMap http://www.networldmap.com/TryIt.htm

NetWorldMap lets users enter IP addresses, then returns information about where that server is located at the city level. It does this by gathering vast amounts of data from volunteers who, for several years, have been entering information about the physical locations of their addresses. Of course, this opens NetWorldMap up to abuse, but the information provided by volunteers is crosschecked for accuracy. The site now claims that "currently it can only locate about 97.8% of the Internet's address space," which is a vast improvement over its earlier claims. While Networldmap is far from the perfect way to geolocate addresses, it is a valuable tool.

GeoTool http://www.rleeden.no-ip.com/geotool.php

This site also uses the same free Maxmind city database and Google maps to locate and show one IP address at a time. GeoTool adds information about the IP address. Note the GeoTool Spy option: this will show who is using the tool at the moment (including you).

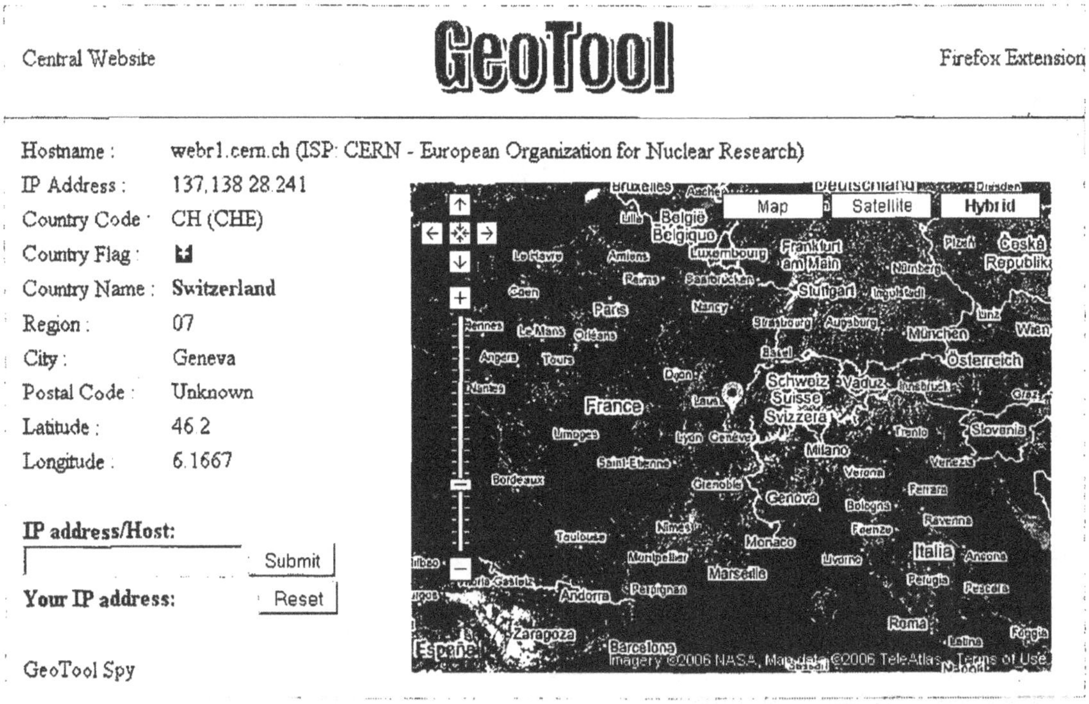
Central Website
GeoTool
Firefox Extension
Hostname : webr1.cern.ch (ISP: CERN - European Organization for Nuclear Research)
IP Address : 137.138.28.241
Country Code : CH (CHE)
Country Flag :
Country Name : Switzerland
Region : 07
City : Geneva
Postal Code : Unknown
Latitude : 46.2
Longitude : 6.1667
IP address/Host:
Submit
Your IP address:
Reset
GeoTool Spy
Map
Satellite
Hybrid
France
Paris
Schweiz
Suisse
Svizzera
Milano
Italia
Marseille
Monaco
Andorra
Barcelona
Zaragoza
Genova
Roma
Österreich
Stuttgart
Frankfurt am Main
Luxembourg

---

158 Geoselect Frequently Asked Questions, Geobytes.com,<http://www.geobytes.com/FAQ.htm> (14 November 2006).

HostIP.info http://www.hostip.info/

HostIP is a community effort to build a non-commercial database of geolocated IP addresses. It uses two sources of information to generate its geolocation tool: people identifying their city as associated with an IP address and automatic traceroutes.

HuntIP http://www.huntip.com/Tools/mapips.php

HuntIP is a set of tools designed to help sysadmins do a number of things, including plot multiple IP addresses on a map. Keep in mind the caveat from the site's creator: "This site should be used as a tool for investigative purposes only. The information provided here may not be correct and should not be trusted." To perform IP geolocation, ***HuntIP uses the free GeoLite data provided by Maxmind***, which also sells more accurate IP geolocation data at its website. Because Maxmind's free site limits queries to 25 per day per incoming IP address, I expect HuntIP operates under the same restrictions. HuntIP uses a Google Map mashup to plot IP address locations. Here I have plotted three IP addresses on the Google Map image just to give you an idea of the results you might get. In each case, the IP addresses were located in the appropriate country and very close to the actual physical location within the country. If the locations had been in countries for which the Google Maps provides greater resolution, that would have been even better.

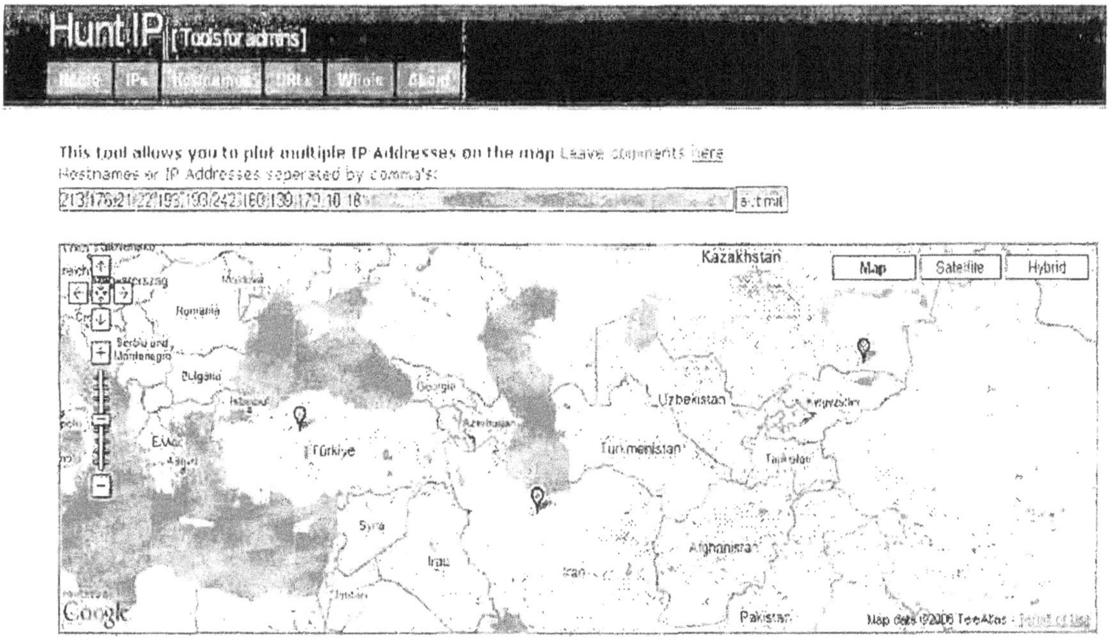
HuntIP
[Tools for admins]
This tool allows you to plot multiple IP Addresses on the map
Kazakhstan
Map
Satellite
Hybrid
Romania
Uzbekistan
Türkiye
Turkmenistan
Syria
Afghanistan
Pakistan
Google

IP2Location http://www.location.com.my/free.asp

This is a very good tool, but unfortunately, users are limited to 20 free lookups per day. However, the next two sites discussed below use this same data and the searches do not appear to be cumulative, in one case, and do not appear to be limited in the other. The results offer the country, city, region, flag, and associated ISP. IP2Location gathers geolocation of IP data using its own proprietary means and claims 95% accuracy, but I can't find any independent verification of this assertion. The company is based in Malaysia.

AJAX Powered IP to Location http://www.seomoz.org/ip2loc/ip2loc.php

SEOmoz.org has a nice little tool based on MaxMind's Geolite Data that maps IP addresses to Google maps. While users must consider that the geolocations are not entirely accurate, compare this tool with the GeoTool above for the same address, and you will see that the SEOmoz application is much more detailed. The street address information below is culled from the RIPE Whois database.

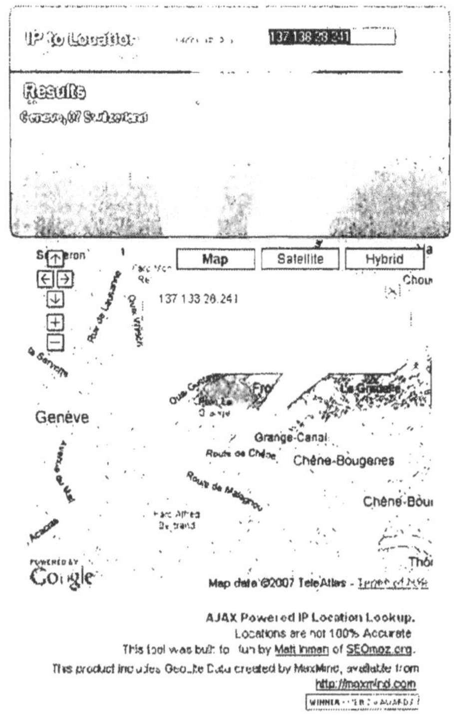
Map
Satellite
Hybrid
137.133.28.241
Genève
Grange-Canal
Chêne-Bougeries
AJAX Powered IP Location Lookup.
Locations are not 100% Accurate

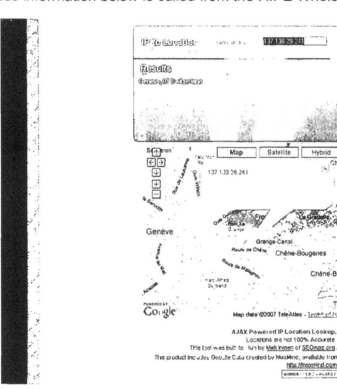

WebHosting.Info http://ip-to-country.webhosting.info/node/view/36

WebHosting.Info only resolves IP addresses to a country, but it appears to do so pretty well. I think the website says it best: "Although not 100% accurate, the IP-to-Country Database is about 98% accurate on country recognition. The main reasons for this lie in the existence of dynamic IP addresses and Internet access through proxy servers. Also, it should be noted that the IP-to-Country Database seeks to indicate the country where resources were first allocated or assigned and are not an authoritative statement of the location in which any specific resource may currently be in use. These cases are very difficult and sometimes impossible to map. However at this moment the IP-to-Country Database is by far the most accurate way to determine the location of Internet users in real-time." *[emphasis added]* Ignore the username/password boxes and look for the demo query box on the right-hand side of the page.

GeoIP Country Lookup http://www.maxmind.com/app/lookup

GeoIP City Lookup http://www.maxmind.com/app/lookup_city

GeoIP by Maxmind is a product that can be purchased; however, the Maxmind website does offer two free demonstration options, one for country lookup and one for city lookup. The country lookup is limited to 100 queries per month while the city lookup is limited to 25 lookups per day (presumably per visitor's IP address). GeoIP claims to "use a number of Internet mapping tools to identify and correct IP addresses where the end-user location does not match the ISP location on the Whois record," but they do not go into detail about exactly how they do this. They do claim they are 95 percent accurate in geolocating IP addresses.

NetGeo http://www.caida.org/tools/utilities/netgeo/

NetGeo, from the Cooperative Association for Internet Data Analysis (CAIDA), takes a completely different approach and one, frankly, that is less reliable. NetGeo correlates both IP address and Autonomous System (AS) numbers to the three major Whois databases at ARIN, APNIC, and RIPE and returns latitude/longitude data for the city, state (or province, district, etc.), and country from the text of the Whois record. The problem is, of course, that there is no guarantee that the physical location of the server is the same as the physical location registered in the Whois database. The site warns users it has not been updated and may give "wildly inaccurate" results. I believe them.

Finding ISPs & Internet Access Points

Thanks to the expansion of the Internet, it has become relatively easy to track down ISPs (Internet Service Providers) and/or Internet access points (IAPs) for any country with public Internet access. The first thing you need to understand is that ***there is no such thing as a complete list of ISPs anywhere***. Furthermore, even a cursory look at websites providing lists of ISPs are either heavily weighted toward or entirely about US ISPs. How do you go about finding ISPs in the rest of the world?

The best information about non-US ISPs, not surprisingly, requires payment, sometimes big payment. I highly recommend you check with your organization's library or other reference resource to see what premium (fee for service) resources they offer. For example, **TeleGeography's Global Internet Geography** is probably the single best source of information on ISPs around the world, but it is far from cheap. Registered users can view free samples of the report, but to view the complete provider list, read the detailed profiles on many of the providers, and see the myriad country profiles, you must have access to the complete report.

Global Internet Geography from TeleGeography
http://www.telegeography.com/products/gig/index.php

One good way to find freely available information about non-US ISPs is to look at organizations or associations, such as the European ISP Association. This page provides links to its members' sites in individual European countries, from which you can quickly dig down and find ISPs that are in turn members of that country's association.

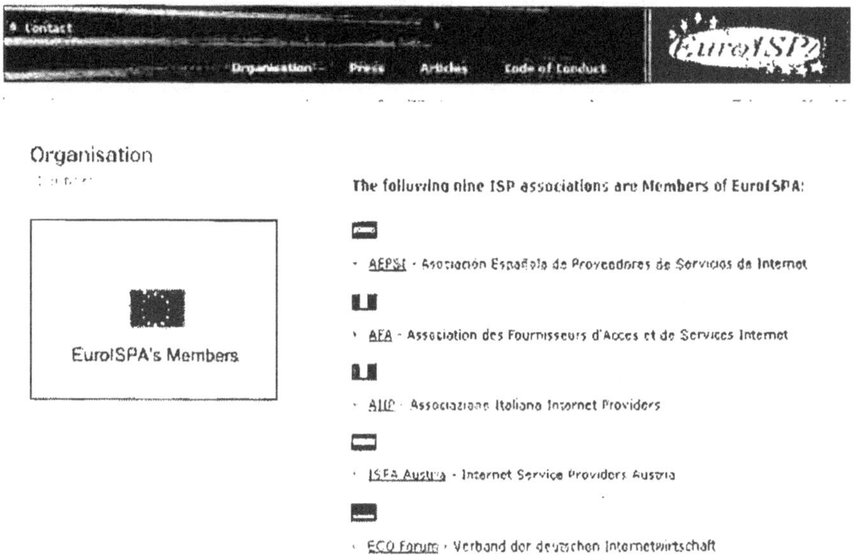
Contact
Organisation
Press
Articles
Code of Conduct
EuroISPA
Organisation
EuroISPA's Members
The following nine ISP associations are Members of EuroISPA:
AEPSI - Asociacion Española de Proveedores de Servicios de Internet
AFA - Association des Fournisseurs d'Acces et de Services Internet
AIIP - Associazione Italiana Internet Providers
ISPA Austria - Internet Service Providers Austria
ECO Forum - Verband der deutschen Internetwirtschaft

EuroISPA's Members http://www.euroispa.org/32.htm

Here are the members of the ISP Association of Ireland

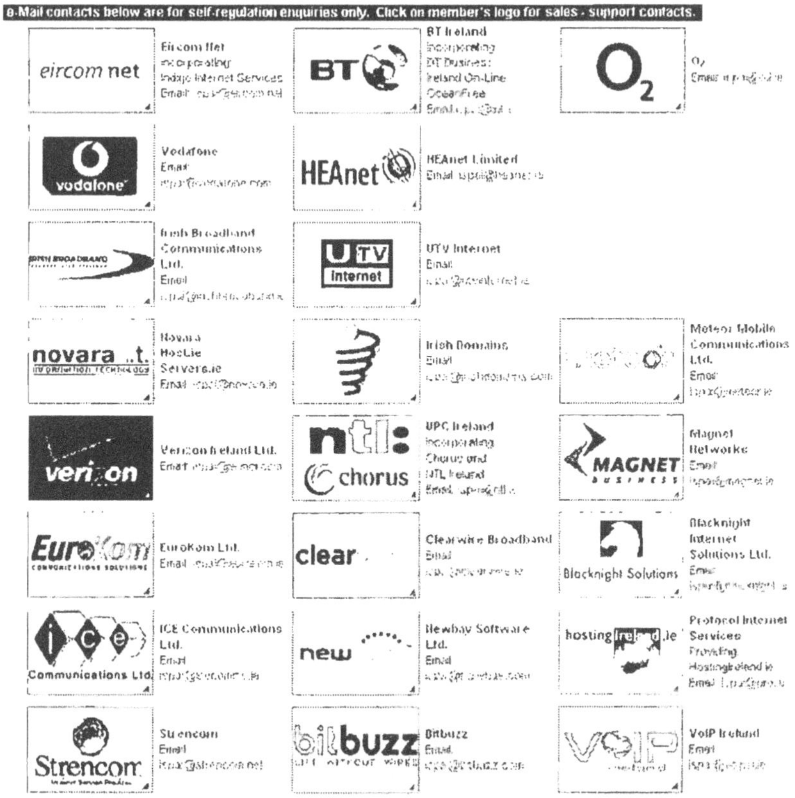
e-Mail contacts below are for self-regulation enquiries only. Click on member's logo for sales - support contacts.
eircom net
Eircom Net incorporating Indigo Internet Services
BT
BT Ireland incorporating BT Business Ireland On-Line OceanFree
O2
Vodafone
HEAnet
HEAnet Limited
Irish Broadband Communications Ltd.
UTV Internet
novara
Novara HostIE Servers.ie
Irish Domains
Meteor Mobile Communications Ltd.
verizon
Verizon Ireland Ltd.
ntl: chorus
UPC Ireland incorporating Chorus and NTL Ireland
MAGNET BUSINESS
Magnet Networks
EuroKom
EuroKom Ltd.
clear
Clearwire Broadband
Blacknight Solutions
Blacknight Internet Solutions Ltd.
ice Communications Ltd
ICE Communications Ltd.
new
Newbay Software Ltd.
hosting ireland .ie
Protocol Internet Services Providing HostingIreland.ie
Strencom
Strencom
bitbuzz
Bitbuzz
VoIP
VoIP Ireland

Other international organizations and associations that list members in their country or region include the following. Keep in mind that **Local Internet Registries** referred to at some of these sites are usually local ISPs.

Asia Pacific Network Information Center (APNIC) Membership List
http://www.apnic.net/member/current-members.html

This is an excellent way to find ISPs in the Asian Pacific region. The list is searchable by country. Below is a snapshot of the ISPs in Afghanistan that are members of APNIC.

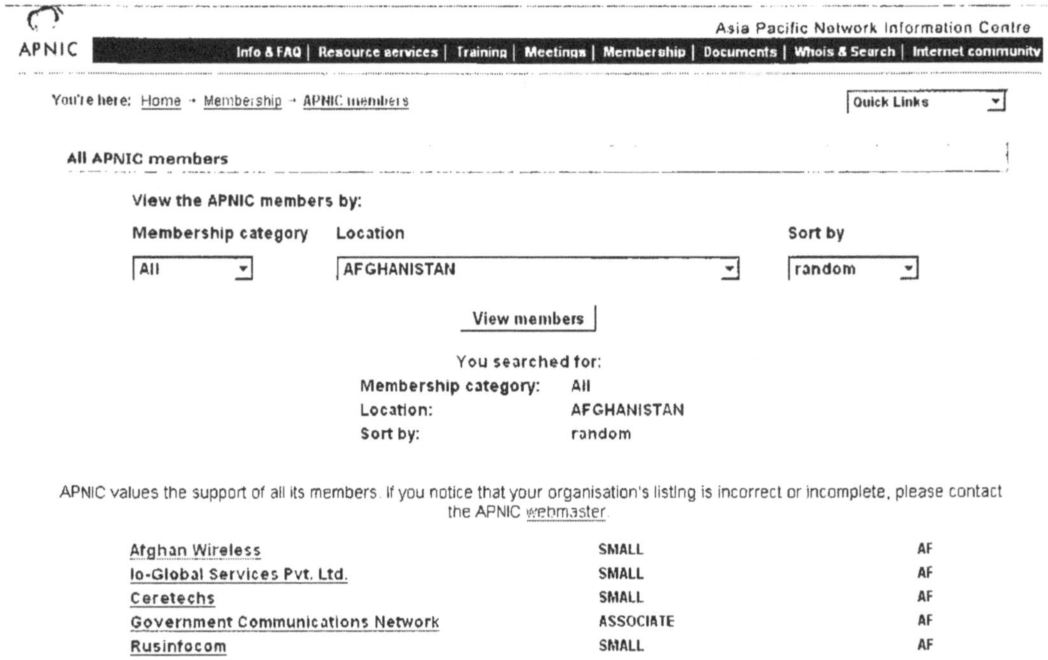
APNIC
Asia Pacific Network Information Centre
Info & FAQ | Resource services | Training | Meetings | Membership | Documents | Whois & Search | Internet community
You're here: Home → Membership → APNIC members
Quick Links
All APNIC members
View the APNIC members by:
Membership category
Location
Sort by
All
AFGHANISTAN
random
View members
You searched for:
Membership category: All
Location: AFGHANISTAN
Sort by: random
APNIC values the support of all its members. If you notice that your organisation's listing is incorrect or incomplete, please contact the APNIC webmaster.
Afghan Wireless SMALL AF
Io-Global Services Pvt. Ltd. SMALL AF
Ceretechs SMALL AF
Government Communications Network ASSOCIATE AF
Rusinfocom SMALL AF

RIPE NCC's Membership List (covers Europe, North Africa, and the Middle East; listed by country code) http://www.ripe.net/membership/indices/

The Internet Service Providers' Association of South Africa
http://www.ispa.org.za/about/memberlist.shtml

Africa Top Level Domains (AFTLD) Members
http://www.aftld.org/html/english/AFTLD_members.html

Internet Service Providers' Association of Nigeria http://www.ispan.org.ng/

Telecommunication Service Provider Association of Kenya (TESPOK): Kenyan ISPs
http://www.tespok.co.ke/ispa.html

Association of African Internet Service Provider Associations
http://www.afrispa.org/founding.htm

The Internet Service Providers' Association of India http://www.ispai.in/
Members are shown in a scroll near the top right of the home page.

The Internet Service Providers' Association of Bangladesh
http://www.ispabd.org/content.php?content.17

Internet Service Providers' Association of Nepal
http://www.ispan.net.np/memlist.php

The Hong Kong Internet Service Providers Association
http://www.hkispa.org.hk/memberlist.htm

Association de Fournisseurs d'Accès et de Services Internet (France)
http://www.afa-france.com/membres.html

Slovene Internet Service Provider Association
http://www.sispa.org/seznam_clani.htm

Nominet UK Internet Names Organization Members
http://www.nominet.org.uk/governance/members/list/

Network Access Points (or Internet Exchanges aka IXs) are also superb sources of information about local ISPs. NAPs or IXs are the junctions where Internet traffic is handed off among different Internet providers and networks. Think of them as performing the same function as airport hubs or highway cloverleaves, where travelers change from one airline or highway to another. Finding IXs is relatively easy because there are a couple of sites that provide links to virtually all the exchanges around the world. The two best free (non-registration) sites are Colosource's Internet Exchange Points and Exchange Points Around the World. If you are able to register at websites, TeleGeography's Internet Exchange Points Directory is free only to registered users.

Telegeography's Internet Exchange Points Directory [Registration Required]
http://www.telegeography.com/products/ix/index.php

Colosource's Internet Exchange Points http://www.colosource.com/ix.asp

Exchange Points Around the World http://www.ep.net/ep-main.html

It is a simple matter to use an IX directory or list to find ISPs. If I am interested in Hong Kong, from Colosource's home page, I scroll down to Asia Internet eXchanges and click on China The Hong Kong Internet eXchange (HKIX), which brings up a page with a link to "connected IAPs":

Licensed Members

The following Internet Access Providers have been directly connected to HKIX:

Internet Access Providers	Link Speed to HKIX	AS Number	PNETS Lic. No.
Asia Netcom Asia Pacific Limited	GE x 2	10026	789
Akamai International B V	10G + GE	20940	1244
AT&T Global Network Services HK Ltd	FE	2687	572
DIN / PCCW Global Limited	GE	9237 & 3491	901
China Internet Corporation	[ATM(155)] + [ATM(155)]	4611	140
ChinaMotion NetCom (Asia) Limited / Wanbao Telecom	FE	7705	1065
China Network Services (HK) Ltd	[20M]	7499	348
China Resources Peoples Telephone Co Ltd	[E1]	9231	648
Citic 1616 Data Limited	10G + GE + GE	17554	712
Cityline (Hong Kong) Ltd	FE + [T1]	9409	379
Communilink Internet Limited	GE	38277	1218
CPCNet Hong Kong Limited	GE + [STM-1 + ATM(155)]	4058	123
The Chinese University of Hong Kong	GE	4641	180
Cable & Wireless Global Network (Hong Kong) Limited	GE	1273	EFTNS(32)
Diyixian com Limited	GE	9584	598
Donghwa Telecom Co. Ltd	GE	9505	1186
Equant Hong Kong Limited	[ATM(155) + E1 x 5]	4862	079
Equinix Hong Kong Limited	GE x 2	17819	756
Era International (HK) Ltd.	GE	24328	1193
ET Net Limited	ATM(16/32) + ATM(16/32)	9906	636
FLAG Telecom Asia Limited	GE + GE	15412	EFTNS(29)
Genuity Hong Kong	[ATM(155)]	202	826
Global Crossing Hong Kong Limited	GE	3549	1139
GlobalNet Communication Limited	FE	17990	873
Google (Hong Kong) Limited	GE	15169	1222
Henderson Data Center Limited	GE	10098	685
Hong Kong Cable Television Limited	10G x 2	9908	FTNS(5)
Hong Kong Broadband Network Ltd	10G x 3	9269	094
hkcolo Limited	GE	23749	842
HKNet Company Ltd	GE	4645	110
The Hongkong & Shanghai Banking Corp Ltd	GE x 2	9221	777
Hutchison Telephone Company Limited	FE x 2	10118	1088
Hutchison MultiMedia Services Ltd	10G + GE x 8	9304 10032	238

The following sites list and link to ISPs around the world, but remember that none of these sites has a complete list of all ISPs everywhere. You will need to look at some or all of these lists, as well as try country-specific resources, if you are want to find a thorough list of ISPs in a particular country. The best list of ISPs is called, not surprisingly, "The List," a site that has been around for years. Click on the highlighted country code for the best list of ISPs in most countries around the world. Other resources vary in quality depending on the country you're researching, so you will probably need to look at all of these sites.

The List http://thelist.internet.com/countrycode.html

NSRC's Connectivity Providers Database http://www.nsrc.org/networkstatus.html

Some of the information at the Network Startup Resource Center is out of date while some has recently been updated; however, the links should help you locate the major ISPs in almost any country.

International Internet Access Providers

http://www.herbison.com/herbison/iap_international_meta_list.html

Herbison's International Internet Access Providers website was last updated in April 2006. While some of the links are out of date, there is so much information at this site it is still useful.

FreedomList http://www.freedomlist.com/find.php3
Free and cheap ISPs by country; fairly well up to date.

African Internet Connectivity
Last updated 2002. http://www3.sn.apc.org/africa/af-isps.htm

Middle East Directory List of ISPs
Good resource for 16 Middle Eastern countries; some links are out of date. http://www.middleeastdirectory.com/me-isps.htm

Major directories can be good sources for ISPs around the world, though they are only starting points and none has a really thorough list of ISPs.

Google Directory
http://directory.google.com/Top/Computers/Internet/Access_Providers/

Yahoo Directory http://dir.yahoo.com/

There are several ways to use Yahoo to find international ISPs. The best is: Business_and_Economy/Business_to_Business/Communications_and_Networking/Internet_and_World_Wide_Web/Network_Service_Providers/Internet_Service_Providers__ISPs_/
This lists ISPs for most countries and all world regions. Also, look at each individual Region and Country to see if Yahoo lists ISPs for that area or country.

Because of the lack of terrestrial broadband capacity in many locations, there is an increasing demand for broadband services via satellite in many parts of the world. These sites will help you locate satellite Internet providers.

Satellite Internet Service Providers for North & South America, Europe, Africa, Asia, Middle East http://www.satsig.net/

Linksat Satellite and Internet Providers (covers most of the globe) http://www.linksat.com/

Satellite Industry Links: Satellite Service Providers (includes but is not limited to Internet service) http://www.satellite-links.co.uk/links/ssp.html

Wireless Internet access points (usually called **hotspots**) have become extremely important in recent years. Many sites have appeared to help people locate these hotspots anywhere in the world. Intel's Mobile Technology Hotspot Finder is among the best of these sites. You can search for hotspots by address, city, state, country/region, distance, business or hotspot name; location type (e.g., an airport); by service provider; or by whether they are free or commercial hotspots, or both. You can also browse hotspots by country. The number of hotspots is listed next to each country's name.

Intel's Mobile Technology Hotspot Finder http://intel.jiwire.com/

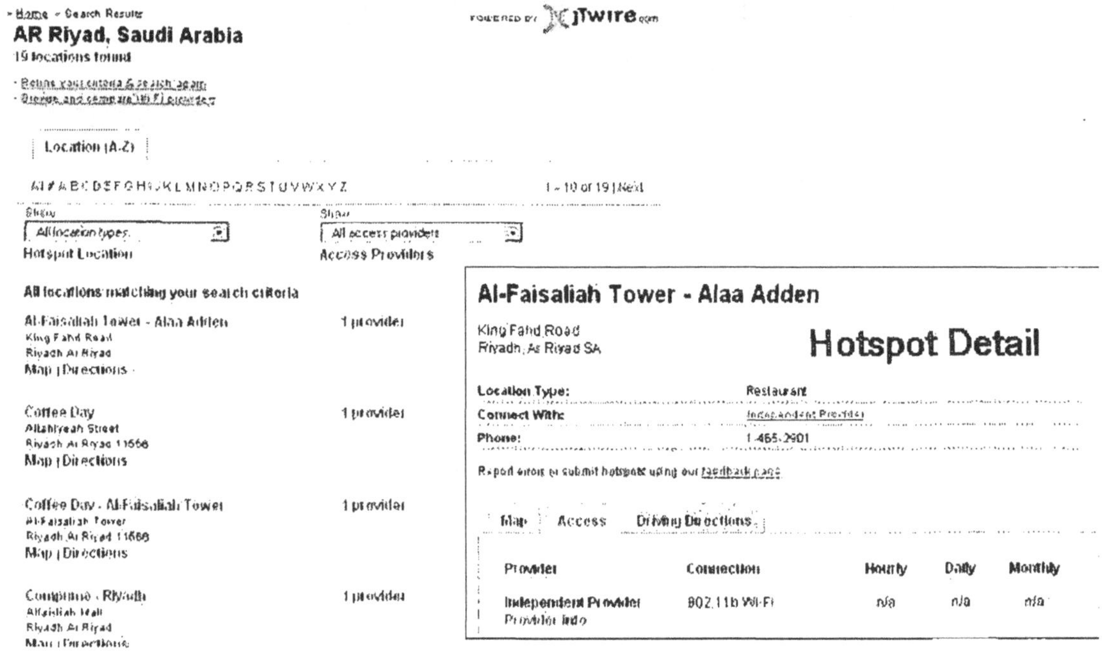
AR Riyad, Saudi Arabia
19 locations found
Location (A-Z)
Hotspot Location
Access Providers
All locations matching your search criteria
Al-Faisaliah Tower - Alaa Adden 1 provider
King Fahd Road
Riyadh Ar Riyad
Map | Directions
Coffee Day 1 provider
Map | Directions
Coffee Day - Al-Faisaliah Tower 1 provider
Map | Directions
1 provider
Al-Faisaliah Tower - Alaa Adden
King Fahd Road
Riyadh, As Riyad SA
Hotspot Detail
Location Type: Restaurant
Connect With:
Phone: 1-465-2901
Map Access Driving Directions
Provider Connection Hourly Daily Monthly
Independent Provider 802.11b Wi-Fi n/a n/a n/a
Provider Info

IPass's Hotspot Finder lets users look for dial-up, ISDN, PHS access points and Ethernet or Wi-Fi Internet access points in any country. For example, it found 72 hotspots in Hokkaido, Japan. Other hotspot finder websites include the following:

iPass Hotspot Finder	http://ipass.jiwire.com/
Wi-Fi Hotspot List	http://www.wi-fihotspotlist.com/
Hotspothaven	http://www.hotspothaven.com/
JiWire Global WiFi Hotspot Finder	http://www.jiwire.com/search-hotspot-locations.htm
WiFinder	http://www.wifinder.com/
WiFi411	http://www.wifi411.com/

Finally, cybercafes remain extremely popular Internet access points in many countries where it may be too expensive to have either a personal computer or an Internet connection in one's home. But they have also become popular with travelers who don't want to lug around a laptop (which is all too easy to steal). More and more people are relying on their neighborhood cybercafes as a cheap, fast, easy way to exchange email and browse the web without the inconvenience of having to carry around a laptop. There are many guides and search engines for finding cybercafes around the world, but do keep in mind that cybercafes come and go very quickly, especially in certain countries. China, for example, closed 8600 unlicensed cybercafes in 2004.

Netcafe Guide http://www.world66.com/netcafeguide

Google Directory: Cybercafes http://directory.google.com/Top/Computers/Internet/Cybercafes/

Cybercaptive Search Engine http://cybercaptive.com/

The country search has been disabled; search by city

Indra's International Cybercafes http://www.indranet.com/potpourri/links/cybercafeint.html

Curious Cat's Cybercafe Connections http://www.curiouscat.com/travel/cybercafe.cfm

Internet Cybercafe Database http://cybercafe.katchup.co.nz/search.asp

A database searchable by country and city that provides results in an easy to read format.

Internet CyberCafe Database

KATCHUP
For Your Own
Web Based Email Address

Further Info

Search

Add Cafe

Email Us

Welcome to the Internet's first free, fully automated and searchable database of Cybercafes and Public Access Internet terminals

Egypt | Cairo | Update

Results for Cairo

Cafe Name	Address	Phone Number	Details
Internet Egypt	Maadi Grand Mall, 3rd Floor, Maadi	202-5184273	Details
Internet Egypt	2 Midan Simon Bolivar, 6th fl. Apt. 40, Garden City	202-3562882	Details
Internet Egypt	Zamalek Sporting Club Passage, Above the Golden Store	3030493	Details
site	30 shehab st. mohandseen	0123428866	Details
site	49 el batal ahmed abd el aziz st. mohandseen	0123428866	Details
Worldwide Internet C@fe	32 Pyramids Street Falco City Main Enterance	202-385 7799	Details
Cyber Cafe	58 Lebanon Street Mohandeseen City		Details
WORLDNET	1 Magd Eleslam St.from Elhegaz St.From Ain Shams St	202-4983474	Details
Forcante	13 El-Thabat St. - Ahmed Orabi - El Mohandessin	0101817722	Details

In short, there are many resources available to help users find Internet access points and sources of all types, although there is no single resource that will do it all for you. And some types of Internet access, especially cybercafes, change so rapidly that it is especially difficult to keep up with the latest information on their locations, requiring users to refresh that data frequently.

Cybergeography, Topology, and Infrastructure

The Internet has created many new ways of seeing, understanding, and knowing the world. It has also created, in a sense, a new world unto itself, a world with its own "geography," which has come to be known as cybergeography: the configuration of the constituent parts of the Internet. The most original and informative of the websites devoted to mapping this new landscape used to be **Cybergeography**, which contains, among other things, cybermaps of many flavors: topology, census, conceptual, historical, etc. Some of these maps are very imaginative and some even display an eerie beauty. However, the site has not been updated since 2004, so it is becoming an archive instead of a library of new information.

Cybergeography Research http://www.cybergeography.org/

For a different view of the Internet, check out the **Internet Traffic Report.** "The Internet Traffic Report monitors the flow of data around the world. It then displays a value between zero and 100. Higher values indicate faster and more reliable connections." The Traffic Report uses "ping" to measure round-trip travel time along major paths on the Internet. It also measures response time, i.e., how long it takes for a piece of data to travel from point A to point B and back (round trip). The Traffic Report also provides data on packet loss, which indicates how reliable the connection is. All this data is available for major routers around the world and displayed graphically.

Internet Traffic Report http://www.internettrafficreport.com/main.htm

Yet another way of visualizing and thereby understanding the Internet is by looking at an **Internet Exchange (IX)** or **Network Access Points (NAPs)**. An IX is where networks and service providers hand off traffic to each other; they function as "hubs" for Internet traffic in very much the way certain airports serve as "switching points" for passengers. And, like airport hubs, they are very crowded, busy places. There are IXs around the world, though obviously many more in congested areas such as the US and Europe. In fact, many countries and even some regions do not have their own IXs, which means they must use exchanges in Europe, Asia, or even the US to route traffic between countries or even within one country, which leads to some very interesting routing patterns.

Internet Exchanges all have websites, and each provides a varying amount of data. The two best free metasites with links to most of the world's IXs are Exchange Point Information and Colosource:

Colosource Internet eXchange Points http://www.colosource.com/ix.asp

Exchange Point Information http://www.ep.net/ep-main.html

Internet Exchanges are invaluable sources of information about how major networks are connected to each other all over the world. Often IX websites will include information about **peering arrangements**. Peering is the "arrangement of traffic exchange between Internet service providers (ISPs). Larger ISPs with their own backbone networks agree to allow traffic from other large ISPs in exchange for traffic on their backbones. They also exchange traffic with smaller ISPs so that they can reach regional end points. Essentially, this is how a number of individual network owners put the Internet together."[159]

In addition to IXs, **Internet backbones**, which are central networks that connect other networks together, are high-interest segments of the Internet. Some of the best known are Worldcom/UUNET, Qwest and KPNQwest in Europe, Genuity, Sprint, AT&T, Cable & Wireless, Savvis, GlobalOne, BELNET, Telia, and Teleglobe.

Internet Backbone Networks
http://www.geog.ucl.ac.uk/casa/martin/atlas/isp_maps.html

Russ Haynal's Major Internet Backbones http://www.navigators.com/isp.html

Boardwatch's Internet Backbone Maps http://www.nthelp.com/maps.htm

BWM's Links to Network Maps
http://www.bandwidthmarket.com/component/option,com_weblinks/catid,74/Itemid,4/

BT Infonet's Network Maps
http://www.bt.infonet.com/services/internet/network_maps.asp

What do backbone maps typically look like? Here is the Ipv6 backbone map for the National Education and Research Network (Rede Nacional de Ensino e Pesquisa – RNP), "the Brazilian infrastructure of advanced network for collaboration and communication in the fields of teaching and research."[160] The interactive web page not only displays the topology seen here, but also includes links with details about the connectivity.

---

[159] "Peering," Whatis.com, <http://whatis.techtarget.com/> (14 November 2006).

[160] RNP Backbone Map, < http://www.rnp.br/en/backbone/index.php> (14 November 2006).

DOCID: 4046925

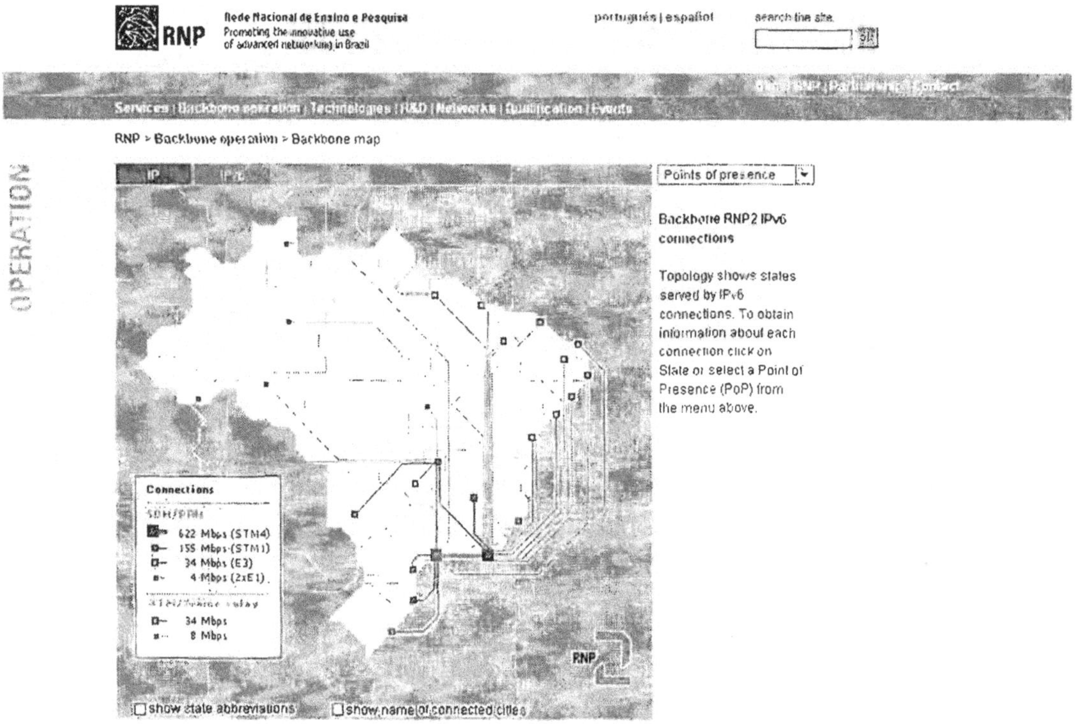
OPERATION
Rede Nacional de Ensino e Pesquisa
RNP
Promoting the innovative use
of advanced networking in Brazil
português | español
search the site
Services | Backbone operation | Technologies | R&D | Networks | Qualification | Events
RNP > Backbone operation > Backbone map
IP
Points of presence
Backbone RNP2 IPv6 connections
Topology shows states served by IPv6 connections. To obtain information about each connection click on State or select a Point of Presence (PoP) from the menu above.
Connections
622 Mbps (STM4)
155 Mbps (STM1)
34 Mbps (E3)
4 Mbps (2xE1)
34 Mbps
8 Mbps
RNP
show state abbreviations
show name of connected cities

Another effort to visualize the topology of the Internet is **Mapnet**, maintained by the Cooperative Association for Internet Data Analysis (CAIDA). Their color map, which runs as a web-based Java applet, allows you to view major providers simultaneously and also to zoom into a specific world region.

Mapnet http://www.caida.org/tools/visualization/mapnet/

Internet Privacy and Security—Making Yourself Less Vulnerable in a Dangerous World

The problems with Internet privacy and security are getting steadily worse, not better, each year as technology to collect information surreptitiously and attack computers at will steadily improves and proliferates. To make matters worse, malicious attacks are not the only or even the most common ways personal data is gathered, stored, and used. Businesses routinely collect information and, in some cases, share it without users' knowledge or consent. By now most Internet users know they are inadvertently giving out information about themselves every time they navigate the Internet. However, it is almost impossible for anyone to have a clear idea exactly what information is being unwittingly provided.

Part of the problem stems from one of the seemingly inviolable rules of progress that applies doubly so to the Internet:

More Convenience = Less Privacy

Major computer companies have recognized this trade-off for years, though rarely will they openly discuss it. In an unusual instance of candor, a Microsoft executive admitted that one of the features in Internet Explorer is a good example of this axiom. Speaking of *userdata persistence* (more on what this is later), Michael Wallent said, "this feature has a trade-off, ***like almost every other feature on the web***—in this case, between functionality and a ***minor***, potential privacy exposure."[161] [emphasis added]

[161] Paul Festa, "IE Feature Can Track Web Surfers Without Warning," *CNET News*, 11 September 2000, <http://news.cnet.com/news/0-1005-200-2751843.html> (14 November 2006).

> **"You have zero privacy anyway. Get over it."**
>
> Scott MacNealy, CEO, Sun Microsystems

Scott MacNealy's now-infamous quote,[162]—"You have zero privacy anyway. Get over it."—may no longer be an overstatement. At the very least, it should serve as a warning to Internet users to be wary of all products and services, but especially new ones that promise to do things faster, better, easier, and cheaper. There is almost always a hidden cost, often in weakened security and compromised privacy.

The costs, however, are not always so "hidden." Many are affecting the bottom line and the budgets of businesses and governments. "***Dealing with viruses, spyware, PC theft and other computer-related crimes costs U.S. businesses a staggering $67.2 billion a year***, according to the FBI. The FBI calculated the price tag by extrapolating results from a survey of 2,066 organizations. The survey...found that 1,324 respondents, or 64 percent, suffered a financial loss from computer security incidents over a 12-month period."[163] Furthermore, as both professionals and individuals become more security savvy, the ***threats become more insidious and therefore harder to detect and protect against***. The "2007 Internet Threat Outlook," a report by software maker CA, predicted that "malware brokers will continue to piece together threats such as Trojan horse viruses, worms and the many forms of spyware to hide their attacks and evade technological defenses employed by both enterprises and consumers. With the level of professionalism rising quickly among the most sophisticated virus distributors, CA predicts that zero-day exploits, drive-by malware downloads and extremely intricate phishing schemes will continue to become more dangerous and harder to detect."[164]

Especially worrisome is the proliferation of bots, the shortened version of 'robot,' which simply refers to any software designed to dig through data. For example, search engines use spider bots to crawl webpages to index them; there are shopping bots that look for the best prices for consumers; bots are at the heart of data mining, the process of finding patterns in enormous amounts of data. But "bad bots" create a virus-like infection under the remote control of a distant computer, network, or individual. This new threat exploits vulnerabilities in security subsystems

---

[162] Polly Sprenger, "Sun on Privacy: Get Over It," Wired, 26 January 1999, <http://www.wired.com/news/politics/0,1283,17538,00.html > (14 November 2006).

[163] Joris Evers, "Computer crime costs $67 billion, FBI says," CNET News.com, 19 January 2006, <http://news.com.com/2102-7349_3-6028946.html?tag=st.util.print> (30 January 2007).

[164] Matt Hines, "CA Predicts More Attacks on Experienced Users," eWeek via Yahoo News, 25 January 2007, <http://news.yahoo.com/s/zd/20070125/tc_zd/199597> (31 January 2007).

and often exploits normally unused ports and channels, permitting the bots to move about on the net unnoticed and undetected. There are now thousands of these bad bots (no one really knows how many) trolling the Internet connected to what is known as a "**botnet**," a kind of underground network of malicious activity. A 2004 study concluded that "two years ago only 200 bot-virus variations existed; today [in 2004] there are about 4,000, according to F-Secure Corp."[165] The security situation is rapidly deteriorating. "David Dagon, a Georgia Institute of Technology researcher who is a co-founder of Damballa, a start-up company focusing on controlling botnets, said the consensus among scientists is that botnet programs are present on about 11 percent of the more than 650 million computers attached to the Internet."[166] None of this is new; botnets have been around for a long time. "What is new is the vastly escalating scale of the problem—and the precision with which some of the programs can scan computers for specific information, like corporate and personal data, to drain money from online bank accounts and stock brokerages."[167]

Security experts believe most spam—in fact more than 80 percent—is now sent by bots.[168] Spam is more than a nuisance. It is the most pervasive and pernicious medium for spreading all sorts of malicious software (malware). To make matters worse, simply using a preview window in an email application may be sufficient to activate scripts sent by spammers, which means many users are unwittingly contributing to the spread of spam and malware. "According to a study by network management firm Sandvine...Trojans and worms with backdoor components such as Migmaf and SoBig have turned infected Windows PCs into drones in vast networks of compromised zombie PCs. Sandvine reckons junk mails created and routed by 'spam Trojans' are clogging ISP mail servers, forcing unplanned network upgrades and stoking antagonism between large and small ISPs."[169] With more and more ISPs pulling the plug on spammers as complaints flood in, spammers are turning to these backdoor means of spreading spam because they are much more efficient, much harder to detect, and much more difficult to stop. "Making things even tougher for IT security administrators in 2007 is the fact that an increasing amount of spam will be image-based, which is more difficult to detect...image-based spam

---

[165] Cassell Bryan-Low, "Virus for Hire: Growing Number of Hackers Attack Web Sites for Cash," *The Wall Street Journal*, pp. A1 & A8, 30 November 2004.

[166] John Markoff, "Attack of the Zombie Computers Is Growing Threat," *The New York Times*, (registration required) 7 January 2007, <http://www.nytimes.com/2007/01/07/technology/07net.html> (16 January 2007).

[167] Markoff.

[168] Markoff.

[169] John Leyden, "Zombie PCs Spew Out 80% of Spam," The Register, 4 June 2004, <http://www.theregister.co.uk/2004/06/04/trojan_spam_study/> (14 November 2006).

accounted for more than 40% of all spam messages generated in the fourth quarter of 2006, compared with less than 5% in the first quarter of 2005."[170]

All it takes is a public willing to open spam email, especially in HTML format, or its (seemingly innocuous) attachments, and there are millions of people still doing this. Spam sent using bots is notoriously difficult to trace because it uses other people's computers to traverse the Internet and, of course, always-on broadband connections only facilitate the movement of bots and spam.

The year 2003 may be remembered in Internet history for reaching one very unfortunate milestone: there were more spam emails than legitimate emails. "In 2003, Brightmail [an anti-spam company] saw spam surpass legitimate email—growing to more than 56% of all Internet email, up from just 40% a year ago."[171] The problem continues to worsen. In 2004, statistics painted a different picture. According to [e-mail security vendor FrontBridge's] figures, spam volume increased two percent, to 87 percent of e-mail, and has continued its growth each month since May of this year [2004]."[172] December 2006 saw a new record: according to one tracking system, ***spam accounted for 94 percent*** of all email that month.[173]

Despite improvements in knowledge and education about computer security risks, too many people still know little or nothing about the vulnerabilities in the tools they use every day, and this ignorance truly is bliss to the bad people wishing to exploit those weaknesses. Therefore, the first essential step in improving your Internet security and privacy is to learn more about basic vulnerabilities, exploits, and ways to protect yourself. Many of the recommendations in this book only need to be implemented one time for the life of a computer. Some, such as keeping basic software up to date, require more diligence. However, all are examples of "good computer hygiene" that will—or should—become second nature over time.

There are new vulnerabilities disclosed literally every week, so this book cannot provide a comprehensive list of problems, flaws, and potential attacks. It can do three things:

✓ Provide general guidance on improving your Internet privacy and security.

---

[170] Paul McDougall, "Organized Malware Factories Threaten Internet Users, Study Says," Information Week, 30 January 2007, <http://www.informationweek.com/story/showArticle.jhtml?articleID=197001739> (31 January 2007).

[171] "Brightmail Reports on Spam Trends of 2003," Networks Unlimited, 27 February 2004, <http://www.netunlim.co.za/news/news18.htm> (link inactive as of November 2005).

[172] Sean Michael Kerner, "The Deadly Duo: Spam and Viruses." ClickZ Stats, 16 November 2004, <http://www.clickz.com/stats/sectors/email/article.php/3433141> (14 November 2006).

[173] Gregg Keizer, "Spam Sets Record, Accounts For 94% Of E-mail," InformationWeek, 10 January 2007 <http://news.yahoo.com/s/cmp/20070111/tc_cmp/196802782>, 23 January 2007.

✓ Describe some known problems and how to cope with them.

✓ Point you to some privacy-related sites that have good information about keeping a low profile and preventing malicious attacks.

Ultimately, each individual must be responsible for staying up to date with the latest news about computer vulnerabilities—*you're your own best line of defense.*

Basics for Improving Your Internet Privacy and Security

The first thing you should do is check one of the sites that lets you see what information is being unwittingly provided about you as you surf. Go to the following sites to see what is known about you as you browse and, in the case of **Shields Up!**, what can be done to your computer while you're on line. As Steve Gibson, the site's creator, explains, "Without your knowledge or explicit permission, the Windows networking technology which connects your computer to the Internet may be offering some or all of your computer's data to the entire world at this very moment!" At his site, Gibson offers very practical ways to protect yourself and your data.

Shields Up! http://www.grc.com/

Junkbusters http://www.junkbusters.com/cgi-bin/privacy

BrowserHawk Browser Analysis http://www.syscape.com/showbrow.aspx

Browser Spy Browser Analysis http://gemal.dk/browserspy/

Russ Haynal's Persona Check http://navigators.com/cgi-bin/navigators/persona.pl

HackerWhacker Free Tools http://whacker4.hackerwhacker.com/freetools.php
especially the Browser Leakage and Quick Scan for open ports

Sygate/Symantec Online Security Services
http://scan.sygate.com/home_homeoffice/sygate/index.jsp
Sygate is now owned by Symantec.

There is some, though not much, "***security through obscurity***." If you have dial-up access to the Internet through a commercial ISP, you likely will be assigned a different ("dynamic") IP address every time you log on, which means it is difficult to link you as an individual customer of a particular ISP to any specific IP number registered to that provider. Furthermore, if you use an ISP with a large geographic coverage area, it *may* be difficult to pin down your location. On the other hand, your

ISP may indicate a very specific location—say, Fairfax County, Virginia—so check your profile (how to do this later). However, just because you're using a dial-up connection doesn't mean you can become complacent. Determined malicious hackers use very sophisticated tools, such as one that automatically dials thousands of random phone numbers until it finds another modem connected to the Internet, maybe your computer modem.

The sheer size of the Internet is also an inhibiting factor in what can be tracked by network administrators. For example, take a look at a ***web statistics*** page at the Department of Pulsar Astrometry of the Pushchino Radio Astronomy Observatory in Moscow:

Pushchino Radio Astronomy Observatory, Russia, Access Statistics: http://psun32.prao.psn.ru/wwwstat.html

0.01	0.01	14425	4	lt.ktu.aitra
0.03	0.06	82579	15	lt.ktu.sc-uni.delta
0.03	0.03	38850	17	lt.mtl.its
0.01	0.02	29682	6	lt.ot.slvie3-a10
0.00	0.01	10854	2	lt.takas.sia.dialup41
0.01	0.02	20170	3	lt.takas.vln.dialup68
0.00	0.01	10890	2	lt.telecom.klp.dialup74
0.01	0.01	12598	4	lu.pt.ppp01-0710-019
0.00	0.00	3197	2	lu.pt.ppp01-0710-065
0.01	0.01	8808	6	lv.alise.gw
0.00	0.00	1574	1	lv.gov.vid.proxy
0.01	0.01	11833	4	lv.lu.fmf.cs.pc06
0.00	0.01	10854	2	lv.riga.dialup166
0.00	0.01	10854	2	lv.riga.dialup181
0.01	0.02	21378	5	mil.af.aviano.cits-fw-1
0.01	0.01	14129	4	mil.af.keesler.kee22-200-52
0.01	0.01	12062	4	mil.af.langley.scm.user237066
0.01	0.01	12082	4	mil.af.pope.jason
0.00	0.00	703	1	mil.af.wpafb.px0o
0.01	0.01	15560	5	mil.uscg.gateway-fincen
0.00	0.00	883	2	mil.uscg.gateway-osc
0.00	0.00	703	1	mv.net.dhivehinet.engine3
0.00	0.00	703	1	mx.com.pvnet.pppd23
0.01	0.08	99068	5	mx.com.spin.blaster37
0.04	0.04	48281	19	mx.inaoep.pactli
0.01	0.01	16789	4	mx.itesm.mty.matematicas
0.00	0.01	14270	2	mx.net.telmex.tntleon1-1-157

Network administrators use these statistics to glean general information about where visitors to their website are coming from, peak activity times, and which internal urls are visited most frequently. Of special interest are accesses by client domain. Most accesses at the observatory, not surprisingly, are from other computers at the site, but if you scroll down the list, many international sites, including .edu, .gov and .mil, also appear. While access to servers from commercial US accounts is generally too commonplace to provide much useful information, access from .gov or .mil accounts show up quite prominently on these statistical listings. Also, generally only older or superficial statistics tend to be available to the public; more recent statistics, which tend to be very detailed, usually require a password.

In addition to unscrupulous people trying to get into your computer, a somewhat less threatening but nonetheless worrisome possibility is that a network administrator at the website you are visiting may be able to tell the following about you:

- Who your provider is.
- Where your provider is located.
- What site you last visited.
- If you link to a site from a search engine, the query you ran.
- What browser software you are using.
- Your email address.

On the other hand, you may *not* be giving out all this information. Specifically, you should make sure your browser does not provide the "HTTP_From" or "REMOTE_USER" variables, both of which give away information about your email address and other indications of your identity. Also ensure that the "REMOTE_IDENT" variable is not being disclosed (more than likely, it is not). How do you know if you are providing these variables? Go to the **Junkbusters** site listed above and it will let you know.

JUNK*BUSTERS* Alert on Web Privacy

You can be tracked from your mouse clicks

Most people surf the net under the illusion that nobody will ever know what they look at. We want you to know what companies find out about you when you visit their web sites.

Your browser assembles each page by making "HTTP requests" for its text and graphics parts from one or more web sites. These sites may not have been named in the link you clicked on: two banner ads on the same page can come from different companies. Your browser gives all of them a lot of information you might prefer to keep private. Most sites store these details indefinitely.

How they know where you came from

The "HTTP Referer" tells them what led you to the request.
In this case it was **not provided**.

- If you use a search engine to find a site, the entire query you typed is typically handed to the sites you then click on.
- If you clicked on a banner advertisement, the URL may contain coded data used to target specific ads at you. (Before clicking on an ad, look at the URL displayed for it. Codes and long addresses suggest that your mouse clicks are being tracked.)
- If the URL you clicked on was in one of your private files, such as your email reader may use, the full file name is still handed over to the web site. It may contain information about you such as an indication of your name or email address, the email program you are using, and the structure of your file space.

Junkbusters http://www.junkbusters.com/cgi-bin/privacy

Increase Your Knowledge

Most computer users think they are much safer on line than they actually are according to a survey of 329 computer users by America Online and the National Cyber Security Alliance (NCSA) during 2004.[174] To conduct the survey, AOL and the NCSA sent technicians to 329 homes to inspect users' computers. Here's what the survey found:

- Four out of five users had spyware and/or adware on their computers, and most did not know this software was running on their computers.
- Nearly two-thirds had been infected by a virus at one time (and this is just the number who *knew about an infection*).
- 85 percent had anti-virus software, but more than half hadn't updated it in a week or more.
- Two-thirds of users did not have any type of firewall protection.
- Nearly three in five users did not know the difference between a firewall and antivirus software.
- 38 percent of wireless users had not bothered to encrypt their networks.

Users are endangering not only their own privacy and security, including any and all financial and personal data stored on a computer, but *they are putting everyone else at risk*. The proliferation of spyware opens the gates to intruders who can potentially gain control of individual computers. When networked together, this system of personal computers can form what is usually called a "zombie army" of PCs that can be used to attack other networks. Like it or not, each individual is responsible for his or her own computer privacy and security, so please pay attention to the basics to protect yourself and others. Here are the minimum steps you need to take:

- Keep your system patched and regularly update all security software.
- Install, routinely run, and UPDATE an anti-virus program (at least once a week).
- In general, do not open email attachments.
- Install and use firewall software and/or hardware (make sure settings are restrictive).

---

[174] America Online and National CyberSecurity Alliance, "AOL/NCSA Online Safety Study," Staysafeonline.org, October 2004, <http://www.staysafeonline.org/pdf/safety_study_2005.pdf> [PDF] (14 November 2006).

- Use strong passwords and change them regularly.
- Do not download and install programs indiscriminately (read user agreements).
- Install, routinely run, and update spyware software (at least one and preferably two).
- Configure your Internet browser(s) to maximize security.
- If you have a wireless network, use strong encryption.

For general information and help with personal privacy and security concerns, visit websites such as are NCSA's Stay Safe Online, CERT's Home Network Security, About's Network Security, and Microsoft's page on security and privacy for home users to learn more about vulnerabilities and how to protect yourself from many dangers on the Internet. All these sites not only warn you about the problems but also do an excellent job of telling you how to fix vulnerabilities.

About's Network Security	http://netsecurity.about.com/
CERT's Home Network Security	http://www.cert.org/tech_tips/home_networks.html
Get Safe Online	http://www.getsafeonline.org/
Microsoft Security & Privacy for Home Users	http://www.microsoft.com/athome/security/default.mspx
NSA's Security Recommendation Guides	http://www.nsa.gov/snac/
NCSA's Stay Safe Online	http://www.staysafeonline.info/
Surf the Net Safely	http://surfthenetsafely.com/

and for Macintosh users...

SecureMac.com	http://www.securemac.com/

The Perils and Pitfalls of Wireless Internet

I believe I need to interject a few comments about **wireless Internet connectivity** here. If you use wireless connections either in your home or on the road, I urge you in the strongest terms to be extremely careful with wireless connectivity. The *New York Time*'s technology columnist David Pogue wrote an interesting piece on how his eyes were opened by a stark demonstration of the insecurity of a public WiFi connection.[175] In fact, if you use WiFi without encryption, expect that anyone and everyone can read everything you read and write, and track every move you make.

You also need to be aware of the WiFi "evil twin" scenario, an attack that is remarkable both for its simplicity and its effectiveness. Here's how it works. The bad guy takes his laptop to a popular coffee shop where lots of people like to use the Internet while enjoying a cup o' joe. The bad guy has set up his computer to transmit a signal that turns his laptop into an Internet gateway or access point, one that looks and sounds remarkably legitimate. Here you come, mocha frappuccino in hand; you open your laptop, start searching for a local WiFi connection, and—bingo—in addition to that coffee shop's fee-for-service Mobile Hotspot, there is a second option Cheap & Friendly Mobile Hotspot or maybe even a Free Mobile Hotspot. If you are like most people, you might well log into the cheap or free service, assuming they are legitimate WiFi hotspots. And what happens if you do log into an evil twin WiFi access point? The bad guy will have software on his computer to capture every keystroke you make, so whatever you have entered once you've logged in, he now owns. And if you used a credit card to log into the cheap WiFi hotspot, the bad guy now has that. Even if you sent any encrypted data, such as a password, that's still probably not a problem for the bad guy because he also undoubtedly has software to break that, too.

The problem is obvious: you don't want to fall prey to this evil twin attack, but how to avoid it and still use WiFi hotspots? Here are good suggestions from PCWorld Magazine:

> "Check Your Wi-Fi Settings: Many laptops are set to constantly search and log on to the nearest hotspot. While this option might seem convenient, it does not allow you to monitor which hotspots you are logging on to and determine if they are legitimate. Turning off this option will prevent your computer from logging on to a hotspot without your knowledge.
>
> Pay Attention to Dialog Boxes: Pop-up warnings are there for a reason—to protect you. If you are lucky enough to have not clicked the "never show this again" option, make sure you read these warnings carefully before agreeing to send information.
>
> Use One of Your Credit Cards on the Web Only: Open a credit card account that is used solely for the purposes of shopping on the Web. Ideally, you should be able to access account records online so you don't have to wait for monthly statements to monitor any activity. "Be prepared to close that account on short notice if it's been compromised," says Schiller.
>
> Conduct Private Business in Private: "Maybe you don't need to move money around or check your bank statements when you are connected to a public

---

[175] David Pogue, "How Secure is Your WiFi Connection," Pogue's Posts, *The New York Times*, 4 January 2007, <http://pogue.blogs.nytimes.com/2007/01/04/04pogue-email/>, 16 January 2007.

> hotspot that you're not really familiar with," says Schiller. If you restrict your public surfing to Web pages you don't mind a stranger reading along with you, there is little an evil twin attacker can do to harm you."[176]

I recommend that you embark on the installation of a home wireless network with trepidation and care. Two good starting places for learning more about secure wireless networking are Tony Bradley's "Introduction to Wireless Network Security" and Brian Livingston's "Wi-Finally: Wireless Security That Actually Works."

Introduction to Wireless Network Security
http://netsecurity.about.com/od/hackertools/a/aa072004b.htm

Brian Livingston, "Wi-Finally: Wireless Security That Actually Works"
http://www.windowssecrets.com/comp/050526/ - story1

"Law #10: Technology is not a panacea."[177]

Web Tip

Virtually all Microsoft products come with all the doors open and unlocked, figuratively speaking. You must take upon yourself to find the open doors, shut them, and lock them tight.

[176] Erin Biba, "Does Your Wi-Fi Hotspot Have an Evil Twin? Identity thieves are going wireless in their quest to steal your personal info," Medill News Service, PCWorld, 15 March 2005, <http://www.pcworld.com/news/article/0,aid,120054,00.asp > (16 January 2007).

Browser Concerns

Using Internet Explorer's Privacy and Security Controls

One of the biggest underlying problems vis-à-vis Internet security is that virtually all Microsoft products,[178] including Internet Explorer, come with all the doors open and unlocked, figuratively speaking. You must take upon yourself to find the open doors, shut them, and lock them tight. The guiding principle for browser security is to place high restrictions on all web sites *by default*, while giving trusted sites only limited security restrictions. This will allow trusted sites to function with limited or no problems.

Because of the many changes occurring with Microsoft products, including the release of Internet Explorer 7 in October 2006 and the Vista operating system in January 2007, as well as the growth in popularity of the Firefox browser, I am no longer focusing on instructions for specific software. Instead, I will discuss the broad issues surrounding browser privacy and security and point you to sites where you can learn the details of securing your own particular browser and other software.

In August 2005, Microsoft released an upgrade to IE version 6 that was only available to users of Windows XP SP2. For more information on IE6 for Windows XP SP2, I recommend these sites to readers who still use IE6:

Windows XP Service Pack 2: What's New for Internet Explorer and Outlook Express
http://www.microsoft.com/windowsxp/sp2/ieoeoverview.mspx

Comparison of the Internet Explorer Security Zones in Windows XP Service Pack 2
http://surfthenetsafely.com/ieseczone5.htm

Then, on November 1, 2006, Microsoft began offering Internet Explorer 7, since renamed Windows Internet Explorer, as a high-priority update via Windows Automatic Updates. Microsoft is no longer updating its browser for any operating systems other than XP Service Pack 2, Windows Server 2003, and Windows Vista.

---

[177] "The Ten Immutable Laws of Security," Microsoft Security Essays, <http://www.microsoft.com/technet/treeview/default.asp?url=/technet/columns/security/essays/10imlaws.asp > (14 November 2006).

[178] The advent of Windows XP Service Pack 2 in 2004 addressed some of these "open door" privacy and security issues, but certainly not all of them. The safest rule is never to assume any product is secure and always read the instructions on how to implement higher levels of privacy and security. A good starting place for Windows security help is Microsoft Technet Security. <http://www.microsoft.com/technet/security/default.mspx> (14 November 2006).

For help with IE7's security and privacy settings, look at the Microsoft website and these other sites, but remember, the IE browser in Windows XP is not the same as the browser that came with Vista. Microsoft offers a set of four steps to improve your online security and privacy. While it is incomplete, it is a good starting place for strengthening your Microsoft-based browser and email security.

Microsoft IE7: Dynamic Security Protection
http://www.microsoft.com/windows/products/winfamily/ie/features.mspx

Microsoft: Improve the Safety of Your Browsing and E-Mail Activities
http://www.microsoft.com/athome/security/online/browsing_safety.mspx

How to surf more safely with Internet Explorer 7
http://www.helpwithwindows.com/techfiles/ie7-surf-safe.html

Brian Livingston, Windows Secrets, IE7 Needs Tweaking for Safety
http://windowssecrets.com/comp/061026/ - story1

Diana Huggins, IE 7.0's Internet Options Privacy and Security Settings
http://www.lockergnome.com/nexus/windows/2007/01/22/ie-70s-internet-options-security-settings/
http://www.lockergnome.com/nexus/windows/2007/01/23/ie-70s-internet-options-privacy-settings-part-i/
Be sure to look at Part II as well.

Marc Liron, Microsoft MVP on Internet Explorer 7
http://www.updatexp.com/internet-explorer-7-download.html

Deb Shinder, Tech Republic, "10 things you should know about Internet Explorer 7 Security" http://articles.techrepublic.com.com/5100-1009_11-6130844.html

Surf the Web Safely: Make IE7 Safer http://surfthenetsafely.com/ieseczone8.htm

Kim Komando's Firefox 2 and IE7's Security Settings
http://www.komando.com/tips/index.aspx?id=2523

One of Internet Explorer's best security features is its **Trusted Sites**. The Trusted Sites option is an excellent way to give some websites more privileges while keeping most sites at higher security settings. Put only sites you absolutely trust, e.g., your bank, in your "trusted sites" zone and keep all others at the highest security settings.

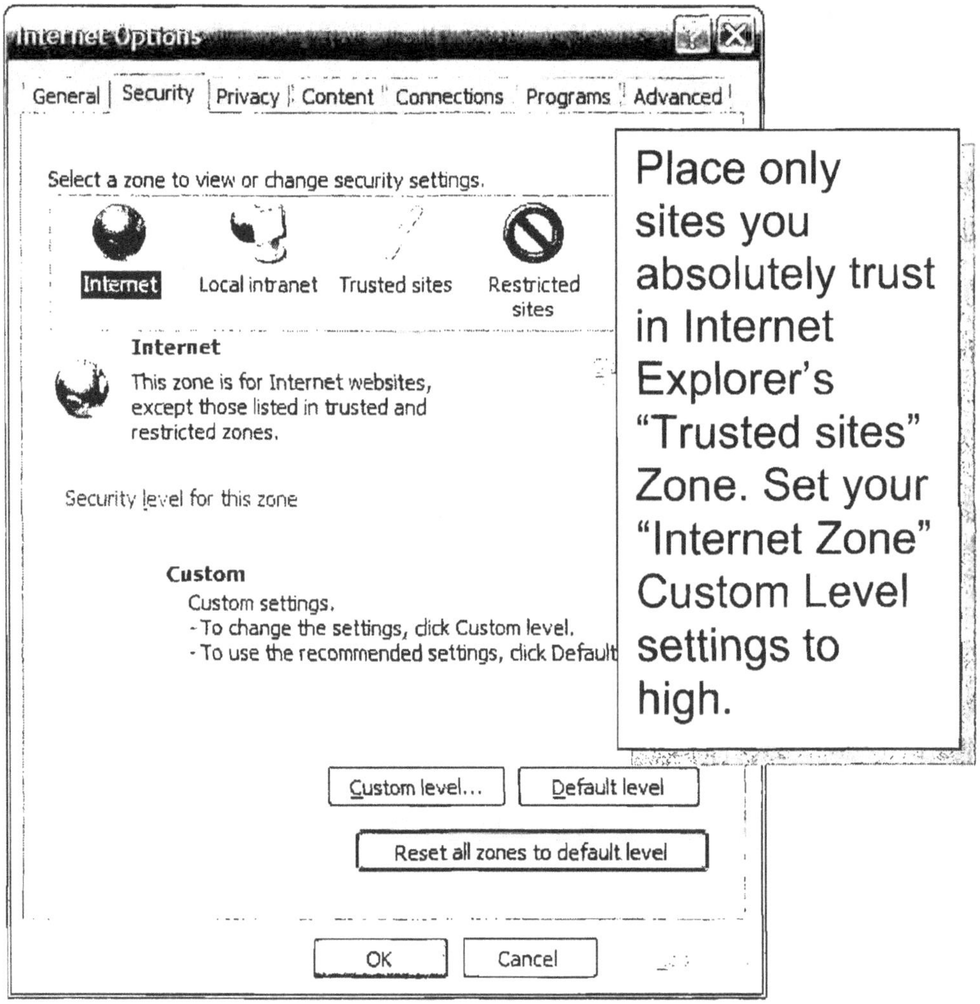
Internet Options
General
Security
Privacy
Content
Connections
Programs
Advanced
Select a zone to view or change security settings.
Internet
Local intranet
Trusted sites
Restricted sites
Internet
This zone is for Internet websites, except those listed in trusted and restricted zones.
Security level for this zone
Custom
Custom settings.
- To change the settings, click Custom level.
- To use the recommended settings, click Default
Custom level...
Default level
Reset all zones to default level
OK
Cancel
Place only sites you absolutely trust in Internet Explorer's "Trusted sites" Zone. Set your "Internet Zone" Custom Level settings to high.

While there are various differences among Microsoft Internet Explorer browser versions, these are generally accepted as safe settings for the **Internet Zone** using the **Custom Level**:

Tools | Internet Options | Security | Internet Zone | Custom Level

- ActiveX Controls and plugins
  - Download signed ActiveX controls **[Prompt or Disable]**
  - Download unsigned ActiveX controls **[Disable]**
  - Initialize and script ActiveX controls not marked as safe **[Disable]**
  - Run ActiveX controls and plug-ins **[Disable]**
  - Script ActiveX controls marked safe for scripting **[Prompt or Disable]**

- Downloads
  - File Download **[Enable]**
  - Font Download **[Prompt]**
- Microsoft VM
  - Java permissions **[High Safety]**
- Miscellaneous
  - Access data sources across domains **[Disable]**
  - Allow META REFRESH **[Enable]**
  - Display mixed content **[Prompt]**
  - Don't prompt for client certificate selection… **[Disable]**
  - Drag and drop or copy and paste files **[Enable or Prompt]**
  - Installation of desktop items **[Disable]**
  - Launching programs and files in an IFRAME **[Disable]**
  - Navigate sub-frames across different domains **[Enable or Prompt]**
  - Software channel permissions **[High Safety]**
  - Submit nonencrypted form data **[Enable]**
  - Userdata persistence **[Disable]**
- Scripting
  - Active scripting **[Disable]**
  - Allow paste operations via script **[Disable]**
  - Scripting of Java applets **[Disable]**
- User Authentication: **Automatic logon only in Intranet zone**

As a general rule, <u>do not rely upon sliders to determine your security settings</u>. These settings <u>will</u> affect your browsing. Some websites require ActiveX or scripting. If you want to run ActiveX or scripts on any website, you can either turn this feature on temporarily or add the site to the **Trusted sites zone**, though I would be very, very careful about which sites you add.

You can add Web sites by selecting the **Trusted sites** icon, and pressing the **Sites** button. The default setting only lets you add secure sites (sites using https); however, if you uncheck the **Require server verification (https:) for all sites in this zone**, you can add any site.

Web Tip

Have you suppressed popups only to continue to see them? Here's why: in Netscape, there is a default list of "exceptions" for sites whose popups are allowed. To view this list and suppress popups from these sites in Netscape 7.x, Edit | Preferences | Privacy & Security | Popup Window Controls | Suppress Popups | Exceptions | Remove All

Firefox, IE 6 for Windows XP SP2, and IE7 block all popups by default but permit users to allow popups from specific websites. Older versions of MSIE do not have a popup blocker option.

Manage ActiveX, Java, & JavaScript

Many security vulnerabilities exploit these applications and most privacy/security experts recommend disabling them. Malicious hackers can use ActiveX, Java, or JavaScript to upload files and run them on your computer when you simply visit a web page that has been cracked or created by malicious hackers. "In January 1997 members of the Hamburg-based Chaos Computer Club staged an electronic break-in on German national television. Using an ActiveX control, they made unauthorized bank transfers through Intuit's Quicken without a personal identification number. The demonstration sought to prove that executable content, particularly Microsoft's ActiveX, isn't secure."[179] In fact, *any ActiveX control downloaded over the web might be a Trojan horse or a virus.*

---

[179] "Preventing Possible Web Intrusions," Smartcomputing.com, Vol 8, Issue 4, April 2000, <http://www.smartcomputing.com/editorial/article.asp?article=articles%2Farchive%2Fg0804%2F37g04%2F37g04%2Easp> (14 November 2006).

More recently, malicious users have found devilishly clever ways to use ActiveX, Java, and JavaScript to hijack browsers, or to be more precise, to hijack Internet Explorer. The least innocuous form of browser hijacking involves changing the browser's home page and favorites, but most browser hijackers do a lot more, from creating endless pop-up windows to taking complete control of your browser. Browser hijacking software also usually includes some form of spyware to monitor and report your Internet activity. Worse, they are notoriously difficult to remove.

ActiveX has been implicated in the surreptitious installation of software known as drive-by downloads. **Drive-by downloads**[180] occur when a user simply visits a website or views an HTML email. These sites exploit a vulnerability in Internet Explorer's ActiveX to download, install, and run software on an unsuspecting user's computer without his knowledge or consent. This type of software can also be very difficult to remove.

Keep in mind that, by default *active scripting is enabled by default in Internet Explorer*! The problem with simply disabling these controls is that you will encounter difficulties viewing some webpages. Experiment with turning them off or, in the case of MSIE, having your browser "Prompt" you and see what happens.

Here are recommendations for increased security settings in IE's **Internet Zone** (remember, you can put sites where you need to use these controls into your **Trusted Sites Zone**):

Tools | Internet Options | Security | Internet Zone | Custom Level

- ActiveX Controls and plugins
    - Download signed ActiveX controls **[Prompt or Disable]**
    - Download unsigned ActiveX controls **[Disable]**
    - Initialize and script ActiveX controls not marked as safe **[Disable]**
    - Run ActiveX controls and plug-ins **[Disable]**
    - Script ActiveX controls marked safe for scripting **[Prompt or Disable]**

---

180 "A **drive-by download** is a program that is automatically downloaded to your computer, often without your consent or even your knowledge. Unlike a pop-up download, which asks for assent (albeit in a calculated manner likely to lead to a "yes"), a drive-by download is carried out invisibly to the user: it can be initiated by simply visiting a Web site or viewing an HTML e-mail message. Frequently, a drive-by download is installed along with another application. For example, a file sharing program might include downloads for a spyware program that tracks and reports user information for targeted marketing purposes, and an adware program that generates pop-up advertisements using that information. If your computer's security settings are lax, it may be possible for drive-by downloads to occur without any action on your part." "Drive-by Download," SearchSMB.com, <http://searchsmb.techtarget.com/sDefinition/0,,sid44_gci887624,00.html> (14 November 2006).

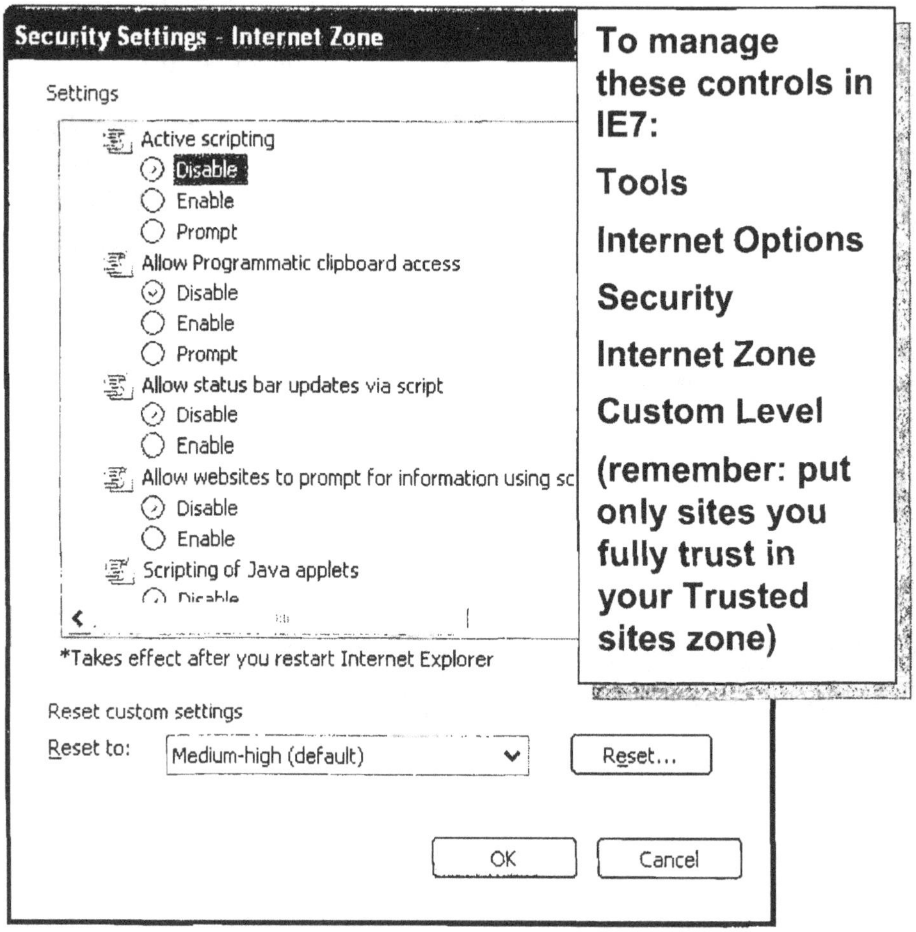
Security Settings - Internet Zone
Settings
Active scripting
Disable
Enable
Prompt
Allow Programmatic clipboard access
Disable
Enable
Prompt
Allow status bar updates via script
Disable
Enable
Allow websites to prompt for information using sc
Disable
Enable
Scripting of Java applets
*Takes effect after you restart Internet Explorer
Reset custom settings
Reset to:
Medium-high (default)
Reset...
OK
Cancel
To manage these controls in IE7:
Tools
Internet Options
Security
Internet Zone
Custom Level
(remember: put only sites you fully trust in your Trusted sites zone)

What's the best way to avoid browser hijacking and drive-by downloads? "First and foremost simply, stop using Internet Explorer. If you use Mozilla browsers (Netscape and Firefox) or Opera, you are immune to all known browser hijackers. You are immune for two reasons. First, most people use Internet Explorer, so most malicious code is custom built to exploit it. Second, Opera's and Mozilla's programmers take security very seriously and have made these browsers very secure. It is not possible to install software from a web site using these browsers without at least seeing a prompt of some sort asking permission." This is the advice of most Internet security experts.[181]

---

[181] Mike Helan, "Prevent Browser Hijacking," *SpywareInfo.com*, 23 March 2004 (Updated 12 January 2005), <http://www.spywareinfo.com/articles/hijacked/prevent.php> (article no longer available).

Despite changes in IE7, Microsoft's browser still relies heavily on ActiveX controls, which are often exploited by browser hijackers. In fact, *PC World* lists IE as the number one Internet threat of 2007 because of IE's "reliance on Microsoft's ActiveX technology, which allows Web sites to run executable programs on your PC via your browser."[182] Although Firefox is becoming a more tempting target for malicious hackers, IE remains the target of choice for now both because it is the most popular browser and because of its dependence on ActiveX.

Disable Autocomplete for Forms and Names/Passwords on Forms

This is another case where placing convenience ahead of security could cost you dearly. You do not want passwords or forms saved to the browser so that someone else might use them for some nefarious purpose. Passwords should not be saved unencrypted or without strong protection anywhere at any time. Many online stores will ask you if you would like to save your credit card information for future use. ***Do not allow websites to save your credit card number***. Make a habit of entering such personal and financial data each time it is needed and only for that transaction.

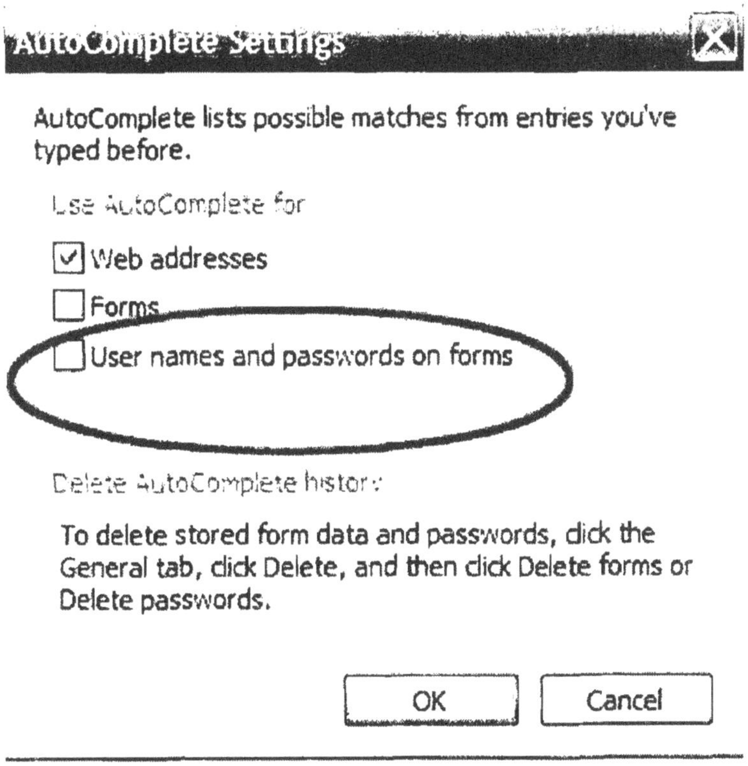
AutoComplete Settings
AutoComplete lists possible matches from entries you've typed before.
Use AutoComplete for
Web addresses
Forms
User names and passwords on forms
Delete AutoComplete history
To delete stored form data and passwords, click the General tab, click Delete, and then click Delete forms or Delete passwords.
OK
Cancel

---

[182] Scott Spanbauer, "Thwart the Three Biggest Internet Threats of 2007," 24 January 2007, <http://www.pcworld.com/printable/article/id,128538/printable.html> (31 January 2007).

In Internet Explorer:

Tools | Internet Options | Content | AutoComplete Settings

Uncheck the second and third boxes (*forms* and *user names and passwords on forms*).

Be sure to Delete AutoComplete History (in IE7) or Clear Forms and Clear Passwords (in IE6) to remove any stored data.

Firefox & the "Clear Private Data" Option

One of the many nice features of Mozilla's Firefox browser is the "Clear Private Data" option. First, why did Mozilla include this option and why would you want to consider using it? A lot of websites imply this option is just for people doing things on a shared computer that they don't want others to know about, but that is an extremely narrow understanding of why it is important to remove certain personal data from your browser and computer. After logging off a secure website, especially your bank's or credit card's site, have you ever gotten a message like this?

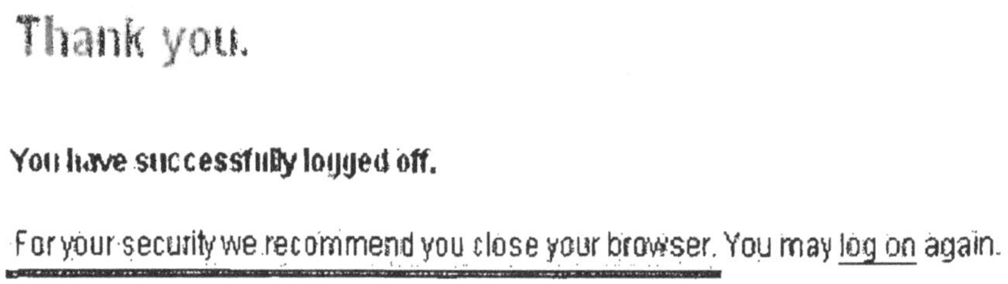
Thank you.
You have successfully logged off.
For your security we recommend you close your browser. You may log on again.

Why is a secure site recommending you close your browser? Browsers do a very good job of keeping track of when and where you have been on the Internet and, in some cases, what you have been doing. Browsers are usually set to save or cache the pages you visit so that the next time you visit that page, you will not have to wait for the server to send the page to your browser. While this makes your surfing faster, it also causes a small but real vulnerability by creating a record of your browsing history. In the case of a site where you have entered information such as a password, a credit card number, your Social Security Number, etc., that information filled into a form on the web may remain in your cache and therefore on your computer. While it is unlikely a malicious user is going to get that information, it is not impossible. Also, if you ever use a computer that others can access, remember, they can get that information, so why not take a few steps to improve your privacy and security?

You can manually clear your browser each time you use it, and as I explained in the previous section, current versions of Internet Explorer offer an option to clear the browser cache automatically when you close the browser. However, this only addresses the cache issue. Firefox provides an easy way to delete stored information of various types either as you are working online or whenever you close your browser. The *default* setting of the "Clear private data" option in Firefox will delete:

- your browsing history.
- form data (this could include credit card numbers).
- your download history, cache, and authenticated sessions (the kind of sessions likely to include passwords).

The "Clear private data" option leaves saved passwords and cookies alone unless you tell it otherwise. I doubt if many users want to preclude the use of cookies altogether; cookies placed on your computer from the originating site (that means the site you are visiting) rarely present a problem and are a genuine help in many instances (for example, cookies let you set preferences at a search engine so you don't have to reset them every time you go there). By now most people know they probably do not want third party cookies, that is, cookies placed on your computer by some uninvited third party, such as from a banner ad. You can tell Firefox (and other browsers, too) to save cookies for the originating site only and leave the "Cookies" option unchecked on the "Clear private data" form. Also keep in mind that Firefox lets you browse, search, and delete individual (or all) cookies using the "View cookies" option.

Saving passwords is a more controversial subject. My preference is not to let any browser save any of my passwords, but many experts think having Firefox manage them for you (with a very strong Master Password) is actually safer than writing them down. I disagree; most people don't break into a house in order to break into a computer. One important caveat: ***do not use the Firefox Remember passwords' option on a laptop***. That way, if someone steals your laptop and accesses the account, the thief will not have access to every saved password you have stored. It is too easy to get at the Firefox Master Password, which in turn will unlock every password saved by Firefox on that laptop.

If you want Firefox to remember your passwords, here is the safest way to do it:

Select Tools | Options | Privacy | Passwords and check "Remember Passwords"

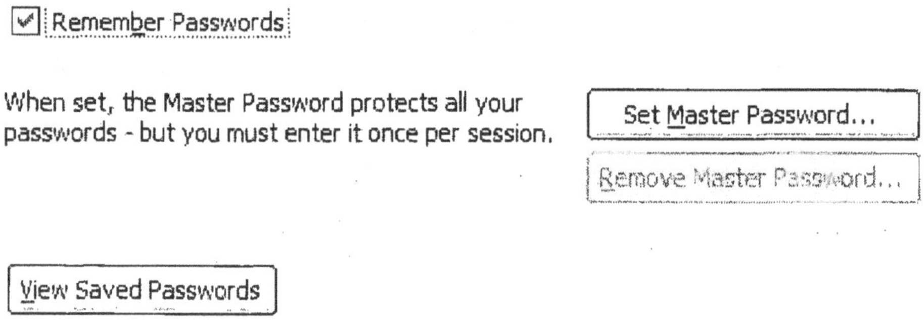
Remember Passwords
When set, the Master Password protects all your passwords - but you must enter it once per session.
Set Master Password...
Remove Master Password...
View Saved Passwords

Now when you enter a password, Firefox will automatically ask you the following question:

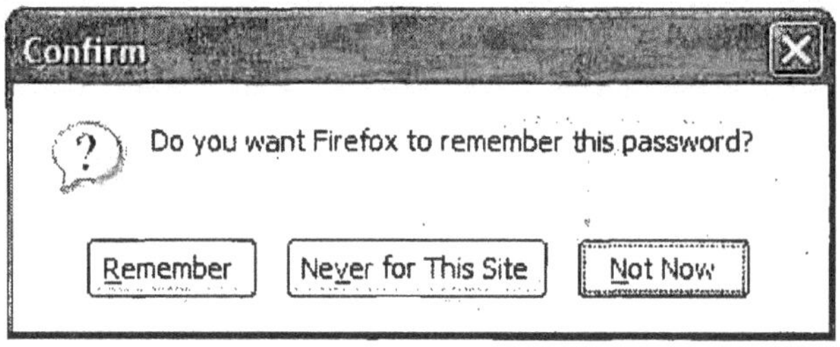
Confirm
Do you want Firefox to remember this password?
Remember
Never for This Site
Not Now

If you are going to allow Firefox to save your passwords, then I believe you must also set a Master Password (and do not forget it). Otherwise, anyone with access to your computer can view your saved passwords.[183] The Master Password will be required for your saved passwords to be loaded. Firefox will prompt you to enter it once per session when it is needed. You can also manage saved passwords and delete individual passwords by clicking the "View Saved Passwords" button. Keep in mind that the Master Password is required in order to change or remove the Master Password, which is why it is important not to forget it (of course, there are hacks for resetting it if you have forgotten it). With a Master Password in place, in principle no one can see your passwords unless he also has your Master Password. However, the Firefox Master Password is not going to stop a knowledgeable malicious hacker with full access to your computer from getting the Master Password and, indeed, all your passwords, but if someone has full file-level access to your computer, you are already in deep trouble.

Finally, I recommend using the Firefox feature that will clear your personal information automatically:

Select Tools | Options | Privacy | Settings and check "Clear private data when closing Firefox," then click OK twice.

---

[183] Some people have discovered, usually the hard way, that viewing someone else's Tools | Options | Privacy | View Saved Passwords | Passwords Never Saved may reveal some rather shocking facts about sites the other person has visited. This only works if Firefox is set to Remember Passwords and the user chose "Never for this site."

These are not my preferences; I like to keep my Browsing History and I have Firefox Clear private data when closing.

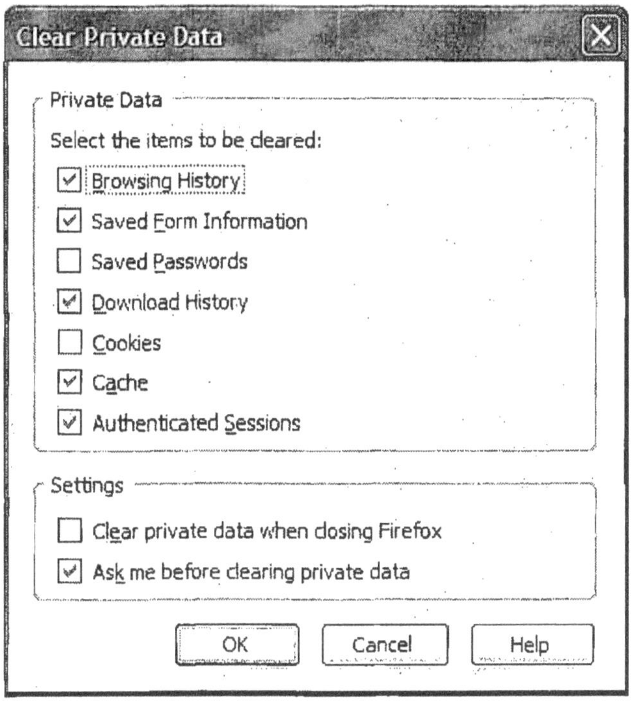
Clear Private Data
Private Data
Select the items to be cleared:
Browsing History
Saved Form Information
Saved Passwords
Download History
Cookies
Cache
Authenticated Sessions
Settings
Clear private data when closing Firefox
Ask me before clearing private data
OK
Cancel
Help

You can also use the Clear Private Data option without closing the browser when you log out of a site where you entered data such as a password, credit card number, etc., (clearing Authenticated Sessions).

Select Tools | Clear Private Data or use the keystroke combination Ctrl+Shift+Del.

- One final comment: all these types of measures are useful and worth taking, especially since most Internet users do nothing to protect their privacy and security. But please do not be lulled into a false sense of security because a proficient malicious hacker can make hash out of all basic computer security measures.

Disable "Userdata Persistence"

Userdata Persistence is a feature in Internet Explorer that lets websites "remember" information you enter, such as search queries. Userdata persistence is an XML-based storage methodology for saving large amounts of user data. If you use Internet Explorer, have you ever come back days or weeks later and find that, as you type your query, your earlier queries suddenly appear in or below the query box? This is user data persistence at work. Many people manage cookies, but few realize they also need to disable this "feature" in MSIE. It is easy to disable: In MSIE 5 & 6:

Tools | Internet Options | Security | Custom Level | Miscellaneous

About two-thirds the way down you will see "userdata persistence." Select *Disable*.

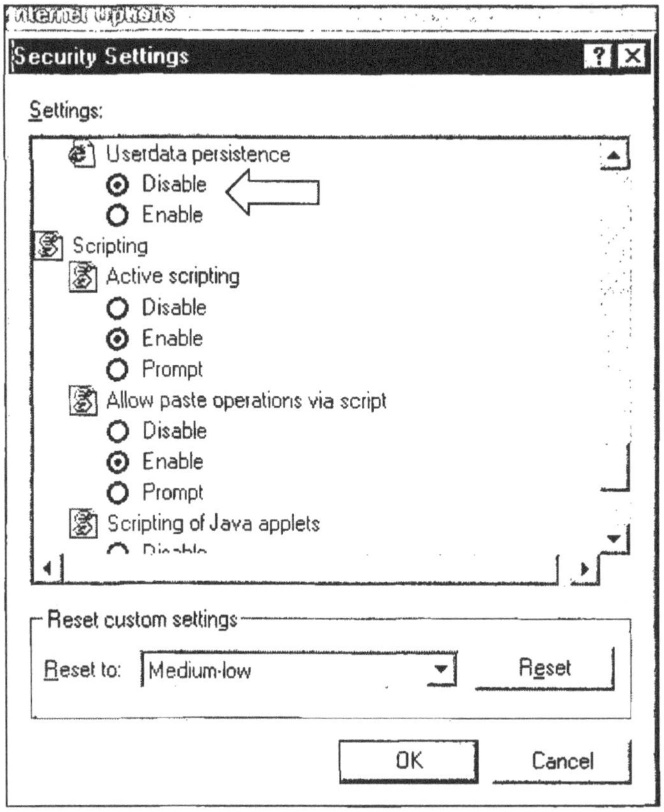
Security Settings
Settings:
Userdata persistence
Disable
Enable
Scripting
Active scripting
Disable
Enable
Prompt
Allow paste operations via script
Disable
Enable
Prompt
Scripting of Java applets
Reset custom settings
Reset to:
Medium-low
Reset
OK
Cancel

You will also need to <u>clear your cache</u> to clean out old user data.

<u>Manage Your Cookies</u>

Cookies are the source of a great deal of discussion and consternation among Internet users. What exactly are cookies? They are text files placed on a user's hard disk (yes, your computer) by a website. Each cookie has two parts: a name and a value. The value can contain a lot of data, such as the address of the server that set the cookie, an expiration date, and (possibly) usernames and passwords. Before you panic, this data is usually encrypted, and cookies were designed so only the site that placed them on your computer in the first place could read them. This being said, nothing is 100 percent secure and it is not a good idea to accept all cookies willy-nilly.

You have many options for handling cookies. Most browsers offer the option never to accept cookies, but many sites require them and turning them all off is generally not a realistic option for most Internet users. One compromise is to accept only those cookies that are sent back to their originating server, which means you won't allow third-party cookies (this, however, will not stop cookies from pop-up windows). Blocking third-party cookies also prevents web bugs from linking together information in your email with the web sites you visit. Newer browser versions have more cookie options with promises for even greater user control in future releases. Check your browser(s) preferences/options to see what choices you have. Also check out:

To manage cookies in IE 6/7:
Tools | Internet Options | Privacy

Internet Explorer 6 and 7 have enhanced cookie management, including a ***privacy settings slider*** with six settings: Block All Cookies, High, Medium High, Medium (default level), Low, and Accept All Cookies. The default setting is Medium. You can override automatic cookie handling for all websites by clicking **Advanced** on the **Privacy** tab. You can also choose to ***always accept*** or ***always reject cookies from specific sites***. I recommend you block all cookies from sites from which you know you never want to accept a cookie, e.g., *doubleclick.net.*

Better still, get a good "cookie crumbler," that is, software designed to handle cookies automatically and intelligently. Cookie Central reviews a number of cookie management programs.

For details on handling cookies in **Internet Explorer 6/7**, I recommend Surf the Net Safely's "Advanced Cookie Management in Internet Explorer 6 and 7, <http://surfthenetsafely.com/cookie_advanced.htm>.

Firefox/Mozilla offers flexible cookie controls, and there is a very good tutorial on using Firefox's cookie management options at:

Firefox's Cookie Options
http://mozilla.gunnars.net/firefox_help_firefox_cookie_tutorial.html

Microsoft's Help Safeguard Your Privacy on the Web (for IE6, but most still applies to IE7) http://www.microsoft.com/windows/ie/using/howto/privacy/config.mspx

Cookie Central's Reviews of Cookie Management Software
http://www.cookiecentral.com/files.htm

Junkbusters Cookie Page http://www.junkbusters.com/ht/en/cookies.html

Disable or Defeat the HTTP-Referrer

The **"http referrer" variable** (often misspelled "referer"[184]) may be a serious concern. This variable lets a site you are visiting know which site you just came from (which site referred you to them). Usually, the value of the "referrer" field is the url of the page you last visited. The problem is that ***the http-referrer variable gives out more information***. If you use a search engine to find a site and then click on that site, the http-referrer will provide the ***entire query*** you used to find the site! Furthermore, it is possible that other sensitive types of information, such as username, password, email address, or even a credit card number, could be sent as part of an http-referrer variable.[185]

There are ways around this problem. Here are three solutions:

1. **Don't click on a link from a search engine**; instead right-click on the link and select Copy Link Location (Mozilla) or Copy Shortcut (IE6); paste the link in the address window, and go to the link from the new browser window. Your query will not be provided. Remember: you must copy the link to the address bar; it is ***not*** sufficient to right-click and "open in new window" or "open in new tab."

2. Use a **browser-based service** that blocks the http-referrer, such as Webwasher or Guidescope, both of which are *free to individual users*, or any number of products that can be purchased for this purpose.

Webwasher[186]
http://www.cyberguard.com/products/webwasher/webwasher_products/classic/index.html

Guidescope http://www.guidescope.com/home/

3. **Disable the http-referrer in Netscape 7 and Firefox** (you cannot do this in Internet Explorer). In the Address/Location bar, type *about:config* and find *network.http.sendRefererHeader*. This variable can be set to 0, 1, or 2:

2—default; send referrer for all requests

1—do not send referrer for images

---

[184] The actual "referrer" code uses the incorrect spelling "referer," which may say something about the spelling skills of programmers.

[185] For an excellent overview of the legitimate use of and problems with the http-referrer, see, Lincoln D. Stein, "Referer Refresher, " *New Architect*, September 1998, <http://www.webtechniques.com/archives/1998/09/webm/> (14 November 2006).

[186] Webwasher Classic is now owned by Cyberguard and is still free, though the company does request a donation.

0—do not send referrer for anything

Open the menu by right-clicking and selecting *Modify*. Then change the numerical value from the default 2 to 0.

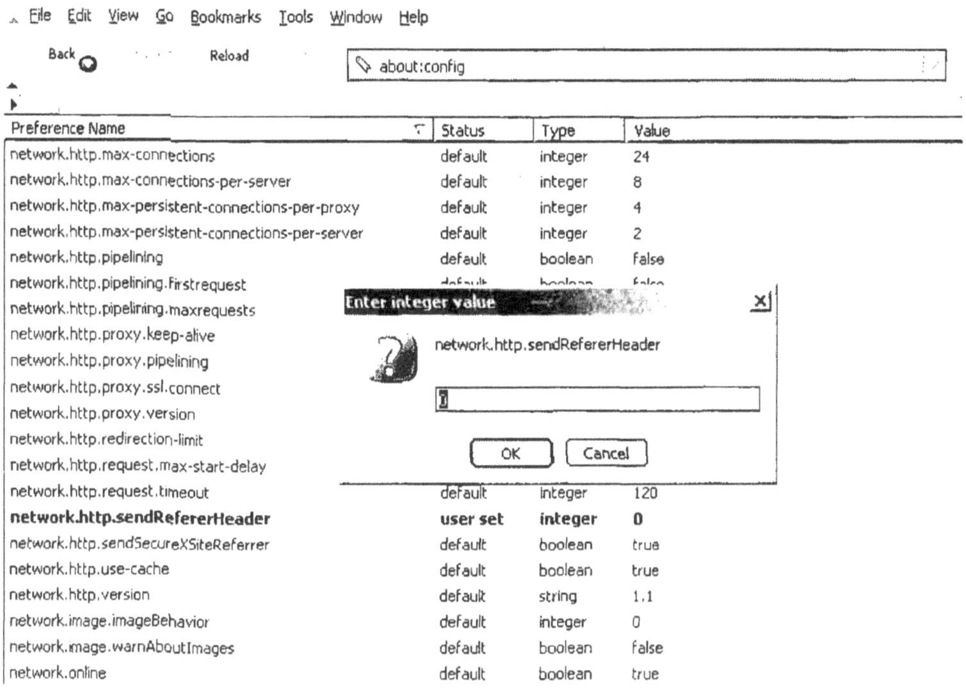
File Edit View Go Bookmarks Tools Window Help
Back
Reload
about:config
Preference Name
Status
Type
Value
network.http.max-connections default integer 24
network.http.max-connections-per-server default integer 8
network.http.max-persistent-connections-per-proxy default integer 4
network.http.max-persistent-connections-per-server default integer 2
network.http.pipelining default boolean false
network.http.pipelining.firstrequest
network.http.pipelining.maxrequests
network.http.proxy.keep-alive
network.http.proxy.pipelining
network.http.proxy.ssl.connect
network.http.proxy.version
network.http.redirection-limit
network.http.request.max-start-delay
network.http.request.timeout default integer 120
network.http.sendRefererHeader user set integer 0
network.http.sendSecureXSiteReferrer default boolean true
network.http.use-cache default boolean true
network.http.version default string 1.1
network.image.imageBehavior default integer 0
network.image.warnAboutImages default boolean false
network.online default boolean true
Enter integer value
network.http.sendRefererHeader
OK
Cancel

Clear Your Cache

Things happen on your computer in the background as you browse the Internet, some of which can affect your privacy and security. **Caching** is the process whereby your browser tries to make your journey on the information highway faster by saving copies of webpages as you visit them. Then, if you decide to go back to a recently visited webpage, instead of having to reload the page from the Internet, your browser simply serves up the stored or "cached" copy from your computer. This is fine until you realize that the cache is a record of your web browsing and, potentially, certain information you may have entered at a website, including passwords or credit card numbers. The safest thing to do is to clear the cache each time you end an Internet session.

Internet Explorer makes this very easy to do. In fact, ***you can tell the browser to clear the cache every time you close it***. To clear the cache manually in Internet Explorer:

Tools | Internet Options | Temporary Internet Files | Delete Files

To have Internet Explorer automatically clear the cache each time you close the browser:

Tools | Internet Options | Advanced | Security | Empty Temporary Internet Folder When Browser is Closed

---

Delete Your History Files

Browsers keep detailed lists of everywhere you have been on the Internet for a variable length of time, depending on the settings you have chosen. These are known as history lists and many people do not like to retain them because they give anyone with access to your computer a clear picture of your browsing habits. Also, there have been a number of malicious exploits that have gained access to users' history lists, which is at the very least an invasion of privacy. You have several options for handling history files. The simplest is not to keep a history list by ***setting the number of days to keep a history list to zero***.

To manage your history file in Internet Explorer:

Tools | Internet Options | General | History | Clear History

Tools | Internet Options | General | History | Days to keep pages in history = 0

To manage your history file in Firefox:

Tools | Options | Privacy | Clear Browsing History Now

Tools | Options | Privacy | Remember Visited Pages for the last 0 days.

Set Up Different Browsers for Different Purposes

There are many ways to handle privacy and security concerns on the Internet and no one way is right for everyone. Some people have one computer (usually an old one that is not connected to any others) they use strictly for Internet browsing and shopping, while all personal and financial information is stored on other computers perhaps not even connected to the Internet at all. Another less expensive way to minimize privacy and security vulnerabilities is to set up one browser (say, Internet Explorer) to run at the highest security settings and use it for all Internet browsing. Then the other browser (say, Firefox) could be set to accept cookies, run scripts, etc., so that it could be used for such things as shopping and financial transactions. The point is to think about your personal privacy and security because the risks are real and growing.

Check Your Browser's Security

After you have made all the recommended changes to enhance your browser's privacy and security, it is a good idea to check your browser(s) for vulnerabilities. There is a free browser checkup run by a reputable company, the Belgian security firm ScanIT, which tests for system vulnerability against a range of 22 simulated attacks. The test works with any browser, including Internet Explorer, Mozilla-based browsers, and Opera. It also appears the test is not operating system dependent because it runs on Linux. Highly recommended.

ScanIt's Browser Security Check http://www.scanit.be/bcheck

Email Concerns

Email is without a doubt the biggest source of security vulnerabilities on the Internet. All the qualities that make email so attractive to users—its speed, ease of use, inexpensiveness, and almost universal presence—also make it the perfect medium for spreading malicious software (malware). All the major virus, worm, and Trojan horse attacks have employed email to infiltrate networks worldwide. Therefore, email security is of the utmost importance to every user.

I recommend these two excellent webpages on email security that deal with all of the issues addressed in section and more.

Security Focus: "Securing Privacy: E-mail Issues" http://www.securityfocus.com/infocus/1579

A Quick Guide to Email Security http://www.zzee.com/enh/email_security.html

Move Outlook and Outlook Express to the Restricted Zone[187]

One of the conveniences and one of the weaknesses of Outlook and Outlook Express are their intimate relationship with the Internet Explorer browser. If you use either Microsoft product for email, it is critical that you make sure they are moved to the **Restricted sites zone** of the Internet from their default location in the Internet Zone. Why? Because malware is often spread via email, so you need to be sure your security settings for your email reader are set very high.

By default, the Restricted sites zone is assigned the High security level. If you assign a site to the Restricted sites zone, it will be allowed to perform only minimal, very safe operations. However, I recommend you do not rely upon the zone slider being set to High; instead choose the **Custom** option for manual settings. It is not hard to do. Make sure your Restricted Zone settings are set to disable all Java, JavaScript, and ActiveX controls because these are the most frequent sources of security problems in email.

These are the generally accepted settings for the Restricted sites zone.

[187] Windows XP Service Pack 2 includes security upgrades to Outlook 2003 that are not covered here. Please see "Microsoft Outlook 2003 Security Tips" for more information. <http://security.fnal.gov/handouts/Outlook_2003_Handout.pdf > [PDF] (14 November 2006).

In Internet Explorer 6:

Tools | Internet Options | Security
Select: Restricted sites zone

- ActiveX Controls and plugins
  - Download signed ActiveX controls **[Disable]**
  - Download unsigned ActiveX controls **[Disable]**
  - Initialize and script ActiveX controls not marked as safe **[Disable]**
  - Run ActiveX controls and plug-ins **[Disable]**
  - Script ActiveX controls marked safe for scripting **[Disable]**
- Downloads
  - File Download **[Disable]**
  - Font Download **[Disable]**
- Microsoft VM
  - Java permissions **[Disable Java]**
- Miscellaneous
  - Access data sources across domains **[Disable]**
  - Allow META REFRESH **[Disable]**
  - Display mixed content **[Prompt]**
  - Don't prompt for client certificate selection... **[Disable]**
  - Drag and drop or copy and paste files **[Prompt or Disable]**
  - Installation of desktop items **[Disable]**
  - Launching programs and files in an IFRAME **[Disable]**
  - Navigate sub-frames across different domains **[Disable]**
  - Software channel permissions **[High Safety]**
  - Submit nonencrypted form data **[Prompt]**
  - Userdata persistence **[Disable]**
- Scripting
  - Active scripting **[Disable]**
  - Allow paste operations via script **[Disable]**
  - Scripting of Java applets **[Disable]**
- User Authentication: Prompt for user name and password

Once you have finished selecting the Restricted site zone settings, you are not finished yet. You must add your email reader (Outlook or Outlook Express) to the Restricted Zone.

Open Outlook Express or Outlook
Select: Tools | Options | Security
Select: Restricted Zone

For more details see:

About.com Email Help Center http://antivirus.about.com/library/bloutlook.htm

---

Don't Open Email Attachments

I can't say *never* open any email attachments because there are times when you trust the user and are expecting a document via email. However, do not open email or attachments from unknown or even questionable sources. If you don't know the person who is sending you an email, do not open the email or any file attached to it. Even if you do know the sender, be very careful about opening the email and attachment (people sometimes unwittingly spread malware). If the mail appears to be from someone you know, still be careful, especially if it has a suspicious subject line (e.g. "I love you" or "look at this!") or if it seems odd (e.g., it was sent in the middle of the night). It may not actually be from the person you know but may be using a "spoofed" or fake email address using your friend's identity. Also be especially wary if you receive multiple copies of the same message from any source because they are likely to be spam.

The best thing to do with suspicious email is to delete the entire message, including any attachment, and empty your email reader's trash. If you really must open a file from an unknown source, save it first and virus scan the file. However, you need to know there is still a risk because no virus scanning software can detect every piece of malware.

"Finally, remember that even friends and family may accidentally send you a virus or the e-mail may have been sent from their machines without their knowledge. Such was the case with the "I Love You" or "Love Bug" virus that spread to millions of people in 2001. When in doubt, delete!"[188]

Stop "Email Wiretapping" by Disabling JavaScript in Your Email

A malicious user could insert hidden JavaScript code into an HTML email message and send it to another person's email reader that has both JavaScript and HTML enabled. Then if that unsuspecting person forwards the email message to others, the JavaScript, using a web bug or hidden form, surreptitiously sends a copy of the forwarded email back to the original sender, who can retrieve and read the forwarded message. This is a great method for spammers to harvest email addresses. Turning off JavaScript in email offers some measure of protection for

---

188 Awareness and Outreach Task Force, "Report to the National Cyber Security Task Force," 18 March 2004, < http://www.educause.edu/ir/library/pdf/SEC0403.pdf > [PDF], Top 10 Cyber Security Tips, p. 25, (1 February 2007).

you, but if you reply to or forward an email to a person with a JavaScript-enabled email program, that person is vulnerable.

JavaScript is disabled in **Microsoft Outlook** and **Outlook Express** by adding them to the Restricted Zone where the settings disable all Java, JavaScript, and ActiveX controls (do not just set the zone to "high"; you must choose the *Custom* option and do this manually). For detailed information on Email Wiretapping, see:

About's Email Wiretapping Article
http://antivirus.about.com/library/weekly/aa020501a.htm?once=true&

See What's Arriving in Your Email

Malicious hackers can easily disguise malicious file types sent via email using what are called "double extension" files. For example, you may think you've received a harmless graphic file in *.gif* or *.jpg* format when in fact the file is something else altogether—such as an executable or a visual basic script—and opening it can infect your computer. The reason you are being fooled is that Windows' default setting hides certain file types that are "known" to the operating system, so what you see is *prettypicture.gif* when the full file name is *prettypicture.gif.vbs*.

To see all file extensions, you need to make a simple change to Windows itself (*not* to your browser or email tool). To enable **show all files**:

Windows 2000

- Open **My Computer**.
- Select the **Tools** menu and click **Folder Options**.
- Select the **View** Tab.
- Under the **Hidden files and folders** heading select **Show hidden files and folders**.
- Uncheck the **Hide protected operating system files (recommended)** option.
- Click **Yes** to confirm.
- Click **OK**.

Windows XP

- Click **Start**.
- Open **My Computer**.
- Select the **Tools** menu and click **Folder Options**.
- Select the **View** Tab.

- Under the **Hidden files and folders** heading select **Show hidden files and folders**.
- Uncheck the **Hide protected operating system files (recommended)** option.
- Click **Yes** to confirm.
- Click **OK**.

HTML & Email: Two Things That Do Not Belong Together

One of the worst practices to gain widespread acceptance on the Internet is HTML email. Surprised? It sounds like a nice idea (I mean, don't HTML messages look a lot nicer than text?), but in reality it is the source of lots of problems. HTML was created for web browsers and webpages, and that's where it belongs. But somewhere along the way, someone got the "bright" idea that HTML would make pretty email messages, complete with graphics and scripts and all those things that go into webpages. Unfortunately, all the qualities that make HTML appealing and flexible also make it vulnerable and have created huge problems with email. Here are some (not all) of the major problems with HTML in email:

1. HTML often contains executable code, such as JavaScript, Java, or ActiveX, which can automatically do a number of things on your computer *without your doing anything* to activate it and without your knowledge or consent.

2. Email programs (such as Outlook and Netscape Messenger) often have bugs that have been exploited by email worms and viruses that include automatic execution of attachments, buffer overflows, etc. While the bugs have been systematically patched by their manufacturers, the fact is that many people do not install patches and new exploits come along all the time. HTML facilitates the spread of malicious software.

3. Macromedia Flash is a browser plug-in that "interprets" code, so it could be used to execute malicious code or initiate buffer overflows from a fancy HTML email message.

4. Web bugs (invisible clear images imbedded in HTML email) are used routinely both by advertisers and spammers to track who reads (that means OPENS) their email messages. When you VIEW the message, the web bug (image) is downloaded and a unique ID is sent back to the spammer/advertiser. Now he knows your email address is alive and well and ready to receive more spam! Some email readers prohibit the display of remote graphics in HTML email by default; these include but are not limited to Google's Gmail, Yahoo Mail, Mozilla Thunderbird, and Opera. Outlook 2003 with Windows XP SP2 adds anti-phishing functionality, displaying all junk email in plain text format and removing the ability to click on URLs in the junk email folder and on other suspicious messages.

A spreading threat involves something called image spam. **Image spam** uses HTML code to display the email message, so spam filters cannot detect the spam because there is no text. Some estimates place the amount of image spam at "15-25% of all spam sent in the first half of 2006."[189] Clever image spam emails appear to be plain text messages to the casual observer, but in fact the entire message is nothing but an image. While image spam at present appears to be mainly a nuisance, it has the potential to become a threat as malicious hackers figure out ways to exploit it.

What can you do to protect yourself and others. First, *never send HTML formatted email*. Period. It's easy to **select the format for your outgoing email:** [190]

Outlook Express 6:

Tools | Options | Send | Mail Sending Format
select Plain Text

Outlook (most versions):

Tools | Options | Mail Format | Message Format | Choose a format for outgoing mail | Send in this message format:
select Plain Text from pull-down menu

The next part is more complicated. Of the current versions of Microsoft's Outlook and Outlook Express, only Outlook 2003 and Outlook Express 6 give users the ability to disable HTML in messages *received*. This is a huge problem and one that a lot of users have solved either by switching to another email client, such as Eudora, or by installing a program, such as noHTML. As of now, if you use Outlook[191] or Netscape Messenger for email, you run the risk of falling victim to all the many perils of HTML email.

Outlook 2003: to disable HTML in messages you *receive*

Tools | Options | Preferences | Email Options

---

[189] Mike Chapple, "Battling Image Spam," Search Security.com, 15 August 2006, <http://searchsecurity.techtarget.com/tip/0,289483,sid14_gci1210679,00.html> (1 February 2007).

[190] If you run a different email package, please check its help files for information on how to disable HTML.

[191] There is a way to disable HTML in Outlook 2000 using Tools | Macros | Visual Basic Editor, but it's too complicated for my taste. However, for the less faint of heart, here's where you can get the instructions (*no guarantees on this one*...I've not tried it): "How to Disable HTML Email in Microsoft Outlook," Ostrosoft, <http://www.ostrosoft.com/vb/disable_html_email.asp> (14 November 2006).

check *Read all messages in plain text*

Outlook Express 6: to disable HTML in messages you *receive*

Tools | Options | Read

check *Read all messages in plain text*

Changing this option just changes how messages are displayed, not how they are stored. When a message appears containing HTML or RTF (Rich Text Format), an option will appear in the message allowing you to view the selected e-mail in HTML or RTF format. Also, even in plain-text mode, some URLs still show up as hyperlinks.

Disable the Preview Pane

There are some other things you can do to make your email more secure. One is to ***disable the Preview Pane***, which is a feature in Outlook, Outlook Express, and some other email readers that shows the contents of an email message before the user opens it. The Preview Pane actually opens the email, even if you don't intend to do so. Some of the scripts that malicious users send via email can activate automatically simply when the email message appears in the Preview Pane. Also, web bugs are activated when the message is previewed before opening it, so disabling the Preview Pane is a fairly good way to stop web bugs from acting as the little "homing beacons" they are. Of course, all this presupposes that you will NOT OPEN the message but will delete it unopened and then empty your "deleted messages" folder.

To disable the Preview Pane in:

Outlook Express:

View | Layout deselect Preview Pane (do NOT use the preview pane)

Outlook:

In the *Inbox*: View | deselect Preview Pane (it is a toggle between seeing and not seeing it)

Don't Become "Phish" Food

While "phishing" (or carding) is a scam that has been around for years, it has become an enormous problem recently. Phishing is the use of "spoofed" emails and fraudulent websites that appear to be authentic emails from and links to legitimate company websites designed to lure an unsuspecting user to a fake website where the user will be prompted to enter personal information. Phishing emails have tricked many hapless customers of reputable companies into providing personal data, such as user names, passwords, account numbers, social security numbers, etc. How does phishing succeed? These particular kinds of scam emails, which are criminal in nature, are very professional-looking and use the real companies' logos and, so it seems, web addresses to lure a user to a fraudulent website. Phishing attacks sometimes employ very convincing image spam to trick users. Even the link looks valid to the average user. Phishers are reportedly able to convince up to five percent of recipients to respond to them, and it doesn't take many successful phishing scams to pay big dividends for the criminals behind them.

One celebrated case of phishing involved Citibank. Here's how it worked. Let's say you are a Citibank customer and you get an email "from Citibank" (the email has the Citibank logo and looks as though it came from Citibank). One of the numerous fake Citibank emails says, "We encountered a billing error when attempting to renew your Citibank online banking services." The email then goes on to detail member information from the "Citibank" database and says, "Please take a moment to update your credit card information by clicking here and submitting your information." The email ends with the warning that if you do not take this action, "your service will be terminated!"

On the face of it, the "Citibank" link in the email may look completely legitimate:

https://www.citibank.com/signin/citifi/scripts/user_setup.jsp

However, such links are fake. If the user "mouses over" (moves the mouse over) the link, he will see this:

http://www.citibank.com:ac=8tcBs829uY3T23ue76Hg@FaStWay2StUIpqwrCh7L09j

Now this link might be legitimate, too, except that everything between the *http://* and the at sign (@) in this url is irrelevant, so the real url in this link is what follows the at sign. Not exactly a Citibank website!

As consumers became more cautious and aware of these scams, new "bait" appeared in phishing scams that can fool even savvy Internet users. This attack uses a custom JavaScript to replace the Address or Location bar at the top of a web browser with a fake that is so good that it's almost undetectable. Here's how the attack works.

- Customer receives a forged but very legitimate-looking email from a bank or business with whom he may have a relationship (account, credit card, etc.).
- Email says customer must verify his email address and includes a link inside the email to a website.
- User clicks on the link in the email and the browser opens what appears to be the company's webpage but is in fact a fake website.
- The fraudulent site automatically detects the user's browser (the attack is not browser dependent) and runs custom JavaScript code that removes the real address bar and replaces it with a fake address bar at the top of the browser window. The copy is exact. It has the Address field, it displays a url that appears to be a secure link to the real company website (e.g. "https://"), and it has the *Go* button on the right-hand side. Unlike earlier, less sophisticated phishing attacks that create static (fake) Address bar images, this is a live piece of JavaScript code.
- Even if the user right-clicks on the webpage to View Source, the real source code is not shown; in order to see the real source code, the user must use the View | (Page) Source pulldown menu at the top of the browser to see the real HTML source code.
- The active JavaScript address bar could permit what is known as a "man in the middle" attack, i.e., every subsequent website the user visits after this one could send any information the user enters (passwords, credit card numbers, etc.) to the "phisherman" until the browser is closed.

In short, there are very few clues as to the fraudulent nature of this particularly dastardly phishing scam, but they are important ones:

- Even though the fake page shows the "https://" in the address bar, there is no corresponding Secure Sockets Layer (SSL) padlock at the bottom of the browser.
- If the user types a new url into the Address bar, the browser will continue to display the same fake "Welcome" message.
- The real url appears very briefly while the user is redirected to the fake site.

Take a look at this actual example of a fraudulent webpage used in a real phishing scam. You can see how it would be hard for the user to detect this is a fake.

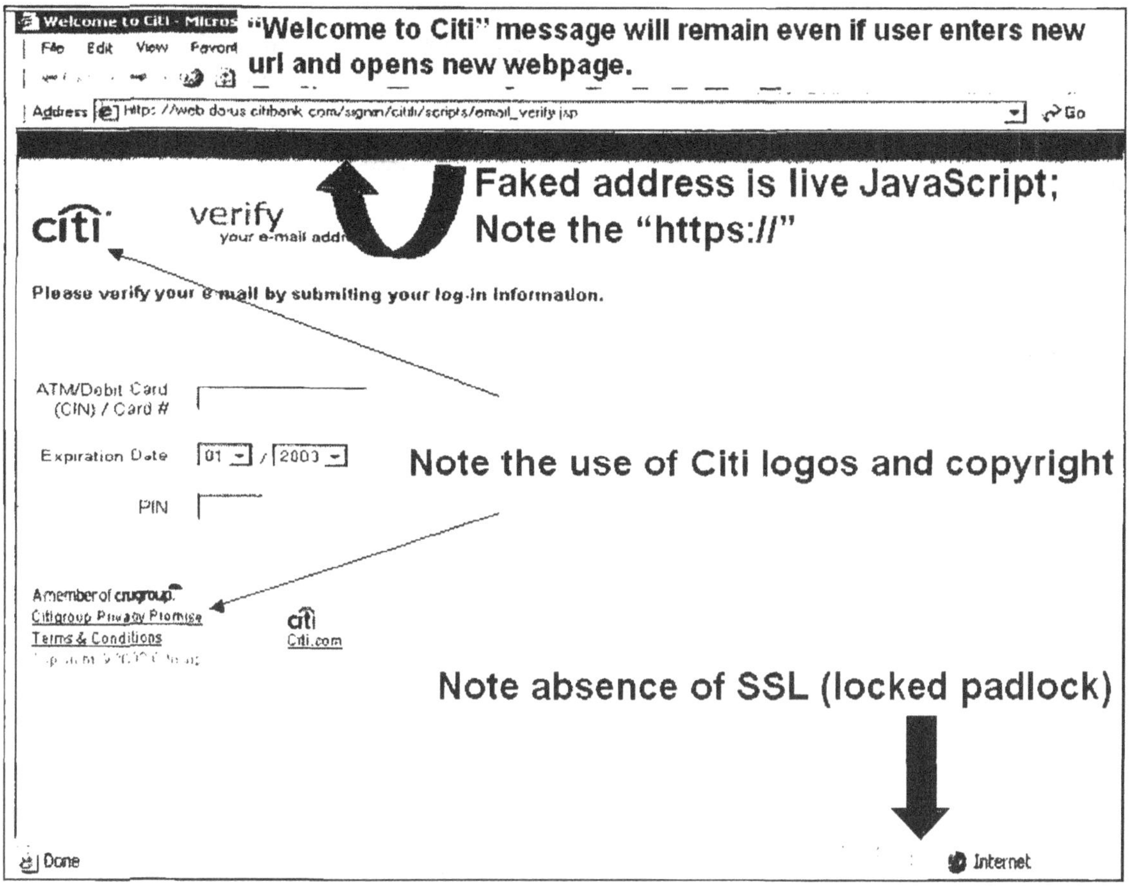

"Welcome to Citi" message will remain even if user enters new url and opens new webpage.
Faked address is live JavaScript; Note the "https://"
Please verify your e-mail by submiting your log-in information.
ATM/Debit Card (CIN) / Card #
Expiration Date
PIN
Note the use of Citi logos and copyright
Citigroup Privacy Promise
Terms & Conditions
Citi.com
Note absence of SSL (locked padlock)
Done
Internet

There is a worrisome refinement of the traditional phishing attack that gained a lot of attention beginning in 2005. **Spear phishing** is exactly what it sounds like: precisely targeted phishing attacks that try to lure users to provide personal data by cleverly conceived social engineering strategies. Instead of the blanket approach of sending thousands (millions) of emails blindly, spear phishing carefully selects its audience and targets these users with very legitimate-sounding emails. For example, one spear phishing attack targeted students and faculty at the University of Kentucky. Spear phishing emails typically appear to be coming from a trusted source: your company's HR or IT department or your own little credit union. Also, a spear phishing attack may try to sound as if your security is at stake, e.g., you have been locked out of your account because of unsuccessful attempts to break into it, and in order to unlock your account you will need to reenter your personal information.

Fake spear phishing emails have even been used to educate people about the dangers of spear phishing. "In June 2004, more than 500 cadets at West Point received an email from Col. Robert Melville notifying them of a problem with their grade report and ordering them to click on a link to verify that the grades were correct. More than 80% of the students dutifully followed the instructions. But there is no Col. Robert Melville at West Point. Aaron Ferguson, a computer-security expert with the National Security Agency who teaches at West Point crafted the email. The gullible cadets received a 'gotcha' email, alerting them they could easily have downloaded spyware, 'Trojans' or other malicious programs and suggesting they be more careful in the future. Mr. Ferguson, who runs similar exercises each semester, said many cadets have been victimized by real online frauds."[192]

The problem with this approach is that it can undermine company trust. Who would ever trust another company email after being caught in a fake spear phishing attack? The fact remains, users must be extraordinarily vigilant and never provide personal information that is solicited by anyone without taking steps to verify the authenticity of that request. Sometimes a phone call is the best way to ensure that an email request from the HR person for your personal data really came from that department and not from a spear phisher.

How do you protect yourself from more and more sophisticated phishing scams?

- <u>never, ever under any circumstances click on a link in an unsolicited email</u>, especially one that asks you to click on the link to confirm or update personal or financial information.
- instead, type the address directly into the browser yourself and then check to see if that company has any security alerts about phishing scams.
- always make sure that the SSL is enabled before entering any personal or financial data; the browser will show a locked padlock: or
- learn how to view and interpret the message source code of an email message; when in doubt about the true source, assume the worst.
- stay on top of the news about scams; frequent websites such as the one run by the Anti-Phishing Working Group.
- when in doubt, contact the source by telephone to make sure the request is legitimate.

For anyone concerned about phishing attacks (and that should be all of us), there are several free online tools to help you tell if a url in an email or on a webpage is legitimate (that is, is it what it says it is, or is it something entirely different?). These "url decrypters" are designed to reveal the real addresses of obfuscated urls. Nothing could be simpler to use: just copy the obfuscated url from an email or from a

---

[192] David Bank, "'Spear Phishing' Tests Educate People About Online Scams," *The Wall Street Journal Online*, 17 August 2005, <http://online.wsj.com/public/article/SB112424042313615131-z_8jLB2WkfcVtgdAWf6LRh733sq_20060817.html?mod=blogs> (14 November 2006).

webpage and paste it into the query box, hit return, and the hidden address will be revealed.

Sites to De-obfuscate URLs

URL Decrypter	http://www.cyber-junkie.com/tools/urldecrypter.shtml
Un-Obfuscating URLs	http://www.wilsonmar.com/1tcpaddr.htm

You must be very careful to avoid becoming "phish food" because the scams are increasingly sophisticated and hard to detect. Banks, lending institutions, insurance companies, and ***legitimate account holders*** of any kind (eBay, PayPal, Amazon, etc.) ***never send requests for account information via email***. If you are in doubt about any request for information via email, ***do not click on the link in the email***. Instead, open your browser, type the url of the company's home page into the browser's address bar and go to the site that way. Then you can log into your account and see if there is really a need for you to do anything. You can also use an online tool to de-obfuscate urls to determine the real address of any url. Phishing is a form of the con game discussed later.

Another potentially dangerous type of phishing scam involves fraudulent e-commerce websites that lure searchers to their sites, which present malware disguised as legitimate-looking images of a product supposedly for sale. The "image" is in fact a self-extracting zip (compressed) file that installs a Trojan horse on the user's computer, usually in order to steal personal and financial data. Be wary of any site that asks you to "click here to download images." This is an especially difficult scam to detect because many legitimate sites offer users the option to download image files (though usually not zipped files). The phishing sites purportedly are offering very inexpensive products, so if an offer looks too good to be true or if it looks in any way "phishy," it's best to avoid it.

A new type of attack gained prominence in 2006: "voice phishing" or **vishing**. Vishing is a type of phishing scam that uses VoIP (voice over Internet Protocol) phone numbers to trick users into providing their private information. Unlike traditional telephone numbers, it is relatively easy to get a VoIP number anonymously. "That makes it easier for scammers to carry out these vishing scams. In some ways, vishing may be even more dangerous than phishing scams, because consumers are used to entering private information into automated phone systems."[193]

Vishing indicates that as consumers wise up to scams such as phishing, bad people come up with creative new ways to separate you from your money (and sometimes your identity). One reason it's so easy to use a vishing scam is that some

---

193 Issue #189, Scambusters.org, 26 July 2006, <http://www.scambusters.org/vishing.html> (12 December 2006).

companies, notably Skype, allow customers to pick both their area code and prefix, which means a call can appear to be coming from a very specific entity, such as your bank. The simple solution for customers is not to respond either to automatic emails (aka spam or phishing scams) or to automatic phone messages asking you to call a number. If you are in doubt about the legitimacy of any email or phone call, call your bank or credit card company at their main number and ask if there is a problem with your account. ***Good rule of thumb: Initiate, do not respond.***

How Not to Get Hooked by a "Phishing" Scam
http://www.ftc.gov/bcp/edu/pubs/consumer/alerts/alt127.htm

The Anti-Phishing Working Group http://www.antiphishing.org/

Phishtank (known and suspected phishing sites) http://www.phishtank.com/

PayPal's Protect Yourself from Fraudulent Emails
https://www.paypal.com/cgi-bin/webscr?cmd=xpt/general/SecuritySpoof-outside

---

Protect Yourself from "Pharming" Attacks

Not content with trying to lure victims to fraudulent websites using phony links in email messages, malicious users have devised an even more insidious trick to redirect users to fake websites. These scams have been dubbed **pharming**,[194] and the potential for the trouble they could cause is just becoming apparent. Basically, a pharming attack involves redirecting web users from a legitimate site by any number of dirty tricks. Usually the attacker exploits a browser vulnerability, such as what has been happening since late 2004 when the security company Secunia began identifying vulnerabilities in Internet Explorer, Opera, all the Mozilla-based browsers, and a number of other browsers that permit an attacker to inject content into a legitimate website, for example, by inserting the attacker's content into a popup at someone else's website. All these attacks are described as "spoofing" attacks, i.e., fooling users into believing they are at a legitimate website when in fact they are at a fake or spoofed site instead. Secunia provides details of these many vulnerabilities and demonstrations of whether your browser is vulnerable at its website.

Secunia's Advisories: Dialog Origin Vulnerability Test
http://secunia.com/multiple_browsers_dialog_origin_vulnerability_test/

It gets worse. In January 2005 a pharming attack successfully diverted all email and web traffic from the New York ISP Panix. According to a statement from Panix, "The

[194] This term may create confusion because there is already a use of the neologism pharming, i.e., "The production of pharmaceuticals from genetically altered plants or animals."

ownership of panix.com was moved to a company in Australia, the actual DNS records were moved to a company in the United Kingdom, and Panix.com's mail has been redirected to yet another company in Canada." How was this accomplished? According to Ed Ravin, systems administrator at Panix, "Our registrar, Dotster, told us that according to their system, the domain had not been transferred, even though the global registry was pointing at Melbourne IT. Something went wrong with the Internet registry system at the highest levels." This particular pharming attack involved a **domain hijack**, but it's not the latest type of possible pharming attack.

The newest browser vulnerability could enable even more sinister and harder to detect pharming attacks primarily because it is not a true vulnerability but rather simply an unintended side effect of a new browser feature designed to implement **International Domain Names (IDN)**.

This pharming attack does not involve a domain hijack. Rather, it is a spoofing attack that works by displaying fake addresses (urls) in the browser's address bar, the status bar, the hyperlinks, and even in the SSL Certificate. It is almost impossible to detect with the naked eye. The problem stems from the implementation of IDN, the standard that allows users to register domain names in different languages and different encodings. The flaw was first reported at ShmooCon, a hacking/computer security convention held in Washington, D.C., in January 2005. The Shmoo Group issued an advisory along with a demonstration of the attack using the domain for PayPal, in which they substituted an alternate Unicode character for the first "a." The address looks like the real PayPal url—http://www.paypal.com—but with a slightly smaller "a." With the implementation of IDN, there are now a huge number of ways to display domain names, many of which look very much like the original Latin character set.

The vulnerability affects IE7 (but not IE6 because IDN was not implemented before version 7). Firefox 1.0.6, Firefox 1.5 beta, Netscape 8.0.3.3, and Mozilla 1.7.11. The Firefox 1.5 release of November 2005 corrected the problem, so be sure you are using version 1.5 *or later* if you use Firefox. Previous versions of these browsers may also be affected. Mozilla released a self-installing patch that disables the International Domain Name (IDN) processing that makes the vulnerability possible.

Mozilla 1.7.12 http://www.mozilla.org/products/mozilla1.x/

Firefox http://www.mozilla.com/firefox/

"The State of Homograph Attacks," by Eric Johanson, The Shmoo Group, 31 Jan 2005 http://www.shmoo.com/idn/homograph.txt

Secunia's Multiple Browsers IDN Spoofing Test http://secunia.com/multiple_browsers_idn_spoofing_test/

If you use a Mozilla-based browser or simply don't want to install the patch, there is a very simple workaround that negates the vulnerability:

- in the browser address bar, enter *about:config*
- scroll down to or search for the parameter *network.enableIDN*
- right-click on that parameter and select *Modify*
- change the value from *true* to *false*

Here are other suggestions for preventing this and other pharming/phishing attacks from being successful:

- never follow hyperlinks from HTML-formatted emails (in fact, don't accept HTML email in the first place); this is especially important in the case of emails from banks; and from companies such as Amazon, eBay, or PayPal; credit card companies, etc.
- do not click on hyperlinks from a website if you have any doubt about the site's integrity. You can always type the url into the address bar to ensure you go to the real website.

Go Offline to Read Your Email

You can go offline to read your email once you have downloaded it. You can tell Outlook Express to "Work Offline." Working offline in Outlook is more complicated, so I cannot recommend it. Also, if you are using a firewall program like Zone Alarm, it's easy to go offline. Just "lock" your Internet connection or block your email client's access to the Internet while you go through the junk email. There is no way the evil little web bugs can phone home to the mothership while your Internet connection is blocked or inactive. Then you can safely delete the messages (and *empty your deleted items folder*) before reconnecting. Of course, you will not be able to see any images or read any HTML emails that require access to a website, but you probably don't want to read these anyway because the ones that require access are likely spam or worse.

<u>To work offline in Outlook Express:</u>

in the Inbox: File | Work Offline

Try to Avoid Being Joe Jobbed

This may be hard to avoid. It's one of the oldest tricks around. Joe jobbing is an email spoof that sends out huge volumes of spam that appear to be from someone other than the actual sender. It got its name from its first known victim, Joe Doll, who offered free webpages to anyone who agreed to his rules of netiquette. In 1996 one of his free page users started sending newsgroup and email spam in violation of Joe's rules. When Joe terminated the user's free account, the spammer retaliated with forged messages that appeared to be from Joe Doll. The angry recipients of the spam that appeared to be from Joe in turn retaliated by attacking Doll's website, shutting it down for 10 days.

Because Joe jobbing is so easy to accomplish—sometimes nothing more than changing the *Reply-to* address is required—it's very hard to prevent. The best way to avoid being Joe jobbed is to follow the general rules for spam avoidance (and we all know how well these work). However, Joe jobbing tends to involve retaliation and is personal whereas spam is about as impersonal and universal as anything can get, so most people will be victims of the latter but not the former. Still, these are wise precautions for avoiding both the Joe job and spam.

- Don't unsubscribe from anything. Unsubscribing lets spammers know they have a valid email address.
- Don't open web-based emails as it also alerts spammers to a valid address.
- Don't open spam; simply opening spam may activate a script or web bug that alerts a spammer to a valid email address.
- Don't send and receive HTML email; it may contain code that alerts a spammer to a valid email address.
- Do not sign Guestbooks or, if you must, use a disposable email address, such as a Hotmail or Yahoo email account.
- Do not post your email address on a website. Email spiders can easily find and harvest your email address for spammers.
- Be very careful about signing up for anything free that requires your email address, especially newsletters.
- If you have ever posted to a newsgroup using your real email address, it's gone. Spammers have it. Get a new address.

For even more ways spammers gather email addresses and ways to avoid being harvested, see:

How Spammers Get Your Email Address
http://www.junk-mail.org.uk/articles/spam.html

First Spam, Now Spim

You thought spam was bad, but now there is a torrent of what has been dubbed "spim" or unwanted messages sent to instant messaging programs. According to a report from the technology market research company Radicati Group, spim tripled in 2004, growing to 1.2 billion spims sent, 70 percent of which are pornographic. While the number of spim messages is small compared to the estimated 35 billion spam messages in 2004, spim is growing at a rate of three times that of spam. Spim is also more intrusive than spam because spim messages pop up on a user's computer screen when he is logged into his IM program, making them very hard to ignore.

While many IM users employ "buddy lists" to limit whose messages they can receive, spimmers have developed clever ways to get around this restriction by illegally "borrowing" identities or by persuading users to add them to their buddy list by posing as someone they are not. Experience shows that people are much more likely to click on spim messages than to open and/or respond to spam, in part because spim is not as well known and in part because it appears to be from a friend.

Celeste Biever, "Spam Being Rapidly Outpaced by 'Spim'," *NewScientist.com*, 26 March 2004,<http://www.newscientist.com/news/news.jsp?id=ns99994822> (1 February 2007).

Microsoft and Windows Concerns

The computer security company Symantec reported in 2006 that "home users now comprise 86 percent of all targeted attacks against computers,"[195] in large part because most home users do not take even the most rudimentary steps to secure their own computers. There is no such thing as a "secure" computer that is connected to the Internet. If you **never** connect your computer to the Internet—and by that I mean not for one minute ever—and never install any new software on your computer, you do not need to worry about computer security. Otherwise, you need to be concerned. I agree with Eric Vaughan of Tweakhound's assessment of the current state of computing:

> "1. There is no such thing as a secure OS (operating system), or web browser. If you want true security (read something like this somewhere at some time); *disconnect your network card, turn off/unplug your computer, take out the hard drive and smash it to bits, take computer to a construction site and ask the bulldozer operator to run over it.* [emphasis added]
>
> 2. In the real world, Windows operating systems are less secure than the newest versions of Linux (distro) and Mac OS X. We'll leave the argument over why that is and the advantages of one OS over another to internet forums/discussion boards.
>
> 3. A fully patched Windows XP and to a lesser degree Windows 2000 are the only non-server Microsoft OS's that are even remotely secure. If you care about security you shouldn't be running any other Microsoft OS's. If you have machines on your home network that run anything less than a fully patched XP, 2k, Linux (distro), OS X then the security of any machine on your network is lessened."[196]

To make matters worse, most home users are running Windows XP Home Edition. "Windows XP Home has too many major security flaws (e.g., in XP Home every default account has superuser privileges and cannot belong to any domain) to enable it to achieve even a baseline level of security."[197] However, there are specific

---

[195] Jay Wrolstad, "Hackers Targeting Home Computer Users," Newsfactor.com, 25 September 2006, <http://news.yahoo.com/s/nf/20060925/tc_nf/46488> (article no longer available).

[196] Eric Vaughan, "Securing Windows XP," Version 2 BETA, Tweakhound.com, 30 September 2005, <http://www.tweakhound.com/xp/security/page_1.htm> (14 November 2006).

[197] "Checklist for Securing Windows XP Systems," Lawrence Berkeley National Laboratory, <http://www.lbl.gov/cyber/systems/wxp-security-checklist.html> (14 November 2006).

steps you can take to improve your home computer security. It is important to keep in mind that every computer, like every person, is *unique*, which means I cannot cover every possible configuration that might occur. However, there are numerous excellent websites that discuss how to enhance security on a home computer and/or network, and I will point you to those sites.

Some of the best sites for home computer and network security for Windows' user are the following:

Tweakhound's Securing Windows XP
http://www.tweakhound.com/xp/security/page_1.htm

Fred Langa's 5 Essential Steps To PC Security
http://www.informationweek.com/shared/printableArticle.jhtml?articleID=177100010

NIST's Guidance for Securing Windows XP Home Edition
http://csrc.nist.gov/itsec/guidance_WinXP_Home.html

CERT's Home Network Security http://www.cert.org/tech_tips/home_networks.html

Gary Kessler's Protecting Home Computers and Networks
http://www.garykessler.net/library/protecting_home_systems.html

University of Cambridge's Securing Windows XP Home Edition for Stand Alone Use
http://www-tus.csx.cam.ac.uk/pc_support/WinXP/collegehome.html

Lawrence Berkeley Lab's Checklist for Securing Windows XP **PRO**
http://www.lbl.gov/ITSD/Security/systems/wxp-security-checklist.html

Windows XP Security Checklist
http://labmice.techtarget.com/articles/winxpsecuritychecklist.htm

Tom-Cat.com's Secure Your Home Computer v.2.22
http://www.tom-cat.com/security.html

Download Operating System, Browser, & Other Software Updates Regularly

If you have a slow Internet connection, this is a painful process, but it is necessary. Many updates are in fact security patches in response to reported vulnerabilities. You should also be aware that the patches are not always explicitly described as fixing a security flaw. Updates are *not* the same thing as new version releases. New versions often (dare I say usually?) have new vulnerabilities, so the best advice is to wait until a new version has been around for a while before downloading it. Be sure to check Microsoft's Security page frequently for news, updates, and patches. Also, don't forget other software, such as Microsoft Office, which needs to be patched and updated separately.

If you are unlucky enough to be one of the many people who installed a Microsoft patch only to discover it caused problems with your computer, this article could come in handy.

Gregg Keizer, "How To Uninstall A Microsoft Patch," TechWeb News, 21 April 2006, http://www.techweb.com/wire/186500738 (31 October 2006).

Important: If you are using a router that offers additional ActiveX filtering, you will no longer be able to run Microsoft Updates with the filter enabled. You must disable (remove the check beside) any ActiveX filter on your router in order to update Microsoft products.

Microsoft Security Home Page http://www.microsoft.com/security/

Microsoft Internet Explorer Security Updates http://www.microsoft.com/windows/ie/downloads/default.asp

Microsoft Office Download Center http://office.microsoft.com/downloads/

Microsoft Windows Update Page http://update.microsoft.com/windowsupdate/

Turn Off File Sharing in Windows

You may or may not have file and print sharing enabled on your Windows computer. One of the changes in Windows XP Service Pack 2 (SP2) is that it includes the Windows Firewall, which is enabled by default in both the Home and Pro editions. And ***by default Windows Firewall blocks printer and file sharing***, which is the appropriate setting for most home users. Unless you need it, and you probably don't on your home computer, leave file and print sharing disabled.

However, if you turn off the Windows Firewall in order to use a better firewall, you may need to disable file and print sharing manually in order to thwart such cracking programs as "***ShareSniffer***,"[198] which is designed to find computers with file sharing enabled, access all the files on the hard drive, and perhaps modify or delete them. In any event, check your Windows' settings and make sure file and print sharing is disabled.

Windows XP. To disable file and print sharing in Windows XP:

1. Click the Start button in the lower left corner of the desktop.
2. Click Settings, then click Control Panel.
3. In the Control Panel, click Network Connections.
4. In the Network Connections window, right-click on the appropriate connection, then select Properties.
5. Uncheck the File and Printer Sharing for Microsoft Networks check box.
6. Click OK, then close the Control Panel window.

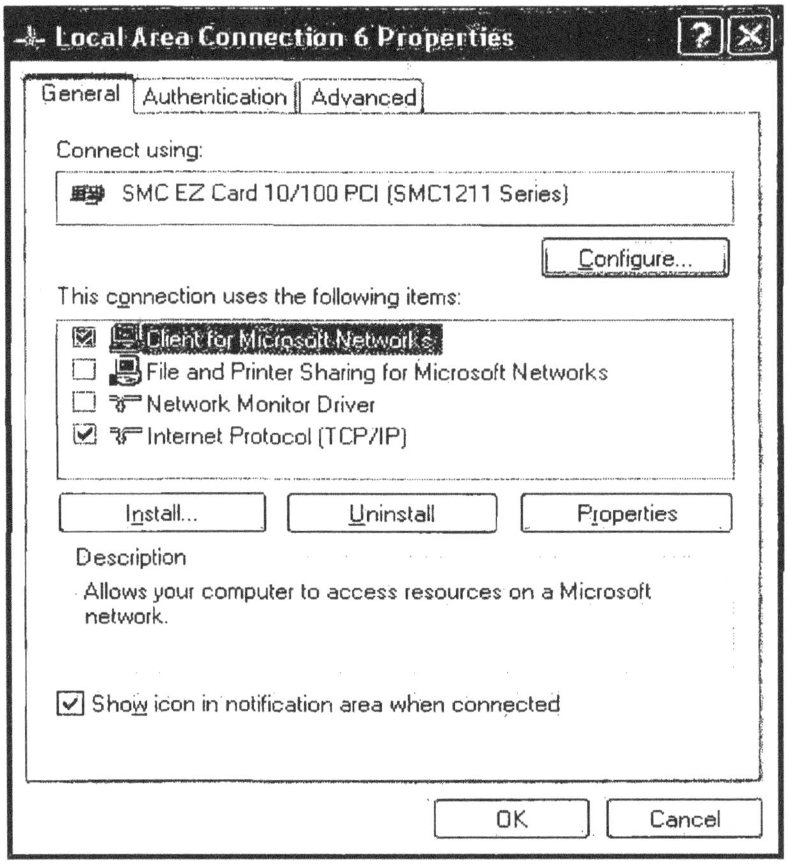
Local Area Connection 6 Properties
General
Authentication
Advanced
Connect using:
SMC EZ Card 10/100 PCI (SMC1211 Series)
Configure...
This connection uses the following items:
Client for Microsoft Networks
File and Printer Sharing for Microsoft Networks
Network Monitor Driver
Internet Protocol (TCP/IP)
Install...
Uninstall
Properties
Description
Allows your computer to access resources on a Microsoft network.
Show icon in notification area when connected
OK
Cancel

[198] For more information on Sharesniffer, see Robyn Weisman, "New Hackerware Makes Everyone a Hacker," Newsfactor Network, 6 March 2001, <http://www.newsfactor.com/perl/story/7906.html> (14 November 2006).

Windows 2000. To disable file and print sharing in Windows 2000:

1. Click the Start button in the lower left corner of the desktop.
2. Click Settings, then click Control Panel.
3. In the Control Panel, click Network and Dial-up Connections.
4. In the Network and Dial-up Connections window, right-click on the appropriate connection, then select Properties.
5. In the Connection Properties window, click the Networking tab.
6. Uncheck the File and Printer Sharing for Microsoft Networks check box.
7. Click OK, then close the Control Panel window.

You may want to go one step further and unbind file and print sharing from TCP/IP, which (yet again) is the default setting for Windows. What does this mean? Simply, "the same capability that allows peer-to-peer networking and file sharing on your home/office LAN is available to anyone on the Internet!"[199] Instructions for unbinding print and file sharing from TCP/IP, which will still permit a local area network to share printers and files using a different protocol known as NetBEUI, are available at security expert Gary Kessler's website.

Securing Windows XP http://www.tweakhound.com/xp/security/page_1.htm

Protecting Home Computers and Networks
http://www.garykessler.net/library/protecting_home_systems.html

---

Disable Visual Basic Script in Windows

The infamous "Love Bug" worm exploited vulnerabilities in Windows Visual Basic Script via email. This is more than a browser problem because VBS (sometimes dubbed the "Virus Building System") is part of Windows, not the browser. There are several methods for thwarting potential VBS attacks. These sites all provide methods for preventing visual basic scripts from running automatically without your knowledge or consent. In addition, the first site offers detailed information about vulnerabilities associated with VBS.

How to Disable VBS http://www.cvm.uiuc.edu/net/virus/outlook.html

Disable Windows Scripting Host http://www.sophos.com/support/faqs/wsh.html

Remove Windows Scripting Host http://www.f-secure.com/virus-info/u-vbs/

---

[199] Gary Kessler, "Protecting Home Computers and Networks," Gary Kessler.net, November 2002, <http://www.garykessler.net/library/protecting_home_systems.html> (14 November 2006).

Know What Your Computer is Loading (Check Your Start-Up Applications)

It seems that most programs today think they are important enough to start automatically each time you reboot your computer. That is, the default installation on most programs tends to add them to your Windows start-up list, so every time you start your computer, these programs are running whether you want them to or not. The problem with this is, at the very least, they are an unnecessary drain upon memory and other system resources and, at worst, some of these unknown programs may in fact be spyware or even viruses or Trojans that add any number of different entries to start-up.

Fortunately, most Windows operating systems (95/98/Me/XP) come with a handy System Configuration Utility called **MSCONFIG** that lets users identify start-up applications. The exception is Windows 2000, which does not come with msconfig. Before using this tool, I recommend you visit these excellent websites devoted to helping users demystify applications that run at start-up and explain which can be removed from start-up without danger. The sites also provide an exhaustive list of programs potentially residing in a computer's start-up list.

Greatis Start Up Application Database http://www.greatis.com/regrun3appdatabase.htm

Pacman's Start-Up Applications http://www.pacs-portal.co.uk/startup_index.htm
or http://www.sysinfo.org/startupinfo.html

Know What Your Computer is Running

What exactly are those invisible programs running in the background on your computer using up system resources? Can you remove them safely or are they necessary? Are they spyware or Trojans undermining your privacy and security or maybe just useless junk clogging up the works? Or are they programs vital to keeping your operating system operating. It is very hard to tell because the names of so many of these programs are unrevealing, but there are several websites that help de-obfuscate these processes, tell you which ones you need, and recommend removal procedures when appropriate. However, it is very important to be careful about removing or disabling programs because many illegitimate programs have names that are almost—or in some cases are—identical to valid programs precisely to confuse users.

In order to see the processes running on your computer, the traditional method in Windows is to use Ctrl+Alt+Del to activate the Task Manager and view the Process

List, but in Windows 2000 and Windows XP, you can right-click on the task bar and select Task Manager.

Process ID maintains a large database of processes that might show up on the Process List. Process ID explains each process, its function, the associated program, and whether or not it is legitimate or malware. Process ID does not tell you how to remove unwanted or dangerous processes, but does refer you to free software designed to eliminate these types of threats.

The **Answers That Work** website provides a comprehensive and easily understandable database of most programs that any Windows user might see in his Task Manager. In addition to identifying the process, the site makes sensible recommendations about how to handle unnecessary or malicious processes. The site is selling a product, but you can handle most of the recommended removals by using the Start Up utility in MSCONFIG (above).

The Process Library will tell you exactly what the processes are, which ones must run, which ones can be safely disabled, and which ones are known threats. The Process Library is searchable by process name or alphabetically browsable. There is also a comprehensive DLL library. Both illicit processes and DLLs are identified as to the type of threat or problem (virus, Trojan, or spyware).

Process Library is also very good at explaining the nature of the problem and when a threat may be easily confused with a legitimate process or DLL. See, for example, the entry for *rundll32.exe*, which is a legitimate process on most Windows operating systems but may indicate a virus on Windows 2000 and XP. Do not, however, confuse *rundll.exe* with *rundll32.exe* or *rundll16.exe*...see, it is confusing. The problem with this site is that it, too, is selling something. When you do find a real threat or problem and click on the remove option, you are taken to a site selling a product to remove the process or DLL. However, Process Library is very good at identifying the many processes running on your computer.

Many of these problems can be avoided in the first place by keeping your virus scanning software up to date or, in the event you do get a virus, using that software to remove it. A very good site for help with removing a variety of types of malware—viruses, browser hijackers, exploits, Trojans, spyware—is PC Hell (motto: *You've Been Here Before But Now You're Just Visiting*). PC Hell doesn't try to sell you anything, just help save you from your current damnable situation, so to speak. So, once you have learned about your problem, it's worth a trip to PC Hell to see if there is a way out (sometimes, however, there is no exit).

Process ID	http://www.processid.com/
Answers That Work	http://www.answersthatwork.com/Tasklist_pages/tasklist.htm
Process Library	http://www.processlibrary.com/
PC Hell	http://www.pchell.com/

Shoot the Messenger!

With the release of Service Pack 2 for Windows XP, Microsoft finally shut one of the many wide open, unlocked "doors" in one of its operating systems by disabling Windows Messenger Service as the default setting. Unfortunately, Windows Messenger Service remains a problem for other operating systems. First, it is important to understand that ***Windows Messenger Service is something entirely different from instant messaging services and turning it off will not affect IM in any way***. Messenger is primarily used by network administrators to send administrative alerts to network users or, for example, to let a user know when a print job on a network printer is complete. However, most home users are not networked and never need or want Messenger. The problem is that Messenger comes enabled by default on most Windows operating systems and is, in fact, automatically launched whenever a user boots his computer. This may not sound too bad, except that the ever-enterprising spammers and malicious hackers of the world found a way to exploit the darned thing. The spammers found they could flood users with pop-up messages using Messenger and, worse, malicious hackers found a way to use a buffer overflow in Messenger to install and run malicious code on a victim's computer.

If you use a Windows operating system other than Windows XP/SP2 or Vista, I recommend you turn off Messenger Service—that is, if you can. Users of Windows 2000 systems can disable Windows Messenger Service. However, ***Windows Messenger Service cannot be disabled on Windows 98 or ME***. For Windows 2000 users, it is easy to disable Windows Messenger and, if needed, turn it back on by reversing these steps:

Windows 2000

Click Start | Settings | Control Panel | Administrative Tools | Services

Scroll down and highlight "Messenger"

Right-click the highlighted line and choose Properties.

Click the STOP button.

Select Disable or Manual in the Startup Type scroll bar

Click OK

User Profiles and the RunAs Command in Windows XP

One of the best features of Windows XP, even in the Home Edition, is user profile administration and the *RunAs* command. While these options existed in Windows 2000/NT, Windows XP was the first Microsoft operating system to make these very important computer management and security features easily accessible and configurable for the home user. Although Windows XP Home Edition offers limited user and profile management when compared to the Professional Edition, it does introduce the concept of the Administrator versus the user as part of its **user accounts**. You should set up different types of accounts on your computer(s) running Windows XP Home Edition. Here's why and how.

Windows XP automatically creates certain built-in groups when it is installed. In Windows XP Home Edition, you belong to one of two broad types of "Groups": either Administrator or User. Belonging to a group gives a user rights and abilities to perform various tasks on the computer. Unfortunately, in Windows XP Home Edition ***by default, all user accounts have administrative privileges and no password***. This is a potentially serious security vulnerability that should be remedied right away. If you always use your computer as the Administrator, it means that, if you encounter a virus, a Trojan horse, or a worm while you are logged on as Administrator, your entire system could be compromised because the Administrator has full control over every aspect of the computer. When you are logged on as Administrator, ***every program you run has unlimited access to your computer***. If malware finds its way to one of those programs, it also gains unlimited access. However, if you create user accounts and normally log in as a user and not as the Administrator, any malware you encounter will be limited in the amount and kind of damage it can do to your computer.

Here is how to set up user accounts in Windows XP Home Edition.[200]

[200] Windows XP Professional has additional user categories, including Power User, that are absent from the Home Edition. If you have Windows XP Pro at home, you have more options for how to administer your computers and your network.

- Logon as Administrator
- Start | Settings | Control Panel | User Accounts

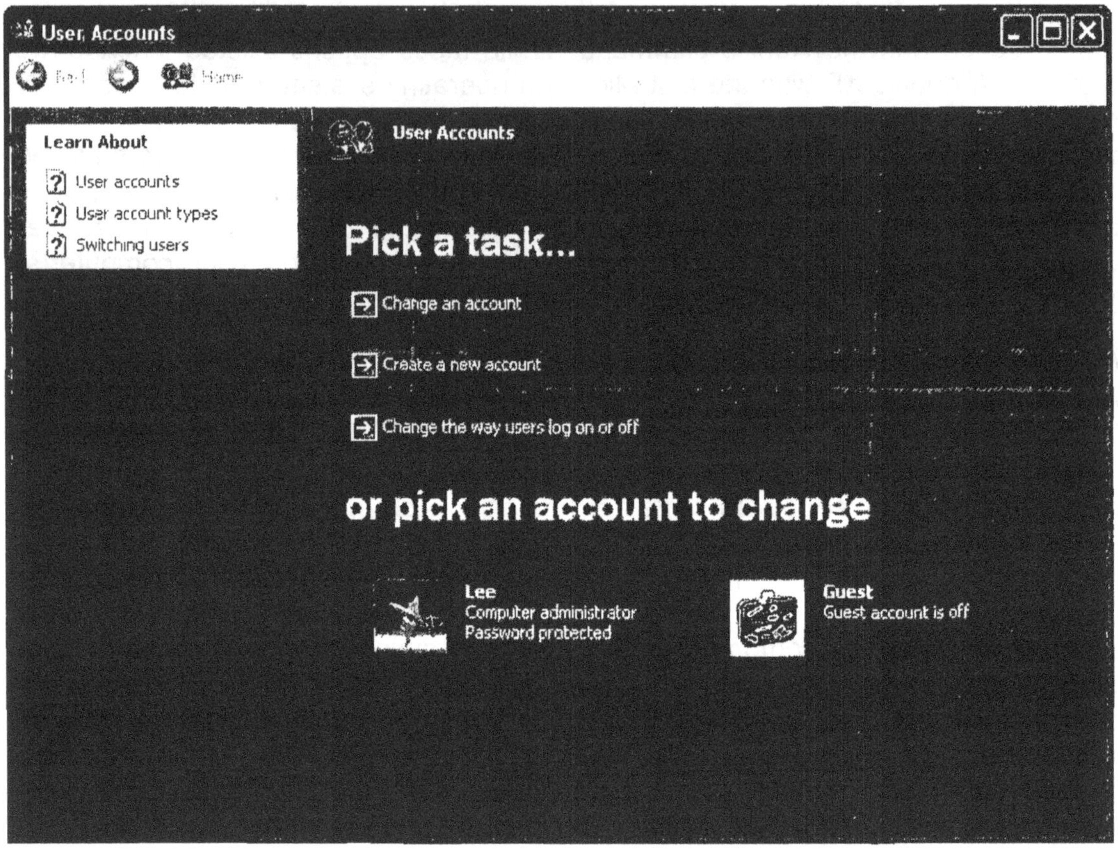
User Accounts
Back
Home
Learn About
User accounts
User account types
Switching users
User Accounts
Pick a task...
Change an account
Create a new account
Change the way users log on or off
or pick an account to change
Lee
Computer administrator
Password protected
Guest
Guest account is off

From here, it is a simple matter to set up and change user accounts and account types. Create a "Computer administrator" account for yourself with a strong password. Then create a new account for yourself and each user of the computer as a Limited user. Make sure each Limited user account also is password protected. Remember, user names are not case sensitive but passwords are.

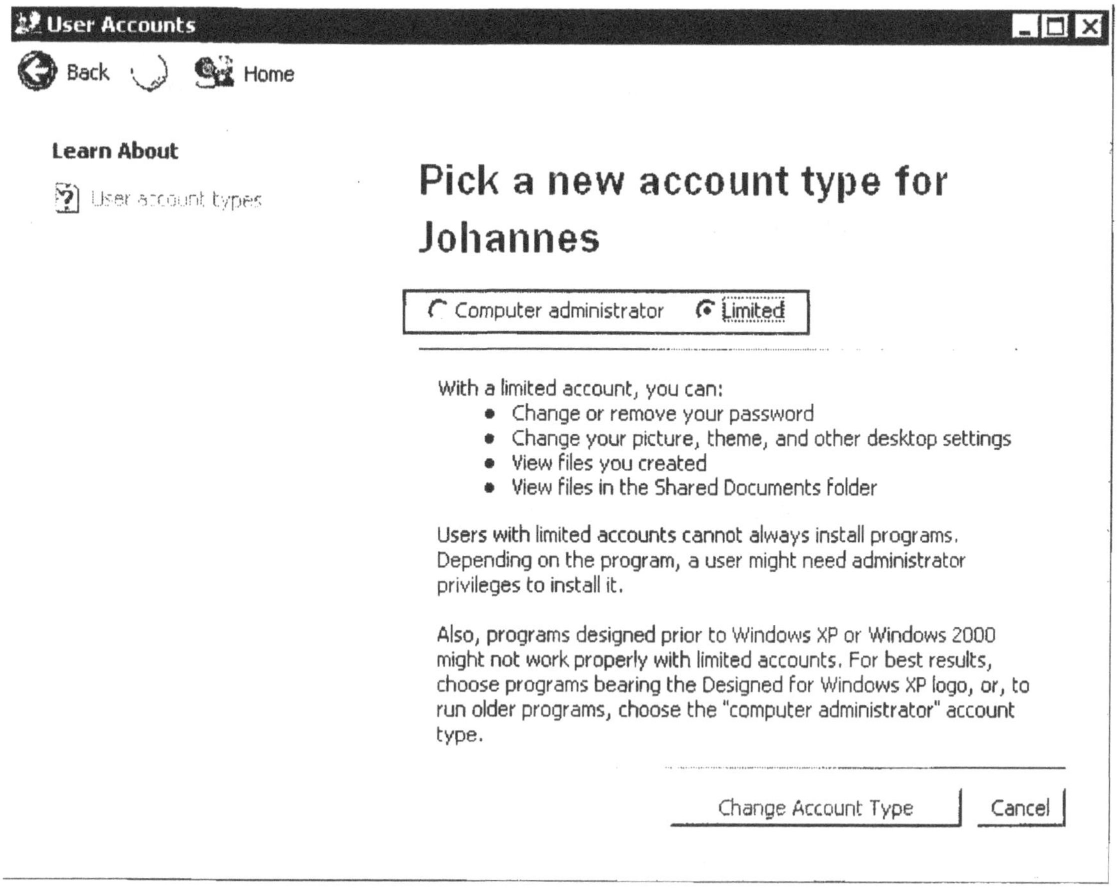
User Accounts
Back
Home
Learn About
User account types
Pick a new account type for Johannes
Computer administrator
Limited
With a limited account, you can:
Change or remove your password
Change your picture, theme, and other desktop settings
View files you created
View files in the Shared Documents folder
Users with limited accounts cannot always install programs. Depending on the program, a user might need administrator privileges to install it.
Also, programs designed prior to Windows XP or Windows 2000 might not work properly with limited accounts. For best results, choose programs bearing the Designed for Windows XP logo, or, to run older programs, choose the "computer administrator" account type.
Change Account Type
Cancel

As you can see, Limited users are just that: strictly limited to what they can and cannot do on a computer. For the most part, logging in as a limited user should cause no problems in using applications on the computer. Email, web browsing, and instant messaging do not require administrative privileges, and are common avenues for malicious code to attack end users' systems. However, certain actions—such things as installing software, creating new network connections, or even running certain programs—require you to access them as the Administrator. There are two simple ways to accomplish this. First, you can always switch from Limited user to Administrator:

- Start | Log Off
- at this point, a new screen will appear; select **Switch User** and logon as the Administrator.

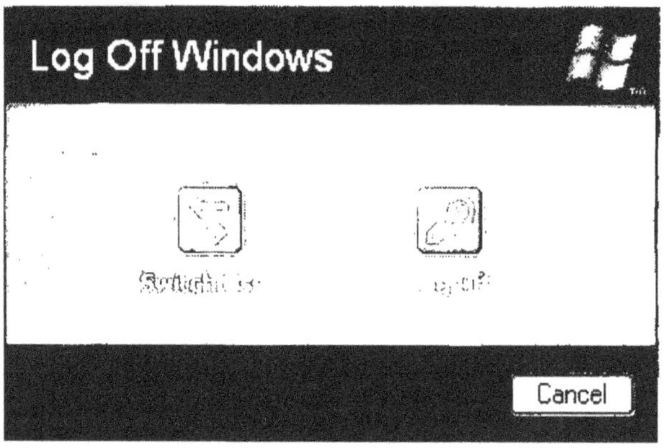
Log Off Windows
Cancel

Fast User Switching should be enabled on Windows XP Home Edition by default, but just in case it isn't here is how to enable it:

- To Enable or Disable Fast User Switching:

    1. Start | Settings | Control Panel | User Accounts
    2. Pick a Task | **Change the way users log on or off**
    3. On the **Select logon and logoff options** page, check **Use the Welcome screen** and **Use Fast User Switching**

The second and, to my mind, much easier way to "be" the Administrator temporarily is to use the **RunAs** command. To sign on as Administrator using the RunAs command, simply right-click on a shortcut and select RunAs. When you right-click on a shortcut or application, you will see this dialog box, which gives you the option to run this specific program as the Administrator. As long as you know the user name and password, you can sign on as the Administrator or as any other user. This is an invaluable tool because a number of programs simply will not run for Limited Users. Keep in mind, however, that the RunAs command gives any Limited User the power of the administrator, so only permit a trusted user to use the RunAs command. That means if you don't trust your teenager to use RunAs responsibly, do not give him or her the administrator password. In this case, Windows XP Professional is a better choice because it gives you more user options.

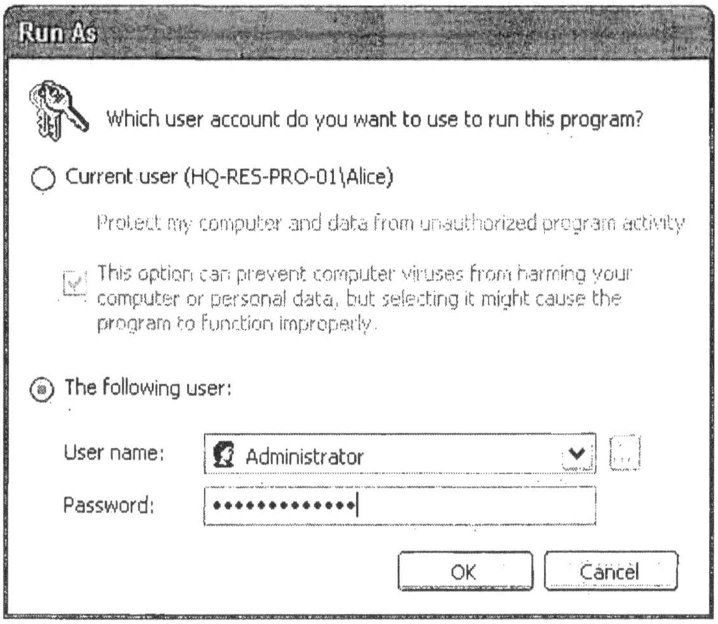
Run As
Which user account do you want to use to run this program?
Current user (HQ-RES-PRO-01\Alice)
Protect my computer and data from unauthorized program activity
This option can prevent computer viruses from harming your computer or personal data, but selecting it might cause the program to function improperly.
The following user:
User name:
Administrator
Password:
OK
Cancel

There is one more user account type that needs attention in Windows XP Home Edition: the **Guest account**. Guest accounts have been notorious gateways for malicious hackers to break into computers. Unfortunately, in the Home Edition you cannot (or, rather, *should* not) disable the Guest account ("disabling" the account from the Control Panel simply removes the Guest account from the Fast User Switching system). According to the Microsoft website, "You can use the **User Accounts** tool in Control Panel to turn off the Guest account. When you turn off the Guest account, you remove the Guest account from the **Fast User Switching** welcome screen. However, the Guest account is not disabled. We do not recommend that you disable the Guest account. If you disable the Guest account, you may not be able to access network resources. Additionally, you cannot access resources on a local computer from another computer on the network."[201] Okay, so do not try to disable the Guest account in Windows XP Home Edition. What can you do to minimize the risk posed by the Guest account? At this time, the best work-around is to ***assign the Guest account a very strong password***.

Sounds simple enough, doesn't it? Yet for some reason I really cannot comprehend, Microsoft failed to include an option to add a password to the Guest account in Windows XP Home Edition. However, all is not lost; you can still create a password for the Guest account very simply.

[201] "Description of the Guest User Account in Windows XP," Microsoft.com, <http://support.microsoft.com/default.aspx?scid=kb;en-us;300489> (14 November 2006).

- Logon as Administrator.
- Open a Command Prompt (Start | Settings | Accessories | Command Prompt).
- Type **net user guest *password*** (replace the word *password* with your new Guest password and make sure it is a strong password because no Guest password is better than a weak one.)

In summary, as Aaron Margosis advocates in his excellent "non-admin" blog, "do your everyday computing as a Limited user and log on as Administrator only when it is absolutely necessary, such as when installing new software or hardware, or changing security settings." Words to live by. For more detailed information about administering accounts, securing Windows XP Home Edition, and using RunAs on Windows XP Home Edition, refer to these links:

5 Steps to Secure Windows XP Home
http://netsecurity.about.com/cs/windowsxp/a/aa042204_2.htm

Non-Admin Blog, Aaron Margosis' Weblog
http://blogs.msdn.com/aaron_margosis/archive/2005/04/18/TableOfContents.aspx

"RunAs" basic (and intermediate) topics, Aaron Margosis' Weblog
http://blogs.msdn.com/aaron_margosis/archive/2004/06/23/163229.aspx

---

Encrypt Files in Windows

One of the basic privacy and security functions some versions of Windows offer is easy to use and provides a better degree of protection for files on your personal computer. However, not all Windows versions have this feature. ***The Windows operating systems that offer Microsoft's Encrypting File System (EFS) are XP Professional (another reason to go with Pro over the Home edition) and Windows 2000, beginning with Service Pack 2.*** Since most readers are probably using Windows XP, I will only discuss this operating system.

Microsoft provides clear instructions on how to encrypt a file in Windows XP Professional; keep in mind you can either encrypt a single file or a file and its parent folder.

How to Encrypt a File

You can encrypt files only on volumes that are formatted with the NTFS file system. To encrypt a file:

1. Click **Start**, point to **All Programs**, point to **Accessories**, and then click **Windows Explorer**.
2. Locate the file that you want, right-click the file, and then click **Properties**.
3. On the **General** tab, click **Advanced**.
4. Under **Compress or Encrypt attributes**, select the **Encrypt contents to secure data** check box, and then click **OK**.
5. Click **OK**. If the file is located in an unencrypted folder, you receive an **Encryption Warning** dialog box. Use one of the following steps:
   - If you want to encrypt only the file, click **Encrypt the file only**, and then click **OK**.
   - If you want to encrypt the file and the folder in which it is located, click **Encrypt the file and the parent folder**, and then click **OK**.

If another user attempts to open an encrypted file, that user is unable to do so. For example, if another user attempts to open an encrypted Microsoft Word document, that user receives a message similar to:

> Word cannot open the document: *username* does not have access privileges (*drive*:*filename*.doc)

If another user attempts to copy or move an encrypted document to another location on the hard disk, the following message appears:

> Error Copying File or Folder
> Cannot copy *filename*: Access is denied.
> Make sure the disk is not full or write-protected and that the file is not currently in use.

Back to the top

Troubleshooting

- You cannot encrypt files or folders on a volume that uses the FAT file system.

  You must store the files or folders that you want to encrypt on NTFS volumes.
- You cannot store encrypted files or folders on a remote server that is not trusted for delegation.

issues, and more.

- Newsgroups
  Pose a question to other users. Discussion groups and Forums about specific Microsoft products, technologies, and services.

Page Tools

Print this page
E-mail this page
Microsoft Worldwide
Save to My Support Favorites
Go to My Support Favorites
Sign In

Notice that ***only the user who encrypted the file or folder can open, copy, or move that file or folder***. If you keep information such as your passwords, financial information, etc., on your computer, especially if that computer is connected to the Internet, you should encrypt those files. In addition to adding a password to a sensitive Microsoft Office files, it is also a very good idea to encrypt those files as well.

How to Encrypt a File in Windows XP	http://support.microsoft.com/kb/307877
How to Encrypt a Folder in Windows XP	http://support.microsoft.com/kb/308989

For those who really want the nitty gritty on the EFS:

Windows XP Professional Resource Kit, Using Encrypting File System
http://www.microsoft.com/technet/prodtechnol/winxppro/reskit/c18621675.mspx#EVD

Do Not Save Encrypted Pages to Disk

Internet Explorer uses caching to save website information as you browse in order to allow faster access to pages you frequently visit. The actual copies of webpages are stored in the Temporary Internet Files folder on your hard drive. Normally, this process is a benefit to users, but there is one circumstance in which you do not want pages saved. If encrypted webpages are cached, the copies saved to your hard drive are not encrypted and can be read by someone who might gain access to your computer using malicious software such as a Trojan horse or virus (or even someone with physical access to your computer).

To prevent this from happening, select:

Tools | Internet Options | Advanced | Security

- Check box next to "Do Not Save Encrypted Pages to Disk"

Next, you need to erase any encrypted pages that might have already been saved to disk.

Tools | Internet Options | General | Temporary Internet Files | Delete Files

In pop-up message that says "Delete all offline content," click OK

Handle Microsoft Files Safely

It can be risky to open certain Microsoft file types, especially those you may encounter on the Internet or in email, because of the potential for infection via what are known as macro viruses. **Macro viruses** exploit an application such as Word or Excel (which use little programs called macros) to infect a document and then spread the infection to other computers and networks. One of the dangers with macro viruses is that they do not infect programs, so you do not have to run an executable file to become infected. All you need to do is to open an infected Word, Excel, Access, or PowerPoint file to activate the virus.

However, there are some simple precautions you should take to avoid the risk of infection. After all, the awful Melissa virus of 1999 was a Word 97 and Word 2000 macro virus, and it spread like crazy around the world very quickly as an email attachment. There was another major outbreak of Word macro viruses in 2006, so the problem is still very much with us. As more search engines make it possible to search for non-HTML file formats, including all Microsoft file types, it is vital to take steps to protect yourself and your employer from potentially damaging viruses that could lurk in these types of files.

There are several ways to handle the problem of macro viruses and prevent both infection and spread of these nuisances:

- One of the safest and easiest ways is to use **Google** or **Yahoo** to locate the web page with the link to the file you wish to view, then select *view as html* or *view as text*. These options will permit you to see the file (whether it is a .doc, .xls, .ppt, .ps, etc.) as an HTML file or a text file (in the case of Postscript files in Google) with no fear of viruses.

➢ However, this solution will not work in every situation. There is an alternative available to users with access to Keyview Pro viewers. These viewers should be able to handle most file types you will encounter and handle them safely because the viewers do not run the underlying program and thus cannot execute a virus. The viewers also permit printing and some other functions. Documents should be saved to your computer's desktop; then right-click on the document to select the "View with" option. DO NOT DOUBLECLICK to open or you will execute the underlying program and possibly a virus as well. Here are the available viewers and the types of files they handle (this is not a complete list):

<u>Keyview Pro</u>
Microsoft Word
Microsoft Excel
Microsoft PowerPoint
Applix Words
Corel WordPerfect
Corel Presentations
Corel Quattro Pro
Lotus Freelance Graphics
Lotus 1-2-3
Lotus Word Pro
XyWrite for Windows
Enhanced Metafile (EMF) (KeyView Pro 32-bit only)

<u>Adobe Acrobat</u>
PDF
FrameMaker

<u>GSView/Ghostscript</u> (GSView is a Windows GUI for Ghostscript)
Postscript
PDF

➢ Did you know that Microsoft offers free viewers for Word documents, Excel spreadsheets and other applications as well, including PowerPoint and Access files? This freeware lets you open, view, and print all Microsoft Office files without concerns about macro viruses because the viewers cannot run macros. The free viewers are built to automatically configure themselves for use with both Mozilla and Internet Explorer. They are available at:

All Microsoft Office Viewers http://www.microsoft.com/office/000/viewers.asp

➢ As an additional precaution, make sure all your Microsoft applications have <u>macros security settings at high</u>. For example, in Word 2000:

1. open Tools | Macro | Security

2. select Security Level High

3. make sure there are no Trusted Sources

Doing this will ensure that no macros can run on your computer in Word because Word will not execute any macros at all with these settings.

- Configure your virus scanning software to perform an automatic virus scan of ALL downloaded files. Ensure that your virus scanning software ***scans all downloaded files***, not just executables.

An excellent guide to home network security is available on line from the CERT Coordination Center.

Home Network Security from the CERT Coordination Center
http://www.cert.org/tech_tips/home_networks.html

"Law #3: If a bad guy has unrestricted physical access to your computer, it's not your computer anymore."

Handle with Care: More Privacy and Security Concerns

Use Anti-Virus Software and Keep It Up To Date

I expect most readers have anti-virus software on their computers, but having it and using it properly are not the same thing. Make sure you configure the software to maximize protection of your computer, especially against email-borne malware, run a full system scan at least once a week, and keep your virus definitions up to date. Most anti-virus software now offers automatic updates and scans, which can relieve users of some of the burden of remembering these tasks. There are new viruses—not to mention Trojan horses and other nasty invaders—unleashed via the Internet every day. The Kaspersky Lab's **VirusList** encyclopedia contains more than 30,000 entries. No anti-virus software is a guarantee against infection, but not using and updating it is akin to leaving your car door unlocked and the keys in the ignition.

There are several good free anti-virus packages, and AOL began offering "free" virus scanning software from Kaspersky during 2006. However, I would be careful about the AOL package, which is only free for 30 days; after that, there is a $50 per year subscription. I recommend reading Fred Langa's article[202] about the AOL offer before making a decision.

About's Free Antivirus Software Reviews
http://antivirus.about.com/od/freeantivirussoftware/Free_Antivirus_Software.htm

VirusList Virus Encyclopedia http://www.viruslist.com/en/viruses/encyclopedia

"Law #8: An out of date virus scanner is only marginally better than no virus scanner at all."

[202] Fred Langa, "Should you use AOL's free antivirus?" Windows Secrets and Langalist, 7 December 2006, <http://windowssecrets.com/comp/061207/#langa0> (12 December 2006).

Make Sure You Are Not Inadvertently Running "Spyware"

Spyware is often distinguished from "adware," that is, advertising supported software, which was designed to help shareware authors make money. There are a few examples of "good" adware, software that you can get for free if you are willing to put up with sponsored ads each time you use it. ***Good adware explicitly asks you if you are willing to accept the ads*** in exchange for the program and also promises not to share or sell any information it collects about your browsing habits.

Spyware, on the other hand, rarely asks for your permission to do what it was created to do. An exception would be something like the Google Toolbar, which offers an option to turn off data collection and, even if it is enabled, does not share its tracking data with anyone else. Spyware by definition contains some sort of tracking software that regularly tries to "phone home" via your Internet connection to report data about your browsing habits, virtually never with your explicit permission. Most spyware then sells your personal information or, worse, exploits it to attack you. To make matters worse, it is now so hard to detect spyware that even the most sophisticated users often do not realize they have been infected.[203]

Here are several ways to avoid spyware: do not download shareware or freeware, such as Kazaa, Quickclick, WebHancer, CuteFTP, etc. However, most people are going to download software at some point. If you do, try to make sure it doesn't include spyware by visiting a website that lists known spyware, such as those listed below. Be aware, however, that more and more spyware is not actively installed by users but is downloaded, installed, and run on computers using nefarious techniques such as drive-by downloads, which exploit browser features such as ActiveX.

There is software available to check your system on a regular basis for spyware. Sadly, not all such software does what it claims; instead, there are unscrupulous people who are offering "spyware detection" software that is itself spyware. Do not download any antispyware software without checking it out beforehand. There is even a website devoted to finding and exposing bogus antispyware products. Spyware Warrior Rogue/Suspect Anti-Spyware Products maintains a long and growing list of these untrustworthy products.

While you can buy good antispyware software, some of the best is available for free. Ad-Aware SE Personal Edition and Spybot Search & Destroy are excellent free utilities that detect and remove spyware. Microsoft offers its own free antispyware

---

[203] Leslie Walker, "Theft You Don't Even See," *Washington Post*, 1 September 2005, <http://www.washingtonpost.com/wp-dyn/content/article/2005/08/31/AR2005083102486_pf.html> (14 November 2006).

software, Windows Defender. As of 2007, Windows Defender is only available for use on Windows XP, SP2 and Windows Server 2003, which means Windows Defender is no longer supports Windows 2000.

Most experts agree that ***there is no single product that can detect all spyware***. If I were only going to use one antispyware product on the Windows XP operating system, I would choose Windows Defender for several reasons: it has a very high detection rate; it is easy to configure; it will run automatically on a schedule; it automatically updates its detection rules; and, of course, it is free. In addition, Microsoft products tend to work very well on Windows computers.

Free Antispyware Products

Ad-Aware Spyware Checker
http://www.lavasoftusa.com/products/ad-aware_se_personal.php

Windows Defender
http://www.microsoft.com/athome/security/spyware/software/default.mspx

Spybot Search & Destroy http://www.safer-networking.org/en/home/index.html

Antispyware Guides & Articles

11 Signs of Spyware
http://www.pcmag.com/article2/0%2C1759%2C1522648%2C00.asp

Anti-Spyware Guide http://www.firewallguide.com/spyware.htm

Monitoring Software on Your Computer: Spyware, Adware, and Other Software, Staff Report, Federal Trade Commission, March 2005 **[PDF]**
http://www.ftc.gov/os/2005/03/050307spywarerpt.pdf

PC Hell Spyware Removal Help http://www.pchell.com/support/spyware.shtml

Spychecker http://www.spychecker.com/

Spyware Guide http://www.spywareguide.com/product_list_full.php

Spyware Warrior Rogue/Suspect Anti-Spyware Products
http://www.spywarewarrior.com/rogue_anti-spyware.htm

Spyware Watch http://www.spyware.co.uk/

Stop Internet Abuse http://www.celticsurf.net/webscape/abuse.html

"Law #2: If a bad guy can alter the operating system on your computer, it's not your computer anymore."

Install Software and Hardware Firewalls

Whether you are using an always-on connection such as cable or DSL, or you are accessing the Internet via a dial-up connection, you need to install at least a software firewall and, I believe, a hardware firewall as well. Firewalls, while not foolproof, are the home user's best protection against Trojan horses and spyware. Both types of malware are a huge threat to the Internet community because they are insidious, hard to detect, and harder still to remove. The best advice about Trojan horses and spyware is don't get them in the first place, and firewalls remain the best defense against these types of malicious software.

Software firewalls[204] can be purchased for a relatively low price or, even better, some of the best are free. Check Firewallguide's Personal Firewall Reviews for some options.

However, while all Internet users need a software firewall, anyone with cable, DSL, or satellite Internet access needs a hardware firewall, too. The bad news is that "true" hardware firewalls are still fairly expensive and hard to configure. The good news is that there is a very inexpensive alternative for the home user that offers similar basic protection: a cable/DSL router. As with a hardware firewall, routers use Network Address Translation or NAT to hide your computer's Internet address from the bad guys. The firewall—and not your computer—becomes your connection to the Internet, making it harder for malicious hackers to see your computer, much less scan or attack it. In addition to NAT, firewalls (and good home routers) also use something called Stateful Packet Inspection (SPI) to let through only those Internet connections you request and block connections that are trying to break into your computer.

Make sure the router you purchase offers SPI and good advanced control settings. And, please, ***change your router password as soon as you install it!*** Malicious hackers know all the default logins and passwords for every router ever made. For example, check this site (just one of many):

Default Password List http://phenoelit.darklab.org/cgi-bin/display.pl

It is important to understand that while a good home router will help protect your computer from attacks, it is not impervious. Nothing really is, but for a home user, you are going to be much more secure with software and hardware firewalls than the vast majority of users who don't do anything to protect themselves. However, in order to get the most good out of these products, you must configure them properly.

---

[204] The firewall that is part of Windows XP (including the improved firewall in Service Pack 2) does not provide "extrusion protection," i.e., it only detects incoming data, not data that might flow from your computer. Do not rely solely on the XP firewall.

I have compiled some of the most useful websites for learning about firewalls and routers here:

Firewallguide's Personal Firewall Reviews http://www.firewallguide.com/software.htm

Firewallguide: Wired Routers http://www.firewallguide.com/hardware.htm

Firewall Forensics http://www.linuxsecurity.com/resource_files/firewalls/firewall-seen.html

Firewall Q&A http://www.vicomsoft.com/knowledge/reference/firewalls1.html

Gibson Research's Firewall Page http://grc.com/su-firewalls.htm

HomeNetHelp's Broadband Router Guide http://www.homenethelp.com/router-guide/index.asp

Home Network Router Security Secrets http://www.informit.com/articles/printerfriendly.asp?p=461084&rl=1

How Firewalls Work http://www.howstuffworks.com/firewall.htm

Internet Firewall FAQ http://www.interhack.net/pubs/fwfaq/

Introduction to Firewalls http://netsecurity.about.com/od/hackertools/a/aa072004.htm

10 Steps To Make Your Firewall More Secure http://www.itsecurity.com/features/more-secure-firewall-012207/

Free Software Firewalls for Windows

Free Personal Firewall Software http://netsecurity.about.com/od/personalfirewalls/a/aafreefirewall.htm

Sunbelt Kerio Firewall http://www.sunbelt-software.com/kerio.cfm
Full version free for 30 days, then reverts to basic version.

Comodo Free Personal Firewall http://www.personalfirewall.comodo.com/

Zone Alarm http://www.zonelabs.com/

Test Your Online Security

So you installed firewall software and perhaps even hardware protection in the form of a router and you're feeling pretty smug. Before you get too comfortable, you should test your firewall to make sure it is doing the job it should be. My favorite set of tests is **Sygate/Symantec's Online Services**, which puts your computer through a whole range of scans to test its vulnerability to attack. I also recommend you run Steve Gibson's **Internet Vulnerability Profiling** at his **Shields Up!** website. This is what you want to see for every test you run at Shields Up:

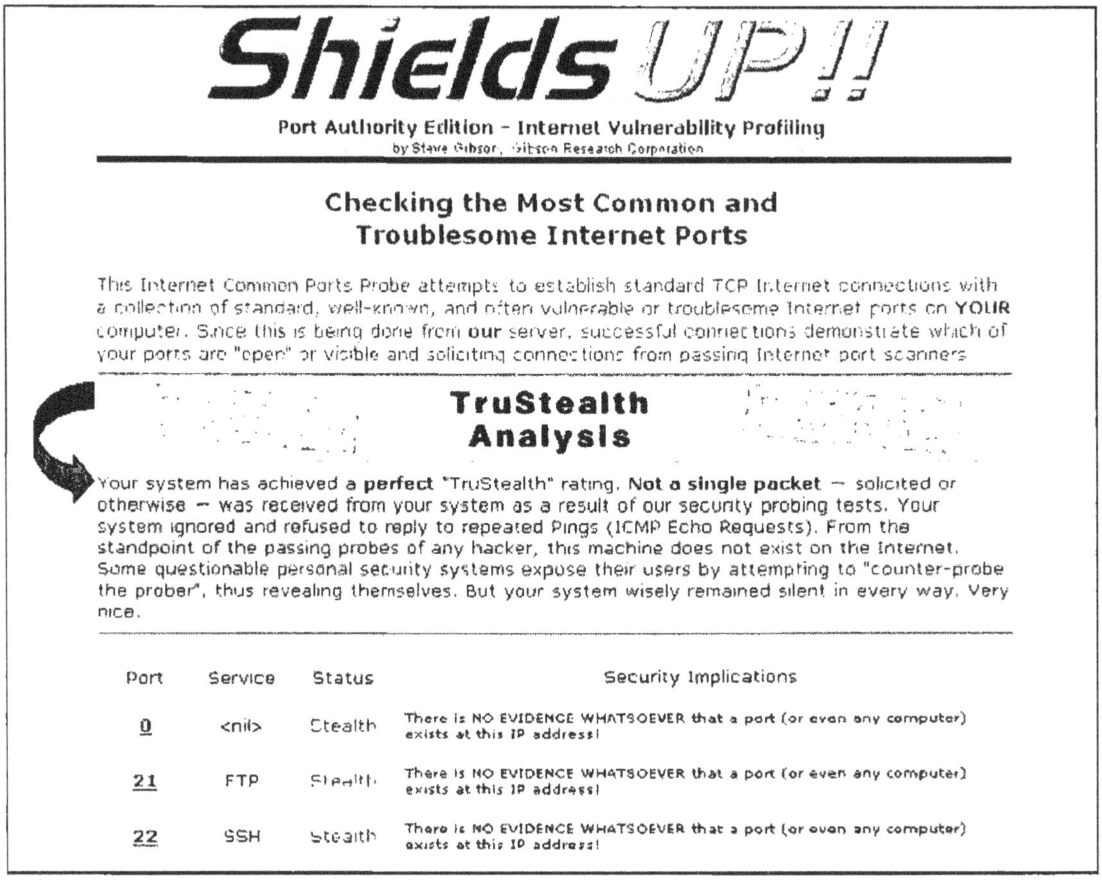

Shields UP!!

Port Authority Edition – Internet Vulnerability Profiling

by Steve Gibson, Gibson Research Corporation

Checking the Most Common and Troublesome Internet Ports

This Internet Common Ports Probe attempts to establish standard TCP Internet connections with a collection of standard, well-known, and often vulnerable or troublesome Internet ports on **YOUR** computer. Since this is being done from **our** server, successful connections demonstrate which of your ports are "open" or visible and soliciting connections from passing Internet port scanners

TruStealth Analysis

Your system has achieved a **perfect** "TruStealth" rating. **Not a single packet** – solicited or otherwise – was received from your system as a result of our security probing tests. Your system ignored and refused to reply to repeated Pings (ICMP Echo Requests). From the standpoint of the passing probes of any hacker, this machine does not exist on the Internet. Some questionable personal security systems expose their users by attempting to "counter-probe the prober", thus revealing themselves. But your system wisely remained silent in every way. Very nice.

Port	Service	Status	Security Implications
0	<nil>	Stealth	There is NO EVIDENCE WHATSOEVER that a port (or even any computer) exists at this IP address!
21	FTP	Stealth	There is NO EVIDENCE WHATSOEVER that a port (or even any computer) exists at this IP address!
22	SSH	Stealth	There is NO EVIDENCE WHATSOEVER that a port (or even any computer) exists at this IP address!

If your computer does not pass every test, find out why and fix it. Unfortunately, every computer, like every person, is unique, so one solution definitely does not fit all. However, with a little patience and some trial and error, most vulnerabilities can be eliminated. Steve Gibson's website is especially useful in helping home users diagnose and correct computer security and privacy related problems.

Gibson's Shields Up Internet Vulnerability Profiling https://grc.com/x/ne.dll?bh0bkyd2

Sygate/Symantec Online Security Services
http://scan.sygate.com/home_homeoffice/sygate/index.jsp
Sygate is now owned by Symantec.

HackerWatch Port Scan and Simple Probe http://www.hackerwatch.org/probe/

HackerWhacker Free Tools http://whacker4.hackerwhacker.com/freetools.php
especially the Browser Leakage and Quick Scan for open ports

DSL Reports http://www.dslreports.com/tools/

PC Flank Advanced Port Scan http://www.pcflank.com/scanner1.htm

Planet Security Port Scans http://www.planet-security.net/index.php?xid=%F7%04T%BDP%92nD

Firewall Guide: Firewall Testing http://www.firewallguide.com/test.htm

Most personal firewalls do a decent job of blocking intruders from gaining external access a computer (i.e., *intrusion detection*). However, many of these same programs (most notably the Windows XP firewall) fail to catch applications residing on a computer that access the Internet without your knowledge or consent (i.e., *internal extrusion*). Why? Often these personal firewall packages come pre-programmed to allow some applications to pass through them without the user's knowledge. Also, it's quite easy for a malicious person to simulate a preapproved application and fool a computer into "phoning home." All that is required is to rename the malware with a commonly used file name, such as *iexplore.exe*, which is usually allowed free access to the Internet, and the attacker has opened a back door into your computer.

Check to see if your firewall passes the "leak" test by downloading Gibson's tiny Leak Test application or try one of these online firewall testers. If your firewall is properly configured (meaning you do not let programs—especially browsers—access the Internet without your permission), your firewall will pass all three leak tests. If it doesn't, you need to reconfigure your software.

Gibson Research's Firewall Leaktest	http://grc.com/lt/leaktest.htm
PCFlank Firewall Leaktest	http://www.pcflank.com/pcflankleaktest.htm
Tooleaky	http://tooleaky.zensoft.com/
Firewall Leak Tester	http://www.firewallleaktester.com/index.html

---

Don't Fall for the Con

Never download software or open and/or run an email attachment unless you are absolutely sure you know what it is. It used to be known as a con job and the person who committed this type of fraud a con artist. Then in the computer hacker world, the con became "social engineering, one of the most pernicious ways malware is spread. Social engineering is a con game designed to trick users into violating normal security procedures. One famous example involves a malicious user sending email that looks as though it is from a trusted source, such as "Microsoft Corporate Security Center," warning you to install the attached "fix" to a vulnerability or to go to a certain website to download a file. That "fix" is in fact a virus or some other piece of malware. Read Microsoft's policies on software distribution (they *never* distribute software directly via email) and all Microsoft downloads are from their site <http://www.microsoft.com/>.

Microsoft Policies on Software Distribution
http://www.microsoft.com/technet/security/bulletin/info/swdist.mspx

Although it is not new, another type of con called "pretexting"[205] made headlines during 2006 when some executives at Hewlett Packard got into serious trouble because of this method of obtaining information. The HP execs weren't after financial data but telephone records, and at that time it was not clear if pretexting to obtain phone records was illegal or not in the US. HP admits that it hired a firm to investigate board leaks to the press. The firm HP engaged to look into the leaks in turn hired private investigators who impersonated HP board members to get phone records belonging to at least nine reporters and one HP board member.

This is just the most high-profile complaint about the ready availability of personal records obtained by "data brokers." You need to be aware that pretexting is a widespread tactic, and the ***laws governing fraudulently obtaining non-financial personal records and information are murky at best.*** Frankly, there isn't much you can do to protect yourself from a clever and determined con artist who is going after your phone records at your phone company. The best ways to combat pretexting are laws that make pretexting a crime and companies that train their employees better.

Understand Website Certificates

If you are concerned about phishing attacks and other social engineering scams, you have probably been advised to make sure the site you are visiting has a valid site certificate. And then you probably scratched your head and wondered, "how the heck can I tell if that certificate is valid or not?"

First, it is important to understand what a site certificate is and what it does for the site and for you. Any website that wants a secure connection must use encryption. In order to use encryption over the Internet, the website owner must obtain a site certificate. There are, then, two parties involved in verifying the validity of a certificate: the website owner and the trusted certificate authority. At present, your browser is probably set to recognize more than 100 trusted certificate authorities, but not all of these have the same strictness about ensuring the validity and security of

[205] The earliest use I have found of the term 'pretexting' to mean obtaining private or confidential information by pretending to be someone who has a legitimate right to or need for that information is 1980: Fair Financial Information Practices Act: Hearings Before the Subcommittee on Consumer Affairs by the United States Senate Committee on Banking, Housing, and Urban Affairs, Subcommittee on Consumer Affairs.

their data. You can check the validity of the site certificate by clicking on the locked padlock, but clever malicious hackers know how to create a fake padlock that appears to provide valid site certificate data. A more reliable way to verify a certificate is to view the webpage's *Page Info* in Mozilla or, in Internet Explorer, to right click on the webpage and select *Properties* to see, first, the general information about the page security and then, by clicking on *Certificates*, the actual certificate information such as from the PayPal website:

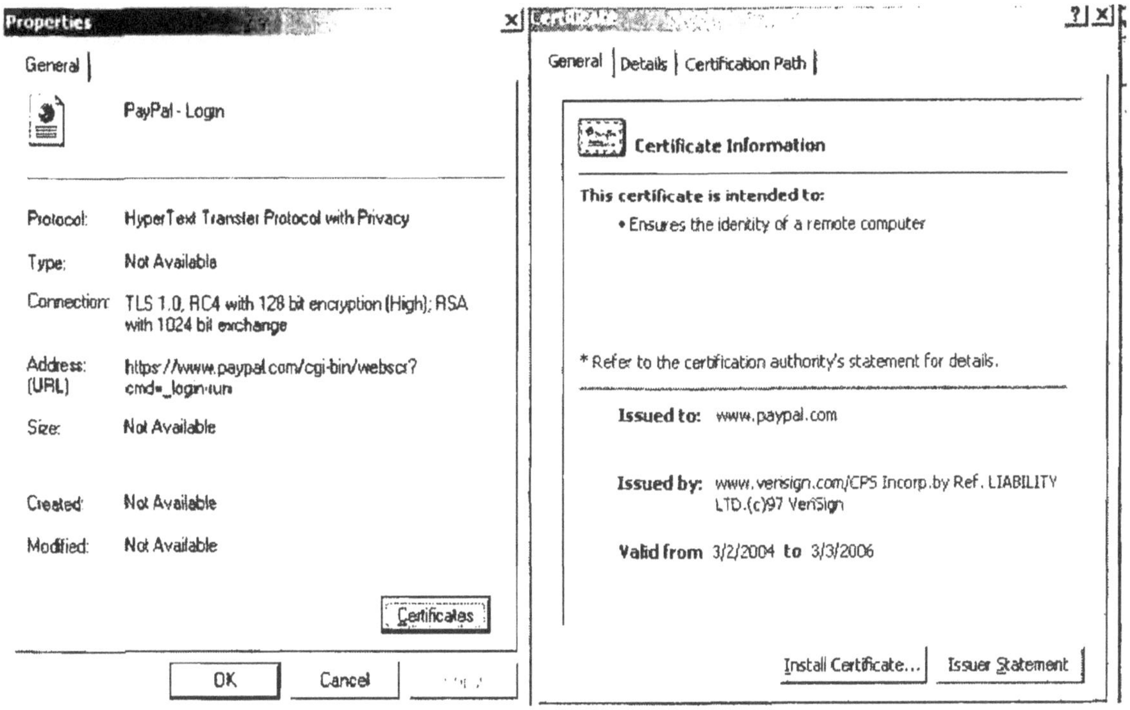
Properties
General
PayPal - Login
Protocol: HyperText Transfer Protocol with Privacy
Type: Not Available
Connection: TLS 1.0, RC4 with 128 bit encryption (High); RSA with 1024 bit exchange
Address: (URL) https://www.paypal.com/cgi-bin/webscr?cmd=_login-run
Size: Not Available
Created: Not Available
Modified: Not Available
Certificates
OK
Cancel
General
Details
Certification Path
Certificate Information
This certificate is intended to:
• Ensures the identity of a remote computer
* Refer to the certification authority's statement for details.
Issued to: www.paypal.com
Issued by: www.verisign.com/CPS Incorp.by Ref. LIABILITY LTD.(c)97 VeriSign
Valid from 3/2/2004 to 3/3/2006
Install Certificate...
Issuer Statement

In viewing the certificate information you should make sure the ***trusted certificate authority is legitimate***. If you do not recognize the name, check your browser's list of certificate authorities:

- in Firefox: Tools | Options | Advanced | View Certificates | Authorities
- in Netscape: Edit | Preferences | Privacy & Security | Certificates | Manage Certificates | Authorities
- in Internet Explorer: Tools | Internet Options | Content | Certificates | Trusted Root Certification Authorities

The certificate should have been issued to the website owner. If the ***name on the certificate*** does not match the name you expected, do not trust it.

Also look at the certificate's ***expiration date*** to make sure it has not expired.

For more information on how to understand site certificates, the US CERT site has a new Cyber Security Tip addressing this topic.

Understanding Web Site Certificates http://www.us-certgov/cas/tips/ST05-010.html

---

Watch Out for "Web Bugs"

"Web bugs" are virtually invisible 1-pixel images that act as electronic tags to help websites and advertisers track users' movements across the Internet. "Also called a 'Web beacon,' 'pixel tag,' 'clear GIF' and 'invisible GIF,' it is a method for passing information from the user's computer to a third party Web site. Used in conjunction with cookies, Web bugs enable information to be gathered and tracked in the stateless environment of the Internet."[206] At present, there is no sure way to counteract all web bugs, but products are becoming available to let you "see" web bugs, block them, or remove them. All the products designed to handle web bugs must be downloaded and installed. Only products with free versions are listed here.

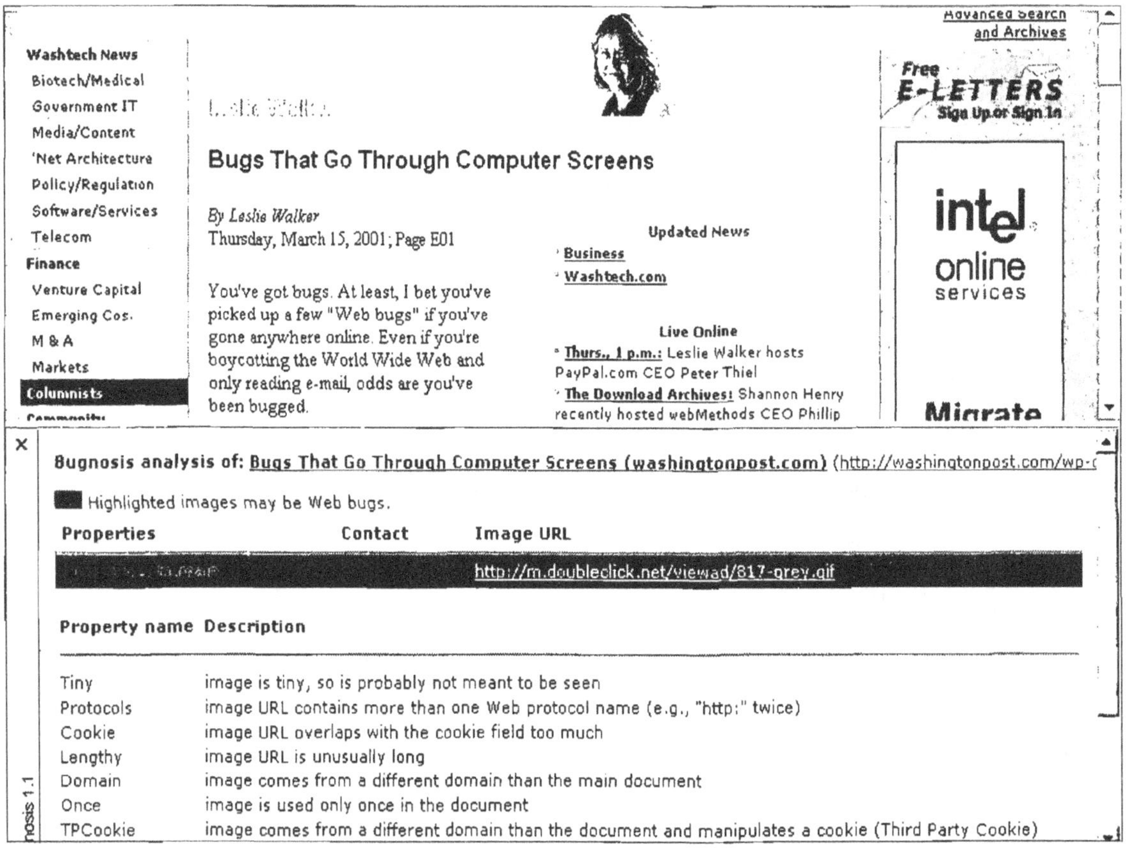
Advanced Search and Archives
Washtech News
Biotech/Medical
Government IT
Media/Content
'Net Architecture
Policy/Regulation
Software/Services
Telecom
Finance
Venture Capital
Emerging Cos.
M & A
Markets
Columnists
Free E-LETTERS Sign Up or Sign In
Bugs That Go Through Computer Screens
By Leslie Walker
Thursday, March 15, 2001; Page E01
You've got bugs. At least, I bet you've picked up a few "Web bugs" if you've gone anywhere online. Even if you're boycotting the World Wide Web and only reading e-mail, odds are you've been bugged.
Updated News
Business
Washtech.com
Live Online
Thurs., 1 p.m.: Leslie Walker hosts PayPal.com CEO Peter Thiel
The Download Archives: Shannon Henry recently hosted webMethods CEO Phillip
intel online services
Bugnosis analysis of: Bugs That Go Through Computer Screens (washingtonpost.com)
Highlighted images may be Web bugs.
Properties
Contact
Image URL
http://m.doubleclick.net/viewad/817-grey.gif
Property name Description
Tiny image is tiny, so is probably not meant to be seen
Protocols image URL contains more than one Web protocol name (e.g., "http:" twice)
Cookie image URL overlaps with the cookie field too much
Lengthy image URL is unusually long
Domain image comes from a different domain than the main document
Once image is used only once in the document
TPCookie image comes from a different domain than the document and manipulates a cookie (Third Party Cookie)

Bugnosis does not block or clear web bugs, but it will certainly make you want to fumigate your computer by letting you see just how many of these pests are infesting the websites you visit, but it only works with Internet Explorer. For a detailed explanation of how web bugs work and how they are used, see the **Web Bug FAQ** provided by the Privacy Foundation.

Bugnosis http://www.bugnosis.org/

Web Bug FAQ http://www.bugnosis.org/faq.html

Guidescope http://www.guidescope.com/home/

WebWasher[207]
http://www.cyberguard.com/products/webwasher/webwasher_products/classic/index.html

Find and Remove Trojan Horses[208]

There are few things worse that can happen to your computer than to become infested with a Trojan horse in general or a RAT in particular. A RAT is a special form of a Trojan: the Remote Access Trojan, which is malicious software that runs invisibly on a computer and permits an intruder to access and control that computer remotely. The reason Trojans and RATs are so pernicious, dangerous, and infuriating is that ***they are difficult to detect and harder still to exterminate***. The best defense, not surprisingly, is a good offense: don't get a Trojan in the first place. So how do most people get Trojans on their computers? There are many ways, but the most common are unwittingly installing them in games or other software, or by opening email attachments.

As if this isn't bad enough, in September 2005, F-Secure identified a new Trojan horse that moves from mobile phones to computers. It appears to be a pirated version of a mobile phone game users can download from the web; the malware installs itself and runs on a PC when a user transfers data from his mobile phone to his computer. The Trojan also infects the phone. While this vulnerability is rated as a

[206] "Web bug," *Computer Desktop Encyclopedia*. Computer Language Company Inc., 2005, *Answers.com*, <http://www.answers.com/topic/web-bug > (14 November 2006).

[207] "Cyberguard has changed the license for Webwasher Classic to Donationware and asks you to make a donation before downloading Webwasher Classic." However, the donation is voluntary.

[208] A Trojan horse is "a program in which malicious or harmful code is contained inside apparently harmless programming or data in such a way that it can get control and do its chosen form of damage." Trojan horses are often used in what are known as "zombie" distributed denial of service attacks in which attackers place Trojans on many computers, then use them as part of a concerted attack, flooding a website or server with so much data it is effectively shut down. Many people have Trojan horses on their computers without knowing it. "Trojan horse," SearchSecurity.com, <http://searchsecurity.techtarget.com/sDefinition/0,,sid14_gci213221,00.html> (14 November 2006).

"low" threat, it is significant because it marks the first time that malware has successfully infected both mobile devices and Windows-based computers.[209]

Trojans are hard to detect because they often use what are called "binder programs" to link them with a legitimate program so that the Trojan will execute in the background at the same time that the legitimate program runs, making the Trojan invisible to the victim.

How can you tell if you have a Trojan on your computer? Some of the telltale signs are unexplained slow performance, a CD tray that mysteriously opens and closes randomly, inexplicable error messages, strange screen images, or the computer automatically rebooting itself. These are by no means the only symptoms and, in fact, there may be no symptoms at all.

Once the Trojan has started to run, it may communicate with its home base via email, by contacting a hidden Internet chat channel, or by using a predefined TCP port, providing the attacker with the computer's IP address. Once activated, the Trojan can then be instructed to do many things, such as formatting a hard drive, sending back financial data, attacking another computer, or participating in a Distributed Denial of Service (aka "zombie") attack against a website. It gets worse. ***Trojans may have the ability to capture keystrokes***, meaning they can gather absolutely any data on a victim's computer, including passwords, credit card numbers, personal communications, files—anything you have, they have. Anything you do, they can do. Anything you see, they can see.

So how do you find and eradicate these vile vermin? First, understand that although good virus scanning software *may* detect and remove many Trojans, ***typical anti-virus scanners may not detect Trojans***. That's because Trojans use techniques to hide themselves. How then can you find out if you have a Trojan? A major clue to a Trojan infection is an unexpected open IP port, especially if the port number matches a known Trojan port. How do you find out which IP ports are open on your computer? It's easy: use the ***netstat*** utility that comes with many operating systems, including Windows. Here's how on a Windows computer:

1. disconnect the computer from the Internet
2. using Task List, close all programs that connect to the Internet (e.g., email, IM)
3. close all open programs running in the system tray

---

[209] Robert McMillan, "Mobile Trojan Horse Trots onto PCs," IDG News Service, PCWorld.com, 22 September 2005, <http://www.pcworld.com/news/article/0,aid,122658,00.asp> (14 November 2006).

4. open a DOS command tool and type netstat –a 15 or select Start | Run | netstat –a 15

Netstat will display all the active and listening IP ports on your computer refreshing every 15 seconds. What you are looking for is ***suspicious port activity***. For example, if port 31337 is active, there is a good chance you have the Back Orifice Trojan on your computer. Also, look for unknown FTP server processes (port 21) or web servers (port 80) that show up using netstat. But remember, you ***must*** disconnect from the Internet and shut down all programs that might use the Internet to get an accurate reading.

Or you can try a free online Trojan scanner such as the one available from PCFlank.com or WindowSecurity.com (below). While a negative report is no guarantee you do not have a Trojan horse, a positive test means you need to take action to remove this infection.

What should you do if you think you have a Trojan on your computer? I strongly recommend that you not start deleting software indiscriminately because something you don't recognize may in fact be a piece of vital software! Instead, if something suspicious shows up in your netstat investigation, now is the time to get some good Trojan-detection and removal software. Below are some sites that will help you locate legitimate anti-Trojan software and provide other advice on how to prevent and remediate infection.

What if you ultimately discover that your computer is infested with a Trojan? Even after you have successfully removed the malware, this may not be the end. How long was the Trojan on your system? What kind of information did it collect and forward? It is probably prudent (if inconvenient) to change all your passwords and even get new credit cards if you have used them on that computer just to be on the safe side. If you do such things as stock trading on your computer, you should probably assume your account has been compromised. In fact, assume everything on your computer has been compromised and treat the invasion as if a thief broke into your house and lived in it for months without your knowledge.

As you can see, Trojan horses are bad, really bad. Again, it is best to avoid them, and the single best defense is not to be promiscuous when it comes to downloading software and opening email attachments. The second best defense is a good firewall. But keep in mind that it is up to you to set the firewall options at a high level of protection to ensure that no Trojan can "phone home" without your permission.

List of Trojan Ports	http://secured.orcon.net.nz/portlist_list.html
Onctek's Trojan Port List	http://www.onctek.com/trojanports.html
Anti-Trojan Software Reviews	http://www.anti-trojan-software-reviews.com/
Anti-Trojan.org	http://www.anti-trojan.org/
Anti-Trojan Guide from Firewall Guide	http://www.firewallguide.com/anti-trojan.htm

PCFlank's Trojan Test Page	http://www.pcflank.com/trojans_test1.htm
WindowSecurity.com TrojanScan	http://www.windowsecurity.com/trojanscan/

Use Good Passwords

Enterprising malicious hackers and thieves are now using sophisticated programs to break passwords. Take a look at this screen shot of just one website offering Windows password crackers:

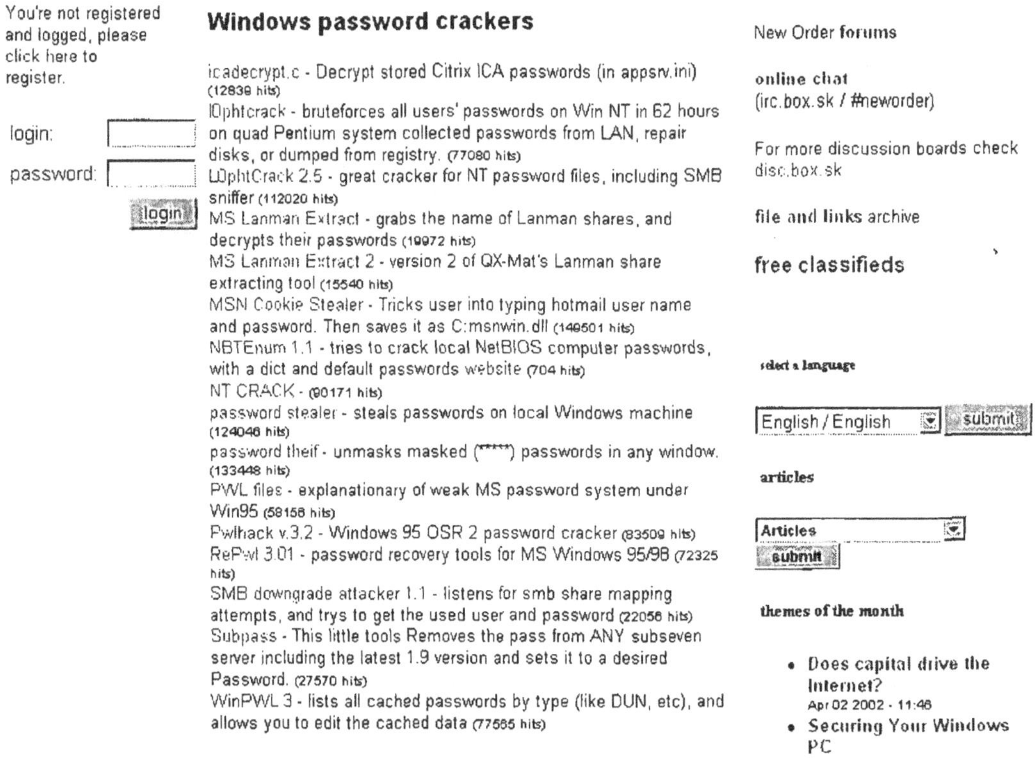
You're not registered and logged, please click here to register.
login:
password:
login

Windows password crackers

icadecrypt.c - Decrypt stored Citrix ICA passwords (in appsrv.ini) (12839 hits)
l0phtcrack - bruteforces all users' passwords on Win NT in 62 hours on quad Pentium system collected passwords from LAN, repair disks, or dumped from registry. (77080 hits)
L0phtCrack 2.5 - great cracker for NT password files, including SMB sniffer (112020 hits)
MS Lanman Extract - grabs the name of Lanman shares, and decrypts their passwords (19972 hits)
MS Lanman Extract 2 - version 2 of QX-Mat's Lanman share extracting tool (15540 hits)
MSN Cookie Stealer - Tricks user into typing hotmail user name and password. Then saves it as C:msnwin.dll (149501 hits)
NBTEnum 1.1 - tries to crack local NetBIOS computer passwords, with a dict and default passwords website (704 hits)
NT CRACK - (90171 hits)
password stealer - steals passwords on local Windows machine (124046 hits)
password theif - unmasks masked (*****) passwords in any window. (133448 hits)
PWL files - explanationary of weak MS password system under Win95 (58158 hits)
Pwlhack v.3.2 - Windows 95 OSR 2 password cracker (83509 hits)
RePwl 3.01 - password recovery tools for MS Windows 95/98 (72325 hits)
SMB downgrade attacker 1.1 - listens for smb share mapping attempts, and trys to get the used user and password (22056 hits)
Subpass - This little tools Removes the pass from ANY subseven server including the latest 1.9 version and sets it to a desired Password. (27570 hits)
WinPWL 3 - lists all cached passwords by type (like DUN, etc), and allows you to edit the cached data (77585 hits)

New Order forums
online chat
(irc.box.sk / #neworder)
For more discussion boards check disc.box.sk
file and links archive
free classifieds
select a language
English / English
submit
articles
Articles
submit
themes of the month
- Does capital drive the Internet? Apr 02 2002 - 11:46
- Securing Your Windows PC

While there is no guaranteed protection against a determined malicious hacker, following these basic rules probably will help protect you and not following them is an invitation to disaster:

- Never use a real word in *any* language (too easy for dictionary attacks to break).
- Never use just letters.
- Make it *at least* 8 characters long.
- Include both upper and lower case letters.

- Include numbers.
- Include special characters.

For a good article on how easy seemingly "good" passwords can be broken and how to pick a strong and memorable password, see Fred Langa's "How to Build Better Passwords" in *Information Week*.

The Simplest Security: A Guide To Better Password Practices
http://www.securityfocus.com/infocus/1537

Microsoft: How to Create Stronger Passwords
http://www.microsoft.com/security/articles/password.asp

Password Security Guide http://www.umich.edu/~policies/pw-security.html

Fred Langa: How to Build Better Passwords
http://www.informationweek.com/story/showArticle.jhtml?articleID=164303537

"Law #5: Weak passwords trump strong security."

Use Desktop Tools with Care

The past few years we have witnessed an explosion in new tools that can be downloaded for free and, in many cases, integrated into the user's browser or operating system. The highest profile of these applications was desktop search. Microsoft, Yahoo, and Ask all have some version of desktop search and there are other smaller companies such as Copernic, X1 Technologies, and Blinkx offering desktop search technology as well. However, Google's product garnered the most attention and generated the greatest controversy. According to Google, its Google Desktop is an "application that provides full text search over your email, computer files, music, photos, chats and web pages that you've viewed." Google Desktop now also indexes the entire content of PDF files and the metadata of multimedia files. In August 2005 Google introduced Google Desktop 2 in beta and dropped "Search" from its name because it does much more than just search. According to Google, "Google Desktop [2.0] doesn't just help you search your computer; it also helps you gather new information from the web with Sidebar, a new desktop feature that shows you your new email, weather and stock information, personalized news and RSS/Atom feeds, and more."[210]

What are the privacy and security concerns surrounding desktop search tools? I think Wendy Boswell, the editor of About.com's *Web Search Guide*, sums up the current state of affairs not only with Google Desktop but with all the major desktop search tools when she writes, "In a very small nutshell, the trouble with Google's Desktop Search is that when you are hooked up to a network of other computers, there are holes in Google's Desktop Search that exploit already known holes in Internet Explorer, and these two just basically open up your computer to any malicious hacker that feels like a bit of snooping."[211] Boswell points out that she uses Google Desktop Search on her own computer, but only because her computer is not networked to any others and she is has anti-virus/security/firewall protection, another backup firewall, and a broadband firewall router. And, I would add, I suspect she knows a lot more than the average user about personal computer security.

The fundamental issue with all the desktop search applications is a familiar one: balancing a very useful tool with a potential loss of privacy. "Desktop search undermines your personal security. Every time you use it, your life's an open book. Or, in this case, an open hard drive."[212] It is precisely the power and scope of desktop search tools that make them so potentially dangerous. Unlike kludgy old Microsoft Windows Explorer, which can take many minutes to search a large hard drive, desktop search tools index a hard drive upon installation and catalog the results to make retrieval very quick, usually within seconds. And desktop search tools can and do find pretty much everything on your computer, even the cache of web pages where you might have entered credit card information, for example. Which helps explain why putting desktop search tools on networked computers may not a good idea at this time. In fact, many organizations have banned the installation and use of Google Desktop, but some have discovered it came preloaded on new computers, such as one state agency that found it preinstalled on its new Dell desktops.[213]

Google's Desktop 2.0 addressed some of these security issues. Google Desktop no longer indexes or stores secure web pages or password-protected files, and the index can be encrypted. The corporate version also allows network administrators to

---

[210] "About Google Desktop," Google.com, <http://desktop.google.com/about.html > (14 November 2006).

[211] Wendy Boswell, "Are You Using Google Desktop Search?", About.com, 20 January 2005, <http://websearch.about.com/b/a/140602.htm?nl=1> (14 November 2006).

[212] David Sheets, "Desktop Search Threatens Your Privacy," *St. Louis Post-Dispatch*, 21 January 2005, <http://www.stltoday.com/techtalk> (article no longer available).

[213] C.J. Kelly, "Google Desktop - Yet Another Security Frightener," Computerworld, 28 December 2006, <http://www.techworld.com/features/index.cfm?featureID=3066&printerfriendly=1> (5 February 2007).

restrict the indexing of specific files. Nonetheless, users who have registered with Google—for example, Gmail account holders—should have more concerns because of the potential for Google to "connect the dots" and create a detailed profile of its registered users.[214]

Google Desktop is not alone in creating concern for security experts. All desktop search tools are inherently problematic, but Microsoft's desktop search tool is probably the most worrisome because it launches ActiveX in Internet Explorer, and ActiveX controls are among the most notoriously vulnerable applications on the web. Neither Microsoft nor Yahoo integrates web and local desktop search as Google does (yet). However, users can limit the Google Desktop to searching the hard drive, disabling the web search feature and thus gaining a measure of security. To do so, users need to make a decision during the setup. At the end of the setup process, Google Desktop asks you to enable or disable "Advanced Features." Enabling Advanced Features "sends Google non-personal data about how you're using the program, along with reports if it ever crashes. It also sends information about the websites you visit so that Sidebar can show personalized info, such as personalized news. Analyzing this data from many users helps our engineers better understand how people actually use Google Desktop and therefore how we can improve it. If you don't want Google Desktop to send this information, simply uncheck the Advanced Features checkbox. Desktop will immediately stop sending any of this non-personal information to Google."[215] You should also uncheck the option to keep your local files and cached web pages permanently out of your Google web search results; this option is under "Google Integration" in the Preferences window.

Search expert Danny Sullivan offers a very good and measured assessment of desktop search, in particular Google Desktop, in which he offers sensible advice for keeping your data safe and private while still enjoying the benefits of desktop search.

Danny Sullivan, "A Closer Look at Privacy and Desktop Search," SearchEngineWatch.com, 14 October 2004, http://searchenginewatch.com/sereport/article.php/3421621

---

[214] Elinor Mills, "Google Balances Privacy, Reach," *CNET News*, 14 July 2005, <http://news.com.com/Google+balances+privacy%2C+reach/2100-1032_3-5787483.html> (14 November 2006).

[215] Google Desktop Features, <http://desktop.google.com/features.html#senddata> (14 November 2006).

Protect Yourself from Search Engine Leaks

In late July 2006, AOL published a list of 20 to 36 million search inquiries collected over a three-month period that included identification numbers for 658,000 unnamed users at their now defunct Research website <http://research.aol.com/>. It didn't take long for some fairly bright researchers to piece together some of the information and come up with real people whose queries were released. This was possible largely because AOL kept individual user's queries together in order to show the pattern of a person's searches over a period of time. "Searches by individual users are grouped together, often forming small profiles of a user's habits and interests. The files include the date and time of each inquiry and the address of the Web site the user chose to visit after searching."[216]

Why would AOL do such a thing in the first place? AOL's intention was to provide useful data to researchers performing "search research." However, the data turned out to be more "helpful" than AOL intended. If you think about it, how much effort does it take to figure out a specific user's name and location if you have three months of his or her searches? And since all the queries also included a date/time stamp and the link to the site they visited from AOL, there are other ways a site manager could use site logs to put together a profile on someone. What some truly enterprising person or group could do with this data is limited only by their imagination. Once the news came out that individuals could be identified from the database, AOL took the data off its website, but of course it was too late. Sites mirroring the database immediately popped up.

The lessons to be drawn from this episode are too many to name, but at the very least we know that what we like to think of as privacy is largely an illusion and what seems like an innocent act of "openness" and "sharing" can backfire in the worst possible way. What can you do to protect yourself against disclosures such as the one described above or from inadvertent leaks of search engine data? I have repeatedly warned people about using search services that require you to log into the site. AOL, Google, Live, and Yahoo all offer such services, which illustrate my rule of thumb: anything that adds convenience brings with it some degradation of privacy and/or security. The fact is that ***you are personally identifiable if you have an account with a search engine site***.

But what is the risk that you can be identified from your searches if you do not have an account at a search site? In light of the AOL incident, *Wired* updated a January 2006 article on this topic, and some of the points they make are as follows:

---

[216] Saul Hansell, "AOL Removes Search Data on Group of Web Users," New York Times, 8 August 2006, <http://www.nytimes.com/2006/08/08/business/media/08aol.html> (archived article requires payment).

"How does a search engine tie a search to a user?

If you have never logged in to a search engine's site, or a sister service like Google's Gmail offering, the company probably doesn't know your name. But it connects your searches through a cookie, which has a unique identifying number. Using its cookies, Google will remember all searches from your browser. It might also link searches by a user's internet protocol address.

How long do cookies last?

It varies, but 30 years is about average. AOL drops a cookie in your browser that will expire in 2034. Yahoo used to set a six-month cookie but now its tracker expires in 2037. A new cookie from Google expires in 2036.

What if you sign in to a service?

If you sign in on AOL, Google or Yahoo's personalized homepage, the companies can then correlate your search history with any other information, such as your name, that you give them. If you use their e-mail or calendar offerings, the companies can tie your searches to your correspondence and life activities. Together these can provide a more complete understanding of your life than many of your friends or family members have.

Why should anyone worry about this leak or bother to disguise their search history?

Some people simply don't like the idea of their search history being tied to their personal lives. Some people check to see if their Social Security or credit card numbers are on the internet by searching for them. Ironically, for more than a few AOL users, the leak of the search terms means that this sensitive information is now on the web."[217]

One of the things the *Wired* article recommends is cookie management. The problem is that unless you routinely refuse all cookies, it is very difficult to avoid some risk of identification, however small that risk may be. Using the Internet without using any cookies is not a realistic option for most of us most of the time, so we have to find a reasonable balance between no cookie use and wide open acceptance of all cookies. Luckily, browsers have gotten much better in the way they permit users to manage cookies. Refer to the section on Managing Your Cookies for details on how to minimize problems with cookies. The *Wired* article also mentions more sophisticated options for protecting your privacy, such as anonymizers and proxy services. None of these comes without a downside or is a guarantee of privacy.

The best approach is to be prudent by limiting your use of cookies via browser settings and/or third-party software to "crunch" cookies. Also, ***never search for personal data, such as your social security or credit card number at any site***

[217] Ryan Singel, "FAQ: AOL's Search Gaffe and You," Wired, 11 August 2006, <http://www.wired.com/news/politics/privacy/1,71579-2.html> (14 November 2006).

where you are registered or logged in, e.g., if you use personalized Google, AOL, Yahoo, Live, etc. If you do, you can be sure there is a record of that search. If you want to run these types of searches, the best thing to do is to block cookies for that search session, then clean out your browser's cache. That way, your search will not be stored anywhere and there will be no "cookie trail" at any site.

A number of articles recently have touted **IxQuick**, a metasearch engine, as an alternative search engine because IxQuick does not keep records of searchers' IP addresses. According to the company, "We have a program running which opens the log files and deletes the user IP addresses and overwrites them...[and] the company removes the unique ID from Ixquick.com's cookies."[218] Of course, you still must place your trust in this Amsterdam-based company not to change their policy or make a mistake. Another option to consider is Clusty, a superb search service based on Vivisimo's technology. Clusty says, "We at Clusty don't track you. Our toolbar doesn't track you. We don't want to know your email address." <http://clusty.com/privacy>

IxQuick	http://ixquick.com/
Clusty	http://clusty.com/

Finally, I also want to mention an article that includes more drastic measures one can take to keep searches private. The focus of the article is Google, but many of the suggestions work with other search engines. *I am not recommending or endorsing any of the software mentioned in the article*, but I thought you should know of other options.

Amit Agarwal
"How to Stop Google from Recording Your Search Habits"
Digital Inspiration, 13 August 2006,
http://labnol.blogspot.com/2006/08/how-to-stop-google-from-recording-your.html

---

Think Twice Before Registering at Search Sites

During the summer of 2005 Google became upset over an article[219] in CNET News demonstrating how much information the author could find about Google CEO Eric Schmidt using—you guessed it—Google. All the information the CNET reporter

---

[218] Declan McCullagh, "FAQ: Protecting Yourself from Search Engines," CNET News, 9 August 2006, <http://news.com.com/2102-1025_3-6103486.html?tag=st.util.print> (14 November 2006).

[219] Elinor Mills, "Google Balances Privacy, Reach," CNET News, 14 July 2005, <http://news.com.com/Google+balances+privacy%2C+reach/2100-1032_3-5787483.html> (14 November 2006).

found was from publicly available sources only. While that is interesting and not surprising, far more intriguing are the observations in the article about what she could have found had the reporter had access to Google's databases.

> "Assuming Schmidt uses his company's services, someone with access to Google's databases could find out what he writes in his e-mails and to whom he sends them, where he shops online or even what restaurants he's located via online maps. Like so many other Google users, his virtual life has been meticulously recorded."[220]

It's not just Google, of course, that collects personal data from ***registered users***. Yahoo, Live Search, A9, and other search services offering registration, online businesses, etc., also collect personal information when you register with them. But Google has so much of the current market share they are the highest profile company in terms of privacy concerns. "Kevin Bankston, staff attorney at the Electronic Frontier Foundation, said Google is amassing data that could create some of the most detailed individual profiles ever devised."[221] How does this happen?

> "As is typical for search engines, Google retains log files that record search terms used, Web sites visited and the Internet Protocol address and browser type of the computer for every single search conducted through its Web site. *[comment: this is true of any website you visit: any site can gather limited, non-personally identifying information that is readily available from the browser.]*
>
> In addition, search engines are collecting personally identifiable information in order to offer certain services. For instance, Gmail asks for name and e-mail address. By comparison, Yahoo's registration also asks for address, phone number, birth date, gender and occupation and may ask for home address and Social Security number for financial services."[222]

The danger lies in the ability to put together all these pieces of data to create a personal profile: "If search history, e-mail and registration information were combined, a company could see intimate details about a person's health, sex life, religion, financial status and buying preferences."[223] ***Simply using Google or any other search engine to search poses little privacy risk*** because of the sheer volume of traffic at these sites and the lack of any personal data about the searcher. ***The real privacy concerns arise when someone is a registered user*** at a site such as Google, Yahoo, AOL, Live Search, or A9. In theory, the information collected and stored about a user could enable someone to put together a remarkably thorough profile of that individual user.

---

[220] Mills.

[221] Mills.

[222] Mills.

[223] Mills.

Both the original CNET article and the Newsfactor article[224] make a good case for why users should either not register at sites such as Google, Yahoo, AOL, Live Search, and A9. However, if you do register, then you should consider using one browser for web searches and another for services such as the search engine's email, toolbar, instant messaging, etc. While there are no known abuses of this information as of now, who knows what the future holds or, worse, what could happen if unscrupulous persons got their hands on this data. This is something to keep in mind, especially when using search engines in the workplace.

Take Care with ZabaSearch

A new people search service called ZabaSearch opened during 2005 and caused an immediate firestorm. This was somewhat surprising given that it is only one more among many such sites offering personal data, but ZabaSearch has been the catalyst for a lot of anger and frustration about our ever-shrinking privacy. One reason ZabaSearch garnered so much attention is because it is offering some of its tantalizing data for free, unlike most services that charge for the same information. But the main reason ZabaSearch captured so much attention is it is the focus of one of those panicky emails warning people about its dangers. While the essence of the email is true, it is misleading because it encourages people to think ZabaSearch is something new, special, or unique. If one were truly cynical, one might even suspect ZabaSearch of being behind those spam mailings as a way of getting people to ask to have their data removed.

I need to emphasize this: ***do not try to have your data removed from ZabaSearch***. ZabaSearch says:

> "If you are interested in creating, editing or deleting records, please submit a valid e-mail address below and we will send you specific instructions on how to do that. Please make sure you can receive e-mail from the ZabaSearch.com domain to insure you receive our reply."[225]

People who have tried to remove their information from ZabaSearch have discovered that ZabaSearch demands they provide even more detailed information about themselves than ZabaSearch already has access to (purportedly on the grounds that they have to ensure you are really who you claim to be). ZabaSearch does not view itself as responsible for the information it provides because it does not own that information. ***All of ZabaSearch's data comes from public databases***

---

[224] Jack M. Germain, "Google Has Your Data: Should You Be Afraid?" Newsfactor Network, 17 August 2005, <http://www.newsfactor.com/story.xhtml?story_id=37466> (15 November 2006).

[225] ZabaTools, ZabaSearch.com, <http://www.zabasearch.com/thankyou.php> (14 November 2006).

maintained by such entities as state, local, and even the US government. Most of this type of data simply cannot be removed from the public record.

If you think we can stop companies like ZabaSearch, think again. As attorney Anita Ramasastry, points out, "[I]n a recent court case, the First Amendment has been held to allow publication *even when it predictably will threaten the safety of particular individuals.* Threats themselves can be made criminal, consistent with the First Amendment. But when information is not itself a threat—but does pose one—courts have recently tended to allow the information to be published, even on the Internet."[226] [*emphasis added*] Ramasastry goes on to say that, in her opinion, sites providing this detailed kind of personal information should be regulated. However, at present only medical records are afforded the kind of legal protection many people would like to see extended to other types of information, e.g., bankruptcy records, divorce data, real estate transactions. As of now, this information is fair game, our privacy is under assault, and the balance of power is on the side of the First Amendment: "...when constitutions do protect privacy, they typically protect it against invasion by the government—not by other citizens. Meanwhile on the other side of the balance, the First Amendment protects a person's right to speak and publish information, absent a compelling governmental interest in silence. So while *privacy rights don't help those who find themselves the subject of digital dossiers,* free speech rights do help the dossier-makers."[227] This is a difficult issue and one the Founders could hardly have imagined because the concepts of things like computers, the Internet, and online identity theft were simply unimaginable for them.

Can You Opt Out of Online Directories?

Many people are interested in (in some cases, desperate to) get their personal information out of the many online directories that now brazenly sport that data. The Privacy Rights Clearinghouse offers a very useful webpage on this subject, including a handy chart of the major "data vendors" who do and who do not offer opt out provisions. The prospect of getting your personal information out of the many databases is daunting and some of the procedures are highly dubious. For example, to get your data out of PeopleFinders, you are required to provide the following information:

> Complete Social Security number, First name, Last name, Middle initial, Aliases and A.K.A.'s, Complete current address, Complete former addresses going back

[226] Anita Ramasastry, "Can We Stop ZabaSearch—and Similar Personal Information Search Engines?: When Data Democratization Verges on Privacy Invasion," FindLaw.com, 12 May 2005, <http://writ.news.findlaw.com/ramasastry/20050512.html> (14 November 2006).

[227] Ramasastry.

20 years , Date of Birth - including month, day, and year. Include print out of info. to be removed.

If you actually provide this much detailed data, you may be opening yourself up to identity theft. Furthermore, the Privacy Rights' page identifies twenty data vendors who offer opt out policies and fifteen that do not. All the vendors who allow users to try to remove personal information have their own procedures and requirements, and even if you diligently follow all these steps and these vendors really do remove the data, this still leaves many more vendors who will not remove your data as well as new vendors, unknown vendors, and foreign vendors. However, that's not the worst of it: "Opting out may prove to be a fruitless venture since often online vendors will simply repopulate the data when they obtain their next download of information from the source. According to People Data, their information is refreshed every three to four months. Your only option would be to check back and go through the opt-out process again if you find your information has been reposted."[228] ***Unless and until there is a way to get personal information out of public databases, requesting online data brokers to remove your information is probably counterproductive.***

In short, trying to keep your personal data private will quickly turn into a full-time job, you almost certainly will not fully succeed, and you will have to keep asking to have your data removed over and over again. So what are we to do? If you are a victim of domestic violence, stalking, or some other such crime, it is worth your time and energy to try to keep your personal information off the Internet and out of these databases. For the rest of us, prevention is the best approach. Guard your "holy trinity" of personal data—name/date of birth, address, and Social Security Number. Be especially leery of providing your Social Security Number. Most companies want your business, and if you refuse to provide an SSN, they probably will still do business with you rather than lose a customer. For now, it appears we are going to have to live with the uneasy balance between privacy and the free flow of information.

Understand the Pros and Cons of an Anonymizing Proxy

If you are truly concerned about revealing anything about yourself as you surf the web, consider using an anonymizing proxy. A proxy is an agent that interfaces between you and the Internet. Most proxies strip out all references to your IP address, your location, your email, types of software you are using, and the previously visited page (http-referrer). Some, such as **Anonymizer**, also let you block cookies and disable scripts, both of which can potentially be used to track your

---

[228] "Online Data Vendors: How Consumers Can Opt Out of Directory Assistance and Non-public Information," Privacy Rights Clearinghouse, February 2006, <http://www.privacyrights.org/ar/infobrokers.htm> (12 September 2006).

movements on the web or disclose information about you. ***One of the big drawbacks with many proxy services is that you may be identified as using an anonymizing proxy, which could "flag" you as someone to watch. Also, keep in mind that you are not anonymous to the proxy provider.***

Most anonymizing services are strictly "http" proxies, which means they only give you "anonymity" when browsing webpages, which is all you need most of the time. My experience with proxies is that they *probably will slow you down*. Several years ago there were documented problems with anonymizers that allowed websites to view your real IP address. These bugs have largely been fixed but if you are using any of these services, *be sure to turn off JavaScript, Java, and ActiveX controls in your browser*. Check privacy guru Richard Smith's Computerbytesman page to test any anonymizing service for leaks.

Finally, anonymizing proxies may create a *false sense of security* that in itself can be dangerous. One experimental Trojan horse program, Setiri, actually disguises itself as Internet Explorer, connects to a website via Anonymizer.com, and uses Anonymizer to execute commands from the victim's computer. Once connected the Trojan can download programs, such as keystroke monitoring software, and steal any data on that computer, sending it via Anonymizer so it cannot be traced.[229] While the Setiri Trojan does not exploit a flaw in Anonymizer, it does point to how malicious users can turn good things to evil purposes.

Warning: Never use an anonymizing proxy that requires registration to use a **free** service! Some proxies have been associated with people and organizations that want to gather information about users.

InfoAnarchy's Anonymous Web Searching http://www.infoanarchy.org/en/Anonymous_Web_Surfing

Free Web Anonymizer Services http://www.cexx.org/anony.htm

Web Anonymizing Services http://www.computerbytesman.com/anon/index.htm

Test Page for Web Anonymizing Services http://www.computerbytesman.com/anon/test.htm

"Law #9: Absolute anonymity isn't practical, in real life or on the web."

---

[229] Kim Zetter, "Trojan Horse Technology Exploits IE," PCWorld.com, 5 August 2002, <http://www.pcworld.com/news/article/0,aid,103620,tk,wb081202x,00.asp> (14 November 2006).

Convert with Caution

As part of its initiative to enhance software security and share this information with users, the National Security Agency's Information Assurance Directorate published a new guide in December 2005: "Redacting with Confidence: How to Safely Publish Sanitized Reports Converted from Word to PDF." This is a very important issue because failure to redact documents properly—whether they are declassified government documents, court records, proprietary company documents—can lead not just to embarrassment but also to very serious security violations and potential risks to individuals. I call your attention to the very sad case in May 2005 in which an improperly prepared PDF document about the killing of the Italian intelligence agent Nicola Calipari in Iraq was quickly discovered and exploited by the press worldwide. Not only was classified information leaked to the world, but the lives of those whose identities were revealed were also put in jeopardy by the improper method of removing data from a MS Word file and converting it to PDF. This is an important guide and I urge you to keep a copy for yourself and your organization.

"Redacting with Confidence: How to Safely Publish Sanitized Reports Converted from Word to PDF"
Architectures and Applications Division of the Systems and Network Attack Center (SNAC)
Information Assurance Directorate, National Security Agency
last updated 2 February 2006
http://www.nsa.gov/snac/index.cfm?MenuID=scg10.3.1

For details on the Calipari incident and the ensuing disclosure of classified information, I recommend an article from the Times Online (UK).[230]

Your privacy and security are only as good as the weakest link using your computer (a spouse, a child, or your teenager's friends...)

---

[230] Simon Freeman, "Italy Releases Report into Death of Security Agent," Times Online, 2 May 2005, <http://www.timesonline.co.uk/article/0,,7374-1594880,00.html> (14 November 2006).

Always Put Privacy and Security Before Convenience

Remember the quote from Scott MacNealy? It is tempting to store credit card and password information on your hard drive or let a site retain your credit card number or log you in automatically. I highly recommend you eschew these conveniences and force yourself to ***enter sensitive information every time you need to use it and only when it is absolutely necessary***. Do not volunteer information about yourself and only fill in the required boxes on forms. An enterprising thief can break into your computer, steal the contents of it, and get out without your ever knowing he was there. Also, if you don't store credit card information at websites, that data won't be sitting in a database potentially waiting to be stolen. Every time you do something new or different on the Internet or your computer, ask yourself if it could potentially compromise your privacy or security, then decide if the benefits outweigh the risks before proceeding.

"Law #1: If a bad guy can persuade you to run his program on your computer, it's not your computer anymore."

General Security & Privacy Resources

The best defenses against the many dangers lurking on the Internet are awareness and information. Because security and privacy threats are so pervasive and increasing in number and potency, staying on top of threats and means of protection is crucial.

Steve Gibson, rightly famous for his Shields Up! website and free software (e.g., "UnPlug n' Pray"), launched a new service with TechTV's Leo Laporte in 2005. Every Thursday afternoon they create a 20-25 minute audio column about personal computer security called "Security Now!" The topics covered include personal passwords (a must read), NAT routers as firewalls (another must read), "HoneyMonkeys" (no, I'm not making that up), unbreakable WiFi security, and bad WiFi security. The audio broadcasts are archived in several formats, including a text file, a PDF version, and an HTML webpage. There is also an option to receive an email reminder whenever the page is updated. Gibson has the ability to cut through the jargon to explain these topics clearly and to offer practical advice on how to handle personal computer security issues.

Security Now! http://www.grc.com/securitynow.htm

The following are a few more of the many excellent sites providing news, information, and advice on Internet privacy and security.

Center for Privacy and Technology Ten Ways to Protect Privacy Online
http://www.cdt.org/privacy/guide/basic/topten.html

EPIC Online Guide to Privacy Resources
http://www.epic.org/privacy/privacy_resources_faq.html

Georgi Guninski Security Research http://www.guninski.com/

Security Focus http://www.securityfocus.com/

Yahoo News Computer Security
http://fullcoverage.yahoo.com/fc/Tech/Computer_Security

Yahoo News Cybercrime and Internet Fraud
http://news.yahoo.com/fc/Tech/Cybercrime_and_Internet_Fraud/

Yahoo News Internet Privacy http://news.yahoo.com/fc/Tech/Internet_Privacy/

Conclusion

The overall implications of the Internet for how we work and how we play are just beginning to be discussed and understood. The Internet is changing, or at the very least touching, people's lives in ways we have not imagined. I close with an example of the reach of the web. My 97-year-old aunt in South Carolina had a bit part in an obscure movie in 1989. Despite the fact that the movie has been largely forgotten, my aunt has an "Actress Filmography" in the Internet Movie Database. She, of course, was unaware of her Internet presence and was both thrilled and more than a little shocked to find that even she was "in cyberspace."

The point, of course, is that no one is out of reach of this powerful, invasive technology. We change the world with our technology and we, in turn, are altered by that same technology. It remains to be seen where our technology leads us, whether into an "endless frontier"[231] or, more ominously, into a "cemetery of dead ideas."[232]

[231] Vannevar Bush, *Science: The Endless Frontier,* Washington, D.C.: United States Government Printing Office, 1945.

[232] Miguel de Unamuno, *The Tragic Sense of Life,* Princeton: Princeton University Press, 1990. (November 2005), p. 100.

Web Sites by Type

General Purpose Search Engines

A9	http://a9.com/
Ask	http://www.ask.com/
Exalead	http://www.exalead.com/search
Gigablast	http://www.gigablast.com/
Google	http://www.google.com/
Live Search	http://www.live.com/
Yahoo	http://search.yahoo.com/

Directories

Best of the Web	http://botw.org/default.aspx
Galaxy	http://www.galaxy.com/
Google Directory	http://directory.google.com/
Open Directory	http://dmoz.org/
Yahoo Directory	http://dir.yahoo.com/

Metasearch Sites

Open Directory's List of Metasearch Sites	http://dmoz.org/Computers/Internet/Searching/Metasearch/
Clusty	http://clusty.com/
Dogpile	http://www.dogpile.com/
Ithaki	http://www.ithaki.net/indexu.htm
IxQuick	http://www.ixquick.com/
Jux2	http://www.jux2.com/
Mamma	http://www.mamma.com/
Metacrawler	http://www.metacrawler.com/
The Pandia Metasearch Engine	http://www.pandia.com/metasearch/index.html

Search.com http://www.search.com/

Surfwax http://www.surfwax.com/

Megasearch Sites

All Search Engines http://www.allsearchengines.com/

Find It Quick http://www.quickfindit.com/Search_Engines/

Search—22 http://www.search-22.com/

SearchEzee http://www.searchezee.com/search.shtml

Internet Guides and Tutorials

BrightPlanet's Guide to Effective Searching of the Internet http://www.brightplanet.com/deepcontent/tutorials/search/index.asp

Finding Information on the Internet: A Tutorial http://www.lib.berkeley.edu/TeachingLib/Guides/Internet/FindInfo.html

Internet Tutorials from University of Albany Libraries http://www.internettutorials.net/

Internet Scout Report http://scout.wisc.edu/Projects/PastProjects/toolkit/searching/index.html

Intute: Virtual Training Suite http://www.vts.intute.ac.uk/

Pandia's Goalgetter http://www.pandia.com/goalgetter/index.html

Phil Bradley's Searching the Internet http://www.philb.com/searchindex.htm

Search Engine Watch Tutorials (old but still useful) http://www.searchenginewatch.com/resources/article.php/2156611

Web Search Guide http://www.websearchguide.ca/tutorials/tocfram.htm

Google Help & Tools

Google Help http://www.google.com/help/features.html

Google Guides http://www.google.com/press/guides.html

Google Book Search http://books.google.com/

Google Language Tools http://www.google.com/language_tools

Google Scholar http://scholar.google.com/

Google International Sites http://www.google.com/language_tools

Google Blog Search http://blogsearch.google.com/

Google Patent Search http://www.google.com/patents

Google Directory	http://directory.google.com/
Google SMS	http://www.google.com/sms/
Google Scholar	http://scholar.google.com/
Google Trends	http://www.google.com/trends
Google Find Related Images	http://blog.outer-court.com/related/
Simply Google	http://www.usabilityviews.com/simply_google.htm
Google Rankings	http://www.googlerankings.com/kdindex.php
Google Compare	http://oy-oy.eu/google/world/

Specialized Search Tools

Answers.com	http://www.answers.com/
Babelplex	http://www.babelplex.com/
Fagan Finder Search by File Type	http://www.faganfinder.com/filetype/
Google Trends	http://www.google.com/trends
Neighborsearch	http://www.blog.outer-court.com/neighborsearch/
OAIster	http://www.oaister.org/
Searchroller	http://www.researchbuzz.org/2004/10/new_yahoo_hack_searchroller_fo.shtml
The Wayback Machine	http://web.archive.org/
Yahoo Proximity Search	http://www.researchbuzz.org/2004/10/ynaps_yahoo_nonapi_proximity_s.shtml

Custom Search Engines

Gigablast's Custom Topic Search	http://www.gigablast.com/cts.html
Google Custom Search Engine	http://www.google.com/coop/cse/overview
Yahoo Search Builder	http://builder.search.yahoo.com/
Windows Live Search Macros	http://search.live.com/macros/default.aspx
Rollyo	http://rollyo.com/
Eurekster's Swicki	http://swicki.eurekster.com/
PSS	http://www.pssdir.com/
Alexa Web Search Platform	http://websearch.alexa.com/welcome.html

Subject Guides, Virtual Libraries, and Reference Desks

About	http://www.about.com/
Encyclopedia.com	http://www.encyclopedia.com/
Encyclopedia Britannica[233]	http://www.britannica.com/
Hotsheet	http://www.hotsheet.com/
INFOMINE	http://infomine.ucr.edu/
Information Please	http://www.infoplease.com/
Intute (formerly RDN)	http://www.intute.ac.uk/
The Internet Public Library	http://www.ipl.org/
Librarians' Index to the Internet	http://lii.org/
The Library Spot	http://www.libraryspot.com/
Martindale's The Reference Desk	http://www.martindalecenter.com/
My Virtual Reference Desk	http://www.refdesk.com/
Pinakes Subject Gateway[234]	http://www.hw.ac.uk/libWWW/irn/pinakes/pinakes.html
Wikipedia	http://en.wikipedia.org/
WWW Virtual Library	http://vlib.org/Overview.html
Yahoo Reference	http://education.yahoo.com/reference/

Wikipedia and Wikipedia Search

Wikipedia	http://en.wikipedia.org/
Search Web Links at Wikipedia	http://en.wikipedia.org/w/index.php?title=Special%3ALinksearch
Clusty's Wikipedia Search (English only)	http://wiki.clusty.com/
FUTEF (Beta)	http://futef.com/
Qwika	http://www.qwika.com/
LuMriX	http://wiki.lumrix.net/
Wikiseek	http://wikiseek.com/
WikiWax	http://www.wikiwax.com/

---

[233] Although full-text articles require a paid subscription to *Encyclopedia Britannica*, the site is still a useful starting place for research and includes free access to the *Britannica Concise Encyclopedia*.

[234] Pinakes is the gateway to EEVL and dozens of other equally valuable specialized research sites.

Best Mapping Sites

Ask Maps http://maps.ask.com/maps

France Telecom's Pages Jaunes http://photos.pagesjaunes.fr/

Google Earth (must be downloaded) http://earth.google.com/

Google Maps http://maps.google.com/

Map24 http://www.map24.com/

MapQuest http://www.mapquest.com/

Maporama http://www.maporama.com/share/

Mappy's Aerial Photos http://www.mappy.com/ (select Maps | Aerial Photos)

Multimap (excellent source of maps worldwide) http://www.multimap.com/

Spain's Callejero Fotographico http://www.qdq.com/indexfotos.asp

Mappy (Europe) http://www.mappy.com/

ViaMichelin (Europe, US, Canada)
http://www.viamichelin.com/viamichelin/gbr/dyn/controller/Maps

Windows Live Local/Virtual Earth http://local.live.com/

Windows Live Spaces/Virtual Earth http://virtualearth.spaces.live.com/

Yahoo Maps http://maps.yahoo.com/

Best Map MetaIndices

About's Maps http://geography.about.com/science/geography/msub1.htm

Martindale's "Virtual" Geoscience Center
http://www.martindalecenter.com/GradGeoscience_5_GG.html

Odden's The Fascinating World of Maps and Map-Making
http://oddens.geog.uu.nl/index.html

Perry-Castaneda Library Map Collection at the University of Texas Austin
http://www.lib.utexas.edu/Libs/PCL/Map_collection/map_sites/map_sites.html

ReisWijs Route Planner Metasite
http://www.reiswijs.co.uk/routeplanner/routeplanner.html

Book Search

A9 (select "books by Amazon") http://www.a9.com/

Amazon (search "Books") http://www.amazon.com/

Google Book Search http://books.google.com/

Live Book Search (Beta) http://books.live.com/

Metasearch for Books http://kokogiak.com/booksearch/

The "Invisible Web"

A9 http://www.a9.com/

Aardvark Asian Databases http://www.aardvarknet.info/user/subject26/index.cfm?all=All

Amazon http://www.amazon.com/

Answers http://www.answers.com/

BUBL Catalog http://www.bubl.ac.uk/link/

The Complete Planet http://www.completeplanet.com/

Deep Web Research http://www.deepwebresearch.com/

Infomine http://infomine.ucr.edu/

Intute (formerly RDN) http://www.intute.ac.uk/

Pinakes Subject Launchpad http://www.hw.ac.uk/libWWW/irn/pinakes/pinakes.html

Research Beyond Google: 119 Authoritative, Invisible, and Comprehensive Resources http://oedb.org/library/college-basics/research-beyond-google

The Wayback Machine http://web.archive.org/

Scholarly Search

Answers.com http://www.answers.com/

Citeseer☆ http://citeseer.ist.psu.edu/

CiteULike http://www.citeulike.org/

Cornell University's arXiv.org http://arxiv.org/

Foreign Doctoral Dissertations http://www.crl.edu/content.asp?l1=5&l2=23&l3=44&l4=25

Google Scholar http://scholar.google.com/

Ingenta Connect http://www.ingentaconnect.com/

Infomine's Electronic Journals Search http://infomine.ucr.edu/cgi-bin/search?ejournal

ISI Highly Cited http://isihighlycited.com/

Live Academic http://academic.live.com/

OAIster http://www.oaister.org/

Research Now	http://researchnow.bepress.com/
Scholar Universe	http://www.scholaruniverse.com/index.jsp
Science Direct (select Abstract Databases tab)	http://www.sciencedirect.com/
Scirus	http://www.scirus.com/srsapp/
Wiley InterScience Journal Search	http://www3.interscience.wiley.com/

Browser-Related Pages

Microsoft Internet Explorer	http://www.microsoft.com/windows/ie/default.htm
Mozilla Firefox	http://www.mozilla.com/firefox/
Netscape 7.1 Streamline	http://sillydog.org/narchive/sd/71.html
Netscape Archive (7.1 or 7.2)	http://browser.netscape.com/ns8/download/archive.jsp

Search News and Blogs

Google Operating System	http://googlesystem.blogspot.com/
John Battelle's Searchblog	http://battellemedia.com/
Live Search Weblog	http://blogs.msdn.com/msnsearch/default.aspx
Official Google Blog	http://googleblog.blogspot.com/
Pandia Search Central	http://pandia.com/
Philipp Lenssen's Google Blogoscoped	http://blog.outer-court.com/
Phil Bradley's Weblog	http://philbradley.typepad.com/phil_bradleys_weblog/
Research Buzz	http://www.researchbuzz.com/
Resource Shelf	http://www.resourceshelf.com/
Search Day	http://searchenginewatch.com/searchday/
Search Engine Showdown	http://www.searchengineshowdown.com/
Search Engine Watch	http://searchenginewatch.com/
Search Engine Watch Web Searching Tips	http://www.searchenginewatch.com/facts/index.html
Web Master World	http://www.webmasterworld.com/
Web Search Guide	http://www.websearchguide.ca/
Search Engine Watch Blog	http://blog.searchenginewatch.com/blog/
Yahoo Search Blog	http://www.ysearchblog.com/

Links to Online Dictionaries

Foreignword http://www.foreignword.com/Tools/dictsrch.htm

Language Automation's Glossaries http://www.rahul.net/lai/glossaries.html

Martindale's Language and Translation Center http://www.martindalecenter.com/Language.html

Paderborn University List of Dictionaries http://www-math.uni-paderborn.de/dictionaries/Dictionaries.html

Word2Word http://www.word2word.com/dictionary.html

yourDictionary http://www.yourdictionary.com/

Online Multilingual Dictionaries

Digital Dictionaries of South Asia http://dsal.uchicago.edu/dictionaries/

Eurodicautom* http://europa.eu.int/eurodicautom/Controller

Foreignword http://www.foreignword.com/Tools/dictsrch.htm

Language to Language http://www.langtolang.com/

Logos * http://www.logos.it/lang/transl_en.html

OneLook Dictionaries http://www.onelook.com/

Online Dictionary http://www.online-dictionary.biz/
English↔French, German, Spanish, Italian, Japanese, Chinese, Russian

Papillon Project http://www.papillon-dictionary.org/Home.po
English↔Estonian, German, French, Japanese, Vietnamese, Korean, Malay, Chinese

FreeDict http://www.freedict.com/

Travlang's Translating Dictionaries http://dictionaries.travlang.com/

UltraLingua http://www.ultralingua.net/
English↔German, French, Spanish, Italian, Portuguese, Esperanto, Latin

Word Reference http://www.wordreference.com/

Online Text Translators

AjaxTrans http://ajax.parish.ath.cx/translator/

Babelfish from Yahoo http://babelfish.yahoo.com/

FreeTranslation** http://www.FreeTranslation.com/

Foreignword http://foreignword.com/Tools/transnow.htm

Google Language Tools http://www.google.com/language_tools

InterTran**	http://www.tranexp.com/win/itserver.htm
Mezzofanti Translations	http://www.mezzofanti.org/translation/
PhraseBase	http://www.phrasebase.com/english/phrases/
PopJisyo (Asian languages)	http://www.online-dictionary.biz/
PROMT**	http://www.translate.ru/eng/text.asp
Reverso**	http://www.reverso.net/text_translation.asp
VoyCabulary	http://www.voycabulary.com/
WorldLingo**	http://www.worldlingo.com/products_services/worldlingo_translator.html
yourDictionary	http://www.yourdictionary.com/diction1.html#translate

Online Web Page Translators

Ajeeb! Arabic ↔ English+	http://tarjim.ajeeb.com/ajeeb/default.asp?lang=1
Babelfish from Yahoo	http://babelfish.yahoo.com/
Google Language Tools	http://www.google.com/language_tools
InterTran**	http://www.tranexp.com/win/itserver.htm
PROMT	http://www.translate.ru/eng/srvurl.asp
Reverso**	http://www.reverso.net/url_translation.asp
Systran	http://www.systransoft.com/
VoyCabulary	http://www.voycabulary.com/
WorldLingo**	http://www.worldlingo.com/en/websites/url_translator.html

+ Requires free registration
* Translates to/from multiple languages at once
** Site offers virtual keyboard or special characters for non-English translations

Finding International Search Engines

All Search Engines.com	http://www.allsearchengines.com/foreign.html
Beaucoup!	http://www.beaucoup.com/
European Search Engines	http://www.netmasters.co.uk/european_search_engines/
FetchFido European Search Engines	http://homepage.ntlworld.com/fetchfido2/interface/search_engines_european.htm
FetchFido World Search Engines	http://homepage.ntlworld.com/fetchfido2/interface/search_engines_worldwide.htm
FinderSeeker	http://www.finderseeker.com/

Google International Sites http://www.google.com/language_tools

Infisource Foreign Language Search Engines http://www.infinisource.com/search-engines.html#foreign

International Search Engines http://www.arnoldit.com/lists/intlsearch.asp

ISEDB Local and Regional Search Engines http://www.isedb.com/html/Internet_Search_Engines/Local_and_Regional_Search_Engines/

ISEDB Local and Regional Directories http://www.isedb.com/html/Web_Directories/Local_and_Regional_Directories/

Phil Bradley's Country Based Search Engines http://www.philb.com/countryse.htm

Regional and Special Search Engines http://www.ntu.edu.sg/lib/search/specialframe.htm

Search Engine Colossus http://www.searchenginecolossus.com/

Search Engine Guide http://www.searchengineguide.com/pages/Regional/

Search Engine Index http://www.search-engine-index.co.uk/Regional_Search/

Search Engines 2 http://www.search-engines-2.com/

Search Engines Worldwide (2003) http://home.inter.net/takakuwa/search/

Ultimate Search Engines Links Page http://www.searchenginelinks.co.uk/

Yahoo International http://world.yahoo.com/

Finding Email Directories

Email-Directory.com http://www.email-directory.com/

Nedsite http://www.nedsite.nl/search/people.htm#email

International Email Lookup Tools

Addresses.com http://www.allemailaddresses.com/

Infospace Email Lookup http://www.infospace.com/home/white-pages/email-search

Infospace Reverse Email Lookup http://www.infospace.com/home/white-pages/reverse-email

Look4U http://www.look4u.com/english/

MESA MetaEmailSearchAgent http://mesa.rrzn.uni-hannover.de/

Peoplesearch Reverse Email Search http://peoplesearch.net/peoplesearch/peoplesearch_reverse_email_address.html

World Email Directory http://www.worldemail.com/freemail.htm

Email Megadirectories

Freeality Email Lookup http://www.freeality.com/findet.htm

Infospace International Directories http://www.infospace.com/intl/int.html

MESA MetaEmailSearchAgent http://mesa.rrzn.uni-hannover.de/

Nedsite http://www.nedsite.nl/search/people.htm#email

Peoplesearch
http://peoplesearch.net/peoplesearch/peoplesearch_reverse_email_address.html

Infobel Email Lookup http://www.infobel.com/teldir/teldir.asp?page=/eng/more/email

Tools for International Telephone Lookups

AnyWho International http://www.anywho.com/international.html

AOL International Directories http://www.aol.com/netfind/international.html

EscapeArtist Telephone Search Engine
http://www.escapeartist.com/global/telephone.htm

Global Yellow Pages http://www.globalyp.com/world.htm

Infobel http://www.infobel.com/World/default.asp

Infobel's Telephone Directories on the Web http://www.infobel.com/teldir/

Infospace International Directories http://www.infospace.com/intl/int.html

International White & Yellow Pages http://www.wayp.com/

Nedsite http://www.nedsite.nl/search/people.htm#telephone

Phonebook of the World http://www.phonebookoftheworld.com/

SBN International Yellow Pages http://www.sbn.com/international/international.asp

Specialty Telephone Lookups

ACR's International Calling Codes by country
http://www.the-acr.com/codes/cntrycd.htm

ACR's International Calling Codes listed numerically
http://www.the-acr.com/codes/cntryno.htm

Americom's International Decoder http://decoder.americom.com/

International Dialing Codes http://kropla.com/dialcode.htm

International City Codes http://www.numberingplans.com/kropla/

World Telephone Numbering Guide http://www.wtng.info/index.html

Online Video Search

AOL Video Search http://search.aol.com/aolcom/videohome

BBC Video http://news.bbc.co.uk/

Blinkx http://www.blinkx.tv/

CBS News Video Search http://www.cbsnews.com/sections/i_video/main500251.shtml

CNN Video Homepage http://www.cnn.com/video/

CNN Video Almanac http://www.cnn.com/resources/video.almanac/

C-SPAN http://www.c-span.org/

C-SPAN Store http://www.c-spanstore.org/shop/

Google Video http://video.google.com/

IFILM http://www.ifilm.com/

MSN Video http://video.msn.com/

Pixsy http://pixsy.com/

Reuters Video http://today.reuters.com/tv

RocketInfo http://www.rocketnews.com/ [select the VIDEO tab]

RooTV http://www.rootv.com/

Searchforvideo http://www.searchforvideo.com/home/index.html

Searchforvideo IM Service http://www.searchforvideo.com/misc/im.jsp

Searchforvideo Reel Time Lens http://www.searchforvideo.com/misc/reel.jsp

Sky News Video http://www.sky.com/skynews/video

TVEyes http://tveyes.com/

Yahoo Video Search http://video.yahoo.com/

Yahoo News Video http://news.yahoo.com/video

YouTube http://www.youtube.com/

Podcasting

Blinkx http://www.blinkx.tv/

Odeo http://odeo.com/

Podcast Alley http://www.podcastalley.com/

Podcast.net http://www.podcast.net/

Podscope http://www.podscope.com/

Podzinger	http://www.podzinger.com/
Yahoo Podcast Search (Beta)	http://podcasts.yahoo.com/

Podcast Directories

iPodder	http://www.ipodder.org/directory/4/podcasts
Podcast Directory	http://www.podcastdirectory.com/
Podfeed	http://www.podfeed.net/
Podcasting Station	http://www.podcasting-station.com/
Podcast Shuffle	http://www.podcastshuffle.com/

Newsgroups & Mailing Lists

Google Groups	http://groups.google.com/
BoardReader	http://www.boardreader.com/
BoardTracker	http://www.boardtracker.com/
Omgili	http://www.omgili.com/
Yahoo Groups	http://groups.yahoo.com/
Yahoo Member Directory	http://members.yahoo.com/
CataList	http://www.lsoft.com/lists/listref.html
Tile.net	http://www.tile.net/

Weblog Search

Blogdigger	http://www.blogdigger.com/
Blog Search Engine	http://www.blogsearchengine.com/
Blogwise	http://www.blogwise.com/
Bloogz	http://www.bloogz.com/
Clusty Blog Metasearch	http://blogs.clusty.com/
Daypop	http://www.daypop.com/
Feedster	http://www.feedster.com/
Google Blogsearch	http://blogsearch.google.com/
IceRocket	http://blogs.icerocket.com/
Sphere	http://www.sphere.com/
Technorati ☆	http://www.technorati.com/

General News Sources

ABYZ Newslinks	http://www.abyznewslinks.com/
Guardian's World News Guide	http://www.guardian.co.uk/worldnewsguide/
HeadlineSpot	http://www.headlinespot.com/
Kiosken	http://www.esperanto.se/kiosk/engindex.html
Metagrid (newspapers & magazines)	http://www.metagrid.com/
NewsCentral (online newspaper links)	http://www.all-links.com/newscentral/
NewsDirectory	http://newsdirectory.com/
Newslink	http://newslink.org/
Online Newspapers	http://www.onlinenewspapers.com//index.htm
RefDesk (My Virtual Newspaper)	http://www.refdesk.com/papmain.html

News Search Services

Google News	http://news.google.com/
Google News Archive	http://news.google.com/archivesearch
HavenWorks	http://havenworks.com/news/search/
JournalismNet	http://www.journalismnet.com/
MSN Newsbot	http://newsbot.msnbc.msn.com/
NewsNow	http://www.newsnow.co.uk/
Pandia Newsfinder	http://www.pandia.com/news/
Topix.net	http://www.topix.net/
Worldnews	http://www.wn.com/
Yahoo News	http://news.yahoo.com/

Technology News on the Web

Newsfactor Network	http://www.newsfactor.com/
TechNews.com	http://www.washingtonpost.com/wp-dyn/technology/
TechWeb	http://www.techweb.com/
The Register	http://theregister.co.uk/
Wired News	http://www.wired.com/
ZDNet News	http://zdnet.com.com/

Telecommunications on the Web

Analysys Telecoms Virtual Library http://www.analysys.com/vlib

Computer and Communication Entry Page http://www.cmpcmm.com/cc

Goodman's Bookmarks http://www.gbmarks.com/

IT Landscape in Nations Around the World
http://www.american.edu/academic.depts/ksb/mogit/country.html

Lido Telecom Web Central http://www.telecomwebcentral.com/secure/links/

Bandwidth Market Telecom Links
http://www.bandwidthmarket.com/component/option,com_weblinks/Itemid,4/

World Wide Web Telecommunication Center
http://home.planet.nl/~wvhwvh/teletop.htm

Researching PTTs & Telecom Operators Around the World

Country Index for Major PTTs, PTOs, and Major Service Providers
http://home.planet.nl/~wvhwvh/countidx.htm

Goodman's International Telecom Companies
http://www.gbmarks.com/html/international.html

IT Landscape in Nations Around the World
http://www.american.edu/academic.depts/ksb/mogit/country.html

ITU's Global Directory of Regulators (select ***Regulators*** for PTTs)
http://www.itu.int/cgi-bin/htsh/mm/scripts/mm.search

World Wide Web Telecommunication Resource Center
http://home.planet.nl/~wvhwvh/teletop.htm

Radio, Television, and Satellite Broadcasting

Radio Locator http://www.radio-locator.com/

Radio, TV, and Satellite Links http://www.liensutiles.org/sat.htm

Live Radio http://www.live-radio.net/info.shtml

Radio Station World http://radiostationworld.com/default.asp

Mike's Radio World http://www.mikesradioworld.com/

vTuner http://www.vtuner.com/

USC Satellite Database
http://www.ucsusa.org/global_security/space_weapons/satellite_database.html

Heaven's Above Satellite Database http://www.heavens-above.com/selectsat.asp

SatcoDX Satellite Chart http://www.satcodx.com/eng/

NASA's J-Track Satellite Tracking http://science.nasa.gov/RealTime/JTrack/

Small Satellites Home Page http://centaur.sstl.co.uk/SSHP/

Search for People

Biography Center http://www.biography-center.com/

Biography Reference Center from MacGill University http://www.library.mcgill.ca/refshelf/biograph.htm

Chinese Biographical Database http://www.lcsc.edu/cbiouser/

Wolfram's Science World Biography http://scienceworld.wolfram.com/biography/

ISI Highly Cited Researchers http://www.isihighlycited.com/

Google Groups http://groups.google.com/

Yahoo Member Directory http://members.yahoo.com/

ICQ User Directory http://people.icq.com/whitepages/

Forbes People Lists http://www.forbes.com/lists/

Search the SEC's Edgar Database http://searchwww.sec.gov/EDGARFSClient/jsp/EDGAR_MainAccess.jsp

SurfWax SEC Search http://lookahead.surfwax.com/edgar/

EdgarScan Advanced Search http://edgarscan.pwcglobal.com/servlets/advancedsearch

Deadline Online's People Finders http://www.deadlineonline.com/peoplefinders.html

Langenberg.com Person Finder http://person.langenberg.com/

LexNotes Telephone and Email Directories http://www.lexnotes.com/sources/people/fonemail.shtml

Pandia People Search http://www.pandia.com/people/

People Search Engines http://www.people-search-engines.com/

People Search Links http://www.peoplesearchlinks.com/

People Search Sites http://www.nettrace.com.au/resource/search/people.html

Power Reporting People Finders http://powerreporting.com/category/People_finders

Public Record Finder Outside the US http://www.publicrecordfinder.com/outside_usa.html

Searchbug People Finder http://www.searchbug.com/peoplefinder/

Search Systems Free Public Records Database http://www.searchsystems.net

The Virtual Chase People Finder Guide http://www.virtualchase.com/people/

The Virtual Chase Finding People Guide
http://www.virtualchase.com/topics/people_finder_index.shtml

The Virtual Gumshoe http://www.virtualgumshoe.com/

ZoomInfo http://www.zoominfo.com/

Landings Certified Pilots Database
http://www.landings.com/_landings/pages/search/certs-pilot.html

The Virtual Chase Criminal Records
http://www.virtualchase.com/resources/criminal_records.html

CrimeNet http://www.crimenet.com.au/

The Black Book Online http://www.crimetime.com/online.htm

NameBase http://www.namebase.org/

Business Search & Research

10K Wizard http://www.10kwizard.com/

Annual Reports from Report Gallery http://www.reportgallery.com/

Arab Data Net http://www.arabdatanet.com/

Business.com http://www.business.com/

Business Information on the Internet http://www.rba.co.uk/sources/index.htm

Corporate Information http://www.corporateinformation.com/

Search the SEC's Edgar Database
http://www.sec.gov/edgar/searchedgar/webusers.htm

EdgarScan Advanced Search
http://edgarscan.pwcglobal.com/servlets/advancedsearch

Free Reports for Top 20 European Companies
http://amadeus.bvdep.com/amadeus/top20/_top20.htm

Global Edge International Business Research (Michigan State University)
http://globaledge.msu.edu/ibrd/ibrd.asp

Hoovers* http://www.hoovers.com/

Kompass* http://www.kompass.com/

MacRae's Blue Book http://www.macraesbluebook.com/

MacRae's EuroPages Search
http://www.europages.net/co_branding/macraesbluebook/home-en.html

Market Access and Compliance http://www.mac.doc.gov/

MSN Money's Key Developments http://news.moneycentral.msn.com/ticker/sigdev.asp

PRNewswire http://www.prnewswire.com/

Researching Businesses and Non-Profits on the Web http://www.ojr.org/ojr/technology/1028068074.php

Researching Companies Online http://www.learnwebskills.com/company/

The Scannery http://www.thescannery.com/

SEDAR http://www.sedar.com/

ThomasGlobal* http://www.thomasglobal.com/

Virtual Business Information Center http://www.vbic.umd.edu/

Virtual International Business and Economic Sources http://library.uncc.edu/display/?dept=reference&format=open&page=68

Yahoo Finance Press Releases http://biz.yahoo.com/prnews/

*Full access requires subscription, but limited information is free.

Researching Countries

Aardvark: Asian Resources for Librarians http://www.aardvarknet.info/user/aardvarkwelcome/

Academic Info http://www.academicinfo.net/

Admi.net http://admi.net/world/

BBC Country Profiles http://news.bbc.co.uk/1/hi/country_profiles/

BUBL Country List http://www.bubl.ac.uk/link/countries.html

Bucknell University's Russian Studies http://www.departments.bucknell.edu/russian/

The Economist Country Briefings http://www.economist.com/countries/

Google Directory Country Index http://directory.google.com/Top/Regional/Countries/

Library of Congress Country Studies http://lcweb2.loc.gov/frd/cs/cshome.html

Middle East and Jewish Studies http://www.columbia.edu/cu/lweb/indiv/mideast/cuvlm/

NationMaster http://www.nationmaster.com/

Northwestern University Library Foreign Governments http://www.library.northwestern.edu/govpub/resource/internat/foreign.html

Northwestern University Library International Governmental Organizations http://www.library.northwestern.edu/govpub/resource/internat/igo.html

The Organization for Economic Co-operation and Development http://www.oecd.org/

Unrepresented Nations and Peoples Organization (UNPO) http://www.unpo.org/

WWW Virtual Library http://vlib.org/Regional

Yahoo Countries http://dir.yahoo.com/regional/countries/index.html

Researching Governments, Political Parties, and Politicians

Council of the Baltic Sea States http://www.cbss.st/

East & Southeast Asia: An Annotated Directory of Internet Resources
http://newton.uor.edu/Departments&Programs/AsianStudiesDept/index.html

European Countries http://europa.eu/abc/european_countries/index_en.htm

Foreign Government Resources on the Web
http://www.lib.umich.edu/govdocs/foreign.html

Global Edge http://globaledge.msu.edu/

InterParliamentary Union http://www.ipu.org/english/home.htm

Northwestern University's Foreign Governments
http://www.library.northwestern.edu/govpub/resource/internat/foreign.html

Northwestern University's International Governmental Organizations
http://www.library.northwestern.edu/govpub/resource/internat/igo.html

Political Resources on the Net http://www.politicalresources.net/

Political Resources on the Net: Unrepresented People
http://www.politicalresources.net/int6.htm

Political Database of the Americas http://www.georgetown.edu/pdba

Current Rulers Worldwide http://www.terra.es/personal2/monolith/

Rulers of the World http://rulers.org/

Finding Foreign Ministries

Library of Congress: Portals to the World
http://www.loc.gov/rr/international/portals.html

Ministries of Foreign Affairs from Lawresearch
http://www.lawresearch.com/v10/global/ciministries.htm

Stefano Baldi's Ministries of Foreign Affairs Online
http://hostings.diplomacy.edu/baldi/mofa.htm

US Institute of Peace Library Foreign Affairs Ministries on the Web
http://www.usip.org/library/formin.html

Finding Embassies

Embassies & Consulates Worldwide http://www.mypage.bluewin.ch/caccia/

Embassy.org http://www.embassy.org/

Embassy World http://www.embassyworld.com/

Latin American Embassies http://www-personal.si.umich.edu/~rlwls/embajadas.html

Library of Congress: Portals to the World http://www.loc.gov/rr/international/portals.html

Tagish Worldwide Embassies http://www2.tagish.co.uk/Links/embassy1b.nsf/

Yahoo Embassies and Consulates http://dir.yahoo.com/Government/Embassies_and_Consulates/

Internet Surveys and Statistics

Clickz Stats http://www.clickz.com/stats/

Cyberatlas http://cyberatlas.internet.com/

Global Reach's Global Internet Statistics by Language http://www.glreach.com/globstats/

Internet Traffic Report http://www.internettrafficreport.com/main.htm

Netcraft http://news.netcraft.com/

Network Wizards Domain Survey http://www.isc.org/ds

Zooknic Internet Statistics http://www.zooknic.com/

ICANN and the Regional Internet Registries (aka NICs)

ICANN http://www.icann.org

AfriNIC http://www.afrinic.net/

APNIC http://www.apnic.net

ARIN http://www.arin.net

European Registry of Internet Domain Names (EURid) http://www.eurid.eu/

LACNIC http://lacnic.net/en

RIPE http://www.ripe.net

Ipv6

Ipv6 Information Page http://www.ipv6.org/

Domain Name Resources

Ins and Outs of DNS http://www.webmonkey.com/webmonkey/02/31/index3a.html

DNS for Rocket Scientists http://newweb.zytrax.com/books/dns/ch1/

The Domain Name Service http://www.scit.wlv.ac.uk/~jphb/comms/dns.html

DNS and BIND, 3rd Edition, O'Reilly Online Catalog
http://www.oreilly.com/catalog/dns3/chapter/ch02.html

Domain Name Registries Around the World http://www.norid.no/domreg.html

IANA's Contact List for TLD Administrators http://www.iana.org/cctld/cctld-whois.htm

InterNIC FAQ on New Top-level Domains http://www.internic.net/faqs/new-tlds.html

Whois Data Problem Report System http://wdprs.internic.net/

Yahoo's Computers and Internet Domain Name Registration
http://dir.yahoo.com/Computers_and_Internet/Internet/Domain_Name_Registration/→
Top_Level_Domains__TLDs_/Registry_Operators/International_Country_Codes/

NSLookup Tools

AnalogX http://www.analogx.com/contents/dnsdig.htm

Check DNS http://www.checkdns.net/quickcheck.aspx

DNS Stuff* http://www.dnsstuff.com/

Eye-Net Consulting* http://www.enc.com.au/itools/

Infobear http://www.infobear.com/nslookup.shtml

Multiple NSLookup http://www.bankes.com/nslookup.htm

SmartWhois NSLookup http://swhois.net/

Squish DNS Lookup http://www.squish.net/dnscheck/

WebReference NsLookup Gateway
http://www.webreference.com/cgi-bin/nslookup.cgi

ZoneEdit NSLookup http://www.zoneedit.com/lookup.html?ad=goto&kw=nslookup

*These sites provide Ipv6 lookups in addition to Ipv4.

Whois Queries

APNIC Whois lookups http://www.apnic.net/search/index.html

APNIC Whois help http://www.apnic.net/db/search/all-options.html

ARIN http://www.arin.net/whois/index.html

ARIN Whois help	http://www.arin.net/tools/whois_help.html
AfriNIC Whois	http://www.afrinic.net/cgi-bin/whois
AfriNIC User Manual	http://www.afrinic.net/docs/db/afsup-dbgs200501.htm
EURid Whois	http://www.whois.eu/whois/GetDomainStatus.htm
LACNIC Whois	http://lacnic.net/cgi-bin/lacnic/whois
RIPE	http://www.ripe.net/perl/whois/
RIPE Reference Manual	http://www.ripe.net/ripe/docs/databaseref-manual.html
RIPE Whois help	http://www.ripe.net/ripencc/pub-services/db/whois/whoishelp.html

Domain Queries

Allwhois	http://www.allwhois.com/
CheckDNS☆	http://www.checkdns.net/quickcheck.aspx
Checkdomains	http://www.checkdomain.com/
CoolWhois	http://www.coolwhois.com/
DNS411	http://dns411.com/
Domain Dossier	http://centralops.net/co/DomainDossier.vbs.asp
Domainsearch	http://www.domainsearch.com/
Domainsurfer	http://www.domainsurfer.com/
Domain Tools	http://www.domaintools.com/
Domain Tools Whois Source	http://whois.domaintools.com/
DrWhois	http://www.drwhois.com/
EasyWhois	http://www.easywhois.com/
Geektools	http://www.geektools.com/whois.php
IP-Plus	http://www.ip-plus.ch/tools/whois_set.en.html
MSV.DK Network Whois	http://msv.dk/ms593.aspx
Multiple DNS Lookup Engine	http://www.bankes.com/nslookup.htm
Namedroppers	http://www.namedroppers.com/
Netcraft	http://news.netcraft.com/
Network-Tools	http://network-tools.com/
Whois.net	http://www.whois.net/
Whois at Webhosting.info	http://whois.webhosting.info/
Whoix?	http://www.whoix.com/

Whoix? Advanced Search	http://www.whoix.com/advdomsearch.html
Xwhois	http://www.xwhois.com/

Internet Utilities and Tools for Windows

All-Nettools.com	http://www.all-nettools.com/tools1.htm
Centralops	http://centralops.net/co/body.asp
Domtools.com	http://www.domtools.com/domtools/
Internet Query Tools	http://www.demon.net/external/
iTools Internet Tools	http://www.itools.com/internet/
Logbud Online Tools ☆	http://www.logbud.com/
Network-Tools	http://www.network-tools.com/
RodentNet Ad Hoc IP Tools	http://tatumweb.com/iptools.htm

Traceroute Tutorials

Mapping Where the Data Flows	http://www.isoc.org/oti/articles/0200/dodge.html
Traceroute Tutorial	http://www.exit109.com/~jeremy/news/providers/traceroute.html
Russ Haynal's Traceroute Overview	http://navigators.com/traceroute.html

Traceroute for Windows

All Nettools.com	http://www.all-nettools.com/toolbox
Cogentco	http://www.cogentco.com/htdocs/glass.php
Geektools Traceroute	http://www.geektools.com/traceroute.php
IP-Plus Traceroute Servers	http://www.ip-plus.ch/tools/traceroute.en.html
Logbud Online Tools ☆	http://www.logbud.com/
Multiple Traceroute Gateway	http://www.tracert.com/cgi-bin/trace.pl
New York Internet Traceroute Links	http://www.nyi.net/traceroute.html
Opus One Traceroute	http://www.opus1.com/www/traceroute.html
SixXs IPv4 and IPv6 Traceroute	http://www.sixxs.net/tools/traceroute/
Traceroute.org	http://www.traceroute.org/
Tracerouters Around the World	http://tracerouters.nielssen.com/
BGPNet IPv4 Wiki	http://www.bgp4.net/tr

BGPNet IPv6 Wiki http://www.bgp4.net/tr6

More Traceroute Tools

Airport & City Code Database http://www.airportcitycodes.com/aaa/CCDBFrame.html

World Airport Codes http://www.world-airport-codes.com/

Airlines of the Web Airport Codes http://flyaow.com/airportcode.htm

Sarangworld Traceroute Project Known Hostname Codes http://www.sarangworld.com/TRACEROUTE/showdb-2.php3

IP Geolocation Tools

DNS Stuff's Version of IP2Location http://www.dnsstuff.com/

Geobytes' IP Locator http://www.geobytes.com/IpLocator.htm

GeoIP http://www.maxmind.com/app/lookup

HuntIP http://www.huntip.com/Tools/mapips.php

IP2Location http://www.location.com.my/free.asp

IPAddressGuide.com http://www.internetipaddress.com/ip2location.aspx

NetGeo http://www.caida.org/tools/utilities/netgeo/

NetWorldMap's Geolocation Tool http://www.networldmap.com/TryIt.htm

WebHosting.Info http://ip-to-country.webhosting.info/node/view/36

Finding ISPs and Email Providers Around the World

The List http://thelist.internet.com/countrycode.html

NSRC's Connectivity Providers Database http://www.nsrc.org/networkstatus.html

International Internet Access Providers http://www.herbison.com/herbison/iap_international_meta_list.html

FreedomList http://www.freedomlist.com/find.php3

African Internet Connectivity http://www3.sn.apc.org/africa/af-isps.htm

Middle East Directory List of ISPs http://www.middleeastdirectory.com/me-isps.htm

Satellite Internet Service Providers for North & South America, Europe, Africa, Asia, Middle East http://www.satsig.net/

Linksat Satellite and Internet Providers http://www.linksat.com/

Satellite Industry Links: Satellite Service Providers http://www.satellite-links.co.uk/links/ssp.html

ISP Directories

Google Directory http://directory.google.com/Top/Computers/Internet/Access_Providers/

Yahoo http://dir.yahoo.com/

There are several ways to use Yahoo to find international ISPs. The best is: Business_and_Economy→Business_to_Business→Communications_and_Networking→Internet_and_World_Wide_Web→By_Region

WiFi Hotspot Finders

Hotspothaven http://www.hotspothaven.com/

Intel's Mobile Technology Hotspot Finder http://intel.jiwire.com/

iPass Hotspot Finder http://ipass.jiwire.com/

JiWire Global WiFi Hotspot Finder http://www.jiwire.com/search-hotspot-locations.htm

WiFinder http://www.wifinder.com/

WiFi411 http://www.wifi411.com/

Wi-Fi Hotspot List http://www.wi-fihotspotlist.com/

Cybercafe Finders

Curious Cat Cybercafe Connections http://www.curiouscat.com/travel/cybercafe.cfm

Cybercaptive Search Engine http://cybercaptive.com/
The country search is disabled; search by city

Google Directory: Cybercafes http://directory.google.com/Top/Computers/Internet/Cybercafes/

Indra's International Cybercafes http://www.indranet.com/potpourri/links/cybercafeint.html

Internet Cybercafe Database http://cybercafe.katchup.co.nz/search.asp

Netcafe Guide http://www.world66.com/netcafeguide

Internet Exchanges and Backbone Networks

Colosource Internet eXchange Points http://www.colosource.com/ix.asp

Exchange Point Information http://www.ep.net/ep-main.html

Boardwatch's Internet Backbone Maps http://www.nthelp.com/maps.htm

BT Infonet's Network Maps
http://www.bt.infonet.com/services/internet/network_maps.asp

BWM's Links to Network Maps
http://www.bandwidthmarket.com/component/option,com_weblinks/catid,74/Itemid,4/

Russ Haynal's Major Internet Backbones http://www.navigators.com/isp.html

Check Your Internet Profile and Vulnerability

Shields Up! http://www.grc.com/

Junkbusters http://www.junkbusters.com/cgi-bin/privacy

BrowserHawk Browser Analysis http://www.syscape.com/showbrow.aspx

Browser Spy Browser Analysis http://gemal.dk/browserspy/

Russ Haynal's Persona Check http://navigators.com/cgi-bin/navigators/persona.pl

HackerWhacker Free Tools http://whacker4.hackerwhacker.com/freetools.php
especially the Browser Leakage and Quick Scan for open ports

Sygate/Symantec Online Security Services
http://scan.sygate.com/home_homeoffice/sygate/index.jsp

Improving Your General Computer & Network Security

About's Network Security http://netsecurity.about.com/

CERT's Home Network Security http://www.cert.org/tech_tips/home_networks.html

Get Safe Online http://www.getsafeonline.org/

Microsoft Security & Privacy for Home Users
http://www.microsoft.com/athome/security/default.mspx

NSA's Security Recommendation Guides http://www.nsa.gov/snac/

NCSA's Stay Safe Online http://www.staysafeonline.info/

Surf the Net Safely http://surfthenetsafely.com/

SecureMac.com http://www.securemac.com/

Securing Home Computers and Networks

Tweakhound's Securing Windows XP
http://www.tweakhound.com/xp/security/page_1.htm

Fred Langa's 5 Essential Steps To PC Security
http://www.informationweek.com/shared/printableArticle.jhtml?articleID=177100010

NIST's Guidance for Securing Windows XP Home Edition
http://csrc.nist.gov/itsec/guidance_WinXP_Home.html

CERT's Home Network Security http://www.cert.org/tech_tips/home_networks.html

Gary Kessler's Protecting Home Computers and Networks
http://www.garykessler.net/library/protecting_home_systems.html

University of Cambridge's Securing Windows XP Home Edition for Stand Alone Use
http://www-tus.csx.cam.ac.uk/pc_support/WinXP/collegehome.html

Lawrence Berkeley Lab's Checklist for Securing Windows XP **PRO**
http://www.lbl.gov/ITSD/Security/systems/wxp-security-checklist.html

Windows XP Security Checklist
http://labmice.techtarget.com/articles/winxpsecuritychecklist.htm

Tom-Cat.com's Secure Your Home Computer v.2.22
http://www.tom-cat.com/security.html

Browser Setup & Testing

ScanIt's Browser Security Check http://www.scanit.be/bcheck

Microsoft Security Home Page http://www.microsoft.com/security/

Microsoft Internet Explorer Security Updates
http://www.microsoft.com/windows/ie/downloads/default.asp

Microsoft Office Download Center http://office.microsoft.com/downloads/

Microsoft Windows Update Page http://update.microsoft.com/windowsupdate/

Microsoft IE7: Dynamic Security Protection
http://www.microsoft.com/windows/products/winfamily/ie/features.mspx

Microsoft: Improve the Safety of Your Browsing and E-Mail Activities
http://www.microsoft.com/athome/security/online/browsing_safety.mspx

How to surf more safely with Internet Explorer 7
http://www.helpwithwindows.com/techfiles/ie7-surf-safe.html

Marc Liron, Microsoft MVP on Internet Explorer 7
http://www.updatexp.com/internet-explorer-7-download.html

Brian Livingston, Windows Secrets, IE7 Needs Tweaking for Safety
http://windowssecrets.com/comp/061026/ - story1

Diana Huggins, IE 7.0's Internet Options Privacy and Security Settings
http://www.lockergnome.com/nexus/windows/2007/01/22/ie-70s-internet-options-security-settings/
http://www.lockergnome.com/nexus/windows/2007/01/23/ie-70s-internet-options-privacy-settings-part-i/
Be sure to look at Part II as well.

Deb Shinder, Tech Republic, "10 things you should know about Internet Explorer 7 Security" http://articles.techrepublic.com.com/5100-1009_11-6130844.html

Surf the Web Safely: Make IE7 Safer http://surfthenetsafely.com/ieseczone8.htm

Kim Komando's Firefox 2 and IE7's Security Settings
http://www.komando.com/tips/index.aspx?id=2523

Cookies

Firefox's Cookie Options
http://mozilla.gunnars.net/firefox_help_firefox_cookie_tutorial.html

Microsoft's Help Safeguard Your Privacy on the Web (for IE6, but most still applies to IE7) http://www.microsoft.com/windows/ie/using/howto/privacy/config.mspx

Cookie Central's Reviews of Cookie Management Software
http://www.cookiecentral.com/files.htm

Junkbusters Cookie Page http://www.junkbusters.com/ht/en/cookies.html

Microsoft Security

Microsoft Internet Explorer Security Updates
http://www.microsoft.com/windows/ie/downloads/default.asp

Microsoft Office Download Center http://office.microsoft.com/downloads/

All Microsoft Office Viewers http://www.microsoft.com/office/000/viewers.asp

Microsoft Policies on Software Distribution
http://www.microsoft.com/technet/security/topics/policy/swdist.mspx

Microsoft Security Home Page http://www.microsoft.com/security/

Microsoft Windows Update Page http://windowsupdate.microsoft.com/

5 Steps to Secure Windows XP Home
http://netsecurity.about.com/cs/windowsxp/a/aa042204_2.htm

Non-Admin Blog, Aaron Margosis' Weblog
http://blogs.msdn.com/aaron_margosis/archive/2005/04/18/TableOfContents.aspx

"RunAs" basic (and intermediate) topics, Aaron Margosis' Weblog
http://blogs.msdn.com/aaron_margosis/archive/2004/06/23/163229.aspx

Email Security

About.com Email Help Center http://antivirus.about.com/library/bloutlook.htm

About's Email Wiretapping Article http://antivirus.about.com/library/weekly/aa020501a.htm?once=true&

How Spammers Get Your Email Address http://www.junk-mail.org.uk/public/articles/spam.html

A Quick Guide to Email Security http://www.zzee.com/enh/email_security.html

Security Focus: "Securing Privacy: E-mail Issues" http://www.securityfocus.com/infocus/1579

Anti-Phishing

How Not to Get Hooked by a "Phishing" Scam http://www.ftc.gov/bcp/edu/pubs/consumer/alerts/alt127.htm

The Anti-Phishing Working Group http://www.antiphishing.org/

Phishtank (known and suspected phishing sites) http://www.phishtank.com/

PayPal's Protect Yourself from Fraudulent Emails https://www.paypal.com/cgi-bin/webscr?cmd=xpt/general/SecuritySpoof-outside

URL Decrypter http://www.cyber-junkie.com/tools/urldecrypter.shtml

Un-Obfuscating URLs http://www.wilsonmar.com/1tcpaddr.htm

Disabling Visual Basic Script

How to Disable VBS http://www.cvm.uiuc.edu/net/virus/outlook.html

Disable Windows Scripting Host http://www.sophos.com/support/faqs/wsh.html

Remove Windows Scripting Host http://www.f-secure.com/virus-info/u-vbs/

See What Your Computer is Loading and Running

Greatis Start Up Application Database http://www.greatis.com/regrun3appdatabase.htm

Pacman's Start-Up Applications http://www.pacs-portal.co.uk/startup_index.htm
or http://www.sysinfo.org/startupinfo.html

Process ID http://www.processid.com/

Answers That Work http://www.answersthatwork.com/Tasklist_pages/tasklist.htm

Process Library http://www.processlibrary.com/

PC Hell http://www.pchell.com/

Encrypt Data in Windows XP

How to Encrypt a File in Windows XP http://support.microsoft.com/kb/307877

How to Encrypt a Folder in Windows XP http://support.microsoft.com/kb/308989

Anti-Virus Information

About's Free Antivirus Software Reviews
http://antivirus.about.com/od/freeantivirussoftware/Free_Antivirus_Software.htm

VirusList Virus Encyclopedia http://www.viruslist.com/en/viruses/encyclopedia

Spyware Checkers and Information

Free Antispyware Products

Ad-Aware Spyware Checker
http://www.lavasoftusa.com/products/ad-aware_se_personal.php

Windows Defender
http://www.microsoft.com/athome/security/spyware/software/default.mspx

Spybot Search & Destroy http://www.safer-networking.org/en/home/index.html

Antispyware Guides & Articles

11 Signs of Spyware
http://www.pcmag.com/article2/0%2C1759%2C1522648%2C00.asp

Anti-Spyware Guide http://www.firewallguide.com/spyware.htm

Monitoring Software on Your Computer: Spyware, Adware, and Other Software, Staff Report, Federal Trade Commission, March 2005 **[PDF]**
http://www.ftc.gov/os/2005/03/050307spywarerpt.pdf

PC Hell Spyware Removal Help http://www.pchell.com/support/spyware.shtml

Spychecker http://www.spychecker.com/

Spyware Guide http://www.spywareguide.com/product_list_full.php

Spyware Warrior Rogue/Suspect Anti-Spyware Products
http://www.spywarewarrior.com/rogue_anti-spyware.htm

Spyware Watch http://www.spyware.co.uk/

Stop Internet Abuse http://www.celticsurf.net/webscape/abuse.html

Web Bugs

Bugnosis http://www.bugnosis.org/

Web Bug FAQ http://www.bugnosis.org/faq.html

Guidescope http://www.guidescope.com/home/

WebWasher[235]
http://www.cyberguard.com/products/webwasher/webwasher_products/classic/index.html

Trojan Horse Prevention, Detection, and Removal

List of Trojan Ports http://secured.orcon.net.nz/portlist_list.html

Onctek's Trojan Port List http://www.onctek.com/trojanports.html

Anti-Trojan Software Reviews http://www.anti-trojan-software-reviews.com/

Anti-Trojan.org http://www.anti-trojan.org/

Anti-Trojan Guide from Firewall Guide http://www.firewallguide.com/anti-trojan.htm

PCFlank's Trojan Test Page http://www.pcflank.com/trojans_test1.htm

WindowSecurity.com TrojanScan http://www.windowsecurity.com/trojanscan/

Passwords

The Simplest Security: A Guide To Better Password Practices
http://www.securityfocus.com/infocus/1537

Microsoft: How to Create Stronger Passwords
http://www.microsoft.com/security/articles/password.asp

Password Security Guide http://www.umich.edu/~policies/pw-security.html

Fred Langa: How to Build Better Passwords
http://www.informationweek.com/story/showArticle.jhtml?articleID=164303537

Firewall Information

Firewallguide's Personal Firewall Review http://www.firewallguide.com/software.htm

Firewallguide: Wired Routers http://www.firewallguide.com/hardware.htm

Firewall Forensics
http://www.linuxsecurity.com/resource_files/firewalls/firewall-seen.html

---

[235] "Cyberguard has changed the license for Webwasher Classic to Donationware and asks you to make a donation before downloading Webwasher Classic." However, the donation is voluntary.

Firewall Q&A http://www.vicomsoft.com/knowledge/reference/firewalls1.html

Free Personal Firewall Software http://netsecurity.about.com/od/personalfirewalls/a/aafreefirewall.htm

Gibson Research's Firewall Page http://grc.com/su-firewalls.htm

HomeNetHelp's Broadband Router Guide http://www.homenethelp.com/router-guide/index.asp

How Firewalls Work http://www.howstuffworks.com/firewall.htm

Internet Firewall FAQ http://www.interhack.net/pubs/fwfaq/

Introduction to Firewalls http://netsecurity.about.com/od/hackertools/a/aa072004.htm

Free Software Firewalls for Windows

Securepoint http://www.securepoint.cc/products_pcfirewall_en.html

Sygate Personal Firewall http://smb.sygate.com/products/spf_standard.htm

Zone Alarm http://www.zonelabs.com/

Firewall Leak Tests

Gibson Research's Firewall Leaktest http://grc.com/lt/leaktest.htm

PCFlank Firewall Leaktest http://www.pcflank.com/pcflankleaktest.htm

Tooleaky http://tooleaky.zensoft.com/

Firehole http://keir.net/firehole.html

Firewall Leak Tester (Test Results) http://www.firewallleaktester.com/index.html

Anonymizing Proxies

InfoAnarchy's Anonymous Web Searching http://www.infoanarchy.org/en/Anonymous_Web_Surfing

Free Web Anonymizer Services http://www.cexx.org/anony.htm

Web Anonymizing Services http://www.computerbytesman.com/anon/index.htm

Test Page for Web Anonymizing Services http://www.computerbytesman.com/anon/test.htm

Internet Security and Privacy News and Information

Center for Privacy and Technology Ten Ways to Protect Privacy Online http://www.cdt.org/privacy/guide/basic/topten.html

EPIC Online Guide to Privacy Resources http://www.epic.org/privacy/privacy_resources_faq.html

Georgi Guninski Security Research http://www.guninski.com/

Security Focus http://www.securityfocus.com/

Security Now! http://www.grc.com/securitynow.htm

Yahoo News Computer Security http://fullcoverage.yahoo.com/fc/Tech/Computer_Security

Yahoo News Cybercrime and Internet Fraud http://news.yahoo.com/fc/Tech/Cybercrime_and_Internet_Fraud/

Yahoo News Internet Privacy http://news.yahoo.com/fc/Tech/Internet_Privacy/

Notes

Notes

Notes

DOCID: 4046925

UTW

31228

www.ingramcontent.com/pod-product-compliance
Lightning Source LLC
Chambersburg PA
CBHW060919060326
40690CB00041B/2724
* 9 7 8 0 9 8 4 2 8 4 4 9 8 *